Genetic Testing

Contents

8 Prenatal Screening and Diagnosis   163

9   Genetics of Common Neurological Disorders   219

10   Newborn and Carrier Screening   238

11   Susceptibility Testing   268

12   Test Samples and Laboratory Protocols   292

Contributors

AUTHORS AND EDITORS

Neil F. Sharpe
Neil F. Sharpe's particular interests focus on the legal, ethical, social, and psychological aspects of medical genetics and genetic counseling. Neil was a practicing lawyer, completed a Master of Laws degree, and subsequently received training in medical genetics and genetic counseling. He serves as a consultant in health policy and the development of appropriate legal, medical, and counseling standards of care for the delivery of genetic testing services

Ronald F. Carter, PhD, FCCMG, FACMG
Director, Genetic Services, Hamilton Health Sciences
Professor, Department of Pathology and Molecular Medicine, McMaster University
Dr. Carter completed a post-doctoral research fellowship at the Banting Institute, University of Toronto, and fellowships in laboratory cytogenetics and molecular genetics at the Hospital for Sick Children, Toronto. He is a former president of the Canadian College of Medical Genetics, and a founding fellow of the American College of Medical Genetics. Dr. Carter was appointed to direct a regional cytogenetics laboratory in 1991, and now directs a medical genetic service with a catchment of over 2 million people. He has particular interests in cancer genetics, teaching and curriculum development for medical genetics, and policy development and implementation for new genetic services.

CONTRIBUTORS

S Annie Adams, MS, Mayo Clinic Cytogenetics Laboratory, Department of Laboratory Medicine and Pathology, Rochester, New York

Riyana Babul-Hirji, MSc, CGC, Genetic Counsellor, Lecturer, Hospital for Sick Children, Lecturer, Department of Molecular Genetics and Microbiology, University of Toronto, Canada

Jon Beckwith, PhD, Department of Microbiology and Molecular Genetics, Harvard Medical School, Boston, Massachusetts

Peter J Bridge, PhD, FCCMG, FACMG, Associate Professor, Molecular Diagnostic Laboratory, Alberta Children's Hospital, Calgary, Alberta, and Department of Medical Genetics, University of Calgary, Alberta, Canada

CH Browner, Phd, Professor, Center for Culture and Health, Department of Psychiatry and Biobehavioral Sciences University of California, Los Angeles, California

Nancy P Callanan, MS, CGC, Director, Genetic Counseling Program, Graduate School, University of North Carolina at Greensboro, North Carolina

David Castle, PhD, Associate Professor, Department of Philosophy, University of Guelph, Ontario, Canada

Timothy Caulfield, LLB, LLM, Professor and Canada Research Chair in Health Law and Policy, Health Law Institute, Faculty of Law, University of Alberta, Edmonton, Alberta, Canada

David Chitayat, MD, FABMG, FACMG, FCCMG, FRCPC Head, The Prenatal Diagnosis and Medical Genetics Program, Professor, Departments of Medical Genetics and Microbiology, Pediatrics, Obstetrics and Gynecology and Laboratory Medicine and Pathobiology, Medical Director, The MSc Program in Genetic Counselling, University of Toronto, Ontario, Canada

Edwin H Cook, Jr, MD, Professor of Psychiatry, Pediatrics, and Human Genetics, University of Chicago, Illinois

Dena S Davis, JD, PhD, Professor, Cleveland-Marshall College of Law, Cleveland, Ohio

Stephen H. Dinwiddie, MD, Professor of Psychiatry, University of Chicago, Illinois

Daniel L Van Dyke, PhD, Mayo Clinic Cytogenetics Laboratory, Department of Laboratory Medicine and Pathology, Rochester, New York

MJ Esplen, PhD, NCIC Clinician Scientist, Head, Program of Psychosocial and Psychotherapy Research in Cancer Genetics, Associate Professor, Department of Psychiatry, Faculty of Medicine, University of Toronto, Ontario, Canada

Lisa Geller PhD, JD, Wilmer Cutler Pickering Hale and Dorr LLP, Boston, Massachusetts

Elliot S. Gershon, MD, Professor of Psychiatry and Human Genetics, University of Chicago, Illinois

Béatrice Godard, PhD, Programmes de Bioéthique, Université de Montréal, Quebec, Canada

E Richard Gold, LLB, LLM, SJD, Associate Professor and Director, Centre for Intellectual Property Policy, McGill University, Montreal, Quebec, Canada

Mark A Hall, JD, Professor of Law and Public Health, Wake Forest University Medical School, Department of Public Health Sciences, Winston-Salem, North Carolina

Jinger G Hoop, MD, Fellow in Psychiatric Genetics and Senior Fellow in Medical Ethics, University of Chicago, Illinois

Rhett P Ketterling, MD, Mayo Clinic Cytogenetics Laboratory, Department of Laboratory Medicine and Pathology, Rochester, New York

Bartha-Maria Knoppers, PhD, LLD, Professor, Faculté de droit Université de Montréal, Quebec, Canada

Mireille Lacroix, BSS, LLB, LLM, Research Associate, Centre de recherche en droit public, Université de Montréal, Quebec, Canada

Bonnie S LeRoy, MS, ABMG, ABGC, Department of Genetics, Cell Biology and Development and the Institute of Human Genetics, University of Minnesota, Minneapolis, Minnesota

Andrew R MacRae, PhD, FCACB, Director, The Research Institute at Lakeridge Health, Assistant Professor, Department of Pathobiology and Laboratory Medicine, Faculty of Medicine, University of Toronto, Ontario, Canada

Allyn McConkie-Rosell, PhD, CGC, Assistant Professor, Division of Medical Genetics, Duke University Medical Center, Durham, North Carolina

Jon F Merz, JD, PhD, Associate Professor, Department of Medical Ethics, University of Pennsylvania, Philadelphia, Pennsylvania

Roxanne Mykitiuk, BA, LLB LLM, Associate Professor of Law, Osgoode Hall Law School, York University, Toronto, Ontario, Canada

Julianne M O'Daniel, MS, CGC, Assistant Director, Educational and Training Programs Institute for Genome Sciences and Policy, Duke University, Durham, North Carolina

Jillian S Parboosingh, PhD, FCCMG, Assistant Professor, Molecular Diagnostic Laboratory, Alberta Children's Hospital, Calgary, Alberta, and Department of Medical Genetics, University of Calgary, Alberta, Canada

H Mabel Preloran, PhD, Center for Culture and Health, Department of Psychiatry and Biobehavioral Sciences University of California, Los Angeles, California

C Anthony Rupar, PhD, FCCMG, Associate Professor Biochemistry, Pediatrics, and Pathology, Chairman, Division of Clinical Biochemistry, Medical Leader, Molecular Pathology, Child and Parent Resource Institute, University of Western Ontario and the London Health Sciences Centre, London, Ontario, Canada

Maren T Scheuner, MD, MPH, Visiting Associate Professor, UCLA School of Public Health, Department of Health Services, ATPM Fellow, CDC Office of Genomics and Disease Prevention, Los Angeles, California

Lorraine Sheremeta, Research Associate, Health Law Institute, Faculty of Law, University of Alberta, Edmantan, Alberta, Canada

Cheryl Shuman, MS, CGC, Associate Professor Division Clinical and Metabolic Genetics, Medical Genetics and Microbiology, Hospital for Sick Children, Assistant Professor, Department of Molecular Genetics and Microbiology, University of Toronto, Ontario, Canada

Anne Summers, MD, FRCPC, Director of Genetics, North York General Hospital, Toronto, Ontario, Canada

Stephanie Turnham, BA (honors), LLB candidate, class of 2006, Osgoode Hall Law School, York University, Toronto, Canada

Erik C Thorland, PhD, Mayo Clinic Cytogenetics Laboratory, Department of Laboratory Medicine and Pathology, Rochester, New York

Foreword
Advances in Genetics Reach the Clinic: Challenges for Health Care Providers and Patients

Our understanding of genetic mechanisms of disease has improved so rapidly in recent years that concepts and applications to medicine have moved well beyond the knowledge base of most health care providers and their patients. From the discovery of the correct number of human chromosomes in 1956 [Tjio and Levan, 1956] to the complete sequencing of the human genome in 2003, technical advances in the analysis and manipulation of DNA have rapidly permitted an increasingly accurate understanding of the effects of genetic sequence variations on the human condition.

Only a few years ago, medical genetic researchers used to struggle to catalog thousands of syndromes based upon clinical presentations and tireless examinations of families and their histories [see McKusick, 1990, for discussion]; a minority of diagnoses could be confirmed by testing, and useful treatment was not possible for even half of all genetic conditions. Now, thousands of genes have been cloned, diagnostic tests are available for over 1000 different genetic diseases, accurate predictions for clinical outcome and recurrence risks are commonly available for many syndromes, therapies with molecular levels of specificity of action are becoming available, and the complex interactions of environmental and genetic components of common adult disorders are increasingly clear.

What was once a quiet little corner of medicine has transformed into a cornerstone of all medical care. The "new genetics" is a focus of vigorous research and high technological competition, the cause of both uneasiness and expectation for patients, a bountiful source of ethical, legal, and social dilemmas, and a minefield of unknowns for many health care practitioners. In the United States, the societal impacts of these issues were extensively reviewed twice within the past 15 years [Andrews et al., 1994; Holtzman et al., 1997]. More recently, a uniquely comprehensive and authoritative inquiry commissioned by the Australian government has issued a landmark report, *Essentially Yours* [Australian Law Reform Commission, 2003] that is discussed in more detail in Appendix 1 by Professor D. Weisbrot. The office of the Surgeon General of the United States also engaged public interest recently by launching "Family History Day" on Thanksgiving 2004 [www.surgeongeneral.gov]. The Surgeon General's initiative has encouraged families to discuss and document their medical family histories, and the website can be used to construct pedigrees easily and accurately.

The scope of practice for genetics is extending into all medical disciplines, and with less than 90 out of 104,000 residency slots in the United States (0.085 percent of the total) allocated for medical geneticists, there are not nearly enough medical geneticists for the work that needs to be done. Accordingly, other health care professionals must invoke genetics into their level of care. Genetics is quickly moving into the mainstream, propelling advances in prenatal screening, newborn screening, neonatology, pediatrics, oncology, behavioral medicine, and adult medicine. Everything from infertility to psychiatry is being recast in pathogenetic principles founded on genetic paradigms, and nowhere is this more evident than in our understanding of the complex, multifactorial diseases that commonly affect adults: cardiovascular disease, neurodegenerative conditions, and cancer.

The combination of epidemiology, information technology, and large-scale molecular diagnostics has permitted increasingly accurate descriptions of the interactions of environmental and genetic variables affecting morbidity and mortality, and these advances, in turn, are informing a beckoning era of better clinical management through the related developments of pharmacogenomics and proteomics. Even while it remains to be seen how much of this technology we can afford to provide in health care systems around the world, we all look forward to the potential of new diagnostic and therapeutic techniques already demonstrated in such sentinel examples as imatinib (in the treatment of chronic myelogenous leukemia) and diagnostic microarrays (e.g., for developmental genetics and the prognostic evaluation of breast cancer).

Even though many promising advances have yet to make it to the clinic, the sheer rapidity of advances in medical genetics makes it very difficult for nongeneticist health care providers to keep up. Increasingly, genetics is more and more a part of medical care for all health care professionals, yet surveys have repeatedly shown that many health care providers are aware that they have very little knowledge of even basic genetic principles. "Should I be looking for a genetic component of this disease?" "What is the most recent understanding for the disease?" "How much of this is relevant to practice?" "How to know when to refer?" "Where to get testing, and what do the results mean?" "Who else in a family is at risk?" "What about ethical and legal issues, patient privacy, access to genetic information?" "Where do I find the answers?"

This book is intended to be a source of answers for medical students, residents, and nongeneticist health care providers. In keeping with current pedagogical trends away from knowledge-based curricula ("Flexnerian" medical education) and toward competency-based approaches [Carraccio et al., 2002], we have not attempted to compile a definitive knowledge base for teaching medical genetics. We assume that the knowledge base for medical genetics will continue to evolve rapidly.

Rather, we are presenting concepts, issues, and professional skills that will illustrate to the reader how to continue to provide better care for patients with genetic diseases. The information is intended to be quickly accessible at two levels: important concepts and sources of details. Thus, topics have generally been compiled in a format that frequently includes:

- Introductions of topics
- Illustrative case scenarios
- Core concepts
- Resources for self-directed learning
- Detailed, practical discussions written by experts in the field
- Sources of professionally reviewed, consistently updated information

This book contains practical information for health care professionals of all types who need to provide counseling about genetic tests. Numerous common clinical scenarios are discussed with concise and accurate descriptions of essential information, accompanied by further discussion and evolving reference resources for those who wish to delve into further detail. We have provided an overview of both medical and legal perspectives on many issues, and, in addition, we refer extensively to other sources of accurate information throughout this book.

We emphasize that this book is *not* intended to suggest practice guidelines or appropriate standards of care; our purpose is only to provide an accessible resource that presents principles, concepts, and illustrated discussions in an accessible clinical context as an aid to the consideration of genetics in medicine.

It is timely that the American Academy of Family Physicians chose genetics for the 2005 "Annual Clinical Focus" (www.aafp.org/acfgenomics.xml). Professional organizations are responding to the need to provide continuing education in this area, and excellent materials for continuing education in medical genetics are becoming available from a wide variety of sources (e.g., the Foundation for Medical Practice Education, www.fmpe.org/index.asp). The authors of this book hope that health care professionals will also find this format useful in approaching unfamiliar clinical territory in medical genetics.

Chapter 1

Genetic Counseling and the Physician–Patient Relationship

OVERVIEW

Medicine's ethical maxim, *primum non nocere*—above all do no harm—and its underlying values and objectives of treatment and cure, traditionally have been regarded as the fundamental role and value of medicine in society [Gadow, 1984; Jonsen et al., 1998]. Within the context of genetic testing, the goals and values adopted in the practice of counseling need to reflect a variety of applications and settings. The lack of effective treatment and cure for many genetic diseases and disorders has triggered significant debate concerning what objectives and values constitute an appropriate model of care. Accordingly, it is appropriate to consider how the traditional medical concepts of "benefit," "harm," and the physician–patient relationship have been interpreted within the context of genetic testing and to look at practical steps that can be taken to ensure that the patient benefits from counseling.

INTRODUCTION

Genetics has moved beyond the scope of individual families and quiet conversations about which uncle might have had what problem. With widespread uptake of prenatal screening, expansion of newborn screening programs, the advent of conclusive laboratory diagnostics for many developmental disorders of childhood, predictive testing and risk assessment for disorders of adult onset, new reproductive technologies, cancer diagnostics, the widely promoted promises of biotechnology, and the immediate accessibility of Internet-based information, patients have moved from passive to active users of genetic services. As genetic components of common diseases become

Genetic Testing: Care, Consent, and Liability, by Neil F. Sharpe and Ronald F. Carter
© 2006 John Wiley & Sons, Inc.

recognizable, quantifiable, and predictable, risk modification techniques will be extended to whole populations. We are all now the patient. Governments, professionals, private enterprise, and the lay press have responded to this situation, and the result is more information (accurate or not) from more sources (biased or not) available to more patients (receptive or not) every day.

To return to the clinic, this movement of genetics into the mainstream of health care and public debate has changed the expectations of our patients. However, for many situations we still face an unfortunate technological lag between our theoretical knowledge of a given genetic disease and our systemic ability to provide effective therapy. Thus, we not only frequently have a knowledgeable patient but often also a frustrated one. Genetic counseling is by definition an interactive process. More than ever, health care professionals will have to assess how much the patient can comprehend about the situation at hand and be skilled at both extracting and providing critical information that will benefit the patient in decisions about care. Clearly, communication skills and the fundamental nature of the relationship between the patient and the provider are very important.

Why has the patient sought professional services? While concern and worry about a perceived medical problem are a common motivation to seek medical attention, the specific components of a patient's expectations might become particularly important if there is no effective treatment for the disease in question. Discussion can clarify what the patient expects to learn, uncover the patient's psychosocial status and cultural and family context, correct misconceptions, assess concerns, and provide an opportunity to discuss the nature, objectives, risks, and limitations of genetic testing. The patient needs a realistic understanding of the implications posed by a positive, negative, or inconclusive result. A failure to review these issues may contribute to patient misunderstanding and dissatisfaction, which are both critical factors in a patient's decision to seek redress through malpractice litigation.

CORE CONCEPTS

The patient–health care professional relationship in genetic counseling is:

- Interactive
- Respectful of patient autonomy (nondirectiveness and incorporation of patient values)
- Client centered
- Founded on the goals of achieving supportive patient education, enabling an informed decision, providing accurate risk assessment and risk perception, facilitating decisions, attending to psychosocial needs for the patient and the family, empowering the patient, and facilitating future care

Health care professionals need to adapt their counseling techniques and goals to a variety of applications and indications, including type of practice (clinical setting) and type of indication [diagnostic confirmation, at-risk testing (carrier status), prenatal screening, population screening, presymptomatic testing, susceptibility testing].

Patient motivation to participate in and act upon genetic counseling is influenced by the patient's perception of risk and effectiveness of clinical management (treatment). Patients accrue knowledge from many different sources. Patients may access information that is of dubious origin, authenticity, or benefit. Patients may not be able to critically appraise the clinical relevance of the information they obtain and may either seek guidance about their information or use it as an opinion in contrast to the genetic counseling they receive. The patient's beliefs, knowledge, and cultural values may conflict with the goals and expectations of a nondirective counseling process. Health care professionals may need to prepare the patient for counseling about test results.

Genetic counseling is often provided at a time of emotional stress. The timing and setting for counseling may prevent the patient from understanding what is said. The process of risk assessment, understanding risk, and modifying perception of risk may be difficult for patients to utilize (see also Chapter 14, Test Results: Communication and Counseling). The psychological status of the patient needs to be assessed so that effective communication can ensue (see also Chapter 2, Communication).

Health care practitioners need to understand the limitations and applications of genetic testing and predictions of clinical significance and be able to accurately convey that information to the patient (see also Chapter 7, Informed Consent). However, health care professionals who have not received specialized training in genetics may be uncertain as to the nature and extent of their responsibilities [Veach, 2004].

Patients may make management decisions based upon their own frame of reference. The patient should expect the freedom of informed choice in regard to management decisions, and the health care provider has to accept that some patients will make choices that do not appear to be based upon the information provided.

TRUST, GENETIC TESTING, AND THE PHYSICIAN–PATIENT RELATIONSHIP

When a patient is faced with a medical illness or problem, the patient asks the physician to find an appropriate solution, usually in terms of treatment and cure. In seeking the physician's aid, the patient must provide the physician with a certain degree of trust. The patient must have confidence that the physician will be able to resolve the medical difficulty, will exercise good judgment, and will utilize medical skills and knowledge with due care, diligence, and caution [Sharpe, 1997; Mechanic, 1998]. In times of emergency, uncertainty, and anxiety, the patient may be especially dependent upon the physician to come to her or his aid. Due to their medical knowledge and skills, physicians perceive themselves as trustworthy and assume that because their patients trust and have confidence in them, a reasonable patient will, or should, adopt the recommended course of action because the physician has determined that it provides the best opportunity for therapeutic benefit and the prevention of medical harm [DuBose, 1995; Sharpe, 1997].

"Trust," within the context of this type of physician–patient relationship, reflects medicine's ethical maxim, *primum non nocere*—above all do no harm—and its underlying values of treatment and cure. Given these perceptions and values, the physician–patient relationship traditionally was perceived as one that reflected an inherent disparity in medical knowledge and expertise. Accordingly, the nature of the relationship was deemed to be paternalistic, the patient dependent and passive, the physician paternal, active [Katz, 1984; Apfel and Fisher, 1984; Gillon, 1986], and best able to determine "what constitutes well-being" [Katz, 1977, 1993; Infelfinger, 1980; Kinsella, 1988].

The traditional therapeutic model of care, first proposed by René Descartes in the seventeenth century, was based on the idea that the human body was a machine and functioned pursuant to discernible mechanical principles. Descartes separated the soul—the patient's beliefs, needs, fears, and values—from the bodily processes and argued that such emotional concerns had little in common with physical health and, therefore, the care and cure of bodily problems [Carter, 1983]. Late into the twentieth century, medical and social commentators argued that the Cartesian biomedical model remained the dominant therapeutic model, exemplifying medicine's traditional values of "the relief of pain, the prevention of disability, and the postponement of death by the application of theoretical knowledge incorporated in medical science" [Seldin, 1981; Krauser, 1989]. This included the operative assumption that disease can be fully accounted for by "deviations in measurable biological variables," without reference to the "social, psychological, and other behavioral dimensions of illness" [Engel, 1991].

However, many of these operative assumptions, and their underlying values, increasingly were questioned and challenged [Lynch, 1985; Englehardt, 1986; Gillon, 1986]. In the last quarter of the twentieth century, it was argued that the physician–patient relationship had become less paternalistic with increased shared decision making, as patients took on a great responsibility and demanded more autonomy in medical decision making [Balint and Shelton, 1996; Charles et al., 1997, 1999; but see Zupancic et al., 2002]. In 2003, a World Health Association survey of 3707 patients and doctors in the United States, United Kingdom, Canada, Germany, South Africa, and Japan reported that only about 20 percent of those surveyed defined the physician–patient relationship as paternalistic or authoritarian [Pincock, 2003]. Health Maintenance Organizations (HMOs) and the advent of "managed care," however, continue to have a dramatic impact in the United States. One study reported that nearly one in three doctors in the United States withheld information from patients about medical services because they were not covered by health insurance plans [Wynia et al., 2003; Gottlieb, 2003; Mechanic, 1998].

Many types of genetic tests may not clearly promote diagnosis, treatment, and cure. They may only provide information about a medical condition that is likely to occur at some time in the future. Testing may fail to predict how severe the medical condition may be, when it will occur, or even, due to reduced penetrance, that it will occur [Andrews and Zuiker, 2003]. These limitations can give rise to complex and troubling ethical and moral issues. By way of example, should a health care professional recommend adoption or reproductive technology in order to prevent the transmission of a genetic disorder [Shaw, 1987]? If so, what type of genetic disease or disorder justifies a recommendation of adoption, or reproductive technology, or termination of a pregnancy; what is the appropriate recommendation, *if any*, in the case of hereditary deafness [Taneja et al., 2004] or achondroplasia [Green, 1997] or mental conditions [Andre et al., 2000; compare to Vehmas, 2002] (see also section by Mykitiuk et al. in Chapter 8, Prenatal and Neonatal Screening)?

Because of these limitations, physicians can be confounded by patient anguish, anxiety, and despair. In these circumstances, the patient may feel helpless and alone, and the physician also may experience a sense of powerlessness. In such a context, the *traditional* perception of medical trust is reduced to an illusion [DuBose, 1995].

If health care professionals are to provide benefit and prevent harm to the patient before, during, and after genetic testing, they will need to change the way they perceive the nature of their relationship with the patient and how they can best respond to the patient's pleas for assistance and relief [Maley and NGSC, 1994; Sharpe, 1997]. Health care professionals need to focus on a patient's communicative, emotional, and psychological needs in order to assist the patient to understand, adjust to, and cope with the implications posed by genetic testing and test information [Ad Hoc Committee on Genetic Testing, 1975; Sandhaus et al., 2001]. In short, health care professionals have to apply a therapeutic model of care that encompasses a more "human vision" as opposed to a purely Cartesian "medical vision" [DuBose, 1995; and see also section by Mykitiuk et al. in Chapter 8, Prenatal and Neonatal Screening].

GENETIC COUNSELING AND THE PHYSICIAN–PATIENT RELATIONSHIP

Genetic counseling represents a significant departure from other forms of medical care in not giving overt advice concerning decision making by a patient. Given the lack of effective treatment and cure for many genetic diseases and disorders, the pivotal factor in decision making can be the nature of the genetic test information and its impact on the patient and the family, including—as will be discussed in later chapters—the patient's objectives, concerns, values, and beliefs, as well as the potential for insurance and employment discrimination, stigmatization, and the psychological implications [Wiggins et al., 1992; Tibben et al., 1993a; Michie et al., 1997a; Michie and Marteau, 1999b; Codori et al., 1999; Glanz et al., 1999; Kirschner et al., 2000; Bassett et al., 2001; Hadley et al., 2003; Keller et al., 2004].

Whether seen from the perspective of good patient care and/or reducing the risk of patient misunderstanding, dissatisfaction, and allegations of malpractice, genetic counseling and the specialized practice of genetic counselors, should be considered a critical, if not mandatory, component of the testing process. Counseling:

- Provides improvement in risk estimation and comprehension [Lerman et al., 1997; Burke et al., 2000; Edwards et al., 2000].
- Promotes discussion of risk management options [Johnson et al., 2002].
- Facilitates quality assurance [Gibons, 2004], better understanding, more informed decision making, and helps to reduce exposure to malpractice allegations.

Effective counseling helps the patient and family to: (1) comprehend the medical facts including the diagnosis, probable course of the disorder, and the available management, (2) appreciate the way heredity contributes to the disorder and the risk of recurrence in specified relatives, (3) understand the alternatives for dealing with the risk of recurrence, (4) understand, cope with, and adapt to the emotional and psychological aspects of testing and test results; (5) choose a course of action that seems to them appropriate in view of their risk, their family goals, and their ethical and religious standards; and (6) make the best possible adjustment to the disorder in an affected family member and/or to the risk of recurrence of that disorder [Ad Hoc Committee on Genetic Counseling, 1975; Royal Commission, 1993; Burgess et al., 1998].

NONDIRECTIVENESS

A comprehensive review of practitioner surveys and human/medical genetics journal and text literature indicates that a significant majority of geneticists and counselors subscribe to the ethical value of *nondirective* counseling [Ad Hoc Committee on Genetic Counseling, 1975; Kessler, 1979; Fletcher et al., 1985; Wertz and Fletcher, 1988; Wertz et al., 1990; De Dinechin et al., 1993; Committee on Assessing Genetic Risk, 1994; Benkendorf et al., 1997; Biesecker, 2001; Kessler, 2001; but see Yarborough et al., 1989; Clarke, 1991; Royal Commission, 1993; Bernhardt, 1997; Williams et al., 2002a; Weil, 2003]. A prevailing value of nondirective counseling is to acknowledge that the patient has the right to autonomous decision making that seems appropriate to the patient in view of her or his particular values, objectives, and ethical and religious standards [Ad Hoc Committee on Genetic Counseling, 1975; Kessler, 1979; Emery, 1984; Royal Commission, 1993; Andrews, 1997]. For example, the code of ethics for the National Society of Genetic Counselors states that counselors "should strive to enable their clients to make informed independent decisions, free of coercion, by providing or illuminating the necessary facts and clarifying the alternatives and anticipated consequences" [NSGC, 1992]. The value of patient autonomy is to be supported even where the health care professional personally may disagree with the patient's decision (with regard to genetic testing of children under the age of 18 years for adult-onset disorders, professional organizations have stated that testing may be discouraged in absence of potential benefit; however, each case should be considered on an individual basis [Bloch and Hayden, 1990; Committee of IHA and WFN, 1990; Sharpe, 1993; Wertz et al., 1994; ASHG/ACMG Report, 1995; AMA,

2003; Bioethics Committee (CPS), 2003; WHO, 2003; but also see Duncan, 2004] (see also, Chapter 11, Susceptibility Testing).

However, as discussed in the following section by Nancy Callanan and Bonnie LeRoy, important distinctions are made between directing the *process* compared to the *outcome* of genetic counseling [Bartels et al., 1997; Kessler, 1997a; Biesecker, 2001], especially in those scenarios where patients may not have sufficient information [Bower et al., 2002], may perceive or retain information in a way not intended by the provider, and may wish to share the provider's experiences and expertise [Royal Commission, 1993; Michie et al., 1997b, 1997c, 1998; Bernhardt et al., 2000; Bower et al., 2002]. The Canadian College for Medical Geneticists [CCMG, 2004b] professional accreditation standard for a clinical geneticist acknowledges these potential scenarios, stating that when providing genetic counseling:

> *The physician should be able to be sympathetic and empathetic, remain objective and impartial, delineate his/her own ethical standards and appreciate those of the patient, appreciate the general mores of the culture of the patient, be non-directive in most instances, but be prepared to advise in certain situations and provide psychological support, either personally or through referral.*

With regard to outcomes, nondirectiveness may be found in conflict with those health care professionals who do not feel that medical decision making about outcomes should be the sole prerogative of the patient [Royal Commission, 1993; Geller et al., 1996; Quill and Brody, 1996; Andrews, 1997; see also section by Mykitiuk et al. in Chapter 8, Prenatal and Neonatal Screening]. Some have argued that nondirectiveness may be perceived as representing a neglect of the necessary role of accountability or responsibility in autonomous decision making [Yarborough et al., 1989; Pauker and Kassirer, 1987; Pauker and Pauker, 1987]. Although patient autonomy in decision making is acknowledged, the health care professional may have a significantly different perception from the patient about the impact of genetic disabilities [Royal Commission, 1993; Ormond et al., 2003] and may wish to make recommendations on the basis of those perceptions and experiences [Williams et al., 2002a; also see Geller et al., 1993; an American court has stated that where a patient selects a treatment from a group of reasonable alternatives and the physician regards the choice as inappropriate, the physician may withdraw from further care without liability for failure to provide treatment provided that the physician has reasonable assurances that treatment and care will continue: *McLaughlin v. Hellbusch*, 1999; see also section by Mykitiuk et al. in Chapter 8, Prenatal and Neonatal Screening]. For example, if a patient has been advised of a significant risk that her child will be born with Down syndrome, what course of action provides the most benefit and prevents harm? The health care professional may argue that the option with the greatest benefit will have the least burden [Pauker and Kassirer, 1987] such as the monetary cost of care and the avoidance of the birth of a genetically afflicted child [Pauker and Pauker, 1987]. If the patient, however, due to her attitudes toward Down syndrome decides to proceed with the pregnancy, the health care professional may perceive this to be the selection of the option with the lowest benefit. The patient's choice, and underlying values and beliefs, therefore, may be deemed by the health care professional to be nondeliberative and irrational [Pitz, 1987]. Studies indicate, however, that the health care professionals may be equally susceptible to such "inconsistencies" [Pitz, 1987; Michie, 1997d; Michie and Marteau, 1999b]. One health care professional, on the basis of experience, may be of the opinion that a child afflicted with a genetic disorder can enjoy a full and satisfying life; another health care professional may believe that the financial and emotional costs of the same disorder justifies a recommendation of abortion [Abramsky et al., 2001].

Health care professionals who have not received specialized training in genetics, when confronted with situations where any course of action may result in harm to the patient (e.g., the decision whether to terminate a pregnancy or to give birth to a child with a severe genetic condition) may be uncertain as to the nature and extent of their roles and responsibilities [Veach, 2004]. How then, within the context of genetic testing, is medicine's traditional ethical maxim—*primum non nocere*, above all do no harm—to be defined and from whose perspective? For example, in prenatal diagnosis, no desirable option may exist, and any course of action, whether the termination of a pregnancy or the birth of an afflicted child, can result in human anguish and tragedy [Tymstra, 1991; Grant, 2000].

Martha Nussbaum, in *The Fragility of Goodness: Luck and Ethics in Greek Tragedy and Philosophy*, discusses scenarios where any course of action can result in harm (Agamemnon was told by the gods that if his daughter was not offered for sacrifice, the expedition to Troy would be becalmed, and everyone, including his daughter, would die):

> *Agamemnon is allowed to choose; that is to say, he knows what he is doing; he is neither ignorant of the situation nor physically compelled; nothing forces him to choose one course of action over another. But he is under necessity in that his alternatives include no desirable options. There appears to be no incompatibility between choice and necessity here—unless one takes the ascription of choice to imply that the agent is free to do anything at all. On the contrary, the situation seems to describe precisely a kind of interaction between external constraint and personal choice that is found to one degree or another in any ordinary situation. For a choice is always a choice among possible alternatives; and it is a rare agent for whom everything is possible. The special agony of this situation is that none of the possibilities is even harmless [Nussbaum, 1986].*

The issue of choice with regard to outcomes in such clinical scenarios is not necessarily a matter of harm or benefit or what the physician may perceive to be "rational" or "irrational" decision making [Emery, 1984; Royal Commission, 1993; Andrews, 1997]. The issue is one of autonomy.

Rather than perceiving the physician–patient relationship as a paternalistic one that reflects an inherent disparity in knowledge and expertise [Katz, 1984], within the context of genetic counseling, the role of the health care professional is to weigh and balance personal observation, human and medical genetics knowledge, skills, and experience against available options and their respective risks and benefits. The patient's role is to describe and evaluate symptoms, detail a family medical history, and discuss expectations, objectives, concerns, beliefs, and values. The result of these respective contributions is an informed choice by the patient [Hartlaub et al., 1993; Sharpe; 1994a, 1997; Jonsen et al., 1998; but see Quill and Brody, 1996; see also section by Mykitiuk et al. in Chapter 8, Prenatal and Neonatal Screening].

Nondirectiveness represents the acknowledgment of the complex and troubling challenges posed by the processes of genetic testing and counseling. Of practical importance is the fact that many geneticists and counselors subscribe to the associated values and objectives. As will be discussed, this lends both insight and guidance to the manner in which professional guidelines, professional accreditation practice norms, and specialty program content for genetic testing services are likely to be interpreted by geneticists, counselors, health care professionals, and a court of law.

Conflict in decision making is resolved through recognition of the patient's right of autonomy in decision making, to choose the course of action that seems appropriate to the patient in view of his or her values, objectives, and ethical and religious standards [Ad Hoc Committee on Genetic Counsling, 1975; Kessler 1979; Emery, 1984; Royal Commission, 1993; Andrews, 1997]. It is the task of the health care professional and the patient to willingly remain exposed to such risks, for these are the risks inherent in any autonomous decision [Nussbaum, 1986].

WEB RESOURCES

All sites accessed December 15, 2004.

Continuing Medical Education: Genetics

American Medical Association: Genetics in Clinical Practice: A Team Approach, an Interactive Medical Laboratory Virtual Clinic. This CD-ROM is now available at no cost to physicians and other health care professionals that have an interest in clinical genetics: http://www.ama-assn.org/ama/pub/article/1615-8311.html

National Coalition for Health Professional Education in Genetics: http:// www.nchpeg.org

National Center for Biotechnology Information: Genes and disease: a collection of articles that discuss genes and the diseases that they cause: http://www.ncbi.nlm.nih.gov/books/bv.fcgi?rid=gnd

National Institutes of Health: National Library of Medicine: http://www.nlm.nih.gov/pubs/cbm/health_risk_communication.html

National Information Resources on Human Genetics and Ethics: http://www.georgetown.edu/research/nrcbl/nirehg/

National Society of Genetic Counselors (1992) National Society of Genetic Counselors code of ethics. *J Genet Couns.* 1(1):41–43: http://www.nsgc.org/about/code_of_ethics.asp

Genetic Counseling Organizations

American Board of Genetic Counseling: http://genetics.faseb.org/genetics/abgc_diplomates.html

National Society of Genetic Counselors: http://www.nsgc.org/resourcelink.asp

How to Find a Genetic Counselor (U.S.): http://www.genetichealth.com/Resources_How_to_Find_a_Genetic_Counselor.shtml

Canadian Association of Genetic Counsellors: http://www.cagc-accg.ca/ and http://www.nsgc.org/consumer/faq_consumers.asp

National Cancer Institute Cancer Genetics Services Directory: http://www.cancer.gov/search/genetics_services/

American Board of Genetic Counselors: http://genetics.faseb.org/genetics/abgc_diplomates.html

American Board of Genetic Counseling: http://www.abgc.net/

Genetic Counseling Resources, Human Genome Project: http://www.ornl.gov/sci/techresources/Human_Genome/medicine/genecounseling.shtml

University of Kansas Medical Center: http://www.kumc.edu/gec/prof/gc.html

Glossary of Terms

University of Kansas Medical Center: http://www.kumc.edu/gec/glossnew.html

National Human Genome Research Institute: http://www.ornl.gov/sci/techresources/Human_Genome/glossary/

National Human Genome Research Institute: Glossary in Spanish: http://www.genome.gov/sglossary.cfm

"Talking" Glossary of Genetic Terms: http://www.genome.gov/10002096

Laboratories

Association of Public Health Laboratories: Newborn Screening and Genetics: http://www.aphl.org/Newborn_Screening_Genetics/index.cfm

GeneTests: A publicly funded medical genetics information resource developed for physicians, other health care providers, and researchers, available at no cost to all interested persons: http://www.genetests.org/

Canadian College of Medical Geneticists: Cytogenetics: http://www.hrsrh.on.ca/genetics/CanCyt/

Directory of Medical Cytogenetic Laboratories in Canada: http://www.hrsrh.on.ca/genetics/canlabs.htm

Professional Organizations

American Board of Genetic Counselors: http://genetics.faseb.org/genetics/abgc_diplomates.html

American Board of Genetic Counseling: http://www.abgc.net/

American Board of Medical Genetics: http://www.abmg.org/

American College of Medical Genetics: http://www.acmg.net/

American Society of Human Genetics: http://genetics.faseb.org/genetics/ashg/ashgmenu.htm

Association of Genetic Technologists: http://www.agt-info.org/

Australasia: The Human Genetics Societies of Australasia: http://www.hgsa.com.au/

Canadian College of Medical Genetics: http://ccmg.medical.org/

Council of Medical Genetics Organizations: http://genetics.faseb.org/genetics/ashg/comgo.htm

European Society of Human Genetics: http://www.eshg.org/

Genetics Society of America: http://www.genetics-gsa.org/

International Federation of Human Genetics Societies: http://www.ifhgs.org/

Latin American Network of Human Genetics Societies: http://www.relagh.ufrgs.br/

International Society for Nurses in Genetics: http://www.isong.org/

National Society of Genetic Counselors: http://www.nsgc.org/

National Coalition for Health Professional Education in Genetics: http://www.nchpeg.org/

National Society of Genetic Counselors: http://www.nsgc.org/resourcelink.asp

International Federation of Human Genetics Societies: http://genetics.faseb.org/genetics/ifhgs/

Coalition of State Genetics Coordinators (CSGC): http://www.stategenetics coordinators.org/

GENETIC COUNSELING APPROACH TO GENETIC TESTING

Nancy P. Callanan, Genetic Counseling Program, Graduate School, University of North Carolina at Greensboro, Greensboro, North Carolina
Bonnie S. LeRoy, Department of Genetics, Cell Biology and Development and the Institute of Human Genetics, University of Minnesota.

INTRODUCTION TO GENETIC COUNSELING

Historically, genetic counseling has been defined as a communication process that deals with the human problems associated with the occurrence or risk of occurrence of a genetic disorder in a family [American Society of Human Genetics, 1975]. This often quoted definition emphasizes the interactive nature of genetic counseling, as a process performed by appropriately trained individuals, which encompasses both education and counseling, and is performed in the context of respect for individual autonomy in patient decision making. Respect for individual patient autonomy continues to be a central guiding ethical principle in contemporary genetic counseling and is a basic tenet for the professional code of ethics adopted by the National Society of Genetic Counselors [NSGC, 1992]. The tradition of nondirectiveness as a means of respecting patient autonomy has formed the framework for the values of the profession and the practice of genetic counseling. The genetic counselor aims to appreciate the values of the patient and incorporate those values into the counseling and facilitated decision-making components that are vital to effective services.

In this chapter, we will present a brief overview of the history of the profession and discuss the evolving tradition of nondirectiveness. Some of the primary factors that have contributed to the shift in responsibility to primary health care providers for identifying patients at risk, providing genetics education, genetic testing, and referral to genetic specialists will also be explored. Additionally, the major categories of genetic testing and issues raised by each will be discussed.

HISTORICAL PERSPECTIVE AND THE ROLE OF NONDIRECTIVENESS IN GENETIC COUNSELING

Sorenson [1993] summarized the history of "genetic counseling" by describing three phases, with phase A representing the late 1800s to the late 1930s, phase B, the late 1930s to the late 1960s, and phase C the late 1960s and beyond. Phase A constituted the major period of eugenics in this country during which applied human genetics was practiced as part of a social movement. The belief that heredity contributed not only to medical but also to social conditions, such as poverty, crime, and mental illness, and enthusiasm over the possibility that genetics might be used to improve the human condition led, in the United States, to the development of institutions where data on human traits was collected, to state laws that mandated involuntary sterilization of individuals determined to be "mentally defective," and federal laws to limit immigration by various "inferior" ethnic groups [Walker, 1998]. In phase B applied human genetics moved out of the organizational and institutional context of eugenics and into academia, practiced initially by Ph.D. academic-based geneticists, followed by the emergence of academic physician-geneticists, and ultimately to the establishment of clinical genetics departments of medical schools [Sorenson, 1993]. In 1947, the term *genetic counseling* was introduced by Sheldon Reed, who proposed that in order to be effective, genetic counselors must have the knowledge and ability to provide genetic information in the context of respect for the sensitivities, attitudes, and reactions of clients [Reed, 1955]. Recognizing the complexities of reproductive genetic decision making, and anxious to avoid an association with early eugenic programs, many clinical geneticists of the time also advocated a nondirective type of counseling [Resta, 1997]. Phase C was characterized by rapid advances in the field of human genetics, leading to further development of clinical genetics as a medical specialty and the emergence of masters-level genetic counselors [Sorenson, 1993; Walker, 1998].

The value of informed reproductive decision making and nondirective counseling assumed major roles in the development of genetic counseling in the United States [Sorenson, 1993]. As the profession evolved, a nondirective counseling approach based on Carl Roger's client-centered counseling model emerged as a good fit for genetic counseling [Fine, 1993]. Although the principle of respect for client autonomy remains a core value in the practice of genetic counseling, several have challenged the use of the term *nondirective* as it has been applied to the practice of genetic counseling. Narrow definitions of the term are problematic as they imply value neutrality that is both ethically insufficient and not achievable [Weil, 2003].

In describing nondirectiveness as "procedures aimed at promoting the autonomy and self-directedness of the client" Kessler [1997a] not only provided a more broad definition of nondirectiveness but also described the quality counseling skills that are needed in order to achieve this goal within the context of genetic counseling. A survey of genetic counselors [Bartels et al., 1997] was undertaken to assess how they defined nondirectiveness, its value to their practice, and whether there were circumstances in which they employed a directive approach. The results suggest that nondirectiveness is defined in a consistent way and is highly valued by genetic counselors; however, important distinctions between directing the *process* and the *outcome* of genetic counseling were apparent as respondents cited as examples of when a more directive approach was utilized. Clients rely on genetic counselors to share their expertise and experience and to provide guidance about the process. A totally value-neutral stance is neither achievable nor appropriate. Even within the context of nondirectiveness, it is appropriate for genetic counselors to fully utilize their knowledge of medical genetics, psychological, and ethical issues in the process of supporting client autonomy by promoting active, self-confident decision. Further work in describing an empirically established model that articulates the process of genetic counseling is still needed [McCarthy et al., 2002].

As primary health care providers are forced to address their patient's questions about genetics, it is important to remain cognizant of the values central to the practice of genetic counseling. These values have guided the provision of service for more than 30 years and remain constant because they respect the ability of the patient to ultimately make best decision. Primary care providers must work in collaboration with genetic specialists to develop models for providing genetic risk assessment, patient education, and testing for their patients that incorporate the ethical principles that continue to serve as the underpinning of traditional genetic counseling.

PRACTICE OF GENETIC COUNSELING TODAY

The practice of genetic counseling has changed dramatically since its inception. The only tools available to genetic counselors years ago were the statistics drawn from observations of affected families that indicated the likelihood that someone might inherit or pass on a gene that could cause a disease. Direct gene testing did not exist.

For instance, in reproductive genetic counseling, the only options available to most individuals at risk for having a child with a genetic birth defect involved decisions about whether or not to have children in light of the known risks and, after conception, whether to continue or terminate a pregnancy known to be or suspected to be affected. Today, couples may consider sophisticated prenatal testing and reproductive options such as egg donation, preimplantation genetic diagnosis, surrogate motherhood, and more. Advances in genetic research now allow for over 1000 tests available for diagnosis, risk assessment, prenatal, and predictive testing [GeneTests, 2004].

The popular media present information about new genetic tests and technologies to the public almost daily. Individuals and families are encouraged to seek genetic counseling and testing, and primary care providers are encouraged to offer genetic tests to their patients. Some laboratories are currently participating in direct-to-consumer marketing. These efforts are expected to result in an increased demand from patients for tests such as susceptibility testing for hereditary breast and ovarian cancer, the genes *BRCA1* and *2*. There are concerns that this type of marketing will adversely affect the care a patient receives in that patients may not be making fully informed decisions [Hull and Prasad, 2001a, 2001b]. The demand for genetic counseling services likely will increase in response to advances in genetics research. These advances are expected to lead to genetic population screening to better predict the risk for developing many common conditions such as diabetes, heart disease, and cancer [Collins and McKusick, 2001]. These are the disorders that will affect almost every family. These are also the disorders that result in the majority of health care costs today.

GENETIC COUNSELOR GOALS

Although there is a dearth of research investigating the model or models of practice for genetic counseling, the literature addressing this issue is relatively consistent with regards to genetic counselor goals. Studies suggest that giving information is not the only important factor in meeting the patient's needs but that patients expect, welcome, and need supportive counseling at a genetic counseling visit. A study examining genetic counseling sessions [Matloff, 1994] found that the content of prenatal counseling sessions varied significantly among their sample. However, the majority of counselors listed patient education and informed decision making as their primary goals in the session. Bernhardt et al. [2000] interviewed 16 genetic counselors and 19 patients about genetic counseling goals. They identified the following as the major goals:

1. Increase in patient knowledge and understanding
2. Accurate risk assessment
3. Facilitated decision making
4. Patient support and anticipatory guidance
5. Alleviate guilt
6. Empower patients to feel in control
7. Make the necessary referrals

The patients in this study perceived the benefits of genetic counseling as including:

1. Acquisition of information and increased knowledge
2. Immediate psychosocial support
3. Long-term psychosocial support
4. Anticipatory guidance
5. Facilitation of family communication
6. Assistance with decision making

Both counselors and patients considered the nature of the interpersonal interaction of utmost importance. A survey [Lobb et al., 2001] of 29 Australian genetic counselors and clinical geneticists, working in cancer genetics, identified 5 counselor goals:

1. Assessing patient needs and concerns
2. Providing technical genetic information
3. Conducting an individual risk assessment in the context of supportive interaction
4. Discussing the pros and cons of genetic testing
5. Developing a follow-up plan

These findings may have limited applicability to U.S. genetic counseling, and they are further limited to one type of genetic counseling—cancer genetics, but working within the context of a supportive interaction was found to be a basic underlying goal. In a longitudinal study of 43 families who received genetic counseling, Skirton interviewed each family prior to counseling and twice after counseling, at 2 to 4 weeks and 6 months, respectively [Skirton, 2001]. The outcome of genetic counseling was influenced by four factors: the patients' need for certainty, the quality of their relationships with the genetic counselor, the integration of lay and scientific explanations, and the patient's psychological adaptation. The results of this study support the notion that genetic counselor behaviors play a major role in producing effective genetic counseling outcomes. Although all of these studies show variation in the way genetic counseling is practiced, the basic underlying goals are constant:

- An informed patient empowered to make autonomous decisions
- Respect for the values and decisions made by the patient
- Attending to psychosocial needs of the patient
- Providing supportive counseling and appropriate follow-up care

IMPORTANT CULTURAL CONSIDERATIONS IN GENETIC COUNSELING

Cultural differences often challenge a practitioner's ability to provide effective medical care, but in the genetic counseling setting, these differences can become a true obstacle. In some cases, the history of genetics with respect to instances of blatant discrimination remains in the collective memories of people whose ancestors were victims and impedes building the trust that is necessary to provide good medical care. Differences in language, ways of communicating, worldviews, and religious beliefs can hinder the process of communicating accurate information and examining the associated psychosocial issues that are fundamental to effective genetic counseling. In addition, there are often major differences in perceptions about the cause of a disorder and the consequent burden that that disorder places on the family. The genetic counselor cannot rely on her or his worldview for guidance. For example, a survey [Greeson et al., 2001] about the perception of disability in the Somali immigrant population in a city in the upper Midwest reported that participants were constant in their belief that disabilities and birth defects are considered a gift from God not to be questioned in their culture. Families who had a child with a birth defect were special because they were "chosen by God," and God knew that they could care for this specific child. It was obvious that traditional genetic counseling would not be useful in this population. Practitioners would need to change their focus in order to be effective. Participants were very interested in learning about anything that would improve the health of a child (or any individual) but not in anything that would predict or prevent what God had planned [Greeson et al., 2001].

In many cultures, the traditional nondirective approach and concern for promoting patient autonomy may directly conflict with the expectation that health care providers will tell the patient what to do in order to best deal with the disease. Patients often have complete confidence that their health care provider will make the best decision for them, and, by not doing so, the patient may loose confidence in the abilities of health care provider. Lewis [2002] expresses concern about disparities in the genetic services received by members of culturally diverse groups, in particular, ethnic and racial minority groups. He refers to two genetic counseling models that have been proposed by Kessler [1997b]—a teaching model and a counseling model. Lewis states that although these models fail to explicitly address issues of culture, the counseling model may best support multicultural approaches to genetic counseling. Ota Wang [1993] argues that genetic counselors must develop a foundation of cultural knowledge, awareness of self and others, and *specific counseling skills*. These studies emphasize the importance of cultural influences in providing effective genetic counseling.

FACTORS THAT PROMOTE EFFECTIVE GENETIC COUNSELING

Although there are few studies examining what constitutes effective genetic counseling, attending to certain issues often contributes to a better outcome. Careful attention to patient expectations, perceptions, and needs is fundamental to effective genetic counseling. Many studies have examined patient expectations of genetic counseling, patient risk perception, motivations for wanting genetic counseling and genetic testing, and patient needs. However, these studies find that patients often do not know what to expect from genetic counseling or how to directly communicate their needs [Hallowell et al., 1997]. Most patients could benefit from precounseling education.

One strategy suggested in the literature for dealing with this issue is to send patients written information about genetic counseling prior to their appointment. If patients and families know what to expect, they will come to the session more prepared to communicate their needs.

For many patients, the perceived burden of the disease is a major factor in the decision to seek genetic counseling and genetic testing. Many people have a difficult time relating risk to their own situation [Ravine et al., 1991]. Patients who are unaware of their risk or do not clearly comprehend their risk in relation to their own health usually do not see a benefit in genetic counseling services. When patients gain a better understanding of their risk and appreciate the options available to *modify* that risk, they are more likely to benefit from genetics services.

It is critical that the genetic counselor provide accurate and current information, but this is only one facet of effective genetic counseling. Equally essential for providing effective genetic counseling service is the counselor's ability to ascertain the patient's motivation for wanting testing. In some cases, the genetic test will not answer the patient's questions or provide the patient with a level of certainty about the future. In other cases, patients may come in for testing in response to the urging of a third party such as a family member. Genetic counselors also need to explore the patient's perception of risk and burden of the disease, assist the patient in identifying the specific issues that are important to their own health and family situation, and possess the skills needed to provide supportive counseling.

GENETICS IN MAINSTREAM MEDICAL CARE: THE EMERGING MILIEU

The steady surfacing of genetics in mainstream medicine has been recognized for quite some time. The National Coalition for Health Professional Education in Genetics (NCHPEG) was established in 1996 by the American Medical Association, the American Nurses Association, and the National Human Genome Research Institute. The major focus of NCHPEG is "a national effort to promote health professional education and access to information about advances in human genetics." Over 100 health professional organizations, consumer and volunteer groups, government agencies, private industry, managed care organizations, and genetics professional societies are members of NCHPEG (http://www.nchpeg.org/). In 2001, NCHPEG published the document, *Core Competencies in Genetics Essential for All Health-Care Professionals.* This publication delineates the essential information needed by health care professionals as they encounter the integration of genetics into all aspects of medical practice.

Genetic counselors also are rapidly becoming practitioners in mainstream medicine. They now provide services in multiple settings at infertility clinics, cancer centers, psychiatric clinics, neurology and cardiology clinics, and many more. In such settings, patients requesting genetic counseling may be considering presymptomatic diagnosis for a later-onset condition such as Huntington disease or testing for predisposition to a familial cancer or more information about the role that inheritance plays in heart disease. The near future of genetic counseling might involve gene testing to see if an existing condition, such as clinical depression, some forms of heart disease, or Alzheimer disease, has a genetic basis in a specific individual or family. Information from this type of testing likely will generate a more accurate risk assessment specific to an individual. The arena of pharmacogenetics will emerge as an increasing role in identifying genetic factors that affect response to drugs used for common ailments [Begley, 1999]. It will be possible to identify some individuals who are likely to respond poorly to a medication. These individuals may seek genetic counseling in some form to better understand the implications of these test results for themselves and other family members.

One of the major differences of genetic medicine when compared to traditional medical care is the impact that genetic information has on the whole family. Genetics means families and often the practitioner must work within the family system in order to be most helpful. When one individual is diagnosed with a genetic disease or found to carry a gene mutation, this information has direct implications for parents, brothers, sisters, children, and other family members. The strain on a marriage and the effects on family dynamics can be overwhelming. In a study looking at the effects of predictive testing for Huntington disease on marital relationships, researchers found that *partners* of at-risk individuals demonstrated significantly higher levels of depression than those individuals at risk [Quaid and Wesson, 1995]. A genetic condition has a wide range of affects on multiple family members and the resulting psychological repercussions often disrupt relationships. Exploring family issues with the patient prior to testing is a foremost element of successful genetic counseling.

Genetic information learned through a risk assessment and genetic testing can be devastating to an individual. The consequent emotional reactions often interfere with a complete understanding of the technical information needed by the patient to make decisions and cope with the disease. Common reactions include denial, anger, fear, despair, guilt and shame, sadness, and grief [Djurdjinovic, 1998]. Although these reactions occur when dealing with patients in many areas of medicine, guilt and shame play a particularly prominent role in genetic counseling. Patients may feel personally responsible for passing a gene on to a child or to be ashamed of the history of genetic disease in their family, believing that their family is defective. The diagnosis may mean a surprising and unwanted change in the patient's life and affect the patient's perception of self [Delaporte, 1996]. The reactions are often emotional and not cognitivly based in fact. Once again these studies emphasize the importance of attending to the psychosocial needs of patients and families in the genetic counseling setting especially when the patient is considering a genetic test.

CATEGORIES OF GENETIC TESTING: ISSUES TO CONSIDER

The issues that surround genetic testing relate to the indication for doing the test. In considering these issues, genetic counselors distinguish genetic *testing* from genetic *screening*. The term *genetic testing* describes evaluation of individuals who have an increased risk for a specific genetic condition based on family history or clinical symptoms. This can include *diagnostic* testing, or testing that is performed to confirm a clinical diagnosis of a genetic condition or syndrome, for example, performing a cytogenetic analysis to confirm the diagnosis of Down syndrome in an infant with features of this condition. As such, the use of cytogenetic or molecular genetic tests within the context of a diagnostic evaluation is not unlike other forms of laboratory testing.

Issues to consider in the use of diagnostic genetic testing include the following:

- The use of genetic testing to establish or confirm a diagnosis is consistent with providing the best medical care for patients. As with similar medical recommendations, a nondirective approach is not required.

- Diagnostic genetic testing of a patient often results in information concerning potential risks, either for the same diagnosis or for reproductive risks for other relatives. Therefore, it is important to discuss these potential implications prior to testing and to encourage patients to share relevant information with at-risk relatives. Strategies for providing information to patients and their relatives that is relevant and comprehensible must be employed.

- In some situations a health care provider may face situations in which a patient is unwilling to inform relatives about their genetic risks. The responsibility to protect the confidentiality of a patient's medical information would prevent a health care provider from disclosing the information directly to at-risk relatives. However, in rare situations in which the risk is great, and effective medical interventions are available, it has been argued that limited disclosure may be acceptable [ASHG Social Issues Subcommittee on Familial Disclosure, 1998; Offit, 2004; also see Chapter 15, Confidentiality and Recall].

Genetic testing also is used to determine the genetic status of an individual who is at risk for carrying a gene mutation for a specific disease or genetic condition based on their family history, for example, a testing for cystic fibrosis (CF) mutations in a woman whose brother is affected with this condition. In this situation, *carrier* testing is accompanied by a genetic risk assessment based on pedigree analysis and should be preceded by a discussion of the risk, the availability of testing, the potential benefits and limitations of testing, and the implications of testing with regards to reproductive planning.

Issues to consider in the use of carrier testing of individuals with a positive family history for a specific disorder include:

- It is important to obtain accurate information about the family history as well as confirmation of the diagnosis in the affected relative prior to testing [Uhlmann, 1998].

- Accurate testing may require information about the specific genetic mutation identified in the affected relative. Therefore, cooperation of relatives is needed in order to provide accurate testing for your patient. If such information is not available, then the limitations of testing must be communicated to the patient.

- Carrier testing should be preceded by a discussion of the genetic risk, as well as the potential implications of the test result, including reproductive risks and options.
- Counseling strategies should be employed to assist patients in evaluating the pros and cons of testing in their situation in light of their own values, beliefs, and goals [McCarthy Veach et al., 2003].
- Carrier testing for familial genetic conditions in minors is generally discouraged [ASHG/ACMG, 1995; Davis, 1998].

Prenatal diagnosis is genetic testing performed during a pregnancy for individuals who are at risk for a specific condition due to family history, known carrier status, or prior reproductive history, for example, using amniocentesis to obtain a sample for biochemical or molecular testing for Hurler syndrome in a woman whose first child was affected by this disorder.

Issues to consider in the use of prenatal genetic diagnosis include:

- A careful assessment of the reproductive risks, availability of testing, evaluation of the accuracy, sensitivity, and potential limitations of testing should also precede prenatal diagnosis [Walker, 1996].
- Informed consent also should be obtained prior to prenatal diagnostic testing. This should include a full discussion of any risks to the procedures as well as the limitations of testing.
- Genetic counseling strategies that support patient autonomy and facilitate patient decision making are crucial prior to prenatal diagnosis [McCarthy Veach et al., 2003].

The term *genetic screening* is used to describe testing for genetic disorders in specific groups of people independent of a family history or clinical features. Criteria for genetic screening can include factors such as age, gender, or ethnic background [McCabe and McCabe, 2002]. Examples of genetic screening include newborn screening for the diagnosis for conditions such as phenylketonuria (PKU), prenatal screening by evaluation of maternal serum markers for open neural tube defects, Down syndrome and other chromosome disorders, and carrier screening for conditions such as cystic fibrosis or Tay-Sachs disease in individuals who belong to populations known to be at increased risk for these disorders. Offering prenatal testing for Down syndrome and other chromosome disorders to women over the age of 35 is also a form of genetic screening. It is anticipated that genetic screening for common disorders, such as Type II diabetes will be developed within the predictable future [Khoury et al., 2003].

Issues to consider in the use of genetic screening tests include:

- It is important to educate patients about the difference between screening tests and diagnostic tests as well as the possibility of false-negative and false-positive results.

- Participation in genetic screening programs should be voluntary and accompanied by education and informed consent prior to screening. Individuals considering screening tests should be informed about the potential risks of screening, which include the possibility for stigmatization and for genetic discrimination in life or health insurance and/or employment.
- Screening programs should make genetic counseling available to individuals who are diagnosed with or found to be carriers of genetic disorders or susceptibilities.
- As with diagnostic genetic testing, the results of genetic screening tests frequently have implications for other relatives.

The term *predictive genetic testing* is used to describe genetic testing in healthy individuals for genetic disorders with late onset. This includes *presymptomatic testing*, for example, testing for Huntington disease in individuals at risk for this condition prior to the onset of symptoms. It also includes *susceptibility testing*, or testing for genetic changes that predispose an individual for certain health conditions. Testing for *BRCA1* or *BRCA2* mutations in a woman with a family history of breast cancer is an example of susceptibility testing. The distinction between presymptomatic and susceptibility testing is important because the former implies certainty of developing the specific condition at some point in the future, while the latter identifies individuals who have an increased risk for developing a specific condition.

Issues to consider in the use of predictive genetic testing include:

- Careful psychosocial assessment and exploration of the reasons for seeking testing should precede any type of predictive testing. Alternatives to testing should be discussed, and counseling strategies to help patients evaluate the pros and cons of testing in the context of their own values, needs, and beliefs should be employed [Armstrong et al., 2002].
- Follow-up with regard to medical management issues and assessment of the patient's emotional response to testing is important in any type of predictive testing, regardless of the results.
- Testing of minors for adult-onset conditions is generally discouraged, with the exception of situations in which there is a clear medical benefit to testing [ASHG/ACMG, 1995]. Counseling strategies to help parents explore their rationale for seeking predictive testing in their children is important.
- As with any type of genetic testing, confidentiality of test results is required. However, the results of testing in one patient often have implications for other relatives. Patients should be encouraged to share information about potential risks with their relatives, and health care professionals should be prepared to facilitate this process.
- It is important to note the difference between predictive testing and susceptibility testing. Although there is a lot of overlap of the issues to consider for both types of testing, it is especially important in susceptibility testing to ensure that patients are aware of the limitations of the test results.

As illustrated in the examples provide above, there are important issues that surround all types of genetic testing and screening. These relate directly to the core values and goals of genetic counseling: promoting patient autonomy by providing information, sharing expertise and support, and employing counseling strategies to facilitate decisions that reflect the patient's own values and beliefs. While there is considerable overlap in issues, by focusing on the specific context in which testing is offered or performed, the practitioner can be better prepared to anticipate issues and approach patients appropriately.

SUMMARY

Genetic counseling is an exceptionally complicated process that is fundamentally different from other areas of medical practice in that it is almost always necessary to consider not only the needs of the patient but the impact on the patient's family or prospective family. Moreover, deep-seated personal and cultural values influence the decisions patients make about genetic testing. Many genetic tests generate information that changes self-perception, alters family relationships, and compels patients to adjust their life choices. Given the potential for serious consequences resulting from genetic information and genetic testing, it is essential that the provider possess the skills needed to address all of these issues with patients and help patients find the decision that is right in their situation.

Genetic counseling has supported a nondirective counseling approach as a means of promoting patient autonomous decision making. Nondirectiveness has been a way of thinking about an approach that centers the counseling on the values of the patient and not those of the counselor. The profession of genetic counseling has evolved over the past 30 years in response to the extensive developments in our understanding of the role of genetics in disease and the increase in available technologies. With this evolution, challenges to the actual significance of the term *nondirectiveness* have emerged. Genetic counselors now have a history of practice such that practitioners are able to share their wealth of experience and recognize that appropriate professional advice and guidance is not synonymous with coercion. Nonetheless, the core goals of striving to understand the world of the patient as he or she sees and experiences it and of focusing decisions around the values of the patient remains central to genetic counseling. Over the years, Dr. Seymour Kessler has constantly reminded us of what is most important in the practice of genetic counseling through his multitude of publications and invited speaking engagements. Over and over he tells us that the critical question is not, "Am I being directive with this patient?" but rather, "How can I best help this patient." Resta and Kessler [2004] summarize the basic counseling skills required to provide good genetic counseling:

- The ability to understand the psychological needs of others
- The ability to understand the psychological meaning of client's behaviors
- The ability to communicate that understanding in ways that leave clients emotionally enriched, psychologically stronger, and more competent to deal with their own lives

We still have much to learn about the ways in which advances in genetics is impacting traditional medical practice. Moreover, it continues to be important to study the means by which patients and their families cope with genetic information and how all health care practitioners can best help their patients. Nonetheless, it is obvious that the genetic counselor in practice today needs to not only know how to obtain, comprehend, and communicate the most current technical information about the disease affecting a patient and/or their family but, in addition, possess the ability to empathize with patient's situation, facilitate informed decisions, and help patients and families cope with their lives. As other health care professions enter the arena of providing genetic information to patients, they need to hold themselves to these same standards.

Chapter 2

Communication

OVERVIEW

Communication is the essence of genetic testing and counseling [Royal Commission, 1993; Committee on Assessing Genetic Risk, 1994; Sharpe, 1994a; Roter, 2004]. Ineffective communication:

- Reduces the accuracy of a diagnosis [Woolf et al., 2004; this study reports 80 percent of error chains were initiated by informational or personal miscommunication].
- Affects the estimation of risks of occurrence or recurrence.
- Impairs a patient's ability to understand and to retain information.
- Is a primary factor in medical malpractice actions.

Although many physician–patient encounters involve communication, studies have questioned the effectiveness of these communicative relationships. The health care professional (the "counselor" in this context) needs to allow the patient time to provide information; good communication depends on the transfer of information in both directions.

- A 1984 study of a university medical clinic reported that only 23 percent of patients were allowed to complete an opening statement of concern, that patients were interrupted on an average of 18 seconds after they began to speak, and that only one of the 52 interrupted statements was completed during the interview [Beckman and Frankel, 1984].
- In 1999, a follow-up study reported that although physicians solicited patient concerns in 75.4 percent of the interviews reviewed, only 28 percent of the patients' initial statements of concerned were completed. Physicians redirected these statements after an average of 23.1 seconds, even though those patients who were allowed to complete their statements only took 6 seconds longer on

Genetic Testing: Care, Consent, and Liability, by Neil F. Sharpe and Ronald F. Carter
© 2006 John Wiley & Sons, Inc.

average. The authors concluded that the interruption of patients' statements of concerns could impair potentially important data [Marvel et al., 1999; Suchman et al., 1997; Waitzkin, 1985].

Counselors need to recognize when difficulties in communication arise with patients and the potential for tragic outcomes [Geller et al., 1997; Lerman et al., 1997; Miesfeldt et al., 2000; Carroll et al., 2000; Dillard et al., 2000; Marteau and Dormandy, 2001; Lobb et al., 2002]. Even the simplest of test results can provide room for misunderstanding. For example, a patient went to an ob/gyn clinic to obtain a pregnancy test [*Duplan v. Harper*, 1999; Bernhardt et al., 1998]:

- Because her job environment put her at increased risk of becoming infected with cytomegalovirus (CMV), the patient wished to be tested to determine whether she was immune to CMV. She and her husband agreed to terminate the pregnancy if she was not immune rather than take the risk of having a child with CMV-induced birth defects.

- The patient took the CMV test. The test results indicated that she was pregnant and had an ongoing primary CMV infection that posed a significant risk of severe birth defects.

- The physician instructed a nurse to notify the patient of the results of the CMV test. The nurse contacted the patient by phone and told her that the results were "positive."

- The patient was uncertain whether a positive result meant that she was immune to CMV or that she was infected. She called the nurse back and asked for clarification of the test results.

- The nurse incorrectly told the patient that a positive test result meant that she was immune to CMV.

- Based on this erroneous information, the patient decided not to terminate the pregnancy. She gave birth to a son, Zachary, who was born with CMV-induced birth defects including hearing loss, delay in development and loss of certain fine and gross motor skills, mental retardation, microcephaly, and nystagmus.

The patient was given the test result in an ambiguous, misleading manner; this error was repeated even when the patient made it clear that she did not understand the clinical significance of the results as reported. Patient understanding and recall of genetic test information can be affected by:

- The terminology, format, and context in which the information is presented

- The timing, manner, and method of communication, including the health care professional's tone of voice and facial expressions

- Emotional and psychological responses

- Personal experiences, values, and beliefs on the part of the health care professional as well as the patient [Abramsky et al., 2001]

Common errors in terminology, format, and content include the use of:

- Scientific jargon
- Imprecise words
- Attempts to mitigate the impact of bad news by using evasive or less severe terms
- Failure to recognize and respond to language and cultural barriers
- Too much or too little detail

The health care professional needs to be aware of these potential difficulties and to ensure that the appropriate information is provided in clear, precise terminology in a language, format, rate, and intellectual level that the patient can understand in her or his own terms. The counselor can assess the rate of progress by asking patients to summarize what they have been told.

The timing, manner, and method of communication can be a significant factor in successful communication. Patients will seek further clues about the clinical significance of a result from the counselor's body language, facial expression, and demeanor. When facing a situation that is not clear cut, for example, an inconclusive lab result, the patient will look for clues to good or bad outcomes that the counselor is not verbally stating. Even in the context of informed choice, patients often seek guidance from heath care professionals and ask questions with the intention of seeking some clue from the counselor about the "true" significance of the results and what the counselor would do if placed in the same situation.

The personal beliefs and values of the counselor can have a significant impact on the outcome of counseling. Consider, for example, a published account of two different approaches to counseling for a common cytogenetic abnormality in a prenatal test situation [Abramsky et al., 2001]:

One health professional considered 47,XXX to be "as devastating as Down syndrome" and said "that there was a possibility of mental retardation, intelligence down, stunted."

Another health professional who considered 47,XXX to be a very mild condition told parents that: "She would be a perfectly normal baby and she would go to a school, a normal school, and she would grow up normally and that she had an extra chromosome . . . [I] explained to them that the child would look just like every other child and that she was a normal child and that her intelligence might not be quite as great as you would expect for her parents but that she would still cope in a normal school."

There are obviously substantial differences in counseling that originate in personal opinion. Personal opinions can vary greatly in situations where laboratory results are inconclusive (e.g., the interpretation of a molecular sequence variant that is of unproven clinical significance) or the abnormality is well described but there may be substantial variation in clinical outcomes (e.g., Turner syndrome, Klinefelter syndrome). Counselors need to present an accurate, informed viewpoint that is based on a reasonable and unbiased assessment of published and personal experience.

In summary, effective communication of genetic test information represents a two-way dialog of exploration, explanation, and empathy [Kessler, 1979; Sharpe,

1994a; Tudor and Dieppe, 1996; Edwards et al., 2002]. An empathetic approach [Bellett and Maloney, 1991; Zinn, 1993; Suchman et al., 1997], with a focus on the concerns of the patient and patient's family, has been associated with:

- More effective communication [Suchman et al., 1993; Ambady et al., 2002]
- Improved patient comprehension and reduction of dissatisfaction and allegations of malpractice [Adamson et al., 1989; Levinson and Roter, 1995; Beckman et al., 1994; Levinson et al., 1997; Levinson, 1997; Bogardus et al., 1999; Blackston et al., 2002; Ambady et al., 2002].

As will be discussed in later chapters, medical genetics and genetic counseling studies recommend:

- Using language that is "easily understandable" [European Commission, 2004] with an avoidance of technical jargon
- Taking into account the timing, tone, manner, and place of all discussions, including emotional, psychological, linguistic, cultural, and socioeconomic differences
- Communications to other health care professionals need to be tailored to ensure appropriate understanding including the implications for care and follow-up [Sandhaus et al., 2001; Statham et al., 2003]

NEED TO KNOW

Health care professionals may presume that patients understand the risk information presented, however, this may not always be the case [Axworthy et al., 1996; Marteau et al., 1999a; Bogardus et al., 1999; Lloyd 2001], especially if health care professionals themselves are uncertain as to the nature and implications posed by test results [Sandhaus et al., 2001; McGovern et al., 2003b; Goos et al., 2004].

Accurate test interpretation and risk perception appear to be difficult for both providers and patients to achieve. For example:

- A survey of commercial genetic testing for familial adenomatous polyposis [Giardiello et al., 1997] reported that doctors misinterpreted test results in 31.6 percent of the cases and only 18.6 percent of patients received genetic counseling before testing.
- A survey of neonatal screening for cystic fibrosis (CF) reported that after the test results had been communicated, although 88.3 percent of parents understood that their child was a carrier for CF, 15.4 percent of parents were unsure whether being a carrier could cause illness, 12.4 percent of parents were unsure whether at least one parent was a carrier of the CF gene, and only 57.percent of parents knew there was a 1 in 4 chance that if their child reproduced with another carrier of the CF gene, a child could be born with CF [Ciske et al., 2001].
- Another study reported that out of 353 parents who were assessed for knowledge about test results for CF 18 months after screening had been offered, 11

of 47 carriers "falsely believed they were only very likely to be carriers, while nearly a third of test-negative individuals falsely believed they were definitely not carriers" [Gordon et al., 2003].

Physicians may:

- Underestimate the patient's needs [Bertakis et al., 1991; Street, 1992].
- Not take into account a patient's emotional and psychological responses [Shapiro et al., 1992; Roberts et al., 1994].
- Provide too little information [Waitzkin, 1985; Michie et al., 1997a,b; Mechanic, 1998].

These limitations are discussed in this and later chapters.

Standards

Professional accreditation standards for medical genetics and genetic counseling reflect the critical role of communication. The American Board of Medical Genetics, Inc. [ABMG, 2004] requires skills to:

- Elicit from the patient or family the information necessary to reach an appropriate conclusion.
- Transmit pertinent information in a way that is comprehensible to the individual or family.

The American Board of Genetic Counseling [ABGC, 2004] requires the ability to:

- Communicate information to their clients in an understandable manner.

The Canadian College for Medical Geneticists [CCMG, 2004b] requires that clinical genetics have the ability to:

- Communicate appropriately with the patient/family, both verbally and in follow-up letters.

An open question is: How are these recommended norms of practice to be interpreted by health care professionals who lack specialized training in genetics yet who elect to provide genetic testing services including the communication of genetic test information? Although these are accreditation standards for medical specialties, courts of law may regard accreditation standards, as well as the number of reports in leading human/medical genetics and medical journals, as reliable guidelines to desired norms of practice [Capron, 1979; *Pedroza v. Bryant*, 1984; *Menzel v. Morse*, 1985; Havighurst, 1991; Jutras, 1993; *Kramer v. Milner*, 1994; *ter Neuzen v. Korn*, 1993; but note *Helling v. Carey*, 1974; Campbell and Glass, 2000].

It may be argued that obligations to take into account linguistic and cultural differences as well as the timing, tone, and manner of the communication already are a part of a physician's obligations to the patient [Caulfield, 2001; Sharpe, 1997]. For example, as will be discussed with regard to informed consent, courts have recognized that health care professionals have a reasonable obligation to ensure that discussions regarding informed consent occur at a suitable time and place, in a manner appropriate to the communicative needs of the patient, and to take into account the patient's emotional state of mind including the potential impact on the understanding and retention of information. It is reasonable to conclude that the same obligations are likely to be applied to the discussion of sensitive, and often complex and ambiguous, genetic test information.

WATCH OUT FOR

Obligations

A health care professional has an obligation to accurately interpret and communicate genetic test results, to advise patients of the implications posed by the test results, and to discuss treatment options [Andrews, 1992; Sharpe, 1996; Welch and Burke, 1998]. This may prove challenging:

- Genetic test results may be perceived differently than nongenetic test results [Emery, 2001]. As will be discussed in later chapters, they can have significant implications for the patient and the family, including potential insurance and employment discrimination, stigmatization, and familial difficulties.

- Will the health care professional have appropriate communication skills to discuss the implications of the information that often may be complex and ambiguous, to assess the patient's understanding [Bramwell and Carter, 2001], and to elicit a patient's preferences as to decision making [Say and Thomson, 2003]?

- Patients may be uncertain as to the meaning of, and implications posed by, terms such as *autosomal recessive* and *autosomal dominant*.

- The health care professional may lack appropriate genetic counseling skills to appropriately interpret the probability of an outcome [Bramwell and Carter, 2001; Freedman et al., 2003; a survey of U.S. physicians reported that only 29 percent felt qualified to provide genetic counseling to patients with regard to cancer susceptibility].

- Patients' desire for information, and involvement in decision making, may be greater than expected by the health care professional [Coulter, 1997; Guadagnoli and Ward, 1998; Helmes et al., 2002; but see Blenkinsopp et al., 1998; Robinson and Thompson, 2001].

- Although some patients may be of the opinion that they do not have preexisting knowledge to make decisions [Royal Commission, 1993; Committee on Assessing Genetic Risk, 1994], they still may wish to communicate their preferences [Coulter et al., 1999; Bowling and Ebrahim, 2001; Say and Thomson, 2003].

Methods

To facilitate understanding, test information should be presented in:

- Simple, commonly understood terms [Edwards et al., 1996; Cosmides and Tooby, 1996; European Commission, 2004].
- The language the patient best understands.
- An empathetic approach with a focus on the patient's, and family's, concerns. Patients often wish to discuss the psychological, socioeconomic, and familial concerns [Geller et al., 1998; Marteau and Croyle, 1998; Menahem, 1998; McConkie-Rosell et al., 2001; Esplen et al., 2001; Lion et al., 2002; Speice et al., 2002; Parsons et al., 2003]; however, the health care provider may be more focused on factual test results [Levinson and Roter, 1997; Michie et al., 1997b; Roter et al., 1997; Levinson et al., 2000; Carroll et al., 2000; Phillips et al., 2000; McConkie-Rosell et al., 2001; Bensing et al., 2003].
- A variety of formats including the use of graphs and charts [Kessler and Levine, 1987; Marteau, 1989; Shiloh and Sagi, 1989; Redelmeier et al., 1993; Grimes and Snively, 1999; Sandhus et al., 2001; Edwards et al., 2001, 2002; Woloshin et al., 2002].

Terms, Tone, and Follow-up

Health care professionals may overestimate the burden and negative effects of a genetic disease or disorder [Blaymore et al., 1996; Bogardus et al., 1999; Kirschner et al., 2000; Abramsky et al., 2001]. A patient's perception and understanding can be influenced by the health care professional's tone of voice and facial expressions [Shapiro et al., 1992; Ambady et al., 2002].

Health care professionals need to understand that communication of test information only by telephone and/or by letter [Tluczek et al., 1991; Phelps et al., 2004] and/or electronic means [Cepelewicz et al., 1998; Hodge et al., 1999] without access to face-to-face discussion [AMA Report on Scientific Affairs, 1999] can increase the risk of misunderstanding. Discussion about risk management options [Lerman and Croyle, 1994; Kash, 1995; Lerman et al., 1995, 1997; Burke et al., 1997] with referral for genetic counseling is recommended.

Patients may be uncertain as to the meaning and implications posed by terms such as *rare* and *carrier* and may wish to discuss when is an appropriate time, if at all, to discuss and disclose carrier status with their children [Fanos and Johnson 1995; McConkie-Rosell et al., 1999, 2002; Michie and Marteau, 1999(a); Ciske et al., 2001; Tercyak et al., 2001b; Edwards et al., 2002].

Follow-up visits and a written letter [McConkie-Rosell et al., 1995; Michie et al., 1997c; Hallowell and Murton, 1998; Lobb et al., 2004] summarizing the genetic counseling, the test results, and their significance, helps to facilitate understanding.

CULTURE AND COMMUNICATION IN THE REALM OF FETAL DIAGNOSIS: UNIQUE CONSIDERATIONS FOR LATINO PATIENTS*

C. H. Browner and H. Mabel Preloran Center for Culture and Health, Department of Psychiatry and Biobehavioral Sciences, University of California, Los Angeles, Los Angeles, California

Amniocentesis and other prenatal genetic tests have become well-established features of modern prenatal care but place a considerable decision-making burden on the expectant mothers who are offered them. The genetic and emotional issues involved are complex and the "best" course of action sometimes ambiguous. It is often assumed that ignorance, poverty, and illiteracy, more acceptably termed *cultural* factors, lead women from certain ethnic minority backgrounds to turn down amniocentesis at higher rates than ostensibly less "ignorant" women from European American backgrounds. However, the extent to which *both* amniocentesis acceptance and refusal by ethnic minority women is influenced by miscommunication between providers and patients—and the extent to which such miscommunication stems from *cultural factors*—has rarely been explored [Resta, 1999]. We focus on Latinos because they represent a large and rapidly growing population in the United States (38.8 million as of July 2002, with a growth rate of 9.8 percent), and because they face a higher risk of neural tube defects than most other U.S. groups [Stierman, 1995]. There have been few studies of why Latinos are more likely to decline prenatal testing and why they accept when they do. This section is intended to highlight some of the ways communication and miscommunication may impact decision making.

Most babies born in the United States today are screened for the most commonly occurring birth anomalies, including neural tube defects and certain chromosomal disorders [ACOG, 1996]. In California, where our studies were conducted, women who screen positive are referred to a state-certified prenatal diagnosis center where they are offered a consultation with a certified genetic counselor and additional testing, typically high-resolution ultrasound, and if indicated, an amniocentesis. (Prenatal genetic services tend to be organized differently in different parts of the United States [see, e.g., Rapp 1999; Hunt et al., 2004].) Prenatal genetic consultations in California typically follow a fairly standard protocol that includes obtaining the woman's reproductive and family medical history, describing her options for additional testing, reporting the mathematical probability of a fetal anomaly based on the screening test result, and outlining the risks associated with the amniocentesis procedure. The consultation can also include a discussion of the benefits of reassurance and preparedness that fetal diagnostic testing can provide, the woman's right to accept or decline it, and her right to terminate the pregnancy if an anomaly is found.

* Funding was provided in part by the National Center for Human Genome Research (1R01 HG00138401), the Russell Sage Foundation, UCMEXUS, the UCLA Center for the Study of Women, and the UCLA Center for Culture and Health. Thanks to Maria Casado for unfailing patience and invaluable assistance and Richard M. Rosenthal for sharpening the prose.

Our data come from an ethnographic study of amniocentesis decision making by women from Latino backgrounds who were referred for prenatal diagnosis because they had screened positive for a birth anomaly [Preloran et al., 2001]. The study population consisted of 156 women and 128 male partners who completed a face-to-face, semistructured interview. In addition, we analyzed sociodemographic and other background data from 25 women the clinics classified as "no shows" because they did not return for a follow-up visit and a second chance to be offered amniocentesis (their behavior represented de facto refusal, although we did not have the chance to ask them about their reasons). We also interviewed 60 medical personnel including genetic counselors, geneticists, family practitioners, ob/gyns, perinatologists, nurses, health educators, and the translators whose work included offering prenatal genetic information or services to pregnant Latinas in southern California. We systematically observed 73 clinical encounters where such information or services were offered or provided and 12 prenatal education classes at primary or prenatal care facilities and recorded detailed notes based on our observations.

The cases we offer highlight some of the complexities associated with offering prenatal genetic information and services; the unit of analysis is the actual interaction between the patient, her clinicians, and any others present. We focused on two broad questions in developing our analyses: (1) What factors contributed to the clients' decisions to accept or decline amniocentesis? (2) Were the reasons clients gave for their decisions based on information received directly from a clinician or from other sources? In our examples of amniocentesis refusal based on miscommunication, our intent is not to imply that refusal is wrong, nor that women who relied on biased or incorrect information from male partners or other family members were necessarily victims of undue or unwelcome family pressure. Most such women said they wanted their relatives' help in interpreting the information they received. Likewise, the sources of misinformation we describe should not necessarily be considered the sole—or even the most important—factor in the women's decisions. Like any other, clinical communication is an inherently fragile endeavor [Mattingly and Lawlor, 2001]. Although counselors may genuinely wish to help patients make informed decisions, the priority biomedicine puts on neutrality may indeed deter them from directly challenging patients' views that they know or believe to be wrong.

MISCOMMUNICATION AND PRENATAL GENETIC CARE

There was a distinct and significant connection between women's understandings of the significance of having screened positive and their amniocentesis decision. Whereas most that declined attributed their positive test result to concrete but transitory physical or mental conditions, most who said yes attributed it to mere chance [Browner and Preloran, 2000b]; only a tiny proportion thought there was actually anything wrong with the pregnancy. The explanations for screening positive given by those who accepted were usually a reiteration of the medical information communicated during the genetic consultation; in contrast, the reasons given by those who refused diverged much more widely. This led us to consider the possibility that miscommunication may have influenced the women's decision processes.

It was no surprise to find that misunderstanding appeared to play a role in some women's amniocentesis refusals. Some genetic counselors expressed frustration at what they felt was their inability to convey their message to certain patients, a situation they attributed to patients' low levels of education or ethnic backgrounds. At the same time, we observed many instances of successful communication between prenatal genetic service providers and clients. Counselors almost invariably followed the standard protocol regardless of clients' educational or ethnic background. They detailed the advantages and disadvantages of the amniocentesis procedure and sought to make it clear that any decision (i.e., to accept or decline the test, to continue or terminate the pregnancy) would be the patient's. Most counselors talked sensitively about the often difficult issue of abortion as well as other matters such as caring for a child born with a disability, as well as religious, economic, and practical issues. *Mutuality*, in which the clinician provides expertise and shares the responsibility of medical decision making with the patient [Roter and Hall, 1992], was the approach most counselors sought to emulate.

Most women in our study understood that they had been referred for genetic counseling because certain substances measured by a blood test were either too high or too low. But we also found that clients' interpretations of what this meant *for them* often differed from those of the genetic counselor. To cast some light on the origins and characteristics of these misunderstandings, we have selected several cases that exemplify the recurrent sources of miscommunication we observed. The cases are offered next.

Idioms and Jargon

Health care providers, no less than other professionals, become fluent in a professional discourse. Steeped in specialized terminology and jargon, it may be inevitable that clinicians occasionally lose sight of the fact that terms and usages familiar to them can have different meanings for patients and other laypersons. Prenatal care providers are particularly well practiced in discourse aimed at reassurance intended to manage the anxiety of women undergoing testing. Yet communication is compromised if patients misunderstand either the idioms or the reassurances at the core of health care providers' professional language. In our research we found these kinds of misunderstandings could sometimes be traced back to information clients were given even before their screening test, when they attended prenatal education classes given as part of the State of California Expanded AFP program.

These prenatal care instructors, usually nurses or health educators, are acutely aware of the anxiety most pregnant women experience if they screen positive and have learned to underscore the fact that screening positive does not *necessarily* mean there is a problem with the pregnancy. Sometimes, however, even the most conscientious of these efforts can backfire. In one class we observed, participants were told that they would be offered a "standard" test designed to measure, among other things, a "protein" produced by the fetus that is also present in the mother's blood. Normal amounts, said the instructor, indicate a "very good chance" that "the baby is healthy." She then continued: "Knowing that everything is normal is a great relief, and don't be alarmed if your test comes out positive. We will call you if it's positive. You will be informed if your baby's protein came out high or low. That means that your baby has low or high protein compared with other babies that were born healthy. That is all." Before dismissing the class she repeated: "I see that most of you have signed the consent form for the test. Good! You will be called only if your test is positive. If we don't call you, don't worry . . . And if we call you, please, you don't have to worry either. We may offer you other tests that are more complete or more accurate. Because, you know, this test [XAFP], the one that some of you are going to have today, doesn't say much. It's like a red flag that indicates 'be careful, pay attention.' So this is why you might be offered other tests that will reassure you that everything is fine" [Browner et al., 2003, p. 1937].

In this example, the educator did not distinguish between the fetal "protein" for which clients would be tested and the popular meaning of protein, as found in food, to maintain energy. Drawing on the more conventional meaning of protein, many clients attributed their own "low" fetal protein test results to weakness brought about by a poor diet [Browner and Preloran, 2000b] and believed that by changing their eating habits, their pregnancies would no longer be at risk. The educator also failed to explain that a "positive" medical test indicates possible fetal abnormalities. As we will see in the following examples, these misunderstandings combined with providers' repeated efforts to reassure patients that the screening test was "standard," "painless," and "nothing to worry about," may have made a positive test result seem less important. Convinced there was little need to worry, some saw no compelling reason to undergo amniocentesis, which in the minds of many was an exceedingly risky procedure [Browner et al., 1999].

Suggestions

In addition to misconstruing clinicians' reassurances, we found it not uncommon for clients to misconstrue the nondirective nature of genetic counseling. Accustomed to receiving prescriptive medical advice and carrying out "doctor's orders," these clients took the genetic counselor's reluctance to directly recommend an intervention as a sign that the intervention was not truly needed, rather than it being a standard part of prenatal care [Williams et al., 2002a; ACOG, 1996]. For example, one participant we call Maria, age 24, and with no disabilities in her family, was living on a small income (less than $20,000 a year for a family of four). She had first come to the United States 6 years prior to the interview when she was 18 and pregnant by another partner. This was Maria's third pregnancy. She had completed elementary school, worked as a babysitter, and believed that she was going to have a healthy baby because "it moves happily like my first born." The following comes from our face-to-face interview and explores Maria's reasons for declining amniocentesis.

When we asked whether she had thought things over before deciding to decline, she replied that there had been no need because *she already knew* she did not want the test. When called and invited to attend a genetic consultation, she told the "nurse" on the phone that she was not interested in an amniocentesis, but the nurse had insisted that even if she planned to refuse the test, she would be better off attending the consultation "for the [good of the] baby [because] the blood was a little low [and it was better] to have an ultrasound first, and later the test of the needle [amniocentesis]."

Asked what she thought having her "blood a little low" meant, Maria explained she had been told it was from "the baby's protein . . . because I haven't been eating well." She said she was told she should eat more and that she could have an amniocentesis. When we asked directly why she had declined the amniocentesis, she explained that the clinicians "did not pressure me at all" ("*No me obligaron para nada*"), which she interpreted to mean that they themselves saw no real need for the test. From Maria's perspective, the offer was quite casual: "They said, 'I suggest you consider it, but it is up to you to decide.' They did not recommend it at all; it was only a suggestion. I'm sure I could catch up with my weight if I make an extra effort to eat for two."

In our larger sample, we found that women like Maria usually drew on two types of information when trying to make sense of a positive screening test result: the prenatal classes they had attended and personal experiences of friends and family [Browner and Preloran, 2000b; Mittman et al., 1998]. Both sources led some to decline testing. The medical setting, in its effort to be nondirective and highly reassuring, sometimes resulted in recommendations that were perceived as only casual suggestions. In the home and community setting, the ubiquity of what some counselors referred to as amniocentesis "horror stories," which magnified risk of fetal harm, also inhibited test acceptance. Information drawn from these two sources led some to conclude that amniocentesis was not sufficiently justified in their own cases [Browner et al., 2003, p. 1939].

Cultural Sensitivity

California genetic counselors who typically deal with an ethnically diverse client population generally recognize that they must be sensitive to their clients' cultural backgrounds. Occasionally, however, a counselor's efforts to be culturally sensitive can inadvertently interfere with direct and open communication and, in the process, jeopardize informed choice. Our observations reveal that many counselors are reluctant to challenge the mistaken beliefs of their clients when they appear to be rooted in the client's ethnic or cultural background. The following excerpts come from observations of a formal genetic consultation and that of a more casual encounter between a patient and her genetic counselor.

Genetic Consultation

The participants, 24-year-old Lidia and her husband Rodrigo, 25, have lived in California for 3 years. She is a homemaker and he works in building construction; this is their second pregnancy. Their family medical history is uneventful, except that one of Rodrigo's brothers was apparently diagnosed with spina bifida as a child in Mexico, although the couple reported that he showed no long-term effects. Rodrigo also confessed to some skepticism about U.S. medical care because he has noticed fewer "mongoloid" and paralyzed children in Mexico than in the United States. During the 35-minute consultation, the genetic counselor avoided discussing what she deemed to be her client's ethnic beliefs because she feared she would jeopardize rapport. She eventually timidly suggested that they check the brother's diagnosis with his mother and even offered to talk to the Mexican doctor in an effort to further clarify the brother's diagnosis. (The couple did not take the counselor up on her offer.)

After an inconclusive ultrasound, the counselor told the couple she was offering amniocentesis mainly to rule out the possibility of spina bifida. Rodrigo responded that it was unnecessary because 20 years ago, his mother's doctor had told her the baby's problem would abate if she took "very good care of herself." Rodrigo proudly explained that was just what she did "and my brother walked normally like you and me." The counselor tried to clarify Rodrigo's misunderstanding by explaining that the severity of the condition depends on the location of the spinal defect and suggested that an older diagnosis could be less accurate. But her explanations were insufficient for the couple to reverse their decision.

Opportunistic Observation

Elena is 20 years old with no disabilities in her family. This is her first pregnancy. She completed secondary school in Mexico, was living with her grandmother, and had no plans to establish a domestic relationship with the child's father. Elena came alone to the genetic consultation and remained mostly silent throughout. She asked for a few minutes alone when it was over, to which the counselor replied, "Take your time." Returning to the waiting room, Elena asked the receptionist for the location of a pay phone and, when she returned a short time later, asked to see the counselor again. She told her that she had discussed the test with her grandmother who had convinced her to agree because "it is the only way to be 99 percent sure that my baby will be 100 percent healthy." Neither then, nor subsequently, did the genetic counselor explain that a negative amniocentesis was not a guarantee that the baby would be "100 percent healthy." The counselor had only said that the test is "99 percent accurate," which Elena reinterpreted to mean she could be "99 percent certain" that the fetus would be fine.

The above are examples of amniocentesis decisions based on clients' misinformation not clarified by clinicians. In the first, the counselor did not try to dispel the mistaken notion that spina bifida may be reversed during pregnancy if the woman takes good care of herself and did not emphasize the increased risk associated with her client's family history and AFP positive test result. In the second, the counselor did not clarify that a normal amniocentesis did not 100 percent guarantee a perfectly healthy baby.

Regarding the first case, when we subsequently asked the counselor to explain her reticence, she said that when clients bring contradictory "ethnic" data to a genetic consultation, she seeks a balance between "respect" for their cultural beliefs and her professional objective to convey the fact that there may be genetic or developmental explanations for the condition in question. In the second case, the counselor said, "I know she agreed for the wrong reason. I know it was because her grandmother talked her into doing it; but I didn't want to intervene. First, I know that among Latinos, family relations are sacred and felt it was not my place to contradict the grandmother. Second, I found no harm in the family advice." These two providers and others like them were uncomfortable with directly challenging clients' cultural beliefs because they assumed it could offend their clients. But, as we later argue, avoiding open discussion of these very issues seems to lead to a serious gap in communication and possible consequences for informed decision making.

TRANSLATION AND SECOND-HAND INFORMATION

Multilingualism is a stark and increasing reality in many U.S. medical settings, and its challenges can take on special salience during pregnancy, when complex information is communicated and important decisions must be considered (see also discussion on linguistic differences in Chapter 7, Informed Consent). Since little is known about the impact of translators on amniocentesis—or other types of medical decisions—we focused attention on this issue. Most U.S. medical settings offer translation services, either through an on-site interpreter (who may or may not have formal training and/or primary job responsibilities such as clerical or nursing) or a telephone translation service. Patients with limited English also may be accompanied by a family member or friend who ostensibly speaks English better than the patient. Either alternative can present its own difficulties, as will be seen below in the examples offered to illustrate how a translator can, in fact, impede comprehension and/or compromise informed decision making.

On-Site Translators

Our data derive from two sources. Part I is from observation of a genetic consultation with Rosalia, in her midtwenties, with no family history of disabilities. Part II comes from our face-to-face interview with Rosalia, immediately after the genetic consultation. Rosalia had resided in the United States for 3 years prior to the interview and suffered continuing economic setbacks. She had completed high school in her native country, was married to a man 20 years her senior, but attended genetic consultation alone. Although she said she had been studying English, at the time of the interview, she spoke it poorly and requested the help of an interpreter. A secretary was called away from her usual duties to translate (but was not unaccustomed to providing such service in that setting).

Genetic Consultation

Assisted by the translator, the genetic counselor asks Rosalia about her family's medical history. The interpreter literally translates the conversation. Rosalia reports no problems on her father's side but indicates that her younger half-sister's (with her mother's second husband) legs were "semiparalyzed" as a child. Hearing this, the counselor probes to learn more about the half-sister, but her questions, when translated literally (e.g., "*estaba retardada?*"), sound harsh and crude, and Rosalia appears discomfited. To each question about her sister's condition, she repeatedly answers that she is walking perfectly well now. Eventually, the counselor turns to the translator to discuss how the questions can be asked without further upsetting the client. This exchange is not translated and throughout the continuing consultation, the translator looks mainly at the counselor, only once making eye contact with the client. At one point, Rosalia repeats that her sister has improved—as if she wants to convince the genetic counselor that her sister's medical condition should not be of concern: "She is walking well now," Rosalia asserts, "She only has to use special shoes." But the interpreter says only: "She said her sister is O.K. now." As we later show, the client seems to notice that her words are not being faithfully conveyed, enlarging her sense that what she says is being disregarded.

The counselor then explains that the ultrasound showed the pregnancy to be dated correctly and she would therefore like Rosalia to consider amniocentesis. Rosalia appears doubtful. She says she would "perhaps prefer to consider" the test later because right now she is not feeling all that well, adding that she recently had the flu and, in fact, had been eating poorly when she had the triple marker screening. The counselor explains that "the flu" would not cause a positive screening test and urges Rosalia to decide as soon as possible because her pregnancy is very advanced. When Rosalia does not reply, the counselor begins to describe the amniocentesis procedure, adding that it allows the doctors to analyze the chromosomes for the presence of Down syndrome and spina bifida. She then asks Rosalia if she remembers what they had previously discussed about these conditions. Rosalia shakes her head. In response, the counselor asks if Rosalia wants to review the information again, but she quickly says no. The counselor repeats that Rosalia can have the test today and avoid having to make another special trip. The patient responds: "If you want to do it because of my sister, she is well now. My mother says she was a little behind (*quedada*) when she went to school, but she is walking well now. She only has to use special shoes." The interpreter again translates this by saying: "She said her sister is O.K. now." Instead of responding, the counselor sighs and changes the subject, suggesting that Rosalia use the office phone to talk over the situation with her husband.

Interview after Consultation

When we ask Rosalia if she would be willing to discuss her genetic consultation, she readily agrees. When we ask if it was difficult for her to understand the information the counselor conveyed, she says it was, but adds that she disliked the way that the counselor and the translator interacted with her: "They were talking to each other, not to me ... I told her about my sister, she only needs special shoes. She [the translator] didn't say 'shoes'; I think she didn't say anything [about that]." Rosalia said that she knew the English word for shoes from her English as a second language classes and complained that they were "very impolite" and excluded her from the interaction. She adds that she much prefers her neighborhood clinic where everyone speaks Spanish and is friendlier and less intimidating. Despite our best efforts to ascertain how well Rosalia felt the information about fetal diagnosis was conveyed, she repeatedly shifts back to the antagonistic emotions she experienced [Browner et al., 2003, pp. 1940–1941].

Family and Friends as Translators

These data come from interviews with two couples: one who accepted and one who declined amniocentesis. Both husbands attended the genetic consultation and served as their wives' translators. Although both couples described their amniocentesis decisions as jointly made, in each case, the men shaped the information they gave to their wives to achieve their own desired outcome.

Elisa, 29, is a housewife married to an insurance salesman. Because her English is rather poor, she wanted her husband to translate the genetic consultation. She explained that she felt comfortable in his doing so because he had been very involved in the pregnancy, accompanying her to all her prenatal visits including the genetic consultation offered after she had screened positive. Regarding the amniocentesis offer, she said, "At the [genetics] clinic, he explained all the pros and cons of the test and later [after the genetic consultation] when we were relaxed at home, I could see things more clearly." There, she said, they had talked "about all the advantages of the test," [the most important being] the opportunity to be reassured," something Elisa said her husband told her the counselor had mentioned several times during the consultation. She added that her husband already knew the advantages of "being prepared" because as an insurance agent, he witnessed many families struggle "to pay medical bills for chronically ill patients." Elisa said that talking with her husband enabled her to see that the "small risk" of amniocentesis was worth taking.

Lia, 28, was employed part-time and married to Pedro, a full-time college student. Both had been raised Catholic, left the Church, and had recently returned to become active adherents to their faith. They were thrilled by the prospect they would soon become parents, having unsuccessfully tried for several years to conceive. Both were fearful they might loose the pregnancy, but Pedro seemed the more anxious of the two about the miscarriage risk associated with amniocentesis. He indicated he would abide by his wife's decision about amniocentesis but also told us he was committed "to helping her to decide by considering all the facts, especially that she might lose this baby." He said he advised her to take time and think "[because] now that we belong to the Church, we can't do whatever we want and in the Church, they say one could kill his own child with that test. I made her realize that that is the truth. If not, look at what happened to my friend [his co-worker's wife who had a miscarriage that he attributed to an amniocentesis]." As the interview was ending, Pedro spontaneously returned to their amniocentesis decision: "For me it represents a lot of responsibility. I am alone, I lost my entire family, I only have her [Lia] and now this child. [For me] it was love at first sight, and when I learned about that test, I was afraid. She was also. My co-worker told me that when that happened to them [miscarriage], his wife cried and cried—so, I told Lia better not to have the test, because if she loses the baby, what is she going to do? She struggled years to get pregnant."

While, as Roter and Hall [1992] observe, power differentials render client–clinician relationships inherently problematic, these differentials are intensified when actors lack a common language. With these examples, we demonstrate sources of miscommunication that go beyond simple errors in translating words. Skilled interpreting is a creative technique demanding the communication of ideas and concepts as well as the accurate translation of words. Listeners depend upon pauses, word emphasis, eye contact, and a trusting attitude to understand the message. In the first example, those conditions were not fully met. On the one hand, the Spanish-speaking secretary was insensitive to the harsh connotations of a literal translation. On the other hand, she lost the trust of the patient when her translations were not faithful or comprehensive enough. In our second and third examples, women received genetic information through their partners, who clearly had their own agendas. Because research on the use of friends and family as translators in medical encounters is virtually nonexistent, we have included these examples to highlight some of the issues.

CONFIDENCE AND TRUST

Although the relationship between counselor and client is a professional one, it is not wholly impersonal. The degree to which a client can trust and feel comfortable with the counselor can have significant bearing on the effectiveness of communication between them [Rapp, 1999]. Anecdotal data from genetic counselors [Alvarado, 1999; Caldwell, 2000; Tatsugawa, 1998] suggests that minorities in general, and Latinos in particular, may be highly skeptical of the motivation for offering genetic tests and the results that follow. They are unlikely to air that skepticism or engage in a full and frank discussion of the implications of the tests offered unless they trust the counselor. But women from minority backgrounds may find it difficult to develop a trusting relationship with health service providers of different cultural backgrounds, especially ones with whom they have had no prior relationship [Browner et al., 2003, p. 1941]. The following cases of two women, one who declined and the other who accepted amniocentesis, helps illustrate the significance of trust in medical decision making.

Woman Who Declined Amniocentesis

Rosa was 27 and had lived in the United States for the past 2 years. She said she and her family had all been quite fortunate with regard to their health but described her background as unusual in that she had been "touched by adversity." She confided that she had been abused as a child, did not complete elementary school, ran away from home, landed in a relationship at age 22 with a man who tried to kill her, and 2 years later became pregnant—but lost the baby at birth. When she decided "to cross the border" in search of a better life, it was also to get away from abusive relationships. One year later, she found a new partner in the United States and again became pregnant. The partner was currently under arrest, and Rosa was living with a friend who participated in the genetic consultation and took care of her "out of charity." The friend was also instrumental in motivating Rosa to seek prenatal care at a free clinic close to their home. There, she was offered triple-marker screening and, when she screened positive, additional testing including amniocentesis, which she declined on the spot.

Asked why, Rosa explained that she intended "to have this baby because it is the only family I have" and that she found the risk of amniocentesis unacceptable. Besides, she added, she does not understand why she was offered "the needle test." She proceeded to ask the ethnographer's opinion and, when told that the test is offered for diagnostic purposes, she insisted that in her case, there was no need. She added that in the neighborhood clinic where she attended prenatal education classes, she was informed that "the baby's protein was low, but all other things were fine." In addition, she trusted the doctor who saw her there because he was "a good doctor . . . he helped my neighbor with her baby. I can trust him." Probing about her experience at the genetics clinic, Rosa expressed doubts: "They said the baby could be mongoloid. But how could she [genetic counselor] know only because of that [measurement of AFP]?" When the ethnographer asked whether Rosa had voiced her concern to the counselor, she said no "[because] she doesn't listen to things. I told her about the [neighborhood] clinic [where I was told my baby was fine] and I told her about my [financial] problems, but she didn't pay attention. [In contrast] in the [free] clinic, each time I go there, they check me and [they check] my baby [too]. . . . They said the baby is fine; nobody there told me I needed to have the needle test."

Woman Who Accepted Amniocentesis

Ana, a 28-year-old clinic clerk, is pressed into duty to translate for Rocio, the patient, a 45-year-old woman with a history of hereditary family illness; one daughter, born with Down syndrome, died at age 8 [Browner and Preloran, 2004]. Rocio attributed her daughter's medical problems to an injection, which the genetic counselor said that she doubted. Although Rocio came to the genetics consultation mentally prepared to accept amniocentesis, she changed her mind twice before ultimately agreeing to have the test. This case is a clear example of how the dynamics of ambivalence, trust, and compliance interact in the context of a decision about fetal diagnosis.

During our subsequent interview, we ask Rocio the standardized question: "If doctors recommend a test, such as the amniocentesis, do you think a person should agree because doctors know more than the patients?" At first she categorically answers, "No," but a bit later softens her response adding, "Well, as much as we can, we should follow the doctor's advice," quickly adding "[agreeing is acceptable] but we have to be careful, because sometimes they [doctors] lie because they don't know 100 percent what is going on [referring to her inconclusive ultrasound]."

Asked about the genetic consultation, Rocio explained that she had problems communicating with "the blonde" [the genetic counselor] to clarify her doubts and concerns. She found her "too critical": "If she does not believe in me, I won't believe in her;" and she found her words of comfort false: "If everything is going to be O.K. why would they send me to do that [amniocentesis]?" She said she much preferred the directness of the translator with whom she "could talk." Asked if her preference was due to the fact that they could converse in Spanish, Rocio said it went beyond that. She said that Ana did not challenge her explanation of the cause of her daughter's problem and openly told her that things could go wrong such as miscarriage, but she should agree to amniocentesis if she wanted to be reassured. Rocio said that Ana's "frankness inspired my confidence" (*esa franqueza de ella me dió confianza*). She added that being so indecisive, she felt relieved and supported when Ana "pushed her a little" to agree to amniocentesis if she wanted to "stop worrying."

These two cases illuminate how the dimension of "trust" can shape the course of a prenatal genetic consultation. Rosa felt that what she had told the genetic counselor about her diet and financial circumstances was dismissed and she reciprocated in kind: "Well, if she doesn't believe me, I don't believe her." Rocio used virtually the same words when describing her own encounter with "the blonde" [genetic counselor], whom she found critical and skeptical: "If she does not believe in me, I won't believe in her." Establishing trusting relationships may be particularly challenging in cross-cultural clinical encounters. In Rosa's case, she gave more credence to clinicians she felt she could trust, such as her neighbor's doctor at the free clinic close to home, than the suggestions of the anonymous genetic counselor. In Rocio's case, although a typical genetic consultation usually lasts about 45 minutes, hers spanned almost 5 hours. During that time, we often observed Ana listening patiently to Rocio, comforting her during the ultrasound procedure, and bringing her orange juice while she was waiting to see the genetic counselor for a second time. Ana's open, warm, and friendly style of communication won Rocio's trust.

DISCUSSION: COMMUNICATION, MISCOMMUNICATION, AND GENETIC COUNSELING

The traditional doctor–patient relationship is rapidly changing, no less in the field of genetics than elsewhere in medicine; the goal today is to forge partnerships aimed at better enabling patients to make informed decisions. But, acting as partners demands mutual knowledge, trust, and unerring communications that are sometimes difficult to achieve, especially when interactions involve more than two parties.

The genetic counselors in our study were genuine advocates for their clients' views and sought to take into account their clients' values about abortion, their socioeconomic situation, their need for concrete medical explanations, and even their capacity to handle anxiety. Indeed, the counselors in our "cultural sensitivity" examples were, if anything, guilty of displaying *too much* deference to their clients' beliefs.

Most clients rated their genetic consultations very highly, even when an interpreter was required [Browner et al., 1999]. The open dissatisfaction one participant expressed (in the "they're talking to themselves" example) was almost an exception. Women assisted with translation by friends and family also reported they were generally satisfied with the experience. Yet we found clear evidence of misunderstandings, even among women who reported satisfaction. This was most evident when it came to women's assessments of risk and choice and their understandings of the nature of genetic conditions and the meaning of fetal "protein." Genetic testing can also create new ambiguities and uncertainties [Lock, 1998] and may challenge patients to generate creative responses. Many of the women in our study who refused amniocentesis used the opportunity of the positive screening result to forge their own strategies for dealing with the anxieties produced by the positive screen [Browner and Preloran, 2000a].

Prenatal instructors aim to inform women about the benefits and limitations of screening tests. They adopt a discourse of reassurance to reduce anxiety but, by doing so, can also contribute to misunderstandings. Instructors frequently repeated assurances such as, "This test [XAFP] is only a red flag; it doesn't mean too much." Such statements led some women who screened positive to believe they were either at a very low risk or no risk at all. In their own minds, they assumed that prompt remedial actions—usually modification of diet and lifestyle aimed at minimizing stress—would control or eliminate any risks indicated by the test.

For many women, the refusal of amniocentesis was based on perceptions reinforced by the genetic counseling that followed their prenatal classes. Counseling suffered from the fact that patients and health providers, both attuned to a certain model of medical practice, tend to expect a definitive diagnosis and prescription for treatment and cure. Patients do not, however, know or perhaps comprehend that in the field of genetics direct recommendations are usually avoided. Thus, when counselors refrain from making suggestions or recommendations and remain neutral, miscommunication may well occur [Browner et al., 2003].

Our evidence suggests that many of those who declined amniocentesis believed that physical or psychological conditions were responsible for their positive screening result. Clients who confused fetal protein with standard protein, as the word is commonly understood, were especially prone to believe that simple dietary changes could improve their condition. This misinterpretation allowed women to feel more in control and made the risk (*peligro*) much less threatening. This type of misunderstanding is not exclusive to women of Mexican origin; researchers working with other ethnic groups have reported similar findings [Cohen et al., 1998]. Neither is it only manifest among women who refused the test. Some women who accepted amniocentesis also expressed doubts about the biomedical explanation that screening positive was a random event. These women usually looked for more concrete explanations for the test result and sometimes blamed such factors as their spouse's alcohol or drug use for the threat of a genetic abnormality.

If misunderstanding has its consolations, genetic counselors face the difficult task of removing these misunderstandings and possibly depriving clients of the benefits they provide. As we saw in the excess of cultural sensitivity example, while counselors may be aware their clients are emotionally invested in mistaken beliefs, they are sometimes reluctant to address them for fear of undermining trust and rapport. But this is to confuse the means with the end: Rapport is but an aid to open communication, and it should not be compromised in order to create or maintain rapport. In other cases, counselors were quick to point out to clients that their lay theories, preconceptions, or expectations were mistaken or irrelevant. But as we saw in the trust example, clients of Mexican origin, many with limited or no prior knowledge of genetic testing, seem to not be predisposed to accept the guidance of genetic counselors over advice from other, more familiar sources. Counselors thus find themselves in the difficult position of having to earn the right to have their counsel heard.

Our examples show that clients who do not feel their views have been heard and respected are much less likely to listen to and respect the counselor's guidance. Miscommunication is often the result of asymmetric communication: The client is asked to accept what the counselor has to say before the counselor has heard the client. In one of our confidence and trust examples, the client complained that the counselor "doesn't listen to things"; in the suggestion example, the client mentioned that the counselor "didn't pay much attention." On the other hand, in Rocio's case, another of the confidence and trust examples, she accepted in large part due to the "little push" by the translator who inspired her "confidence"; in the "second-hand information" example, the client complained that her translator disregarded what she had to say. All of the foregoing suggests that the best way for counselors to improve comprehension and win their clients' ears and understanding is to give a better listen to their clients' words.

CONCLUSION

Our research on communication between genetic counselors and pregnant Latinas and their partners revealed predictable sources of miscommunication, including the use of idioms and jargon, inadequate translation, counselors' efforts to maintain neutrality (the edict of nondirectiveness), and a reluctance to directly contradict what they feel are clients' erroneous worldviews. Cutting-edge diagnostic technologies, as much in the prenatal domain as elsewhere in medicine, have challenged the long-established hierarchical relationship between patient and clinician. Providers are expected to share their knowledge and expertise with clients and they no longer have the final word in patient care. Free and open communication, as well as an awareness of the many opportunities for miscommunication will be key to the success of this newly emerging medical paradigm. Since "informed" consent is increasingly recognized as a basic right and goal in medical care, such consent is best achieved when patients have a good understanding of the means and ends of the care they are being offered. This analysis is offered in the spirit of moving closer toward those objectives.

COMMUNICATION: CLINICAL DIAGNOSIS

Riyana Babul-Hirji, Genetic Counseler, Lecturer, Hospital for Sick Children, Department of Molecular Genetics and Microbiology, University of Toronto, Toronto, Canada
Cheryl Shuman, MS, CGC, Assoc. Professor Division Clinical & Metabolic Genetics, Medical Genetics & Microbiology, University of Toronto, Toronto, Canada.

HISTORY

Individual III:1 in the pedigree is a 4-year-old-boy of Chinese origin who has been referred to the Genetics Clinic to confirm a suspected diagnosis of neurofibromatosis

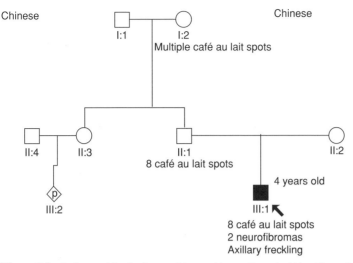

Figure 2.1   Pedigree of family discussed in case history. Numerical identifiers refer to individuals discussed in the text. Arrow indicates proband.

type 1 (NF1). He presents with eight café au lait spots, two neurofibromas, and axillary freckling. Based on his clinical presentation, a diagnosis of NF1 is confirmed. Examination of his parents reveals that his father (II:1) has eight café au lait spots and reportedly the paternal grandmother (I:2) has multiple café au lait spots. II:1's sister (II:3) is currently pregnant. (see Fig. 2.1.)

Background and Diagnostic Criteria for NF1

Neurofibromatosis type 1 is one of the most common dominantly inherited conditions with a prevalence at birth of 1 in 3000. The clinical manifestations are extremely variable, and the diagnosis is based on the presence of two or more of the following clinical findings [diagnostic criteria established by National Institutes of Health (NIH) consensus conference, 1987]:

- Six or more café au lait spots over 5 mm in greatest diameter in prepubertal individuals and over 15 mm in greatest diameter in postpubertal individuals
- Two or more neurofibromas of any type
- Intertriginous freckling (axillary or inguinal)
- Optic glioma
- Two or more Lisch nodules
- Distinctive osseous lesion (e.g., sphenoid dysplasia)
- First-degree relative with NF1

Molecular Genetics

Located on chromosome 17q11, *NF1* is the only gene currently known to be associated with this condition. Genetic testing is challenging as many mutations occur in this gene due to its large size. The detection rate is dependent upon the testing methodology, which may include protein truncation, Fluorescence in situ hybridization (FISH) and/or cytogenetic analysis. Linkage analysis can be considered if there are two or more available family members with confirmed clinical diagnoses and a mutation has not been identified [Friedman, 2004].

Issues

This case scenario highlights a number of issues related to communication of a clinical diagnosis and genetic testing. Some of these issues are listed below in italics and are explored in more detail in this chapter. The reader may wish to (1) consider the potential issues in this scenario that could hinder the communication process and (2) how such issues could be addressed.

- This family has recently emigrated from China and has very limited facility with English and therefore *a medically trained interpreter* should be requested for the appointment.
- Discussion of diagnosis, prognosis, mode of inheritance, and recurrence risk for NF1 should be provided at *an appropriate level* and could be supplemented by the use of *educational tools or aids*.
- As part of the *discussion regarding the availability of genetic testing* for NF1, the family should be aware of potential benefits of pursuing testing as well as the limitations. If DNA (deoxyribonucleic acid) testing is pursued, discussion should include testing approaches, sensitivity and specificity, and relevance to and implications for other family members.
- The family may be very concerned about protecting their privacy regarding the diagnosis in keeping with their *cultural values*.
- The clinician should discuss with the family the importance of *informing other family members* about the family history of NF1, especially II:1's sister who is currently pregnant. As *family dynamics* may influence their willingness to communicate this information to others, the clinician should explore this with the family.
- The counseling session should incorporate *psychosocial and social issues* relevant to this family, for example, guilt, anger, isolation, and stigmatization.

INTRODUCTION

As the case history illustrates, effective communication in the context of genetic testing is critical to empowering patients so that they might make informed and autonomous decisions in keeping with their value and belief systems. Clinicians should strive to promote an active dialog and to create an environment that is supportive and empathetic. However, several factors, beyond the control of the health care provider, may influence the communication process including cultural background, intellectual capacity, language facility, family dynamics, social support, and cognitive reasoning styles and abilities as well as coping mechanisms. This section will provide a concise overview of select factors and tools that may influence the communication process as related to genetic testing.

Genetic testing can be utilized for various reasons: to determine carrier status, to confirm a diagnosis, to recommend appropriate therapies and/or preventative behaviors, and lastly for presymptomatic genetic testing for which medical management may or may not be altered [Collins, 2003]. Currently, consumers of genetic testing include adults and capable minors for specific genetic conditions, women who are pregnant for prenatal diagnosis, and parents for newborn screening as well as for their children who would medically benefit from genetic testing. However, with the mapping of the human genome and the search to find genes causing or involved in susceptibility to common diseases such as heart disease, diabetes, and asthma, future consumers will include virtually anyone and everyone.

The process of genetic testing can be complicated and the results received difficult to interpret. The intricacies of testing can make the communication of such information to patients that much more challenging. Genetic counseling is the practice of helping individuals and families understand the medical, psychological, social, and reproductive implications of genetic and congenital conditions. Elements of the practice include: assessment of the chance for recurrence or occurrence of a condition; education about inheritance, testing options, medical management, prevention, social support, and research; and counseling to help clients adapt to the choices and to the psychological, familial, and social issues that stem from the risk or condition in the family [Working Group, 2004]. Counseling in the realm of genetic testing would include imparting relevant medical and genetic information as well as discussing the availability of the test, examining the need for genetic testing, eliciting the reasons for wanting the test, discussing the available testing modalities, their limitations and their complexities, and discussing the possible misconceptions about the test [Helmes et al., 2002]. This may involve a vast amount of information that could make it overwhelming for both the client and the clinician. Therefore, it is important to balance communicating critical information with the client's expectations and processing abilities [McCarthy Veach, 2003].

Counseling is recommended both prior to and potentially after results are received and should incorporate a discussion of how such testing would impact not only the individual seeking testing but also other family members. If one is to make an informed decision about genetic testing, then the information should be presented in a comprehensible, caring, and culturally sensitive manner. This implies that to communicate effectively, one must search for cues that will provide some insight into the client's perspective and motivations as well as to engage the client and his or her family members in active dialog. This lays the foundation for espousing a trusting relationship with the client.

CLIENT CHARACTERISTICS: CULTURE AND INTELLECT

In developing and practicing communication strategies that are culturally sensitive, a number of factors need to be assessed at the outset of the appointment. These include race, religion, ethnicity, socioeconomic status, and gender. Any assumptions about the clients should be avoided, especially those based solely upon appearance. Acculturation and assimilation into Western culture may have taken place or an adherence to traditional customs and values may be retained. A culturally sensitive clinician will strive to understand the reference frame of the client by applying interview skills in addition to exploring his or her own personal assumptions and beliefs [Baker, 1998; Sue and Sue, 1990; Sue, Arrendondo and McDavis, 1995]. Incorporation of culturally sensitive approaches and counseling tools when presenting information promotes a sense of trust in the medical system and engagement in the counseling process. It also allows the information to resonate with the target population [Baty et al., 2003].

Throughout the interview process, the clinician should also seek evidence of intellectual functioning that may be elucidated through active dialog together with detailed history taking. Care must be taken to deliver the information at an appropriate level in order to empower clients so that they can process such information for decision making. This can be especially challenging when the client(s) has mental retardation. In such situations, "the focus needs to be psychosocial, whether or not any informational goals are realized" [Finucane, 1998 p. 63].

EDUCATIONAL TOOLS AND FACILITATORS

Counseling regarding genetic testing involves a significant educative component. Individuals have different learning styles and abilities that influence how they comprehend information. It is not feasible to assess each patient's learning style prior to engaging in a discussion about genetic testing. However, the delivery and processing of highly complex genetic information can be facilitated by incorporating educational tools. Such tools will not address all aspects of learning but may serve to enhance the provision of genetic information. Currently, educational tools include pictorial aids either published [e.g., Counselling Aids for Geneticists, Sweet et al., 1989] or created by the clinician for the individual patient session, videos, audio-guided workbooks, and computer-based programs [O'Connor et al., 1999; Green et al., 2001a]. The choice of educational tool(s) is dependent on a number of factors, including the clinical setting (e.g., an obstetric practice may use videos to describe prenatal testing options), the condition for which genetic testing is being requested (e.g., cartoon depicting CGG repeats for fragile X syndrome), and client preferences (visual learner vs. verbal learner). Although educational aids are effective in enhancing the information-giving component of a session, the aids currently available do not meet the needs of individuals who are intellectually challenged, of low literacy, who are economically disadvantaged or of specific ethnic groups [Baty et al., 2003].

Videos and computer-based programs have been shown to be effective alternatives to the traditional face-to-face interactions in providing the educative component for specific patient populations. Ongoing evaluation of the efficacy of one such educational tool, a computer-based program regarding genetic testing for breast cancer, revealed that it was more effective than standard genetic counseling for increasing knowledge, especially among women at low risk of carrying *BRCA1* or *BRCA2* mutations. This computer-based decision aid facilitated the educative component by allowing participants to proceed at their own learning pace and in privacy [Green et al., 2001a]. However, genetic counseling was valued to address personal concerns and psychosocial issues as well as to discuss personalized risk assessments and decision making [Green et al., 2001b, 2004].

The near future may well see at least a proportion of clients arriving at a session to discuss genetic testing with a copy of their family history, prepared by a computer-based pedigree program available on the Internet; one such program can be found on the website for the American Society of Human Genetics (http://genetics .faseb.org/genetics/ashg/ashgmenu.htm). In addition, they may have the factual knowledge base and be aware of genetic testing issues specific to their condition as a result of having used a computer-based decision aid. Such well-informed clients may make it easier for the health care provider to convey critical aspects of genetic testing. However, some clients may present with a skewed information base or may, in fact, be misinformed. This can be challenging as the health care providers must adhere to a professional standard of communicating relevant genetic information. Some clients, feeling that they are already well informed, may be resistant to engaging in further discussion and may view the clinician as an obstacle to accessing genetic information about themselves. Such situations highlight the importance of contracting and establishing a mutually agreed upon agenda at the outset of the session.

Essential for facilitating the communication process are medically trained interpreters for clients who do not speak English and when the health care providers themselves are not bilingual or multilingual. In such situations, the health care providers rely upon interpreters to convey the information, questions, and concerns to the clients and vice versa. The clinician should be especially vigilant in assessing client demeanor and comprehension by observing visual cues and listening to interpreter–client dialog for both tone and unusually lengthy discourse. In addition, the clinician may direct the clients to relevant videos, brochures, and/or pamphlets that have been translated to serve specific patient populations. Other aids such as telephone-based interpreting services (e.g., AT&T Language Line) and computer-based translating programs for summary letters may also be useful, but the latter should be used with discretion as the translation may not accurately reflect the intended content.

Summary letters for clients, in general, are commonly utilized in the provision of genetic care and serve to document the information provided. These letters, which may include relevant medical information and recommendations as well as genetic test results, should be accurate and written at an appropriate level of understanding. Consideration should be given to avoid potentially stigmatizing phrasing and medical jargon. Moreover, summary letters should not disclose confidential information about other family members, such as genetic test results, unless appropriate consents have been obtained. Specific guidelines have been developed for formatting client letters, as well as how to best present medical and genetic information with sample patient letters, in an excellent article by Baker et al. [2002]. There is evidence that written summaries enhance patient understanding and information recall as well as information dissemination to other family members [Hallowell and Murton, 1998].

DECISION MAKING: CONSIDERATIONS AND APPROACHES

As mentioned earlier, genetic testing not only impacts the individual but potentially their family members as well. Family systems literature describes the family as a complex and interactive social system and emphasizes the need to understand family structure, family change and development, and ways in which families interface with larger systems such as the medical system [Peterson et al., 2003]. Thus, genetic counseling should also include exploring the role of family members, as well as their interpersonal relationships and communication styles. The individual who openly communicates with his or her family may have better adaptation skills to new and different situations due to ongoing emotional support. Other individuals may not communicate relevant information to other family members, either as a conscious (e.g., do not want to elicit sympathy) or unconscious (e.g., denial) decision [McCarthy Veach, 2003]. This leads to ethical and legal questions when other family members may be directly impacted by the information that the client is resistant to sharing. Also complicated is the situation where the defined risk pertains to other family members in attendance who have conflicting needs and cognitive styles. In this circumstance, questions such as "Who is the patient?" need to be entertained.

Understandably a significant amount of information is provided to the client(s) in order to appropriately frame the option of genetic testing. How individuals process this information and ultimately make decisions is influenced by a number of factors, including client affect, learning styles, gender, cultural and spiritual background, socioeconomic status, family and support networks, personal experiences, and risk perception. All of these factors are complex and interdependent. Risk perception itself, is a multifaceted concept and will be explored in greater detail in Chapter 14. A number of different strategies to communicate risk can be utilized either individually or in combination, in keeping with the client's cognitive processing style. Briefly, risks may be presented in a probability-based approach with numerical data or in a contextualized approach with personally relevant background and potential outcomes [Julian-Reynier et al., 2003]. These should be complementary and adapted to the client's cognitive needs, however, the difficulty lies in accurately assessing those needs.

Another tool potentially useful in risk communication especially when one of the goals includes the motivation of behavior change is "fear appeal," a form of behavior modification [Sweet et al., 2003]. This approach has been used effectively in the advertising realm and is now being studied for applicability in certain subspecialty areas of genetic counseling, such as those involving cancer genetics. It has been suggested that direct terminology rather than couching the message (e.g., cancer risk assessment rather than genetic counseling for family history of cancer), with the immediate provision of medical management options and personalization of message are likely to be more effective for adopting better health behaviors [Sweet et al., 2003]. This communication strategy would not be applicable to all areas of genetic testing but rather could be incorporated for those conditions where altered medical management has been shown to improve health outcomes. The clinician should be cautious in applying and incorporating fear appeal when communicating with clients as the intent is not to incite fear. If this strategy is used appropriately, however, it should empower clients to have more control of their health management.

Individuals make decisions after having received information that is thorough, factual, and accurate. If the communication process is effective, then clients may make decisions that are informed and autonomous. Informed choice encompasses three main characteristics: "(1) the decision is based on relevant, high quality information, (2) reflects the values of the person making the decision and (3) is behaviorally implemented" [Wang et al., 2004 p. 1433]. In order to empower clients in their decision making regarding genetic testing, clinicians should have a basic understanding of the processes or strategies involved in decision making.

Decision-making processes can be either analytic or automatic [Wang et al., 2004; Broadstock and Michie, 2000]. Analytic processes imply that clients make rational, cerebral choices in an attempt to attain the very best outcome. Anticipatory guidance and consideration of the pros and cons of genetic testing are just two approaches that may be helpful for analytic decision makers [McCarthy Veach, 2003]. Automatic processes are more instinctual in nature with the purpose being to reach a satisfactory outcome [Wang et al., 2004]. It is not known whether such processes or strategies actually facilitate the decision-making process for clients and further study is indicated. In addition, certain clients may have a preferred decision-making style: (1) autonomous, (2) in consultation with the clinician, or (3) delegated to their clinician. Regardless of the client's preferred decision-making style, the clinician should endeavor to incorporate a shared decision-making approach [Helmes et al., 2002; Burke et al., 2001] in examining and discussing the need for genetic testing with the client. This shared decision-making approach supports open dialog between the client and clinician in order to promote ongoing engagement in their genetic care.

Finally a discussion on communication would not be complete without mention of a term commonly referred to in the genetic counseling literature—*nondirectiveness*. In the 1950s, Carl Rogers incorporated nondirectiveness in his client-centered psychotherapeutic theory and intervention [Rogers, 1951]. This approach was later adopted by the genetic counseling profession in an attempt to dissociate itself from the eugenic movement [Bartels et al., 1993; Weil, 2003]. Although the application of nondirectiveness is not consistent among genetic practitioners, its basic premise is to avoid any semblance of coercion, either verbal or nonverbal. Inherent in this approach is effective communication of balanced, value-neutral information in a supportive environment to promote autonomous decision making in keeping with the client's value and belief systems [Weil, 2000]. Recently, the practice of nondirectiveness has been questioned in a number of genetic settings including the realm of genetic testing. When a genetic test is requested by a client and there is no medical or genetic indication for such testing, is it ethically sound for the clinician to remain nondirective? Similarly, when there is an evidence-based medical protocol that will clearly improve health outcome, is there any rationale for remaining nondirective? Promotion of client autonomy through active engagement and open dialog together with further exploration of psychosocial issues in a caring manner are being considered as more appropriate guiding principles of genetic counseling, rather than the previously ascribed term of nondirectiveness [Weil, 2003].

As technology advances and the integral role of genetics in common disorders become more widely recognized, the demand for genetic testing will continue to increase. The complexity of issues inherent in genetic testing necessitates an approach by clinicians that integrates the clients' psychological and social situations together with the educative component. The potential for ineffective communication ultimately leading to psychological harm is very real, whereas open and effective communication can serve to help the client cope more effectively, not only emotionally but also in the decision-making process [Weil, 2000]. As such, the basic premise of genetic care should rest on the communication of balanced information (including facts, alternatives, anticipated consequences), devoid of coercion, in a supportive environment and with emphasis on clients' psychological and social dynamics [Weil, 2003].

Chapter 3

Psychological Aspects

OVERVIEW

Counseling and testing for genetic disease are likely to raise psychosocial concerns for patients. The fact that many clinical situations revolve around estimations of risk of disease onset, rather than disease onset itself, immediately introduces uncertainties and complexities related to the discussion of possibilities. Another major complicating factor is the essential hereditary mechanism of disease transmission, often leading to an obligatory concern and involvement for family members in counseling situations where privacy may be urgently desired by the patients. This chapter will describe factors contributing to psychological distress during genetic counseling and testing, suggest techniques for recognizing signs of distress, and outline approaches to avoiding harmful outcomes associated with a patient's inability to cope.

INTRODUCTION

Genetic disease and testing can trigger a host of emotional and psychological responses that impair the communication, understanding, and retention of information. Patients can be under a significant level of stress during counseling, even if they do not show it. This is especially true when the counseling concerns serious outcomes or where the prognosis is difficult.

- Patients may consciously or subconsciously choose to focus only on specific components of the information, or discount information that they cannot deal with, or be unable to comprehend little if any of the information at the time.

- They may perceive and retain significant details in a way that is unintended or unexpected by the health care provider.

Counselors and physicians need to ensure that the patient accurately can recover the information provided during counseling. Inadequate assessment and support during counseling, particularly during periods of significant stress, can result in an inaccurate

Genetic Testing: Care, Consent, and Liability, by Neil F. Sharpe and Ronald F. Carter
© 2006 John Wiley & Sons, Inc.

and/or incomplete family history, inappropriate testing, inaccurate diagnosis, patient misunderstanding of test results, patients who over- or underestimate the seriousness of the situation, and/or inappropriate follow-up and aftercare (including loss of continuity of care). This chapter illustrates clinical objectives and practices that promote effective counseling with the individual, couple, and family.

Scenario 1

A couple is considering having a second child. The woman requests an appointment with her physician because her husband has been told recently that his father has Huntington disease. The couple's first child is a healthy 3-year-old son, who is also at the appointment. They want to know if the husband can be tested. The woman is the primary provider in the family; the husband is currently not employed and looking after the son at home.

Issues Raised by This Scenario

Huntington disease is frequently diagnosed in a family at a point of adult onset, and persons at risk for inheriting a mutation already may have children. In this situation, the family is likely to be still struggling with the impact of the recent diagnosis, which will heighten their anxiety. The parents will be concerned not only for the husband's own status but also for the son and their hopes for more children. Because the age of onset for Huntington typically decreases with succeeding generations, the couple already may have formed an opinion that the husband is affected; the reason why the husband is unemployed may turn out to be important. If the mother already was pregnant, these issues would be even more acute and anxiety provoking. This is a situation where the patients are likely to be highly anxious and to have developed preconceived ideas; the counselor will need to address that anxiety and determine whether the patients are able to participate in the development of appropriate plan for care and comprehend an accurate risk assessment.

Scenario 2

A severely deformed infant is born. Investigation reveals that the deformities are caused by a familial chromosomal anomaly; the father has a reciprocal chromosome translocation, which the baby inherited in an unbalanced fashion. Appropriate counseling is given, other family members are notified, and relatives of the father undergo counseling and karyotyping. The father's sister requests an appointment with the geneticist and brings along her 15-year-old daughter. Both are found to carry the translocation. It becomes apparent that the daughter has left home for an ongoing relationship and already has had one miscarriage. Both mother and daughter want to know whether she can have more children. Further discussion reveals that the mother strongly believes that intervention is unnecessary because God will take care of them.

Given the chromosome breakpoints of the translocation and the empirical risks associated with the known occurrence of abnormal livebirth, the counselor explains that there is a significant increase in risk for miscarriages and stillbirths, and that the risk for an abnormal liveborn baby is probably in the range of 20 to 30 percent. Also, the counselor carefully explains in simple terms the options for prenatal diagnosis and assures the daughter that these options should be offered to her for any pregnancy. The mother repeatedly tells the daughter during the counseling not to worry about any testing because everything will turn out fine. The appointment lasts for more than an hour, and the counselor concludes by speaking directly to the daughter, assuring her that: (1) she still has at least a 70 percent chance for a normal infant with each pregnancy; (2) she should be offered prenatal counseling and diagnosis for any pregnancies, and (3) she has the right to choose her options for management of any pregnancies. The counselor sends a consult letter to the family physician and another letter to the family, carefully outlining the risks and options for future pregnancies. One year later, the daughter presents in late-stage labor at a hospital and delivers a severely malformed infant. When asked by the neonatologist about what happened during her pregnancy, she said she had been told not to have children by the genetic clinic.

Issues Raised in This Scenario

This is a situation where the counseling information did not provide sufficient support and understanding to the daughter. Whatever she apparently retained as information did not reflect the counseling provided. This occurred despite the care taken by the counselor to provide accurate information in regard to her risks and options and the appropriate letters written to both physician and patient. At the time of counseling, the geneticist was worried that communication was hindered by the mother's insistence on the protection provided by religious belief and her repeated instructions to the daughter to disregard the pessimistic predictions. The daughter's actions following the counseling suggested that the counseling experience probably instilled a sense of fear or defensiveness from medical care. In this scenario, the combination of the age of the daughter, the influence of her mother, the family's personal and religious beliefs, and anxiety associated with medical intervention, all apparently combined to leave the daughter with the erroneous impression that she was forbidden to try to have children.

NEED TO KNOW

- Potential for psychological harm
- Potential for psychological benefit
- Anxiety and perceived risk: pretest status of patient
- Recognizing psychological stress
- Coping with psychological stress: improving outcomes

Potential for Psychological Harm

The possibility of psychological harm is a major concern during the process of genetic counseling and testing.

- Much of the psychological stress associated with genetic counseling is associated with concerns for future onset of genetic disease for the patient or a family member.
- Feelings of guilt, loss, and doubt can arise if the patient learns of a mutation that was passed on to other family members, decides to terminate a pregnancy, or chooses to have prophylactic surgery.
- The uncertainties often implied in evaluations of risk of disease also can cause confusion and anxiety, or the patient may not be able to comprehend and benefit from the significance of a result because of psychological barriers.
- In situations where the perceived risk has become assimilated into the patient's perception of self, a test result that changes that risk can cause doubt and conflict.
- Psychological harm may arise secondarily if members of the family are subjected to stresses arising from harmful reactions of others: friends, relatives, insurance companies, employers, and health care professionals.
- Religious and cultural backgrounds can strongly modify a person's ability to make use of genetic counseling and testing.

The impact of these stresses is often increased by their frequent coincidence with situations that already are difficult: family planning decisions, pregnancy, a diagnosis of cancer, or death of a relative.

Potential for Psychological Benefit

Patients and their families can derive psychological benefit from genetic counseling and testing; it is this expectation of benefit that is part of the motivation for patients in many cases.

- A positive test result can bring feelings of relief, explanation, or justification for previously unexplained concerns.
- Accurate identification of a causal genetic mutation can reduce anxiety in the longer term and improve psychosocial status through improved clinical management and a sense of control and empowerment.
- Test results that reduce the risk of disease frequently result in an immediate reduction in anxiety and a sense of relief.
- Negative test results usually are assumed not to trigger harmful effects; however, it should be noted that a sense of *survivor guilt* may occur in family members who are spared the inheritance of a mutation that affects close relatives.

- Patients often derive a positive sense of altruism and involvement in health care research by participating in genetic counseling and testing. They may recognize the potential benefit to their children and other family members, and they may feel satisfaction and accomplishment in helping to provide information that will be helpful to others.

Anxiety and Perceived Risk: Pretest Status of Patient

Anxiety influences the patient's motivation and degree of compliance with genetic counseling and testing. Low levels of anxiety can be associated with failure to seek counseling, while higher levels of anxiety can increase the patient's motivation to seek and complete the process. Precounseling anxiety also can represent a barrier to the patient, leading to feelings of fatalism, denial, or avoidance.

Anxiety levels typically vary during the process of counseling; they are affected by:

- Precounseling status of the patient
- Perceived level of risk
- Perceived severity of outcomes
- Process of counseling
- Waiting periods
- Provision of test results and associated counseling in regard to actual risk
- Long-term management of patient care

The pretest psychosocial status of the patient is a major predictor of the patient's reaction to testing and ability to cope with outcomes. Because testing may increase patient anxiety, it is important to assess the patient's initial status for possible barriers to appropriate recognition of risk and response to counseling.

If the patient is distressed or highly anxious, preoccupied by an inaccurate or exaggerated perception of risk, depressed or guilt-ridden, fatalistic about the expected onset of disease, convinced that a particular type of result will be revealed (either positive or negative), or strongly misinformed about the actual nature and cause of the situation, the counselor needs to be able to recognize and address these issues before testing is initiated. Failure to respond to a poor pretest psychosocial status increases the possibility of adverse psychological reactions following testing.

Recognizing Psychological Stress

The counselor needs to have a clear perception of the patient's ability to make an informed choice in regard to testing and to respond appropriately to the test result. Pretesting assessment, therefore, needs to include discussions of the patient's:

- Concerns and motivations for seeking counseling and testing
- Prior experiences, family relationships, personal health, and lifestyle

- Religious and cultural influences
- Comprehension of potential implications and outcomes of testing
- Expectations from counseling, particularly the usefulness of the test result and any implications for family planning for health care
- Preparation for testing and receiving a result

In situations where concerns arise, it may be useful to apply a standardized approach to measuring psychosocial status. Various screening instruments can be used [Derogatis, 1993; Radloff, 1997; Zigmond and Snaith, 1983; Beck and Beamsderfer, 1978] to obtain an objective assessment for the patient. Clinical psychologists and social workers can provide very useful support to the counseling process and have become an important component of clinical genetic teams. Their skills are invaluable in assessment and modification of an individual's psychosocial status.

Coping with Stress Induced by Genetic Counseling and Testing

The degree of management and intervention required during and after genetic counseling will vary with the patient's status.

- In the pretest phase, support and expert assistance may be required for modification of the patient's anxiety.
- Following a positive test result, health care providers have to be alert for evidence of increased anxiety, depression, isolation, guilt, disbelief, denial, and other signs that the patient will have difficulty comprehending and coping with the test result.
- Sometimes the testing protocol provides an inconclusive report; either the genetic status of the patient cannot be ascertained by currently available methods or a genetic variant of uncertain clinical significance is identified. Uncertainty and confusion may result. The counselor needs to be able to understand and to effectively communicate the clinical significance of such results accurately to the patient.

Following provision of a test result, the patient is often faced with the need to make choices on future care. Sometimes the choices must be made rapidly, as in the decision to continue or terminate a pregnancy, or to have bilateral mastectomy rather than a lumpectomy for a breast tumor. The counselor needs to be able to recognize how much information the patient can understand, communicate it effectively, and ensure that the patient's psychosocial status permits the patient to choose the most appropriate clinical action. The manner in which results are communicated to the patient must be matched to the patient's needs.

Results usually become part of a larger management plan that often includes decisions on referrals to other specialists, family planning, prenatal diagnosis, communication with other family members, and improved management of affected family members. Health care providers have an important role in ensuring the continued functional use of the information by the patient during this process.

Although it is clear that the passage of time itself can improve a patient's anxiety level and psychosocial status, some patients may experience persistent problems, and these need to be recognized and appropriately addressed. Patients who are known to be coping poorly need to be followed and reassessed; this takes time and continuing involvement in the delivery of care. More than one type of professional assistance may be required.

Physicians, genetic counselors, social workers, and psychologists each have their own roles in the care of a patient. Patients may form strong attachments to counselors as part of their coping strategy, but sometimes the patient's need for help is beyond the scope of the counseling resources. The support of other team members, family members, friends, and the community is important, and when it appears that a patient may need intervention and skilled assistance, referral to expert help is critical. For patients and care providers in remote communities, models for shared care may be useful. Teleconferencing also appears to be particularly suited to this application because clinical genetics practice consists primarily of exchanges of information; consultations with relatively few family members present the least technical challenge [Gattas et al., 2001].

Patients are using Internet-based resources for information and support. Patient groups and associations also provide practical advice, access to resources, enthusiastic support, a sense of community unique to affected persons, advocacy, and increased societal awareness. All of these sources of aid and support can contribute to the continuing improvement of quality of life for the patient.

WATCH OUT FOR

Factors Affecting Perception

The perception and interpretation of risk and test results can be affected by personal experiences and values on the part of health care professionals as well as patients [Lloyd, 2001; Doust and del Mar, 2004; Abramsky et al., 2001]. A person whose mother died of breast cancer at 40 years of age may perceive risk quite differently from another whose mother was diagnosed with breast cancer but survived [Shapiro et al., 1992; Kelly, 1992; Thirlaway and Farrowfield, 1993; Audrain et al., 1995; Edwards et al., 2002].

- The emotional and psychological responses of the patient may be more critical to risk perception than the actual facts (diagnosis, risk information, prognosis) [Lippman-Hand and Fraser, 1979; Wertz et al., 1986; Lerman et al., 1995; Hoskins et al., 1995; Phillips, 1999; Press et al., 2001; Horowitz et al., 2001; Lodder et al., 2001; Peskin et al., 2001].

- Providing detailed information at the same time as a discussion of diagnostic information may be confusing and not be understood by the patient due to resulting stress and psychological responses [Lerman et al., 1995; Menahem, 1998; Hagerty et al., 2004].

- A negative test result may not reassure patients who continue to believe that they are at risk [Huggins et al., 1992; Lynch and Watson, 1992; Michie et al., 2003].

- Psychological difficulties have been associated with patient delay and a lack of compliance with surveillance and management options [Metcalfe et al., 2000; Kash et al., 2000].

Health care professionals should be prepared to provide emotional and psychological support especially in the case of:

- False-positive and false-negative results [Milunsky et al., 1989a; Tluczek et al., 1991; *Humana of Kentucky, Inc. v. McKee*, 1992; Hall et al., 2000; Waisbren et al., 2003].

- Potentially life threatening genetic diseases or conditions [Frets et al., 1991; Sharpe, 1994b; Lerman et al., 1995; Hoskins et al., 1995; Epstein et al., 1998; Marteau and Croyle, 1998; Kash et al., 2000; Collins et al., 2000; Peskin et al., 2001; Horowitz et al., 2001; Edwards et al., 2002; Metcalfe et al., 2002; Waisbren et al., 2004].

Standards and Interpretation

Professional accreditation standards for medical genetics and genetic counseling require the ability to recognize and to help patients with their emotional and psychological needs. The American Board of Medical Genetics, Inc. [ABMG, 2004] standard for a clinical geneticist and Ph.D. medical geneticist requires skills to:

- Help families and individuals recognize and cope with their emotional and psychological needs.
- Recognize situations requiring psychiatric referral.

The American Board of Genetic Counseling [ABGC, 2004] requires the ability to:

- Assess psychosocial factors, recognizing social, educational, and cultural issues.

The Canadian College for Medical Geneticists [CCMG, 2004b] standard for a clinical geneticist requires that clinical genetics have the ability to:

- Recognize the psychological implications of a genetic problem in the family.
- In providing genetic counseling, "the physician should be able to be sympathetic and empathetic, remain objective and impartial, delineate his or her own ethical standards and appreciate those of the patient, appreciate the general mores of the culture of the patient, be non-directive in most instances, but be prepared to advise in certain situations and provide psychological support, either personally or through referral."

An open question with regard to the psychological aspects of genetic testing and counseling is: What standards of care will be applicable to health care professionals who lack specialized training in genetics yet who elect to provide genetic testing services including communicating, and counseling for, test information? The clarity of desired clinical norms and objectives, such as "provide psychological support," does not necessarily imply a similar clarity with regard to what constitute appropriate protocols. This issue may turn on the particular nature of the patient's emotional and psychological or psychiatric needs [e.g., Sharpe, 1994b]. Could a health care professional be held to the standard of care applicable to a psychologist or a psychiatrist [Sharpe 1994b; Kapp, 2000]? Although, from both a clinical and legal perspective, the issue of how these standards will be interpreted continues to evolve, courts have held that a physician who knows or should know that a patient's condition is beyond the physician's professional skills and knowledge has a duty to refer [e.g., *Naccarato v. Grob*, 1970]. If the health care professional does not do this and undertakes to treat the patient, the physician may be held to the same standard of care applicable to the specialty [*Larsen v. Yelle*, 1976; Sharpe, 1994b; Kapp, 2000].

PSYCHOLOGICAL ASPECTS OF GENETIC TESTING FOR ADULT-ONSET HEREDITARY DISORDERS

M. J. Esplen, Department of Psychiatry, Faculty of Medicine, University of Toronto, Toronto, Ontario, Canada

INTRODUCTION

The Human Genome Project is a fast-developing area of health research. New gene-based diagnostic tests, such as genetic testing, are currently being performed, leading the way to revolutionary treatment and prevention options. In the past, most genetic tests were used to determine genetic risk within the context of reproductive decision making, were for rare disorders, and were provided by a few specialists. To this area of practice has been added counseling and testing for adult-onset hereditary disorders (AOHD), where individuals may be provided with information about their own risk of future disease. The Health Canada Working Group on Public and Professional Educational Requirements Relating to Genetic Testing of Late Onset Disease 2000, defines AOHD as "disorders in which there are heritable factors and that normally become symptomatic in adult life. These diseases include, but are not limited to, cancers (e.g., breast, ovarian and colorectal cancer) and some neurodegenerative diseases, such as Alzheimer's disease and Huntington disease." As new gene loci are identified, genetic testing for other common disorders, such as heart disease and diabetes, may become more widespread. The resulting increased demand for testing may necessitate involvement of new groups of health professionals, including primary care physicians and nurses.

The benefits of risk evaluation are readily apparent. For example, for the cancer syndromes, the potential to accurately distinguish carriers from noncarriers allows for more precise targeting of early detection and prevention strategies. For Huntington disease, the potential benefits of genetic testing include increased certainty and an improved basis for reproductive decision making [Meiser et al., 2000]. Knowing that one is at substantial risk for a disease may help people avoid or "prepare" for it. There may be steps one can take to prevent disease onset, and there may be important personal issues that need to be resolved before illness develops. However, the identified advantages of testing must be considered alongside the potential for harm. A number of psychosocial challenges can arise for an individual undergoing genetic testing, suggesting the need to integrate psychological care into genetics services.

A brief overview of the process of genetic testing in AOHD will be presented, with reference to its phases and corresponding psychological aspects. Published findings on the psychological impact of genetic testing for adult-onset disorders, focusing chiefly on Huntington disease and the hereditary cancer syndromes, for which most data are available, will be incorporated, followed by a section on implications for clinical practice.

PROCESS OF GENETIC TESTING

Initial studies on testing for Huntington disease and cancer have included standardized protocols involving one or more counseling sessions by a genetic counselor or geneticist to help people decide whether or not to proceed with testing. The offer of the test is generally separated in time from the taking of a blood sample to conduct the DNA (deoxyribonucleic acid) test. During the pretest counseling stage, the focus is on collecting a family history, assessing eligibility and risk level, and providing information on the pros, cons, and potential meaning of a positive test result. In addition, at this stage of the counseling family implications are discussed. Following this session, there is a waiting period at which time DNA testing is conducted. A posttest counseling session is then provided at which time the disclosure of the genetic test result occurs. During this session the impact of the genetic test result is fully explored. Emphasis is placed on assisting the individual to comprehend the meaning of the genetic test result in terms of its associated risk for disease and its implications. The individual's immediate reactions to the genetic test result are explored in order to gauge the person's sense of his or her assimilation of the information and in relation to his or her expectations of the test result. Specific surveillance and prevention recommendations are presented, which often include the scheduling of follow-up appointments for screening tests and for consultations with specialists to facilitate medical decision making concerning prevention options (e.g., prophylactic surgery). During the disclosure session, the implications for the testing of family members are also discussed and the individual is given the opportunity to explore potential reactions and issues concerning his or her role in the dissemination of genetic information to relatives. Some programs build in a follow-up in-person or telephone session within

BOX 3.1 *Psychosocial Aspects of Genetic Testing*

Pretest Counseling Phase

- Assessment of interest in genetic testing, attitudes, beliefs, motivations
- Screening for individuals at psychological risk or experiencing psychological distress

Disclosure of Test Result Phase

- Management of emotional responses evoked by risk and complex genetic information
- Screening of individuals at psychological risk
- Screening of individuals with current adjustment difficulties
- Assessment of potential family dynamic issues/responses
- Supporting medical decision making

Follow-up Phase

- Management of emotional responses due to ongoing medical decision making or procedures
- Supporting efforts at dissemination of genetic information to family
- Facilitating individual and family adjustment
- Supporting other family members through genetic testing process

one week to one month in order to assess the individual's adjustment, perceptions around surveillance and prevention recommendations, and to inquire about any additional support needs concerning dissemination, decision making, or adaptation to the genetic test result. The following sections deal with each phase of the genetic testing process and potential associated psychosocial issues (Box 3.1).

PRETEST GENETIC COUNSELING PHASE

At the pretest genetic counseling stage, an important psychological aspect is the assessment of the motivation and expectations of the individual prior to the taking of a blood sample. A primary motivation for genetic testing is to gain certainty around risk and information concerning surveillance or preventive measures for a disease [Esplen et al., 2001; Biesecker et al., 2000]. In general, uptake rates for genetic tests tend to be higher when there are effective ways of treating or preventing the condition. If little can be offered, fewer individuals want information about their risk status. To exemplify, the uptake rate for DNA predictive testing is approximately 10 to 15 percent for Huntington disease for which there is no treatment; for breast cancer for which there is some possibility of prevention and treatment, it is approximately 50 to 70 percent; and it is approximately 80 percent for familial adenomatous polyposis for which there is effective treatment [Marteau and Croyle, 1998; Bloch et al.,1989; Meiser et al., 2000]. Expressed interest does not necessarily translate into uptake of a new genetic service. For example, interest in genetic testing for Huntington disease was as high as 75 percent, but only 15 to 20 percent actually took the test when it was implemented and offered [Bloch et al., 1992; Wiggins et al., 1992; Babul et al., 1993; Binedell et al., 1998].

Motivations for pursuing testing are personal, and an individual brings to the consultation expectations, some of which are based on obtained knowledge of genetics and associated health risks, and others based on the prior history [Stiefel et al., 1997]. The past history includes prior experiences with the heritable illness in the family, the awareness of multiple diagnoses in the family, the possible experience of prior losses of relatives to a disease, and the disruptions associated with the illness experience [Esplen et al., in press; Wellisch et al., 1991; Wellisch and Lindberg, 2001]. These experiences often contribute to a person's sense of his or her personal risk for the disease under consideration during genetic testing [Kash and Lerman, 1998].

One of the most consistent predictors of interest in undergoing genetic testing is perceived risk, regardless of an individual's actual risk level for disease [Bowen et al., 1999; Kelly, 1987; Codori et al., 1999]. Individuals who feel at risk for disease due to family experience with illness or a cognitive style that involves the monitoring of health threat information are often interested in attending a center for genetic detection [Rees et al., 2001; Leventhal et al., 1995; Kash and Lerman, 1998]. These individuals frequently carry representations and emotions in regard to their destiny [Stiefel et al., 1997]. They may believe, for example, that developing the disease is inevitable, citing perceptions of themselves as being at "high" risk or has having a 100 percent chance of developing the disease in consideration. An individual may believe that he or she is at high risk for disease when he or she is not, and in some cases the family history may not meet eligibility criteria for genetic testing. Women with one first-degree relative with breast cancer frequently overestimate their risk for cancer [Black et al., 1995; Durfy et al., 1999; Esplen et al., 2000]. Risk perception is important to evaluate at the pretest counseling phase of genetic testing, as elevated risk perceptions are associated with high levels of psychosocial distress or anxiety [Kash and Lerman, 1998]. Overestimations of risk are often persistent, profound, and difficult to modify [Kash et al., 1992; Lerman et al., 1995; Kash and Lerman, 1998], as individuals who experience elevated risk perceptions feel vulnerable to disease and often experience intrusive thoughts about the disease threat.

Factors that have been identified as being associated with elevated perceptions of risk include having a family history of disease, beliefs about the disease and risks, previous loss of a family member as a result of the disease, overidentification with a family member who has had the disease, and media representation [Kash and Lerman, 1998]. These factors are based on personal experiences and mental representations that formed through the person's prior history and are frequently in contrast to the "rational" approach taken during genetic counseling.

Genetic and risk information tends to focus on probabilities, risk factors, and categories, which are based on factual information derived from the family history—the number of family members with the disease, age of onset, and closeness of the affected relatives to the individual considering testing [Stiefel et al., 1997]. Professionals must bear in mind that the framing of disease risk has the potential to further heighten psychological arousal that may already be present in specific populations [Bottorff et al., 1998; Huibers and Van't Spijker, 1998; Kelly, 1987; Croyle and Lerman, 1999]. During genetic counseling, much emphasis is spent on making sure that the individual comprehends the complexity of the information, the nature of heritable disease, transmission patterns, and risk implications. A common response in attending to difficulty in a person's comprehension is to provide additional information or to simplify its presentation. However, this information must be presented while considering the context of a person's complex psychosocial history. Emotional reactions may pose barriers to the ability to assimilate and comprehend risk information, and, therefore, individuals may not be reassured by the provision of accurate objective risk information, or even a test result indicating the absence of a genetic mutation. Individuals may have to reconstruct a balance between thoughts and images evoked by expectations and the objective information obtained during a counseling session. Such a reconstruction often requires time, and in some circumstances psychological assistance. Stiefel et al. [1997] have described this process as involving the meeting of two worlds where "new objective and scientific knowledge from the outside" encounters "old, subjective and often irrational knowledge from inside."

Added psychosocial intervention might be required in order to deal with the past experiences (e.g., loss) that are generating emotional responses [Lerman et al., 1998b; Evers-Kiebooms et al., 2000; Esplen et al., 2000; Lindberg and Wellisch, 2004]. Cancer-specific anxiety and unresolved grief patterns, for example, have been noted among women who have lost mothers to breast cancer [Esplen and Hunter, 2002; Wellisch et al., 1991, 1992]. Interventions, such as individual psychotherapy or a support group, that facilitate the resolution of the past experiences can be implemented to assist an individual to fully appreciate the complex genetic and risk information and, therefore, optimally use this medical technology.

The interplay between personal perceptions of risk and emotional response is also important in considering disease screening behaviors. Studies on surveillance indicate that a minimum level of distress and anxiety is necessary to facilitate attendance to a screening program. However, if anxiety exceeds a certain level, it can have a negative influence on adherence to surveillance programs [Kash et al., 1992; Lerman et al., 1993; Vernon, 1999; Black et al., 1995; Codori, 1997]. Some individuals feeling a threat of disease may adopt a denial coping strategy, resulting in noncompliance or delay [Codori, 1997; Vernon, 1999; Kash et al., 1992]. Similarly, the adoption of lifestyle changes and other health behaviors are influenced by levels of anxiety and concern. The seeking out of additional information or health services to alleviate concerns about risk is a common response to elevated risk perception.

Additional commonly cited personal motivations for testing include the following: to participate in research, to learn about risk for offspring, to explore further surveillance options, and to make child-bearing and marital decisions [Esplen et al., 2001; Lerman et al., 1994b, 1996b; Press et al., 2001; Kinney et al., 2001; Meissen et al., 1991]. Gathering information on the motivations for testing is useful in order to plan ahead for the disclosure of a test result. A person's expectations and rationale for attending a clinic may mediate the adjustment to a test result. For example, individuals highly focused on offspring may experience high levels of concern and worry at the pretest stage that will need to be reassessed following the disclosure of a positive test result, given its associated implications for offspring risk.

Typically, family members who have been previously diagnosed with the disease under consideration for testing are the first to be offered a genetic test. While they may have adjusted to a diagnosis and have undergone previous treatments, the experience of genetic testing places an additional burden. These family members are the initial ones to assist in the collection of the family history and medical information and are given the responsibility to inform other members who may be eligible for testing once a mutation is discovered. The first family contact may also be female, as women are more likely than men to undergo testing [Tibben et al., 1993b; Evans et al., 1997; Press et al., 2000]. Observed gender differences may be linked to differences in knowledge about health threats and the differences between men and women in the way they cope with adverse information about their health. Some evidence suggests that men may be more likely than women to engage in "minimization" and/or denial [Bekker et al., 1993].

As with any medical diagnostic test, a waiting period for the test result exists. Periods during which a state of uncertainty exists tend to be associated with elevations in psychological distress. Individuals awaiting a genetic test result may experience greater distress than those who have the knowledge that they carry a genetic mutation [Broadstock et al., 2000; Esplen et al., 2001]. These reports suggest that a reduction in uncertainty among those testing positive may account for the observed benefit. It is also important to note that some individuals will decline genetic testing at the pretest counseling stage. Reported reasons for declining testing include concerns about stigma and insurance and feeling unable to emotionally deal with a positive genetic test result [Kash and Lerman, 1998; Codori et al., 1999; Decruyenaere et al., 1997]. Some individuals who opt out of testing may experience significant levels of distress [Lerman et al., 1998b, 1999]. One study found among individuals with high baseline cancer-related anxiety, who declined testing, that depression rates were significantly higher at one month follow-up than those who were tested [Lerman et al., 1998b]. In another study, individuals who opted to decline testing for heritable nonpolyposis colorectal cancer had higher rates of depression [Lerman et al., 1999]. These individuals who may be at high risk for disease may need to be carefully attended to, as they may be potentially avoiding important surveillance and health behavior modifications that may lower their risk. Therefore, reasons why an individual declines a genetic test needs careful exploration, as declining testing or the receipt of a test result may be an indication of significant psychological distress; however, for others it may reflect only a mature understanding of current limitations of testing and management [Lim et al., 2004].

POSTTEST COUNSELING PHASE: DISCLOSURE OF THE GENETIC TEST RESULT

The psychological aspects of the phase where disclosure of the test result occurs include the different aspects of communication and information and the management of difficult emotional states evoked by the information [Stiefel et al., 1997; Decruyenaere et al., 2000]. In a genetic consultation, disease may be absent and the information provided to the individual can be very complex. The information includes risk estimates and probabilities that may lead to confusion or be difficult to comprehend. The nature of the information varies from that of providing a diagnosis of disease, which is experienced as being more concrete, typically involving being informed whether or not a disorder is either absent or present [Stiefel et al., 1997]. In addition, while the diagnosis of disease is an existential threat to the individual, genetic information is also a threat to family members. This can arouse not only fears about developing the disease but feelings of anger (toward those who transmitted the disease) and guilt (about the potential transmission of the genes) to offspring [Lerman et al., 1994b; Murakami et al., 2004]. These emotional reactions have been documented in the literature and have potential for influencing the person's well-being and relationships with family members. Such dynamics can also interfere with the dissemination of genetic information to family members who may be at risk [Appleton et al., 2000].

Individuals may also wrongly assume that an illness with a genetic cause is inevitable and not treatable. For example, in one study individuals were asked to imagine that they had been tested by their general practitioner and found to have an increased risk of heart disease [Marteau and Senior, 1997]. One-half of the participants were told that this increased susceptibility had been determined by a genetic test and the other half were not told the type of test. When risk was determined by a genetic test, heart disease was seen as less preventable. The extent to which individuals consider a condition to be preventable is an important predictor of whether they follow advice on how to reduce the risk of developing the condition or ameliorate the condition once it has developed [Skinner, 1996]. This may mean that if individuals do not consider genetic tests in the same light as tests for biological risk factors of disease they will not be motivated to change their behavior. In fact, there is some evidence suggesting that genetic information does not always lead to behavioral change. One study found among women carriers in families with hereditary breast–ovarian cancer that 68 percent were adherent to mammography recommendations before *BRCA1/2* testing and 68 percent reported adherence 1 year after receiving positive test results [Lerman et al., 2000].

TEST RESULT INDICATING PRESENCE OF GENETIC MUTATION

Studies indicate that individuals who learn that they carry a genetic mutation for disease (e.g., Huntington disease, hereditary colorectal cancer, breast and ovarian cancer) do not receive the beneficial decreases in distress observed in those who test negative [Broadstock et al., 2000; Lerman et al., 1996a; Croyle et al., 1997; Schwartz et al., 2002]. Among those with elevated distress, approximately 16 to 25 percent undergoing risk counseling demonstrate levels consistent with a clinical diagnosis of depression or anxiety [Coyne et al., 2000; Lodder et al., 2001; Wellisch and Lindberg, 2001; Matthews et al., 2002; Vernon et al., 1997]. Studies that have utilized standardized disease-specific measures of distress (e.g., cancer worry scale) have demonstrated higher rates [Kash, 1999; Trask et al., 2001; Coyne et al., 2003]. However, by

1-year follow-up, psychological functioning levels among the majority of individuals testing positive is similar to those receiving negative test results [Almqvist, 1997; Schwartz et al., 2002], suggesting the effectiveness of genetic counseling in facilitating adjustment. Reactions to a positive test result understandably include a feeling of vulnerability to the onset of a disease. Common responses among individuals include the sense of "feeling like a walking time bomb" and a belief that "it is just a matter of time before the disease occurs."

Qualitative and quantitative descriptive studies have attempted to identify predictors of distress. Individuals who have elevations in distress before they undergo testing, or who have a premorbid psychiatric history (e.g., substance abuse, depression), are at increased risk for an adverse psychological outcome after testing [Thewes et al., 2003; Murakami et al., 2004; Van Oostrom et al., 2003; DudokdeWit et al., 1997; Broadstock et al., 2000; Meiser et al., 2000; Coyne et al., 2000]. In fact, a person's psychosocial functioning level at the time of the pretest counseling phase is a better indication of how he or she will react than the nature of the test result [Broadstock et al., 2000; Marteau and Croyle, 1998; Codori, 1997; Croyle et al., 1997], underlining the potential benefit for psychosocial screening early on in the testing process. Additional factors associated with distress following the disclosure of a test result include the level of penetrance associated with the gene mutation (probability of developing the disease), the perception of control over the disease (including the number of prevention/treatment options) [Codori et al., 1997; DudokedeWit et al., 1998], and the perception of the immediacy of risk (proximity in age to perceived disease onset) [Codori et al., 1997; Meiser et al., 2000]. Having younger children at the time of pretest and a history of previous losses to cancer also may play a role [Van Oostrom et al., 2003; Esplen et al., 2003a]. The expectation of having a negative test result and coping style (e.g., anxious preoccupied), as well as having a lower level of social support, are associated with greater distress [Vernon et al., 1997].

Despite having experience with a diagnosis, individuals who have dealt with a disorder and who are notified that they carry a genetic mutation may underestimate their emotional reactions following receipt of their results [Dorval et al., 2000]. While these individuals are aware that they did not intentionally pass along a gene mutation and its associated risk for disease, a common and understandable reaction includes feelings of guilt or worry concerning offspring [Lerman et al., 1998a; Lerman and Croyle, 1996; Dorval et al., 2000; Esplen et al., 2004]. Additional stressors include feeling at increased risk for an additional diagnosis of a disease associated with the genetic mutation.

Any test result that brings forward serious and difficult decisional options, anticipated or not, can pose additional psychological burden. Individuals who are notified that they carry a genetic mutation are often confronted with ongoing issues and decisions, which can lead to a reduced sense of control, as well as feelings of isolation. Examples of such decisions include decision making around current prevention and treatment options (e.g., increased surveillance, prophylactic surgery, chemoprevention), test result notification to extended family members and offspring, and relationship decisions. (e.g., marriage, childbearing) [Eisen et al., 2000; Lloyd et al., 2000; Lerman and Croyle, 1996; Esplen et al., 2004]. In some cases a person may feel "ambushed" by unanticipated emotional reactions and options. For example, in the case of *BRCA1/2* mutation testing, women may have been previously unaware that they are at increased risk for ovarian cancer in addition to their breast cancer risk [Esplen et al., in 2004]. These women are confronted with the additional medical decision making around prophylactic oophorectomy to reduce their ovarian cancer risk. Decision making around prophylactic surgery frequently involves the consideration of conflicting medical information, which can lead to confusion and a sense of feeling overwhelmed [Esplen et al., 2004]. For individuals who undergo medical procedures, such as prophylactic mastectomy to reduce a breast cancer risk, or an oophorectomy to lower ovarian cancer risk, ongoing stress will be experienced. Although women report satisfaction and the benefit of a decreased sense of vulnerability and concern about a cancer risk following prophylactic surgery [Metcalfe et al., 2004b; Van Oostrom et al., 2003; Elit et al., 2001], reports also indicate negative impact on psychological functioning, body image, sexual functioning and self-concept [Lloyd et al., 2000; Elit et al., 2001]. Individuals, in their decision making, must weigh these potential costs against the health benefits.

Not surprisingly, the familial context of the genetic test results may play an important role in an individual's psychological adjustment. For example, for an individual testing positive, indicating the presence of a gene mutation, in the context of other siblings testing either negative or having "inconclusive" results, he or she may experience higher levels of distress [Smith et al., 1999]. Such an emotional response may occur as a result of feeling isolated. In contrast, individuals who receive the news that they carry a genetic mutation within the context of having siblings who share a similar test result, may experience less elevation in distress [Smith et al., 1999]. Some individuals describe feeling closer to family members who carry the same genetic mutation. For example, a woman who carried a genetic mutation for hereditary colorectal cancer found that she and her father (who carried the same gene mutation) had numerous discussions around health, risk, and the implications of genetic testing on their life goals. These discussions were described as directly contributing to increased feelings of closeness through the sharing of personal information and mutual support. [Esplen et al., study in progress].

Long-term follow-up data on psychological morbidity are available on individuals who have undergone predictive testing for Huntington disease and breast cancer. To date, evidence suggests that most carriers of genetic mutations do not become more distressed over time [Almqvist et al., 1997, 1999; Broadstock et al., 2000; Schwartz et al., 2002]. For some individuals, however, ongoing family and personal reactions can occur over the long term, with individuals experiencing emotional distress [Esplen et al., in press; Lindberg and Wellisch, 2004] and changes in relationships [Van Oostrom et al., 2003; Almqvist et al., 1999; Tibben et al., 1997; Meiser et al., 2000; Smith et al., 1999]. Spouses can also have difficulty adjusting and have demonstrated high levels of distress and a sense of loss and concern for their partner [Williams et al., 2000].

While genetic testing can present challenges for individuals, the perceived benefits of testing are important in counteracting the emotional costs. Most individuals identify specific benefits to testing [Struewing et al., 1995; Cappelli et al., 1999; Lim, et al., 2004]. Individuals who obtain increased certainty concerning their disease risk often experience beneficial feelings of relief, knowing that they are being monitored closely. Individuals frequently note that their deceased or older relatives were not afforded such opportunities as intense surveillance and the chance to participate in trials of risk-reducing interventions. Others indicate a sense of feeling "empowered" and of resilience through the prior experience of dealing with diagnoses in the family. Individuals often describe "knowing their body well" and believe that through self and medical monitoring they will detect the disorder early and through treatment survive a disease [Esplen et al., 2003b; Lim et al., 2004]. Others express hope for the development of future medical technology that will address disease risk and effectively treat illness.

Additional quality of life benefits to genetic testing include the opportunity for an individual to examine his or her health behaviors and life decisions [Wiggins et al., 1992]. The existential threat associated with being at increased risk for disease can be a powerful motivator, enabling an individual to adopt healthier lifestyle behaviors (e.g., changes in diet, exercise habits, smoking cessation). The experience of feeling "life is vulnerable" and the knowledge that one may develop a disease at an earlier life stage often results in an individual examining many aspects of his or her life in an honest and authentic manner. For example, individuals may become more aware of personal life choices or values, resulting in a greater need to connect in relationships, and spend more time in desired activities or with valued others [Esplen et al., 2004; Lim et al., 2004]. Often such reactions occur naturally; however, for others, they can occur through psychological and behavioral interventions that provide the opportunity to examine the full range of their reactions to a genetic test. These direct or indirect benefits from the genetic testing experience are life enhancing in that they can lead to improvements in health outcomes and quality of life. Additional reported benefits include using genetic information to make career choices, plans for retirement, plans to obtain life insurance, and for medical decision making [Cappelli et al., 2001; Broadstock et al., 2000]. One final important benefit to note is the feeling of altruism, through the opportunity to assist in health research and to provide information that may assist future generations (including one's own relatives) to deal with, or fight disease [Esplen et al., 2004]. Opportunities to support others have direct benefit in buffering the impact of a stress response and facilitate active coping [Esplen et al., 2004].

NEGATIVE TEST RESULT INDICATING ABSENCE OF GENETIC MUTATION

The news that one does not carry a genetic mutation is generally associated with positive emotional responses, including improved psychological functioning, decreases in worry about disease risk, and a sense of relief [Lerman et al., 1996a; Lerman and Croyle, 1996; Marteau and Croyle, 1998; Codori, 1997; Croyle et al., 1997]. However, there is evidence that testing negative can result in some individuals experiencing difficulty in adjusting to their risk status, especially when they have integrated a sense of being at increased risk into their self-concept [Marteau and Senior, 1997; Van Oostrom et al., 2003; Tibben et al., 1993b; Lim et al., 2004]. Individuals who initially thought that they were at increased risk for a disease may have experienced a sense of merger or strong identification with the relative affected with the disease. Individuals told that they do not carry a gene mutation can experience a sense of loss, in the form of a loss of identity formed around a connection to the affected parent [Esplen et al., 2004]. For example, one woman who lost her mother to breast cancer when she was a child and who learned that she did not carry a *BRCA1* or *2* mutation, described feeling less connected to her mother following notification of her test result. This woman described a sense of understanding her mother's suffering, knowing her deceased mother, and feeling closer to her through her sense of her genetic identity [Esplen et al., 2004]. Her unanticipated test result challenged this area of her self-identity and needed to be integrated into her sense of self. The assimilation of new genetic information occurred through a professionally led support group where she had ample opportunity to explore her reactions, process feelings of grief, and gain perspective on the impact of her mother's death.

As noted above, adjustment to genetic testing can vary dependent upon the test results of other family members. For example, when an individual tests negative, indicating the absence of a genetic mutation while siblings are notified that they are carriers, he or she may experience feelings of guilt that he or she was spared the legacy of the disease risk in the family, a response often referred to in the literature as *survivor guilt* [Smith et al., 1999; Murakami et al., 2004; Lim et al., 2004]. Such a feeling is experienced in conflict with the sense of relief that usually accompanies a favorable test result.

RECEIVING AN AMBIGUOUS GENETIC TEST RESULT

It is important to note that apart from genetic test results that are able to provide conclusive evidence of the presence or absence of a known genetic mutation in the family, there is also the possibility of receiving ambiguous results. If a genetic muta-tion is not identified in a family, it may be due to the limitations of genetic technology or the presence of unidentified genes, rather than ruling out a hereditary condition altogether. Such genetic information, often referred to as an *inconclusive* result may have important implications for interpretation and adjustment. While individuals fre-quently experience relief that they do not carry an identifiable mutation, there is the potential for confusion and misinterpretation, which can cause concern to clinicians who fear that the individual will be falsely reassured, believing that he or she has tested "negative" and is therefore not at risk for a disease. The potential for oversimplifying such a test result highlights the need for careful attention in assessing the individual's comprehension of an inconclusive test result and follow-up, particularly in relation to surveillance and preventive options. Individuals who receive inconclusive test results will need to continue screening and monitoring. Ambiguous test results or information that can be misunderstood has implications for adjustment. Individuals may feel that they need to keep checking in to see if more information becomes available through research in genetics. They may not experience the reduction in uncertainty around their risk afforded by those who receive negative or positive test results and may continue to experience elevated risk perceptions and worry around their risk. In addition, an individual who misunderstands his or her test result may communicate inaccurate information to family members. To date, there is little empirical information on the adjustment of individuals who receive ambiguous test results; however, one recent study on women undergoing testing for *BRCA1/2* found that women were not falsely reassured by their inconclusive test result, suggesting the benefit of genetic counseling in facilitating comprehension and screening adherence [Dorval et al., 2003].

IMPACT OF GENETIC TESTING ON THE FAMILY

The genetic test result has implications beyond that of the individual, and often the family is considered the unit of focus during genetic counseling. Relationships among siblings, parents and offspring can be complicated by the different test results that individuals receive. For example, some of those found not to carry the gene for Huntington disease felt rejected by their families when they were found to no longer have one of the key bonds that had previously tied them together: being at risk for Huntington disease [Tibben et al., 1999]. Partners must also adjust to a genetic test result. One study found that the partners of those who tested positive for Huntington disease experienced more posttest distress and poorer quality of life than their tested partners [Sobel and Cowan, 2000]. In addition, for a parent who carries a genetic mutation, he or she is posed with the dilemma of "when" and "how "to inform adolescent or adult offspring. Women with *BRCA1/2* mutations in a group support program reported that one of the greatest challenges they experienced was how to inform an adult daughter of a mutation in the family. Mothers frequently experience the dilemma of wanting an adult daughter to have close surveillance on the one hand, while on the other feeling a need to maintain her sense of invulnerability, typically experienced during youth, and its associated sense of freedom in having a whole future ahead [Esplen et al., 2004]. Even when parents decide to delay telling younger offspring about a genetic mutation until a later developmental stage, they may need to deal with informing children about medical appointments or procedures. For example, a woman with the *BRCA1* mutation who chooses to undergo surgery to remove her ovaries or breasts must find a way to communicate to her children the necessity for a surgical procedure, while reassuring them that she is not currently ill or dealing with a disease.

Families also have challenges around the notification to family members who they may not know well or who they fear burdening with unwanted or unexpected risk knowledge [Evers-Kiebooms et al., 2000; Lerman et al., 1998a; Lim et al., 2004]. For example, individuals may feel they are "opening a can of worms" or feel ill-equipped to discuss a sensitive topic when relationships are strained or unformed. In contrast, some family members feel strong pressure to encourage siblings or other family members to be tested, even when these individuals may not be interested, thereby posing relationship challenges. Within a family, it is not uncommon for individuals with different coping styles to express conflicting views on genetic testing. For example, some individuals prefer to know up front what they are dealing with in terms of their risk for disease, while others wish to take a "wait and see" approach and simply maintain screening efforts, preferring to avoid the confirmation of risk through genetic testing. These potential family communication challenges may require guidance or support by health care professionals in order to minimize the potential for additional stress and to facilitate family adjustment.

SOCIETAL AND ETHICAL ISSUES LINKED TO PSYCHOSOCIAL OUTCOMES

Individuals and their families can face a number of societal and ethical challenges that may impact on quality of life and lead to potential distress. These issues are now well cited in the literature [Kash and Lerman, 1998; Kash, 1995] and include the following: issues around maintaining confidentiality and privacy (e.g., Who will have access to genetic test results?), the potential for discrimination (health, medical, or life insurance, employment opportunities, the linkage of specific cultural groups to a genetic mutation), disclosure issues associated with familial information (e.g., the encouragement to disclose information concerning a genetic mutation to a family member versus the relative's right "not to know"), and interest in prenatal testing for AOHD, which is currently discouraged.

Cultural factors may also be relevant in the consideration of the provision of genetic testing [Meiser et al., 2001]. To date, epidemiological investigations and studies of clinic samples have tended to include Caucasians and individuals from higher socioeconomic levels, who appear to be most interested in seeking out testing. Different cultures may have specific beliefs and attitudes concerning genetic testing [Meiser et al., 2001]. African American women have been shown to have lower levels of knowledge about breast cancer genetics [Hughes et al., 1997] and were less likely to have heard or read about genetic testing for inherited disease, even after controlling for educational level [Durfy et al., 1999; Hughes et al., 1997]. In addition, African American women have been shown to be more likely than women in the majority culture to agree that health care providers should be able to disclose the results of genetic tests to spouses or immediate family without written consent, and that parents should be able to decide whether to test their minor children for genetic susceptibility to breast or ovarian cancer [Benkendorf et al., 1997].

Culture can play a role also in preventive treatments. For example, in a support group for women with *BRCA 1/2* mutations [Esplen et al., 2004], one immigrant woman from a Middle Eastern community described challenges around her options for prophylactic surgery. As she listened in the group to other women expressing personal views and choices, she became aware that such discussions were resulting in her experiencing emotional reactions of fear and decisional conflict. She described to the group that she personally felt that the removal of her ovaries was the best medical option for her in managing her ovarian cancer risk, given that she had completed child-bearing. Her mother had died of ovarian cancer in her thirties and the woman described fearing a similar fate. Although she had previously encountered some difficulty in convincing her husband about the option, he came to understand the potential benefits of surgery and her need to consider the option. However, a greater challenge for this woman was her cultural and religious beliefs. She explained to the group that unless she could find a male leader within her religion who would permit such an action, she would be unable to consider it. She described the role of the female within her culture and beliefs that medical procedures had to be approved by the male leaders of her religious faith.

Cultural aspects of genetics is the focus of an emerging body of research. Investigations are needed on the psychosocial impact of specific cultural groups, particularly given the role that culture plays in health outcomes [Meiser et al., 2001]. Studies are underway and are specifically interested in investigating any concerns that these groups may have in being identified and their levels of interest, in order to develop culturally sensitive approaches to implementing genetic testing [Schwartz et al., 2000].

CLINICAL IMPLICATIONS

The implementation of genetic testing should include the integration of a complementary psychological framework to guarantee optimal conditions for information provision and to enhance adjustment throughout the genetic testing process [Stiefel et al., 1997; Decruyenaere et al., 2000]. Such a framework should include all psychological aspects surrounding the genetic testing phases, including education around the pros and cons of genetic testing, assessing motivations and attitudes, and the facilitation of coping with a test result and any associated medical decision making. Most individuals will adapt over time, following genetic counseling and demonstrate comprehension around their risk and options [Van Oostrom et al., 2003; Broadstock et al., 2000; Schwartz et al., 2002; Braithwaite et al., 2004]. However, others will exhibit inaccurate risk perception and continued elevated distress following genetic counseling [Watson et al., 1999; Brain et al., 2000; Lerman et al., 1995, 1996a; Michie et al., 1997b; Esplen et al., 2000]. Persons undergoing genetic testing who are at risk for adjustment difficulties are in some ways similar in characteristics to those who develop psychological disorders in other health areas. For example, in the field of oncology, individuals with a prior psychiatric history, such as depression or substance abuse, those who rely on maladaptive coping strategies, and who have unstable relationships and difficult social conditions are at increased risk for developing psychiatric complications following diagnosis and treatment [Fallowfield et al., 1990].

The management of knowledge provision around future disease onset can implement psychosocial support in a fashion that addresses the anticipation of problems before they occur. An important task is to screen individuals with psychological distress or disorders and those who may be "at risk" for adjustment difficulties (see Box 3.2). Identifying individuals with known vulnerability risk factors can facilitate the provision or organization of required additional support during the various stages of genetic testing. Screening instruments can be an efficient way of identifying psychosocial issues and difficulties since a balance should be found between the time in

BOX 3.2 *Factors Associated with Psychological Risk or Adjustment Difficulties*

1. Sociodemographic Age/Developmental Level/Proximity to Age of Affected Family Member

2. Gender

3. Culture/ethnicity

4. Socioeconomic status

5. Having young children

6. Medical
   a. Penetrance
   b. Severity/nature of disease
   c. Prevention options and risk-reducing procedures

7. Psychosocial
   a. Loss of relative to disease (especially parent)
   b. Care-giving of family member with disease
   c. Prior history of additional life losses/trauma
   d. Premorbid psychological history/condition
   e. Current level of psychological functioning (e.g., depression, anxiety, disease-specific worry)
   f. Current life stressors (e.g., job stress, divorce)
   g. Expectation of receiving a negative test result
   h. Coping style (e.g., anxious preoccupied, health monitoring)
   i. Social support level (low level)

clinic allocated for interviews and that allowed for treatment and support [Thewes et al., 2003]. Research efforts have focused on the screening of anxiety and depression among medical populations, which are the most frequent symptoms of adjustment and affective disorders. For these conditions, the few studies in other health areas that have assessed the performance of screening methods report good rates of sensitivity and specificity [Razavi et al., 1990, 1992]. Descriptive studies on genetic risk populations have utilized common standardized screening instruments, for example, the Brief Symptom Inventory [Derogatis, 1993], Center for Epidemiology Depression Scale [Radloff, 1977], the Hospital Anxiety and Depression Scale [Zigmond and Snaith, 1983], and the Beck Depression Inventory (BDI) [Beck and Beamesderfer, 1978].

However, as demonstrated above, there are important relevant psychosocial issues that need to be considered in the provision of genetic testing that will not be captured by existing standardized screening instruments. Development and testing of screening instruments that incorporate specific relevant risk factors are needed in future research. Issues such as the identification with the member of the family who has had the disease, prior unresolved grief as a result of the loss of an affected parent, and the guilt associated with the transmission of a gene mutation are specific to the context of genetic testing. In addition, in screening individuals who may be at risk psychologically, it is important to note that many will not suffer from a psychiatric disorder per se but will face a complex problem that leads to a variety of thoughts, feelings, and behaviors that need to be understood within the context of a person's history. These individuals are often described as the "worried well" and may benefit if the psychological aspects are taken into consideration proactively during the provision of information.

Genetic health care providers will have the skill to provide psychosocial support to the majority of individuals undergoing testing. However, additional psychological intervention is warranted in cases where persistent adjustment difficulties exist. Psychiatrists, psychologists, and other mental health professionals are increasingly being asked to function as consultants and as members of multidisciplinary teams. For some individuals at psychological risk the genetic testing process will require vigilant, neutral, and competent attention that gives way to adaptation and integration of this new information. In genetic detection, as in any other high-tech medical approaches, the presence of a psychosocial clinician team member helps to guarantee that a neutral reflection on these technologies and options occurs and allows for a view "from outside." Mental health professionals are able to provide comprehensive biopsychosocial assessments, play an active role in developing treatment plans, particularly regarding ongoing medical decision making or procedures, organize the management of psychotropic medications if required, facilitate referrals to more specialized psychological services, and can assist in the management of co-morbid disorders.

Areas where universal access is challenging include geographic areas, such as isolated communities and specific target populations such as particular cultural groups. Most provinces already face difficulties in providing specialized health services to individuals living in more isolated communities. While recruitment of specialists to work in genetics will remain an important component of any overall solution, there is also a need to develop alternate models of care that are based in primary care and use psychological resources as efficiently as possible. Models that provide direct and indirect consultation, teaching, and skill development, as well as ongoing care can be implemented. One such innovative model is that of the "shared care" model, where specialized mental health professionals provide ongoing support, education, and consultation to medical teams [Kates et al., 1997a, 1997b]. Functions that lend themselves well to the shared care model include early detection and the initiation of treatment, ongoing monitoring, and mental health education [Kates et al., 1997a, 1997b]. Shared care can lead to improved patient outcomes and quality of life and a more efficient use of resources [Kates et al., 1997a].

It may be helpful to think of distress or adjustment difficulties along a continuum. Different levels of intervention can be applied depending on where an individual is along a continuum of distress (see Box 3.3). For example, for individuals who experience adjustment difficulties (e.g., having a low level of worry, concern) that do not interfere with daily functioning and educational needs that require minimum support, Internet support services and pamphlets may address their specific needs. Individuals with moderate levels of distress (e.g., worry or anxiety that can impede coping activity, sleep, adequate decision making) may benefit from additional one-on-one

BOX 3.3 *Continuum of Distress*

Low Level of Distress →	*Moderate Level of Distress* →	*High Level of Distress*
Educational pamphlets, CDs, Internet support	Cognitive-behavioral (e.g., stress management, coping strategies for living with uncertainty)	Individual psychotherapy/ supportive counseling
	Interactive CD-ROMs	Professionally led support groups
	Manual and computer-based decisional aids	Family counseling
	Telephone counseling and follow-up counseling	Psychotropic medication
	Peer support (one-on-one group)	

sessions with a genetic counselor or a mental health professional, CD-ROM and other decisional aids, peer support programs (including one-on-one and group) and cognitive-behavioral interventions designed to manage anxiety and to assist in living with uncertainty or decision making. Individuals who demonstrate high levels of distress (e.g., depression or anxiety levels that interfere with daily functioning, decision making, self-care) or who express feelings of hopelessness and suicidal ideation will require specialized services offered by a skilled mental health professional. Potential interventions for this group might include one-on-one psychotherapy sessions to facilitate coping or unresolved grief issues, psychotropic medication, professionally led psychotherapy and support groups, and long-term follow-up.

Supporting the genetic health care professional as the primary provider and organizer of psychosocial care has a number of advantages. He or she has a continuing relationship with the patient and family, knowledge of the individual's family and of the physical and social environment in which the patient lives, an understanding of coexisting general medical problems, screening guidelines for surveillance and available community resources, and of providers who are involved with the patient (e.g., other medical specialists). These advantages put the genetic health care provider in an excellent position to identify and treat mental health issues at an early stage, assist individuals and families in decision making and adjustment, and to coordinate health and mental health services as the individual and his or her family may require.

SUMMARY

Information resulting from the Human Genome Project will continue to change the way individuals, families, and society view health and disease and its management. Knowledge resulting from this new and evolving genetic information has the potential to provide many benefits but can also lead to increased emotional distress, misperceptions of illness, denial, stigmatization, and other suboptimal health outcomes.

Genetic knowledge in and of itself does not translate directly into desired health behavior changes. Rather, evidence is emerging to suggest that psychosocial, emotional, and family context factors play an important role in the understanding of genetic information, health behavior change, appropriate utilization of the health care system, and quality of life.

Adverse psychological and behavioral reactions will be less common when genetic testing is provided with an integrated complimentary psychological framework. Programs should aim to present clear information and emotional support throughout the genetic testing process. Built-in follow-up support, both to the individual and the family, should be considered an important component to facilitate ongoing monitoring and assistance with any additional challenges as they emerge. Such an approach facilitates a biopsychosocial model of care, providing a comprehensive range of preventive health care, while enhancing the well-being of individuals and their families, and the optimal use of this medical technology.

Chapter 4

Duty of Care

OVERVIEW

Due to a shortage of qualified geneticists and genetic counselors, expanding test technology capacities—for example, tandem mass spectrometry (MS/MS)—and growing public demand, genetic testing services are likely to be provided by primary service physicians and other related medical specialties [Task Force, 1997; Rothstein and Hoffman, 1999]. Health care professionals increasingly will need to have the practical ability to provide appropriate care [Burke, 2004].

Health care professionals may not understand that the model of care applicable to genetic testing and counseling services, although analogous to the standard of care generally applicable to physicians, includes several distinct and important obligations [Capron, 1979; Andrews, 1991, 1992; Reilly, 1993; Sharpe, 1994a, 1994b, 1996; Reutenauer, 2000]. These include the obligation to undertake all reasonable procedures to facilitate effective communication and the obligation to take into account the psychological aspects of genetic testing [ACMG, 2004; ABGC, 2004; CCMG, 2004b].

A health care professional who agrees to provide genetic testing services will be expected to perform in a manner consistent with a reasonable standard of care or be held responsible for injury caused by a breach of this standard. It is unclear how a court of law will evaluate the actions of a health care professional who provided genetic testing services but who does not have specialized training in human/medical genetics and genetic counseling [Sharpe, 1996; Kapp, 2000].

It must be emphasized that it is *not* the purpose of this book to identify and/or to define what constitutes appropriate standards of care and the resulting legal duties of care for genetic testing and counseling. Although these chapters provide an extensive survey of medical and human/medical genetics literature, much of this literature flows from research trials and protocols. Recommendations with regard to what constitute appropriate protocols for the delivery of genetic testing services continue to evolve. Accordingly, unless otherwise noted, the discussions reviewed throughout this book

Genetic Testing: Care, Consent, and Liability, by Neil F. Sharpe and Ronald F. Carter
© 2006 John Wiley & Sons, Inc.

have not been formally recognized or endorsed as standards of care, although they can provide important information about potential shortcomings in the delivery of genetic testing services [Rowley, 1993].

With respect to the legal aspects of genetic testing, like it or not, the practical reality is that decisions of courts of law can have a major impact on health care [e.g., ACOG, 2004]; that impact must be taken into account in discussing how genetic testing will be applied in health care. The following reviews the traditional medical malpractice principles for negligence. With this framework in place, it will be possible to sketch in those *conceptual* guidelines of how a court of law is likely to interpret a health care professional's duty of care as it relates to the processes of genetic testing and counseling [Capron, 1979; Andrews, 1991, 1992; Reilly, 1993; Sharpe, 1996; Reutenauer, 2000; Andrews and Zuiker, 2003].

Given that genetic testing technology and services continue to evolve and new protocols and obligations continue to be debated—for example, the proposed duty to recontact former patients reviewed in Chapter 15—at the time of this writing, *any attempt to identify what obligations constitute an appropriate legal duty of care would be premature and potentially misleading.* Accordingly, any legal analysis will be general in nature, designed only to provide an overview of the potential issues and the manner in which courts of law may evaluate the nature and scope of a health care professional's legal duty of care to the patient. Due to the relative lack of legal precedent in the context of genetic testing, both American and Canadian case law have been cited to provide an overview of the manner in which various jurisdictions have considered analogous issues and their implications, especially with regard to the law of medical malpractice.

Several limitations must be noted. The legal commentary in this book is not intended to serve as: (a) a legal brief or opinion or (b) a comprehensive analysis of malpractice principles and case law. Rather, the legal discussion is employed as a heuristic device to provide a framework for the analysis and evaluation of these issues as well as to stimulate further discussion and debate. A case law decision is binding only within the jurisdiction in which the court rendered the decision. Current case law and legislative requirements should be consulted for the jurisdiction in which the reader practices. Although many of the cases cited in these pages have considered conventional medical scenarios, the potential impact of genetic testing and counseling makes it reasonable to extend a discussion of these obligations to providers of genetic testing services.

NEED TO KNOW

Standard of Care

A physician who agrees to provide services to a patient must possess and use the same degree of learning and skill as that on an average practitioner in similar cases; that is, the physician's conduct must be at least the equal of the prevailing standard of care [*Pike v. Honsinger*, 1890; *Morlino v. Medical Center of Ocean County*, 1998; *Wilson v. Swanson*, 1956].

This includes the obligation to keep up-to-date and abreast of significant developments [*Harbeson v. Parke-Davis Inc.*, 1983; *Nowatske v. Osterloh*, 1996]. Common usage, however, may not always serve as a reliable indicator of what constitutes good medical practice [*Morgan v. Sheppard*, 1963; *Lundahl v. Rockford Memorial Hospital*, 1968; *Toth v. Community Hospital*, 1968; *Helling v. Carey*, 1974; Bovbjerg, 1975; *White v. Turner*, 1981]. To hold that it does may sanction the unnecessary application of medical procedures irrespective of a patient's particular needs [Capron, 1979; Gilfix, 1984; *Adair v. Weinberg*, 1995; *Nowatske v. Osterloh*, 1996; *Crits v. Sylvester*, 1956].

Accordingly, a physician not only must meet the prevailing standard of care but must also fulfill a separate duty to exercise good judgment, particularly with regard to the application of technologies and procedures [*Pike v. Honsinger*, 1898; Capron, 1979; *Jones v. Karraker*, 1982; *Nowatske v. Osterloh*, 1996; *Poole v. Morgan*, 1987].

Courts of law have recognized that a physician's obligation to a patient requires that the physician exercise a reasonable degree of care toward the patient, and this may be adjusted to take into consideration the particular facts and exigencies of a medical situation [*Hood v. Phillips*, 1977; *Phillips v. Good Samaritan Hospital*, 1979; *Rodych v. Krasey*, 1971], such as the nature of treatment provided in an emergency clinic.

> *The duty of care . . . takes two forms: (a) a duty to render a quality of care consonant with the level of medical and practical knowledge the physician may reasonably be expected to possess and the medical judgment he may be expected to exercise, and (b) a duty based upon the adept use of such medical facilities, services, equipment and options as are reasonably available.* [*Hall v. Hilbun*, 1985]

A physician who knows, or should know, that the patient's situation is beyond the professional's knowledge, competence, and skills, has a duty to refer. If a physician does not do this and undertakes to care for the patient, and/or the physician professes to be a specialist, and/or seeks to provide the services of a specialist, the physician may be held liable to the standard of care applicable to the specialty [Capron, 1979; *Shilkret v. Annapolis Emergency Hospital Ass'n*, 1975; *Larsen v. Yelle*, 1976; *Hall v. Hilbun*, 1985; *Crits v. Sylvester*, 1956; *Villeneuve v. Sisters of Joseph*; 1975; but note *Johnson v. Agoncillo*, 1994; *McKeachie v. Alvarez*, 1970; *MacDonald v. York Hospital*, 1973].

Generally speaking, a standard of care will not be adjusted in deference to a lack of experience or training [*Poole v. Morgan*, 1987; *Yepremian v. Scarborough General Hospital*, 1978; *Wills v. Saunders*, 1989]. In the United States, a specialist generally will be held to a national standard of care compared to the standard applicable in a particular locality [*Shilkret v. Annapolis Emergency Hospital Ass'n*, 1975; *Hall v. Hilbun*, 1985; LeBlang, 1994; *Jordan v. Bogner*, 1993].

WATCH OUT FOR

Duty of Care

When a physician treats a patient, a *physician–patient* relationship is created. Due to a physician's medical knowledge and skills, the patient must place trust and confidence in the physician. Certain duties arise from this "special" relationship, including the duties to act with good faith and to exercise a reasonable degree of care toward the patient [Torres and Wagner, 1993; Moore v. Webb, 1961]. These obligations are termed the physician's *duty of care*. When one speaks of this duty of care, one also speaks of its constituent elements; these generally include the duty to attend, the duty to diagnose, the duty to treat, the duty to refer, the duty to monitor, and the duty to provide appropriate aftercare.

To create the physician–patient relationship, generally speaking, the physician must offer, directly or indirectly, to see or to counsel the patient, must have made some rudimentary evaluation of the patient's complaint or condition, and the patient must rely on that evaluation and/or advice and/or offer. A physician need not have direct contact with the patient in order for such a relationship to be created; however, a physician's express or implied consent to advise or to treat the patient generally is required [e.g., *Irvin v. Smith*, 2001]. For example, American courts have held that a physician–patient relationship existed when a pathologist examined a tissue sample [*Walters v. Rinker*, 1988; Spielberg, 1999] and that a physician–patient relationship can be created with a telephone call [Torres and Wagner, 1993; *Miller v. Sullivan*, 1995] where the intention is to seek medical advice [but note *Weaver by Weaver v. University of Mich. Bd. of Regents*, 1993; a telephone call made to schedule an appointment did not create a physician–patient relationship where there was no ongoing relationship and the patient did not seek or obtain medical advice]. It has been argued that this could be extended to include consultation by e-mail [Eysenbach and Diepgen, 1998; Kuszler, 1999].

Where the physician does not have direct contact with the patient, if the physician has examined the patient's medical records and has provided advice with respect to the patient's condition, this may be sufficient to give rise to a physician–patient relationship [*Molloy v. Meier, Backus, et al.*, 2004; however, an examination on behalf of an employer or insurer may not create a physician–patient relationship e.g., *Violandi v. New York*, 1992].

Genetic counseling, as defined within the context of providing genetic information about a present or pending medical condition, has been recognized as constituting "treatment" sufficient to create a physician–patient relationship [*Pratt v. University of Minnesota Affiliated Hospitals and Clinics*, 1987; Reilly, 1977; Capron, 1979; Weaver, 1997]. The case of *Pratt v. University of Minnesota Affiliated Hospitals and Clinics* concerned an allegation for "negligent non-disclosure" during genetic counseling. Parents of an afflicted child wanted to know whether the condition "was genetic in origin," and whether future offspring would have an increased risk. The consulting physicians reviewed the parents' respective medical histories, took blood samples, and performed a "chromosome study." Although the physicians were unable to diagnose the child's anomalies, both consulting physicians were aware of the possibility that the disorder may be an autosomal-recessive condition. However, neither physician felt that "the possibility was significant enough" to inform the parents. At trial, the physicians argued that genetic counseling did not constitute treatment and therefore the doctrine of negligent nondisclosure did not apply. The trail court held that it was vital for patients in the situation of having to make a choice to be able to base their decision on as much information as possible. "This is especially true in the area of genetic counseling." The court concluded that the physicians had replied to the parents' request for information "in order to make a major health care decision. In providing that information, we conclude that the respondents [the physicians] did render treatment . . . " Medical malpractice encompasses errors or commission and omission by health care professionals that fall below the generally accepted, or appropriate, standard of care [e.g., *Bardessono v. Michels*, 1970; *Davis v. Patel*, 2001].

If a health care professional's conduct complies with the average standard of care, civil liability for a patient's injury generally may be avoided if the conduct in question is deemed an error in medical judgment. The law does not expect a standard of perfection [*Wainwright v. Leary*, 1993: "The physician will not be held to a standard of perfection nor evaluated with benefit of hindsight"]. No human being is infallible, and even the most skilled health care professional may not accurately diagnose the true nature of a disease and/or condition [Shaw, 1986; *Moore v. Preventive Medicine Medical Group Inc.*, 1986; *Christensen v. Munsen*, 1994; *Wilson v. Swanson*, 1956; *ter Neuzen v. Korn*, 1993]. In evaluating this issue, a court of law generally will distinguish between an "error in judgment" and a failure to exercise her or his best judgment. A physician may be held liable if an error of commission or omission shows an absence of reasonable skill, care, and judgment falling below that of the generally accepted standard of care [*Riggins v. Mauriello*, 1992; *Lapointe v. Hôpital le Gardeur*, 1992].

American and Canadian jurisdictions generally will regard expert medical evidence, professional standards of practice, as well as the number of reports in leading medical and human/medical genetics journals, to be reliable guidelines to what constitutes a generally accepted standard of care [*Darling v. Charleston Community Memorial Hospital*, 1965; *Pedroza v. Bryant*, 1984; *Menzel v. Morse*, 1985; Havighurst, 1991; *Stanley v. McCarver*, 2003; Jutras, 1993; *ter Neuzen v. Korn*, 1993; but note *Helling v. Carey*, 1974; *Kramer v. Milner*, 1994; Campbell and Glass, 2000].

To be successful in a medical malpractice action, the plaintiff-patient must establish:

- A legally recognized duty of care was owed by the defendant–health care professional to the plaintiff-patient.
- The defendant–health care professional breached this duty of care by failing to meet the required standard of care.
- The plaintiff-patient has suffered a legally cognizable, compensable, injury.
- The injury was cause by the defendant–health care professional's breach of the duty of care.

With the above framework in place, it is possible to sketch in those components that may be described as *conceptual* guidelines for the duty of care as it relates to the processes of genetic testing and counseling. As will be discussed in subsequent chapters, these include the obligations, among others, to:

- Take a detailed, accurate family history.
- Refer for genetic counseling.

- Undertake all reasonable procedures to facilitate communication.
- Advise a patient of all appropriate diagnostic genetic testing services, and to provide, or to refer the patient for, testing.
- Make a timely, accurate diagnosis.
- Secure an informed voluntary consent.
- Employ competent laboratory facilities.
- Appropriately interpret and to communicate genetic test results and risk information, including the implications for the patient and the family.
- Promptly disclose all test results.
- Provide appropriate emotional and psychological support.
- Provide appropriate monitoring, follow-up, and aftercare.

It must be emphasized that the *conceptual* components of the duty of care discussed in this book have not been imposed by a court of law. They have been developed by geneticists, counselors, and physicians to promote further discussion and debate in terms of what constitutes good patient care and how the health care professional can best respond to the needs of patients [Andrews, 1991, 1992; Sharpe, 1996; Reutenauer, 2000; Andrews and Zuiker, 2003].

Chapter 5

Family History

OVERVIEW

An accurate family medical history is an essential requirement for evaluation of a patient, making a diagnosis, and assessing risk. As genetic components of a broader spectrum of disease are recognized, and options for management and treatment of genetic disease improve, the need for accurate and comprehensive family histories has increased [CDC, 2003, 2004; NCHPEG, 2004; U.S. Department of Health and Human Services, 2004]. Family histories provide clues to basic questions: whether or not the disorder appears to be of genetic origin, the pattern of hereditary genetic transmission, strategies for laboratory testing to confirm a diagnosis, information as to possible environmental origins (e.g., exposure to teratogens and mutagens [Thompson and Thompson, 2001]), and the risks of occurrence and recurrence among family members.

INTRODUCTION

The ability to elicit and construct a comprehensive family history is commonly recognized as an essential skill for geneticists and genetic counselors. Both the type of information to be solicited and the need for proper interpretation in the relevant context are important. The specific expectations identified by professional bodies include [ABMG, 2004; ABGC, 2004; CCMG, 2004b]:

- The American Board of Medical Genetics, Inc. [ABMG, 2004], accreditation standards for clinical, and Ph.D., geneticists requires skills to:
  - Elicit from the patient or family the information necessary to reach an appropriate conclusion
  - The ability to elicit and interpret individual and family histories

Genetic Testing: Care, Consent, and Liability, by Neil F. Sharpe and Ronald F. Carter
© 2006 John Wiley & Sons, Inc.

- The American Board of Genetic Counseling [ABGC, 2004] accreditation standards for counselors requires the ability to:
  - Elicit and interpret individual, family, medical, developmental, and reproductive histories
- The Canadian College of Medical Geneticists [CCMG, 2004b] requires that clinical geneticists have the ability to:
  - Elicit a comprehensive medical history and an appropriate family history

In soliciting a history, it is important to allow the patient sufficient time to identify the actual point of concern in an opening statement. Health care professionals, however, may impair the elicitation of information by rushing or interrupting the patient:

  - A 1984 study of a university medical clinic reported that only 23 percent of patients were allowed to complete an opening statement of concern, that patients were interrupted on an average of 18 seconds after they began to speak, and that only one of the 52 interrupted statements was completed during the interview [Beckman and Frankel, 1984].
  - In 1999, a follow-up study reported that although physicians solicited patient concerns in 75.4 percent of the interviews reviewed, only 28 percent of the patients' initial statements of concerns were completed. Physicians redirected these statements after an average of 23.1 seconds even though those patients who were allowed to complete their statements only took 6 seconds longer on average. The authors concluded that the interruption of patients' statements of concerns could impair potentially important data [Marvel et al., 1999; Suchman et al., 1997; Waitzkin, 1985].
  - An American study of a community family practice, surveying 4454 patient visits to 138 family physicians, reported that the average duration of family history discussions was 2.5 minutes [Acheson et al., 2000].

Allowing patients to complete their monolog and to fully share their interests and concerns:

- May require little additional time [Marvel et al., 1999; Rabinowitz et al., 2004].
- Can result in better information, improved diagnosis, and increased patient satisfaction [Haidet and Paterniti, 2003; also see AAMC, 1999].

Any omission in a medical record, or in the taking of a medical history, can be held against the physician in a malpractice action [*Siemieniec v. Lutheran General Hospital*, 1987; *Rietze v. Bruser*, 1979; Hirsh, 1995; Haydon, 1999], especially if it impedes the physician, or other professionals, from later consulting and/or making a diagnosis and/or making decisions affecting the health care of the patient [*Stokes v. Dailey*, 1957; Grover et al., 2004]. The common situation of prenatal counseling provides an example. The father or mother of a fetus may unwittingly carry a chromosomal abnormality, such as a reciprocal translocation, that places the fetus at significantly increased risk for developmental anomalies if inherited in unbalanced fashion; in this situation, sometimes the only clues to the presence of risk are the outcomes of previous pregnancies in the family history, and a standardized pattern of careful, open-ended questions can ensure that a personal or family history of recurrent miscarriages, stillbirth(s), neonatal deaths, or babies born with developmental anomalies is not missed. Recognition of the possible risk will then lead to parental karyotyping, identification of the unsuspected translocation, and clear delineation of the risks of an abnormal outcome and the options for prenatal diagnosis.

Health care professionals who elect to provide genetic testing services must be aware of the potential complexity of taking, and counseling for, an appropriate family history [Pender, 1998; CDC, 2003; AMA, 2004]. The family history may contain clues to the presence of hereditary risk that only become evident once an accurate and detailed three-generation pedigree is constructed, especially in regard to hereditary cancers [Stopfer, 2000]. For example, an inquiry into any past cases of cancer in a family [*Fernandez v. U.S.*, 1980] can be critical to early diagnosis and treatment [*Beckcom v. U.S.*, 1984]. Cancer, however, is a common disorder, and separating familial from nonfamilial cancers can be problematic. An individual may have a family history of breast cancer, but only 5 to 10 percent of all breast cancer cases are associated with mutations in the *BRCA1* and *BRCA2* genes. Concern has been expressed that family histories for genetic testing for cancer may not be sufficiently accurate or comprehensive [Lynch, 2002]. Identifying incidences of cancer spanning three generations can indicate risks that otherwise may not be identified. There is legal precedent that a physician has a duty to inquire into the family history of cancer and that three generations would be an appropriate genealogical range to be investigated [*Fernandez v. U.S.*, 1980; Severin, 1999]. Malpractice allegations may increase because of inattention to what constitutes an appropriate family history for types of hereditary cancers [Lynch, 2002].

> In reviewing medical records during the past 40 years, we have found that in most cases there is not sufficient information about family history to identify a hereditary cancer syndrome, should it be present in the family. This is a pity when one considers the lifesaving potential of a well-orchestrated family history... "Our family history is loaded with people who have died early in life of colorectal cancer; yet no one has ever told me that I was at high cancer risk." [Lynch, 2002]

Negligence claims for not meeting the standard of care with respect to genetic hereditary conditions are among the most commonly filed, including claims that a health care professional did not take an appropriate history and/or failed to diagnose a patient by making use of the patient's genetic hereditary history [Flamm, 1986; Severin, 1999]. An example is the cause of action for "wrongful birth," recognized by a majority of American jurisdictions, which refers to a claim alleging that the parents would have avoided conception or terminated the pregnancy but for the negligence of the health care professionals responsible for informing the parents about the likelihood of giving birth to a mentally and/or physically handicapped child [Botkin, 2003; e.g., *Siemieniec v. Lutheran General Hospital*, 1987; and see Chapter 6, Referral and Diagnosis]. Examples include where the health care professional knew or should have known of the risk because a previous child had been born with a genetic disease or disorder or because of a woman's advanced maternal age [*Schroeder v. Perkel*, 1981], or because of one or both parents' ethnic and racial background [*Naccash v. Burger*, 1982: failure to offer carrier screening for Tay Sachs; but note *Munro v.* Regent of the University of California, 1989].

Clinical Scenario

A woman presents for genetic counseling due to late maternal age. Counseling is provided and options for testing are discussed. The woman chooses to undergo amniocentesis, and the result is 46,XY (normal male). The pregnancy proceeds uneventfully. At birth, a male infant with multiple and severe anomalies is born. The attending neonatologist submits a peripheral blood specimen for confirmation of the normal karyotype. A very small deletion of the short arm of chromosome 1 is identified. This deletion is associated with severely abnormal development and poor outcomes (deletion 1p36 syndrome [e.g., Riegel et al., 2004]). The baby boy dies shortly after birth. A more detailed review of the family history reveals the following findings: The father had a nephew who was institutionalized with multiple anomalies, and the father had a previous partner who suffered a miscarriage and a stillbirth. The parents are karyotyped, and the father is found to have a subtle reciprocal translocation involving chromosomes 1 and 12; the infant boy has inherited the abnormal chromosome 1 from the translocation.

Issues Raised by This scenario

The woman was appropriately referred for a recognized increase in risk associated with her age. However, the history elicited failed to disclose the known abnormal outcomes associated with multiple pregnancies. Chromosomal anomalies are a common cause of miscarriage and stillbirths [Gardner and Sutherland, 2004]; approximately 6 percent of stillbirths are caused by chromosomal anomalies, and the nephew with developmental anomalies severe enough to cause institutionalization merited further investigation. Knowledge of the stillbirth and/or the institutionalized relative would suggest that karyotyping of the parents should be requested. Karyotyping on peripheral blood lymphocytes provides the highest level of banding resolution (chromosomal detail), and if the translocation is present in a normal parent, it will be present in a balanced form, that is, both abnormal partner chromosomes in the translocation will be present. This increases the chance of detecting subtle rearrangements. The abnormality was missed in the amniocentesis because the technical quality of amniotic fluid specimens is usually poorer than blood specimens, and the deletion would be too small to detect unless special staining methods specifically targeting that area of chromosome 1 were used. This situation could have been prevented by a simple question such as: "Do you know of anyone in your family who had difficulty getting pregnant, lost a pregnancy, lost a baby, or had a baby with problems such as anomalies?" The answers would have triggered the request for parental karyotypes, the translocation would have been recognized, and the amniocentesis investigation would have specifically focused on the chromosomes of interest and found the problem. (*See also Chapter 12, Scenario 1 for further discussion of laboratory perspective.*)

NEED TO KNOW

Family Pedigree

Health care professionals who elect to provide genetic testing services should have the ability to construct a family *pedigree*, a graphic representation of the family history information—a family "medical" tree—usually for at least three generations [Bennett et al., 1995; Bender, 1998; CDC, 2003; ASHG, 2004]. This may necessitate contacting family members and other physicians for medical records because sometimes the details of a specific diagnosis or laboratory test provide critical differences in the interpretation of a risk. An organized approach is required. At the least, this represents an issue of time and effort on the part of the patient and the clinician for contacting relatives, obtaining consents for release of information, and ensuring timely receipt of documents or messages, all of which need to be documented in the patient chart. Moreover, a patient may not wish to notify other family members out of concern for privacy, emotions, family considerations [McConkie-Rosell et al., 1995; Richards, 1999; Metcalfe et al., 2000], and potential discrimination and stigmatization [Fanos and Johnson, 1995; Metcalfe et al., 2000; Dugan et al., 2003; Gordon et al., 2003; Kass et al., 2004]. When critical information is not available, it is important to revise the clinical impression and risk estimates in the light of the known information while acknowledging the potential risk associated with unconfirmed aspects of the history.

Constructing an accurate family history for prenatal care raises additional issues. A common consideration is the possibility that the mother has had previous pregnancies about which she has not informed her partner. If the partner is present during the counseling session, it is important to ask the mother to volunteer her obstetrical history, even if it is clearly stated in documentation already available to the interviewer; if necessary, discrepancies can be checked later with the mother. The possibility of nonpaternity is another potential problem, and not only because the results of some genetic test protocols will clearly indicate when it is a consideration. Given that anecdotal evidence from diagnostic molecular testing suggests that, depending upon sociocultural background, up to about 10 percent of pregnancies may involve false paternity, it should be noted that the chance of nonpaternity in a given pregnancy in some circumstances may be substantially higher than the genetic risk addressed by testing. Further, false paternity presents an awkward counseling issue that has to be handled with tact and care.

WATCH OUT FOR

Inaccurate or Incomplete Information

Medical history information may be incomplete or inaccurate [Redelmeier et al., 2001], and histories may have to be cautiously interpreted [Rasmussen et al., 1990; Novakovic et al., 1996]. Patients may not have all of these details and may have inaccurate recollections. A study of patients registered in an English general practice reported that 53 (44.9 percent) patients said that they were not aware of diseases running in their family; 42 (38.2 percent) listed one or more diseases that they thought did run in their family [Rose et al., 1999]. Specific types of cancer, such as breast cancer, may be more accurately recalled than other types [Ziogas and Anton-Culver, 2003].

Although the answers given by a patient may be incorrect, in the absence of clinical evidence to the contrary, a physician can be entitled to rely on them [e.g., *MacKey v. Greenview Hospital*, 1979; *Leadbetter v. Brand*, 1980]. A physician may be exonerated or a patient found liable for contributory negligence where the patient *knowingly* provides insufficient or false information [*Rostron v. Klein, Lankenav v. Dutton*, 1986] or does not report medical ailments, symptoms, and complaints [*Guimond v. Laberge.* 1956; Mayer, 1992].

Courts of law have recognized that a physician has a duty to secure the best factual data, to consider the history given by the patient [*Walstad v. University of Minnesota Hospitals*, 1971; *Haughian v. Paine*, 1987] and to make use of all reasonable scientific means and facilities to appropriately follow up and respond to such information [*Forrestal v. Magendantz*, 1988; *Humana of Kentucky, Inc. v. McKee*, 1992; *Dale v. Munthali*, 1978; *Rietze v. Bruser*, 1979].

- A couple had consulted with a physician regarding the risk that a prospective child would be born a hemophiliac. The couple was concerned because two of the wife's cousins had the disease. The couple was advised that the risk was "very low." The child was born with hemophilia [*Siemieniec v. Lutheran General Hospital*, 1987]. The evidence established that the physician had failed to adequately inquire into the patient's background, had misrepresented the risks, and did not advise that some forms of hemophilia (at the time) were not test detectable.

- A physician not only failed to take a complete medical history but also prevented the patient from giving such information by refusing to let her explain her past history and advising that he would ask the questions. As a result, the physician did not discover that the patient had undergone two previous biopsies. The patient later died of cancer [*Beckcom v. U.S.*, 1984].

- Parents sought advice about a prospective child. The physician, an obstetrician, took a medical history during which time the women advised that her father had a "skin condition." The physician did not inquire about the nature of the skin condition and did not make any further inquiries to discover the nature of the condition and whether it might be harmful if a child was born. The father had been treated since birth for a skin condition caused by a relatively mild form of neurofibromatosis. A child was later born with a severe form of neurofibramotosis [*Ellis v. Sherman*, 1986].

Prenatal and Neonatal Testing

Prenatal and neonatal genetic testing [U.S. National Screening Status Report, 2004] can pose complex and troubling medical, legal, and ethical dilemmas [Harmon, 2004a,

2004b]. Health care professionals have an obligation to ensure that genetic tests are provided in appropriate cases. Mandatory testing for specific conditions are required with the nature and number varying from jurisdiction to jurisdiction [U.S. National Screening Status Report, 2004]. Additional tests may be required depending on the patient's ethnicity and conditions that run in his or her family.

Will health care professionals who do not have specialized training in genetics and genetic counseling have the knowledge, skills, and ability to make such clinical evaluations and decisions [Stopfer, 2000]?

- Surveys indicate that health care professionals may be uncertain as to the nature, reliability, and meaning of genetic tests and may not appropriately advise patients about the limitations, risks, and implications associated with genetic testing [Hayflick et al., 1998; Hunter et al., 1998; Cho et al., 1999; Wolpert et al., 2000; Escher and Sappino, 2000].

- A health care professional who knows, or should know, that a patient's situation is beyond the health care professional's skills and competence has a duty to refer for genetic counseling [Reilly, 1995; Ang and Garber, 2001; Stopfer, 2000]. Health care professionals need to keep up to date, especially with regard to the latest information about genetic testing technologies in order to know when it is appropriate to refer the patient for testing and counseling.

Keeping up to date with significant advances, having sufficient knowledge of the nature and limitations of genetic tests, and knowing when patients should be referred to genetic counselors and specialists can be a daunting task. These difficulties are compounded with the rapid proliferation of tests delivered by commercial organizations [Andrews, 1997] and with the potential for direct-to-consumer advertising [Hull and Prasad, 2001a,b; Caulfield, 2001; Gollust et al., 2002; Tsao, 2004].

At the time of writing, more than 1000 genetic tests are reported to be available for clinical or research purposes—http://www.genetests.org—including prenatal screening for deafness dwarfism, and skin disease, with more than 100 prenatal/neonatal tests added in 2003 alone [Harmon, 2004a]. Health care professionals may withhold information about the availability of specific genetic tests [Harmon, 2004b] out of concern that the tests are not clinically valid and do not have appropriate clinical utility [Collins, 1996; American Society of Human Genetics, 1996; Task Force, 1997; Welch and Burke, 1998; Yang et al., 2000; Caulfield, 2000; Khoury et al., 2000; Burgess, 2001; Haga et al., 2003; Khoury et al., 2004; and please see Chapter 7, Informed Consent, and Chapter 8, Prenatal and Neonatal Screening].

Prenatal genetic tests for non-life-threatening, developmental disabilities conditions such as deafness [Harmon, 2004b] can pose complex and troubling ethical and legal dilemmas. What if the patient requests testing for a developmental disability condition such as deafness either to consider terminating the pregnancy [Harmon, 2004b; and see Chapter 8, Prenatal and Neonatal Testing] or to ensure that the child is born with a condition "just like" the parents [Green, 1997]? What if the health care professional decides to withhold the information based solely on the fact that the tests are not covered by the patients' health insurance plans [Gottlieb, 2003; Wynia et al., 2003]?

Generally speaking, the nature and scope of a health care professional's obligation is determined by the prevailing standard of care. A health care professional, however, must not only meet the prevailing standard of care but also fulfill a separate duty to exercise good judgment—as discussed early in Chapter 5—particularly with the application of new technologies and procedures.

Although a genetic disease or condition may be comparatively rare in terms of relative population risk, a patient's ethnic background, family history, situation (e.g., workplace environment with exposure to teratogenic agents—drugs, alcohol, environmental agents, infections—associated with birth defects) [Thompson and Thompson, 2001] can indicate that a specific test may be appropriate, such as Tay-Sachs disease for patients with an Ashkenazi Jewish background. In situations where a health care professional knows or should know of these potential factors, or the patient makes a specific request, the professional arguably has a reasonable obligation to take an adequate and appropriate family history and to refer the patient for all appropriate and reasonably available diagnostic tests, notwithstanding that the test in question—such as the case with fragile X—has not been formally recognized as a standard of care by professional medical genetic and/or medical organizations yet is clinically valid with demonstrated clinical utility [but see *Munro v. Regent of the University of California*, 1989].

What if the patient requests testing even though the family history, ethnic background, or situation does not indicate that the test would be applicable? Although the health care professional may seek to rely on a family history to help determine this issue, this arguably is based on the operative assumption that the medical history will be adequate and accurate, when histories are often incomplete. A child may be born with a genetic condition like fragile X even though there were no indications of related problems in past generations. Although a clear majority of geneticists and counselors support the value of patient autonomy in decision making (as discussed in Chapter 2), these situations will have to be considered on a case-by-case basis, and referral for genetic counseling is recommended. The health care professional will need to know about the nature and scope of the test, its limitations and risks, whether the laboratory performing the test is accredited, the status of the test in terms of its clinical utility and validity, the potential impact on the patient and family including emotional and psychological responses, and the potential risk of stigmatization, familial unrest, and discrimination [Harmon, 2004a].

Family History in Adult-Onset Disorders

As with prenatal and developmental genetics, the family history provides a critical source of information for identifying potential risk factors for pediatric and common adult-onset disorders. A complete three-generation family history can be used to determine the presence of genetic risk, the apparent pattern of heredity, the possible presence of common environmental risk factors, and which family members are at risk.

For pediatric disorders, much of the focus is on considerations similar to prenatal counseling: knowledge of abnormal outcomes for pregnancies, stillbirths, neonatal outcomes, developmental anomalies, syndromes, and confirmed diagnoses of genetic disorders. Clinical examinations, detailed review of the medical history, and/or genetic testing for children in question may be required.

However, because common adult disorders usually arise from interactions of genetic and environmental mechanisms, interpretation of the family history may not be straightforward. Additionally, even in families with hereditary onset of common diseases, sporadic cases may also occur and therefore obscure the true hereditary pattern of disease in the family. Accordingly, the family history assessment can help place the patient's risk of genetic disease into a category [e.g., low, intermediate, or high (see discussion by Scheuner in the following section)]. Risk assessments can be based upon disease-specific criteria or practical guidelines for testing [e.g., Taylor, 2001], and for some diseases there are even specific risk calculation models available to facilitate identification of putatively affected families [e.g., for hereditary breast/ovarian cancer or colorectal cancer: Claus et al., 1993, 1994; St John et al., 1993; Srivastava et al., 2001]. Using models for assessing the likelihood of a positive result, threshold criteria can be used to determine which patients should be offered testing. This type of approach can minimize unnecessary testing and serve to reassure patients who are truly at low genetic risk for onset of a common disease, while permitting early recognition, testing, and recommendation of management strategies for patients at intermediate or high risk.

IMPORTANCE OF FAMILY HISTORY IN APPROACHES TO COMMON CHRONIC DISEASES OF ADULTHOOD

Maren T. Scheuner, UCLA School of Public Health, Department of Health Services, CDC Office of Genomics and Disease Prevention, Los Angeles, California

OVERVIEW

Common chronic diseases such as coronary heart disease, stroke, diabetes, and cancer are caused by interaction of genetic and nongenetic risk factors. Recognizing genetic factors that contribute to common chronic diseases can help identify individuals with increased risk, especially at early ages of onset. Understanding genetic susceptibilities can also help determine which diseases might develop in an individual given a particular exposure or behavior. Family history collection and interpretation is the most practical screening strategy for identifying individuals with a genetic susceptibility to many common chronic diseases. It represents complex interactions of genetic, environmental, cultural, and behavioral factors shared by family members.

Family history characteristics that increase disease risk include: early age at diagnosis, two or more close relatives affected with a disease or a related condition, a single family member with two or more related diagnoses, multifocal or bilateral disease, and occurrence of disease in the less often affected sex. By recognizing the magnitude of risk associated with these familial characteristics, stratification into different familial risk groups (e.g., low, intermediate, and high) is possible, which can guide risk-specific recommendations for disease management and prevention.

Clinical genetic evaluation should be considered for individuals with a high familial risk for common chronic diseases or when a Mendelian disorder is suspected. The process includes (1) genetic counseling and education, (2) risk assessment and diagnosis using personal and family medical history, physical examination, and genetic testing, and (3) risk-appropriate recommendations for management and prevention.

INTRODUCTION

Common disease genetics is the study of genetic aspects of diseases that are of major public health concern. Coronary heart disease, stroke, cancer, and diabetes [American Heart Association, 2002; American Cancer Society, 2003] are diseases of affluence that have increased in prevalence with cultural factors such as tobacco use, physical inactivity, excess caloric intake, processed foods, environmental irradiation, and pollution. These conditions develop over decades of life and usually are diagnosed in adulthood. Early disease detection and prevention are possible because of the chronic nature of these disorders.

Common chronic diseases are complex genetic conditions because they arise due to an interplay between genetic and nongenetic/environmental factors. In some individuals, the interaction of multiple genetic factors explains most of the disease susceptibility (i.e., polygenic inheritance), whereas in others, environmental factors predominate. However, in most, disease is due to the interaction of genetic and environmental risk factors (i.e., multifactorial inheritance). Only rarely do common chronic diseases of adulthood occur primarily as the manifestation of a single gene mutation. However, even in these cases, there is substantial clinical heterogeneity [Scheuner et al., 2004].

Recent improvements in our ability to identify variable genetic susceptibilities to common chronic diseases will lead to improved strategies for diagnosis, treatment, and prevention. With increasing longevity and pervasive environmental risk factors, our genetic susceptibilities are readily revealed. This is especially true for most individuals diagnosed with a common chronic disease at an early age (about 10 to 20 years earlier than the typical age of diagnosis). In addition, because a variety of chronic diseases can be linked to each environmental risk factor (e.g., coronary artery disease, lung cancer, and emphysema are all associated with smoking), genetic susceptibilities become paramount in determining which individuals have the greatest risk for developing a specific disease given a particular exposure or lifestyle.

GENETIC RISK ASSESSMENT STRATEGIES FOR INDIVIDUALS

Strategies for assessing genetic susceptibility to common disease can utilize personal and family history collection and interpretation, phenotypic assessment of biochemical and physiological traits, deoxyribonucleic (DNA)-based testing. The approach to individualized risk assessment may combine each of these sources of information.

REVIEW OF FAMILY HISTORY

Family history collection and interpretation is currently the most practical screening strategy for identifying individuals with a genetic susceptibility to many common chronic diseases [Scheuner et al., 1997]. It represents complex interactions of genetic, environmental, cultural, and behavioral factors shared by family members. For many common diseases, a positive family history is quantitatively significant with relative risks ranging from two to five times those of the general population, and this risk generally increases with an increasing number of affected relatives and earlier ages of disease onset (Table 5.1). Clues to increased risk include early age at diagnosis, two or more close relatives affected with a disease or a related condition, a single family member with two or more related diagnoses, multifocal or bilateral disease, and occurrence of disease in the less often affected sex. By recognizing the magnitude of risk associated with these familial characteristics, stratification into different familial risk groups (e.g., low, intermediate, and high) is possible (Table 5.2), which can guide risk-specific recommendations for disease management and prevention (Fig. 5.1). Referral for genetic evaluation by a geneticist or other specialist should be considered for individuals with high familial risk.

Table 5.1   Risk Estimates Due to Family History for Selected Common Chronic Diseases of Adulthood

Disease	Risk due to family history
Coronary heart disease	OR = 2.0 (one first-degree relative) [5] [Ciruzzi et al., 1997]
	OR = 5.4 (two or more first-degree relatives) [6] Silberberg et al., 1998
Type 2 diabetes	RR = 2.4 (mother) [7] Klein et al., 1996
	RR = 4.0 (maternal and paternal relatives) [8] [Bjornholt et al., 2000]
Osteoporosis	OR = 2.0 (female first-degree relative) [9] Keen et al., 1999
	RR = 2.4 (father) [10] [Fox et al., 1998]
Asthma	OR = 3.0 (mother) [11] [Tariq et al., 1998]
	RR = 7.0 (father and mother) [12]
Breast cancer	RR = 2.1 (one first-degree relative) [13] Pharoah et al., 1997
	RR = 3.9 (three or more first-degree relatives) [14] [Collaborative Group, 2001]
Colorectal cancer	OR = 1.7 (one first-degree relative) [15] [Fuchs et al., 1994]
	OR = 4.9 (two first-degree relatives) [16] [Sandhu et al., 2001]
Prostate cancer	RR = 3.2 (one first-degree relative) [17] [Cerhan et al., 1999]
	RR = 11.0 (three first-degree relatives) [18] [Steinberg et al., 1990]

OR: odds ratio
RR: relative risk

Table 5.2 Suggested Guidelines for Risk Stratification Based on Family Medical History of Common Chronic Diseases of Adulthood[a]

Low familial risk

1. No affected relatives
2. Unknown family medical history (includes adopted person with unknown family medical history)
3. Only one affected second-degree relative with any age of disease onset from one or both sides of the pedigree if the disease is not sex-limited
4. If the disease is sex-limited, only one affected second-degree relative with any age of disease onset from the parental lineage of the more often affected sex
5. If the disease is sex-limited, only one affected second-degree relative with a late or unknown age of disease onset from the parental lineage of the less often affected sex

Intermediate familial risk

1. Only personal history with a later age of disease onset
2. Only one affected first-degree relative with late or unknown age of disease onset
3. Only two affected second-degree relatives from one lineage with a late or unknown age of disease onset if the disease is not sex limited
4. If the disease is sex limited, only one affected second-degree relative with an early age of disease onset from the parental lineage of the less often affected sex

High familial risk

1. Personal history with an early age of disease onset
2. Personal history with a late or unknown age of disease onset and personal history of a related condition at any age of onset
3. Personal history with a late or unknown age of disease onset and one affected first- or second-degree relative with an early age of disease onset, or one affected first- or second-degree relative with a late or unknown age of disease onset and a related condition in the same relative
4. One affected first-degree relative with an early age of disease onset
5. Two affected first-degree relatives at any age of disease onset
6. One affected first-degree relative with a late or unknown age of disease onset and one affected second-degree relative with an early age of disease onset, or one affected second-degree relative with a late or unknown age of disease onset and a related condition in the same relative
7. Two affected second-degree relatives from the same lineage with at least one having an early age of disease onset, or at least one having late or unknown age of disease onset and a related condition in the same relative
8. Three or more affected first- and/or second-degree relatives from one lineage with any age of disease onset
9. One first- or second-degree relative with an early age of disease onset and two first- and/or second-degree relatives from the same lineage with related conditions
10. Intermediate familial risk on both sides of the pedigree

Source: Adapted from Scheuner et al. 1997.

[a] Early age of disease onset refers to disease that occurs about 10 to 20 years earlier than typical. Examples of sex-limited diseases include coronary heart disease, osteoporosis, breast cancer, ovarian cancer, and prostate cancer. Examples of related conditions include coronary heart disease, diabetes, and stroke or colorectal cancer endometrial cancer, and ovarian cancer.

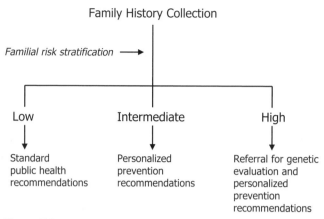

Figure 5.1 Family history collection followed by risk stratification that recognizes family history characteristics that increase disease risk (e.g., early age at diagnosis, two or more close relatives affected with a disease or a related condition, a single family member with two or more related diagnoses, multifocal or bilateral disease, and occurrence of disease in the less often affected sex) can guide risk-specific recommendations for disease management and prevention. Standard public health messages would be appropriate for individuals with a low familial risk. Personalized prevention recommendations should be provided to individuals with intermediate and high familial risk, such as lifestyle modifications, earlier and more frequent screening, and use of chemoprevention when available. Referral for genetic evaluation by a geneticist or other specialist should be considered for individuals with high familial risk.

Review of the family history is important because it is a prevalent and relatively accurate predictor of risk for several chronic conditions, including many forms of cancer, coronary heart disease, stroke, and diabetes. About 43 percent of healthy, young adults will have a family history for one of these disorders. Depending upon the specific disorder, approximately 5 to 15 percent will have a moderate familial risk (about two to three times the population risk), and 1 to 10 percent will have a familial risk level that is high enough to approach risks associated with Mendelian disorders [Scheuner et al., 1997]. Most sensitivity values for a positive family history of these diseases in a first-degree relative range from 70 to 85 percent and specificity is usually 90 percent or greater [Hunt et al., 1986; Acton et al., 1989; Kee et al.,1993; Bensen et al., 1999; Glazier et al., 2002] Overall, the available studies suggest that a positive family history can generally be used with a high degree of confidence for the identification of individuals at increased risk for developing many common chronic diseases.

MOLECULAR TESTING

Although numerous susceptibility alleles for common chronic diseases have been discovered, there has been limited progress in the discovery of genes for non-Mendelian forms of these diseases that have meaningful clinical relevance [Kip et al., 2002]. Before testing for low-risk susceptibility genes have widespread clinical application, we will need to have a better understanding of the prevalence and penetrance of these genotypes, as well as the effect of other genes and environmental factors on their expression. Furthermore, the clinical utility of DNA-based testing for disease susceptibility compared to other risk assessment strategies, including familial risk assessment and assessment of biochemical risk factors, must be proven.

Currently, genetic testing for chronic disease susceptibility is generally only available for rare Mendelian disorders. Personal and family history characteristics are crucial for identifying Mendelian disorders [Scheuner et al., 2004]; therefore, genetic testing will likely remain a clinical intervention based on familial characteristics for many years to come. Thus, use of family history is central to providing access to genetic testing services that are currently available, and it is likely the paradigm of familial risk assessment will inform future genetic testing of less penetrant susceptibility alleles.

CLINICAL GENETIC EVALUATION

Clinical genetic evaluation for common disease should be performed for individuals with a high familial risk or when a Mendelian disorder is suspected. Genetic evaluation is comprised of several components. The process includes: (1) genetic counseling and education, (2) risk assessment and diagnosis using personal and family medical history, physical examination, and genetic testing, and (3) recommendations for management and prevention options appropriate for a genetic risk.

GOALS OF GENETIC COUNSELING AND EDUCATION

An important goal of genetic evaluation for common chronic diseases is the development of individualized preventive strategies based on the genetic risk assessment, the patient's personal medical history, lifestyle, and preferences. Genetic counseling is critical for delineating a patient's motivation and likely responses to learning of a genetic risk. Through genetic consultation, patients will be educated about the role of behavioral and genetic risk factors for disease, the mode of inheritance of genetic risk factors, and the options for prevention and risk factor modification.

This is a necessary process for individuals at risk for common chronic diseases since there is evidence that awareness of increased risk due to genetic or familial factors does not automatically translate to spontaneous improvement in lifestyle choices [West et al., 2003; Tavani et al., 2004]. For example, the occurrence of a heart attack or stroke in an immediate family member did not lead to self-initiated, sustained change in modifiable risk factors in young adults [West et al., 2003], and among low-income, rural African American women who had not had a recent mammogram, knowledge of family history of breast cancer was not associated with perceived risk or screening [Tavani et al., 2004]. These results argue that counseling and education are needed to actively intervene in people with a family history of common chronic disease, where the opportunities for prevention are substantial [Tavani et al., 2004; Slattery et al., 2003; Lerman et al., 1996a].

The process of genetic counseling ensures the opportunity to provide informed consent, including discussion of the potential benefits, risks, and limitations of genetic risk assessment. Knowledge of genetic susceptibility to a common chronic disease has the potential to improve diagnosis, management, and prevention efforts. Psychological benefits can also result from knowledge of a genetic risk [Tonin et al., 1996; Croyle et al., 1997]. Confirming a suspected genetic risk can be empowering and may relieve anxiety related to not knowing, while excluding a familial susceptibility can also reassure individuals. On the other hand, potentially harmful psychological effects, such as increased anxiety, can result from knowing of a genetic risk, particularly if there are no proven interventions available for management or prevention, or if such interventions are inaccessible or unacceptable (e.g., prophylactic oophorectomy for a woman who has not completed childbearing). Family dynamics may change from knowledge of a genetic risk for disease. For example, a parent may feel guilt about passing on a disease predisposition, or a sibling for whom genetic susceptibility has been excluded may experience survivor guilt if the susceptibility is identified in another sibling. Family members may experience loss of privacy when asked to share their medical history and medical records, and if labeled as having a genetic risk for disease, family, friends, or society may stigmatize them. There is also the possibility of misuse of genetic risk information by third parties such as employers, educators, and insurers that could exclude individuals from employment or education opportunities or from obtaining health, life, or disability insurance, although the evidence of genetic discrimination against otherwise healthy individuals is minimal [Billings et al., 1992; Geller et al., 1996; Epps, 2003; please see Chapter 7, Informed Consent Genetic Discrimination; J Beckwith, L Geller]. These potential harms should be considered and weighed against the potential benefits when providing genetic risk assessment.

APPROACH TO GENETIC COUNSELING AND EDUCATION

Pedigree Analysis

Pedigree analysis is typically the first step in genetic risk assessment and diagnosis. The pedigree structure usually includes all first- and second-degree relatives spanning three to four generations. Demographic information for each family member is documented, which typically includes each relative's name and current age or age at death. Medical history is documented for each family member including age at diagnosis, cause of death if deceased, and known interventions or procedures, which can help clarify a diagnosis. For example, questioning regarding coronary artery bypass surgery, angioplasty, heart transplant or pacemaker placement may help clarify a relative's diagnosis of heart disease. Information also is collected regarding important risk factors for a disease, such as use of hormone replacement therapy or chest irradiation for breast cancer, and smoking, asbestos exposure, and coal mining for lung cancer. Medical records are reviewed when possible to verify the medical history of each family member or at least those who are critical to the genetic risk assessment and diagnosis. The family history should include ethnicity and country of origin of grandparents since certain conditions might be more prevalent in certain ethnic groups. For example, the prevalence of insulin resistance is high among individuals of Native American admixture [Arnoff et al., 1997] and Asian Indian origin [Sharp et al., 1987], and there are common *BRCA1* and *BRCA2* gene founder mutations in Ashkenazi Jewish families with breast and ovarian cancer [Struewing et al., 1997].

Once this information is collected, pedigree analysis is performed to determine the most likely mode of inheritance (i.e., Mendelian versus multifactorial) and the risk of disease to the patient and to unaffected relatives based on their position in the pedigree. Pattern recognition is used to elucidate the differential diagnosis [Scheuner et al., 2004]. For example, when considering an inherited form of breast cancer, there are at least seven different Mendelian disorders to consider, including hereditary site-specific breast cancer, hereditary breast–ovarian cancer syndrome, Li–Fraumeni syndrome, Cowden syndrome, Peutz–Jeghers syndrome, hereditary nonpolyposis colon cancer, and ataxia telangiectasia [Scheuner et al., 2004; Sharp et al., 1987]. The types of cancers and other conditions reported help to distinguish each of these syndromes (Fig. 5.2). Mutations in different genes underlie the susceptibility in these syndromes, and genetic testing can help to confirm a suspected diagnosis.

For pedigrees consistent with multifactorial inheritance or that lack convincing evidence of Mendelian inheritance, quantitative risk assessment (and see Chapter 13, Risk Assessment) can be performed for specific conditions using disease-specific mathematical models or published estimates [Amos et al., 1992; Claus et al., 1993, 1994; St John et al., 1993]. For example, a woman's absolute risk for breast cancer by age 80 or in the next 10 years can be provided based on the family history of breast or ovarian cancer in first and- second-degree relatives and their age at diagnosis [Claus et al., 1993, 1994], and she can contrast this to the population risk of breast cancer by age 80 or in the next 10 years. Because family history is often the most significant predictor of risk, these risk estimation models provide good estimates of risk. However, they do have limitations, and they should not be the only means for risk assessment if they do not account for significant environmental exposures or behaviors that might influence disease risk.

Personal History

In addition to review of past medical history and medical records for confirmation, assessment of signs and symptoms of the disease of concern should be performed to more accurately assess risk for the patient. For example, when evaluating a genetic risk for heart disease, review of systems should include questions regarding angina, shortness of breath, dyspnea on exertion, paroxysmal nocturnal dyspnea, pedal edema, claudication, and exercise tolerance. In the case of risk assessment for colorectal cancer, questions should be asked regarding frequency of bowel movements, caliber of the stool, color of the stool, and presence of blood in the bowel movement. If symptoms are present, confirmatory testing should be recommended, such as exercise treadmill testing or echocardiogram to evaluate cardiovascular symptoms or colonoscopy to assess change in bowel habits or blood in the stool.

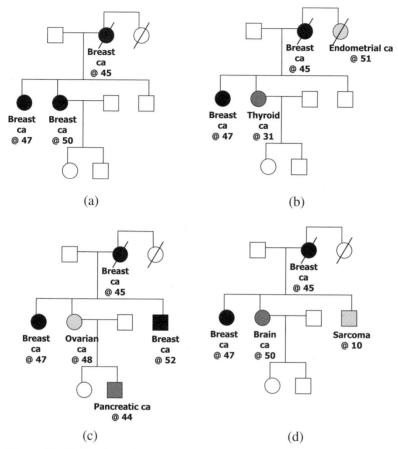

Figure 5.2   Each pedigree shown has a high familial risk for breast cancer. However, by recognition of the patterns of cancer in the family, a more accurate diagnosis can be made. Pedigree (*a*), which features early, onset breast cancer and no other cancers, is most consistent with hereditary site-specific breast cancer, which is often due to *BRCA1* or *BRCA2* gene mutations. Pedigree (*b*) features early-onset breast, thyroid, and endometrial cancer and is most consistent with Cowden syndrome due to *PTEN* gene mutations. Pedigree (*c*) features early-onset breast and ovarian cancer and is most consistent with hereditary breast–ovarian cancer syndrome, which is almost always due to *BRCA1* or *BRCA2* gene mutations. In this case, a *BRCA2* gene mutation is likely given the family history of male breast cancer and pancreatic cancer. The history of early-onset breast cancer, brain tumor, and childhood sarcoma in pedigree (*d*) is most consistent with Li–Fraumeni syndrome due to *TP53* gene mutations. Multiple primary cancers are common among individuals with Li–Fraumeni syndrome.

Physical Examination

Physical examination should be performed for signs of the disease of concern as well as to identify characteristic manifestations of Mendelian forms of a disease. For example, an evaluation for cardiovascular risk should include auscultation of the heart, lungs and major vessels in the neck, abdomen, and groin, and palpation of the aorta and distal pulses. Any abnormalities can be followed up with additional studies, such as ultrasound. Blood pressure in the upper and lower extremities can identify hypertension, and these measurements can be used to calculate the ankle/brachial blood pressure index (ABI). Values <0.9 are correlated with atherosclerosis. Weight and height should be obtained and body mass index should be calculated to identify overweight and obese patients, and follow-up measurements can help monitor diet and exercise interventions. Waist circumference should be obtained, as increased values are associated with the metabolic syndrome [Expert Panel on Detection, Evaluation, and Treatment of High Blood Cholesterol in Adults, 2001], a common cause of cardiovascular disease [Park et al., 2003].

Evaluation of possible lipid disorders should include examination of the eyes, assessing corneal arcus, and lipemia retinalis. Examination of the skin should in-clude assessment for xanthelasma and tendonous xanthomas. Physical signs of Mendelian disorders that feature cardiovascular disease should be assessed, such as dolichostenomelia and arachnodactyly associated with Marfan syndrome, abnor-mal scarring and translucent skin associated with Ehlers–Danlos syndrome type IV, papular skin lesions and plaques in flexural creases and angioid streaks on the retina associated with pseudoxanthoma elasticum, and angiokeratomas (vascular cutaneous lesions) and corneal and lenticular opacities associated with Fabry disease.

GENETIC TESTING

Identification of appropriate tests that may be helpful in confirming a genetic diagnosis and refining genetic risk is the initial step in the genetic testing process. Genetic tests for assessing common diseases usually include DNA-based tests and biochemical tests. DNA-based testing is typically performed when assessing a Mendelian form of cancer susceptibility, whereas biochemical testing is more often available to assess a susceptibility to cardiovascular disease or diabetes [Scheuner et al., 2004].

Most commercial laboratories performing DNA-based testing utilize PCR-based (polymerase chain reaction) methods to amplify patient DNA, followed by sequencing or other methodologies. While sequencing is often considered the "gold standard" for identification of a mutation, even sequencing may miss rearrangements or large deletions of DNA or mutations in regulatory regions [Yan et al., 2000; Gad et al., 2001]. Rarely, errors may occur with sample handling, contamination by airborne particles in the laboratory, or failure of the PCR technique.

When considering genetic testing, clinicians should choose a testing strategy that will be the most informative for a patient and their family members. Testing of all family members will be more conclusive and clinically relevant if a causal genetic mu-tation is clearly identified in the family. Ideally, genetic testing should begin with an affected family member, which provides the best opportunity to identify the specific genetic determinant(s) of disease susceptibility and to recommend appropriate man-agement. If an abnormality or abnormalities are identified, then at-risk relatives can be tested for those familial factors. Normal results can provide reassurance regarding disease risk due to those factors; however, there will always be a background or population risk for disease development. At-risk relatives with abnormal results can begin surveillance and prevention activities appropriate for their risk.

It is not always possible to test an affected family member, as they are often deceased and therefore unable to provide a sample, or not interested in testing, or unable to participate due to financial reasons. For deceased family members, archived pathology specimens may be available and provide usable DNA. If all efforts are unsuccessful, unaffected family members may participate in testing. Identification of a genetic risk factor or factors may help in planning appropriate screening and preventive strategies. Normal test results, however, cannot be entirely reassuring since the available clinical testing is not comprehensive. A normal test result can exclude the genetic risk factors that have been tested but not the possibility of an inherited susceptibility. In such a case, medical management would typically depend upon the empiric risks associated with the family history. To further clarify the situation, testing additional family members, both affected and unaffected, may be helpful.

Once a genetic testing strategy has been formulated, laboratories that can provide the testing are identified. Obtaining, processing, and shipping the sample may also be necessary, as well as facilitation of the billing process (providing the most appropriate diagnostic codes and documenting medical necessity for testing when requested). Issues to consider when choosing a laboratory include methodology, analytic sensitivity and specificity, technical support, cost, and turn-around time. Understanding which methodology is used is important since standards for many genetic tests currently do not exist. There is great variability in testing procedures for many conditions. As a result, clinicians need to be personally familiar with a given laboratory's protocol and the test limitations and should ensure that the test protocol applied is appropriate for the clinical issue in question.

An individual's genetic test results should be interpreted within the context of her or his family history. The interpretation should incorporate a discussion of the impact of the results on risk assessment, diagnosis, and plans for management and prevention. Family members who might benefit from risk assessment and genetic testing should also be identified, and the clinician should facilitate communication between the index case and at-risk relatives as well as provide referrals to genetics professionals or other specialists for relatives. This facilitation can take the form of providing an open letter to family members for circulation by the patient or by offering help in referral for care.

Interpretation of genetic test results for disease susceptibility alleles also should consider the variable effects of an allele. For example, polymorphisms in the methylenetetrahydrofolate reductase (MTHFR) gene are associated with cardiovascular disease, neural tube defects, and unexplained recurrent embryo losses in early pregnancy [Isotalo et al., 2000; Eikelboom et al., 1999]. The apolipoprotein E genotypes contribute to serum cholesterol levels and have been associated with cardiovascular disease risk as well as risk for Alzheimer disease [Wilson et al., 1996; Farrer et al., 1997]. Recognizing these associations has important implications regarding informed consent when offering specific tests for the purpose of refining a diagnosis or for presymptomatic detection of disease.

MANAGEMENT AND PREVENTION

Identifying individuals with genetic susceptibilities to common chronic diseases may have profound implications not only for their health but the health of their family. Knowledge of a genetic risk can influence the clinical management and prevention of a disease. Prevention strategies include targeted lifestyle changes, screening at earlier ages, more frequently and with more intensive methods than used for average-risk individuals, use of chemoprevention, and for those at highest risk, prophylactic procedures and surgeries (see Table 5.3 for examples).

Screening and prevention guidelines are available for many chronic disorders [Scheuner et al., 2004; Expert Panel on Detection, Evaluation, and Treatment of High Blood Cholesterol in Adults, 2001; Smith et al., 2002; Diabetes Prevention Program Research Group, 2002; Straus et al., 2002], and data are accumulating regarding the effectiveness of these strategies in high-risk individuals [Jarvinen et al., 2000; Rebbeck et al., 1999, 2002; Narod et al., 1998, 2000]. Because many behavioral risk factors for disease aggregate in families, a family-based approach to risk factor modification ought to be an effective strategy, and this has been demonstrated in a few studies [Knutsen and Knutsen, 1991; No authors listed, 1994; Pyke et al., 1997].

Table 5.3   Examples of Potential Benefits, Risks, and Limitations of Knowing of a Genetic Susceptibility

Potential benefits	Potential risks and limitations
Improve diagnosis, prognosis, and risk assessment.	False reassurance of no disease risk if genetic susceptibility is excluded.
Improve management (including response to therapy).	Increase anxiety if no interventions are available or if available interventions are inaccessible or unacceptable.
Enhance opportunities for early detection and prevention.	Change in family dynamics such as guilt for transmitting a genetic susceptibility or survivor guilt for not having inherited a genetic susceptibility. Or possible coercion by family members to pursue genetic risk assessment/testing to inform testing for the family.
Psychological benefits such as feeling of empowerment (better than not knowing), and reassurance if genetic risk is excluded.	Loss of privacy among family members asked to share medical information.
Provide important health information to family members at risk.	Labeling/stigmatization by family, friends, and society if genetic susceptibility is confirmed. Use of genetic information to deny opportunities such as employment or education or access to insurance (health, life, disability).

Table 5.4 Examples of Early Detection and Prevention Strategies According to Familial Risk Level

Colorectal cancer	
	Low Familial Risk. Beginning at age 50, annual Focal Occult Blood Test (FOBT) and/or flexible sigmoidoscopy every 5 years, or double-contrast barium enema every 5 to 10 years, or colonoscopy every 10 years [52–55]. [Winawer et al., 1997, Rex et al., 2000; Smith et al., 2002; United States Preventive Services Task Force, 2002a]
	Intermediate Familial Risk. Beginning at age 40, annual FOBT and/or flexible sigmoidoscopy every 5 years, or double-contrast barium enema every 5 to 10 years, or colonoscopy every 10 years [52, 53]. Winawer et al., 1997; Rex et al., 2000
	High Familial Risk. Beginning at age 40 or 10 years before the earliest age at diagnosis in the family, colonoscopy every 3 to 5 years [53]. Rex et al., 2002. Consider daily use of aspirin [Baron et al., 2003; Sandeer, 2003] [56, 57], and daily intake of supplemental folic acid 400 mcg [58] [Giovaunueci et al., 1998] and calcium 1200 mg [59, [Holt et al.,1998, Baron et al., 1999] 60].
Coronary heart disease	
	Low Familial Risk. Beginning at age 45 for women and at age 35 for men, screen for lipid disorders every 5 years or shorter intervals for people with lipid levels close to those warranting therapy [61]. [United States Preventive Service Task Force, 2001]
	Intermediate Familial Risk. Beginning at age 20 for women and men, screen for lipid disorders every 5 years or shorter intervals for people with lipid levels close to those warranting therapy [61][b]. [United States Preventive Service Task Force, 2001] Consider use of aspirin daily or every other day [62][b]. US Preventive Services Task Force, 2002b,
	High Familial Risk. Beginning at age 20 for women and men, screen for lipid disorders every 5 years or shorter intervals for people with lipid levels close to those warranting therapy [61][b]. [United States Preventive Service Task Force, 2001] Assess emerging risk factors [e.g., homocysteine, lipoprotein(a), C-reactive protein, low-density lipoprotein (LDL) cholesterol density] Scheuner, 2003 [63]. Consider early detection strategies [e.g., carotid artery duplex scanning to measure intima-media thickness, ankle-brachial index, electron beam computed tomography (CT) to detect coronary artery calcification] [Scheuner, 2003; Smith GC et al., 2000] [63, 64]. Consider use of aspirin daily or every other day [62][b] USPSTF, 2002b.

[a]For all individuals, regardless of familial risk level, assess relevant exposures, lifestyle choices, and habits and recommend changes if indicated. Consider referral for a genetic evaluation for individuals with high familial risk.

[b]Based on United States Preventive Services Task Force recommendations for individuals with increased coronary heart disease risk.

Lifestyle changes such as dietary modification, weight control, and smoking cessation are likely to be more effective when delivered to the family than to an individual because family members can influence each other and provide ongoing support to one another.

The plan for management and prevention should be tailored to an individual's genetic risk while taking into consideration their personal medical history, family history, habits, and preferences. Other disease risks that may influence decision making regarding options for risk reduction should be identified as part of the management and prevention plan. For example, use of tamoxifen for breast cancer prevention might be contraindicated given a personal or family history of thrombophilia. In addition, as with any treatment plan, potential risks or complications that may result from the recommended management and prevention strategies must be recognized.

Finally, the recommendation for management and prevention options specific to an individual's genetic risk should be communicated in writing so the patient and referring physician can incorporate these recommendations into a plan for future health management. Plans for continued contact or follow-up with a geneticist may be appropriate to review the individualized plan for management of genetic risk, and updated personal and family history information can be obtained with revision of the management and prevention plan as indicated. New information and technology available for genetic risk assessment or management and prevention of disease can also be discussed.

SUMMARY

Genetics play an important role in the development, progression, and response to therapy of many common chronic diseases of adulthood. Family history collection and interpretation is the best strategy for initial identification of genetically susceptible individuals. Individuals can be assigned different levels of familial risk through recognition of family history characteristics associated with increased risk and genetic susceptibility. Personalized recommendations for disease management and prevention can be made according to the level of familial risk, including referral for genetic evaluation. This personalized approach helps prioritize lifestyle choices and behaviors based on an individual's genetic susceptibility, and, as a result, it has the potential to improve health promotion and disease prevention efforts.

Acknowledgments

This work was supported by a Career Development Award sponsored by the Centers for Disease Control and Prevention and the Association of Teachers of Preventive Medicine, Cooperative Agreement U50/CCU300860.

Chapter 6

Referral and Diagnosis

OVERVIEW

Case Scenario

As a result of investigations relating to the birth of a child with spina bifida, cytogenetic testing is done and a familial reciprocal translocation is identified by coincidence. Other family members are tested, and appropriate counseling is given with regard to the risks and options for prenatal diagnosis. One family member who is shown to carry the translocation is a 12-year-old boy. Years later, the boy is a married and his wife is referred to an obstetrician because she is 11 weeks pregnant. During the referral visit, the wife mentions that her husband knows he has an abnormal chromosome, but she does not know exactly what kind of abnormality. The obstetrician comments to her that her baby will be fine.

Routine care is provided, including a dating ultrasound, but the patient is not referred for genetic counseling nor is she offered prenatal testing. Late in the pregnancy, she is referred for an ultrasound because of some bleeding. Multiple anomalies are discovered on the ultrasound. The woman delivers an infant with severe developmental anomalies (such children can continue to live for years and require expert, constant, special care). Karyotyping is requested. The father is confirmed to have a reciprocal translocation, and the infant's anomalies are found to be due to an unbalanced karyotype resulting from the presence of one derivative chromosome from the translocation. The parents sue the obstetrician for failure to provide adequate care. The case is settled out of court.

Issues Raised in This Scenario

In this situation, a clinically significant chromosome abnormality had been discovered fortuitously in the proper investigation of an unrelated medical problem. Appropriate steps were taken: Other family members were tested, carriers were identified, and counseling was given in regard to the risks for miscarriages and abnormal livebirths.

Genetic Testing: Care, Consent, and Liability, by Neil F. Sharpe and Ronald F. Carter
© 2006 John Wiley & Sons, Inc.

However, because one of the family members was 12 years of age, he was too young to accurately remember the details of the counseling when it was needed years later. Nevertheless, it was clearly established that he knew that he had *some* kind of chromosome anomaly, that his wife knew, and that his wife raised that issue with the obstetrician at a time when prenatal counseling, testing, and subsequent therapeutic abortion could be offered if desired. The obstetrician did not seek confirmation of the type of anomaly. Even if records from the initial consultation were not available (and they were), it would be reasonable to request karyotyping of the parents in order to clarify exactly what type of risk the suspected chromosome anomaly actually presented. The obstetrician should have requested the records and ordered karyotypes for comparison and updating. These steps would have provided evidence of significant risk to the fetus. The risk would be clarified, the options offered, and the patient would have been able to make an informed choice. Even if the patient declined testing, or declined to terminate a pregnancy when the abnormal fetal karyotype was reported, the appropriate actions would have been taken.

NEED TO KNOW

Approaches to Diagnosis of Genetic Disease

Obligations of the Clinician

In the past, many genetic disorders could not be conclusively identified because a relevant test was not available. Clinical genetics essentially consisted of careful observation, diagnostic acumen, and considered judgment. Clinicians collected family histories, charted physical and developmental anomalies, and poured over pedigrees to determine patterns of inheritance and categorize their findings into syndromes and diseases. It was recognized that more than one genetic mechanism could cause the same apparent clinical outcomes and also that one genetic mutation might cause a variety of different outcomes. Treatments were generally ineffective, and recurrence risks were hard to estimate. Without the benefit of conclusive testing, the patterns of clinical care for genetic diseases were derived from years of experience with painstakingly accrued empirical evidence.

The advent of biochemical tests, cytogenetics, and molecular diagnostics has transformed the diagnosis and management of genetic disease. Ultimately, genetic testing will enable patients to undertake preventive measures and to modify behaviors in order to reduce the risk, and effects, of genetic disease. Even though the clinical utility and validity of many genetic tests remain open to question [Collins, 1996; American Society of Human Genetics, 1996; Task Force, 1997; Welch and Burke, 1998; Yang et al., 2000; Caulfield, 2000; Khoury et al., 2000, 2004; Burgess, 2001; Haga et al., 2003], there has been a rapid surge in the importance of genetic testing.

Health care professionals are obligated to:

- Keep up to date and abreast of significant developments [*Harbeson v. Parke-Davis, Inc.,* 1983]

- Timely inform patients about the availability of diagnostic genetic tests and their associated risks and limitations [*James G. v. Caserta*, 1985; Andrews, 1992; Derrick, 2002]
- Timely and accurately inform a patient about a diagnosis and the known implications, risks, or dangers associated with the result in order to enable the patient to make a decision as to a course of treatment [American Jurisprudence, 2002].

These expectations need to be met in a time where there is consistently rapid improvement in the sophistication, accuracy, and breadth of the diagnostic genetic testing available. How do health care providers keep up? Basically, the options are:

1. Become an expert.
2. Know enough to recognize when an expert is needed (when to refer).
3. Avoid caring for that kind of patient.

We note that option 3 is becoming less feasible as more and more diseases, common and rare, are demonstrated to have at least some recognized pathogenetic component that involves hereditary or acquired genetic mechanisms.

Gaining Expertise in Diagnostic Genetics

Professional accreditation standards for M.D. and Ph.D. geneticists generally require the ability to understand the indications, uses, limitations, interpretation, and significance of specialized laboratory and clinical diagnostic procedures [ABMG, 2004; CCMG, 2004b]. The American Board of Genetic Counseling requires the ability "to identify, coordinate, interpret, and explain genetic laboratory tests and other diagnostic studies" [ABGC, 2004].

Professional accreditation in genetics involves specialty training and certification as a clinical geneticist (or related discipline such as pediatrics), genetic counselor, laboratory scientist, or genetic nurse. Professional organizations such as the American Board of Medical Genetics (ABMG), the American College of Medical Genetics, and the Canadian College of Medical Geneticists (CCMG) are recognized as specialty societies for physicians and scientists. Both the ABMG and the CCMG have credentialing, training, and examination procedures that lead to specialty certifications in genetics for M.D.s and Ph.D.s. Genetic counselors have a similar framework of expectations, leading to certification by the American Board of Genetic Counseling or the Canadian Association of Genetic Counsellors.

However, less than 100 out of 104,000 residency slots in the United States, at the time of this writing, are allocated to clinical genetics (a number that appears to be declining), and there are less than 50 training sites [ABMG, 2004; http://genetics.faseb.org/genetics/abmg/abmgmenu.htm]. Genetic counselor training positions are similarly in short supply. At the time of this writing, in the United States there are only 1075 certified clinical geneticists and 1410 certified genetic counselors [ABMG, 2004]. The option of training and certification as a recognized specialist in genetics represents an option for very few individuals; therefore, a considerable component of the overall burden of care is being provided by individuals who have certified in other disciplines. These individuals need to ensure that they either personally maintain an appropriate level of competence in medical genetics or follow appropriate guidelines for referral to medical geneticists.

WEB RESOURCES

All websites were accessed as of December 15, 2004. For individuals seeking clinically relevant information about medical genetics, numerous resources for continuing education and maintenance of competence, professional statements, and practice guidelines are available. Information from many societies in the United States is accessible from a common website (http://genetics.faseb.org/genetics/), including:

American Society of Human Genetics: http://genetics.faseb.org/genetics/ashg/ashgmenu.htm

American College of Medical Genetics: http://genetics.faseb.org/genetics/abmg/abmgmenu.htm

American Board of Genetic Counseling: http://www.kumc.edu/gec/prof/abgc.html

Association of Professors in Human and Medical Genetics: http://genetics.faseb.org/genetics/aphmg/aphmg1.htm

Human Genome Variation Society: http://www.hgvs.org/

In Canada useful information is available through:

Royal College of Physicians and Surgeons of Canada (RCPSC): http://rcpsc.medical.org/main_e.php

Canadian College of Medical Geneticists: http://ccmg.medical.org/

Canadian Association of Genetic Counsellors: http://www.cagc-accg.ca/

WATCH OUT FOR

Competency and the Obligation to Refer

Surveys indicate that health care professionals who have not received specialized training in genetics may be uncertain as to the nature, reliability, and meaning of genetic tests [Hayflick et al., 1998; Hunter et al., 1998; Cho et al., 1999; Wolpert, 2000; Escher et al., 2000; Goos et al., 2004]. If the patient's condition is beyond the physician's knowledge, ability, or capacity to treat, the physician has a duty to disclose this to the patient and to refer to an appropriate specialist. Diagnostic errors have been cited as the most common underlying cause for medical errors and malpractice claims [Phillips et al., 2004].

Courts of law have recognized that a physician has a duty:

- To secure the best factual data and to make use of all reasonable scientific means and facilities to appropriately follow up and respond to such information [*Forrestal v. Magendantz*, 1988; *Humana of Kentucky, Inc. v. McKee*, 1992; *Dale v. Munthali*, 1978; *Rietze v. Bruser*, 1979; Derrick 2002; *Geler v. Akawie*, 2003]

- To exercise reasonable skill and competence and to refer the patient for appropriate diagnostic procedures [*Munro v. Regent of the University of California*, 1989; Derrick, 2002]

- To exercise good judgment, particularly with regard to the application of new technologies and procedures [*Pike v. Honsinger*, 1898; *Jones v. Karraker*, 1982; *Wilson v. Swanson*, 1956]

An incorrect diagnosis in itself generally is not sufficient to support a finding of liability. In evaluating this issue, a court of law generally will distinguish between an "error in judgment" and a physician's failure to exercise her or his best judgment. A physician may be held liable if the error of commission or omission shows an absence of reasonable skill, care, and judgment falling below that of the generally accepted standard of care [*Riggins v. Mauriello*, 1992; *Lapointe v. Hôpital le Gardeur*, 1992].

To establish a legal cause of action for malpractice, a plaintiff-patient generally must satisfy four elements:

- A legally recognized duty of care was owed by the defendant-physician to the plaintiff-patient.
- The defendant-physician breached this duty of care by failing to meet the required standard of care.
- The plaintiff-patient has suffered a legally cognizable, compensable injury.
- The injury was caused by the defendant-physician's breach of the duty of care.

Prenatal/Neonatal Diagnosis

The field of prenatal/neonatal diagnosis represents a particular focus of concern in regard to competency and the duty to refer. In prenatal medicine, the parents and clinicians have a restricted timeframe for investigating and acting upon screening, diagnostic, and management options. In neonatology, timely and accurate diagnosis of genetic diseases in infants can allow appropriate management, amelioration of clinical signs, and better outcomes [e.g., phenylketonuria (PKU) and medium chain acyl-CoA dehydrogenase (MCAD)] and also the adoption of disease-specific prenatal screening and diagnostic techniques for future pregnancies. Accurate diagnoses are essential because of the implications for lifelong sequelae for the family and the fetus, including associated risks of recurrence for future pregnancies.

A health care professional has an obligation [*Becker v. Schwartz*, 1978; *Robak v. United States*, 1981; *Speck v. Finegold*, 1981; *Eisbrenner v. Stanley*, 1981; *Naccash v. Burger*, 1982; *Pitre v. Opelousas General Hosp.*, 1988; *Geler v. Akawie*, 2003] to provide, or to refer the patient for, appropriate prenatal and neonatal genetic tests and genetic counseling [*Canesi v. Wilson*, 1999; *Geler v. Akawie*, 2003; *Knoppers and Isasi*, 2004; but see *Munro v. Regent of the University of California*, 1989; Reutenauer, 2000; please see Chapter 8, Prenatal and Neonatal Screening].

Malpractice cases concerning genetic testing have been ongoing since the 1960s. Typical causes of action [Andrews and Zuiker, 2003] will allege that a negligent act was committed with respect to:

- Risks that a prospective parent is a carrier of a gene associated with a genetic disease or condition
- Risks that a child may be born afflicted with an inherited genetic disease or condition
- Consequences to the child of inheriting the genetic disease
- Duties to timely and accurately interpret diagnostic results and to timely and effectively communicate test results to the parents

Examples include:

- *Naccarato v. Grob*, 1970: Failure to offer testing for PKU
- *Becker v. Schwartz*, 1978: Failure to advise of the availability of amniocentesis
- *Schroeder v. Perkel*, 1981: Failure to identify cystic fibrosis in a child with the result a second child was born with cystic fibrosis
- *Naccash v. Burger*, 1982: Failure to offer carrier testing for Tay-Sachs disease to a Jewish couple with an Eastern European background
- *Proffitt v. Bartolo*, 1988: Failure to diagnose a mother's rubella and warn her of the risk to a prospective child; the child later was born with congenital rubella syndrome

The early cases gave rise to the controversial causes of action known as *wrongful birth* and *wrongful life*. These claims are based on the obligation of health care professionals to provide timely, appropriate, and accurate information concerning a prospective or existing pregnancy including the availability of testing for possible genetic alternations [Andrews, 1992; *Canesi v. Wilson*, 1999; Wildeman and Downie, 2001; Mandl et al., 2002].

When a physician knows or should know of the existence of an unreasonable risk that a child will be born with a birth defect, he [she] owes a duty to the unconceived child as well as to its parents to exercise reasonable care in warning the potential parents and in assisting them to avoid the conception of the deformed child. The time has come when we can and should say that each person owes a duty to take reasonable care to avoid acts or omissions which he [she] can reasonably foresee would be likely to injure a present or future member of society unless there is some justification or valid explanation for its exclusion . . . The persons at whose disposal society has placed the potent implements of technology owe a heavy moral obligation to use them carefully and to avoid foreseeable harm to present or future generations. [Pitre v. Opelousas General Hosp., 1988]

Wrongful Birth

Wrongful birth refers to a claim brought by parents, "being deprived of the option of . . . making an informed and meaningful decision either to terminate the pregnancy or to give birth to a potentially defective child" [*Liddington v. Burns*, 1995; *Greco v. United States*, 1995]. Although wrongful birth actions are accepted by a majority of American jurisdictions [e.g., see Reutenauer, 2000; in Canada see *Colp v. Ringrose*, 1976; *Krangle (Guardian ad litem of) v. Brisco*, 1997; *Mickle v. Salvation Army Grace Hospital*, 1998; Nelson and Robertson, 2001], the debate is far from over. Courts have overturned existing legal precedents both to deny [*Taylor v. Kurapati*, 1999] and to recognize [*Schirmer v. Mt. Auburn Obstetrics and Gynecological Assoc.*, 2003] a claim for wrongful birth.

Although a court may recognize that an act of negligence has been committed, liability may be denied based on the argument that no legally cognizable injury has been caused by the alleged act of negligence. In 2003, in the case of *Grubbs v. Barbourville Family Clinic*(consolidated with the case of *Bogan v. Altman, McQuire and Pigg, PSC*, 2001), the Kentucky Supreme Court refused to recognize a claim for wrongful birth, stating:

> *Although the parents . . . allege that their injury was in being deprived of accurate medical information that would have led them to seek an abortion, we are unwilling to equate the loss of abortion opportunity resulting in a genetically or congenitally impaired human life, even severely impaired, with a cognizable legal injury.*

However, the Kentucky Supreme Court did recognize that:

> *The Bogans present an interesting issue unrelated to the tort aspects of the claim, yet which merits attention. The Bogens believe that patients should have a breach of contract action against physicians who offered and charged for diagnostic prenatal testing, yet who allegedly did not perform these services properly. Despite our denying the tort claim as a matter of law, a physician who contracts and charges for a service such as a prenatal ultrasound and consequent opinion as to the results of the ultrasound, is liable for any breach of contract in this regard. We do not believe that physicians should be relieved of any proven contractual responsibility to report to patients the accurate results of diagnostic procedures, even if the condition is "incurable". In the absence of such a conclusion, we would be forced to hold that physicians could perform and charge for diagnostic procedures and report whatever they want if the diagnosis is of an incurable condition, and physicians could legally charge and be paid for services they did not perform . . . the case is remanded . . . for adjudication of the claims for pain and suffering and permanent scarring suffered in connection with the caesarean section delivery.*

The child in the case of *Grubbs v. Barbourville Family Clinic* was born with spina bifida. In the case of *Bogan v. Altman, McQuire and Pigg, PSC*, the child was born without any eyes, or brain, an underdeveloped brain stem, a cleft palate, and was unable to speak.

Examples of scenarios that *may* provide a basis for this legal cause of action (and for wrongful life dependent upon the jurisdiction in which the health care professional practices) include:

- Failing to inquire about a patient's background (e.g., family medical history and ethnic ancestry) "to assess what information might be useful to the patient's deliberative process and then to discuss that information" [*Canesi v. Wilson*, 1999]

- Failing to advise of all reasonably available diagnostic procedures [*Haymon v. Wilkerson*, 1987; *Krangle (Guardian ad litem of) v. Brisco*, 1997; *Jones v. Rostvig*, 1999] including all alternative procedures [*Logan v. Greenwich Hosp.*, 1983; Derrick, 2002]

- Failing to offer testing services [e.g., *Naccash v. Burger*, 1982—Tays Sachs disease; *Reed v. Campagnolo*, 1993—neural tube defects]

- Failing to undertake all reasonable procedures to establish a timely diagnosis [*Naccarato v. Grob*, 1970—PKU; *Cherry v. Borsman*, 1990; *Ciceron v. Jamaica Hospital*, 1999—spina bifida; ABMG, 2004; ABGC, 2004; CCMG, 2004; *Costa v. Boyd*, 2003]

- Failing to accurately perform genetic tests [*Curlender v. Bio-Sciences Labs*, 1980—Tay-Sachs; *Nelson v. Krusen*, 1984—Duchenne muscular dystrophy]

- Failing to properly perform or to appropriately interpret prenatal examinations (including taking into account the risk of false positives and false negatives) [*Creason v. Department of Health Services*, 1998; Waisbren et al., 2003], where such failure results in the unnecessary termination of a pregnancy, or the birth of a genetically afflicted child, or the loss of opportunity to terminate a severely affected fetus [*Greco v. United States*, 1995; Mavroforou et al., 2003]

- Failing to advise the parents that they were at risk of conceiving or giving birth to a child with a genetic disorder [*Gildiner v. Thomas Jefferson University Hospital*, 1978]

- Failing to diagnose a first child's medical condition as hereditary (blindness) and but for this negligence the parents would not have had a second child afflicted with the same condition [*Lininger v. Eisenbaum*, 1988]

- Failing to advise of test results indicating the percentage risks that a child may be born with a genetic disease [*McAllister v. Ha.*, 1998—failure to disclose a one in four risk of sickle cell anemia]

- Failing to provide or to refer for genetic counseling [*H. (R.) v. Hunter*, 1996], where the health care professional knew or should have known of the risks associated with a prospective genetic condition. By way of example, these may include the patient's ethnic background [*Naccash v. Burger*, 1982; but see *Munro v. Regent of the University of California*, 1989], advanced maternal age [*Schroeder v. Perkel*, 1981; Botkin, 2003], and/or the fact that a previous child was born with a genetic disease or disorder [*Keel v. Banach*, 1993].

- Failing to timely and accurately interpret and/or to timely report diagnostic test results [e.g., *Bogan v. Altman, McQuire and Pigg, PSC*, 2001; later consolidated with the case of *Grubbs v. Barbourville Family Clinic*, 2003].

Wrongful Life

Wrongful life claims are brought on behalf of a genetically afflicted infant against health care professionals alleging that but for an act of negligence, the child would not have been born [*Turpin v. Sortini*, 1982]. The allegation is not that the health care professional(s) caused the genetic condition or disease but that the negligent act of performing or failing to perform appropriate genetic testing has resulted in the birth of a genetically afflicted child. The plaintiff claims that recovery is required where the defendant's negligence is the proximate cause of the child's need for extraordinary medical care and training.

Wrongful life actions have proven highly controversial [*Goldberg v. Ruskin*, 1986; *Kassama v. Magat*, 2002; in Canada see *Jones v. Rostvig*, 1999, *Mickle v. Salvation Army Grace Hospital*, 1998; but see Shapira, 1998; *Curlender v. Bio-Science Laboratories*, 1980; *Turpin v. Sortini*, 1982]. In the 1967 decision of *Gleitman v. Cosgrove*, it was alleged that the physician had assured the mother that a disease contracted during pregnancy—rubella—would not affect the child. The mother claimed that if she had been adequately informed she would have considered an abortion. The court denied recovery, finding it impossible to "weigh the value of life with impairments against the nonexistence of life itself."

In comparison, the 1980 decision of *Curlender v. Bio-Science Laboratories* concerned a defendant laboratory that had been hired by the plaintiff parents to conduct a genetic test for Tay-Sachs disease. False negative results were given, and the plaintiff child was born with Tay-Sachs. The court allowed the claim for wrongful life on the basis that the child had a right to recover damages for pain and suffering endured during the span of the child's life as opposed to the life span if the child had been born without Tay-Sachs disease.

Wrongful life actions are not recognized in most jurisdictions. The few jurisdictions that have recognized this action [e.g., *Turpin v. Sortini*, 1982; *Harbeson v. Parke-Davis, Inc.*, 1983; *Procanik v. Cillo*, 1984] generally have denied general damages for the suffering of being born in an afflicted condition [Botkin, 2003].

Some courts have sought to limit liability based on what is perceived to be the predictive "quality" of a genetic test [Andrews and Zuiker, 2003]. In the 1989 decision of *Simmons v. W. Covina Medical Clinic*, a California court held that a physician was not liable for failing to offer a genetic test, maternal serum alpha-fetoprotein screening (AFP). The child was born with Down syndrome. The court held that liability could be imposed:

> *Only when there is a reasonable medical probability of predicting the outcome of the pregnancy ... A mere 20 percent chance does not establish a "reasonably probable casual connection" between defendants' negligent failure to provide the test and plaintiff's injuries. A less than 50–50 possibility that defendant's omission caused the harm does not meet the requisite reasonable medical probability test of proximate cause.*

It should be noted that this ruling was made in 1989 and was applied by this court as a benchmark to determine proximate cause in a wrongful life action. In contrast, is the 1997 decision of *Gardner v. Pawliw*. A New Jersey court held that where a diagnostic test may constitute a standard of care but it is unknown whether performing the test will help diagnose or treat a preexisting condition, the plaintiff must demonstrate that the failure to provide the test increased the risk of harm from the preexistent condition. This may be demonstrated, stated the court, even if such tests are helpful in only a small percentage of cases. Arguably, this requirement would be satisfied in scenarios where the failure to provide information about genetic testing services would prevent a patient from undergoing:

- Increased surveillance to promote early detection, prevention strategies [e.g. hereditary breast and ovarian cancer: *Katskee v. Blue Cross/Blue Shield of Nebraska*, 1994]
- Risk management [e.g., Alzheimer disease: Kapp, 2000]
- More informed reproductive decision making including the opportunity to consider preimplantation genetic diagnosis [Verlinsky et al., 2004; Moutou et al., 2004; also see Chapter 8, Prenatal and Neonatal Screening].
- Procedures to prevent transmission of the gene associated with a genetic disease to a child, such as an autosomal-dominant form of early-onset Alzheimer disease [Verlinsky et al., 2002; Bertram and Tanzi, 2004] or Tay-Sachs disease [Hansis and Grifo, 2001].

Wrongful Abortion

A new variation of the wrongful life action, involving genetic testing, is *wrongful abortion*. A woman was told that genetic testing showed that her baby would be born with Down syndrome [*Breyne v. Potter*, 2002]. The woman proceeded to have an abortion. The following day, the physician telephoned the woman, admitted that he had been mistaken and said that the baby actually had Trisomy 21 (Downs Syndrome), a condition that would not cause retardation but could result in developmental delays in learning, speech, and motor skills. The defendant physician argued that he should not be subject to liability because the woman had made an independent decision to have the abortion. The court, however, stated: "Under Dr. Potter's theory, a patient who had her breast amputated unnecessarily after her doctor mistakenly told her she had cancer would have no malpractice claim . . . Patients are entitled to rely on their doctors' diagnoses in deciding a course of treatment."

Adult-Onset Disease: Genetic Susceptibility Testing

Another field of genetics that has generated attention is the use of genetic counseling and testing to identify individuals at increased risk of a future diagnosis of a disease. In this situation, the focus is on the benefit of accurate risk assessment because knowledge of increased risk might lead to the adoption of risk prevention and disease-specific surveillance strategies.

Genetic susceptibility testing can identify individuals who have an increased risk for later developing a genetic disease due to hereditary factors as indicated by a family's medical history. Benefits of susceptibility testing may include improved management (reduced morbidity and/or mortality) through early detection and prevention strategies, risk management [e.g., Alzheimer disease: Kapp, 2000], increased certainty, and more informed reproductive decision making [Verlinsky et al., 2002, 2004; and see Tasca and McClure, 1998].

Failure to take an appropriate family history, to advise a patient of the implications posed by a family history, to provide the patient with appropriate information to make an informed choice [McKinnon et al., 1997; Kapp, 2000; *Canesi v. Wilson*, 1999] and to provide or to refer the patient for genetic susceptibility testing may provide the basis for a legal cause of action for negligence [Severin, 1999].

Health care professionals need to ensure that the patient understands the nature and objectives of the genetic test, its associated benefits, risks, and limitations, and the implications posed by genetic testing and test information [Sharpe 1994a; Geller et al., 1997; Lerman et al., 1997; Kirschner et al., 2000].

If a physician fails to advise a patient of the availability of genetic susceptibility testing and/or fails to timely and accurately interpret and report the test results to the patient, an open question is whether a court will deny liability based on the grounds that because the actual genetic disease has not been actually expressed, it follows that no legally cognizable injury has occurred? In wrongful birth and wrongful life actions, courts have denied recovery based on the argument that no legally cognizable injury has resulted from an act of negligence. In susceptibility testing, this issue remains open to debate, although at least one state initially has refused to recognize genetic susceptibility testing for hereditary breast/ovarian cancer as being "medically necessary" [Klanica, 2004; California Genetic Testing Dispute, no date]. However, a Nebraska case did hold that a genetic predisposition to cancer was covered as an "illness" in an insurance policy, given that "appellant's condition is a deviation from what is considered a normal state arises, in part, from the genetic makeup of the woman. The existence of this unhealthy state results in the woman being substantially at risk of developing cancer. The recommended surgery [prophylactic hysterectomy] is intended to correct the morbid state by reducing or eliminating the risk" [*Katskee v. Blue Cross/Blue Shield of Nebraska*, 1994].

As clinical data relating to the value of risk reduction strategies such as tamoxifen, prophylactic bilateral mastectomy, [for example, see Hartmann et al., 1999; Rebbeck et al., 1999, 2002; recently reviewed by Calderon-Margolit and Paltiel, 2004] or oophorectomy becomes available, clinicians will need to ensure that the potential benefits and risks of such strategies are clearly identified to the patient.

CRITERIA FOR REFERRAL

Physicians have a responsibility to ensure timely referral to an appropriately competent specialist. In medical genetics, criteria for referral and criteria for testing are commonly available. A useful summary list of indications for referral to clinical genetics includes the following criteria [Thompson and Thompson, 2001]:

- A previous child was born with a genetic disorder, genetic disease, birth defect or chromosome abnormality
- Family history of a hereditary condition such as cystic fibrosis, fragile X, diabetes, or cancer
- Advanced maternal age of 35 or over for a pregnancy
- Repeated pregnancy loss, stillbirths, early infants deaths, or infertility
- Consanguinity
- Exposure to teratogens, including alcohol, medications, drug, occupational chemicals
- Children with developmental delay and unusual features
- Individuals of certain ethnic groups associated with an increased risk for specific genetic diseases or disorders
- Individual or family concerns about genetic diseases and disorders including late-onset diseases such as cancer, cardiovascular, or neurological disease
- Follow-up for a positive newborn screening test, such as PKU, or a heterozygote screening test such as Tay-Sachs

For more specific recommendations, various specialty professional bodies and government organizations provide regularly updated websites. Organized by type of practice, examples are presented below (all websites following were accessed December 15, 2004). Two useful resources especially represent authoritative sources of information in regard to genetic diseases. They are:

- OMIM (Online Mendelian Inheritance in Man) is the definitive catalog of all published literature relevant to genetic diseases identified in humans. It is accessed at: http://www.ncbi.nlm.nih.gov/entrez/query.fcgi?db=OMIM.
- Sponsored by the National Institutes of Health (NIH), Genetests is the site for patients and health care professionals. This website offers clinically relevant information summarizing current concepts in the diagnosis and management of an expanding number of genetic diseases, along with updated listings for clinical centers, laboratory services, and research laboratories. The review articles in particular represent valuable sources of information for clinicians. Genetests is accessible at: http://www.geneclinics.org/.

Prenatal/Neonatal

American College of Obstetricians and Gynecologists: Committee opinion on first-trimester prenatal screening methods for chromosome abnormalities, including nuchal fold translucency screening: http://www.dnapolicy.org/FirstTrimesterScreening

National Conference of State Legislatures (U.S.): Newborn Genetic and Metabolic Screening: http://www.ncsl.org/programs/health/genetics/nbs.htm

American Academy of Pediatrics—Section on Birth Defects: Policy Statement and Information on Specific Disorders: http://www.aap.org/healthtopics/birthdefects.cfm

American College of Medical Genetics Foundation: Evaluation of the Newborn with Single or Multiple Congenital Anomalies: A Clinical Guideline: http://www.health.state.ny.us/nysdoh/dpprd/index.htm

Association of Public Health Laboratories: Newborn Screening and Genetics: http://www.aphl.org/Newborn Screening Genetics/index.cfm

Catalogue of Rare Genetic Diseases in Children: http://www.med.nyu.edu/rgdc/disease.htm

Newborn Screening Quality Assurance Program: http://www.cdc.gov/nceh/dls/newborn screening.htm

Frequency of Inherited Disorders Database: http://archive.uwcm.ac.uk/uwcm/mg/fidd/

Genetics and Public Policy Center: Prenatal Genetic Testing: http://www. dnapolicy.org/genetics/prenatal.jhtml

March of Dimes Fact Sheets: These cover a wide range of prenatal and genetic topics in English and Spanish: http://www.marchofdimes.com/professionals/681 1116.asp

March of Dimes Fact Sheets: Newborn Screening: http://www.marchofdimes.com/professionals/580.asp

National Newborn Screening and Genetics Resource Center: http://genes-r-us.uthscsa.edu/

Cancer Genetics

National Cancer Institute: Elements of Cancer Genetics Risk Assessment and Counseling: http://www.cancer.gov/cancertopics/pdq/genetics/risk-assessment-and-counseling/HealthProfessional/page4

U.S. Cancer Genetics Services Directory (a resource for referral of patients to specialists) http://www.cancer.gov/search/genetics services/

Genetic cancer risk assessment and counseling: Recommendations of the National Society of Genetic Counselors (2004) *J Genet Counsel* **13**(2):83–114: http://www.guideline.gov/summary/summary.aspx?viewid=1&docid=5274

Another particularly useful resource is the review document:

Lindor NM, Greene MH: The concise handbook of family cancer syndromes. Mayo Familial Cancer Program. *J Natl Cancer Inst* 1998; 90:1039–1071.

Genetic Test Centers

Centers for Disease Controls: Laboratory Practice Evaluation and Genomics Branch Genetic Testing: http://www.phppo.cdc.gov/dls/genetics/links.aspx

Clinical Laboratory Improvement Amendments: http://www.cms.hhs.gov/clia/

National Institutes of Health: http://www.genetests.org

WATCH OUT FOR

Referral and Primary Care Physicians

Patients value primary care physicians as their first contact care providers and co-ordinators for referral services [Grumbach et al., 1999]. However, although health care professionals may acknowledge the need for referral for genetic testing services [Menasha et al., 2000], professionals who have not been specifically trained in genetics may be uncertain as to the nature, reliability, and criteria of genetic tests [Hayflick et al., 1998; Hunter et al., 1998; Cho et al., 1999; Wolpert, 2000; Escher et al., 2000; Freedman et al., 2003], may not refer the patient for genetic services [Myers et al., 1999; Rose et al., 2001; Mouchawar et al., 2003], and may not have the time [Menasha et al., 2000], ability, or knowledge to appropriately counsel the patient about the meaning and implications posed by often complex, incomplete, and ambiguous test results [Menasha et al., 2000; Freedman et al., 2003]. For example:

- In a study of 15,824 patients seen for genetic counseling (including a three-generation family history and pedigree assessment), over 50 percent were identified as having a clinically significant risk that inferred a recurrence risk and/or additional testing [Cutillo et al., 2002].

- In comparative study, almost 40 percent of 145 high risk obstetric patients referred for genetic counseling were found to have unsuspected, additional risk factors requiring clinical attention [Koscica et al., 2001].

- In a study of the factors determining when family physicians discussed pregnancy risks and referral to clinical genetic centers, 71 percent of referrals in the study were delayed until the patient was already pregnant because the physician was not aware of potential risk factors until that time (i.e., the physician did not utilize family history information to identify potential risk factors prior to pregnancy) [Aalfs et al., 2003]. For the patients who were referred, 40 percent of referral requests were initiated by the patient and only 31 percent by the general practitioner [Aalfs et al., 2003; Albertson et al., 2000].

With respect to cancer genetics, 84 percent of American oncologists considered themselves qualified to recommended genetic testing, but only 40 percent of primary care physicians and 57 percent of tertiary care physicians expressed the same confidence [Freedman et al., 2003]. This lack of confidence appears to be well founded; in a recent study, 75 out of 387 high-risk colorectal cancer patients met guidelines for genetics assessment, but only 13 of 75 (17 percent) were referred by physicians [Grover et al., 2004].

Interestingly, in a recent study examining the merits of patient registry services for families with known genetic syndromes, participant patients actually showed a greater preference for initial communication about risk (i.e., the process of contacting family members unaware of their risk) to occur through a family contact rather than through a registry service or the family physicians [Wright et al., 2002]. However, in this situation, the family members who assumed a role in family communication would not appear to be representative of the majority of genetic patients; these were patients who belonged to families with confirmed genetic diseases and had high knowledge levels in regard to their risk.

A health care professional's obligations with regard to genetic risk, counseling, and genetic testing, potentially affects all primary care providers and specialists alike [Committee on Assessing Genetic Risk, 1994]. The legal implications are significant. A health care professional who knows or should know that the patient's situation is beyond the professional's knowledge, skill, and ability has a duty to disclose this to the patient and to refer in a timely fashion [*Erickson v. Walker*, 1977; Louisell and Williams, 2001] to an appropriate specialty [*Estate of Tranor v. Bloomsburg Hospital*, 1999; *Grubbs ex rel. Grubbs v. Barbourville Family Health Center*, 2003; *Gill Estate v. Marriott*, 1999; *H. (R.) v. Hunter, Rosenbloom, Viner*, 1996].

Examples of potential liability for the failures to advise of genetic risks, of the availability of testing [*Goldberg v. Ruskin*, 1986; *Haymon v. Wilderson*, 1987], and to refer the patient for genetic testing [*Siemieniec v. Lutheran General Hospital*, 1987; *Estate of Tranor v. Bloomsburg Hospital*, 1999; Louisell and Williams, 2001] include:

- Where a previous child was born with a genetic disease, due to the failure to timely diagnosis cystic fibrosis before a second child was born [*Schroeder v. Perkel*, 1981] or the failure to provide genetic counseling and genetic testing to parents when their first two children born with blood disorders—the couple gave birth to a third child with sickle cell disease [*Didato v. Strehler*, 2001].

- Where one or both parents are of an ethnic group associated with a particular genetic disease; for example, people with an Ashkenazi Jewish background and the risk that a child will be born with Tay-Sachs disease.

- Where a prospective mother has contracted an infection, for example, rubella infection (German measles) [*Robak v. United States*, 1981].

- Where a mother was bleeding during pregnancy and despite her inquiries the physicians failed to provide genetic testing; the child was born with Wolf–Hirschhorn syndrome [*Branca v. Miro*, 0735/2001; and see Fortado, 2004].

- Advanced maternal age in the case of pregnancy. The failure to provide information about the risks for Down syndrome to women is reported to be the basis for the single largest number of cases for the legal action of wrongful birth [Botkin, 2003].

- The failure to provide genetic testing for unexplained developmental delay in a child [Fragile X—*Sabeh v. Khosla*, 2003].

The expectations placed on the health care provider relate not only to the need for timely referral but also to appropriate referral; that is, the health care professional needs to refer to a specialist with the appropriate expertise. If the health care professional knows, or has reason to know, that the specialist does not possess the necessary training and skills, the referral itself may be a basis for liability [*Estate of Tranor v. Bloomsburg Hosp.*, 1999].

The referral request needs to provide sufficient information to the specialist. Communications to other health care professionals should be tailored to ensure appropriate understanding and the implications for care and follow-up [Sandhaus et al., 2001; Glabe et al., 2002; Statham et al., 2003]. It is important to ensure that the scope of assessment, communication, and follow-up includes not just the patient but other family members who may be at risk.

Studies have identified significant shortcomings with regard to the referral to, and reports received from, other health care professionals [Forrest et al., 2000; Barnes et al., 2000]. Important information may be omitted including:

- An appropriate patient history
- Recommendations for follow-up and future care
- Clear guidelines for managing patients with positive results

For example:

- An American study reported that 68 percent of specialists had not received information from the PCP (primary care physician) prior to specific referral visits, even though 38 percent of these said that the information would have been helpful. Four weeks after referral visits, 25 percent of the PCPs had not received any information from the specialists [Gandhi et al., 2000; Pantilat et al., 2001].

- An English survey reviewing the communication of prenatal screening and diagnostic results to primary care health professionals reported that 16 percent were not informed of high risk results, 28 percent were not advised about the possibility of abnormalities, 28 percent did not know what follow-up action was appropriate, and only 17 percent reported that the information was sufficient to discuss the issues with patients [Statham et al., 2003].

- In a study reviewing 1856 consecutive referrals made by 142 American pediatricians in a national practice-based research network, pediatricians were less likely to coordinate telephone referrals than office visit referrals and frequently were unaware whether or not referrals had been completed [Glade et al., 2002].

Lack of timeliness of information (i.e., information received prior to the time the patient is seen for follow-up) can impair communication and the quality of care [Forrest et al., 2000; Gandhi et al., 2000; Barnes et al., 2000; Pantilat et al., 2001; Freedman et al., 2003].

GENETIC TESTING AND TORT ACTIONS

Dena S. Davis,
Cleveland–Marshall College of Law, Cleveland, Ohio

INTRODUCTION

Professionals may be held liable for civil actions of *wrongful birth* or *wrongful life*, if, through their negligence, a genetically damaged child is born whose birth or conception would otherwise have been avoided. Parents of children born with birth defects that are caused by the negligence of health professionals can successfully sue for the tort of wrongful birth. The tort of wrongful life stems from the same events, but the action is brought by the child, rather than the parent. Other possible legal actions include the tort of wrongful adoption.

WRONGFUL BIRTH

The classic wrongful birth case goes something like this: A pregnant woman engages the services of an obstetrician. The physician ascertains that the woman is 40 years old but fails to inform her that she is at increased risk of such genetic anomalies as Down syndrome and fails to offer prenatal testing. These omissions amount to a failure to conform to the standard of care. The woman gives birth to a child with Down syndrome. She then sues the physician for the tort (injury) of wrongful birth. If she can convince a judge or jury that, had the physician warned her of the risks, she would have opted for testing and would have terminated a pregnancy where the fetus had Down syndrome, then she can succeed in her claim.

There are many ways in which such a scenario could arise. A physician could negligently fail to diagnose a genetic disease, such as cystic fibrosis, in a first child, and this negligence could be the cause of the parents going ahead with a second pregnancy, which also resulted in a child with cystic fibrosis. A laboratory could be sued for wrongful birth if it mixed up test results, so that parents who are both carriers of a recessive disease such as cystic fibrosis or Tay-Sachs could erroneously believe that they are not at risk and, therefore, fail to take precautions against having an affected child. A physician could erroneously reassure a woman who has been exposed to a teratogen that her fetus is not at risk. As our understanding of medical genetics grows, and as we increase our ability to test for destructive alleles in parents, embryos, and fetuses, the potential for wrongful birth cases increases. Especially where families are not aware that they fall into specific risk groups, this situation highlights the unanswered question of how much the ordinary obstetrician or pediatrician can be expected to know about the most recent advances in human genetics and genetic testing.

The tort of wrongful birth must be distinguished from other actions with which it is sometimes confused. Crucial to the tort of wrongful birth is that the professional is not accused of *causing* the child's injury but rather of causing the birth of an injured child who would otherwise not have been conceived or brought to term. This is also distinct from the tort of wrongful pregnancy, where conception occurs due to, for example, a negligently performed tubal ligation or the substitution of sugar pills for oral contraceptives.

The tort of wrongful life, discussed below, arises from the same event as that of wrongful birth. The difference is that the plaintiff (injured party) in the wrongful life case is the child him or herself, not the parents. The same event can give rise to both types of claims simultaneously.

ELEMENTS OF A TORT CLAIM

If a tort action is to succeed, the plaintiff must be able to show that, more likely than not, the defendant owed the plaintiff a duty of care, the defendant was negligent, the negligence caused the injury, and an injury did in fact exist. A physician, genetic counselor, or any other health professional who enters into a professional relationship with a client owes that client a professional standard of care. Breaching the standard of care constitutes negligence. Care is also owed to others who are also foreseeably at risk from the defendant's actions even if they are not in an immediate relationship with the defendant. For example, a physician who negligently fails to treat a married person's sexually transmitted disease would be liable in tort to the patient and also to the patient's spouse if the disease were transmitted.

The standard of care can be evidenced by statements and policies of the relevant professional organizations or entities such as the Centers for Disease Control, manufacturers and distributors of tests (in their accompanying material), educational training, and accepted practice within the professional's peer community.

There must be a causal link between the defendant's negligence and the plaintiff's injury. The plaintiff must be able to show that "but for" the defendant's negligence the harm would not have occurred. Thus, if the parents in the Down syndrome example would not have taken the test even if informed, or would not have terminated the pregnancy if the fetus was known to have Down syndrome, they cannot argue that the physician's failure was the cause of the birth of their damaged child. However, had they consulted with the physician before becoming pregnant, and the physician had failed to warn them of the risk, they might be able to prove that, rather than take the risk, they might have chosen not to conceive in the first place.

Finally, the parents must be able to show that they indeed were injured. When genetic anomalies are substantial in their impact upon a child and a family, this is not a problem. Despite the fact that many couples do not terminate pregnancies in which the fetus is known to have Down syndrome, Down syndrome has substantial emotional, practical, and financial impact upon any family. Many people do decide to terminate. The serious nature of Down syndrome is one of the reasons why it is standard practice to offer prospective parents prenatal testing to avoid its occurrence. Controversy can arise when parents complain of an "injury" that courts refuse to accept as serious enough to offset the positive aspects of the birth of a child. Parents who claim to be injured because they were not given the opportunity to abort the birth of a child who has a minor, correctable problem, such as cleft lip, may find their arguments falling on deaf ears.

There is a rough symmetry, even a circularity, between what parents and professionals consider substantial injuries and what is mandated by "standard of care." If physicians see a problem as serious enough to engender routine testing, then that standard can suggest to parents that the problem is, indeed, serious and worth taking drastic steps to avoid. Conversely, societal norms, such as the trend toward smaller families, can drive consumer demand for tests that help avoid the birth of disabled children. Companies that develop and market genetic tests can also affect public perception about what constitutes serious risk. With the completion of the first phase of the Human Genome Project, and with the rise of less invasive and safer methods of testing, our increasing medical sophistication raises important questions about which genetic characteristics are the proper foci of testing.

WRONGFUL LIFE

Wrongful life is a claim brought by the child whose birth comes about as the result of the professional's negligence. Parents may bring a wrongful birth claim on their own behalf, and also bring a wrongful life claim on behalf of their child. Sometimes, parents will resort to a claim for a wrongful life when their own claim for wrongful birth is barred for some reason, such as the statute of limitations.

Wrongful life claims have been barred or rejected in most jurisdictions in which they have arisen. The reasons for this are largely philosophical. In a wrongful *birth* suit, courts may look askance at parents' claims for compensation for the birth of a child who now exists. However, abortion is a legally protected right, and many people do choose to terminate pregnancies for fetal anomalies. Thus, the professional's negligence did in fact deprive the parents of a real opportunity. Because children born with serious problems are likely to bring greater emotional and financial burdens, parents who can argue that they would have terminated such a pregnancy have a logical, compensable claim. In wrongful *life* claims, however, children are asserting that their very existence is an injury to them, that they would have been better off never having existed. This claim presents courts with a number of problems. First, agreeing that someone would have been better off had they never been born seems profoundly derogatory to that person and others similarly handicapped. Second, it seems impossible to calculate what would be appropriate compensation. One purpose of the tort system is to "make victims whole" by economic compensation. In the typical tort claim, plaintiffs show that the defendant's actions put them in a worse position than they would have occupied otherwise, and therefore the defendant should pay to make them whole. Before defendant's negligent surgery, for example, plaintiff had two feet; now the plaintiff has only one foot and is obviously worse off; compensation is required to compensate, as far as money can, for this worsening of plaintiff's condition. How can one understand the claim: "Before defendant's negligence, I did not exist; now I am worse off because I do exist"? As one judge said: "To recognize a right not to be born is to enter an arena in which no one could find his way" [*Gleitman v. Cosgrove*, 1967].

A wrongful life claim sometimes meets with limited success. In 1984 a New Jersey court allowed a child to be compensated for wrongful life, but the court was clearly forced to that conclusion only because the parents' claim for wrongful birth was barred by the statute of limitations. The wrongful life claim was the only way to address the child's extraordinary educational and medical needs. The court was careful to rule that wrongful life claims for "emotional distress or for an impaired childhood" would not be entertained. In France, a successful wrongful life claim brought by a child who was born severely disabled because his mother was counseled to continue a pregnancy after she had been exposed to German measles was followed in 2002 by passage of a law that would "forbid" people "to seek damages simply for having been born" [Daley, 2002].

WRONGFUL ADOPTION

In the tort of *wrongful adoption*, parents sue because an adoption agency, through deception or negligence, failed to warn them that a child they adopted was at high risk for a genetic disease. The parents claim that they would not have adopted the child had they known, or perhaps that they would have adopted the child but would have been better prepared financially and emotionally. Under the general rubric of "wrongful adoption," come cases alleging negligent misrepresentation, negligent nondisclosure, and failure to investigate. Competing interests here include the parents' goal of adopting a healthy child, the agencies' concern not to be considered a guarantor of every adoptive child's health, and the child's interest in being considered adoptable.

The law in this area remains in its infancy. A 1989 Wisconsin court allowed adoptive parents to sue an agency that had lied to them by saying that the child's father had tested negative for Huntington disease, when such a test did not exist at that time. But the court was careful to confine its ruling to the egregious facts of the case, involving "affirmative misrepresentation;" the court did not create a duty on the part of adoption agencies to investigate a child's health or to disclose facts to parents [*Meracle v. Children's Service Society of Wisconsin*, 1989]. More recently, the Massachusetts Supreme Court did hold that an agency has a duty to disclose all known information about prospective adoptees [*Mohr v. Commonwealth*, 1995]. No court has yet laid upon an agency the duty to actually investigate the child's health or to test the child or the biological parents [Trefethen, 2000].

This is an area likely to experience increasing legal and ethical pressure. Would-be parents with fertility problems have an ever larger array of reproductive technology from which to choose, as an alternative to adoption. A growing fascination and popularization of genetics generates concerns on the part of adoptive parents that their new child may posses a genetic "time bomb." This is especially true as mental illnesses such as schizophrenia are increasingly understood as genetic in origin [Belkin, 1999]. Conversely, the heightened ability of parents to avoid at least some genetic risks in their biological children through a variety of diagnostic techniques, raises parental expectations generally and contributes to a risk-averse climate. The joint statement of the American Society of Human Genetics (ASHG) and the American College of Medical Genetics (ACHG) [2000] argues for parity between the genetic testing considered appropriate for any children, and those considered appropriate for prospective adoptees. The ASHG–ACMG considers that testing is only appropriate for diseases that can be prevented or ameliorated by early treatment, or for serious childhood diseases in children for whom there is some health-related indication that the disease may exist. Some critics, however, maintain that the special circumstances of the adoption process, in which the state's goal is to create the best "match" between parent and child, grounds an argument for allowing more extensive genetic testing of prospective adoptees [Jansen and Ross, 2001].

The ASHG–ACMG holds that, just as a birth family cannot be certain that its natural child will be healthy, so the adoptive family can have no such guarantee about their child's future health. However, as birth parents become increasingly instrumental in trying to guarantee a genetically perfect child, the notion of "parity" provides little foundation for geneticists or adoption agencies who are trying to resist testing.

Chapter 7

Informed Consent

OVERVIEW

A distinct genetics model of informed consent has evolved that exemplifies the pretest educational and counseling processes necessary to help the patient to understand, cope with, and adjust to the impact of genetic information [Ad Hoc Committee, 1975; Royal Commission, 1993; Sharpe, 1994a, 1997; Elias and Annas, 1994; Lerman et al., 1997; Geller et al., 1997; Emery, 2001].

Given the lack of treatment and cure for many genetic diseases and disorders, as well as the limitations of genetic testing technologies, it is recommended that *informed choice* should be the objective of the informed consent process in order to emphasize that a test is not being recommended only offered [Dickens, 1985; American College of Medical Genetics, 1999a; European Commission, 2004].

INFORMED CONSENT

Informed consent reflects what has been described as the "special" [*Nocton v. Lord Ashburton*, 1914] relationship between a physician and a patient, where due to the physician's superior medical knowledge and skills, the patient must have trust and confidence in the physician [Berger, 2003].

> *A physician violates his [her] duty to his patient and subjects himself [herself] to liability if he [she] withholds any facts which are necessary to form the basis of an intelligent consent by the patient to the proposed treatment.* [*Salgo v. Leland Stanford Jr. University Board of Trustees*, 1957; Andrews, 1997a].

The most commonly cited rationale for the doctrine of informed consent was that stated by Justice Cardozo [*Schloendorff v. Society of New York Hospital*, 1914; Jonsen

Genetic Testing: Care, Consent, and Liability, by Neil F. Sharpe and Ronald F. Carter
© 2006 John Wiley & Sons, Inc.

et al., 1998]:

> *Every human being of adult years and sound mind has a right to determine what shall be done with his[her] own body; and a surgeon who performs an operation without his [her] patient's consent commits an assault for which he[she] is liable in damages. This is true except in the case of emergency where the patient is unconscious and where it is necessary to operate before consent can be obtained.*

The doctrine of informed consent [*Canterbury v. Spence*, 1972; *Reibl v. Hughes*, 1980] generally requires disclosure of all material medical information for a generally accepted course of action [e.g., *Moore v. Baker*, 1993], including:

- The diagnosis and nature of the patient's condition.
- The nature, objectives, and limitations.
- The expected outcome and the probability of success.
- The attendant risks and benefits.
- All reasonable alternative procedures [*Martin v. Richards*, 1993; in some jurisdictions this can include more hazardous risks: *Gemme v. Goldberg*, 1993] and their respective risks and benefits including the risks associated with not following the recommended medical course of action [*Gates v. Jensen*, 1979; *Truman v. Thomas*, 1980; *Logan v. Greenwich Hospital*, 1983; *Coughlin v. Kuntz*, 1986; *Gemme v. Goldberg*, 1993] or choosing no treatment [*Wecker v. Amend*, 1996]. Some courts have held that alternatives should be disclosed even if they are more hazardous. This may include a duty to disclose subsequently discovered risk information to former patients if the risk of future injury arises from the original physician–patient relationship [*Tresemer v. Barke*, 1978]. Some courts have recognized a duty to advise that another hospital or clinic will be more appropriate to conduct the procedure given the absence of specialized care [*Martin v. Richards*, 1995; CT scan and availability of a neurosurgeon].
- Consent to a procedure that is novel, experimental, and/or is not generally observed or applied may require the full disclosure of all risks no matter how remote the risk of occurrence [*Gaston v. Hunter*, 1978; *Weiss v. Solomon*, 1989; Prillaman, 1990; *Kus v. Sherman* Hospital, 1995].
- An exception to the duty to obtain an informed consent is in an emergency situation where the patient has a life or health threatening condition, the patient is incapable of giving consent, and a consent form cannot be obtained in time from an authorized representative [*Cobbs v. Grant*, 1972].

A legal cause of action generally may arise when a health care professional has failed to disclose all material information to a proposed medical course of action [*Truman v. Thomas*, 1980; but note *Turner v. Children's Hosp. Inc.*, 1991].

The plaintiff-patient will allege that:

- The patient has suffered an injury as a result of undergoing the recommended procedure or treatment.

- The health care professional failed to disclose all material information [*Truman v. Thomas*, 1980], such as an attendant risk concerning the recommended procedure or treatment.
- Had this information been disclosed, the patient would not have granted consent, and the injury would not have occurred.

Informed consent long has been a flash point for controversy [Meisel and Kuczewski, 1996]. Studies indicate that some physicians regard informed consent as an intrusion in the physician–patient relationship that can have a negative impact on patient care [Taylor and Kelner, 1987] and represents "an empty ritual in which patients are presented with complex information that they cannot understand, and that has little impact on their decision-making" [Lidz et al., 1988]. Others value the doctrine of informed consent as a response to, and the recognition of, good patient care, medical ethics, and the principle of patient autonomy [Andrews, 1985; Finkelstein et al., 1993; Jonsen et al., 1998; Wildeman and Downie, 2001].

Courts of laws are not immune to this debate. As discussed in the following, legal standards for informed consent vary significantly from jurisdiction to jurisdiction [Schuck, 1994; Merz, 1993; *Ketchup v. Howard*, 2000] and have been the subject of significant controversy and debate.

LEGAL MODELS OF INFORMED CONSENT

Three distinct standards have evolved to evaluate the duty of disclosure in the United States and Canada for traditional medicine.

- The traditional, *professional standard* inquires what a reasonable medical practitioner would disclose in the same or similar circumstances [e.g., *Aiken v. Clary*, 1965; *Stauffer v. Karabin*, 1971].
- The *reasonable patient standard* [*Canterbury v. Spence*, 1972] is founded upon the principle that "physicians' communications to the patient . . . must be measured by the patient's need and that need is material to the decision." This standard generally inquires what information a reasonable patient in the patient's position would want to know about the risks and limitations. The Supreme Court of Canada has adopted a variation of the reasonable patient standard [*Reibl v. Hughes*, 1980; *Arndt v. Smith*, 1997].
- A handful of American jurisdictions have applied a *patient-oriented, subjective standard* [*Arena v. Gingrich*, 1987; *Cheung v. Cunningham*, 1988].
- U.S. state jurisdictions also may have specific legislative requirements concerning the nature and scope of disclosure as well as the evidentiary standards for causation (an analysis of these requirements is beyond the scope of this discussion).

Although the underlying objective and rationale of the doctrine of informed consent is to protect the patient autonomy in decision making, legal commentators have argued that a court's reliance on expert medical evidence to evaluate what a "reasonable" patient or "reasonable" physician would have decided to do if the medical information had been disclosed, effectively can subordinate an individual patient's beliefs, values, concerns, and objectives to medical values and objectives. Some argue that this has resulted from the courts' underlying assumption that a reasonable patient will be presumed to follow the recommendations of the physician—even where material information has not been disclosed—because these recommendations offer the best prospect for therapeutic benefit and the prevention of bodily harm [Katz, 1977; Robertson, 1991].

> *The "objective" standard of "causality" contradicts the right of each individual to decide what will be done with his [her] body by denying the patient recovery whenever his [her] hypothetical decision is out of step with the judgment of a prudent person.* [Katz, 1977]

> *[T]he growing acceptance by Canadian courts that the greater the confidence and trust which a person has in a physician, the less likely a reasonable person in the patient's position would decline treatment recommended by the physician, even if full disclosure of medical risks were made.* [Robertson, 1991]

GENETICS MODEL OF INFORMED CONSENT

- Effective treatment and cure is not available for many genetic diseases and disorders. Accordingly, the pivotal factor in decision making may be nonmedical factors: the patient's objectives, concerns, values, and beliefs about genetic testing and the potential impact on the patient and the family [Wiggins et al., 1992; Tibben et al., 1993a; Codori et al., 1999; Glanz et al., 1999; Kirschner et al., 2000; Bassett et al., 2001; Marteau and Dormandy, 2001; Kohut et al., 2002; Hadley et al., 2003; Keller et al., 2004].

- Genetic testing technologies can have significant limitations. Molecular and biochemical diagnostics currently are restricted by the relative lack of affordable, accurate, high-throughput test capacities for addressing the clinical demands that arise.

- Genetic testing may be limited to providing information about an increased susceptibility to develop a medical condition at some time in the future. Testing may not be able to predict the severity or prognosis of the condition, when it will occur, or even that it will occur. Because of these limitations, the physician may be confronted by a patient's acute emotional and psychological reactions especially when testing concerns a potentially life-threatening condition.

- Physicians will need to focus on a patient's communicative, emotional, and psychological needs in order to assist the patient to understand, adjust to, and cope with the implications posed by genetic testing and test information [Ad Hoc Committee, 1975; Sharpe 1994a; Geller et al., 1997; Lerman et al., 1997; Michie et al., 1997a; Michie and Marteau, 1999b; American College of Medical Genetics, 1999a; Kirschner et al., 2000; Emery, 2001]. Please see Chapter Three: Psychological Aspects.

A health care professional who agrees to provide genetic testing services will be expected to perform in a manner consistent with a reasonable standard of care. Although these standards continue to evolve, the recommendations reviewed in the following have not been imposed by a court of law. Geneticists, genetic counselors, and physicians have developed these guidelines in response to patient needs and the distinct processes of genetic testing. (Please see Chapter 8, Prenatal and Neonatal Screening.)

> *If the medical profession wishes to take a proactive and constructive approach, rather than a merely reactive one [to genetic testing], it must define clearly the expectations and prerogatives of all physicians. If not, physicians will be obligated to respond to rules and regulations imposed from outside the profession, to which they are unlikely to adhere.* [Taylor and Kelner, 1996]

Disclosure and Consent

Within the context of genetic testing, informed consent is perceived to be:

- An ongoing process of education and counseling in which the test provider, the patient, and, where appropriate, the patient's family, discuss their respective needs, concerns, objectives, and expectations [Sharpe, 1994a; Sharpe, 1997; Geller et al., 1997] including emotional and psychological adjustment to genetic information.

- Given the lack of effective treatment and cure for many genetic diseases and disorders, neither the physician nor the patient may understand nor accept what the other perceives to be the most *medically* prudent, rational, or intelligent decision [Brock and Wartman, 1990]. The implications are significant, calling into question whether legal standards for informed consent for traditional medicine, which arguably support the underlying assumption that the physician rather than the patient can best determines "what constitutes well-being" [Katz, 1977], are appropriate to good patient care within the context of genetic testing.

- The value of *nondirective* disclosure of information, both medical and nonmedical, appropriate to the needs and interests of the patient is advocated [Geller et al., 1997; Sharpe; 1997; Task Force, 1997; Jonsen et al., 1998; European Commission, 2004 Please see Chapter One: Nondisectiveness]. A review of human/medical genetics journals and texts indicates that although recommendations may be made to prevent "substantial harm" [Bartels et al., 1997; Avard and Knoppers, 2001; and see Chapter 1, Genetic Counseling and the Physician–Patient Relationship], the majority of geneticists and counselors subscribe to the values of nondirective counseling with respect to a patient's right of autonomous decision making [Wertz and Fletcher, 1988; Geller et al., 1993; Benkendorf et al., 1997; Burgess et al., 1998; Kessler, 2001; but see Yarborough et al., 1989; Clarke, 1991; Williams et al. 2002a; Weil, 2003].

- Genetic testing should never be imposed [Andrews, 1997]; the emphasis should be on *informed choice* [ACMG, 1999a] and *free choice* [European Commission, 2004].
- There is a clear consensus of professional opinion that "nonmedical" factors should be included in the disclosure of the material risks associated with genetic testing [courts of law vary significantly from jurisdiction to jurisdiction in terms of the nature and scope of disclosure for traditional medical procedures: compare *Arato v. Avedon*, 1993; *Johnson v. Kokemoor*, 1996; Karlin v and Foust, 1999; *Benson v. Massachusetts General Hospital*, 2000].
- These nonmedical risks include:
  - The risk of employment discrimination [Billings, 1992; Wertz, 1999b; Barash, 2000; Von Bergen et al., 2001; Pagnataro, 2001; Burgermeister, 2003; State Genetic. 2005]
  - The risk of insurance discrimination [Lapham et al., 1996; Birmingham, 1997; Matloff et al., 2000; Hall and Rich, 2000a,b; Lemmens, 2000; Pfeffer et al., 2003; Hellman, 2003; Kass et al., 2004; Klanica, 2004; State Genetic, 2005; Knoppers and Joly, 2004; European Commission, 2004; Watson et al., 2004a; but see Stephenson, 1999; Watts, 1999; Armstrong et al., 2003; compare to Geer et al., 2001]
  - The potential emotional and psychological impact of the testing process [see Kessler, in Capron, 1979; Marteau, in Evers-Kiebooms, et al., 1992; Wiggins et al., 1992; Mennie et al., 1993; Hubbard and Lewontin, 1996; Michie and Marteau, 1999b; Meiser and Dunn, 2000; Bish et al., 2002; Lerman et al., 2002; Braithwaite et al., 2004].

Although physicians may be reluctant to disclose information out of concern for potential patient distress [*Cobbs v. Grant*, 1972; *Canterbury v. Spence*, 1972; *Reibl v. Hughes*, 1980; Hebert et al., 1997], the objective of the genetics model of informed consent is to facilitate effective communication, comprehension, and adjustment to the emotional and psychological aspects of genetic testing [Andrews, 1985; Sharpe 1994a, 1997; Geller et al., 1997; ACMG, 1999a]. This will better enable the patient to knowledgeably understand, appreciate, and evaluate the attendant risks and benefits [*Gates v. Jensen*, 1979; English, 2002] essential to making informed choices about one's well-being.

Accordingly, a court of law will need to look beyond what occurred at the time of a patient's signing of an informed consent form [Sharpe, 1994a; Kuczewski and Marshall, 2002] and consider the nature of the communicative-counseling relationship. Delineating a physician's communicative and counseling obligations for the disclosure of genetics information may be perceived to be a departure from what physicians perceive to be their normal standard of care, especially when there is a shortage of qualified personnel. However, if the underlying objective and rationale of the doctrine of the informed consent is to enable the patient to understand and evaluate the risks and benefits of a proposed medical course of action, it is reasonable for a court of law to inquire whether all reasonable procedures were undertaken to facilitate effective communication and comprehension in order to ensure a truly voluntary decision [Katz, 1977; Note, 1981; *Ciarlariello v. Schacter*, 1993; Sharpe, 1994a, 1997; *Descant v. Administrators of Tulane Educational Fund*, 1998; Andrews and Zuiker, 2003] without coercion [NSGC, 1992; Parker and Lidz, 1994].

WATCH OUT FOR

Test Validity and Utility

Genetic tests are available for over 1000 inherited diseases or conditions—http://www.genetests.org—for clinical or research purposes. With the advent of genetic testing delivered by commercial organizations [Andrews, 1997] and the potential for direct-to-consumer advertising [Hull and Prasad, 2001a,b; Caulfield, 2001; Gollust et al., 2002; Tsao, 2004] and testing [McCabe and McCabe, 2004; but see Human Genetics Commission, 2003], genetic testing is increasingly likely to be provided by primary care health care professionals.

A critical issue is whether a genetic test has received appropriate peer review prior to introduction to routine practice [Prence, 1999], to ensure clinical validity and clinical utility [Collins, 1996; American Society of Human Genetics, 1996; Task Force, 1997; Welch and Burke, 1998; Yang et al., 2000; Caulfield, 2001; Khoury et al., 2000, 2004; Burgess, 2001; Haga et al., 2003; Bassett et al., 2004]. For the purposes of this text, these terms are described as:

Clinical Validity   Initial reports associating one or more genetic variants with a particular genetic disease or condition may not be confirmed and may be contradicted by later studies.

Clinical Utility   Does the genetic test provide information that would influence individual prevention or management recommendations?

Out of concern for these factors, the U.S. Department of Health and Human Services has stated [U.S. Dept. of Health and Human Services, 2000]:

The U.S. Department of Health and Human Services (DHHS) recognizes how important it is for the public to understand that while genetic tests can be extremely beneficial, they also can pose medical and psychological risks to individuals and families as well as socioeconomic risks that may affect entire groups and their individual members. As the diagnostic and predictive uses of genetic testing continue to increase, and as the effects of testing on society become clearer, its impact will become broader and ultimately will affect all of our lives. Because the use and ramifications of these tests are not yet fully realized, additional consideration is needed regarding whether current programs for assuring the safety and effectiveness of genetic tests are satisfactory or whether additional oversight measures are needed before such tests are introduced for wide-scale use.

Genetic tests can be categorized by a consideration of clinical validity and the availability of effective interventions. For genetic tests with high clinical validity and no effective treatment, the primary objective of the informed consent process is adequate nondirective counseling about the outcomes and attendant medical and nonmedical risks to ensure an informed, autonomous decision [Andrews, 1997; Lerman et al., 1997]. For tests with limited clinical validity and no effective treatment, and for tests where uncertainty exists concerning both the clinical validity and effectiveness of treatment, it has been argued that recommendations may be made against test use justified on the principle of avoiding harm [Welch and Burke, 1998; Caulfield, 2000; Genewatch, 2001].

Issues concerning clinical utility and validity include:

- Issues of test sensitivity and specificity are often in question. Patents awarded for specific genes may grant exclusive rights to all diagnostic applications. Exclusive licensing raises the issue of quality control; standards of proficiency testing cannot be evaluated because the patent holder may refuse to divulge the precise nature of the testing technology or the mutation panel being used, and the patent holder's methodology cannot be rigorously tested against other laboratory technologies. Physicians should be aware of the technical complexities of DNA (deoxyribonucleic acid) molecular testing technologies in order to explain to patients their uses and respective limitations.

- Does the laboratory have appropriate standards for personnel and quality assurance to ensure the reliability of the testing and to safeguard the rights of patients with regard to confidentiality and providing informed consent [Andrews, 1997; ACMG, 2004; and see Chapter 12, Test Samples and Laboratory Protocols]? Surveys have indicated that diagnostic laboratories can have quality assurance scores that may reflect "suboptimal" laboratory practices [McGovern et al., 1999, 2003a, 2003b; Grody and Pyeritz, 1999; Dequeker et al., 2001; Quillin et al., 2003].

- Will the test be fully informative; that is, does the mutation panel in question include all known clinically significant mutations? If not, of the total number of known clinically significant mutations, how many are included in the mutation panel? What mutations are clinically significant for specific ethnic groups and how many of these have been included for the patient in question? What are the limitations of the test technology [Yan et al., 2000; Gad et al., 2001]? By way of example, for susceptibility testing for hereditary breast and ovarian cancer, among high-risk families a proportion may not have a detectable mutation in *BRCA1* or *BRCA2* due to the presence of an undetected mutation that may only be found by alternative techniques or has not been identified.

- What does a "negative" result mean? Does this conclusively mean that a specific mutation or mutations have not been detected and the patient will not be a carrier or be at risk for the genetic disease, or does the negative result apply only to the mutation panel used for the analysis? If the latter, what is the clinical significance of the mutation panel tested especially given the ethnic background of the patient? Have all clinical significant mutations been identified? Is the patient still subject to the general population risk for the disease as is the case with breast and ovarian cancer?

PROCESS ISSUES

Communication

Communication is the essence of genetic testing and informed consent. Ineffective communication can affect a patient's ability to understand, retain, and adjust to genetic information. These difficulties are compounded by the emotional and psychological impact that genetic testing procedures can have on the patient and the family. Health care professionals need to know how to effectively communicate and to help patients adjust to and to manage sensitive genetic test information [Royal Commission, 1993; Committee on Assessing Genetic Risk, 1994; also see Roter, 2004].

Although many physician–patient encounters involve communication, researchers have questioned the effectiveness of these communicative relationships for informed consent. Physicians may underestimate the patient's needs and provide too little information [Waitzkin, 1985; Mechanic, 1998]. Patients may not be fully informed [Wu and Pearlman, 1988; Searle, 1997; Braddock et al., 1997] and even when they are, patients' comprehension and recollection of information may be restricted.

- In 1999, it was reported that out of a total of 1057 encounters with primary care physicians and general and orthopedic surgeons, 9.0 percent met the study's definition of "completeness for informed decision making," yet only 0.5 percent of "complex decisions" met the same criteria [Braddock et al., 1999; Manthous et al., 2003; Thorevska et al., 2004].

- A study found that although 86 percent of patients stated that physicians had explained the risks of a procedure, 30 percent later were unable to articulate the indications for the procedure, only 57 percent could articulate more than one risk, and only 53 percent reported that physicians had explained alternatives to the procedure [Sulmasy et al., 1994].

Similar difficulties can arise in genetic counseling, where health care providers may not give sufficient attention to major issues on the minds of patients "and may be too strongly wedded to their own agendas so that room for adequate discussion of the counselee's concerns may not be available" [Kessler, 1992].

- A study of 880 genetic counseling sessions reported that in 58 percent of those sessions, the counselor was not aware of the topic the client most wanted to discuss [Kessler, 1992].

- A 1998 survey concluded that information about genetic testing, during the first prenatal visit, was "inadequate for ensuring informed autonomous decision-making" [Bernhardt et al., 1998].

- A survey of commercial Adenomatous Polyposis Coli (APC) gene testing for familial adenomatous polyposis reported that only 18.6 percent of patients received genetic counseling, 16.9 percent had been provided with informed consent, and doctors misinterpreted test results in 31.6 percent of the cases [Giardiello et al., 1997].

These studies illustrate why physicians should be aware of the potential difficulties created by communicative issues associated with genetic testing [Geller et al., 1997; Lerman et al., 1997; Miesfeldt et al., 2000; Marteau and Dormandy, 2001]. From the perspective of good patient care, a health care professional has an obligation to undertake reasonable procedures to facilitate the effectiveness of communication and the informed consent process [Katz, 1977; *Phillips v. Good Samaritan Hospital*, 1979; *Ciarlariello v. Schacter*, 1993; Pape, 1997; *Descant v. Administrators of Tulane Educational Fund*, 1998; Marteau and Dormandy, 2001]. As will be discussed in the following, these procedures include taking into account:

- "Easily understandable language" [European Commission, 2004]
- The timing, manner, and place of all discussions, together with
- Emotional, psychological, linguistic, cultural, and socioeconomic differences that may impair the communication process

The Form

Although health care professionals may question the use and need for consent forms [Jacobson et al., 2001], within the context of genetic testing the informed consent process is neither an abstract administrative requirement [*Descant v. Administrators of Tulane Educational Fund*, 1998] nor a purely legal requirement. Although specific standards for disclosure vary significantly among legal jurisdictions, it is generally recognized that the physician has a reasonable obligation to ensure that the patient understands in her or his own terms the explanation and the nature of the information [Katz, 1977; *Ciarlariello v. Schacter*, 1993; Sharpe, 1994a, 1997; *Descant v. Administrators of Tulane Educational Fund*, 1998].

Informed consent and the use of consent forms represent the critical pretest educational and counseling process through which the health care professional helps the patient to understand, cope with, and adjust to genetic information [Ad Hoc Committee, 1975; Royal Commission, 1993; ACMG, 1999a], while preserving the patient's right of autonomy in making decisions.

- Forms alone do not account for factors that can impair patient understanding such as the timing of the disclosure, language and cultural differences, and a patient's emotional and psychological state of mind [Meade, 1999; Jacobsen et al., 2001]. Accordingly, a court of law will need to look beyond an executed form and consider what type of information was disclosed, when, and under what circumstances [Sharpe, 1994a; English, 2002; Mazur, 2003].
- A consent form for a medical procedure is not conclusive as to the issues of disclosure, understanding, and consent [*Ditto v. McCurdy*, 1997; English, 2002], although the signing of a form may give rise to a "presumption" that consent was granted [e.g., *Crundwell v. Becker*, 1998]. The content of forms for a specific procedure can vary widely [Durfy et al., 1998; English 2002; Manthous et al., 2003] and may not provide adequate disclosure [Durfy et al.,

1998; Hopper et al., 1998; Bottrell et al., 2000; Marteau and Dormandy, 2001; Jacobsen et al., 2001; Stapleton et al. 2002; Thorevska et al., 2004].

- The readability [Murgatroyd and Cooper, 1991; Meade, 1999; Taylor, 1999], educational level [Meade, 1992; Hochhauser, 1999; Paasche-Orlow et al., 2003], and breadth and complexity of a consent form can call into question the ability of the patient to understand the information [Hopper et al., 1998; Briguglio et al., 1995; Pape, 1997; Taylor, 1999].

- Potential areas of confusion and dissatisfaction are created by forms that make extensive use of medical and legal terminology [Baker and Taub, 1983; Lidz et al., 1988; Pape, 1997; *Descant v. Administrators of Tulane Educational Fund*, 1998], are not easily understandable [*Ditto v. McCurdy*, 1997], are not written in the patient's principal language, and are not culturally or literally appropriate [Pape, 1997; Meade et al., 2002].

- The consent form should be specific. A generally worded form may not constitute consent [*LaCaze v. Collier*, 1983; *Pridham v. Nash*, 1987; *Diack v. Bardsley*, 1983]. A separate form should be obtained for *each* procedure [*Kohoutek v. Hafner*, 1986; and see Andrews and Zuiker, 2003, concerning scenarios where more genetic information is generated than expected], especially where the test can be applied for other clinical purposes [Kapp, 2000].

- The patient should not be advised that unless the consent form is signed the procedure will not be provided [English, 2002; Stapleton et al., 2002]. There is American precedent that this may constitute coercion and be considered a breach of the duty of care to the patient [*Brown v. Dahl*, 1985].

Place and Time Allocated for Discussion

Disclosure and discussion need to occur at an appropriate time and place. Communication and comprehension may be impaired and a patient's consent vitiated [*Sard v. Hardy*, 1977; *Hodge v. Lafayette General Hospital*, 1981] in scenarios analogous to the following:

- Immediately prior to the administration of a procedure [*Sard v. Hardy*, 1977; *Coughlin v. Kuntz*, 1986; *Whiteside v. Lukson*, 1997; Done et al., 1996; Erde, 1999]

- In a preop staging area [*Felde v. Vein & Laser Medical Centre*, 2002]

- In a hospital corridor or waiting room [Ives et al., 1979; *Coughlin v. Kuntz*, 1986]

- While the patient is under sedation [*Moore v. Webb*, 1961; *Ferguson v. Hamilton Civic Hospital*, 1983; *Whiteside v. Lukson*, 1997]

- When the patient is suffering from emotional and psychological distress [Kessler, 1987; *Cherewayko v. Grafton*, 1993; *Felde v. Vein & Laser Medical Centre*, 2002]

- By telephone [Tluczek et al., 1991]

Timing of Discussion

Timing may prove a crucial element in effective communication and comprehension.

- Genetic information, on a case-by-case basis, can trigger a host of emotional and psychological responses [Kessler, in Capron, 1979; Tluczek et al., 1991, 2005; Marteau, in Evers-Kiebooms et al., 1992; Lynch et al., 1999; Collins et al., 2000; Tercyak et al., 2001a; Lerman et al., 2002] that have the potential to impair a patient's ability to comprehend and retain information [Kessler, in Capron, 1979; Watson et al., 1992; Kelly, 1992; Mennie et al., 1993; Lerman et al., 1994a]. Although these types of responses may prove to be part of the normal coping process, adequate time must be provided for the patient to proceed through such reactions and to allow time for reflection and consideration [Royal Commission, 1993; ACMG, 1999a; Carroll et al., 2000].
- Premature discussion of genetic information may lead to misunderstanding [Kessler, in Capron, 1979; Kelly, 1992; Bloch et al., 1993], dissatisfaction, rejection of counseling, and aggravate a patient's feelings of anxiety and despair [Alexander, 1990].

In a case alleging a failure to obtain an informed consent, prior to the procedure the physician did not provide any written advice other than a pamphlet. The patient was in considerable distress and described herself as "foggy," "uptight," and "unable to open her mouth without crying." The court, in a finding of liability, stated that the physician should have taken into account the patient's agitated emotional state and its potential impact on the information received [*Cherewayko v. Grafton*, 1993].

Tone and Manner

The attitude of a health care professional and the use of descriptive terms significantly can influence a patient's perception of the nature and implications of the information [Abramsky et al., 2001; Abramsky and Fletcher, 2002; Wertz and Knoppers, 2002].

- Pessimism may decrease the effectiveness of the communication and lead to increased patient anxiety that is associated with delaying in seeking care [Kelly, 1992; Shapiro et al., 1992].
- The choice of words used to describe the level of risk may significantly affect how the patient perceives that condition or risk [Abramsky et al., 2001]. Technical language and medical jargon [Abramsky and Fletcher, 2002; Browner et al., 2003] should be avoided [*Descant v. Administrators of Tulane Educational Fund*, 1998].
- Perception of risk information can vary from patient to patient, with a 50 percent risk perceived by some as high and others as low [Wertz et al., 1986; Kessler and Levine, 1987; Kessler, 1992; Gurm and Litaker, 2000; Roggenbuck et al., 2000].

- Several forms of explanation may have to be provided to ensure the patient understands the full force and implications of the information [*Hodge v. Lafayette General Hospital*, 1981; *Coughlin v. Kuntz*, 1986; Wu and Pearlman, 1988; Gurm and Litaker, 2000], especially with regard to ambiguous and often complex information relating to risk [Grimes and Snively, 1999; Gurm and Litaker, 2000].

- The physician has a reasonable obligation to adequately assess a patient's comprehension [Hartlaud et al., 1993]. Having the patient repeat in her or his own terms the patient's understanding of the information will help to determine whether the patient fully understands the circumstances, risks, and alternative outcomes [Katz, 1977; Note, 1981].

Delegation

Delegation to an intern, nurse, junior health care professional [Mulcahy et al., 1997], or health or office personnel [Lidz et al., 1983; Levine et al., 1995], with whom the patient has had limited if any previous contact [*Whiteside v. Lukson*, 1997], the responsibilities concerning informed consent may call into question the validity of the consent [Teff, 1985].

Cultural and Socioeconomic Differences

Differences in cultural and socioeconomic background [Lum, in Biesecker et al., 1987; Punales-Morejon and Penchaszadeh, in Evers-Kiebooms et al., 1992; Nidorf and Ngo, 1993; Saldov et al., 1998; Penchaszadeh, 2001; Kuczewski and Marshall, 2002; Browner et al., 2003; Karliner et al., 2004; Institute of Medicine, 2004], and preconceived attitudes and values on the part of both the health care professional as well as the patient [Rapp, 1997; Weil, 2001; Furr, 2002; Meade et al., 2003], can significantly impair the communication and consent process (also see Culture and Communication in the Realm of Fetal Diagnosis: Unique Considerations for Latino Patients in Chapter 2).

- Wherever reasonably possible, communication and disclosure should be undertaken by the health care professional with an appropriate knowledge of, and sensitivity to, a patient's particular cultural beliefs, values, and attitudes [Punales-Morejon, 1997; Penchaszadeh, 2001; Browner et al., 2003].

- Culturally sensitive education and counseling are needed to respond appropriately to patient needs [Lerman et al., 1997; Hughes et al., 1997; Glanz et al., 1999; Singer et al., 2004]. Asian ethnic groups, each possessing their own cultural identify, have been perceived as a homogenous group by health care professionals [Wang and Marsh, 1990]. Other studies have indicated that stereotypical assumptions regarding low-income and minority clients can interfere with the ability to establish a meaningful communicative relationship [Corey et al., 1988].

- A health care professional may seek to establish an effective relationship through the use of direct eye contact. However, studies have indicated that some ethnic groups place value on indirectness and nonverbal communication [Corey et al., 1988; Mio and Morris, 1990; Sue, 1990] and dislike probing personal questions and direct eye contact [Mio and Morris, 1990; Sue, 1990].
- Unrealistic self-demands on the part of the health care professional such as believing that one is, or must be, able to communicate with every patient irrespective of cultural and socioeconomic differences [Searle, 1997], may impair communication and comprehension [Kessler, in Capron, 1979; Corey et al, 1988]. A health care professional must take into account professional and personal limitations with appropriate referral wherever required.

Linguistic Differences

To ensure effective and "meaningful" communication [Pape, 1997], the health care professional has a reasonable obligation to ensure that the patient understands what has been discussed in the patient's own terms (also see Culture and Communication in the Realm of Fetal Diagnosis: Unique Considerations for Latino Patients- Translation and Second Hand Information in Chapter 2).

- U.S. Census reports that the United States has an estimated 19 million residents with a limited command of the English language.
- Where it appears that the patient has some difficulty with the language spoken by the doctor, the onus may be placed upon the doctor to show that the patient understood the information and explanation [*Ciarlariello v. Schacter*, 1993].
- If an interpreter is to be used [*Lau v. Nichols*, 1974], the health care provider has a reasonable obligation to ensure that the interpreter has been appropriately trained [Laws et al., 2004] with regard to the use of medical terms, the principles, objectives, and process of informed consent, and the necessity for confidentiality.
- A study reviewed transcribed pediatric encounters in a hospital outpatient clinic in which a Spanish interpreter was used. Out of 474 pages of transcripts, 396 interpreter errors were noted, with an average of 31 translation errors per patient visit. Of these, 19 had clinical significance. The most common error types were omission (52 percent), false fluency (16 percent), substitution (13 percent), editorialization (10 percent), and addition (8 percent) [Flores et al., 2003; Bonacruz and Cooper 2003].
- Errors of potential clinical consequence are more likely to be committed by ad hoc interpreters such as nurses [Elderkin-Thompson et al., 2001] than those committed by hospital interpreters [Lee et al., 2002a,b; Flores et al., 2003].

- Commentators have suggested working with local community groups and colleges to provide and train translators, and to schedule non-English-speaking patients on specific days when translators are available [Flores et al. 2003; Monroe and Shirazian, 2004]. The U.S. Department of Health and Human Services has published guidelines for accommodating those with limited English proficiency [www.hhs.gov/ocr/lep/revisedlep.html].

Pamphlets, Videos, and Computer Aids

Pamphlets, videos, and computer educational programs [Green and Fost, 1997; Jimison et al., 1998; Green et al., 2001a,b] for informed consent may be perceived as necessary solutions in the face of time restraints and a shortage of qualified personnel. However, significant variations in the content of such aids [Cho et al., 1997] raises concerns about the extent to which any one, or a combination, can be relied upon to facilitate education and understanding [Fischhoff et al., 1993; Cho et al., 1997; Meredith and Wood, 1998; Loeben et al., 1998; O'Cathain et al., 2002; O'Cathain and Thomas, 2004] and to what extent this negatively will affect the necessary dialog and discussion between the physician and the patient [Rosoff, 1999].

Questions concerning the effectiveness of these aids include:

- How comprehensive is the content [Loeben et al., 1998]?
- How appropriate is the readability and the use of the terminology and language [Jubelirer, 1991; Fischhoff et al., 1993]?
- Is the grade level appropriate [Briguglio et al., 1995; American College of Medical Genetics, 1999a]?
- Who prepared the aid? Information brochures prepared by commercial testing organizations have been questioned for claims that may exaggerate the clinical validity and utility of a genetic test [Caulfield, 2000; Loeben et al., 1998; Holtzman, 1999].
- What will be the patient's emotional and psychological response to the information, how could these affect comprehension, and how will this be dealt with?
- In the absence of a structured, face to face, interview that discusses the respective objectives, risks, and limitations of a proposed course of action with the information-giver responding to the patient's questions and concerns, will patients feel compelled and obliged to sign a form [Dawes et al., 1993; Shair et al., 1995] raising the issue of perceived coercion [Geller et al., 1996; Task Force, 1997] and questions as to the validity of the consent?

If educational aids are to be used, patients should be provided with face-to-face discussion especially with regard to potential life-threatening conditions [Andrews, 1997; Lerman et al., 1997; Geller et al., 1997; Rosoff, 1999; Carroll et al., 2000;

Green et al., 2001a,b; Keller et al., 2004]. Genetic counseling, and the specialized practice of genetic counsellors, should be considered an integral part of the informed consent process to facilitate effective communication and to respond to a patient's emotional and psychological needs.

WEB RESOURCES

Discrimination, Insurance, and Genetics Legislative Activity

Genetic Laws and Legislative Activity (United States): http://www.ncsl.org/ programs/health/genetics/charts.htm

State Genetic Discrimination in Health Insurance Laws: http://www.ncsl.org/ programs/health/genetics/ndishlth.htm

State Genetic Employment Laws: http://www.ncsl.org/programs/health/genetics/ ndiscrim.htm

State Genetic Nondiscrimination Laws in Life, Disability, and Long-Term Insurance: http://www.ncsl.org/programs/health/ndislife.htm

Genetic Testing Resources

GeneTests: a publicly funded medical genetics information resource developed for physicians, other health care providers, and researchers, available at no cost to all interested persons. http://www.genetests.org/

Association of Public Health Laboratories: Newborn Screening and Genetics: http://www.aphl.org/Newborn_Screening_Genetics/index.cfm

Centers for Disease Control: Genetic Testing: http://www.cdc.gov/genomics/ gTesting.htm

Centers for Disease Control: Regional and State Genetics Directory: http://www .cdc.gov/genomics/links/regional.htm

Directory of Medical Cytogenetic Laboratories in Canada: http://www.hrsrh .on.ca/genetics/canlabs.htm

European Commission Expert Group on the Ethical, Legal, and Social Aspects of Genetic Testing (2004); 25 recommendations on the ethical, legal, and social implications of genetic testing: http://europa.eu.int/comm/research/ conferences/2004/genetic/pdf/ recommendations_en.pdf

Genetics and Public Policy Center: http://www.dnapolicy.org/genetics/testing .jhtml;$sessionid$ETC3MSIAAAYYUCQBAT3RVQQ

National Cancer Institute: Elements of Cancer Risk Assessment and Counseling: Informed Consent: http://www.cancer.gov/templates/doc.aspx?viewid=c0fc1 ac3-607b-44a5-9d24-39b0a2a4703c&version=HealthProfessional&section ID=1& #Section_89

National Conference of State Legislatures (United States): Newborn Genetic and Metabolic Screening: http://www.ncsl.org/programs/health/genetics/nbs.htm

Holtzman, N, and Watson MS, eds. *Promoting Safe and Effective Genetic Testing in the United States: Final Report of the Task Force on Genetic Testing*. Baltimore: Johns Hopkins University Press, 1998: http://www.genome.gov/10001733

Laboratories

Association of Public Health Laboratories: Newborn Screening and Genetics: http://www.aphl.org/Newborn_Screening_Genetics/index.cfm

GeneTests: A publicly funded medical genetics information resource developed for physicians, other health care providers, and researchers, available at no cost to all interested persons: http://www.genetests.org/

Canadian College of Medical Geneticists: Cytogenetics: http://www.hrsrh.on.ca/genetics/CanCyt/

Directory of Medical Cytogenetic Laboratories in Canada: http://www.hrsrh.on.ca/genetics/canlabs.htm

Language

Genetic Information Websites in Spanish for Public Education: http://www.ornl.gov/sci/techresources/Human_Genome/education/spanish. shtml

Law and Legislatures

Council for Responsible Genetics (CRG): Genetics and the Law: http://www.genelaw.info/

Genetics and Public Policy Center: http://www.dnapolicy.org/policy/legalIssues.jhtml

Genetic Laws and Legislative Activity (United States): http://www.ncsl.org/programs/health/genetics/charts.htm

Psychology

American Psychological Association: http://www.apa.org/science/genetics/homepage.html

National Cancer Institute: Elements of Cancer Risk Assessment and Counseling: Psychological Impact of Genetic Information/Test Results on the Family: http://www.cancer.gov/templates/doc.aspx?viewid=c0fc1ac3-607b-44a5-9d24-39b0a2a4703c&version=HealthProfessional&sectionID=1&#Section_108

Patient and Public Education

Medline Plus: Genetic Testing: http://www.nlm.nih.gov/medlineplus/genetic-testing.html

Genetic Disorders: Human Genome Project Information: http://www.ornl.gov/sci/techresources/Human_Genome/medicine/assist.shtml

Genetic Information Websites in Spanish for Public Education: http://www.ornl.gov/sci/techresources/Human_Genome/education/spanish.shtml

Human Genome Project: Exploring Our Molecular Selves: Online Multimedia Education Kit: http://www.genome.gov/Pages/EducationKit/

Medicine and the New Genetics: Human Genome Project Information: http://www.ornl.gov/sci/techresources/Human_Genome/medicine/medicine.shtml

March of Dimes Pregnancy & Newborn Health Education Center: http://www.marchofdimes.com/pnhec/pnhec.asp

Public Genetics Education: *Gene Almanac* (produced by Dolan Learning Center, Cold Springs Harbor, New York): http://www.dnalc.org/

The 21st Century Community Schoolhouse: Websites for Genetic Disorders: http://www.communityschoolhouse.org/websites.geneticdisorders.htm

Understanding Genetic Testing: http://www.pitt.edu/~super1/lecture/lec2631/001.htm

U.S. Department of Energy Office of Science: http://www.doegenomes.org/

Public Health

Genomics for Public Health Practitioners is a 45-minute introductory presentation on genomics and public health. Resource is intended for public health practitioners who have minimal experience in the area of genomics as it pertains to public health: http://www.cdc.gov/genomics/training/GPHP/default.htm

Centers for Disease Control: Public Health Perspective: Family History Tools: http://www.cdc.gov/genomics/info/perspectives/famhistr.htm

Secretary's Advisory Committee on Genetic Testing (United States): http://www4.od.nih.gov/oba/ sacghs/sacghslinks.html

Policy Issues

American College of Medical Genetics Policy Statements: http://genetics.faseb.org/genetics/acmg/pol-menu.htm

American Society of Human Genetics Policy Papers and Reports: http://genetics.faseb.org/genetics/ashg/policy/pol-00.htm

Risk Communication

National Institutes of Health: National Library of Medicine: http://www.nlm.nih.gov/pubs/ cbm/health_risk_communication.html

GENETIC DISCRIMINATION: ANTICIPATING THE CONSEQUENCES OF SCIENTIFIC DISCOVERY

Jon Beckwith, Department of Microbiology and Molecular Genetics, Harvard Medical School, Boston, Massachusetts
Lisa Geller, Wilmer Cutler Pickering Hale and Dorr LLP, Boston, Massachusetts

INTRODUCTION

The Human Genome Project has brought to the fore a long-standing concern over the use of genetic information to discriminate against individuals or groups of individuals. The availability of the human genome map and sequence, along with the powerful tools of modern genetics are generating a vast amount of information about people's genetic makeup. Along with these advances, some of the potential negative consequences of genomic knowledge have become apparent. At the forefront of these is the concern that insurance companies, employers, or other societal institutions will discriminate against those individuals who are known or believed to be at risk for a genetic disease. *Genetic discrimination*, as it has been named by Lawrence Gostin [Gostin, 1991], refers to such situations in which individuals who are not ill are deprived of certain rights or benefits based on genetic information indicating that they are at risk of becoming ill. Someone who is genetically discriminated against may be perfectly healthy but may have an increased likelihood over the average of developing a disabling and/or cost-incurring condition.

But, the history of concern over genetic discrimination does not begin with the Human Genome Project. Before DNA sequence analysis was available to provide direct information on alterations of the genetic material, a variety of other kinds of tests were developed to detect disease-related mutations. These included examination of cells obtained from subjects for the absence of a metabolic enzyme (e.g., galactosemia) or detection of altered forms of a protein (e.g., sickle cell trait and disease). Early in the history of in sickle cell testing, people lost jobs and insurance if they tested positive as carriers for the trait (carriers, termed *heterozygotes*, have only one copy of a sickle cell gene and, therefore, do not suffer from sickle cell anemia, whereas individuals with two copies of the gene, *homozygotes*, get the disease) [Duster, 2003]. If we look at this history retrospectively, African Americans were the canaries in the mine, exhibiting at an early point in genetic research the potential dangers of genetic information. In these cases, the discrimination took place because of mistaken beliefs that sickle cell carriers would become ill despite the fact that it is only those who carry both copies of the sickle cell mutation (homozygotes) that suffer the disease. In addition to medical or biochemical tests, family history often gives important genetic (as well as environmental) information on the possibility of inherited health conditions. For instance, when one parent in a family has developed Huntington disease, each of the children in that family has a 50–50 chance of also suffering from the ultimately disabling and fatal condition.

The economic reasons for genetic discrimination are obvious. Most profit-making institutions seek to anticipate any added costs and take preemptive action to avoid those costs. Health insurance companies that can predict, based on family history, an individual's chances of developing the costly Huntington disease may refuse an application for insurance or offer insurance at a much higher premium to that individual.

Predictive tests or information are a long-standing tool for underwriters, and such uses of genetic information represent only an extension of current insurance underwriting practices [Pokorski, 1992]. An employer, given the same genetic information, may refuse to hire an individual at risk for developing a genetic disorder either because of fears of declining job performance or because of the health costs to the company. In the past, this use of genetic test information has been a minor problem, or at least not in the public sphere, and only revealed in rare instances such as sickle cell. Genetic tests were associated in the public's mind with single-gene conditions such as Tay-Sachs disease (in which an individual born with two copies of the gene usually dies by the age of about 4) and sickle cell anemia, conditions that are relatively rare in the general population or that affect people in a particular ethnic group.

Within the past 15 years, genetic tests have been developed for a host of more common single-gene diseases such as cystic fibrosis, hemochromatosis, and Duchenne muscular dystrophy. In addition, progress has been made in identifying the genes associated with, and in developing tests for, a small number of multifactorial diseases such as Type I diabetes and certain types of cancers. In multifactorial diseases, more than one gene may be involved in creating a risk of getting the disease as may environmental factors. While a surprising genetic complexity has been revealed in the study of many of the more common diseases, such as mental illness, there is still a continual accumulation of an ever-increasing number of tests. For example, GeneTests provides a searchable directory for laboratories that provide genetic tests for a number of diseases (www.genetests.org).

There are now available hundreds of genetic tests for both rare and more common conditions that have a genetic component. The awareness of the potential problems that might be generated by this amassing of genetic information was one of the contributing factors in the establishment of the Human Genome Project's Working Group on Ethical, Legal and Social Implications (ELSI) [Beckwith, 1991]. The ELSI group was established to anticipate any negative social consequences of the Human Genome Project and to propose means of averting those problems. One of its first accomplishments was a report on the impact of the project on health insurance [NIH/DOE, 1993]. In part, the worries expressed in this report were based on an evaluation of insurance industry policy as expressed by its spokespeople. For example, in 1992, Jude Payne, Health Insurance Association of America (HIAA) senior policy analyst and a participant in deliberations of the ELSI Task Force on Genetic Information and Insurance, stated that HIAA would "actively oppose" any attempt to restrict the use of genetic information in underwriting [Cox, 1992]. At the same time, HIAA, as an organization, endorsed the use of genetic test results by insurers [Cox, 1992]. A CEO task force representing both the life and health insurance industry stated that "[t]he industry considers genetic information to be as potentially relevant to risk classification as any other medical information" [Brostoff, 1992]. Headlines in the insurance industry's newspaper, *National Underwriter*, reflected this policy: "CEOs: Defend Genetic

Test Use in Underwriting" and "Genetic Tests Become Next Underwriting Frontier" [Brostoff, 1992; Cox, 1992].

The major worry expressed by the ELSI report and many other commentators was that the imminent explosion in genetic tests, catalyzed by the Human Genome Project, would magnify the problem of genetic discrimination. Genome scientists predicted that vast amounts of information about each individual's genetic susceptibilities would become available, and one even suggested that people would be able to carry around a compact disc with all of their personal genetic information on it [Gilbert, 1992]. If that information were to be used by insurance companies in underwriting, it could mean that people would ultimately be paying for insurance according to their set of health problem risks. The people with the fewest genetically predictable health problems would pay the least and those with the most would either not be able to obtain insurance or would pay much higher premiums. With increased genetic information available about everyone, the public conception of health insurance as representing a means of sharing risks would vanish and the consequences of underwriting using predictive information could reach an absurd end point in which only those with minimal health risks would have reasonable access to health insurance.

These worries were amplified by studies that reported cases of genetic discrimination [Billings et al., 1992; Geller et al., 1996; Lapham et al., 1996]. Several surveys were carried out soliciting examples from families with a history of genetic conditions to determine whether, in advance of the mass of genetic information that would eventually be available, hidden genetic discrimination existed. Numerous instances, although anecdotal, provided the not surprising conclusion that genetic information had been used by various societal institutions to deprive individuals of employment, insurance, appropriate education, and the like. The purpose of these studies was to identify situations in which information obtained from genetic tests predictive of disease or from family history was used to discriminate against people before those people showed any of the symptoms of that disease. The largest number of cases were those in which a genetic test (not DNA based) or family history was used in health insurance decisions by insurance agents or their companies. However, people also reported discrimination by employers, educational institutions, the military, and adoption agencies. Because it is essentially impossible to obtain unbiased samples from the relevant populations in such studies, no meaningful statistical information could be obtained from the data. For example, it was not possible to obtain any information quantifying the prevalence of genetic discrimination. Despite the limitations of the studies, it became clear that there were many instances in which genetic information was used to limit people's life options. With the publication of these studies, the media took note and presented many more anecdotal accounts [Hilts, 1993; Saltus, 1994; Schieszer, 2001; Stolberg, 1994]. To choose one of the most egregious examples, the younger brother of a boy with a genetic disease resulting in mental retardation was not promoted in grade school because the teacher believed that his sloppy penmanship was a sign of the disease from which his suffered. In this case, a genetic test used to diagnose a disease in one individual was assumed to provide information about the disease status of his sibling [Geller et al., 1996].

HAS GENETIC DISCRIMINATION INCREASED SINCE INCEPTION OF HUMAN GENOME PROJECT?

In the early years of the Human Genome Project the insurance companies held stead-fast to their right to use genetic information. Yet, at that time, and in the years that followed, there were no indications that the companies were rushing to utilize in-formation obtained from the newly developed DNA genetic tests in underwriting insurance. It is important to note that insurance companies have not relinquished their practice of assigning risk using family medical history, which can reflect ge-netic information, nor is an individual generally permitted to keep secret medi-cal/genetic information if the information is requested by an insurance company. However, in certain states insurance companies are not permitted to request such information.

One study to examine the current practices of insurance companies reached the conclusion that little genetic discrimination was occurring, as the authors found that state insurance commissioners received practically no complaints of such practices, at least in the case of life insurance [McEwen et al., 1992]. The findings of this study may have indicated that insurance companies, despite their insistence that new forms of genetic information should be available to them, were not, in fact, using such information. Alternatively, there may have been individuals who did run into problems with insurance agents but who may not have been aware of the option of using insurance commissioners as an avenue of complaint or did not think it was worth the effort.

Nevertheless, there is no indication that the new genetic tests are leading to a dramatic expansion in cases of genetic discrimination. Admittedly, the absence of evidence is not an argument that the practice does not take place. However, given the amount of publicity that the issue has received, one might have expected that if there had been significant instances of genetic discrimination in recent years, some of these would have surfaced and received attention. Both the media and at least certain groups within the medical community have become sensitive to the issue. Either court cases or media investigations would have attracted publicity.

What factors might explain the apparent absence of this use of DNA-based genetic information? One explanation is that the efforts of the ELSI component of the Human Genome Project have had a restraining effect on the use of genetic information to discriminate. The Human Genome Project was novel among "big science" projects in that it attempted to address potential societal issues that might arise as a result of the project and of other new findings in human genetics. The studies on genetic discrimination supported by ELSI and the ELSI Working Group's own report on health insurance influenced public discussions of the issue of genetic discrimination. The ELSI efforts resulted in extensive publicity for the issue, and newspaper reporters followed up with heart-wrenching tales of individuals unable to obtain insurance for genetic reasons. Public figures worried about the issue. For example, as recently as 2004, Hillary Clinton argued in the *New York Times* Sunday Magazine that "[a]s genetic information allows us to predict illness with greater certainty, it threatens to turn the most susceptible patients into the most vulnerable" [Clinton, 2004]. The media coverage may have restrained insurance companies from pursuing their claimed rights to information for fear they would suffer bad publicity.

A second reason for the apparently low levels of reported genetic discrimination is the significant and continuing media coverage of this issue, along with the ELSI report, which stimulated legislatures in many states to pass laws against genetic discrimination [Hall and Rich, 2000a,b; Yesley, 1997]. While the effectiveness of these laws is questionable, it may be that the environment was such that it became more difficult for employers or insurance companies, in particular, to use test information without repercussions [Hall and Rich, 2000a]. State laws largely prohibit the use of genetic information by providers of health insurance or by employers. In addition, various state and federal laws such as the Health Insurance Portability and Privacy Act help preserve the privacy of health-related information. Finally, the ELSI Working Group was successful in convincing the Equal Employment Opportunity Commission to propose rules for the Americans with Disabilities Act of 1990 that disallowed use of genetic information in hiring practices.

Third, it should be noted that there have been few cases in which any of the antigenetic discrimination laws have been applied to cases in litigation. (This conclusion is based on a search of legal cases in Westlaw.) This could mean either that few instances of discrimination have occurred or that there are such cases but they never reach the courts. One reason the latter may be the situation is that the nature of genetic information is such that people want to keep it secret. It is unlikely that genetic information would be kept private in a court case, and, in fact, the information could become a matter of public record. These facts may be sufficient to deter individuals from using legal means for rectifying an illegal disclosure of their private genetic information. In addition, since threatened legal suits are rarely tried in the courts and are generally settled out of court, the latter is a more likely course of action in the case of suits over genetic information. Also, merely making someone aware that he or she is violating a law prohibiting genetic discrimination may be sufficient to halt the offending disclosures. Thus, the laws prohibiting revelation of genetic information may be effective, not as measured by court cases but by acting as laws should: by preventing the offending behavior in the first place.

Fourth, a reason that no flood of cases reporting genetic discrimination has appeared may have been the lack of availability of genetic information. In the present climate of managed care, there is little incentive for people to have tests for disorders for which they have no symptoms or indications. Furthermore, even if tests were available and insurance companies were permitted to require such tests, they may not be cost effective. Genetic tests are generally costly. Thus, testing for even a few markers for disorders having a genetic component would add significantly to the costs of enrolling an individual in an insurance policy.

In addition, at least certain groups were already aware of the possibility that genetic information, or at least information perceived as being genetic, could be harmful; notably African Americans and Jews. African Americans, who, in addition to the discrimination they have suffered in U.S. society, have had at least historical experience of so-called genetic information being used to justify discrimination against them. The policies of the Nazis in Germany that justified their attempts to exterminate groups such as the Jews and the Romany people because of perceived genetic undesirability also may have predisposed at least certain groups to be more wary of genetic information. As a result of public awareness of the possible negative use of genetic information, people may make a greater effort to hide such information or to avoid seeking genetic information for themselves to avoid any possible negative consequences of revealing that information.

MORE COMPLICATIONS: PREDICTIVE COMPLEXITY AND GENETIC CONDITIONS

When the Human Genome Project began, scientists expressed optimism that the availability of the genome map and sequence would allow the identification of genes involved in the overwhelming majority of genetic conditions. This optimism engendered the image of the predictive "compact disc" that would encode each person's chances of suffering from a host of ailments. This "disc" now appears to be something either unachievable or certainly far off in the future. This less optimistic view of genetic predictability arises from a host of studies that reveal a much greater complexity to genetic diseases than previously imagined [Alper, 2002; Altmuller et al., 2001]. This complexity, in turn, influences our discussions of the use of genetic information by doctors, health insurance providers, and other societal institutions.

Complexity in the interpretation of genetic information occurs at two levels. First, if we look back at the history of successes in identifying genes involved in human diseases, these successes were, and continue to be, largely those in which the diseases were due to mutations in single genes that exhibited classical recessive (e.g., sickle cell anemia) or dominant (Huntington disease) properties. While it has long been understood that there is variation in expression or penetrance of single-gene disease-related mutations, the degree to which that is the case revealed by the new genetic tools available had not been imagined [Dipple and McCabe, 2000]]. For example, before the genome sequence allowed the identification of the mutations that can confer Gaucher's disease, it was thought that people carrying the mutation would likely always suffer from the disease. However, once it was possible to map and sequence the gene, it became clear that a significant proportion of the individuals with the Gaucher's disease mutation were asymptomatic [Sidransky and Ginns, 1993]. Even for diseases with high penetrance, such as cystic fibrosis [Davies, 2004; Desgeorges et al., 1994; Donat et al., 1997; Meschede et al., 1993; Parad, 1996] or Huntington disease [Rubinsztein et al., 1996; Szekely, 2002], the manifestations can vary from no symptoms at all in a very small proportion of individuals to only some of the symptoms to variation in age of onset. These "discrepancies" might be explained by modifier mutations in these individuals' genomes, environmental factors, stochastic events, epigenetic effects, or combinations of these factors. Studies have begun to try to sort out between these [Davies, 2004; US-Venezuela, 2004].

What this information tells us is that the predictive value of genetic tests for single gene disorders may be significantly less than has traditionally been assumed. While it might have seemed that the increasing availability of genetic tests would enhance the ability of insurance companies to perform accurate underwriting for each individual, this aspect of complexity makes this expectation significantly less likely. The predictive accuracy of genetic tests may not be very different from that of other predictive medical tests (e.g., for cholesterol levels).

The second revelation of complexity in genetic studies on human health comes from the attempts to identify genes involved in a host of other diseases and conditions. The success of the single-gene/single-disease model in explaining the diseases mentioned above, along with the heritability analyses for various conditions, provided a rationale for seeking single-gene causes for many more common genetic diseases (cancers, diabetes, mental illnesses, etc.). While this approach may well have been the best first line of attack in trying to understand the genetic contributors to these conditions, the failure of these searches has been a major reason for the increased focus on genetic complexity. Of course, there have been some successes in identifying gene mutations that cause cancers. The *BRCA1* and *BRCA2* gene mutations may contribute to perhaps 5 to 10 percent of breast cancer occurrences [Lichtenstein et al., 2000]. However, inherited genes have not been correlated with the remaining 90 percent or so, and it is not clear whether these cancers are largely due to environment or heredity or a combination. Even among the *BRCA* mutation carriers, as with the single-gene disorders discussed above, there is significant variation in expression and penetrance [Burke and Austin, 2002].

The failure to identify single genes contributing to common diseases is perhaps most striking in the case of mental illnesses. While initial reports suggested that such genes have been found, these have all been discounted [Beckwith in Parens et al., 2005]. A very large number of studies of the genetics of both schizophrenia and bipolar manic depression have not led to the discovery of a single gene that the genetics community accepts as conclusively established. This has led researchers in the field to propose that these diseases are polygenic—due to the combined effect of a number of gene mutations, each of which by itself has small effects. Identifying such mutations becomes a considerably more difficult task than that involved in seeking mutations in single-gene diseases. While a number of creative approaches have been proposed for mapping multiple genes for such diseases and the density of chromosomal markers available to researchers has increased [Hoh and Ott, 2003; Holtzman and Marteau, 2000; Merikangas and Risch, 2003; Patterson et al., 2004; Smith et al., 2004], there are no successes to report yet. It should be pointed out that a major factor making this search so difficult may be the failure to take into account the role of environment in reconsidering these mapping efforts [Beckwith in Parens et al., 2005]. Nevertheless, while it is likely now to be a long-term prospect, these new approaches may eventually yield at least some of the genes correlated with the common human diseases.

The implications of this complexity for genetic discrimination are unclear. On the one hand, the concept of the compact disc appearing in the near future with complete information on future health prospects may be much more limited than at first thought. On the other hand, *incomplete* information where only a subset of genes associated with a disease are identified may invite increased unfairness; there could be an overemphasis on the role of the *identified* genes and thereby the potential for unfair discrimination.

CURRENT LEGAL STATUS

Although there are no studies to date exploring the reasons for the apparently small amounts of genetic discrimination, it seems likely that it is a combination of all of the explanations discussed above ranging from underreporting of genetic discrimination to laws effectively deterring the practice. It is important that society continue to be vigilant about unfair practices that may arise due to the improper use of genetic information. The laws that act to prohibit at least some types of genetic discrimination are not yet tested in court. It is possible that there will be a need for additional legislation to shore up such laws if they are found insufficient to protect people or if they are not deemed properly drafted and to provide consistent law between states. Indeed, in 2002, Tom Daschle, the Senate majority leader at the time said that "[t]he current patchwork of state laws concerning genetic information is confusing, inconsistent and inadequate" [Szekely, 2002]. It is unclear how well laws prohibiting genetic discrimination will hold up in court. The Americans with Disabilities Act (ADA) is a federal law that has been interpreted by the Equal Employment Opportunity Commission (EEOC) to preclude certain types of genetic discrimination in the workplace. In what may be the only example of a genetic discrimination case filed by the EEOC, the EEOC determined that Burlington Northern Santa Fe Railway Co. had conducted genetic tests or sought to test some employees who reported having job-related carpal tunnel syndrome. The testing was performed without the employees' knowledge. The utility of the genetic test for identifying a predisposition to carpel tunnel syndrome was also questionable. Before going to trial, the case settled for $2.2 million in favor of the workers. In spite of this apparent success under the law designed to prevent the improper use of genetic information, there are reasons to be concerned about the effectiveness of the legal options. The ADA itself has been eroded by a number of Supreme Court decisions [e.g., *Chevron U.S.A., Inc. v Echazabal* (226 F.3d 1063), *U.S. Airways, Inc. v. Barnett* (228 F.3d 1105), and *Barnes v. Gorman* (257 F.3d 738)].

The Health Insurance Portability and Accountability Act (HIPAA) is also a federally based source of protection against certain types of genetic discrimination that has not yet been tested in the courts. Although this legislation is not specifically directed to protection of genetic information, its potentially strong protection of all medical information provides a possible model or base upon which to build protection of medical information, including genetic information, from unfair uses.

The existence of all of these laws raises the fundamental question of whether there is justifiable reason to provide special protections for genetic information rather than legislating protection of the privacy of medical information, in general. One scientific argument against making genetics special is that there is currently no good methodology for determining the relative contribution of genetics and environment to a person's health. When particular health effects are shown to be associated with a gene variant, those effects usually cannot be readily parsed between the contribution of the identified gene and the contributions of other genes and environmental factors. What does it mean to say that a condition is genetics in these circumstances?

Further, singling out genetic information for special protection in contrast to other kinds of medical information may unfairly favor those who have the fortune to have had their particular genetic condition well studied and thus identifiable. This can result in an undue bias until such time as all genes and their relative contributions to disease are identified. Even if all of the genetic and environmental factors that play a role in determining the course and severity of a disease are identified, there is no rational reason for protecting genetic information and not such information as nongenetic medical test results or environmental contributors to health. For example, an ironic situation exists under the current laws where an individual who has had a genetic test revealing a high susceptibility to heart disease would be protected against discrimination, while the individual whose cholesterol test reveals a similar high susceptibility would not have the discriminatory use of their test information protected.

The current tendency to provide special protection for genetic information may be due to a number of factors. First, there are deeply rooted cultural impulses reflecting a fear that genetic information is much more intimately connected with a person's identity than other kinds of information related to a person's health status. The public misconception that genetics indicates fixity or unchangeability may create greater fear of a genetic diagnosis. The knowledge that a condition "runs in the family" can bring shame, stigmatization, and be disruptive to family life. While these consequences are not necessarily restricted to genetic conditions, the finding that a condition is inherited appears to have had the deeper social and psychological impact. In addition, unlike at least some environmental factors, genetic contributions to disease are not within an individual's control and so there may be a public perception that it is not fair to penalize an individual for a genetic predisposition.

THE FUTURE

There are no current indications of widespread genetic discrimination. Yet, the principles underlying insurance company underwriting make it likely that the availability of a rapidly growing number of genetic tests for a wide variety of diseases will eventually exacerbate the already existing crisis in the health insurance system. Genetic tests are becoming available not only for the relatively rare single-gene diseases such as cystic fibrosis and Huntington disease but also for multifactorial diseases such as some cancers and some cases of heart disease—conditions that affect millions of people. Although, in most cases, these diseases are not caused by a single gene, insurers might make underwriting decisions on the basis of the results of tests that identify genes increasing the likelihood that a person will develop these diseases. Other types of genetic tests, based on finer and finer genetic distinctions, may allow clinicians to determine which individuals suffering from the same disease will benefit most from particular drugs that might vary greatly in their cost.

Assuming no change in the health care system, if these types of information appear on medical records and are available to health insurance companies, an increasing number of people may lose insurance or not be able to afford the higher rates charged because of their genetic susceptibilities. Ultimately, it is conceivable that the genetic profile could be so well defined that the susceptibility of each individual to almost every disease with a genetic component could be determined. This plethora of information would not only severely impact those individuals found to carry susceptibility genes but would also create a nightmare for insurance companies trying to establish rational underwriting criteria.

An oft-presented reason that the health insurance industry should insist on its ability to obtain genetic test information is potential inequities in access to genetic information. Those people who discover that they carry an altered susceptibility gene may decide not to report this information to their insurance company and buy large health insurance policies to cover the high cost of treating these diseases. Similarly, they may purchase large amounts of life insurance to guarantee financial support for their heirs. The insurance industry fears that, as a result of this practice known as adverse selection, high-risk individuals would buy a disproportionate amount of insurance [Pokorski, 1992]. As insurance industry executives put it, "consumers could use genetic testing to foresee coverage needs and exploit the insurance system" [Brostoff, 1992]. This fear, a potentially real one, is one of the major reasons that insurance companies have still argued for their right of access to genetic test results.

The public concern over genetic discrimination may also be having a negative effect on people's health care. People who would benefit from a genetic test that detects the presence of an altered gene for a disease long before symptoms associated with that disease appear may choose not to submit to a genetic test for fear of losing their insurance. It seems evident that, for an increasing number of diseases, the identification of a specific genetic marker will provide information leading to intervention or treatment. Despite the potential medical benefits of such genetic tests, many genetic counselors, clinical geneticists, and other health care professionals regard it as their obligation to warn their patients who are considering genetic testing that they may be at risk for discrimination [Kolata, 1997]. This issue is a source of great concern to both health care professionals and their patients. There is already some evidence that fewer people than expected are availing themselves of genetic tests [Kolata, 1998; Saltus, 1994; Winik, 1998]. Unfortunately, this type of behavior may prevent the individual from receiving early treatment that might make the consequences of their disorder less onerous. In addition, individuals who do not seek early treatment or prevention may cost society more if their illness, when it becomes fully manifest, is more expensive to treat than if preventive measures had been taken.

We suggest that altogether the potential problems raised by the extraordinary advances in contemporary human genetics require a reconsideration of the best way to protect genetic information. In view of these serious problems, it is not surprising that leaders of the Human Genome Project, public policy analysts, and the general public are all calling for regulation of the use of genetic information [Annas, 1993; Hudson et al., 1995; Karjala, 1992]. The potential for improved health care resulting from advances in human genetics has always been fairly obvious. However, it is now equally obvious that these advances may have deleterious effects on people's health as a result of genetic discrimination or the fear of it. To summarize the problem in its starkest form, presently healthy people with positive genetic tests may not be able to obtain or afford insurance, and people whose health care might benefit from genetic test information may be too fearful of genetic discrimination to be tested at all.

Current advances in genetics make it apparent that there is great complexity both in the number of genes contributing to a disease and the interaction of genes with environmental factors. This understanding means that evaluating risks based on genetic information will be highly complex and peculiar to each individual and will involve more than just the genetic information itself. We have raised the fundamental question of whether there is justifiable reason for providing special protections for genetic information. This is not to suggest that genetic information is not deserving of protection, but rather whether it is fair, from both a scientific and social standpoint, to separate the general privacy of an individual's medical information from his or her genetic information. Even if all of the genetic and environmental factors that play a role in determining the course and severity of a disease are identified, there is no rational reason for protecting the genetic information and not environmental or other medically relevant information. There are many social, economic, and philosophical reasons for proposing that the U.S. health insurance system be changed to one in which everyone can afford health care free of discrimination for any reason. The potential catastrophic impact on this system of the accumulation of genetic information is just one more argument that the current system has unfairness built into it.

Finally, it is worth investigating why the anticipated epidemic of genetic discrimination that was to come from all this new genetic information never materialized. If our evaluation is correct, there have been very few known instances of the use of the new types of genetic information to deprive individuals or families of health insurance or employment. We suggest that the efforts of the ELSI component of the Human Genome Project may have played an important role in preventing such an epidemic. The ELSI Working Group's report on health insurance, ELSI's funding of research into genetic discrimination, its support of efforts to establish privacy policies, and its influence on the EEOC's guidelines pursuant to the Americans with Disabilities Act helped generate public understanding and public debate about the issue. The consequences were increasing media attention to genetic discrimination and the passage of numerous state laws. If this interpretation is correct, it provides strong support for the concept of ELSI-like efforts as an integral part of many scientific endeavors. That is, the principle that anticipation of the potential social and ethical issues arising out of a scientific project should accompany that project's inception may well find strong justification in this and other efforts of the ELSI program.

INSURANCE AND GENETIC DISCRIMINATION

Mark A. Hall, Wake Forest University Medical School, Department of Public Health Sciences
Winston-Salem, North Carolina

INTRODUCTION

Early in the development of medical genetics, strong concerns emerged about the possible adverse uses of predictive genetic information by insurance companies. To address this concern, most states in the United States have adopted legislation limiting health insurers' use of genetic information. This legislation is premised on the belief by the public and the medical genetics community that the threat of discrimination by health insurers is real [Hudson et al., 1995]. This belief also accounts for the universal practice in genetic studies of including insurance discrimination as one of the potential risks to be disclosed in the informed consent process. Despite this seeming consensus over the extent and nature of the problem, however, there have been a few voices of scepticism. Some observers have questioned whether the risk of discrimination by insurers is as great as it is perceived to be, and they have suggested that the campaign to highlight the risk has done disservice to the cause of advancing beneficial uses of genetic technologies [Reilly, 1999; Diver and Cohen, 2001].

To explore this issue, it is first necessary to have a clear understanding of what we mean by *discrimination*. This is a multifaceted term whose connotations range from positive to neutral to pejorative, depending on the social context. For instance, from one point of view, the purpose of insurance is to discriminate, in the sense of accurately identifying, classifying, and pricing different insurance risks. For many types of insurance, it is considered unfair *not* to discriminate [Iuculano, 1987]. From a social perspective, however, accurate differentiation of insurance risk may be unfair if socially disapproved criteria are used or if the effect is destructive of social goals. The best example is the use of racial criteria in insurance underwriting. Even though racial classifications may be actuarially accurate, they are prohibited because of the social stigma that results.

In the genetics context, there appears to be a general consensus that at least some forms of insurance discrimination based on genetic information are potentially unfair. It is considered inappropriate for health insurers to use purely predictive genetic information to deny or limit coverage. There is much greater disagreement, however, over whether predictive genetic information should be used by life insurers or others. Also, the distinction between purely predictive information versus information that helps to diagnosis a current disease is not entirely free from ambiguity. Susceptibility to disease can itself be defined as a disease state for purposes of preventive medical treatment. Therefore, there are several difficult and unresolved aspects of this issue, which this section explores.

HEALTH INSURANCE

According to the best evidence, genetic discrimination by health insurers is rare or nonexistent in the United States. The leading studies that document some discrimination solicited case reports of discrimination from over 1000 genetics professionals, 27,700 affected or at-risk individuals, and most of the major genetic disease support groups [Billings et al., 1992; Geller et al., 1996]. Based on these self-reports, several dozen instances of insurance discrimination were identified, but often the exact types of insurance involved were not specified. Also, it is difficult to know whether the type of adverse insurance consequence experience meets our concept of inappropriate discrimination because case reports are sketchy and often the full account cannot be obtained without extensive investigation of the facts of each case. Insurers often make their decisions for a mixture of reasons, and people rarely are able to determine that a particular factor was determinative in an adverse decision. Also, people who encounter an initial difficulty in seeking insurance are often able to work through the problem by questioning the decision, providing more information, or seeking insurance elsewhere with the help of independent insurance agents. From published accounts, this appears to have happened in a number of these reported cases.

The other study that has had a large impact on perceptions of discrimination was conducted by Lapham et al. [1996], using a similar methodology. Working with genetic support groups, 332 survey responses were solicited from a population of roughly 500,000 people. Twenty-two of these subjects believed that they or a family member had been refused health insurance as a result of a genetic condition in the family. However, these researchers did not seek to limit their focus to asymptomatic genotypes. Instead, they included without differentiation people whose insurance denial was based on the health effects of actual genetic disease.

Other studies, which avoid some of the limitations in the primary studies, conclude that genetic discrimination by health insurers is not widespread. A 1994–1995 survey of 500 primary care physicians found "only a few instances of refusals of employment or life or health insurance" based on presymptomatic genetic information [Wertz, 1997]. A survey of 39 fragile X families in Colorado found that none had their health insurance canceled or their premiums increased after the genetic diagnosis. Although 6 of 23 families (26 percent) reported being turned down when they applied for insurance, they each had an affected individual and so did not present an asymptomatic situation [Wingrove et al., 1996].

Evaluating these studies, the researcher [Reilly, 1999] concluded that "despite a design that strongly favoured the discovery of cases of genetic discrimination, [this research] is remarkable for how little evidence of such practices it discovered." Other efforts to detect genetic discrimination by health insurers have also turned up very little evidence. In the leading study, directed by this author, in-depth interviews were conducted in 7 states with 77 people in the insurance industry (agents, actuaries, underwriters, and regulators). Also, 148 health insurance application forms were collected and analyzed, and 142 insurance agents were contacted with inquiries about purchasing insurance using a scripted scenario that presented a fictitious case of a positive test for "the breast cancer gene." The conclusion from this extensive effort was that health insurers do not inquire about predictive genetic test results, do not have a practice of using this information if they come across it in the medical record, and do not include this type of information in their underwriting guidelines.

The only indication from this study that health insurers may engage in genetic discrimination is that, when presented with a hypothetical about the "breast cancer gene," many insurance representatives thought this might make a small difference in insurance rates, or cause the insurer to exclude that particular condition from coverage (but not to deny coverage altogether). Similarly, 13 percent of agents who were presented this scenario in the scripted inquiry by a purported insurance purchaser indicated this might create some difficulty in obtaining coverage. However, the great majority of insurance agents indicated that having the breast cancer gene would not preclude coverage, which suggests that, at most, people with adverse genetic test results may have to search somewhat more than others before finding a willing insurer, or might have to pay somewhat more for coverage, if they do not obtain coverage as part of an employer group.

Health insurers' lack of interest in purely predictive (presymptomatic) genetic information is based on the fundamental economics of health insurance, which are driven by immediate and short-range projections of medical costs not by long-range forecasts. Health insurance is issued on an annual basis, and the premium is recalculated each year. Under the convention of *open enrolment*, people frequently change their insurance coverage, and changes occur when people move or change jobs. On average, the typical health insurance subscriber stays with the same company only about 3 years. Therefore, it is not cost effective to seek information about what might happen 5, 10, or 15 years down the road, even if the information were highly predictive (which genetic information often is not).

Furthermore, health insurance is not affected by the same degree of adverse selection as is other types of insurance. *Adverse selection* occurs when people are motivated to purchase insurance by private information they have about a large expected expenditure. If undetected by insurers, adverse selection can reduce or undermine the marketability of insurance by forcing the price to levels that lower-risk people are unwilling to pay, or eventually to levels that even high-risk people are unable to pay. Where the potential for adverse selection is serious, insurers might require applicants to undergo testing in order to exclude those at very high risk. This is done, for instance, by testing life insurance applicants for the human immunodeficiency virus (HIV). However, health insurance is not affected by adverse selection of this magnitude. Most people already either have or do not have health insurance by virtue of their jobs, so few people have reason to add or increase coverage based on a single test result. And, unlike life insurance, there is only so much health insurance one can purchase.

The low potential for genetic discrimination is confirmed by examining the structure of the health insurance market in the United States. and the industry's medical underwriting practices. About a quarter of the population is covered by government insurance programs such as Medicare and Medicaid, which take all eligible people on the same terms, without regard to health risk. Among the privately insured population, about three-quarters are covered by employers as part of large groups, which also make no inquiry about health conditions. Everyone in the group is covered on equal terms, and the employer's rates are based on the group's recent medical expenditures.

The only segments of the U.S. population potentially exposed to genetic discrimination are those who purchase individual health coverage outside the workplace, or those who obtain health insurance as part of small employer groups, with fewer than about 20 workers. This is the portion of the market where there is some effort by health insurers to gauge future medical costs based on an assessment of individual health risks. For this segment, a 1991 survey of health insurers found that, among various groupings of insurers, 36 to 64 percent would decline or limit coverage if presymptomatic genetic testing revealed "the likelihood of a serious chronic future disease" [USA Congress, 1992]. However, this assumes that insurers learn about predictive genetic test results. Over half the states now prohibit some or all health insurers from inquiring about, or using, predictive genetic tests results. Even where protective laws do not exist, insurers usually do not ask about future health indicators.

For group sizes in the range of about 10 to 20, health insurers ask only very limited information about major current or recent health conditions. When insurers scrutinize each individual, they almost invariably do so with regard to currently existing conditions that may require treatment in the immediate future. This information is usually collected from discrete questions on the application form, which insurance agents assist applicants in responding to. Because of the expense involved, health insurers rarely obtain complete, or even partial, medical records.

Finally, reports are overstated that the insurance industry maintains a giant repository of personal medical information that can queried when needed. There is a *medical information bureau* (MIB) that life insurers subscribe to in order to verify medical information submitted in application forms. However, this is done primarily by companies that sell life insurance. Even then, the MIB classifies information according to broad diagnostic and treatment categories, not according to the type or source of risk information. Thus, the MIB might record that a cancer screening test was performed, but it would not report that a genetic test was done, nor the results of the test.

On balance (in this author's opinion—Editor's note), there is only a weak basis for serious concern, either now or in the future, that health insurers will engage in widespread genetic discrimination of the form that is considered inappropriate. In a very large country, there are only a handful of reported cases of discrimination, which are not well documented. Regular use of purely predictive genetic information does not make sense for health insurance, considering the fundamental structure and economics of the industry. Were discrimination to occur, it is likely to be sufficiently isolated that a resourceful person could search elsewhere for another willing insurer, perhaps at a somewhat higher price. Some presymptomatic genetic conditions can render people higher insurance risks or uninsurable, but these situations are rare, and, as in the case of Huntington disease, the person is often already considered to be at risk based on family history, before any genetic test is performed. In these situations, genetic testing can provide insurance benefit by determining which family members are free from the mutation.

The greatest risk that health insurers will use genetic information against people is where genetic testing reveals predisposition to previously unsuspected disease that requires immediate treatment, in the form of preventive care or corrective genetic therapy. Then health insurers would likely consider this information relevant for insurance products that are not issued to larger groups. For instance, one study of 128 hemochromatosis patients receiving regular phlebotomies to maintain normal iron stores found that 6 percent had experienced some difficulty related to their condition with obtaining or keeping health insurance [Shaheen et al., 2003]. However, where genetic conditions require immediate treatment, it is debatable whether use of this information meets the accepted definition of unfair discrimination.

LIFE AND OTHER INSURANCE

The risk of genetic discrimination is much greater for life insurance and long-term care (nursing home) insurance. This is especially so for life insurance, where underwriting is all about predicting future, longer-range mortality risk. Unlike health insurance, people do not frequently change life insurers. Instead, they often keep their policies until they die. Moreover, it is possible for people to purchase vast sums of life insurance and to add additional coverage when they learn they have a new mortality risk. And life insurance is most often purchased individually rather than in groups, so it requires risk assessment on an individual basis. For these reasons, life insurers insist that they be told about medical tests that predict future mortality, and they have a strong incentive to require testing for conditions (such as HIV infection) that may prompt people to purchase insurance, especially if the test is relatively inexpensive and accurate. Similar attributes apply to long-term care insurance. This too is usually purchased individually and is often purchased only when there is a perceived need rather than as a matter of course.

Life and other types of insurance also contrast with health insurance, however, in the social consequences of discrimination. In the United States, the ability to purchase private health insurance is a social necessity because, for most people, there is no backstop public program. In contrast, the more limited protections provided by life and disability income insurance also exist in social security programs for which everyone is potentially eligible. Long-term care insurance is a closer social fit with health insurance, but, through Medicaid, these benefits are available even to middle-income people who encounter large, ongoing nursing home expenses.

For these types of insurance, the greater salience of predictive medical information, combined with the lesser social necessity of private coverage, may lead one to question whether it is unfair to use predictive genetic information. Instead, the case appears stronger here for subscribing to the insurance industry's notion of *actuarial fairness*, according to which fairness is defined by actuarial accuracy. People are less inclined to view other types of insurance as an entitlement, and lower-risk people are more likely to take offence at having to pay higher rates in order to make insurance more accessible to people with higher risks. Nevertheless, becoming uninsurable for life, disability, and long-term care insurance is a matter of significant concern to those who contemplate genetic testing.

The extent of actual or potential genetic discrimination by other types of insurers has not been studied as extensively as it has for health insurers. The case report studies described above also include reports of discrimination by life and disability insurers, but it is often difficult to know exactly what type of insurance was involved, and what the actual or purported basis was for discrimination. Therefore, it is difficult to know how to classify or evaluate these reports. Nevertheless, life insurers are now routinely using predictive genetic information in the form of family history [McEwen et al., 1992; Rothstein, 2004]. Also, insurers argue strenuously that they need to be able to consider the results of predictive genetic testing as this technology becomes more commonplace [Pokorski, 1997; Rothstein, 2004].

People with a family history of Huntington disease have difficulty obtaining these types of insurance, and those who test positive for the mutation are uninsurable, even if they are currently well. People with hereditary hemochromatosis have reported some difficulty obtaining life insurance at standard rates even if their elevated iron levels have not yet caused any organ damage or other adverse health consequences

[Alper et al., 1994]. One recent study found this to be the case for 13 percent of 128 patients receiving treatment for the condition [Shaheen et al., 2003]. A survey of 46 life insurers regarding someone at 50 percent risk of having the mutation for hereditary nonpolyposis colorectal cancer found that all but two (96 percent) would issue coverage, but 26 percent of insurers would charge a higher premium, and three would require the applicant to be tested before issuing significant levels of coverage [Rodriguez-Bigas et al., 1998]. For disability insurance, this study found that only one would refuse coverage, but 17 percent would charge a higher premium.

Additional cause for concern exists because the initial history with diagnostic or symptomatic genetic testing indicates that life insurers do not always accurately assess the extent of genetic risk. Early in the diagnosis of genetic disease, some life insurers reportedly confused sickle cell *trait*, which is the carrier status, with sickle cell *disease*. This prompted legislation in a number of states to prohibit discrimination based on the former condition. With regard to hemochromatosis, if the condition has not progressed and is being regularly treated, there appears to be little or no medical basis for classifying someone as a higher mortality risk based on asymptomatic levels of iron overload [Alper et al., 1994]. In short, life insurers do not always live up to their own credo of actuarial fairness.

This is because insurers are limited in their ability to process risk information by the amount and quality of actuarial or medical data that is available. As new areas of potential risk are encountered, it requires a significant investment of resources to standardize actuarial projections or research medical prognosis. If an unfamiliar type of risk is encountered that affects only a small fraction of the population, it may not make good business sense for insurers to do the research and acquire the data necessary to develop refined underwriting policies. The cost-effective response where little business is at stake is sometimes simply to refuse coverage based on the unknown degree of risk. Although these practices may be rational business decisions, they do not always meet a common sense notion of actuarial fairness. Still, they are not necessarily invidious (or prejudiced) forms of discrimination but instead are indications of risk aversion in the face of unfamiliar risks.

As genetic information becomes more available and more thoroughly incorporated into medical practice, it can be expected that competitive pressures in the insurance industry will diminish less accurate calculations of possible risk, especially for more common genotypes. Indeed, some people look to the life insurance industry as a major source for advancing genetic information and testing technology. On balance, one informed industry analyst observes that, with the exception of a few rare conditions such as Huntington disease, "it is becoming increasingly apparent that the extra mortality risk associated with testing positive for predisposition to common diseases will be modest in individuals who are otherwise currently in good health. Most should qualify for coverage at standard if not preferred rates. Happily, this makes good sense both medically and politically" [Nowlan, 2000].

Chapter 8

Prenatal Screening and Diagnosis

OVERVIEW

This chapter discusses genetic indications for prenatal screening and diagnosis and reviews generally accepted testing technologies and guidelines for prenatal screening, prenatal diagnosis, and preimplantation diagnostics. Medical, ethical, legal, and public health perspectives relevant to prenatal screening are reviewed. Expectations of health care providers for patient counseling, provision of testing, and ensuring patient access to testing are discussed.

INTRODUCTION

The goal of adopting a screening process is to identify a selected population of patients who are at increased risk for a specific medical condition, enabling the offer of more diagnostic testing to patients selected from the population by the screening criteria. Screen-positive patients require genetic counseling with regard to their options, which include review and confirmation of screen positivity, diagnostic evaluation, and possible interventions where a diagnosis is confirmed. Screening of pregnancies and newborns for selected genetic and congenital diseases is common practice, and constant advances in the design of screening programs have resulted in broad uptake of expanded screening modalities, improved screening performance, and a reduced burden of morbidity and mortality. In this chapter, reviews are provided of:

- Basic concepts underlying the design and implementation of prenatal screening programs
- Clinical and laboratory criteria of prenatal screening options for neural tube defects and chromosomal trisomies

Genetic Testing: Care, Consent, and Liability, by Neil F. Sharpe and Ronald F. Carter
© 2006 John Wiley & Sons, Inc.

- Ethical and legal considerations in prenatal screening, including preimplantation genetic diagnosis (PGD)
- Obligations of health care providers relating to prenatal and newborn screening

In this discussion, it is useful to keep in mind that prevailing ethical guidelines expect physicians to discuss prenatal testing with "all women who have appropriate indications" [e.g., Society of Obstetricians and Gynecologists of Canada, (SOGC), 1998] and that patients who do not meet established risk thresholds nevertheless can request prenatal diagnosis as long as they are advised of the risk of diagnostic testing (AMA Code of Ethics, discussed further below). These expectations can challenge the ability of both clinicians and health care systems to meet the public's expectations for access to screening and diagnostic services. The following discussions include reviews of the obligations of health care providers and guidelines for counseling relevant to prenatal and newborn screening for genetic diseases.

NEED TO KNOW

Core Concepts

The core concepts relating to the design and implementation of a screening program include:

- Criteria for selection of specified diseases to be screened for.
- Designation of cutoff values (threshold) values for a positive screen result; these values will be determined by the analytical performance of the screening test, the risk of the associated diagnostic procedure, the cost of providing the diagnostic test, and the public's expectations.
- Criteria used to describe the performance of the screening test include: detection rate (DR), false-positive rate (FPR), and the odds of being affected given a positive result (OAPR).
- In prenatal serum screening, results are expressed in relation to median values derived to avoid bias within the reportable range of gestational ages;
- Screen results are reported as likelihoods (risks), which means that most screen-positive pregnancies will result in a normal outcome.

Prenatal Screening

Prenatal screening programs are designed to select for fetuses with an open neural tube defect, trisomy 21, or trisomy 18. Screening options have broadened from the standard *triple screen* of maternal serum α-fetoprotein (AFP), human chorionic gonadotropin (HCG), and unconjugated estriol to include additional serum analytes [pregnancy-associated plasma protein A (PAPP-A), inhibin] and ultrasound assessment of nuchal translucency by specifically defined criteria. Commonly offered screening proto-cols include the triple screen, first-trimester screen, integrated screen, and quadruple screens.

The DR, FPR, and OAPR for each screening protocol vary for each abnormality [neural tube defect (NTD) or trisomy], depending on the analytical criteria included, the timing of the testing, and the operating protocols of the laboratory service. The timing of screening steps extend from first trimester, to second trimester, to integrated first- and second-trimester evaluations. Gestational characteristics (age of gestation, gravidity) and maternal information (weight, ethnicity, diabetes, family history of neural tube defects) represent independent variables that must be accurately reported to the laboratory in order to ensure an accurately estimated risk. It is important to provide the patient with accurate information in regard to the performance character-istics (DR and FPR) and timing of each screening option available to the patient prior to screening.

Patients who receive a positive screen result will experience increased anxiety and require prompt counseling attention. Steps in further evaluation include a repeat assessment for gestational age (preferably by ultrasound dating), review of clinical and gestational variables, and reanalysis of the risk assessment if there is signifi-cant revision of the known variables. Counseling includes review of the options for diagnostic testing should the positive screen result be confirmed after review.

Advances in prenatal screening have rapidly increased the detection rate for affected pregnancies, while simultaneously reducing the number of false-positive results and therefore avoiding the costs, pregnancy losses, and anxieties caused by unnecessary diagnostic evaluations. Reductions in the prevalence of NTD, tri-somy 18, and trisomy 21 have been attributed to the widespread uptake of prenatal screening.

Prenatal Diagnostic Testing

For confirmation of neural tube defects, amniotic fluid acetylcholinesterase (AChE) and AFP concentrations are accepted as the most accurate diagnostic tests (DR greater than 96 percent), but targeted ultrasound examinations can detect a high propor-tion of affected pregnancies (65 percent in unselected patients, 80 to 100 percent in high-risk patients). For chromosome trisomies, cytogenetic analysis of amnio-centesis or chorionic villus specimens provides a conclusive diagnostic test (>99 percent accuracy). The actions of chorionic villus sampling and amniocentesis in-voke risks of approximately 1 and 0.5 percent for causing loss of the pregnancy, respectively.

Expert ultrasound evaluations provide a noninvasive and increasingly accurate method of prenatal diagnosis. Specific signs and measurable criteria have been identi-fied for both screening and diagnostic applications; accuracy of interpretation depends upon operator training and quality of equipment. Standardized training criteria are being developed.

Preimplantation Genetic Diagnosis

Preimplantation genetic diagnosis refers to the application of diagnostic testing to embryos prior to the clinical definition of a pregnancy. The procedures involved are invasive. Cells from embryos are tested before the selected (tested) embryos are implanted into the uterus for continuation of the pregnancy. PGD is costly and has a reported success rate to delivery of livebirth in the range of 20 percent per treatment cycle, with a small proportion of abnormal outcomes. Access to PGD is not universal and the range of genetic diseases that can be diagnostically assessed is limited, but both the success rate and the range of indications appear to be improving rapidly [Kuliev et al., 2005].

In counseling patients about PGD, it is important to accurately convey a realistic assessment of pregnancy success, side effects, the scope and limitations of the prenatal testing to be done, potential risks for abnormal outcomes and long-term sequelae, and the risk of a misdiagnosis.

Access to Prenatal Screening and Diagnosis

In most jurisdictions, patients are offered prenatal screening and diagnosis according to an assessment of risk. Threshold criteria can be applied. It is important that the relevant risk factors are appropriately identified through history taking and medical examination. Criteria for assessing risk include ethnicity, family history, pregnancy history, maternal age, paternal age, and ultrasound and serum screening results.

Obligations of the Health Care Provider in Prenatal Screening and Diagnosis

Clinical guidelines identify a baseline of standard of care for offering screening and diagnostic testing to patients. Clinicians need to be familiar with the prevailing guidelines, identify the options for patients, and discuss the implications of screening and testing to patients. Clinicians are not obligated to provide specific, clinically valid testing services should they lack the relevant skill or knowledge, or if they have a conscientious objection to the procedure. However, in this situation, clinicians may be obligated to inform the patient of the availability of the service elsewhere and (in some jurisdictions) to facilitate referral if desired by the patient. Clinicians have a legal duty to offer tests for which patients might benefit, in accordance with prevailing standards of reasonable care and clinical utility (and see Chapter 6, Referral and Diagnosis).

Societal Issues and Public Policies Relating to Prenatal Screening and Diagnosis

Advances in the capacity of prenatal screening and diagnosis have prompted a variety of ethical debates. Issues include the moral and ethical principles applied in offering these services, the rights of patients to decide on intervention and pregnancy selection based upon test results, and the long-term impacts on society. These issues include:

- Pregnancy terminations and abortion laws
- Use of genetic testing to select for or against a fetal sex for nonmedical reasons
- Testing for susceptibility for a disease or for diagnosis of a late-onset disorder
- Selecting for a disability
- Stigmatization associated with selecting against a disability
- Obligations of clinicians to discuss testing that is illegal or not offered in the area of jurisdiction where care is provided
- Limitations of nondirective prenatal counseling

PRENATAL SCREENING

Andrew R. MacRae, Research Institute at Lakeridge Health, Oshawa, Ontario, Canada and Department of Pathobiology and Laboratory Medicine, Faculty of Medicine, University of Toronto, Ontario, Canada
David Chitayat, Prenatal Diagnosis and Medical Genetics Program, Departments of Medical Genetics and Microbiology, Pediatrics, Obstetrics and Gynecology and Laboratory Medicine and Pathobiology and University of Toronto, Ontario, Canada

SCENARIO

The patient is a 36-year-old G3P2L2 woman of Chinese descent and her husband is 45 years old and of the same descent. The couple was seen at 7 weeks gestation for counseling regarding the patient's age and their prenatal diagnosis options in their current pregnancy.

BASIC CONCEPTS OF PRENATAL SCREENING

A useful definition of screening has been provided by Wald [2001]: *Screening is the systematic application of a test or inquiry, to identify individuals at sufficient risk of a specific disorder, to benefit from further investigation or direct preventive action, among persons who have not sought medical attention because of symptoms of that disorder.*

Thus, the test should be uniformly applied to a population that is without symptoms or indicators for the disorder in order to identify those who are sufficiently at risk of a specifically targeted disorder to justify a diagnostic procedure. The screen also identifies patients who are at *low* risk and thus require no further studies.

In prenatal screening, a *cutoff value* is set to define the pregnant women who are at an increased risk for having a baby with an open neural tube defect (ONTD), Down syndrome (DS), or trisomy 18 (T18) (Fig. 8.1). Those who are at a higher risk

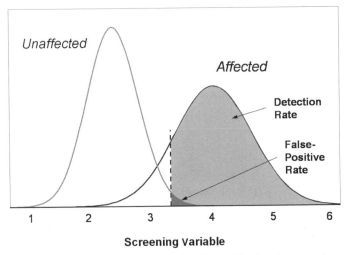

Screening Variable

Figure 8.1   Concept of screening for a specific condition based on a continuous screening variable. A cutoff value, shown by the vertical broken line, defines screen positive (follow-up testing offered) from screen negative (no follow-up testing indicated). Detection is an attribute of the affected population, whereas the false-positive rate is an attribute of the unaffected population.

are called *screen positive*, and those who are not are called *screen negative* for the specific disorder. In selecting a cutoff, consideration should be given to the available financial resources for performing the diagnostic tests, the risk associated with the diagnostic procedure, and the public expectation of the screening service.

A specific prenatal screen is evaluated on the basis of its detection rate (DR) and its false-positive rate (FPR). The DR will determine how many of the patients with the screened condition are screen positive, whereas the FPR will determine the fraction of the unaffected population that is screen positive. Furthermore, in prenatal screening, the positive predictive value of the screen is often expressed as the odds of being affected given a positive result (OAPR). This number shows the prevalence of the target disorder among those who screened positive. Thus, an OAPR of 1 : 100 for ONTD indicates that for every 101 screen-positive patients one will actually have ONTD.

In order to eliminate the need to have normative data for each week of gestation, patient results are expressed as the ratio of the median concentration in unaffected pregnancies at that gestational age. The median value is used instead of the mean value to avoid the influence of substantially elevated results, which would skew the mean in small population samplings. The multiple of the median, or MoM result, is potentially independent of the gestational age, allowing patients of all gestational ages to be assessed together on the basis of their MoM values. In order to achieve this independence of gestational age, the median values must be selected in a manner that avoids any bias throughout the range of reportable gestational ages.

Screening is a process to increase the prevalence of the targeted disorder within the subpopulation that is offered a diagnostic/invasive test. Thus, in circumstances where the a priori risk for having a baby affected with a specific condition is high, prenatal diagnosis screening may not be justified. Thus, if an atrioventricular canal is detected on fetal ultrasound, screening for DS is not justifiable since the risk for having a baby with Down syndrome is already substantially increased.

Following the identification of pregnancies at an increased risk for having fetal open neural tube defects and trisomies 21 and 18, diagnostic tests are offered to determine the fetal karyotype using chorionic villus sampling (CVS) or amniocentesis. The diagnostic test for open neural tube defect is a combination of α-fetoprotein (AFP) level and acetylcholinesterase (AChE) enzyme activity in the amniotic fluid (AF) in combination with second-trimester detailed fetal ultrasound.

Amniocentesis and CVS are associated with a risk of miscarriage usually considered to be 1 : 200 and 1 : 100, respectively. The prenatal screening tests are introduced not only to identify the pregnant women at increased risk but also to lower the number of patients undergoing a diagnostic procedure and, thus, decreasing pregnancy losses.

It is important that patients and health care providers are clear in the understanding that a screening test provides a likelihood of being affected by the disorder and is not a diagnostic test. Most screen-positive patients will not have an affected baby.

Educating the public and counseling patients are part and parcel of every screening program. Only by teaching and counseling the patients/couples will they be able to make informed decisions regarding having the test, interpreting its results, and making a decision regarding diagnostic/invasive tests when needed [Carroll et al., 2000]. Education of the public and health care providers should be a continuous effort to ensure patient and provider knowledge and satisfaction.

Screening for Neural Tube Defects

Neural tube defects (NTDs) are the result of a failure in the process of closure of the neural groove during the third or fourth week of embryonic life. The clinical findings depend on the location of the lesion and the involvement of the neural tissue. Thus, the lesion can be lethal when the anterior part of the tube fails to close, resulting in anencephaly, and can be mild when only the leptomeninges are involved, resulting in meningocele. Abnormalities in closure of the anterior part of the neural tube can result in anencephaly, acrania, exencephaly, encephalocele, and, when the brain tissue is not involved, in meningocele. Abnormalities in closure of the spine can result in myelomeningocele when the spinal cord is involved. It can be open with a thin membranous covering (open spina bifida, OSB) or closed when covered by skin (closed spina bifida). When the spinal cord is not involved, it is called meningocele (which can also be open or closed). In more than 95 percent of spina bifida cases, the defect is open [Milunsky, 1998]. Hydrocephalus is present in most cases with OSB, due to disrupted cerebrospinal fluid (CSF) circulation.

The prevalence of ONTDs varies with geographical region [Greenberg et al., 1983], with an additional socioeconomic factor probably relating to the prepregnancy maternal nutritional status. About 90 percent of the NTDs are multifactorial, and in 97 percent of the cases no family history of this lesion can be identified. Other causes of NTDs are maternal diseases, such as insulin-dependent diabetes mellitus, maternal exposures, such as to valproic acid [Robert and Guibaud, 1982; Guibaud et al., 1993; Lindhout et al., 1992], carbamazepine [Rosa, 1991], and vitamin A [Rappaport and Knapp, 1989].

In some cases, the condition is the result of single-gene disorders such as Walker–Warburg syndrome [Chitayat et al., 1995] and Currarino triad [Belloni et al., 2000], and in 6.5 percent of cases they are associated with chromosome abnormalities [Kennedy et al., 1998].

In North America, the current overall incidence of NTD is between 0.5 and 2.0 per 1000 births. The frequency is lower in the west and increases to the east, with the highest rates occurring in the southeastern United States and in eastern maritime Canada. In Europe, the incidence of NTD is highest in the western United Kingdom, at approximately 2 per 1000 births.

Increased folic acid intake a month before, and during, the first trimester of pregnancy reduces the occurrence and recurrence of ONTD [Milunsky et al., 1989b; Wald et al., 1991c]. In 1996, both the U.S. and Canadian regulatory bodies mandated folic acid fortification of grain products at the level of 0.14 mg folic acid per 100 g. This fortification was associated with a 31 percent decrease in the incidence of OSB and a 16 percent decrease in the incidence of anencephaly, for a 26 percent decrease in ONTD [Williams LJ et al., 2002].

METABOLISM OF AFP AND SCREENING FOR ONTD

Alpha-fetoprotein was first reported as a significant component of the fetal plasma proteins by Bergstrand and Czar in 1956 [Bergstrand and Czar, 1956]. AFP is structurally similar to albumin but has a greater binding of polyunsaturated fatty acids, which may be important in the embryonic cellular development [Benassayag et al., 1999]. However, rare inherited absences of AFP with a normal birth outcome [Greenberg et al., 1992] have been reported.

The AFP is gradually replaced by albumin during the third trimester and in early infancy, and only traces of AFP are present in nonpregnant, healthy adults. However, conditions such as hepatoblastoma and those associated with regeneration of hepatocytes are also associated with an increased serum concentration.

Initially produced in the yolk sac and later in the fetal liver and gut, the AFP reaches the amniotic fluid through the urine and the maternal circulation through the placenta and, by diffusion, through the chorion and amnion. Thus, the AFP level in the fetal plasma and amniotic fluid are maximal at 12 to 13 weeks and 13 to 14 weeks gestation, respectively. The maternal serum concentration of AFP increases through the second trimester and reaches a maximum concentration at 28 to 30 weeks gestation. The fetal plasma to amniotic fluid concentration ratio is 200/1 and the ratio of fetal plasma to maternal plasma is 100,000/1. In ONTD, the AFP is able to pass from fetal plasma and CSF into amniotic fluid, resulting in concentrations 7- to 20-fold higher in amniotic fluid and 4-fold higher in maternal serum.

Brock and Sutcliffe [1972] reported an association in the third trimester between fetal ONTD and high concentrations of α-fetoprotein in the amniotic fluid (AFAFP). Further reports followed and extended this finding into the second trimester, with AFP measurements in both amniotic fluid and maternal serum [Hino et al., 1972; Leek et al., 1973; Brock et al., 1973]. The UK Collaborative Study on Alpha-fetoprotein in Relation to Neural-tube Defects published its findings in 1979 [Second Report, 1979] demonstrating conclusively that AFAFP is elevated during the early second trimester in anencephaly (20-fold) and in OSB (7-fold). By 1981, a second diagnostic test, AChE in amniotic fluid, was reported [Report, 1981], and this combination of tests is currently the "gold standard" method for diagnosis of ONTD.

The first to suggest the use of maternal serum α-fetoprotein (MSAFP) as a screening test for ONTD was Wald et al. [1974], and in 1977, the UK Collaborative Study reported that elevated MSAFP in the early midtrimester of pregnancy could identify 80 percent of OSB and 90 percent of anencephaly cases [Wald et al., 1977] at a false-positive rate of 5 percent. These reports paved the way for the use of MSAFP level as a screening test for fetal ONTD.

Further studies showed that several attributes of the fetomaternal unit influence the concentration of MSAFP:

1. **Gestational Age**   The method of estimating the gestational age in patients is important for the screening performance. Although last menstrual period (LMP) dating is sufficiently accurate to sustain screening, ultrasound biometry improves performance. Ultrasound biometry is considered accurate within 8 percent of the assigned gestational age [Hadlock et al., 1992] or within 9 to 10 days at early midtrimester. The algorithms by Daya [1993] for CRL and the 1982 biparietal diameter (BPD) data of Hadlock [Hadlock et al., 1982] are an example of a pair of biometry algorithms with good concordance that span from 6 to at least 30 weeks. AFP concentrations, particularly in maternal serum, are only informative about the risk or presence of ONTD within a specified period of early midtrimester gestation. The greatest separation from the unaffected population is at the 16th to 18th week of gestation [Wald et al., 1977]. Screening is less effective in week 15 and beyond week 19. AFAFP is best at 16 to 22 weeks gestation.

2. **Racial Origin**   Prenatal screening for fetal ONTD was first developed in largely Caucasian populations. Subsequently, it was found that the black (African American) population has 10 to 22 percent higher concentrations of AFP in both maternal serum and amniotic fluid in pregnant women with and without fetal ONTD [Johnson, 1981; Crandall et al., 1983; Johnson et al., 1990a; Watt et al., 1996].

3. **Maternal Weight**   A relationship between maternal weight and MSAFP was first reported [Haddow et al., 1981] in 1981 and then confirmed in the same year [Wald et al., 1981]. The developing fetus is the source of MSAFP, and the circulating volume of the mother will indirectly affect the MSAFP concentration. Since maternal weight is an easily accessible index of circulating volume, screening programs take the maternal weight into consideration in

calculating the risk for fetal ONTD. Maternal weight has no effect on AFAFP, nor is maternal weight different between OSB and unaffected pregnancies [Johnson et al., 1990b]. There have been a number of reports that an increased risk for neural tube defects is associated with prepregnancy maternal obesity [Waller et al., 1994; Werler et al., 1996; Shaw et al., 1996]. Nondiabetic mothers with a self-reported prepregnancy BMI (body mass index) >30 ("obese") were 3.5 times more likely than average-weight women (BMI 18.5 to 24.9) to have an infant with spina bifida. The relative risks of other disorders were also increased, including omphalocele (3.3-fold), heart defects (2.0-fold), and multiple anomalies (2.0-fold).

4. **Number of Fetuses**   Multiple gestation pregnancies have MSAFP, but not AFAFP, concentrations that are commensurate with the number of developing fetuses [Wald et al., 1979b]. The MSAFP levels in twin pregnancies are, on average, 2.16 times the MSAFP concentration of singleton pregnancies [Canick et al., 1990; Wald et al., 1991b, Neveux et al., 1996]. No adjustment is made to the MSAFP medians, but the MoM cutoff for known twin serum samples is selected at a higher value to maintain approximately the same false-positive rate as in singleton pregnancies.

5. **Insulin-Dependent Diabetes Mellitus**   Pregnant patients with insulin-dependent diabetes mellitus (IDDM) have approximately 20 percent lower concentrations of MSAFP [Wald et al., 1979a; Milunsky et al., 1982]. This factor is taken into consideration when calculating the MSAFP MoM results. No such correction is made in mothers with gestational diabetes, even if the patient is receiving insulin. Although the AFAFP concentrations are also decreased in the presence of diabetes mellitus [Henriques et al., 1993], adjustment is not needed.

6. **Family History**   There is a sevenfold increased risk of having a baby with an NTD with affected first-degree relatives of the developing fetus (e.g., siblings of a previous ONTD-affected pregnancy) over the background prevalence of having a baby with an NTD. This risk is sufficiently high to warrant counseling for diagnostic testing without screening [Milunsky, 1998]. Second-degree relatives have a risk threefold greater than the background prevalence, and this can be factored into a patient's risk calculation [Papp et al., 1987]. More distant relations approach the general population risk.

The prevalence factors of family history, maternal race, diabetes mellitus, and twins are assumed to be independent of each other and, therefore, can be combined directly in any patient with coincident factors.

The performance of the prenatal screen for ONTD using MSAFP, in terms of detection and false-positive rates, depends on several factors, including long-term analytical precision, the gestational age at which patients are screened and the accuracy of its assignment, the MSAFP cutoff value selected for the screen, and the accuracy of elements of clinical information—the confounding variables in the screen. Using a cutoff of 2.2 MoM in a singleton pregnancy detects more than 95 percent of the cases with anencephaly and 80 percent of the cases with ONTD with a FPR of 2 to 3 percent. Factors increasing detection and lowering false-positive rates include the use of consistent ultrasound biometry to assign gestational age, the fraction of the population screened at the optimum 16 to 18 weeks, and the use of maternal weight correction.

Conditions other than ONTD associated with an increased MSAFP include normal variance in a healthy fetus, underestimated gestational age, an undetected twin or multiple pregnancy, a recent or impending fetal demise, the presence of open ventral wall defects detectable by ultrasound (e.g., omphalocele, gastroschisis), and nephritic syndrome. Unexplained high MSAFP is also associated with an increased risk for an intrauterine death, intrauterine growth retardation, and premature delivery [Waller et al., 1996].

A screen positive for ONTD provokes anxiety and thus requires immediate counseling and follow-up once the patient has been notified. Although amniocentesis and determining the levels of AFAFP and AChE remain the best diagnostic procedure, a few less invasive procedures should be implemented initially, including amendment of clinical information. Thus, any clinical information determined to be in error should be corrected (e.g., weight discrepancies over 10 pounds, maternal race, maternal IDDM). A fetal ultrasound is crucial in detecting an error in the gestational age, multiple pregnancies, and an intrauterine death. A new interpretive report should be issued if the clinical data are substantially different. Thus, the gestational age should be amended only if there is a difference of more than 10 days between the gestational age determined by LMP and the one determined by ultrasound [Daya, 1993; Hadlock et al., 1982, 1992]. Amending the gestational age will reduce the false-positive rate from 2.5 to 2.0 percent.

DIAGNOSTIC TESTS FOR ONTD

Detailed Ultrasound Examination in ONTD

Ultrasound diagnosis of meningomyelocele is based on the finding of a cystic mass protruding from the vertebral body and wide separation of the lateral processes of the lamina. However, the cranial sonographic signs are easier to detect. These include the "banana" and "lemon" signs. The banana sign is the result of Chiari malformation, which is the herniation of the cerebellum through the foramen magnum and, thus, obliteration of the cisterna magna. This sign is found in 92 percent of cases [Blumenfeld et al., 1993]. The lemon sign is the scalloping of the frontal bones. Hydrocephalus, the result of obstruction of the fourth ventricle opening, due to the cerebellar herniation, is the third sign, which presents in about 80 percent of cases.

The lesions of OSB are harder to detect, particularly minimal lesions in the caudal spine. Anencephaly is fully detectable with a successful ultrasound examination as an absence of cranium and absence of brain. Encephalocele and cranial meningocele can be detected as a cystic mass protruding from the cranium, and acrania can be detected as the absence of the scalp with the presence of brain tissue.

The general quality of the fetal ultrasound imaging becomes a factor, as does the gestational age when the examination is performed—examinations at 20 to 24 weeks are better than at 16 to 20 weeks. With modern equipment, detection of OSB in high-risk patients with elevated MSAFP ranges from 80 to 100 percent [Wald et al., 1991a]. In the general population, ultrasound as a primary screening test for OSB only detected 68 percent of cases in a recent European study [Boyd et al., 2000]. Although the absence of abnormal findings on ultrasound cannot rule out the presence of OSB, it substantially reduces the likelihood of the disorder.

AFAFP and AChE

The gold standard in the detection of ONTD is based on the measurement of amniotic fluid AFP and AChE [Smith et al., 1979; Report, 1981]. Abnormal AFAFP results are usually defined as ≥ 2.0 MoM (Fig. 8.2). In these patients, AChE can add to the diagnostic certainty and reduce false positives to a dramatically low level. AChE has a low detection rate when the AFAFP is normal, and in this situation AChE testing is not recommended [Wald et al., 1989].

For AFAFP, using a cutoff of 2.0 MoM at 16 to 18 weeks gestation, the detection rate is about 100 percent for anencephaly and 96 percent for an ONTD with an FPR of 2 percent. When the AChE is also positive, the detection rate is about 100 percent for anencephaly and 97 percent for an ONTD with an FPR of <0.1 percent. Conditions other than ONTD that are associated with an elevated AFAFP and/or detectable AChE

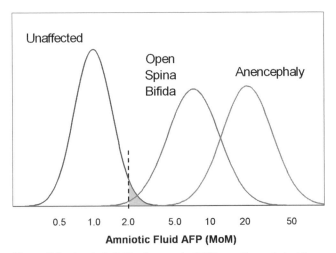

Figure 8.2 Amniotic fluid α-fetoprotein (AFP) as a diagnostic test for open spina bifida and anencephaly. At a cutoff of 2.00 MoM, virtually 100 percent detection is achieved. False positives can be reduced by reflexive testing for acetylcholinesterase (AChE) in fluids with elevated AFP.

include normal variance in a healthy fetus (AFAFP only), fetal blood contamination, open ventral wall defects detectable by ultrasound (gastroschesis and, less commonly, omphalocele), congenital nephrosis (AFAFP only), and fetal demise.

PRENATAL SCREENING FOR DOWN SYNDROME

The first to report Down syndrome as a clinical entity was Langdon Down in 1866 [Down, 1866]. The condition is characterized by mental retardation, characteristic facial features, hypotonia, congenital heart defect, short stature, and higher incidence of congenital heart disease, leukemia, and recurrent infections.

Before the advent of antibiotics, the life expectancy of children with Down syndrome was only 12 years [Wald, 2000]. With the recent advance in the treatment of infections, malignancies, and hypothyroidism, as well as the advance in general and cardiac surgeries, life expectancy has increased. Of those surviving the first 5 years of life, 85 percent will have a life expectancy of at least 30 years, with many living to age 50 [Noble, 1998; Torfs and Christianson, 1998]. In 1959, Down syndrome was found to be the result of an extra copy of chromosome 21 (trisomy 21) [Lejeune et al., 1959], caused mainly by nondisjunction of this chromosome during meiosis I. In most cases the extra chromosome 21 is of maternal origin.

The incidence of Down syndrome in the general population is 1 : 700 live births [Saller and Canick, 1996], and it is the most common form of mental retardation. Down syndrome is the least lethal of all the autosomal trisomies, which accounts for its highest birth prevalence. About 23 percent of trisomy 21 pregnancies are lost between the early second trimester and term [Cuckle et al., 1987; Cuckle, 1999], and the rate of loss between the first trimester and term is about 45 percent [Morris et al., 1999]. In the United States, the prevalence and risks of Down syndrome are usually quoted for the second (or mid) trimester, whereas in Canada and many parts of Europe the risk at term is used.

In 1933, Penrose first published the association between maternal age and the risk for having a baby with Down syndrome [Penrose, 1933]. Amniocentesis, the diagnostic test for fetal chromosome abnormalities, only became available 35 years later [Valenti et al., 1968] and allowed screening for Down syndrome on the basis of maternal age. However, maternal age is a continuous risk variable [Cuckle et al., 1987], a fact often overlooked when screening is based on a threshold set at 35 years of age, with a screen positive being above this age and negative below this age.

Although DS is known to be associated with maternal age, it is paradoxical that most DS babies are born to women younger than 35. Thus, if screening for DS had been based on maternal age, and amniocentesis had been done on all women older than 35, only 30 percent of the babies with Down syndrome would have been detected. To provide a screening test for Down syndrome for the general population of pregnant women and to avoid relying only on maternal age in the determination of this risk, a variety of screening tests have been developed over the last 15 years. In 1984 Merkatz et al. [1984] reported the association between low MSAFP and Down syndrome (Fig. 8.3). In 1987, Bogart et al. [1987] reported the association between high maternal

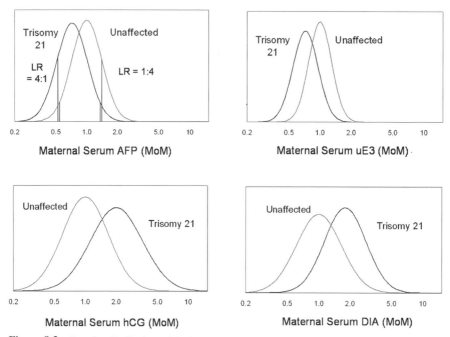

Figure 8.3 Gaussian distributions of the four maternal serum biochemical markers commonly used in second-trimester screening for Down syndrome. Two examples of the calculation of the likelihood ratio (LR) are illustrated for maternal serum AFP, using the relative height of the affected and unaffected distributions at patients' measured MSAFP levels.

serum chorionic gonadotropin (MShCG) concentrations and Down syndrome, and in 1988 [Canick et al., 1988] an association was found between low maternal serum unconjugated estriol (MSuE3) and Down syndrome. The combination of maternal serum AFP, uE3, and hCG became known as the triple test [Wald et al., 1994].

In 1989 and 1990 [Rottem et al., 1989; Szabo et al., 1990], an association was found between the fetal nuchal translucency (NT) measured by ultrasound in the first trimester and Down syndrome. At about the same time, two biochemical markers measured in the first trimester were introduced: the β fraction of hCG [Nebiolo et al., 1990] (Fig. 8.4) and pregnancy-associated plasma protein A (PAPP-A) [Brambati et al., 1994] (Fig. 8.5), and a second-trimester serological marker, dimeric inhibin-A (DIA), was also reported [Van Lith et al., 1992; Wald et al., 1997a]. Combining DIA with the second-trimester triple test yields the quadruple test, which has a better detection rate and a lower false-positive rate (Fig. 8.3). In the mid-1990s, first-trimester screening protocols for Down syndrome were proposed that combined maternal age, NT, and the first-trimester biochemical markers PAPP-A and β-hCG (Brizot et al., 1994, 1995). In 1999, Wald et al. proposed an algorithm combining the best of the first- and second-trimester ultrasound and biochemical markers [Wald et al., 1999], and named this combination the integrated test. The biochemical markers in prenatal screening are either placental derived or fetal derived.

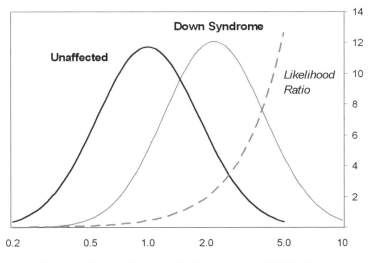

Maternal Serum free β-hCG during week 12 (MoM)

Figure 8.4   Gaussian distributions of maternal serum PAPP-A in the Down syndrome and unaffected populations at 12 weeks gestation. The likelihood ratio for this marker is shown; *decreasing* concentrations are associated with increasing risk.

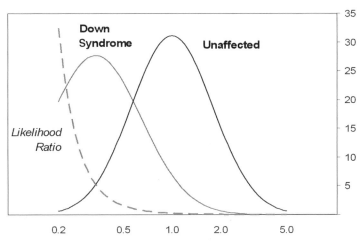

Maternal Serum PAPP-A during week 10 (MoM)

Figure 8.5   Gaussian distributions of maternal serum free β-hCG in the Down syndrome and unaffected populations at 10 weeks gestation. The likelihood ratio for this marker is shown; increasing concentrations are associated with increasing risk. The scale on the right of the figure indicates the relative power of the marker.

PLACENTAL-DERIVED MARKERS

Chorionic Gonadotrophin

The glycoprotein human chorionic gonadotrophin (hCG) is composed of α and β subunits that also exist in serum in an unbound (free) form at low concentrations. It is the β subunit that imparts the hCG's biological specificity. However, the β subunit is biologically active only when bound to the α subunit. Human chorionic gonadotrophin can be measured as either the intact hormone, or the free β-hCG subunit, or the sum of both. *Total hCG* (bound plus free β-hCG subunits) and *intact β-hCG* (bound β-hCG subunits) are probably equally informative in assessing the risk for Down syndrome. Although the free β-hCG concentration seems to be slightly more informative than total or intact hCG [Wald and Hackshaw, 2000], it is less stable in whole-blood specimens and must be separated within 5 hours of collection to run an accurate assay [Knight and Cole, 1991; Muller et al., 1999.] However, the free-ß subunit of hCG [Wald and Hackshaw, 2000] is generally considered to be a better marker than the intact hCG molecule in first-trimester screening.

The median concentrations of hCG reach their peak at 10 to 12 weeks and decrease during the early midtrimester. hCG decreases steeply from 15 to 17 weeks and then very gradually from 17 to 22 weeks. The concentration of free-ß subunit of hCG decreases during the first and second trimester. The gestational age at which all forms of hCG first become informative of the risk of Down syndrome is 10 weeks. Detection with hCG gradually increases in later gestations [Wald et al., 2004].

Dimeric Inhibin A

Inhibin is a glycoprotein hormone composed of an α and a β subunit. The β subunit exists in two closely related forms, A and B, and thus presents as dimeric inhibin A (DIA) and dimeric inhibin B (DIB). The existence of inhibin was predicted in 1932, based on a study that showed an inhibitory effect of testicular vein serum on the anterior pituitary secretion of follicular-stimulating hormone (FSH) [McCullagh, 1932]. In 1975, an ovarian form of inhibin was detected. However, only in 1988 [Vale et al., 1988] were inhibin and its related opposing hormone, activin, characterized.

In 1992, Van Lith first showed the association between fetal Down syndrome and an increase in maternal serum inhibin concentrations in the second trimester [Van Lith et al., 1992]. In 1996, several authors [Cuckle et al., 1996; Wald et al., 1996a; Haddow et al., 1998a] demonstrated that DIA increased the detection rate of Down syndrome fetuses in midtrimester screening for this condition. In 2000, an improved assay format for inhibin A was developed, and the use of this product has been validated in two subsequent case-controlled studies [Knight et al., 2001; MacRae et al., 2001] and in a large observational study [Wald et al., 2003].

Pregnancy-Associated Plasma Protein A

Pregnancy-associated plasma protein A (PAPP-A) is a glycoprotein derived from the placenta. Despite its name, PAPP-A is also present in serum and plasma in nonpregnant women and men [Wald et al., 1992]. During pregnancy, PAPP-A circulates in a covalent complex with the proform of eosinophil major basic protein (proMBP), which is a potential marker in the screening for Down syndrome [Christiansen et al., 1999]. In normal pregnancies, the concentration of PAPP-A rises rapidly during the first trimester; however, in Down syndrome pregnancies, the maternal serum (MS) concentration is less than half that of an unaffected pregnancy [Wald et al., 1992, 1996b; Spencer et al., 1994; Brambati et al., 1994]. PAPP-A in the first trimester is currently the best biochemical marker known for fetal Down syndrome at any gestation and has a higher detection rate when used in the first trimester [Brizot et al., 1994; Haddow et al., 1998b; Cuckle and van Lith, 1999]. Detection with PAPP-A and maternal age (at a fixed 5 percent false-positive rate) decreases from week 10 to week 14 from over 50 percent to less than 30 percent [Bersinger et al., 1994; Cuckle and van Lith, 1999; Cuckle and Arbuzova, 2002; Wald et al., 2004].

FETAL-DERIVED MARKERS

Alpha-fetoprotein

The concentration of AFP in the maternal blood has been found to be low in second-trimester maternal serum in Down syndrome pregnancies. This may be the result of a reduction in the transfer of this protein through the placenta into the maternal circulation in fetal Down syndrome [Newby et al., 1997] and/or a decrease in the AFAFP present in Down syndrome pregnancies [Cuckle et al., 1985; Davis et al., 1985].

Unconjugated Estriol (uE3)

The fetal adrenal gland secretes large amounts of dehydroepiandrosterone sulfate (DHEAS), which is converted to 16α-hydroxy-DHEAS by the fetal liver and to uE3 in the placenta. MSuE3 concentrations rise rapidly (25 percent per week) during the late first trimester and early second trimester of pregnancy. In Down syndrome, the concentrations of uE3 and its DHEAS precursor are lower [Newby et al., 2000], suggesting diminished production.

Fetal Ultrasound Nuchal Translucency

Nuchal translucency (NT) is defined as the maximum thickness of the normally occurring subcutaneous space (translucency) between the fetal skin and the soft tissues overlying the cervical spine, when viewed in the sagittal plane [Pandya et al., 1995c;

Chitty and Pandya, 1997]. A typical NT measurement is in the range of 0.5 to 2 mm [Devine and Malone, 1999], but measurements vary between observers. Increased NT is a nonspecific marker of structural anomalies, in particular cardiac defects, diaphragmatic hernia, fetal akinesia, and abdominal wall defect [Pajkrt et al., 1999; Pandya et al., 1995a; Hyett et al., 1997]. Fetuses with chromosome abnormalities in general, and Down syndrome in particular, have, on average, increased NT measurements [Bronshtein et al., 1989; Cullen et al., 1990; Nicolaides et al., 1992; Pandya et al., 1995c]. However, an increased NT, on its own, is not a fetal abnormality [Snijders and Smith, 2002].

Initial reports regarding the association between fetal aneuploidy and increased NT were based on studies in selected patient populations at increased risk of a variety of chromosome abnormalities, including trisomy 21, 13, 18, and 45,X [Chitty and Pandya, 1997]. Fetuses with chromosome abnormalities, as well as other structural abnormalities, were found to have larger NT. In the mid-1990s, studies in unselected populations were undertaken to assess the use of NT as a screening variable for fetal aneuploidy. Early studies used fixed NT cutoffs ranging from 2.5 to 3.0 mm, with an average of 62 percent detection and a 4.0 percent false-positive rate [Chitty and Pandya, 1997; Pandya et al., 1994, 1995b; Snijders et al., 1996].

It was later recognized that NT measurements increase in association with fetal growth during the 11- to 13-week gestational age [Pandya et al., 1995c; Schuchter et al., 1998]. The NT measurement has been expressed both as the arithmetic difference from the median for a particular gestational age, the approach taken by Nicolaides et al. [2002], and as a ratio of the median in MoM units (Fig. 8.6), the more commonly

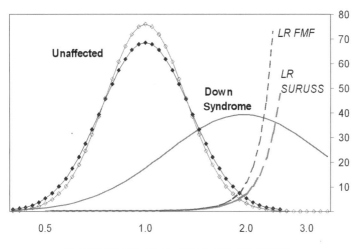

Nuchal Translucency Weeks 10–13 (MoM)

Figure 8.6   Gaussian distributions of nuchal translucency (NT) in the unaffected population at 10 to 13 weeks gestation from two published expert sites, expressed in MoM units: the Foetal Medicine Foundation, UK (open diamonds), the SURUS Study [Wald et al., 2003, closed diamonds]. Both sites are in close agreement on the distribution of this marker in the Down syndrome population.

used method in prenatal risk calculations [Wald and Hackshaw, 1997]. In 1998, Snijders et al. combined the likelihood ratio from the NT measurement and the risk for having a baby with Down syndrome, associated with maternal age [Snijders et al., 1998]. Their study reported a detection rate of Down syndrome of 82 percent with an 8 percent false-positive rate. The Snijders et al. study probably had an ascertainment bias that increased the estimated detection rate [Haddow, 1998]. In a subsequent review of 15 studies that had no identifiable verification bias, Mol et al. [1999] estimated Down syndrome detection to be 63 percent, at a false-positive rate of 5 percent.

By 1998, it was recognized that even experienced sonographers fail to achieve consistent NT measurements without specialized training [Haddow et al., 1998b]. In order to lessen the variation in NT measurements, and to improve the performance of NT as a screening variable, the Foetal Medicine Foundation in London, England, established a training and quality control program to accredit operators who maintain a record of measurement proficiency. This training program has been replicated internationally. Nevertheless, performance is still varied. In Italy, Zoppi et al. [2001] achieved 77 percent detection for a 3 percent false-positive rate, whereas a study in Scotland detected only 54 percent at a 5 percent false-positive rate [Crossley et al., 2002].

To standardize NT measurements, guidelines need to be followed. The NT should be measured with a sagittal plane, so that the measurements will be between the inner surfaces of the two fluid–tissue interfaces, and such that the NT is distinguishable from the space between the back of the fetal neck and the amniotic membranes. Furthermore, the gestational age should be taken into account since the NT increases with gestational age by about 17 percent per week [Schuchter et al., 1998].

SECOND-TRIMESTER DOWN SYNDROME SCREENING

Improvement in the screening for Down syndrome could not come at a better time. In the 1970s, the first method of screening for Down syndrome was maternal age. At that time, only 5 percent of all pregnant women were older than 35, and the detection rate using this method of screening was 25 to 30 percent. Since 1985, when the concept of biochemical screening for Down syndrome was established, women were found to be changing their reproductive plans and conceiving at a later age. Thus, the number of first births per 1000 women 35 to 39 years of age increased by 36 percent between 1991 and 2001. The new methods for screening for Down syndrome enabled a decrease in the number of invasive diagnostic prenatal procedures to determine the fetal karyotype, with a decrease in the incidence of associated pregnancy losses. Furthermore, the screening tests resulted in a decrease in the overall prevalence of Down syndrome among live births by as much as 46 percent [Palomaki et al., 1996].

Second-trimester screening for Down syndrome is performed between 15 and 20 weeks gestation using maternal age (MA) + MSAFP + hCG (total or free β-hCG) (double test), MA + MSAFP + uE3 + hCG (total or free β-hCG) (triple test), or MA + MSAFP + uE3 + hCG (total or free β-hCG) + DIA (quadruple test). With a 5 percent false-positive rate, the double test has a detection rate of approximately 60 percent, the triple test 70 percent, and the quadruple test > 75 percent [Wald and Hackshaw, 2000; Wald et al., 2003].

FIRST-TRIMESTER SCREENING FOR DOWN SYNDROME

First-trimester screening (FTS) for Down syndrome can be performed between 10 and 13 weeks gestation using fetal ultrasound nuchal translucency measurement and maternal PAPP-A and free β-hCG. The PAPP-A loses its discriminatory potential after 14 weeks gestation [Wald et al., 1997b], and NT measurement is less effective after 13 weeks gestation [Roberts et al., 1995].

Overall, NT is the most discriminatory first-trimester marker, although PAPP-A is as informative at 10 weeks gestation [Wald et al., 2004]. At 10 weeks gestation PAPP-A alone has a 58 percent detection rate for a 5 percent false-positive rate, making it the most powerful biochemical marker to date at any gestation. The third best first-trimester marker is free β-hCG and, in contrast to PAPP-A, its discriminatory power increases with each gestational week [Wald et al., 2003]. At a 5 percent false-positive rate, the detection rate of the combined test, which includes MA + NT + PAPP-A + free β-hCG, is 79 to 86 percent [Wapner et al., 2003; Wald et al., 2004].

INTEGRATED FIRST- AND SECOND-TRIMESTER SCREENING

A screening algorithm called the *integrated test* was proposed by Wald et al. in 1999 with the intention of utilizing the potential of the best first- and second-trimester screening markers [Wald et al., 1999]. Rather than choosing markers from just the first or second trimester alone, the integrated test proposed using the best markers of both trimesters. In the integrated test, the first-trimester serum marker PAPP-A and the ultrasound marker NT are measured but not disclosed until the results of the second-trimester AFP, uE3, hCG, and DIA measurements are also available. The integrated test is a second-trimester screen in the sense that the results are not communicated until the second trimester, when a single risk estimate is provided for interpretation. The modeling for the proposed algorithm predicted a detection rate of 90 percent with a 2 percent screen-positive rate [Wald et al., 1999]. In the only observational study reported to date, the performance is very close to what was predicted—a detection rate of 85 percent for a false-positive rate as low as 0.9 percent if the first-trimester tests are performed in the 11th week of gestation [Wald et al., 2003, 2004]. A significant challenge in implementing the integrated test is the required change in ultrasound practice to withhold the results of the NT scan until all the biochemical testing is completed, potentially 2 to 8 weeks later. In practice, there is a tendency to report and act on an increased NT since such elevated NT values are nonspecific and associated with increased risk of other structural problems [Snijders and Smith, 2002]. Perhaps just as concerning is the tendency for women to fail to complete the second-trimester blood testing component of the integrated test because of a verbal reassurance following a normal NT scan.

CONTINGENT SEQUENTIAL SCREENING FOR DOWN SYNDROME

Whereas the integrated test utilizes all available markers uniformly in all patients, most recently interest in prenatal screening has focused on a more selective use of the available markers [Maymon et al., 2004; Wright et al., 2004]. Under this proposal, patients start with the screening tests in the first trimester, and subsequent care options are offered depending on the results of these first-trimester screening tests. Subsequent care options include either invasive diagnostic testing for high-risk patients or further screening in moderate-risk patients or no further testing in very low risk patients as indicated by the results of the first-trimester tests. This approach has the potential benefit of reducing the utilization of second-trimester screening resources while at the same time responding to patient and provider interest in accessing information about the status of the pregnancy as soon as first-trimester tests are completed. Modeling of this approach predicts that the high detection of integrated testing can be retained, with most detection occurring in the first trimester, and with only a very slight increase (0.1 percent) in the overall screen-positive rate [Wright et al., 2004]. Up to 80 percent of women would complete their screening in the first trimester. The modeling is complex, and the practicality and patient acceptance have not been assessed. One issue will be the counseling required for the 20 to 25 percent of women with first-trimester screening results that require further screening in the second trimester.

SCREENING FOR TRISOMY 18

Trisomy 18 is a lethal condition with a prevalence of 1 : 3000 during the first and second trimester and 1 : 8000 at birth. The miscarriage/intrauterine death rate is approximately 70 percent [Hook et al., 1983, 1989]. The condition is characterized by such features as growth retardation, typical facial features, cardiac abnormalities, radial ray defect, and overlapping fingers. Most newborns with this abnormality die within the first year of life.

SECOND-TRIMESTER SCREENING FOR TRISOMY 18

The first prenatal screening marker of any kind, for any fetal trisomy, was a low MSAFP in a case of trisomy 18 [Merkatz et al., 1984]. However, until triple-marker screening for trisomy 21 was implemented, screening for the more rare disorder of trisomy 18 based on MSAFP alone was not beneficial. In trisomy 18, all of the triple markers are low, with uE3 being the most informative (Palomaki et al., 1995). Thus, the triple-marker screen can detect 60 to 80 percent of the cases with trisomy 18 with a very low 0.2 percent false-positive rate [Benn et al., 1999; Hogge et al., 2001]. DIA is not informative of trisomy 18 risk [Lambert-Messerlian et al., 1998].

FIRST-TRIMESTER SCREENING FOR TRISOMY 18

The concentrations of first-trimester PAPP-A and free β-hCG are reduced, and the NT is increased in trisomy 18, sufficient to support a first-trimester screening protocol with a projected detection rate of 89 percent but at the relatively high-false positive rate of 1 percent [Tul et al., 1999].

INTEGRATED BIOCHEMISTRY-ONLY SCREENING FOR TRISOMY 18

Combining the first-trimester PAPP-A marker with second-trimester triple-marker screening can detect 90 percent of cases with trisomy 18 with a 0.1 percent false-positive rate [Palomaki et al., 2003] (Fig. 8.7).

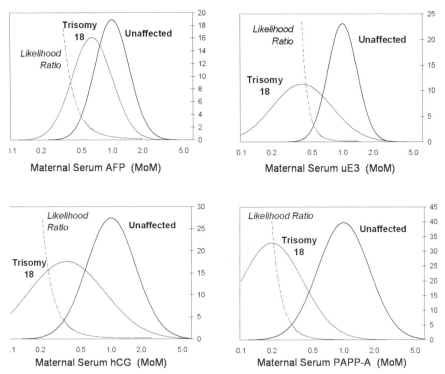

Figure 8.7   Gaussian distributions of the four maternal serum biochemical markers that are informative of the risk of fetal trisomy 18. PAPP-A is measured during the first trimester [Palomaki et al., 2003]. The likelihood ratio for each marker is shown.

SECOND-TRIMESTER ULTRASOUND MARKERS FOR TRISOMY 21

Ultrasound screening for trisomy 21 in the second trimester involves the detection of major or minor findings, which indicate an increased risk for having a baby with Down syndrome. Major findings are structural abnormalities that, in a newborn, will have a surgical, medical, or cosmetic impact. These include abnormalities such as cardiac defects, structural brain abnormalities, and duodenal atresia. The finding of a major abnormality is associated with a substantial increase in the risk for a fetal chromosome abnormality and does not require further risk modification.

Minor abnormalities are findings that, on their own, have no surgical, medical, or cosmetic impact nor cause any harm to the fetus/newborn. Furthermore, they are most frequently detected in normal fetuses. Some are transient and some change with gestational age. However, collectively they are known to be associated with an increased risk for trisomy 21. Thus, in a low-risk population, 11 percent of normal fetuses will have one sonographic marker, whereas the incidence is twice as great in Down syndrome fetuses [Sohl et al., 1999]. The actual sensitivity of a genetic sonogram will depend on various factors including the reasons for referral (a priori prevalence), the markers being sought, whether they are detected in isolation or in combination, and the gestational age.

Sonographic markers may be used in conjunction with second-trimester biochemical screening for trisomy 21, to modify the risk assessment. The postultrasound risk can be estimated by applying specific likelihood ratios, reflecting the strength of individual markers. Markers observed in combination are more informative than isolated findings. In this manner sonographic markers can be an aid in counseling regarding the option for having an invasive, diagnostic test to determine the fetal karyotype by CVS or amniocentesis. A caveat is that there is still very little evidence on the independence of biochemical markers and second-trimester ultrasound markers [Smith-Bindman et al., 2001a]. Although a normal fetal ultrasound (not finding these ultrasound markers) can be used to decrease the assessed risk for having a baby with DS, and may result in a decrease in the number of invasive procedures [Nyberg and Souter, 2003], any practice that diminishes the use of diagnostic testing after a screen will to some extent reduce the detection of that screen.

Selected markers used as part of the risk assessment for trisomy 21 are discussed further below. Some are measurements with a measurement cutoff, such as nuchal fold size. Others are signs that are either present or absent, such as hyperechogenic bowel. Each marker, if detected, is associated with a fixed likelihood ratio.

Increased Nuchal Fold

Nuchal thickening is one of the most sensitive second-trimester markers for trisomy 21. This marker is the measured skin thickness in the posterior aspect of the fetal neck. The association between increased nuchal fold and trisomy 21 was initially reported by Benacerraf et al. [1987]. Using a cutoff of 6 mm between 16 and 24 weeks gestation, they found thickening in more than 40 percent of the fetuses with Down syndrome. Wald et al. reviewed 16 studies that included 390 fetuses with trisomy 21, and the calculated sensitivity (detection rate) was 38 percent and the specificity was 98.7 percent (false-positive rate 1.3 percent) for the diagnosis of trisomy 21 [Wald et al., 1998]. Although nuchal fold thickening is mainly used for screening for trisomy 21, it also can be found in association with single-gene disorders, such as Noonan syndrome, multiple pterygium syndrome, and skeletal dysplasias [Souter et al., 2002; Shipp and Benacerraf, 2002] as well as cardiac abnormalities [Shipp and Benacerraf, 2002].

Hyperechoic Bowel

Hyperechoic bowel is a nonspecific marker most commonly observed in normal fetuses. To minimize subjectivity, an echogenic bowel is identified as a fetal bowel with homogenous areas of echogenicity that are equal to or greater than that of surrounding bone [Sepulveda and Sebire, 2000]. It has been detected in 0.6 to 2.4 percent of all second-trimester fetuses [Bromley et al., 1994; Hill et al., 1994; Nyberg and Souter, 2003] and as an isolated finding in 9 percent of fetuses with chromosomal abnormalities [Yaron et al., 1999; Stocker et al., 2000]. In a study by Deren et al. [1998] that looked at 3838 midtrimester pregnancies, 5 of 34 cases of trisomy 21 had echogenic bowel (sensitivity 15 percent, specificity 98 percent). Sohl et al. [1999] reported their study on 2743 midtrimester pregnancies with ultrasound prior to amniocentesis and found that echogenic gut has a sensitivity of 16.4 percent and specificity of 98.4 percent. As a result, Nyberg and Souter [2003] suggested that the likelihood ratio for this marker is 6.

Echogenic bowel was also found to be associated with an increased risk for congenital infection, intra-amniotic bleeding, congenital malformations of the bowel, cystic fibrosis, and congenital infection [Sepulveda et al., 1996a, 1996b; Sepulveda, 1996; Petrikovsky et al., 1999].

Shortened Limbs

Children with Down syndrome are known to have rhizomelic shortening of the long bones (humerus and femur), which can be detected by ultrasound in some fetuses with trisomy 21 in the second trimester [Nyberg et al., 1993; Bahado-Singh et al., 1996]. A short humerus seems to be more specific for the detection of trisomy 21 than short femur length, but the detection based on humerus length is also less. In a large meta-analysis, Smith-Bindman et al. calculated that short femur length has a sensitivity of 16 percent in the prediction of Down syndrome with a false-positive rate of 4 percent, whereas short humeral length has a sensitivity of 9 percent with a false-positive rate of 3 percent [Smith-Bindman et al., 2001b]. The meta-analysis showed an overall likelihood ratio of 7.5 (95 percent Confidence Interval (CI) 4.7 to 12) for short humerus and a likelihood ratio of 2.7 (95 percent CI 1.2 to 6.0) for short femur. Short femur and humerus can also be associated with skeletal dysplasias and fetal growth restriction [Pilu and Nicolaides, 1999]. The ethnic background and the gender may also influence the results.

Echogenic Intracardiac Focus

Echogenic intracardia focus (EIF) is defined as a focus of echogenicity comparable to bone in the region of the papillary muscle in the right or left ventricle of the fetal heart [Sohl et al., 1999; Winter et al., 2000]. It is found more often in the left than the right ventricle [Wax and Philput, 1998]. The histopathological findings in abortuses with EIF include an isolated intracardiac calcification surrounded by fibrotic tissue. No other changes in the papillary muscle or chordae tendineae were noted, and there was no pathomorphological difference in the calcifications in fetuses with normal and abnormal karyotypes [Tennstedt et al., 2000]. EIF is detected in 3 to 7 percent of karyotypically normal fetuses and in up to 30 percent of trisomy 21 fetuses [Winter et al., 2000]. The prevalence of EIF in unaffected fetuses appears to be significantly higher among Asians [Shipp et al., 2000]. This may have to be taken into consideration when calculating the risk for this ethnic group.

The association with fetal aneuploidy was confirmed by some [Vibhakar et al., 1999; Winter et al., 2000; Bromley et al., 1998] and disputed by others [Achiron et al., 1997]. Bromley et al. [1998] identified differences between right-sided and left-sided EIF regarding the risk for aneuploidy and concluded that right-sided and bilateral EIF had approximately a twofold greater risk for fetal chromosome abnormality compared to left-sided foci. In a meta-analysis of five studies, the likelihood ratio for EIF for trisomy 21 was determined by Smith-Bindman et al. [2001b] to be 2.8, whereas in their review Nyberg and Souter [2003] concluded that the likelihood ratio is 1.8.

Renal Pyelectasis

Isolated pyelectasis (dilation of the fetal renal pelvis) is seen in 2 to 4 percent of fetuses at 16 to 26 weeks gestation [Chudleigh, 2001; Chudleigh et al., 2001] and has been detected as an isolated finding in 2 percent of fetuses with trisomy 21 [Wilson et al., 1997; Benacerraf et al., 1990; Smith-Bindman et al., 2001b]. However, the renal pelvic measurements vary with gestational age and thus the prevalence of this finding changes [Chitty and Altman, 2003]. Using a cutoff of 3 mm, pyelectasis is found in 3 percent of normal fetuses [Nyberg and Souter, 2003] and Snijders et al. [1995] indicated that mild pyelectasis increases the risk of trisomy 21 by 1.6-fold. Wickstrom et al. [1996] found that isolated pyelectasis, defined as renal pelvis of ≥ 4 mm before 33 weeks gestation, and ≥ 7 mm after 33 weeks gestation, is associated with a 3.9-fold increased risk for trisomy 21 and a 3.3-fold increased risk for all chromosome abnormality in the presence of isolated fetal pyelectasis.

Summarizing the literature, Smith-Bindman et al. [2001b] found a likelihood ratio for trisomy 21 of approximately 1.9. However, the 95 percent CI does cross 1 (0.7 to 5.1), indicating lack of significance. Nyberg and Souter [2003] calculated a likelihood ratio of 1.5, based on pyelectasis of ≥ 4 mm. Thus, in the absence of other markers, the risk for trisomy 21 associated with isolated mild pyelectasis may not justify an invasive diagnostic procedure.

Nasal Bone Abnormalities

Nasal hypoplasia has been recognized as a feature of trisomy 21. Following the association between an increased nuchal translucency and trisomy 21 [Bronshtein et al., 1989], it was noted that an absent nasal bone in the first trimester and, perhaps, in the midtrimester, is associated with an increased risk for trisomy 21 [Cicero et al., 2001]. This has led to prenatal evaluation of the nasal bone, which has been shown to be a thin, echogenic line within the bridge of the fetal nose. The fetus is imaged facing the transducer, with the fetal face strictly in the midline. The angle of insonation is 90 degrees, with the longitudinal axis of the nasal bone as the reference line. Calipers are placed at each end of the nasal bone. Absence of the nasal bone or measurements below the 2.5th percentile is considered significant [Cicero et al., 2001; Sonek, 2003; Minderer et al., 2003].

Although not all observers agree [De Biasio and Venturini, 2002], it seems that absent fetal nasal bone could increase the detection rate of prenatal screening tests for Down syndrome and/or reduce the false-positive rate [Cicero et al., 2001, 2003]. However, the authors who presented this test emphasized that ultrasonographers must have "appropriate training and certification in their competence in doing the nasal bone scan" [Cicero et al., 2001, p. 1667]. In fact, it seems that determining the presence or absence of nasal bone and measuring its length is much more difficult than measuring the NT. Furthermore, many women have no access to specialized ultrasound. Thus, a study published by Whittle [2001] showed that in the United Kingdom, where ultrasound markers have been most intensively implemented, only 8 percent of women had access to NT screening, whereas 71 percent of women in the United Kingdom used second-trimester serum screening [Whittle, 2001]. In view of this information, the nasal bone should be considered in the context of research and may only be assessed at tertiary centers.

SUMMARY

In the last 20 years, there has been major research and a fast development in prenatal screening for trisomy 21 and ONTD using maternal serum biochemical markers and ultrasound markers (Fig. 8.8). Screening for fetal chromosome abnormalities started with the option of invasive prenatal diagnostic tests (CVS or amniocentesis) offered to pregnant women older than 35 and continued with the finding that MSAFP used for screening for ONTD can also provide a screen for trisomy 21, when used in conjunction with maternal age. Soon after, additional biochemical markers were found and a second-trimester screen for trisomy 21 using second-trimester triple-marker and, later on, quadruple-marker screening was offered. The path continued with first-trimester screening, including ultrasound measurements of the NT in combination with maternal age and maternal serum PAPP-A and β-hCG, integrated prenatal screening, and, most recently, the suggestion of contingent prenatal screening, offering the second-trimester component of integrated screening only to women with borderline screening test results on first trimester-screening.

The main goal of all screening methods is to maximize detection while minimizing the number of women undergoing invasive diagnostic prenatal tests, thereby

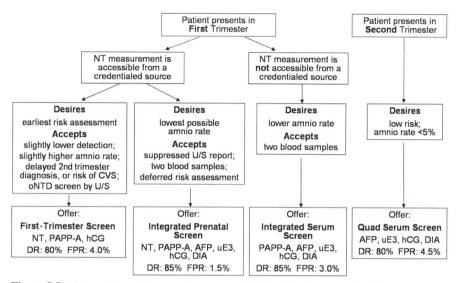

Figure 8.8 Schema of potential prenatal screening algorithms from the perspective of patient choices, with the trade-offs between increased detection, early detection, and frequency of testing. Detection rates (DR) and false-positive rates (FPR) will very with the maternal age of the screened population.

reducing the risk of losing a wanted pregnancy, by offering invasive testing to women at high risk for having the targeted disorder. This can be achieved by having screening tests with a high detection rate and low false-positive rate. By using prenatal screening for trisomy 21 and ONTD, together with the improved ultrasound detection of fetal abnormalities and the primary prevention of NTD by increasing folic acid in the diet, we have been able to decrease substantially the prevalence of ONTDs, trisomy 21, and other chromosomal and nonchromosomal abnormalities.

Currently, intensive research is continuing to find ways to perform fetal chromosome analysis on fetal cells and/or DNA (deoxyribonucleic acid) obtained from maternal blood. If a safe, noninvasive diagnostic test for serious birth defects is found, screening tests and their heroes will pass into history. However, most fetal abnormalities are nonchromosomal and so sonographic detection of abnormalities, as well as the primary prevention of NTD and other abnormalities, will continue to have a major role in the care of pregnancies.

PRENATAL DIAGNOSIS AND PREIMPLANTATION GENETIC DIAGNOSIS: LEGAL AND ETHICAL ISSUES

Roxanne Mykitiuk, B.A., LL.M. Associate Professor of Law, Osgoode Hall Law School
Mireille Lacroix, B.S.S., LL.B., LL.M., Research Associate, Centre de recherche en droit public, Université de Montréal
Stephanie Turnham, B.A.(honours)., (LL.B. candidate, class of 2006, Osgoode Hall Law School)

INTRODUCTION

Prenatal diagnosis (PND) of genetic disorders and fetal anomalies has expanded significantly. Hundreds of conditions can be diagnosed through DNA analysis of fetal cells and ultrasound and maternal serum biochemical screening. The purpose of prenatal diagnosis is to rule out the presence in the fetus of a particular medical condition for which the pregnancy is at an increased risk. This information is provided to the individual or couple to assist in the decision-making process regarding the possible options including: carrying the pregnancy to term, preparing for a difficult delivery, preparing for the birth of a child with genetic anomalies and for special newborn care, or terminating the pregnancy.

Preimplantation genetic diagnosis (PGD) involves the creation of embryos outside the body and their subsequent biopsy in order to test for a genetic disorder. The stated advantage of PGD over prenatal diagnosis or testing is that the genetic diagnosis takes place at a much earlier stage. As a pregnancy has not been established, couples or individuals will not have to consider abortion, which is likely to be a much more stressful and difficult decision than the disposal of affected embryos in their earliest stage of development. It is expected that the range of conditions for which PGD is available will expand as the genes implicated are identified. Prenatal genetic testing and preimplantation diagnosis raise a number of ethical and legal issues that will be discussed following a brief description of PND and PGD techniques.

PRENATAL AND PREIMPLANTATION GENETIC DIAGNOSES TECHNIQUES

Prenatal testing includes prenatal screening, prenatal diagnosis, and preimplantation genetic diagnosis. Screening methods such as determination of maternal age, maternal serum screening, and ultrasound are used to identify women who are at increased risk of having a child with a chromosomal anomaly or a congenital anomaly. If a screening test yields an abnormal result, it can be followed by prenatal diagnosis in order to establish whether the fetus does carry a specific genetic mutation or chromosomal anomaly.

PRENATAL DIAGNOSIS

Techniques currently used for prenatal diagnosis include:

- Amniocentesis
- Chorionic villus sampling (CVS)
- Cordocentesis

Amniocentesis is performed after 14 weeks of gestational age (usually between 15 and 17 weeks) in order to determine fetal karyotype and to detect the presence of molecular and biochemical abnormalities. A needle is inserted into the amniotic sac of a pregnant woman while ultrasound imaging is used to verify the position of the fetus and the location of the placenta. A small quantity of amniotic fluid is aspirated, and its level of α-fetoprotein is measured. The fetal cells contained in the amniotic fluid are also cultured and analyzed using karyotyping, molecular diagnosis, biochemical assays, or direct fluorescent in situ hybridization. Results are highly reliable but are available 1 to 3 weeks after the amniocentesis, at 17 to 20 weeks of gestation. This is regarded by some as the most important drawback of amniocentesis: at this stage, termination of pregnancy—currently the only option to prevent the birth of an affected child—entails a greater physical and emotional risk to the woman than a first-trimester abortion and is difficult to obtain in many regions of Canada and the United States [Childbirth by choice, 2003; Henshaw and Finer, 2003]. However, the risks of the procedure itself are low. The risk of miscarriage due to amniocentesis is 0.5 to 1 percent, while the risk of infection and fetal injury are rare. Minor complications such as leakage of amniotic fluid and bleeding occur in 1 to 5 percent of cases but usually resolve themselves [Chodirker et al., 2001; Jorde et al., 2003].

Chorionic villus sampling is performed earlier, between 10 weeks and 11 weeks 6 days of gestational age. It involves the collection of chorionic tissue either through a transcervical or transabdominal technique and its subsequent analysis. The cells obtained can be analyzed directly or cultured as in amniocentesis. Like amniocentesis, CVS is highly accurate, but it has the advantage of yielding earlier results. The risk of spontaneous pregnancy loss is slightly higher. CVS may also entail a 1 in 3000 risk of facial or limb anomaly due to vascular disruption. In addition, diagnosis may be compromised if the sample has been contaminated with maternal tissue or if there is a discrepancy between the chromosomes in the fetal tissue and the chorionic tissue. These difficulties can, however, be easily circumvented [Chodirker et al., 2001; Jorde et al., 2003].

Cordocentesis is generally performed after 16 weeks of gestation when a rapid diagnosis is required in cases where ultrasound has revealed congenital malformations or growth retardation, to clarify suspected chromosome mosaicism, or to detect viral infections, haematological diseases, inborn errors of metabolism, and maternal or fetal platelet disorders. It involves the insertion of a needle into the umbilical cord vessel with ultrasound guidance and the withdrawal of fetal blood. Cordocentesis has the advantage of allowing direct access to the fetus but entails a slightly higher risk of pregnancy loss, as well as a 5 to 10 percent risk of fetal distress [Chodirker et al., 2001; Jorde et al., 2003; Simpson and Elias, 2003].

In addition to these techniques, prenatal genetic diagnosis can be conducted through fetal tissue sampling. As is the case in amniocentesis and cordocentesis, the procedure is performed under ultrasound guidance. A needle is inserted and fetal tissue such as skin, liver, or fetal urinary tract is collected for analysis. The risks and complications are similar to those linked to cordocentesis. However, the procedure is not commonly practiced. Similarly, the isolation of fetal cells from maternal blood is currently being studied as a technique of prenatal diagnosis. It would have the advantage of noninvasiveness (and therefore extremely low risk) and of being conducted early in the first trimester. It is not yet available in clinical practice [Chodirker et al., 2001; Health Canada 2002; Jorde et al., 2003; Shinya et al., 2004].

PREIMPLANTATION GENETIC DIAGNOSIS

In contrast with prenatal diagnosis, PGD is conducted before a woman becomes clinically pregnant. It is therefore an attractive option for couples at increased risk of having a child with a genetic condition who wish to avoid the possible termination of an affected pregnancy or who oppose abortion for religious or moral reasons [Simpson and Elias, 2003; Sermon et al., 2004; Verlinsky et al., 2004]. Though it has moved from the research setting to clinical application in recent years, PGD is not widely available, and it is considerably more complex (and costly) than traditional PND methods [Health Canada, 2002; Simpson and Elias, 2003]. It involves the creation of embryos through in vitro fertilization and the analysis of only one or a few cells from these embryos (or, in some cases, from the ovum) in order to establish a genetic diagnosis.

Cells can be obtained through biopsy at three different stages: prior to fertilization (biopsy of polar bodies), from blastomeres, and from blastocysts. For polar-body biopsy, an incision is made in the zona pellucida to draw the polar body out of the egg. Analysis of the polar body enables scientists to draw conclusions about the genotype of the oocyte. The utility of this technique is limited to testing for aneuploidy or to cases in which risk is known to originate from the mother's genotype since paternal mutations cannot be evaluated [Jorde et al., 2003]. In the case of blastomere biopsy, one or two cells are removed from the embryo 3 days after fertilization, when the embryo is composed of 6 to 8 cells. A greater number of cells can be obtained for analysis through blastocyst biopsy. This strategy entails removing 10 to 30 cells from the trophoectoderm rather than from the embryo itself. Biopsy is conducted 5 to 6 days after fertilization, when the blastocyst contains over 100 cells [Simpson and Elias, 2003; Sermon et al., 2004].

The cells thus obtained can be analyzed through various methods. Polymerase chain reaction (PCR) is used to amplify the DNA from a single cell and diagnose monogenic disorders such as cystic fibrosis, sickle cell anemia, and Duchenne muscular dystrophy. Specific tests have been developed for approximately 40 monogenic disorders. Though all monogenic disorders could, in theory, be detected through PCR, the cost implications are such that tests exist only for the most common disorders [Sermon et al., 2004; Wells, 2003]. Fluorescence in situ hybridization (FISH) analysis can be used to detect embryo sex and chromosomal aberrations such as aneuploidy, translocations, or other chromosomal rearrangements on selected chromosomes. A new technique called comparative genomic hybridization (CGH) has also been developed to analyze the chromosomal complement of embryos. While FISH can only detect a limited number of chromosomes (between five and nine per cell), comparative genomic hybridization enables scientists to evaluate all chromosomes simultaneously. However, because it is a lengthy process, and because there is only a very short period of time during which PGD can be performed, CGH is not widely used [Sermon et al., 2004; Simpson et al., 2003; Wells, 2003]. Once cell analysis is completed, embryos that are free from the mutations or chromosomal aberrations tested for are implanted into the woman's uterus. In the case of polar-body diagnosis, the selected ova are fertilized and then implanted.

The use of PGD can improve the outcome of a pregnancy for a couple who use in vitro fertilization in order to conceive and for those who are at increased risk of transmitting a genetic disorder to their offspring [Verlinsky et al., 2004]. However, it does not guarantee the birth of a healthy child. Available data from 2001 show that following PGD, pregnancies were established in 16 to 24 percent of treatment cycles [European Society of Human Reproduction and Embryology, 2002; International Working Group on Preimplantation Genetics, 2001] and that a small percentage of the children born as a result of these pregnancies had some form of anomaly [International Working Groups on Preimplantation Genetics, 2001]. There is a slight risk of misdiagnosis in PGD, often due to allelic dropout (a problem associated with PCR amplification failures) or DNA contamination. However, techniques such as multiplex-PCR and amplification of hypervariable fragments of DNA can lower risks of misdiagnosis [Jorde et al., 2003; Sermon et al., 2004; Wells, 2003]. Though PGD is considered a safe procedure, it remains costly and technically challenging [Jorde et al., 2003].

ACCESS TO PND AND PGD

Prenatal genetic testing is not universally accessible. Many factors affect a pregnant woman's ability to access these services, including economics and clinical risk factors.

EVIDENCE-BASED ACCESS: STANDARDS OF USEFULNESS

Whether or not a woman or couple is offered prenatal testing can depend on whether the health provider believes she meets the relevant risk threshold. Evaluating the clinical necessity of intervention is generally considered an appropriate strategy for allocating scarce medical resources [American Medical Association (2003) Code of Ethics, s.E-2.03]. It is also a way to avoid unnecessary invasive tests. Concerns have been raised about recent direct-to-consumer marketing approaches that may cause increased demand on the part of consumers and lead to indiscriminate offers of testing without an appropriate risk assessment [Andrews and Zuiker, 2003; Holtzman and Watson, 1997]. On the other hand, it has recently been argued that the evidence to support the cost utility analysis of the risk thresholds is deficient and that such thresholds can threaten autonomous choice; accordingly, it is argued that all pregnant women should be offered prenatal testing regardless of risk [Harris, 2004]. In any event, risk thresholds remain an important determining factor in access to prenatal testing, and a number of organizations have issued guidelines to assist health care professionals and geneticists in assessing the risk factors. Some risk factors can be identified by asking the right questions, including questions about maternal age, family history of genetic disorders, previous children, ethnic background, and medical and obstetric history [Roop, 2000; SOGC, 1998]. Another way to screen for risk factors is through medical examination, including maternal serum screening and ultrasound. In practice, these various methods are often combined into an overall risk assessment. Evidence-based standards of access arise in the initial screen stage. Screening cannot detect the presence of a disorder; it only identifies women who are at increased risk of having a child with a genetic anomaly.

PRACTICE GUIDELINES

Ethnicity

Certain ethic backgrounds are associated with an increased risk of genetic disorder. For example, Tay-Sachs disease has been linked to Ashkenazi Jews and French Canadians in eastern Quebec; and individuals of other than northern European descent are considered at high risk for thalassemia. It is generally recommended that genetic screening be uniformly offered to patients with high-risk ethnic backgrounds [SOGC, 1998; American Medical Association (2003) Code of Ethics, s.E-2.137].

With respect to Cystic Fibrosis (CF) in particular, a National Institutes of Health (NIH) Consensus Panel originally issued a statement supporting access to CF testing for all pregnant couples, especially those in high-risk populations [NIH, 1997]. Following this, a Steering Committee representing the interests of American College of obstetrician and Gynecologists (ACOG), American College of Medical Genetics (ACMG), and National Human Genome Research Institute (NHGRI) was established to refine the recommendations, leading to creation of general guidelines [Grody et al., 2001], laboratory standards [Richards et al., 2002], and patient education materials [ACOG, 2004]. Overall, the materials recommend offering CF testing to couples in high-risk ethnic groups who are pregnant or planning pregnancy, and making it available along with information about its limitations to couples in other ethnic groups with lower risk. Given that knowledge of one's carrier status can have implications for family planning [Lafayette et al., 1999], it is recommended that members of high-risk ethnic groups be tested for carrier status prior to conception, to allow for genetic counseling at an early stage [Grody et al., 2001; Chodirker et al., 2001]. According to Canadian guidelines, it is preferable for both partners to be tested, but knowledge that the woman is a carrier can still be an indication for prenatal diagnosis [Chodirker et al., 2001].

Family History

It is widely agreed that a recurring genetic disorder in a family's history is "a reasonable indication of risk" [Roop, 2000; Holtzman and Watson, 1997]. According to the AMA Code of Ethics, "prenatal genetic testing is most appropriate for women or couples whose medical histories or family backgrounds indicate an elevated risk of fetal genetic disorders" [American Medical Association Code of Ethics, s.E-2.12]. Physicians should therefore take a full family history, which includes inquiries about siblings, parents, parents' siblings, grandparents, and other extended family members where appropriate. However, family history does have limitations, particularly with respect to reliability and the different modes of inheritance [Holtzman and Watson, 1997]. Canadian practice guidelines appear to envision a spectrum, ranging from cases where prenatal testing must be offered (e.g., where the woman is a carrier for Down syndrome) to cases where further information or consultation with the patient is required (e.g., where one relative is found to have Down syndrome) [Chodirker et al., 2001].

Pregnancy History

Relevant pregnancy history factors include having a previous stillborn birth or a live-born infant with a chromosomal anomaly. Invasive prenatal testing should always be offered in these cases because there is an increased risk of recurrence. Even when the prior infant had a de novo anomaly (i.e., where the parents have normal karyotypes), prenatal testing is offered because of the potential for parental germline mosaicism [Chodirker et al., 2001]. Previous environmental exposure can also be relevant [Holtzman and Watson, 1997].

Maternal Age

It is traditionally recommended that all women over the age of 35 be offered prenatal testing, given the increased risk of fetal aneuploidy [ACOG, 2001]. However, the particular age cutoff has been described as "somewhat arbitrary," and the best practice may be to look at all the factors surrounding a woman's pregnancy [Roop, 2000].

Canadian practice guidelines note that maternal age alone is a relatively poor predictor of fetal anomaly, and it may be inappropriate to rely on it when facilities exist for additional screening methods [Chodirker et al., 2001]. Notably, there is a trend toward women extending their childbearing years, which means that invasive testing would be offered to an increasing number of women, with questionable utility, in the absence of further risk refinement. A recent study found that there was a reduction in the number of women electing to have invasive testing despite the rising maternal age, and the researchers attributed this to a departure from using maternal age alone as a risk assessment tool [Benn et al., 2004].

Advanced paternal age is also associated with an increased risk of genetic anomalies. A frequently used cutoff is 40 years of age at the time of conception, but the ACMG recognizes that risk increases linearly with age. Advanced paternal age is associated with a wide variety of genetic disorders, many of which cannot be detected by ultrasound. Therefore, the ACMG recommends genetic counseling be tailored to the individual needs of couples [ACMG, 1996].

Ultrasound Examination and Biochemical Markers (Maternal Serum Screening)

Genetic assessment is recommended when an ultrasound scan reveals major fetal anomalies. There are also several minor fetal anomalies, or "soft signs," that are statistically linked to chromosomal anomalies, but there is controversy about whether these soft signs have enough positive predictive value to warrant genetic testing [Chodirker et al., 2001].

The maternal age risk factor for Down syndrome can be refined through the use of maternal serum screening for biochemical markers in the first or second trimester. The ACMG recommends that women who screen positive for trisomy 21 or 18 be offered genetic counseling and amniocentesis [Driscoll, 2004]. The ACMG also recommends that women be offered maternal serum screening to identify pregnancies at risk of neural tube defects and anencephaly. In these cases, the risk assessment is refined by factors such as maternal weight, race, and family history, and an elevated result is an indication for genetic counseling and additional testing [Driscoll, 2004]. There are numerous combinations of markers, and second-trimester-screening through the *triple test* or *quad screen* has been found particularly useful in detecting Down syndrome and trisomy 18 [Javitt, 2004]. First-trimester screening methods use a combination of serum screening and ultrasound nuchal fold translucency screening. These methods have comparable detection rates of second-trimester screens and offer the advantage of an earlier test but should not be offered routinely unless there is sufficient ultrasound training and quality control [Driscoll, 2004; Dolan, 2004]. Recent studies have found that *fully integrated screening*, which combines first- and second-trimester screening methods with risk assessment from maternal age, outperforms second-trimester screening alone [Dolan, 2004]. The extent to which these methods are routinely offered will depend on many policy issues. While they offer a superior risk assessment than maternal age alone, there is a concern that some women may rely too heavily on the screen result at the expense of seeking a further diagnostic test. It is also relevant to consider whether a woman who screens positive will be able to access first-trimester diagnostic testing [Javitt, 2004].

Other Medical Factors

Given the variety of genetic disorders that can be detected prenatally, there are other indications of risk that go beyond the scope of this chapter. Health care providers should consult the guidelines and other medical literature for further information. For example, the ACMG has released a number of guidelines with respect to specific disorders, such as fragile X [Maddalena et al., 2001], uniparental disomy [Shaffer et al., 2001], and Prader–Willi and Angelman syndromes [ASHG and ACMG, 1996]. There also are clinical practice guidelines for specific testing techniques, such as FISH [Watson et al., 2000].

OBLIGATIONS OF THE HEALTH CARE PROVIDER

The clinical guidelines are significant for health care providers because they provide a baseline standard of care for offering access to prenatal testing services. Physicians are under both a legal and ethical obligation to ensure that patients have access to necessary or beneficial services. It is therefore necessary for health care providers to be aware of the clinical indications, to identify them in patients, and to discuss the potential implications with the patient.

Under ethical guidelines, physicians are required to discuss prenatal testing with "all women who have appropriate indications" [SOGC, 1998]. The AMA Code of Ethics also notes that women or couples who do not meet the risk threshold may request prenatal diagnosis as long as they are aware of the risks involved [AMA Code of Ethics, s.E-2.12]. However, physicians are not obligated to perform the service if they lack the skills or knowledge or have a conscientious objection, as long as they engage in discussion with the patient or offer a referral. When a physician feels that he or she lacks the necessary skills or competence to perform a genetic test, it is appropriate to offer referral elsewhere, such as to a clinical laboratory [Grody et al., 2001]. When a physician has a conscientious objection to genetic testing or abortion, physicians must at least alert the parents of a potential problem, presumably using the risk factors as guidelines, so that the parents may seek genetic counseling elsewhere [AMA Code of Ethics, s.E-2.12]. Canadian guidelines also explicitly recognize the duty to offer referral to an alternative provider when the opportunity for a full and frank discussion is constrained. "Failure to discuss all options with women at risk or, alternatively, to refer them further, is unethical" [SOGC, 1998].

In conjunction with the ethical duty to make testing available, the courts have recognized that there is a legal duty to offer tests from which patients might benefit [Andrews and Zuiker, 2003]. The above-noted clinical utility guidelines have come up in some cases. For example, a physician may be liable for failure to warn of increased risks due to advanced maternal age [*Becker v. Schwartz*, 1978], failure to sufficiently take into account ethnic background as a risk factor [*Naccash v. Burger*, 1982], failure to diagnose a condition in a previous child that would have provided genetic information relevant to a second child [*Schroeder v. Perkel*, 1981], and failure to offer testing when there is a known family history [*Phillips v. U.S.*, 1981] or a previous child born with a genetic disorder [*Keel v. Banach*, 1993]. Claims in the context of prenatal testing normally arise under *wrongful birth* and *wrongful life* litigation, so the outcome of such cases will depend on the particular facts of each case and the extent to which it fits within the framework of these causes of action.

The threat of legal action can have a significant impact on the physician–patient relationship. Physicians who practice in states that recognize wrongful life and wrongful birth actions are more likely to offer prenatal testing [Roop, 2000]. The threat of malpractice litigation is a daunting prospect for many health care providers, who may feel the need to practice defensive medicine [Mavroforou et al., 2003]. At the same time, it is important to note that case law should not be relied upon as the source for best practice standards, because "case law is primarily reactive." It only "speaks to the minimum standard of professional behaviour," and it is likely that the public and medical community will expect an even higher standard in practice [Botkin, 2003].

There remains a question about whether the obligation to offer a test will depend on its predictive validity. The majority of cases have dealt with single-gene conditions and highly predictive tests, and it may be that a different standard will apply for more complex genetic conditions and tests of lower predictive validity [Andrews and Zuiker, 2003; *Simmons v. W. Covina Medical Clinic*, 1989]. Such an approach may be shortsighted, given that many patients will value the knowledge obtained from such tests, and will want to hold health care professionals accountable for not providing such information [Andrews and Zuiker, 2003; please see, Chapter 6, Referral and Diagnosis, Wrongful Birth].

PUBLIC POLICY CONCERNS

Recent advances in prenatal diagnosis and assisted reproductive technology (ART) have given rise to a number of ethically challenging cases with respect to access rights. While health care providers can refer to a general standard of care where guidelines exist as to clinical risk measurement, there is confusion and debate about the appropriateness of offering testing in other cases. Some countries offer specific guidelines on access, but others, including the United States, have few laws governing ethically complex cases. Instead, it is often left to clinical judgment [Stern et al., 2003; Adams, 2003].

SEX SELECTION

The issue of whether parents ought to be able to use genetic testing to select their fetus on the basis of sex is controversial and can depend on ethical cultural, and religious perspectives. Sex selection can occur through one of three methods: (1) prenatal diagnosis of fetal sex, followed by termination of pregnancy if it is the undesired sex; (2) preimplantation sex diagnosis followed by selection of an embryo of the desired sex; or (3) prefertilization techniques such as sperm separation. However, the latter option is of relative recent development, and its reliability and safety have not yet been proven [Morales et al., 2004].

The primary medical indication for sex selection is to identify fetuses or embryos that may be afflicted with a sex-linked disorder. While the termination of a fetus continues to raise ethical concerns, the use of PGD to prevent transmission of a serious sex-linked disorder is widely considered ethically permissible [ASRM Ethics Committee, 1999; Morales et al., 2004; Csaba and Papp, 2003]. Sex determination may be the only reliable method of identifying sex-linked disorders [Morales et al., 2004].

More ethically problematic cases are those in which prospective parents request sex selection for personal or cultural reasons, such as preferring a child of a certain sex or desiring a sex-balanced family. These motivations raise a host of concerns, including gender bias and discrimination, allocation of scarce medical resources, risk of psychological harm to children, and overall impact on the human sex ratio [ASRM Ethics Committee, 1999]. Sex selection for family balancing purposes is often distinguished from an inherent sex preference. The latter is a particular concern in societies and cultures that have historically shown a preference for male over female children, such as China and India, where screening techniques, abortion, and infanticide have contributed to an imbalanced sex population. While there is no such evidence of sex preference in North America [Royal Commission on New Reproductive Technologies, 1993; Steinbock, 2002], there remains a concern that allowing sex selection of the first child will reinforce negative attitudes and contribute to sex discrimination [Morales et al., 2004]. With respect to family balancing, some have suggested that it is justified as long as one or more children are already born since this would not involve an inherent favoring of one sex over another [Robertson, 2003]. Appropriate standards are difficult to tease out in this area, for ethical doctrines can be invoked on both sides of the argument. For example, it is argued that modern human rights demand that women not be compelled to maintain pregnancies against their will. On the other hand, human rights are also invoked as the basis for prohibiting sex selection, which is held to violate the right to nondiscrimination on the basis of sex [Cook et al., 2001].

Sex selection for nonmedical reasons is not officially regulated in the United States, but it is prohibited in other countries. For example, in March 2004, the Canadian federal government enacted Bill C-6, which provides that:

5.(1) No person shall knowingly:
(e) for the purpose of creating a human being, perform any procedure or provide, prescribe or administer any thing that would ensure or increase the probability that an embryo will be of a particular sex, or that would identify the sex of an in vitro *embryo, except to prevent, diagnose or treat a sex-linked disorder or disease.*

Contravention of this provision carries serious penalties including the possibility of imprisonment. Unfortunately, the provision is not as clear as it could be and leaves loopholes for nonmedical sex selection during prenatal fetal diagnosis.

While it unmistakably prohibits sex selection for nonmedical purposes at the in vitro embryo stage, it leaves loopholes for nonmedical sex selection during prenatal diagnosis. For one thing, the provision refers only to embryos, which are defined under the statute as organisms during the first 8 weeks of pregnancy. It appears that screening the fetus after 8 weeks for the purposes of sex selection is not expressly prohibited. There is also no mention of what happens when sex is determined incidentally (e.g., during ultrasound for another purpose). Is the doctor either permitted or obligated to disclose this information to the patient?

In the United States, the Ethics Committee of the American Society for Reproductive Medicine (ASRM) released a policy statement discouraging the use of PGD for nonmedical reasons, given concerns about the burdens and costs it imposes [ASRM Ethics Committee, 1999]. The AMA Code of Ethics also censures "selection on the basis of non-disease related characteristics" [AMA Code of Ethics, s.E-2.12]. Several world organizations also discourage sex selection for nonmedical reasons [World Health Organization, 1999; International Bioethics Committee of UNESCO, 2003].

Nonetheless, these guidelines are not enforceable as law, and there is evidence that a number of clinics are advertising and offering in vitro fertilization (IVF) services for the purpose of sex selection.

The emergence of preconception selection techniques, which do not involve discarding embryos or fetuses, has added a new dimension to the ethical debate. The Ethics Committee of the ASRM and others have noted that if prefertilization techniques were found to be safe and reliable, it may be ethical to use them for family balancing purposes [ASRM Ethics Committee, 2001; Robertson, 2001].

The sex of a fetus may be determined incidentally to another procedure, such as fetal karyotyping, and a parent may or may not wish to be informed of the sex. As will be discussed below, women have the right to refuse information, just as they have the right to receive it. Therefore, for both legal and ethical reasons, it is best to ask patients in advance whether they want the sex to be disclosed and to allow them to opt not to know the sex of an embryo or fetus [Mavroforou et al., 2003; International Federation of Gynecology and Obstetrics (FIGO), 2004]. Conversely, the ACOG Committee on Ethics recommends that this information not be withheld from a pregnant woman who requests it, given that she has a right to her own medical information [Morales et al., 2004]. If a physician is uncomfortable with the possibility of indirectly contributing to sex selection, he or she ought to clarify the patient's goals in advance and explicitly inform patients of any procedures they are not willing to perform [Morales et al., 2004].

TESTING FOR SUSCEPTIBILITY AND LATE-ONSET CONDITIONS

A further ethical concern is the use of PGD to test for genes that increase one's susceptibility to disease, such as cancer, or late-onset conditions, including Huntington disease and Alzheimer's disease. The two situations are somewhat different: Late-onset conditions have full penetrance and are generally not preventable, while susceptibility genes do not have full penetrance and may be amenable to treatment. However, it has been argued that the ethical issues are similar in each case, given the substantial, nontrivial burden imposed by the diseases, which may serve to justify testing [Sermon et al., 2004; Robertson, 2003]. The discomfort with susceptibility testing relates to the use of PGD to summarily discard embryos that are only associated with an increased risk, not certainty, of disease [Genetics and Public Policy Center, 2004]. Concerns about adult-onset testing include the fact that a child could enjoy years of healthy life and that a treatment or cure may yet be found [Genetics and Public Policy Center, 2004b]. Late-onset testing also has an additional "ethical twist" in that the prospective parents will themselves be carriers of the disease because the genes are dominant. The shortened life span of the parent is a concern because the child may lose the parent while still dependent on him or her [Robertson, 2003; Towner and Loewy, 2002]. However, this may still be an ethically justifiable procedure, given that people with disability have no less interest in reproduction than others, and that denying services in this case could create a slippery slope to other restrictions [Robertson, 2003].

The tests are currently legal in the United States, subject only to gene patent restrictions [Robertson, 2003], but this does not mean that practitioners are obligated to provide the service. It has been argued that the adult-onset risk is not sufficient to warrant this type of testing as a standard of care [Botkin, 2003]. Some ethical committees feel that it is too early to offer a clear opinion on the acceptability of these tests, but an European Society of Human Reproduction and Embryology (ESHRE) Ethics Task Force noted that PGD for adult-onset conditions is acceptable, in spite of limited knowledge about potential therapy in the time gap before onset [Shenfield et al., 2003]. A United Nations Educational, Scientific, and Cultral Organization (UNESCO) committee has suggested that susceptibility testing should be given a low priority and restricted to cases of high risk of severe disease [International Bioethics Committee of UNESCO, 2003]. The National Society of Genetic Counselors (NSGC) has urged caution in the use of adult-onset testing.

Genetic counseling is essential and must include discussion of psychological and social risks and benefits, including the potential for future discrimination and the possibility of disclosing carrier status of other family members. Parents ought to consider whether the decision to test should be reserved for their offspring to make upon reaching adulthood [NSGC Position Statement, 1995].

SELECTING FOR DISABILITY

A difficult issue arises when a woman or couple, who themselves possess a disability or a unique characteristic, wish to have a child with that same characteristic. For example, some deaf couples have expressed a desire to use prenatal testing or PGD to select a child with deafness. For them, deafness is not considered a disability but rather a defining feature of their cultural identity. Similar cases could arise, including dwarves wishing to select a dwarf child or people with an intellectual disability wishing to select a child with the same disability [Savulescu, 2002].

A deaf lesbian couple attracted much criticism in the United States when they attempted to have a deaf child by intentionally using a deaf sperm donor [Savulescu, 2002]. One argument repeatedly raised is that it was a selfish action on the part of the parents. It is argued that intentionally selecting for deafness is not in the best interests of the child because it restricts the child's future options and thereby threatens her autonomy [Davis, 1997]. Even a member of the National Association of the Deaf had trouble understanding the couple's decision because deaf people "don't have as many choices" [Spriggs, 2002].

On behalf of the parents, it has been argued that deaf people can be considered a minority group who merely suffer disadvantage as a result of societal discrimination not of deafness itself. "In this sense, deafness is strictly analogous to blackness" [Levy, 2002]. Second, it is questioned whether a child selected for deafness actually suffers harm as a result. She is not worse off than she would have been—as might be the case if the parents had refused to allow treatment for a correctable condition— because the alternative is that she would never have been born; a different child would have been selected instead. Third, there is an argument based on freedom of choice. Although some members of society may view deafness, or dwarfism, or intellectual disability in negative terms, these value judgments should not be imposed on the couple making the reproductive decision. This argument reminds one of the terrors of the Nazi eugenic program, during which couples deemed "unfit" to create the perfect child were forced to undergo sterilization [Savulescu, 2002]. Indeed, parental choices will influence most of the important aspects of a child's life—from education, to religion, to social interaction—which in turn form the basis for the child's freedom. It is difficult to draw the line between those parental decisions that are the preconditions for the child's freedom of choice, and those which foreclose too many options for the child[Levy, 2002].

Few ethical guidelines have been established in this area. UNESCO's International Bioethics Committee rejects the use of PGD to purposely select a child with a genetic disease, and it includes deafness and dwarfism in its definition of disease. This practice is held to be "unethical because it does not take into account the many lifelong and irreversible damages that will burden the future person" [International Bioethics Committee of UNESCO, 2003].

Potential Stigmatization of Persons with Disabilities

It is important to recognize that for some, the very practice of PND is problematic and raises challenging ethical issues. This means that health care providers and patients can face conflicting messages about the value and acceptability of these tests. The disability rights movement has provided a strong critique of prenatal genetic testing, drawing heavily on concerns about its eugenic implications. Parens and Asch [2003] provide a useful summary of the main arguments. First, it is argued that prenatal testing and selective abortion are against public policy because such practices can lead to stigmatization of individuals living with disability. In particular, there is a worry that the aim of genetic testing is "to eradicate disability and reduce the number of births of genetically disabled individuals" [DeVaro, 1998]. This is said to be harmful in the message it sends about the worth of a life with disability and for ignoring the rich life experiences and useful contributions that disabled individuals can bring to society. In response, proponents of prenatal testing object that the purpose of the procedure is not to eradicate disability but rather to increase the reproductive autonomy of women. Also, prenatal testing does not necessarily entail termination of the fetus; it can provide other useful information and allow for interventions [Chen and Schiffman, 2000; Mahowald, 2003]. However, in practice, there is clearly a tension between the goals of enhancing reproductive choice and preventing disability, particularly given evidence that the success of some genetic screening programs is measured in terms of reduction in disability [Parens and Asch, 2003; Beaudet, 1990], and that some women perceive pressure from the social system and health professionals to undergo testing and to terminate for disability [Fox, 2002; Mahowald, 2003]. Second, there is a concern that the "selective mentality" of prenatal testing, in which parents strive for perfection and lament over disability, fosters a morally deficient view of parenthood [Parens and Asch, 2003]. On the other hand, a distinction can be drawn between parental motivations during prenatal testing of a "potential child" and parental care for a disabled newborn [Mahowald, 2003]. Third, disability rights advocates argue that genetic testing is based on a misunderstanding about what life is like for individuals living with disability. While it is true that some families experience stress and disruption, on average the needs of families with and without disability can be "strikingly similar" [Parens and Asch, 2003].

Some suggest that the practice of selective abortion for disability can be criticized on similar grounds as sex selection. Wolbring [2003] notes, for example, that both practices raise concerns about "significant threats to the well-being of children, the children's sense of self-worth, and the attitude of unconditional acceptance of a new child by parents." Moreover, concerns about stereotypes, discrimination, and oppression of persons of the unwanted sex can apply equally to stereotypes, discrimination, and oppression of individuals with the unwanted disability [Wolbring, 2003; Wong, 2002]. According to Wolbring [2003], the only way to justify disability deselection, while at the same time prohibiting sex selection, is by "arbitrarily" labeling disabilities and diseases as medical problems, thereby applying a different moral standard to them. He suggests that this approach is buttressed by the marginalization of people with disabilities and the corresponding exclusion of disability rights critiques from the bioethics discourse [Wolbring, 2003]. It is therefore important to take seriously the disability rights critique so as to prevent a purely medical model of genetic testing from standing in for a rights-oriented model (please see, Chapter One, Genetic Counseling and the Physician–Patient Relationshi—Trust, Genetic testing, and the Physician–Patient Relationship).

Where Do We Draw the Line?

Parens and Asch [2003] set up a working group to compare and contrast the disability rights critique and the advocacy of prenatal genetic testing and to offer recommendations. Although the group reached consensus on a number of issues, members reached an impasse at the critical question: Is there a helpful and rational way to distinguish between tests that providers should routinely offer and those they should not? According to the group, it comes down to whether a test is judged reasonable or unreasonable under the circumstances, and such line drawing can be an impossible task [Parens and Asch, 2003; International Bioethics Committee of UNESCO, 2003]. It is highly dependent on individual ethical judgment [Garel et al., 2002]. However, many members of society would value some kind of line drawing, whether for ethical reasons [Henn, 2000] or to limit malpractice liability among the medical profession [Botkin, 2003]. Disability groups also fear that a lack of guidelines will encourage more testing and potentially greater intolerance of disability [Parens and Asch, 2003; Wasserman, 2003]. Accordingly, there have been some attempts to draw workable lines between tests or conditions [Murray, 1996; Strong, 1997; Botkin, 2003; Council on Ethical and Judicial Affairs, 1993]. For example, Botkin draws a line between childhood conditions that have significantly adverse effects on the parents ("in terms of heartache, worry, time, effort, and money") and those that do not. Taking a cue from tort law, he suggests that the magnitude of the potential challenge in raising the disabled child can be one criterion for offering prenatal diagnostic testing [Botkin, 2003]. However, the definition of precisely what constitutes a serious parental challenge, and any other potential considerations, will need to be fleshed out with further ethical guidelines.

ABORTION LAW: THE UNITED STATES AND CANADA

The United States

Since the landmark case of *Roe v. Wade*, women in the United States have a constitutional right to abortion. In that case, the U.S. Supreme Court held that the right to privacy included the right to decide to have an abortion. However, this latter right was not absolute; it could be limited by a compelling state interest. The Court went on to devise a rule aiming to balance a woman's right to privacy with the state's interest in protecting her health and its interest in the potentiality of human life. The result was a rule based on a recognition of increasing state interest as the pregnancy progressed. During the first trimester of pregnancy, the woman's right to privacy was prevalent, and the decision to abort was left to the medical judgment of the woman and her physician. After the end of the first trimester, the state could regulate abortion "to the extent that the regulation reasonably relate[d] to the preservation and protection of maternal health." Finally, after the fetus became viable, the state could limit and even proscribe abortion, except where it was necessary to protect the life or health of the mother [*Roe v. Wade*, 1973].

The *Roe v. Wade* decision did not settle the abortion issue in the United States. On the contrary, it opened up a highly polarized debate. States began passing laws restricting access to abortion through a variety of means, including the restriction of funding, the imposing of counseling and waiting periods, the requirement of parental consent for minors, and the prohibition of certain procedures [Greene and Ecker, 2004; Hull and Hoffer, 2001; Peterson, 1996; Partial Birth Abortion Act of 2003; Alan Guttmacher Institute, 2004b]. Some of these laws have been successfully challenged, though many have been upheld by the Supreme Court. The Court has somewhat eroded the Roe decision through its subsequent rulings, but its central finding—that a woman has a right to choose abortion before the fetus becomes viable—still holds [Peterson, 1996]. States are therefore free to prohibit all late-term or postviability abortions, except where the woman's life or health is in danger. Nineteen states have adopted such statutes and 17 others have adopted statutes that do not meet the Supreme Court's criteria because they permit abortion only where there is a threat to the life of the woman or because they prohibit abortion after a specific gestational age [Alan Guttmacher Institute, 2004a]. The problem in the context of prenatal genetic diagnosis is that almost all these states do not recognize an exception in cases of fetal anomaly. Because of the gestational age at which genetic tests can be safely and effectively conducted and the time required to obtain results, some women may be placed in a situation where their state laws prohibit them from choosing to terminate their pregnancy. This situation may become increasingly common as technology enables the medical profession to push the fetal viability limit earlier in the gestation period.

A number of additional barriers limit women's access to abortion in the United States [Henshaw and Finer, 2003]. The first of these is financial: Many states do not provide any public funding for abortions [Peterson, 1996]. Women or couples who do not have private health insurance, therefore, have to pay for the procedure themselves. Second, the number of physicians who provide abortion services is limited. The violence of some elements of the antiabortion movement in the 1990s and the lack of training in medical schools has had a significant impact on the shortage of providers. In many regions, women may have to travel more than 50 or 100 miles to find a physician or a clinic [Finer and Henshaw, 2003; Henshaw and Finer, 2003; Joffe, 2003; Peterson, 1996; Dresser, 1994]. This, compounded with legal barriers such as the obligation imposed on physicians to provide counseling and to require patients to wait 24 hours before the procedure or, alternatively, to evaluate fetal viability, creates a substantial burden for patients who already are in a difficult situation [Henshaw and Finer, 2003].

Canada

A discussion of prenatal diagnosis would not be complete without a discussion of the legal framework governing access to abortion. Until 1988, abortion was regulated as a criminal act in Canada. The *Criminal Code* established a set of rules pursuant to which women could have limited access to therapeutic abortions in Canada. It provided that an abortion only could be performed by a qualified physician in a hospital accredited by the provincial government. In addition, a therapeutic abortion committee had to certify that continuing the pregnancy would or would likely endanger the patient's life or health. The committee was to be composed of three additional physicians from the hospital in which the procedure was to be performed. Abortions that did not meet these criteria carried the possibility of life imprisonment for the physician and 2 years' imprisonment for the female patient [s. Criminal Code: 251].

However, in 1988, following a constitutional challenge, the Supreme Court of Canada struck down section 251 of the *Criminal Code*, stating that it violated the right to security of the person as guaranteed by the *Canadian Charter of Rights and Freedoms* because of the procedural defects its application entailed. One Supreme Court judge also concluded that section 251 violated women's freedom of conscience and religion, as well as their right to liberty [*R. v. Morgentaler*, 1988]. The Court did not conclude that abortion could not be criminalized nor that women had a positive right to access abortion services; these issues remained open. In 1990, the federal government attempted to pass a bill that would have recriminalized abortion, but this effort was defeated in the Senate in 1991 [Bill C-43]. Subsequent federal governments have not attempted to legislate abortion since then. Consequently, abortion is not a crime in Canada if it is performed by a qualified medical practitioner, with the consent of the woman undergoing the procedure [Rodgers, 2002].

This does not mean, however, that abortion is completely unregulated. On the contrary, it is regulated in the same complex manner as all other medical procedures and therefore subject to provincial and territorial laws governing health care professionals, hospitals, and access to health care services [Rodgers, 2002; Farid,

1997]. The provinces and territories have adopted different approaches to the issue of abortion, leading to unequal access to these services across Canada.

For example, the funding of abortion services varies among provinces and territories. Provincial health insurance plans in British Columbia, Alberta, Ontario, Newfoundland, and Labrador cover the full cost of abortions performed in both hospitals and clinics. Those in Saskatchewan, Manitoba, New Brunswick, Prince Edward Island, and the territories fund only abortions performed in hospitals. Québec and Nova Scotia provide full funding for abortions performed in hospitals but only partial funding for those performed in clinics [Childbirth by Choice, 2003].

While some provinces such as British Columbia and Ontario have taken steps to protect women's right to access abortion services without undue interference [Access to Abortion Services Act; *R. v. Lewis*; *Ontario (A.G.) v. Dieleman*], in other jurisdictions legislatures have limited access by imposing funding criteria such as the requirement to obtain certification by a committee of physicians that the abortion is necessary because the health of the woman and/or fetus is at risk [Health Services Payment Regulation; *Morgentaler v. Prince Edward Island* (Minister of Health and Social Services), 1995; Moulton, 2003].

Access to abortion services is also hampered by the lack of availability of services. In 2003, only 17.8 percent of hospitals in Canada were providing abortion services [Canadian Abortion Rights Action League, 2003]. Abortion clinics provide services to a significant number of women, but they are not found in every province or in the territories. Abortion services, whether in hospitals or clinics, are concentrated in large urban centres, leaving many regions underserviced. In addition, abortions are not performed at all in Prince Edward Island or in Nunavut [Childbirth by Choice, 2003]. In many jurisdictions, this lack of services results in long waiting lists and compels women to travel to another region or province to obtain an abortion [Childbirth by Choice, 2003]. This may constitute a significant barrier to access given the costs of travel as well as the delay involved.

This has serious implications for couples who wish to terminate a pregnancy after the prenatal diagnosis of a genetic disorder. Moreover, in many regions of Canada, abortions are performed only up to 12 or 16 weeks gestation. Though some hospitals do perform abortions after 20 weeks in cases of severe disability or risk to the woman's life or health, these are limited in number [Child birth by Choice, 2003]. The delay required to conduct prenatal testing and obtain results combined with the waiting list for abortion services may significantly limit the woman's or couple's choices.

INFORMED CONSENT AND GENETIC COUNSELING

As discussed in Chapter 7, Informed Consent, health care professionals have both a legal and ethical duty to ensure informed consent before providing treatment. In the context of PND and PGD, informed consent recognizes a woman's right to reproductive autonomy, which means both that women have the freedom to choose whether to procreate or not procreate [Strong, 2003], and that they have the right to refuse medically indicated testing [Morales et al., 2004]. Reproductive autonomy is promoted when women or couples are provided information relevant to decisions about whether to continue the pregnancy [Strong, 2003]. The duty demands a full disclosure of information to the patient, including available tests, alternatives, risks and benefits, and outcomes.

Ethical guidelines recognize the importance of ensuring informed consent to prenatal testing [ACOG Code of Ethics, s.I(5), 2003; NSGC Position Statement, 1991; NSGC Code of Ethics, s.II(3), 2003; AMA Code of Ethics, s.E-2.12, 2003]. For example, the ACOG Code of Ethics requires obstetrician-gynecologists to disclose all relevant medical facts, including "alternative modes of treatment and the objectives, risks, benefits, possible complications, and anticipated results of such treatment" [ACOG Code of Ethics, 2003 s.I(5)]. The ACOG also recognizes that informed consent implies not only provision of information, but also free choice and active decision making on the part of the patient; it is "not only a 'permitting' but a 'doing'" [Morales et al., 2004]. Where patients are referred to a clinical laboratory, the responsibility for obtaining informed consent remains with the health professional, though laboratories may be required to document the consent depending on jurisdiction. In turn, laboratories must provide sufficient information to doctors to facilitate the informed consent process [Bradley et al., 2004; Richards et al., 2002; Holtzman and Watson, 1997].

Courts also have noted the importance of providing information relevant to reproductive decisions. As one court noted, "society has a vested interest in reducing and preventing birth defects," and the failure to perform a procedure "which would have yielded information material to the parents' decision whether to abort the fetus, constitutes a breach of . . . duty" [*Blake v. Cruz*, 1984]. Therefore, health care providers are under an obligation to provide information and can face legal liability for failure to enter into a full discussion with the patient. Indeed it has been noted that most medical malpractice suits are based on inadequate communication between health care provider and patient rather than error [Mavroforou et al., 2003]. In the context of prenatal care, the legal duty to provide information has been well-established in wrongful birth cases in which the parents allege that the physician's negligence deprived them of the opportunity to make a reproductive choice regarding conception or termination.

New and difficult questions about informed consent arise in the context of PND and PGD. Prenatal diagnosis is an evolving and complex procedure, characterized by "inevitable ambiguity, uncertainty, and difficult decision making," and involving many ethical dilemmas, social implications, and grave potential for diagnostic errors that are not seen in other kinds of medical intervention [Strauss, 2002]. Perhaps the most unique aspect of genetic testing is that the desirability of treatment depends highly upon moral and ethical judgments; in other words, "there is no single 'right' answer for all women and couples, only answers that are right for the individual woman or couples, based on personal circumstances and values" [Royal Commission on New Reproductive Technologies, 1993]. Respect for an individual or couple's values and beliefs is of "paramount importance" [Holtzman and Watson, 1997]. This may give rise to unique legal and ethical models of informed consent. For example, while Canadian courts typically rely on what a "reasonable" patient or physician would decide, it has been argued that a "genetics model of informed consent" has evolved that is more about incorporating the decision maker's (i.e., the patient's) personal values into the process of decision making [Sharpe, 1997].

The recognition of differing value judgments forms the basis of the principle of nondirectiveness, which is central to genetic counseling. According to this principle, the role of the health professional is:

> ... not to lead clients to make particular decisions or choices (those preferred or recommended by the clinician, the health service or by society) but to help them to make the best decisions for themselves and their families as judged from their own perspectives. [Clarke, 1997]

The AMA Code of Ethics notes that counselors should not substitute their own personal values and moral judgment for that of the prospective parent [AMA Code of Ethics, s.E-2.12, 2003]. It must be remembered that not all women are the same, and that a woman and her physician may have radically different perceptions of risk and burden [Csaba and Papp, 2003]. The counselor or physician therefore should ascertain the patient's personal values and expectations and use them as the framework for the decision-making process. Several reasons for the emphasis on nondirectiveness have been identified, including respect for individual autonomy, a desire to disassociate current genetic practices from the history of eugenics, prevention of overinvolvement by health care providers, and protecting professionals from legal action for medical decisions [Williams et al., 2002]. However, as will be discussed below, there are questions about whether nondirective counseling is possible, or even desirable, in practice.

BARRIERS TO INFORMED CONSENT IN THE PND CONTEXT

Before delving into the requirements for informed consent, it is important to examine the unique barriers to informed consent that arise in the context of prenatal diagnosis. Several studies have unfortunately revealed that many women are not given the tools to exercise an informed choice [Marteau and Dormandy, 2001; Kohut et al., 2002], which no doubt reflects the extent to which health care providers and patients are struggling to grapple with some of these barriers.

First, this area of medicine is highly complex, involving a variety of different screens and tests and a multitude of detectable anomalies. The general public may lack knowledge of the intricate science of genetics [Lanie et al., 2004]. Also, given that prenatal testing involves probabilities instead of certainties, patients can feel frustrated by the ambiguity and vagueness of the genetic predictions when attempting to make an informed decision [Royal Commission on New Reproductive Technologies, 1993]. Discussions of prenatal diagnosis will be unique from other types of medical discussions because many prospective parents will have no first-hand knowledge of the disorders being diagnosed. They therefore will be "almost entirely dependent on their counselors for information about disabilities and may have difficulty imagining the various possibilities and options" [Royal Commission on New Reproductive Technologies, 1993]. There also is a danger that health professionals themselves may lack sufficient knowledge, especially as services are increasingly performed by primary care physicians or obstetricians, rather than geneticists [Ormond et al., 2003; Marteau and Dormandy, 2001; Abramsky et al., 2001; Holtzman and Watson, 1997]. Furthermore, the rapid increase in technology means that an increasing amount of information is available for disclosure. This raises concerns about the detrimental impact of information overload on a patient's ability to understand the process and to make an informed choice.

A second barrier to informed consent is the significant time constraints imposed by the PND context. Not only does the woman or couple face time pressures while trying to process information, but the health care provider also lacks the time to present information adequately [Marteau and Dormandy, 2001; Holtzman and Watson, 1997]. These time pressures become increasingly important when the woman wishes to consider abortion because the longer an abortion is delayed the more risks and trauma it involves for the woman [Royal Commission on New Reproductive Technologies, 1993].

INFORMED CONSENT FOR PND: WHAT TO INCLUDE IN THE DISCUSSION

It is essential to remember that prenatal tests only are *offered* to women and couples. They are not mandatory, and the choice to undergo testing remains with the woman [Kohut et al., 2002]. It is therefore essential that as much information as possible be discussed *prior* to testing, not merely after a diagnosis has been made. It may also be prudent to discuss the genetic counseling service itself, including issues such as confidentiality and nonpaternity, given that many patients will be unfamiliar with the service [Jacobson et al., 2001]. Several elements have been repeatedly identified as central to a well-informed decision with respect to PND.

Information about the Procedures

Under Canadian law, courts tend to impose a very high standard of disclosure for services that affect reproductive capacity because they are generally not life or death situations [Sneiderman et al., 2003]. Ethical guidelines similarly demand a very comprehensive disclosure of information. The necessary information includes available methods of prenatal diagnosis; the difference between screening (e.g., ultrasound) and diagnostic testing (e.g., amniocentesis); risks and benefits of various techniques, including the risk of pregnancy loss; the timing of the procedures; details of the conditions for which procedures screen or test; frequency of abnormal results; accuracy of results, including the implications of false positives and false negatives in screening; the frequency of need for repeat testing; the possibility that abnormalities may go undetected; and, for relatively new and untested procedures, a detailed explanation of their uncertainty and experimental nature [SOGC, 1998; Chodirker at al., 2001].

Overall, there has been a trend toward providing an increasing amount of detailed information, and it has been suggested that this is unlikely to reverse in the future [Mavroforou et al., 2003].

The method of presentation of such complex information can have a significant impact on the patient's reaction and interpretation. Describing every possible risk and consequence in alarming detail can distort the patient's comprehension and impair the informed consent process.

> *It is possible to be frank, honest, and direct without being either grim or frightening. It is possible to discuss risk and still leave hope. Pessimism may decrease the effectiveness of the communication and lead to increased patient anxiety.* [Kelly, 1992]

It is unclear whether physicians are under a duty to provide information about procedures that appear unavailable to the woman. It has been suggested that physicians may now be obliged to inform patients of the inadequacies of the health care system and of the availability of better diagnostic or treatment options elsewhere in Canada or in the United States [Osborne, 2003]. Certainly, if the procedure is unavailable due to lack of hospital resources, then the patient should be informed if it is accessible elsewhere. But if the procedure is unavailable because the cost is too prohibitive or because it is only available in distant countries, "it is not obvious that physicians are legally bound to tantalize" the woman with options beyond her reach [Dickens, 2002]. There may also be no duty to disclose treatment that is not generally recommended [*Munro v. Regent of the University of California*, 1989]. The physician may want to cater the information to options that are within patients' reasonable access. On the other hand, one may want to risk erring on the side of overinforming rather than underinforming because the couple may be willing to make sacrifices the doctor does not consider reasonable or possible [Cook et al., 2001], and doctors may face liability for failing to inform of alternatives that the patient would have considered.

The distinction between prenatal screening and diagnostic testing is a particularly important piece of information that must not be neglected. The increasing "routinization" of ultrasound screening means that it is occurring with greater regularity than diagnostic testing and is more likely to be performed by a general practitioner or obstetrician than a genetic specialist [Suter, 2002]. Unfortunately, studies have documented the poor levels of information conveyed by nongenetic specialists at early stages of pregnancy [Bernhardt et al., 1998; Marteau et al., 1992]. It is essential that women be given the opportunity to opt out, without detriment to their further prenatal care [Mavroforou et al., 2003]. Particular attention must be paid to the woman's potential lack of information about these screening procedures. First, many women are not aware of the full extent of genetic information that can be revealed, and some are not even aware that ultrasound can detect fetal abnormalities at all [Kohut et al., 2002]. Women must be properly informed of this possibility because some may elect not to have the procedure for this reason, or they could at least be better prepared to hear the results. If the woman is not properly informed, and a genetic problem is detected, there is a question about whether this should be disclosed. This is a difficult ethical issue, and the best recommendation is that patients be warned of such possibilities prior to testing so that they can express their wishes with respect to disclosure [Ontario Law Reform Commission, 1996].

Second, patients must be fully informed about the limitations of the screening procedure (the fact that not all anomalies can be detected) and about the risks and benefits involved [Driscoll, 2004; SOGC, 1999; Kohut et al., 2002].

Third, while diagnostic testing can detect the presence of a disability, screening only identifies at-risk individuals. Patients should be informed that a screening test is not a replacement for a diagnostic test, and that detection of a fetal abnormality through screening may necessitate a more invasive diagnostic test [Driscoll, 2004]. Otherwise, women who would not choose diagnostic testing will end up facing a very difficult decision after receiving an abnormal screening result. If they reject the further diagnostic test, "the rest of the pregnancy will be fraught with tension, anxiety and worry." This unnecessary anxiety could have been avoided by refusing the initial screening test [Suter, 2002].

Fourth, it is important to clarify the distinction between probabilities revealed through screening (including the risks of false positives) and confirmatory diagnoses. To enhance understanding of the numbers, evidence shows that quantifying uncertainty in frequencies rather than probabilities, and presenting risks in words rather than numbers is most helpful [Marteau and Dormandy, 2001].

Finally, even women who fully understand this information may face increased anxiety upon a false-positive screening result, feeling that "something must be wrong," and it is recommended that this risk of anxiety be discussed [Suter, 2002].

Information about the Genetic Conditions

It is essential to provide women information not only about the available procedures but also about the conditions that can be detected through testing. First, many patients may not have sufficient understanding of basic genetics, and it is important to clear up misconceptions (e.g., heredity, genetic terminology) [Lanie et al., 2004]. Second, it is important to be as comprehensive as possible about the types of conditions that may be detected to enable an informed choice about whether they want the information disclosed. It must be remembered that although women are entitled to receive information, this does not mean they are under a duty to do so; in other words, they have the "right *not* to know." This is significant in the genetic context because a genetic diagnosis may reveal unwelcome information [Dickens, 2002]. Providing fetal information to a parent who did not request it may cause a number of harms, including forcing the couple to make difficult choices about continuing the pregnancy, changing the parents' attitude toward the pregnancy and the "abnormal" child, and affecting others in the family by revealing genetic information about their own health [Boyle et al., 2003]. As one woman recently stated, "Knowing is not always best" [Kohut et al., 2002]. It is therefore essential to inform patients about the various anomalies that a procedure may detect, to give them an opportunity to (a) accept or reject the procedure as a whole and (b) choose which information they would like disclosed to them should anything be detected [Boyle et al., 2003; FIGO, 2004].

Physicians also should be cautious about attributing a woman's reluctance to discuss fetal anomalies to an informed choice not to know. The reluctance may simply stem from a fear of the unknown, and physicians should attempt to allay this fear [Kohut et al., 2002]. Unfortunately, the breadth of information available means that some information will be conveyed more often than others. For example, it has been found that women are informed less often about the possibility of detecting a severe congenital anomaly (SCA) than the possibility of autosomal trisomies, leading to fear and confusion when SCAs are detected [Petrucelli et al., 1998].

Providing information about conditions is a difficult balancing act. On the one hand, respect for a patient's autonomy demands that counselors convey as much information as is practical to convey, but on the other hand too much information might be confusing or overwhelming for the patient. Because genetic information is framed in language of probabilities and statistics, practitioners must be especially careful not to be misleading and to communicate at a level appropriate to the patient [ABMG, 2004; ABGC, 2004; CCMG, 2004b].

Range of Options Available upon Finding an Abnormal Result

Couples must be informed of the full range of management options available when confronted with an abnormal result. These options include termination of the affected pregnancy or continuing with the pregnancy while preparing for the birth of the child [Chodirker et al., 2001; Strong, 2003]. Studies have shown that information on the first option is given more often than information on the second option, but even that is not always provided [Marteau and Dormandy, 2001].

With respect to abortion, the woman must be informed of the various abortion procedures and their availability, including the time frame during which abortion is legally available, its availability in the local area, and the nearest facilities where it can be obtained [Strong, 2003; Chodirker et al., 2001b]. If a physician has a conscientious objection to abortion, or lacks the necessary resources to perform abortions, this does not justify failing to inform the patient of the option or failing to offer a referral [Strong, 2003; Dickens, 2002; CMA Code of Ethics, 2003; AMA Code of Ethics s.E-2.12, 2003; FIGO, 2004].

Although abortion is to be discussed as an option, women must be advised beforehand that an agreement to terminate a pregnancy if an abnormality is found is *not* a precondition for testing [Chodirker et al., 2001b; SOGC, 1998; FIGO, 1991, 1995]. In other words, "the link between prenatal diagnosis of fetal anomaly and termination of pregnancy is *potential* only" [Bennett, 2001]. This is important because studies show that a decision to undergo prenatal testing is not always linked to the intention to terminate pregnancy [Lafayette et al., 1999]. On the other hand, prenatal testing may not be valuable for women who would not consider an abortion. Counselors should help women avoid the "potentially 'toxic' knowledge" of a fetal anomaly by encouraging them to consider whether abortion is a viable option and what prenatal diagnosis would mean to them [Suter, 2002].

Social Aspects of Disability and Raising a Disabled Child: Stigma, Prejudice, and Access to Resources

It is important to discuss social issues related to disability, including the long-term implications of living with a disease, the prognosis for the disease, and the expected quality of the child's life [Brookes, 2001; Royal Commission on New Reproductive Technologies, 1993]. It can be helpful to describe what the child might look like [Petrucelli et al., 1998]. Unfortunately, health care providers are educated too rarely on the implications of life with disability and may have difficulty conveying such information [Parens and Asch, 2003]. Different health care professionals may assess quality of life in different ways. There is evidence that most health care trainees focus on medical aspects of quality of life, instead of personal and social variables. Genetic counseling students are more likely than other medical students to prioritize the personal and social factors in discussions of quality of life [Ormond et al., 2003]. The perceived seriousness of disability and the negative social predictions can also vary depending on the person [Marteau and Dormandy, 2001]. This may stem from lack of knowledge and experience in dealing with disability. One mother noted that, "without her experience of caring for a child with a genetic condition, the negative picture presented by her medical practitioners in a later pregnancy would have undermined her confidence in her ability to care" [Brookes, 2001]. It is therefore important for health care professionals to receive education and training in order to present an accurate picture of disability. This will necessarily involve sensitivity training and consultation with a variety of groups to achieve a balanced picture of life with disability [Ormond et al., 2003; Marteau and Dormandy, 2001; Parens and Asch, 2003].

Parens and Asch [2003] note that there is some uncertainty about the ideal moment to engage in this discussion. Prior to the screening test, it may be impractical to convey that much information; after a positive screen, parents may be in shock and may feel threatened by such a discussion. Parents might be most willing to gather information after a positive diagnosis, while on the other hand it may be too stressful to process new information at that point. One clear guideline is that if a woman or couple requests information, they should receive it [Parens and Asch, 2003].

Social Pressures to Make the ™Right Decision

Informed choice is threatened by various social pressures that influence a woman or couple's decision in the context of prenatal testing. The very existence of PND makes it possible to view the birth of a disabled child as a woman or couple's "choice," thereby making the woman or couple accountable in the event of discriminatory reactions toward the child and themselves [Brookes, 2001]. Social pressures stem not only from direct stigmatization but also from the lack of economic or social resources for those living with disabilities. It has been recommended that anyone undergoing genetic counseling should be fully informed about the social pressures they may experience, "so that they have an opportunity to consider how such pressures might affect them" [Royal Commission on New Reproductive Technologies, 1993].

At the same time, however, such pressures can originate in the medical community itself. In contrast with the principle of nondirectiveness, studies have found that mothers experience a need to "push against" both overt and covert pressure to undergo invasive PND [Brookes, 2001], and that some are left with the impression that they were not actually given a choice [Kohut et al., 2002]. These pressures are of particular danger when procedures such as ultrasound are offered in a routine "matter of fact" manner. Thus, women ought to be explicitly informed that they have the right to refuse or accept any aspect of the offer [Kohut et al., 2002].

EMOTIONAL AND PSYCHOSOCIAL COUNSELING AS PART OF THE INFORMED CONSENT PROCESS

Another unique aspect of PND relative to other medical services is that it is not concerned with treatment and cure but rather with probabilities and predictions. Given the highly unsatisfactory nature of such information, the anxiety it may cause, and the potential for grave errors, an important aspect of prenatal care in the context of PND is the provision of emotional support. This is based on the ethical principle of beneficence, which requires physicians to reasonably attempt to prevent and remove health-related harms to their patients. When the woman requires help, including emotional support, and the health professional is in a position to offer it, the professional is under a duty to do so [Strong, 2003]. It has been suggested that this type of emotional and psychological counseling imposes upon genetic testing a distinct model of care, based upon a therapeutic *human vision* rather than a mere *medical vision* [Sharpe, 1997]. Not only can emotional distress threaten the patient's well-being, but it can significantly impair his or her ability to comprehend the information. There may be a duty to assess and respond to the patient's emotional and psychological needs to ensure that the communication process is not impaired [Sharpe, 1994b, 1996].

PRE- AND POSTTEST COUNSELING

Before the patient even chooses to undergo testing, counselors may have a duty to inform her of the risk of anxiety subsequent to the test [Suter, 2002]. After testing, it is the responsibility of the health care provider who initiates referral to ensure that, when test results are received, they are reviewed and reported to the woman as quickly as possible [SOGC, 1998]. Because the initial shock of receiving a positive diagnosis may cause "psychological devastation," physicians must respond to the psychological and emotional concerns of the parents, inform them of the potential for mistaken diagnosis, and inform them of the full range of options available [SOGC, 1999]. Although many physicians receive little or no training in giving "bad news" and feel uncomfortable doing so, there is a growing body of literature that can be consulted regarding effective techniques for delivering bad news in these situations [Strong, 2003].

Follow-up Counseling

Following the diagnosis of a severe disorder, or the decision to terminate a pregnancy, the woman or couple may require additional long-term supportive counseling. Follow-up calls and letters can be a valuable source of support [Petrucelli et al., 1998]. Women may feel a strong sense of guilt following an abortion [Royal Commission on New Reproductive Technologies, 1993], and it also marks the beginning of a grieving process for many women [Bennett, 2001; Dallaire and Lortie, 1993]. It is recommended that "special attention" be given to the counseling needs of women and couples in cases involving termination after the discovery of a serious fetal disorder [Royal Commission on New Reproductive Technologies, 1993]. For example, a session should be scheduled to counsel the patient concerning future pregnancies and to assess her emotional needs. Moreover, the woman and her partner may grieve in different ways, creating a sense of isolation from each other, and referral for couple's counseling may be appropriate [Strong, 2003]. Finally, access or referral to self-help groups or associations of people who have had a child with the same disorder may help reduce the feeling of isolation and provide support and helpful advice [Royal Commission on New Reproductive Technologies, 1993; Dallaire and Lortie, 1993; Petrucelli et al., 1998].

Structuring the Genetic Counseling Session

The setting of the genetic counseling session plays an important role in the informed consent process. First, women and couples must be given sufficient time to consider the relative advantages and disadvantages of various procedures [Mavroforou et al., 2003]. Canadian guidelines recommend counseling appointments be scheduled at least one day prior to invasive procedures [Chodirker et al., 2001]. Second, face-to-face encounters may be the best way to communicate information because they allow the counselor to see and respond to the patient's emotional reactions, to gauge her level of understanding based on body language, and to provide helpful visual aids. The physical setting must be conducive to a discussion; namely, private and relatively quiet [Strong, 2003]. However, some studies indicate no large difference between telephone and in-person counseling with respect to patient understanding and anxiety. Indeed, visual privacy of the telephone may be beneficial for open communication [Sangha et al., 2003]. In any event, telephone discussions can be an important adjunct for women who are geographically distant [Chodirker et al., 2001; Sangha et al., 2003]. Third, written brochures or videos should be made available [Holtzman and Watson, 1997] but should not be relied on as the sole means of transmitting information. Concerns have been raised about the quality of these brochures, and whether they may be too complex, technical, or difficult to read [Royal Commission on New Reproductive Technologies, 1993; Marteau and Dormandy, 2001; please see Chapter Seven, Informed Consent, Process Issues]. Canadian guidelines note that if provided, they must be regularly reviewed and updated [SOGC, 1998]. Written summaries of the counseling sessions can also be helpful for the patient [Royal Commission on

New Reproductive Technologies, 1993], although it has been found that counselors are unlikely to use informed consent documents [Jacobson et al., 2001].

INFORMED CONSENT AND PREIMPLANTATION GENETIC DIAGNOSIS

While genetic counseling for PND has been studied in detail, little has been written about the recommended course of action for couples considering PGD, presumably because it is a relatively new and uncommon procedure [Raeburn, 2001]. In this case, the consultation requires discussion of two topics: in vitro fertilization (IVF) and PGD.

There must be a realistic assessment of pregnancy success, side effects, and potential risks. It is also important to explain that in PGD only a limited number of diseases are tested for, meaning that the baby is not guaranteed to be "perfect" [Overton et al., 2001].

It has been suggested that couples be informed of the differing misdiagnosis rates depending on the implantation method, and that they should decide the degree of misdiagnosis that is acceptable. The discussion must also cover the welfare of future children and the needs of the family [Raeburn, 2001].

Couples should receive psychological counseling, involving exploration of their reasons for choosing PGD and their overall feelings. Couples who choose PGD may already have suffered stresses, such as an affected child, repeated miscarriage, previous prenatal diagnosis, or termination [Overton et al., 2001].

Finally, the professionals involved at the referring genetic center and the clinical team from the IVF center should keep in close contact with one another [Raeburn, 2001].

FINAL THOUGHTS: CAN GENETIC COUNSELING BE NONDIRECTIVE?

Although nondirectiveness in genetic counseling has come to be seen as a "universal norm," doubts have been raised about how achievable, how desirable, and how accurate the principle is in practice [Bower et al., 2002; Anderson, 1999]. A recent survey of genetic counselors revealed some of their main concerns. First, the nondirective approach can be confusing for women because they are used to the typical settings in which the health practitioner offers an opinion. Second, some patients are actually seeking direction from practitioners when making important health decisions, to relieve the responsibility from themselves. Third, some women might value the views of others as an information-gathering tool, to test different opinions. Fourth, attempting to give scientific facts and medical knowledge in a value-neutral way is incompatible with the notion of a caring relationship and can undermine open communication. The survey highlighted that counselors did not adopt a uniform strategy for discussions, and much depended on how they framed the information and what information they opted to disclose [Williams et al., 2002].

Aside from the practitioner's behavior, the general move toward routine prenatal testing has an impact on the directiveness of counseling. Both the strong value our society places on knowledge, information, and technology and the increasingly routinized way in which prenatal testing is offered places informed "choice" at risk. In particular, these factors deemphasize the emotional and psychological ramifications of undergoing such testing [Suter, 2002]. Furthermore, genetic counseling and genetic procedures (especially prenatal screening) are performed by a variety of individuals other than genetic counselors, such as obstetricians and midwives. These nongenetic health professionals are less "steeped" in the culture of nondirectiveness and may not understand that genetic testing imposes a unique standard of care for communication [Suter, 2002; Williams et al., 2002; Sharpe, 1994a, 1996]. Evidence does show that the counseling provided by geneticists tends to be more positive and less directive than that provided by other health professionals [Marteau and Dormandy, 2001].

CONFIDENTIALITY

Health care professionals have a legal and ethical duty to protect the confidentiality of personal health information they receive from or about their patients [AMA Code of Medical Ethics; 2004 CMA Code of Ethics, 2004; Feinberg et al., 1984; Marshall and Von Tigerstrom, 2003]. What does this entail for health care professionals who provide prenatal testing and preimplantation genetic diagnosis services? As a general rule, the wishes of patients to keep their information confidential should be respected. Couples or women who choose to undergo prenatal testing or preimplantation diagnosis may wish to keep this a secret. They may feel that their health risks and the risks to their potential children concern only themselves. Alternatively, they may worry that their choice to undergo such tests, to consider a termination of pregnancy or the elimination of embryos, or to voluntarily have a child who will be disabled could be considered morally questionable by others. Parents may also want to keep the results of these tests, and the information it reveals about them, to themselves for a number of reasons, for example, for fear of stigmatization and discrimination or because of feelings of guilt. One parent may oppose the disclosure of prenatal test results to the other parent because he or she believes it is in the other's best interest or because he or she fears abandonment. If a child is born with a genetic disorder, parents may prefer to give an alternative explanation for the child's disability.

Nonetheless, this information could be relevant for a number of third parties, including relatives of the couple and, in particular, those who are of reproductive age or younger (siblings, cousins, nephews and nieces, etc.). Because genetic information is individual and familial, a test result indicating that a fetus carries a genetic mutation or chromosomal aberration could indicate that blood relatives are at increased risk of having a genetic disorder themselves or of having a child with a disorder. The refusal of a patient to share that information creates a dilemma for the health care professional; his or her duty to protect the patient's confidentiality collides with his or her duty to prevent harm [Lucassen and Parker, 2004; Offit et al., 2004]. Various organizations have developed policies and guidelines to assist health care professionals to resolve these issues. The American Society of Human Genetics adopted the position that genetic information "should be protected by the legal and ethical principle of confidentiality" but that disclosure should be permitted in exceptional circumstances, namely when (1) attempts have been made to encourage a patient to disclose the information but have failed; (2) it is highly likely that the nondisclosure of information will lead to a serious and foreseeable harm; (3) the at-risk relatives are identifiable; (4) the harm is preventable or treatable or medically acceptable standards indicate that early monitoring will reduce the genetic risk; and (5) the harm that may result from the failure to disclose outweighs the harm that may result from disclosure [American Society of Human Genetics, 1998]. Though these criteria are broad, they do provide much needed guidance to health care professionals who are confronted with these issues. However, health care professionals should be aware that though a disclosure may be considered ethically justifiable, it may not be legally justifiable. In some settings, the nonconsensual disclosure of a patient's personal information may be authorized only if it is required by law.

Chapter 9

Genetics of Common Neurological Disorders

OVERVIEW

The investigation of genetic mechanisms for neurological disease has been an actively pursued area of genetic research in recent years. Research has focused on:

- Defining syndromes and diseases by reproducible clinical criteria
- Identifying genes responsible for rare, highly penetrant forms of disease, often where evident because of early onset and high risks of recurrence in affected families
- Searching the human genome for genetic loci associated with common types of neurological disease
- Identifying the specific contributions of genetic, epigenetic, and environmental mechanisms in common neurological diseases.

Most of our burden of common neurological disease appears to arise from very complex interactions of many genes, epigenetic mechanisms (e.g., genetic imprinting), and environmental factors. It is apparent that entities that—at the time of this writing—are defined as discretely as possible by clinical criteria have, in fact, multiple different genetic mechanisms of varying expression and penetrance.

The primary goal of research has been to characterize the complexities of these pathogenetic mechanisms, so that improved therapies can be devised. Through this process, some candidate genes for future application of genetic testing have been identified, but such testing appears to be quite premature; neither clinical validity nor the feasibility of therapeutic intervention has been established. Nevertheless, increasingly, direct-to-consumer testing is being offered for a variety of genetic tests that appear to be of dubious value. Clinicians will need to be able to educate their patients about the genetics of common neurological disorders, and both clinicians and

Genetic Testing: Care, Consent, and Liability, by Neil F. Sharpe and Ronald F. Carter
© 2006 John Wiley & Sons, Inc.

patients will have to temper their expectations for the clinical usefulness of genetic testing.

INTRODUCTION

Researchers commonly use several types of studies to tease apart genetic and environmental influences (discussed later in this chapter). Studies of families at high risk for genetic disease are useful for identifying causal genes and mutations and for determining the heritability of disorders. Studies of adopted children help to identify the relative strength of genetic and environmental factors. Twin studies are invaluable in dissection of not only heritability but also shared and nonshared environmental factors.

Because of the complex interactions of multiple genetic and environmental influences in neurological diseases, candidate genetic regions usually are identified by and repeatedly subjected to segregation analyses and whole-genome scanning. One of the most important steps is to demonstrate the involvement of a particular region or gene or allele in multiple different patient groups and by different methodologies. Further, demonstration that the functions, structure, and pattern of tissue expression of the association protein products are consistent with a role in neurological disease is strong supporting evidence of a causal association. At this point in time, numerous candidate genes and alleles have been identified for a wide variety of entities, but far fewer have been consistently shown to have strong, predictable impact in replicated studies.

NEED TO KNOW

Genetics of Neurological Disease

For most categories of neurological disease, clinical presentation can be separated into:

- A minor component of rare, familial cases with autosomal-dominant, autosomal-recessive, or X-linked patterns of inheritance
- A far larger component of disease, commonly prevalent and arising as a result of complex, multifactorial interactions of genetic and environmental causes

The number of genes implicated in specific common behavioral, psychiatric, and neurodegenerative conditions can vary from a few to hundreds, and numerous environmental interactions are known or suspected.

Alzheimer's Disease

Approximately 95 percent of Alzheimer's disease presents as a late-onset, multifactorial disorder. There is clear evidence of a genetic component, but the degree of heritability and the spectrum of associated genes are not clear. The major gene and allele associated with increased risk in Caucasians is *APOE ε-4*. Testing for

APOE ε-4 is currently available from more than 20 laboratories; the test may be useful as a component of diagnostic evaluation for early-onset Alzheimer's disease, but this test is not considered to be highly specific nor highly sensitive, and its application to predictive testing is questionable [Gene Tests, 2004; ACMG, 2004 (http://genetics.faseb.org/genetics/ ashg/policy/pol-21.htm)].

Approximately 5 percent of Alzheimer's disease presents with early onset and an autosomal-dominant pattern of inheritance with high penetrance. Mutations in any one of the genes for amyloid precursor protein (APP), presenilin 1 (*PSEN1*; the most commonly implicated gene) and presenilin 2 (*PSEN2*) are sufficient to cause the disease, and expression may be more severe (earlier age of onset) if the *APOE ε-4* is also inherited. Molecular testing for these genes is available from multiple clinical laboratories, but the applicability of testing to clinical practice is not clear [Goldman and Hou, 2004].

The pathogenetic mechanisms in familial early-onset and common late-onset Alzheimer's disease appear similar. Together, mutations in *APP, PSE1, PSE2*, and *APOE* are estimated to represent about half of the genetic contribution to this multifactorial disease. Details of the molecular genetics of Alzheimer's disease can be found in the literature [e.g., Selkoe, 2001; GeneTests, 2004].

Parkinson's Disease

Parkinson's disease can be caused by exposure to toxins; genetic mechanisms have also been clearly indicated by twin studies, family studies, and segregation analyses. A large number of different genes have been implicated in Parkinson's disease.

Approximately 10 percent of Parkinson's disease presents with relatively early onset and evidence of autosomal-dominant or autosomal-recessive patterns of inheritance. Multiple genes are implicated. Homozygous mutation of parkin (*PARK2*) appears to cause about half of familial early-onset Parkinson's disease.

Approximately 90 percent of Parkinson's disease is common and multifactorial in origin. Multiple genes have been implicated, including heterozygous mutation of parkin (*PARK2*). Details of the molecular genetics of Parkinson's disease can be found in the literature [e.g., Vila and Przedborski, 2004; Huang et al., 2004b].

Autism

Autism is now clearly demonstrated to have a major component of heritability arising from several different genetic mechanisms. Fragile X syndrome and a variety of chromosomal abnormalities have been documented. Familial inheritance with Mendelian patterns have also been shown, and maternal imprinting is implicated in cases associated with duplications of the Prader Willi–Angelman syndrome region on chromosome 15. Genes in numerous different regions of the genome have been implicated, but few appear to account for a significant proportion of the disease. Details of the molecular genetics of autism can be found in the literature [e.g., Veenstra-VanderWeele and Cook, 2004; Muhle et al., 2004].

Epilepsies, Schizophrenia, and Bipolar Disorders

Each of these diseases are known to have multiple clinical variants and strong genetic components. However, teasing apart the genetic contributions to these variable clinical presentations has been challenging; a wide variety of genes—at the time of this writing—tentatively have been implicated, and there is little data that can be applied to routine clinical management outside of the context of research studies at this time. The molecular genetics of these disorders have been reviewed recently [e.g., Godard and Cardinal, 2004; Tan et al., 2004; Shirts et al., 2004; Kato et al., 2001; Tsuang et al., 2004; Maier et al., 2003].

WATCH OUT FOR

Debatable Value of Genetic Testing

The strong interest in mapping out the genetic basis of neurological diseases has extended to the premature application of genetic testing. In a research context, families are identified that fit specific clinical criteria and are demonstrated to have private, familial mutations with high penetrance; in this situation, genetic testing may be of value, but the approach to counseling and testing must be carefully adapted to each family. However, few of the findings to date can be translated to clinical application in the management of common neurological diseases. The sheer number of different genes that have been tentatively identified, the weakness of the evidence for many, the weak penetrance for many alleles and loci, the difficulty in reproducing results in different patient populations, and the complexities of gene–gene and gene–environment interactions together represent daunting barriers to the implementation of interpretable, reliable test protocols. Further, despite the great amount of effort dedicated to clarifying pathogenetic mechanisms, there is little evidence that widespread use of diagnostic or predictive testing can be used to implement effective treatment or prevention at this time.

Despite the lack of convincing indications for testing, an increasing volume of direct-to-consumer and proprietary genetic testing is available to the public [Andrews, 1997; ACMG, 2004; McCabe and McCabe, 2004; but see Human Genetics Commission, 2002]. Such testing raises both the interest and concern of the lay public [e.g., Hull and Prasad, 2001a, 2001b; Caulfield, 2001; Gollust et al., 2002; Tsao, 2004]. Clinicians will need to be able to discriminate between appropriate indications for referral (evidence of familial disease, recognizable syndromes with known genetic mechanisms, access to specialist with appropriate expertise and test capacity) and sporadic, common forms of disease with complex multifactorial inheritance that limit the options for investigation. The clues lie in careful diagnostic assessment (category of disease, age of onset, severity, clinical expression) and review of the family history (evidence of heritability, mode of inheritance, penetrance, ethnicity). The following study provides a detailed review of these concepts.

NEUROGENETICS, BEHAVIOR, AND NEURODEGENERATIVE DISORDERS

Jinger G. Hoop, Edwin H. Cook, Jr., Stephen H. Dinwiddie, and Elliot S. Gershon
University of Chicago, Chicago, Illinois.

INTRODUCTION

Neurogenetics encompasses the study of genetic factors in diseases, disorders, and conditions of the nervous system. As such, it includes attempts to pinpoint rare disease-causing genetic mutations, to identify genetic variants that modify the risk of common neuropsychiatric disorders, to locate the myriad genes thought to influence intelligence and personality, to understand how genes are expressed in the developing and aging brain, and to compare genes among various species to better understand the cognitive, emotional, and behavioral processes that make us human [Christen and Mallet, 2003].

Research in neurogenetics should be of particular interest to genetic counseling professionals because a large proportion of human mortality and disability are attributable to disorders of the nervous system [Christen and Mallet, 2003]. Neuropsychiatric disorders are projected to account for 14.7 percent of the global burden of disease by the year 2020 [Murray and Lopez, 1997], and many of these diseases are poorly understood. For the mental illnesses and common neurodegenerative disorders in particular, neurogenetic research holds great promise for understanding the disease process, developing new treatments, and creating diagnostic and screening instruments [Faraone et al., 1999].

The purview of neurogenetics includes a number of very rare to uncommon conditions such as Huntington disease, Duchenne muscular dystrophy, fragile X syndrome, and Down syndrome. Neurogenetics also includes the common neurodegenerative disorders, mental illnesses, neurologic conditions affecting the central and peripheral nervous system, and behavioral traits such as cognitive ability and aggression. After a brief introduction to the methods of neurogenetics research, the remainder of this chapter will review the current state of knowledge about the genetics of Alzheimer's disease, Parkinson's disease, the epilepsies, autism, schizophrenia, bipolar disorder, and conduct disorder. This is obviously not an exhaustive overview of research in this field, which can be found elsewhere [Rosenberg et al., 2003; Mallet and Christen, 2003], but is intended to illustrate the variety of disorders under study. The chapter will conclude with a discussion of clinical and ethical issues in neurogenetic testing and the role of clinicians as educators and gatekeepers.

RESEARCH METHODS IN NEUROGENETICS

Family, adoption, and twin studies have established the genetic basis of many neuropsychiatric disorders as well as traits such as cognitive ability and personality [Vogel and Motulsky, 1997]. The three methods are complementary: Family studies compare rates of a disease or trait among relatives of affected members and in the general population to determine if the condition runs in families. To find out whether a familial condition is inherited or is caused by some aspect of the family environment, adoption studies compare its prevalence among adopted children and their biologic and adoptive relatives. Twin studies help elucidate the relative importance of genes and environment.

Twin studies examine rates of *concordance* (both members of a set of twins having the same condition or disease) in monozygotic and dizygotic twins. If the etiology of a condition has a genetic component, we will see a higher concordance rate among monozygotic twins, who are genetically identical, than among dizygotic twins, who share only 50 percent of their genes. Twin studies can be used to tease out the contributions of three factors—genes; aspects of the environment that are shared by twins, such as socioeconomic status; and aspects of the environment that are not shared, such as one twin's traumatic injury [Faraone et al., 1999].

Family-based studies are also used to calculate *heritability*, defined as the variance due to genetic factors in a given environment and population [Vogel and Motulsky, 1997]. A heritability of 100 percent indicates that vulnerability to the condition can be explained entirely by genetics; 0 percent means it is due solely to environmental factors [Faraone et al., 1999]. It is important to note that the heritability of a condition may vary across different environmental contexts and among different populations.

Finally, these studies also provide the statistical information needed to perform segregation analyses, mathematical strategies for determining the mode of inheritance of a disease [Faraone et al., 1999]. Neurogenetic diseases include many disorders caused by mutations of a single gene with an X-linked, autosomal-dominant, or autosomal-recessive mode of inheritance. Such diseases are rare in the population. On the other hand, the common neurologic and psychiatric illnesses and many behavioral traits have a complex mode of inheritance. These conditions are produced by at least two and possibly hundreds of genes and/or nongenetic elements such as environmental influences and epigenetic factors. Epigenetic factors are cellular mechanisms that regulate gene expression, such as patterns of DNA (deoxyribonucleic acid) methylation and histone modification (e.g., acetylation).

While genetic epidemiologic studies establish whether or not a condition has a genetic component, molecular research seeks to identify the specific genes involved. For rare, single-gene disorders, researchers typically look for disease-causing genetic mutations by studying affected and unaffected family members in large pedigrees. For common, complex disorders, the search is for *alleles* (variants of a gene) that modify risk. The pool of subjects in these studies may comprise numerous small family groups such as pairs of affected siblings and their parents.

Most contemporary studies in human molecular neurogenetics can be classified as either scans of the entire human genome or more focused examinations of candidate genes. Genome-wide scans are systematic attempts to identify chromosomal regions that may harbor genes involved in a given disease or condition. A linkage-based genome scan uses DNA markers to define regions that are transmitted with the condition within families. An association-based genome scan compares the frequency of alleles at various points on the genome in people with the condition and in a control group. Evidence of linkage or association at a particular *locus* (location on a chromosome) can be followed up with additional association studies to further narrow the region of interest [Nussbaum, 2003].

A candidate-gene study, on the other hand, examines a known gene for association with a condition under study. Typical candidates in neurogenetics are genes for neurotransmitter receptors and ion channels. Candidate-gene studies in neuropsychiatry have yielded some successes, most notably the identification of the amyloid precursor protein gene in Alzheimer's disease [Chartier-Harlin et al., 1991; Goate et al., 1991], but they are hampered by our limited understanding of the pathophysiology of neurogenetic diseases and by the large number of genes expressed in the brain [Faraone et al., 1999].

Single reports of genetic linkage or association may be intriguing, but such findings must be replicated in multiple samples before they can be accepted as conclusive. In the past 20 years, hundreds of mutations associated with rare, single-gene disorders have been identified in this way [Human Genome Project Information, 2004], but progress in finding susceptibility alleles for common, complex diseases has been slower. Because the effect of a given susceptibility allele may be subtle, large samples may be needed to demonstrate statistically significant linkage or association, and conflicting findings in different samples are common [Suarez et al., 1994]. However, in recent years, positive associations of susceptibility alleles for Alzheimer's disease, schizophrenia, and other common neuropsychiatric conditions have been reported and replicated (see discussion below). We may be on the brink of an exciting period for neurogenetics, as medical researchers begin to gain new understanding of the pathophysiology of diseases that have long been poorly understood and open avenues for treatment where none may have existed before.

GENETICS OF ALZHEIMER'S DISEASE

Alzheimer's disease is the most common form of late-life dementia, affecting an estimated 4 million people in the United States [Kawas and Katzman, 1999]. Clinical features include gradually progressive memory loss and cognitive decline leading to severe global dementia and death, usually within about 10 years. Diagnosis can be confirmed by neuropathologic samples showing amyloid plaques and neurofibrillary tangles. Cases are described as early or late onset—depending on whether the disease is manifest before or after age 65—and as sporadic or familial.

In up to 5 percent of cases, the disease is clearly inherited in an autosomal-dominant fashion and onset tends to be early. Three genes have been implicated in this form of the disease: amyloid precursor protein (*APP*) on chromosome 21 [Chartier-Harlin et al., 1991; Goate et al., 1991], presenilin 1 (*PSEN1*) on chromosome 14 [Sherrington et al., 1995], and presenilin 2 (*PSEN2*) on chromosome 1 [Levy-Lahad et al., 1995; Rogaev et al., 1995]. Any one of these is sufficient to cause disease, and penetrance is nearly 100 percent [Rocchi et al., 2003]. Mutations at any of the three loci lead to accumulations of A-β amyloid [Selkoe, 2003]. For individuals with certain mutations of *APP* or *PSEN1*, the age of onset may be decreased by the presence of the ε-4 allele of the apolipoprotein E gene (*APOE*) on chromosome 19 [Rocchi et al., 2003]. The major three alleles of this gene are ε-2, which is the least common, ε-3, the most common, and ε-4 [Sorbi et al., 1995, Pastor et al., 2003].

However, the vast majority of cases of Alzheimer's disease ($>$95 percent) do not fit a single-gene pattern of inheritance [Rocchi et al., 2003] and onset is late [Finckh, 2003]. For this common form of Alzheimer's disease, the precise genetic contribution is unknown but presumed to be significant, based on evidence from epidemiologic studies as well as the observation that sporadic and familial Alzheimer's disease are phenotypically similar. For both early- and late-onset Alzheimer's disease, population-based family studies have demonstrated an increased risk of disease among first-degree family members [Fratiglioni et al., 1993; van Duijn et al., 1993]. Increased risks to relatives have been observed in Caucasian, Caribbean Hispanic, and African-American populations [Devi et al., 2000]. Concordance rates for Alzheimer's disease among monozygotic twins are at least twice as high as among dizygotic twins [Bertoli-Avella et al., 2004].

Among Caucasian patients with late-onset Alzheimer's disease, the *APOE* ε-4 allele confers increased risk of disease as well as decreased age of onset in what has been described as a *dose-dependent* fashion, with homozygotes at significantly higher risk than heterozygotes, whose risk is higher than those without a single copy of the allele [Corder et al., 1993]. In some studies, the ε-2 allele appears to be protective, but evidence is inconsistent [Corder et al., 1994; Talbot et al., 1994; Smith et al., 1994; van Duijn et al., 1995]. The role of *APOE* in modifying risk of Alzheimer's disease in African-Americans is controversial [Farrer et al., 1997, 2000].

Despite the clear evidence of the roles of *APP, PSE1, PSE2*, and *APOE* in Alzheimer's disease, these genes account for only about half of the estimated genetic contribution to the disease [Tang and Gershon, 2003]. Attempts to identify other susceptibility alleles have been difficult, though genome-wide scans have demonstrated linkage on chromosomes 9, 10, and 12, and a number of promising candidate genes have begun to be identified and tested, including genes encoding α-2-macroglobulin, low-density lipoprotein receptor-related protein, and angiotensin I converting enzyme [Finckh, 2003; Rocchi et al., 2003; Kehoe et al., 2003].

GENETICS OF PARKINSON'S DISEASE

Parkinson's disease is the second most common neurodegenerative disorder of human beings. In the United States, the prevalence of the disease is about 1 million individuals [Tanner and Goldman, 1996; Prusiner, 2003]. Symptoms include progressive muscular rigidity, slowed movement, postural instability, and resting tremor. Age is an important risk factor, and most cases are diagnosed after the age of 50. The classic histologic findings of Parkinson's disease are Lewy bodies (cytoplasmic inclusions made of protein aggregates) and a loss of dopaminergic neurons in the substantia nigra pars compacta [Harris and Fahn, 2003].

The role of environmental agents in the etiology of parkinsonism was clearly demonstrated when several cases were traced to exposure to the neurotoxin MPTP

(1-methyl-4-phenyl-1,2,3,6-tetrahydropyridine) [Langston et al., 1983]. The genetic contribution to classic Parkinson's disease has also been well demonstrated. Although early twin studies reported low concordance rates, subsequent ones found significantly higher concordance rates for monozygotic twins than for dizygotic twins [reviewed in Bertoli-Avella et al., 2004]. Family studies showed increased risks to first-degree relatives [Kurz et al., 2003; Sveinbjornsdottir et al., 2000]. Segregation analysis suggested the presence of a major susceptibility gene plus a major dominant gene affecting the age of onset [Maher et al., 2002].

In up to 10 percent of cases, Parkinson's disease follows an autosomal-dominant or autosomal-recessive mode of inheritance [Goldman and Tanner, 1998]. As with Alzheimer's disease, the familial cases have relatively early onset. One form, autosomal-recessive juvenile parkinsonism, is associated with an age of onset before age 40 and a lack of Lewy bodies [Huang, 2004]. The most important gene associated with familial Parkinson's disease is the parkin gene, also called *PARK2*, on chromosome 6. The mode of inheritance is autosomal recessive [Bertoli-Avella et al., 2004]. Since the parkin gene was cloned [Kitada et al., 1998], numerous mutations have been reported. In a large European study, parkin mutations were found in nearly 50 percent of the families with early-onset Parkinson's disease [Lucking et al., 2000]. An autosomal-dominant form of Parkinson's disease is caused by mutations or triplication of the α-synuclein gene on chromosome 4 [Polymeropoulos et al., 1997; Singleton et al., 2003]. Other mutations in familial Parkinson's disease have been identified in the nuclear receptor-related-1 gene (*NR4A2*) on chromosome 2 [Le et al., 2003], the ubiquitin C-terminal hydrolase L1 gene (*UCHL1*) on chromosome 4 [Leroy et al., 1998], *DJ-1* on chromosome 1 [van Duijn et al., 2001], *NF-M* on chromosome 8 [Lavedan et al., 2002], and *PINK1* on chromosome 1 [Valente et al., 2004].

As with Alzheimer's disease, mutations associated with rare forms of Parkinson's disease have been more readily identified than susceptibility alleles for the common, sporadic form of the disorder. However, some genes involved in autosomal-dominant and autosomal-recessive Parkinson's disease are now implicated in the common form. While homozygous parkin mutations cause autosomal-recessive early-onset Parkinson's disease, heterozygous mutations of the parkin gene have been observed in some individuals with late-onset illness [Oliveira et al., 2003]. A common variant of *UCHL1* was significantly and inversely associated with Parkinson's disease in a pooled analysis of 11 studies [Maraganore et al., 2004]. Alpha-synuclein has also been associated with sporadic Parkinson's disease [Farrer, et al., 2001].

Other candidates for susceptibility alleles in Parkinson's disease include mitochondrial genes [van der Walt et al., 2003], genes controlling dopamine synthesis, function, and metabolism [reviewed in Huang et al., 2004b], and genes involved in other neurodegenerative disorders. Mutations in the microtubule-associated protein τ gene (*MAPT*) cause an autosomal-dominant form of the condition frontotemporal dementia and parkinsonism [Hutton, et al., 1998]. *MAPT* resides in a region on chromosome 17 linked to idiopathic Parkinson's disease in Scott and co-workers' genome-wide scan [2001]. A 2004 meta-analysis of *MAPT* studies in Parkinson's disease demonstrated significant association [Healy et al., 2004]. *APOE*, which has been identified as a susceptibility gene for Alzheimer's disease, has also been investigated in Parkinson's disease, with inconsistent results. A meta-analysis of 22 case–control studies concluded that the *APOE ε-2* allele was positively associated with sporadic Parkinson's disease [Huang et al., 2004a].

GENETICS OF THE EPILEPSIES

The epilepsies are neurologic disorders consisting of recurrent seizures, which are themselves caused by abnormal and synchronized cortical neuronal discharges [Dichter and Buchhalter, 2003]. In the United States, the annual incidence of all forms of epilepsy is estimated at 70,000 to 129,000 [Hauser and Hesdorffer, 1990]. The epilepsies can be divided into those causing *generalized* versus *partial* (focal) seizures and syndromes that are *idiopathic* versus *symptomatic*. Idiopathic epilepsy includes syndromes for which the only known or suspected etiologic factor is genetics. Symptomatic, or acquired, epilepsy syndromes may also have a genetic component. Febrile convulsions are seizures accompanied by fever without any known neurologic etiology. They are considered a special syndrome [Commission on Classification and Terminology of the International League Against Epilepsy, 1989].

In the past decade, several genetic mutations that cause human epilepsy have been discovered. Most of the genes code for ion channel proteins and were discovered by linkage analysis in rare families in which epilepsy is inherited in a single-gene fashion, typically autosomal dominant. More recently, linkage and association studies have begun to report putative susceptibility loci or alleles for the more common epilepsy syndromes, which have complex inheritance.

Febrile convulsions are the most common seizure of childhood, affecting about 2 percent of the U.S. population [Dichter and Buchhalter, 2003]. Twin studies have shown concordance rates of 31 to 70 percent in monozygotic twins, compared with 14 to 18 percent in dizygotic twins [Dichter and Buchhalter, 2003]. Segregation analysis supports polygenic as well as autosomal-dominant modes of inheritance [Rich et al., 1987]. Positive associations have been reported with the interleukin-1 β gene (*IL-1B*) on chromosome 2 [Virta et al., 2002]; *CHRNA4* on chromosome 20, which codes for the α-4 subunit of the neuronal nicotinic acetylcholine receptor; and the GABA$_A$ receptor γ-subunit gene *GABRG2* on chromosome 5 [Chou et al., 2003a, 2003b]. Autosomal-dominant febrile seizures have been linked to loci on chromosomes 5, 6, 8, and 19 [reviewed in Berkovic and Scheffer, 2001].

The idiopathic generalized epilepsies (IGEs) are a group of common seizure disorders, comprising childhood absence epilepsy, juvenile absence epilepsy, juvenile myoclonic epilepsy, and epilepsy with generalized tonic-clonic seizures. Concordance rates for monozygotic twins with IGE are 76 percent, versus 33 percent for dizygotic twins [Berkovic et al., 1998]. The mode of inheritance is thought to be complex, including at least two genes [Berkovic and Scheffer, 2001].

Genome-wide scans have suggested linkage for IGE susceptibility at a locus on chromosome 18 [Durner et al., 2001], which was not confirmed [Sander et al., 2002], as well as loci on chromosomes 2, 3, and 14 [Sander et al., 2000b]. Mutations in the chloride-channel gene *CLCN2* in the linked region on chromosome 3 have been noted in some probands [Haug et al., 2003]. The μ-opioid receptor subunit gene has been associated with idiopathic absence epilepsy, and the finding was replicated in a sample of individuals with IGEs [Sander et al., 2000a; Wilkie et al., 2002]. A variant of the *KCNJ10* potassium ion-channel gene was inversely associated with susceptibility to seizures in a case–control study including patients with IGEs or mesial temporal lobe epilepsy [Buono et al., 2004]. Juvenile myoclonic epilepsy has been strongly associated with a transcription factor gene, *BRD2*, on chromosome 6 [Pal et al., 2003].

Generalized epilepsy with febrile seizures plus (GEFS+) is an IGE of great phenotypic variety that follows a pattern of autosomal dominance with incomplete penetrance in some families [Gourfinkel-An et al., 2004]. Affected members of a single family may have febrile seizures, atypically persistent febrile seizures, and/or nonfebrile seizures [Scheffer and Berkovic, 1997]. Mutations in the sodium-channel genes *SCN1A* on chromosome 2 [Escayg et al., 2000], *SCN1B* on chromosome 19 [Wallace et al., 1998], and *SCN2A* on chromosome 2 [Sugawara et al., 2001] have been found in some probands, as have mutations in *GABRG2* [Baulac et al., 2001; Harkin et al., 2002]. *GABRG2* has also been associated with a phenotype consisting of childhood absence epilepsy and febrile seizures [Wallace et al., 2001].

Several other monogenic forms of the IGEs exist, and their causative mutations have begun to be identified. A mutation of the *GABRA1* gene, which encodes the α-*1* subunit of the $GABA_A$ receptor, was discovered in a family with autosomal-dominant juvenile myoclonic epilepsy [Cossette et al., 2002]. Severe myoclonic epilepsy of infancy is associated with de novo mutations of the sodium-channel gene *SCN1A* [Claes et al., 2001]. An autosomal-dominant condition called benign familial neonatal convulsions is characterized by focal or generalized seizures beginning during the first week of life [Dichter and Buchhalter, 2003]. Mutations in two potassium-channel genes, *KCNQ2* on chromosome 20 [Singh et al., 1998] and *KCNQ3* on chromosome 8, have been reported in families with this syndrome [Hirose et al., 2000]. Mutations of *SCN2A* were demonstrated in two families with benign familial neonatal-infantile seizures, a syndrome with a later onset [Heron et al., 2002].

The idiopathic partial epilepsies include benign rolandic epilepsy (BRE), which causes unilateral sensorimotor seizures and centrotemporal spikes on electroencephalogram (EEG) and may account for 25 percent of childhood epilepsies [Dichter and Buchhalter, 2003]. The mode of inheritance was presumed to be complex [Andermann, 1982], and linkage to a locus on chromosome 15 has been reported [Neubauer et al., 1998]. However, a recent small twin study found no concordance in monozygotic and dizygotic twins with BRE, suggesting that the heritability may have been overestimated [Vadlamudi et al., 2004].

A number of putative susceptibility alleles have been associated with temporal-lobe epilepsy, including variants of the prodynorphin gene [Stogmann et al., 2002], *APOE* [Briellman et al., 2000], *IL-1B* [Kanemoto et al., 2000], and *GABRB1* [Gambardella et al., 2003]. An autosomal-dominant form, called lateral temporal-lobe epilepsy with auditory symptoms, has a penetrance of about 70 percent [Dichter and Buchhalter, 2003]. This disorder begins in childhood or young adulthood and consists of rare partial seizures that secondarily generalize. Auditory hallucinations during the seizure occur in some members of every family studied. Mutations in the leucine-rich, glioma-inactivated gene, *LGI1*, on chromosome 10 have been found in some families [Pizzuti et al., 2003].

Autosomal-dominant nocturnal frontal epilepsy is a rare idiopathic partial epilepsy characterized by motoric seizures during sleep, with a typical onset in childhood [Scheffer et al., 1995]. Mutations *CHRNB2* on chromosome 1, which codes for the β-2 subunit of the neuronal nicotinic acetylcholine receptor, and in *CHRNA4* have been found in several families [reviewed in Gourfinkel-An et al., 2004]. Another locus was mapped to chromosome 15 [Phillips et al., 1998], and at least one additional gene is presumed to be involved [Combi et al., 2004].

GENETICS OF AUTISM

Autism is a pervasive developmental disorder characterized by impaired social interaction, stereotyped interests and behaviors, and impaired communication [*Diagnostic and Statistical Manual of Mental Disorders*, 2000]. Mental retardation is common but not necessary for the diagnosis. The term *autistic spectrum disorder* is used to describe a group that includes autism, Asperger syndrome (in which communication is relatively normal in the presence of social and behavioral impairment), and pervasive developmental disorder not otherwise specified in which there are significant deficits in social skills, communication, or behavior but not sufficient to meet criteria for another pervasive developmental disorder. [The other pervasive developmental disorders are childhood disintegrative disorder and Rett syndrome, an X-linked disease caused by mutations in the methyl-CpG-binding protein 2 gene *(MECP2)* [Amir et al., 1999]. In both diseases, children have a period of apparently normal development followed by a devastating loss of skills and mental retardation.]

The population prevalence of autistic disorder has been estimated at 0.1 to 0.2 percent [Chakrabarti and Fombonne, 2001], with boys affected at four times the rate of girls. The relative risk for siblings is 4.5 percent [Jorde et al.,1991]. In twin studies, the concordance rate for autism in monozygotic twins was 60 percent compared with no concordance seen in dizygotic twins, and for autistic spectrum disorder, 92 percent in monozygotic twins and 10 percent in dizygotic twins [Bailey et al., 1995].

Autism is both strongly and heterogeneously genetic. There have been many documented cases caused by chromosomal abnormalities or highly penetrant genetic mutations. But in most cases, the mode of inheritance is presumed to be complex, caused by several genes interacting in a multiplicative fashion and presumably influenced by epigenetic factors.

Cytogenetic studies have demonstrated chromosomal abnormalities in more than 4 percent of autistic individuals [Veenstra-VanderWeele and Cook, 2004]. Almost a third had fragile X syndrome, but abnormalities on every other chromosome were also reported. Interstitial duplication of the chromosome 15q11-q13 region increases susceptibility to autism spectrum disorder considerably, but only if the duplication occurs on the chromosome inherited from the mother [Cook et al., 1997]. The 15q11-q13 region contains a number of imprinted genes, whose expression differs depending on whether they are on the chromosome inherited from the mother or the chromosome inherited from the father. Abnormalities in 15q11-q13 also contribute to the uncommon neurobehavioral disorders Prader–Willi syndrome and Angelman syndrome [Inoue and Lupski, 2003].

Cytogenetic abnormalities may offer clues to the location of genetic mutations or variations associated with autism. Examination of the chromosome 15q11 region identified significant association with *GABRB3*, which encodes the GABA receptor A *β-3* subunit, in two independent samples [Cook et al., 1998; Buxbaum et al., 2002], but this has not been consistently replicated. The observation of a deletion of chromosome Xp22 in three females with autism [Thomas et al., 1999] led to the discovery of mutations in the neuroligin genes *NLGN3* and *NLGN4* in two sets of autistic siblings [Jamain et al., 2003]. Screening in nearly 200 additional autistic subjects failed to find these mutations [Vincent et al., 2004].

Genome-wide linkage scans using relatively small samples have identified several other regions of interest. The highest significance was reported for chromosomes 2 and 3, followed by less significant but replicated findings on 7, 16, and 17 [Barrett et al., 1999; Collaborative Linkage Study of Autism, 2001; Auranen et al., 2002; Yonan et al., 2003]. The region on chromosome 17 linked to autism is also the location of perhaps the strongest candidate gene for autism spectrum disorder, the serotonin transporter gene, *SLC6A4*, which has been associated with autism in several, but not all, studies [reviewed in Veenstra-VanderWeele and Cook, 2004].

GENETICS OF SCHIZOPHRENIA

Schizophrenia is characterized by impaired occupational or social functioning associated with three clusters of symptoms—negative symptoms such as social withdrawal and emotional blunting; positive symptoms such as hallucinations, delusions, and disorganized thought and behavior; and cognitive deficits. The disease is clinically heterogeneous, with variability in both the symptom presentation and the course. Significant gender differences have been noted. Among men, the disease is slightly more prevalent, onset tends to be earlier, premorbid functioning worse, and short-term prognosis worse [*Diagnostic and Statistical Manual of Mental Disorders*, 2000].

The general population prevalence of schizophrenia is 1 percent, and the risk to first-degree relatives is 10 times higher [Levinson and Mowry, 2000]. Twin studies show concordance rates of 44 percent in monozygotic twins versus 10 percent in dizygotic twins [Gottesman and Shields, 1982]. The children of discordant monozygotic twins have equivalent rates of schizophrenia [Gottesman, 1991]. Data from family studies yielded a heritability estimate of about 80 percent [Owen et al., 2003]. The mode of inheritance is complex, involving a combination of genetic and nongenetic factors, such as environmental influences and epigenetic mechanisms [Petronis, 2004].

Molecular genetic studies in schizophrenia include more than 20 whole-genome scans. Badner and Gershon's [2002] meta-analysis of published whole-genome linkage scans in schizophrenia found strong evidence of linkage to regions on chromosomes 8p, 13q, and 22q. A second meta-analysis using a different methodology and overlapping data sets also demonstrated significant linkage to chromosomes 8p and 22q, as well as to regions on chromosomes 1, 2, 3, 5, 6, 11, 14, and 20 [Lewis et al., 2003].

Association studies have begun to yield robust evidence of susceptibility genes for schizophrenia. The three most significant identifications are the neuregulin 1 gene (*NRG1*) on chromosome 8 [Stefansson et al., 2004], the dysbindin gene (*DTNBP1*) on chromosome 6 [Straub et al., 2002], and *G72* on chromosome 13 [Chumakov et al., 2002]. A role for each gene in schizophrenia is biologically plausible, and positive associations for all three have been replicated in independent samples. Other putative schizophrenia susceptibility genes with consistent support for association include the dopamine D_3 receptor gene on chromosome 3 and the serotonin 2_A receptor gene on chromosome 13 [Lohmueller et al., 2003].

The linked region on chromosome 22q has been of particular interest because microdeletions of this area are known to cause velocardiofacial syndrome. This condition (also called DiGeorge, Shprintzen, or 22q11 deletion syndrome) is associated with a range of behavioral and psychiatric abnormalities, including schizophrenia [Shprintzen, 2000]. Candidate genes in the deleted region include *PRODH* [Liu et al., 2002; Li et al., 2004], *COMT* [Shifman et al., 2002], and *ZDHHC8* [Mukai et al., 2004].

GENETICS OF BIPOLAR DISORDER

Bipolar disorder is characterized by recurrent episodes of depression, mania, and mixed-mood states. It has been ranked as the sixth leading cause of disability worldwide [Lopez and Murray, 1998], and rates of completed suicide are higher than for any other psychiatric or medical illness [Goodwin and Jamison, 1990]. Bipolar disorder is part of the spectrum of mood disorders, which includes a mild condition called cyclothymia; unipolar depression, which comprises only depressive episodes; and schizoaffective disorder, bipolar type, which includes psychotic symptoms during periods of normal mood [*Diagnostic and Statistical Manual of Mental Disorders*, 2000].

Epidemiologic research has consistently indicated that there is a strong genetic component to the etiology of bipolar disorder. The population prevalence is 1 to 2 percent [Kessler et al., 1994; Weissman et al., 1991], while Smoller and Finn's review of family studies [2003] estimated the overall risk of bipolar disorder for first-degree relatives at 8.7 percent. Earlier onset of illness appears associated with a further elevated relative risk. Twin studies have generally found higher concordance rates among monozygotic twins than among dizygotic twins, yielding heritability estimates of 60 to 85 percent [Smoller and Finn, 2003]. The mode of inheritance is presumed to be complex, involving multiple genes plus environmental factors [Craddock and Jones, 1999].

The heritability of bipolar disorder, coupled with the enormous burden of the disease, have sparked keen interest in molecular genetic studies. The search for susceptibility genes produced many reports of linkage beginning in the 1980s, although consistent replication proved elusive. Recent studies with more power to detect subtle gene effects include a large genome-wide linkage analysis based on 1163 subjects in 245 families, which found suggestive evidence of linkage on chromosomes 17 and 6 [Dick et al., 2003]. In addition, a meta-analysis of published whole-genome linkage scans found strong evidence of linkage to chromosomes 13 and 22 [Badner and Gershon, 2002] in bipolar disorder and in schizophrenia.

Association studies suggest that a susceptibility allele for bipolar disorder as well as schizophrenia may be in the region of the overlapping *G72* and *G30* genes on chromosome 13 [Hattori et al., 2003; Chumakov et al., 2002]. These findings have been replicated in independent samples for bipolar disorder and early-onset schizophrenia and psychosis [Chen et al., 2004; Addington et al., 2004]. Another strong candidate susceptibility gene is brain-derived neurotrophic factor (*BDNF*), which is thought to be involved in the pathogenesis of mood disorders as well as in treatment response [Green and Craddock, 2003]. The gene resides on chromosome 11, in a region with evidence for linkage to bipolar disorder [Egeland et al., 1987]. Positive associations have been reported in two family-based studies [Sklar et al., 2002; Neves-Pereira et al., 2002].

GENETICS OF CONDUCT DISORDER

Conduct disorder, one of the most common psychiatric diagnoses in children, is a behavioral syndrome consisting of a repeated pattern of violating the rights of others or failure to follow rules [*Diagnostic and Statistical Manual of Mental Disorders*, 2000]. Children with conduct disorder may be aggressive toward people or animals, damage property, lie or steal, and violate rules by running away from home, skipping school, and breaking curfews. They are also at elevated risk of developing antisocial personality disorder in adulthood [Rutter, 1995]. There is greater risk for persistence and severity in the childhood-onset form of conduct disorder compared with the adolescent-onset form.

Estimates of the population prevalence of conduct disorder vary from 1 to 10 percent [*Diagnostic and Statistical Manual of Mental Disorders*, 2000], with higher rates among boys than girls. Family studies suggest that genetic factors play a moderate role in the etiology of the syndrome. Heritability estimates range from 32 to 70 percent in various populations [Dick et al., 2004]. Research into the causes of conduct problems in children has traditionally focused on environmental factors, such as parental neglect. The first published genome-wide linkage analysis for genes influencing conduct disorder did not appear until 2004, in the context of studies of alcoholism [Dick et al., 2004]. The scan found suggestive linkage on chromosomes 2 and 19.

The subtle interplay between genes and environment in this condition has been illustrated by two recent candidate-gene studies of the monamine oxidase A gene (*MAOA*) on the X chromosome. The role of *MAOA* in mediating aggressive behavior has been of interest at least since 1993, when Brunner and colleagues reported that a point mutation in this gene, which resulted in complete MAOA deficiency, co-segregated with severe, impulsive aggression in five Dutch male relatives [Brunner et al., 1993]. Similar highly penetrant variants have not been found in any other families to date.

Caspi and colleagues [2002] found that among adult males who had been maltreated as children, those who had lower-activity *MAOA* alleles were more likely to develop conduct problems than those who had high-activity *MAOA* alleles. The finding was replicated in a 2004 [Foley et al.] study of more than 500 male twins. However, after controlling for the interaction between genotype and maltreatment, the lower-activity *MAOA* alleles were associated with a *reduced* risk of conduct disorder. The study suggests that a single allele may have opposite effects depending on the environment and supports a nondeterministic view of the role of genes in behavior.

EMERGING ISSUES IN NEUROGENETIC TESTING

As research progresses in identifying mutations and susceptibility alleles associated with common neuropsychiatric conditions, we may see a rising public demand for clinical neurogenetic testing. In the United States, commercial laboratories have begun to use direct-to-consumer advertising to promote genetic testing, and these tests as well as the content of the ads are not subject to federal regulation [Human Genome Project information, 2004]. Internet-based companies have begun offering home genetic testing kits with unproven clinical utility as a means to sell nutritional supplements to treat "genetic predisposition to addiction" and other dubious genetic diagnoses [Barrett and Hall, 2003]. This raises a number of concerns about the adverse consequences of genetic testing for what are often dreaded or highly stigmatized conditions. Clinicians will have an increasingly important role to play in educating patients about genetic testing and ensuring that tests are offered in a clinically and ethically appropriate manner.

When considering the ethical issues of genetic testing, it is important to distinguish clinical testing from genotyping in the research setting. Testing that occurs as part of neurogenetic research takes place under protocols designed to ensure that genetic data are obtained, stored, and used in a manner that preserves the participants' confidentiality. Empiric evidence suggests that genotyping as part of a research protocol is without risk. Cubells et al.'s [2004] review of adverse event reports associated with the work of 50 researchers investigating genetic factors in psychiatric and substance abuse disorders found no instance in which data were misused or confidentiality breached.

A useful framework for contemplating the ethical issues in clinical genetic testing was proposed by Burke et al. [2001], who categorized tests according to their clinical validity and the availability of effective treatment. (See Table 9.1.) For tests that have high validity and can be coupled with effective treatment, the major ethical concern is providing access to testing and treatment for all who can benefit. For tests with low validity and no effective treatment, clinicians should recommend against testing according to the principle "do no harm." In the other two categories, the clinical

Table 9.1   Predominant Ethical Concerns in Four Categories of Genetic Tests

	High Validity of Test	Low Validity of Test
Effective treatment available	Ensure access to testing and treatment	Balance clinical benefits against risks of genetic labeling
No effective treatment available	Ensure confidentiality and informed consent	Avoid harm by withholding testing

Adapted from Burke et al. [2001].

benefits of testing must be weighed against the potential risks, including the possibility that testing may cause psychological distress, stigmatization, or discrimination.

To anticipate the issues that may arise in clinical neurogenetic testing, it is important to first consider the validity or predictive value of the tests. As we have seen, many of the common neuropsychiatric disorders are genetically heterogeneous. A small percentage of cases are caused by mutations in a single gene and are inherited in a dominant or recessive pattern, while the majority of cases are characterized by complex inheritance. For the rare cases of single-gene inheritance, the clinical validity of mutation testing will depend upon the penetrance and expressivity of the disorder and upon whether or not all mutations have been discovered and can be evaluated.

For the common forms of disease, however, genetic testing will be less useful, at least in the near term. Unlike mutation testing for a fully penetrant, single-gene disease, testing for a risk allele associated with a complex disorder provides just one piece in a complicated puzzle. The *MAOA* findings in conduct disorder described above [Foley et al., 2004], for example, suggest that a modifying allele may have entirely different effects depending on a feature of the environment. Before susceptibility testing for a complex neurogenetic disorder can yield clinically useful predictions, we would need to identify the major genes involved and understand how the genetic effects interact with each other and with environmental factors.

It is also important to recognize that the major goals of research on common neuropsychiatric diseases have been to understand the pathophysiology of these poorly understood disorders and develop new treatments. Among molecular geneticists, there is a range of opinion about whether genetic susceptibility testing will ever be a clinically useful outcome of research. Most would agree that genetic testing for complex diseases will not be as precise as testing for highly penetrant single-gene diseases. The relative risks associated with each etiologic factor in a complex disorder are likely to have relatively wide confidence intervals that will blur predictions.

APOE testing in Alzheimer's disease is perhaps the best illustration of the potential for genetic susceptibility testing in neuropsychiatry. The *APOE* ε-4 variant has been the most consistently confirmed risk allele among the common neuropsychiatric disorders. Along with age, *APOE4* is the best known predictor of sporadic Alzheimer's disease. Nevertheless, the predictive value of *APOE* testing is low. Because *APOE* is a modifying gene rather than strictly causative, some homozygotes do not develop Alzheimer's disease even after the age of 80, and many Alzheimer's patients do not carry an ε-4 allele [Selkoe, 2003].

Given the low clinical utility of the test, it would clearly be unethical to offer it without fully informing individuals of the test's poor predictive power. In cases in which individuals can be fully informed of the limitations of the test, some clinicians may consider it overly paternalistic not to provide the test to those who believe the results will be useful in life planning. However, several consensus groups have recommended against clinical *APOE* susceptibility testing [reviewed in Post et al., 1997], based on the current lack of therapeutic interventions to prevent dementia, the potential for adverse consequences of learning one's *APOE* status, and the test's poor predictive value.

If an effective intervention for preventing Alzheimer's disease became available, we might assume that *APOE* susceptibility testing would then be indicated. According to Burke et al.'s [2001] schema, in this new situation, the decision to offer the test should be based on an assessment of the potential benefit of the treatment, the potential adverse effects of learning one's *APOE* status, and the harm that could result if the test result were incorrect. If the intervention were safe, cheap, and 100 percent effective, the calculation would appear to be easy. In fact, in that case, it would be preferable to offer the intervention to everyone, without *APOE* testing, since we know that some people without the *APOE4* allele will develop Alzheimer's disease and may be falsely reassured if tested.

Because tests for susceptibility alleles are less predictive than tests for highly penetrant single-gene mutations, they should also have less potential as means of discrimination and stigmatization. Unlike disease-causing genetic mutations, suscep-tibility alleles are highly prevalent in the population. For example, if a disorder with 0.1 percent population prevalence is caused by two genes with multiplicative reces-sive inheritance (affected individuals are homozygous for the risk alleles for both genes), then more than 50 percent of the population would be expected to carry at least one of the risk alleles. Such a high prevalence challenges our notion of what is meant by a *disease gene*. In addition, the risk allele may actually be the ancestral form of the gene. The *APOE ε-4* allele is the form present in the chimpanzee genome, for example, and the other *APOE* alleles can be considered "protective mutations."

Because every human being can be presumed to carry several susceptibility alleles for common disorders, there should be no rational basis for insurance or employment discrimination based solely on the presence of a single risk allele. Realizing that ev-eryone has some genetic vulnerability to neuropsychiatric conditions might also serve to reduce stigmatization, especially if researchers find evidence that risk alleles for these diseases also confer some benefits [Cook, 2000]. We may discover that genetic susceptibility to bipolar disorder also boosts one's creative potential, for example, and risk alleles for conduct disorder encourage individualism and assertiveness.

CONCLUSION

Neurogenetics is a diverse and quickly evolving area of medical research. For common neuropsychiatric disorders, genetic research may offer the best hope of understand-ing the pathophysiology of illness and creating new, rationally based treatments.

Developing diagnostic and screening tests for these diseases may not have been a primary research goal, but susceptibility testing will become technically feasible as more risk alleles are identified. Susceptibility testing may have low predictive power and be of limited clinical utility, except in uncommon cases of families with highly penetrant susceptibility variants. As such tests become available, clinicians will need a sophisticated understanding of genetics and statistics to decide whether it is ethical to offer the tests and, if so, to help patients weigh the risks and benefits.

Chapter 10

Newborn and Carrier Screening

OVERVIEW

Screening newborns and adults is an important consideration for common genetic diseases. For newborn screening, this chapter discusses testing technologies and guidelines (including the next generation of newborn screening techniques such as tandem mass spectrometry), as well as medical, ethical, legal, and public health perspectives. Expectations of health care providers relating to newborn screening are also discussed. This chapter also presents a similar review of considerations relating to carrier screening for children and adults. Screening is now recommended and commonly performed for cystic fibrosis in the United States. Screening is also commonplace for hemoglobinopathies as well as for genetic diseases in the Ashkenazi Jewish population. However, test methodologies may miss some types of mutations, and positive results create patient concern and clinical management issues. Patients may expect testing to be absolutely conclusive. Alternatively, patients may not understand the justification for such testing and feel that it is an unnecessary intrusion. For some forms of testing, although hundreds of gene mutations have been identified (e.g., cystic fibrosis), laboratories only may screen for the most common mutations, with the result that an individual with a *negative* test result could still be a *carrier*. The clinician's approach to screening will depend upon prevailing recommendations as well as the laboratory services available. The increasing breadth and technical capability for carrier screening will place new demands on health care professionals in both primary care and specialist settings.

INTRODUCTION

The goal of adopting a screening process is to identify a selected population of patients who are at increased risk for a specific medical condition, thus enabling the

Genetic Testing: Care, Consent, and Liability, by Neil F. Sharpe and Ronald F. Carter
© 2006 John Wiley & Sons, Inc.

offer of more costly and/or invasive diagnostic testing to patients selected from the population by the screening criteria. Screen-positive patients require counseling in regard to their options, which include review and confirmation of screen positivity, diagnostic evaluation, and possible interventions where a diagnosis is confirmed. Screening of newborns, children, and adults to identify individuals for detection of disease or carrier status is now common practice for selected genetic and congenital diseases, and constant advances in the design of screening programs have resulted in broad uptake of expanded screening modalities, improved screening performance, and a reduced burden of morbidity and mortality. In this chapter, experts provide reviews of:

- Basic concepts underlying the design, implementation, and regulation of newborn screening programs.
- Clinical and laboratory considerations in newborn screening for biochemical disorders such as phenylketonuria (PKU), medium-chain acyl-CoA dehydrogenase (MCAD), and other inborn errors of metabolism.
- Regulatory and administrative issues in newborn screening in North America.
- Principles and practice of screening for detecting carriers of selected genetic diseases.
- Obligations of health care providers relating to screening.

The following discussions include reviews of the obligations of health care providers and suggestions for counseling relevant to newborn and carrier screening for genetic diseases.

NEED TO KNOW

Core Concepts

The core concepts relating to the design and implementation of a newborn screening program include:

- Criteria for selection of specified diseases to be screened for.
- Designation of cut-off (threshold) values for a positive screen result; these values will be determined by the analytical performance of the screening test, the risk of the associated diagnostic procedure, the cost of providing the diagnostic test, and the public's expectations.
- Criteria used to describe the performance of the screening test include: detection rate (DR), false-positive rate (FPR), and the odds of being affected given a positive result (OAPR).

Newborn screening was prompted by the development of an inexpensive, reliable, and convenient method for screening large numbers of specimens (the Guthrie blood spot) and the availability of effective treatments for PKU and congenital hypothyroidism.

Most inborn errors of metabolism are inherited in autosomal-recessive fashion; affected infants are born to normal carrier parents and often do not exhibit detectable clinical symptoms at birth. However, time and rapidity of onset of disease is highly variable.

Early detection of infants affected with MCAD deficiency has become a public health priority, given that the risk of unexpected metabolic crises resulting in severe neurological impairment and/or death can be avoided by simple dietary precautions and parental awareness of risk-inducing situations. The fact that MCAD can be detected by tandem mass spectrometry (TMS) of blood spot acylcarnitines has resulted in the replacement of biochemical newborn screening methodology by TMS. TMS also permits detection of more than 30 other less common disorders. The range of disorders screened in jurisdictions utilizing TMS varies from 3 to over 30. This means that parents need to be educated more carefully and completely about newborn screening and properly prepared for the possible outcomes of screening.

Newborn screening does not provide a diagnosis. Families with babies who are determined to be positive by the adopted screening criteria will require counseling about the significance of the screening result, the options for diagnostic testing, and the options for therapeutic management if a diagnosis is obtained.

Newborn screening is not just a laboratory test. For a newborn screening program to be effective, implementation requires:

- Quality assurance extending from specimen collection to reporting.
- Timely reporting and follow-up of test results; for positive results, this includes notification of the family, primary care physician, and relevant specialists.
- Confirmation by diagnostic testing.
- Clinical management for families with confirmed diagnoses.
- Ongoing evaluation of performance of the screening program.
- Education of public and providers.

Legal Implications of Newborn Screening

Newborn screening programs are usually mandated by legislation and administered as a public health function, but details of implementation vary by jurisdiction, including:

- Specific requirements set out by laws or regulations
- Diseases included in the test panel
- Oversight responsibilities and advisory mechanisms
- Reporting protocols for notification of parents and responsible providers
- Requirements for informed consent and options for exemption
- Policies governing storage and access to blood spots and disclosure of screening data
- Payment and/or source of funding

Criteria for Selection of Disorders in Newborn Screening Programs

The decision to include a specific disease in a newborn screening panel depends upon the cost effectiveness of adding the analyte (includes considerations listed below):

- Performance characteristics of the test protocol (DR, FPR, threshold criteria)
- Clinical burden associated with the disease
- Effectiveness of early diagnosis and intervention
- Implications of false positives for patients and the program

Providers need to be familiar with the particular details of the newborn screening in their area of practice, which are available from the National Newborn Screening and Genetics Resource Center (NNSGRC: http://genes-r-us.uthscsa.edu).

For programs converting to TMS methodology, the number of diseases included in the test panel can be increased at negligible test cost, but the systemic implications relating to detection of positives, intervention, and subsequent long-term impact on health care costs and/or savings all need to be considered. One impact of adoption of TMS for newborn screening has been upwardly revised estimates of the population incidence of the diseases screened, particularly MCAD, which in turn will have an influence on inclusion criteria for other diseases in the future.

NEWBORN SCREENING

C. Anthony Rupar
Child and Parent Resource Institute, University of Western Ontario and the London Health Sciences Centre, London, Ontario, Canada

INTRODUCTION

Newborn screening (NBS) programs constitute the largest volume of all genetic services in most developed countries and are a significant achievement of public health programs. The NBS detection and subsequent treatment of infants with phenylketonuria (PKU) and congenital hypothyroidism has prevented about 40,000 children from being mentally retarded in North America since 1980.

The purpose of population screening has been defined as to "identify individuals at sufficient risk to benefit from further investigation or direct preventive action, amongst persons who have not sought medical attention on account of symptoms of that disorder" [Wald, 1994].

This section discusses newborn screening programs for genetic diseases, especially diseases due to inborn errors of metabolism. However, NBS programs can also include infectious diseases and other disorders such as hearing loss. In many respects the criteria that are applied to include a disorder are similar for both genetic and nongenetic disorders.

BOX 10.1 *Historical Vignette*

Pearl Buck, Nobel Prize (1938), novelist and social activist wrote about her experience as the mother of a child born with PKU in "The Child Who Never Grew" [Buck, 1950].

"I do not know where or at what moment the growth of her intelligence stopped. Nor to this day do we know why it did. There was nothing in my family to make me fear that my child might be one of those who do not grow. Indeed, I was fortunate in my own ancestry on both sides." "At any rate, no young mother could have been less prepared than I for what was to come." "My little daughter's body continued its healthy progress." "She was still beautiful, as she would be to this day were light of the mind behind her features. I think I was the last to perceive that something was wrong. She was my first child, and I had no close comparison to make with others. She was three years old when I first began to wonder." "For at three she did not yet talk." "Yet I can remember my growing uneasiness about my child. She looked so well, her cheeks pink, her hair straight and blond, her eyes the clear blue of health. Why then did speech delay!" "Thus my child was nearly four years old before I discovered for myself that her mind has stopped growing. To all of us there comes the hour of awakening to sad truth. Sometimes the whole awakening comes at once and in a moment. To others, like myself, it came in parts slowly. I was reluctant and unbelieving until the last."

The history of NBS programs is closely intertwined with the development of enabling technologies, concepts of public health medicine and economic resources. Phenylketonuria was the first disorder for which newborn were screened and continues in many respects to be a reference point for the inclusion of other disorders.

The experience described in Box 10.1 echoes that of so many mothers whose children were undiagnosed or diagnosed too late to prevent mental retardation. The publication of the discovery of the cause of PKU by Følling [1934] as described by his son Følling [1994] came too late to be of benefit to Pearl Buck's daughter.

DEVELOPMENT OF NEWBORN SCREENING PROGRAMS

PKU and Congenital Hypothyroidism

Key developments that occurred to enable newborn screening for PKU included the successful treatment of children with PKU [Bickel, 1996] and an inexpensive, reliable, and convenient test for screening large numbers of specimens [Guthrie, 1996]. These advances made it possible to establish policies and standards for newborn screening programs. Similarly, the development of a suitable screening test for congenital hypothyroidism enabled newborn screening of this relatively common (1/3500) and treatable disorder [Dusseault and Laberge, 1973].

Most inborn errors of metabolism including those that are detectable by newborn screening programs are inherited in an autosomal-recessive manner. Typically, there are no dysmorphic features or other symptoms present at birth to alert the parents or physician. Most of the inborn errors of metabolism detected by newborn screening programs are caused by the deficient expression of an enzyme that participates in the metabolism of a small molecule such as an amino acid, fatty acid, or organic acid. These disorders can most commonly be identified by the accumulation of the substrate of the deficient enzyme or some other related metabolite. Occasionally, the concentration of the product of the enzyme reaction is helpful as seen in the measurement of both phenylalanine and tyrosine in the tandem mass spectrometric (TMS) screening for PKU. The range of presentations of disorders that can be detected by newborn screening is from the slow and insidious onset of PKU to acute life-threatening hypoketotic hypoglycemic decompensations seen in medium-chain acyl-CoA dehydrogenase (MCAD) deficiency.

MCAD Deficiency

The role of MCAD deficiency in the development of TMS as newborn screening tool in the 1990s was arguably similar to the role of PKU in developing the Guthrie assay.

The realization that MCAD deficiency can be detected in newborns by TMS of blood dot acylcarnitines stimulated efforts to develop expanded newborn screening programs [Van Hove et al., 1993]. There are many other disorders that occur at much lower incidence than MCAD that are detected by TMS.

MCAD deficiency most frequently presents in infants over 3 months of age with acute episodes of hypoketotic hypoglycemia. The mortality rate of these episodes is about 25 percent. Survivors of acute episodes may have long-term neurological disability. Infants and children with MCAD deficiency are unable to effectively utilize fatty acids for energy through the mitochondrial β-oxidation pathway and typically present with a metabolic decompensation after fasting that may occur after a prodromal illness. However, many children with MCAD deficiency do not experience acute decompensations presumably as a result of genotype–phenotype correlations and the circumstantial nature of precipitating fasts.

The risk of acute decompensation and the sequelae in patients with MCAD deficiency can be significantly reduced with a straightforward management of avoiding fasts and providing glucose during periods of illness. Supplemental carnitine is often provided, although there is a lack of robust evidence for its efficacy. In general, children who do not experience acute decompensations do well [Wilson et al., 1999].

MCAD deficiency meets the criteria of disorders to screen in most NBS programs on the basis of its incidence in many populations of about 1/15,000. Iafolla et al. [1994] studied the natural history of MCAD deficiency.

CONFIRMATION OF POSITIVE SCREENING TEST RESULTS

Newborn screening programs are designed to identify individuals who are at risk to develop disease. All individuals identified as being at risk need to be tested further with definitive diagnostic tests. Most often the diagnostic test uses a different testing method such as quantitative plasma amino acids, urine organic acid analysis, enzyme assay, or mutation detection.

NEWBORN SCREENING POLICIES

Principles of Newborn Screening

Since 1968 there have been several major policy documents on newborn screening [Therrell, 2001]. These include the World Health Organization [1968], Wilson and Jungner [1968], Frankenberg [1974], the National Academy of Sciences [National Research Council, 1975], the American College of Medical Genetics [1997], and the American Academy of Paediatrics [2000]. Wilson and Jungner [1968] delineated the most frequently quoted principles of early disease detection (see Box 10.2).

Frankenberg [1974] concluded that for "certain non-treatable hereditary conditions, screening offers the opportunity for informed decisions as part of family

---

BOX 10.2 *Principles of Early Disease Detection [Wilson and Jungner, 1968]*

1. *The condition sought should be an important health problem.*
2. *There should be an accepted treatment for patients with recognized disease.*
3. *Facilities for diagnosis and treatment should be available.*
4. *There should be a recognizable latent or early symptomatic state.*
5. *There should be a suitable test or examination.*
6. *The test or examination should be acceptable to the population.*
7. *The natural history of the condition, including development from latent to declared disease, should be adequately understood.*
8. *There should be an agreed policy on whom to treat as patients.*
9. *The cost of case finding (including diagnosis and treatment of patients diagnosed) should be economically balanced in relation to possible expenditure on medical care as a whole.*
10. *Case-finding should be a continuing process and not a "once and for all" project.*

planning." This criterion recognizes that it is appropriate to screen untreatable disorders for the benefit of the family, even though the affected screened infant will likely receive little benefit. This is more consistent with a clinical genetics approach to newborn screening than a public health approach. The psychosocial impact on a child and family needs to be considered in assessing newborn screening test that presymptomatically identifies children at risk for an untreatable disorder.

Some criteria espoused by Wilson and Jungner [1968] and others are not universally accepted. Advocacy groups such as the March of Dimes [2004] consider that testing cost is not important in newborn screening policy decisions. The March of Dimes position is that newborn screening for rare diseases should be conducted on every newborn, "as long as its early discovery makes a difference to the child" [Howse and Katz, 2000]. This position is not consistent with selection criteria that include not only costs but also disease incidence.

Systems Approach to Newborn Screening Programs

Effective newborn screening programs are much more than laboratory tests. The principles behind newborn screening imply the need for a systems approach. Therrell [2001] identified six major components to newborn screening programs:

1. Screening, including specimen collection, and testing with quality control
2. Follow-up on test results in a timely manner (especially important for positive screening tests and needs to include family, primary caregiver, and specialist)
3. Confirmatory diagnostic tests
4. Patient management
5. Ongoing evaluation
6. Education at all levels and stages of the program

The American College of Medical Genetics [1997] policy statement on genetic screening describes "screening for genetic disease or genetic predisposition to disease as a unique opportunity to prevent the effects of the disease. Retrieval, diagnosis and intervention before irreversible damage represent goals for an effective genetic screening program." The American College of Medical Genetics guidelines identify four major elements of screening program as:

- The screening program should have a clearly defined purpose.
- A screening program is more than a laboratory test.
- A screening program should be reviewed by the appropriate board.
- The screening program should be evaluated periodically to determine if it is meeting its goals.

LEGAL IMPLICATIONS

Newborn screening programs are most frequently mandated by legislation and organized on a jurisdictional basis. In the United States and Canada programs are the responsibilities of state and provincial jurisdictions. In the United States, state legislation mandates the provision of newborn screening programs with a range of laws or regulations, specific tests, oversight responsibilities, state advisory boards, processes for informing parents, exemptions, storage policies, and use of blood samples and payment for newborn screen procedures.

The selection of disorders to be screened is within the jurisdiction of each state. Historically, all states have screened for PKU and congenital hypothyroidism, but, with the advent of new screening technologies, especially tandem mass spectrometry, there is no longer a uniform standard across the United States. The lack of uniform screening policies across the country has been a source of concern to both service providers and federal legislators [General Accounting Office, 2003].

In Canada newborn screening programs are provincial jurisdictions. In most provinces there is no legislated requirement to provide newborn screening, but the programs are provided universally. As in the United States, there is no national standard as to which disorders are screened.

In the past, courts have determined that health care providers are negligent if mandatory newborn screening for PKU is not performed or if performed and negative results are falsely reported as in the case of *Humana of Kentucky, Inc. v. McKee* [1992]. Before PKU screening was a mandated program, the Michigan Supreme Court established the principle that "the standard of care for a specialist was that of a reasonable specialist practicing medicine in the light of present day scientific knowledge. Locality does not matter" [*Naccarato v. Grob*, 1970].

The expansion of newborn screening programs in the mid-1990s in some but not all jurisdictions may result in similar legal actions. Some states such as Mississippi and Illinois are requiring parents to be informed that there are newborn screening tests that are available in other jurisdictions and from the private sector that are not part of the state-mandated program.

Newborn screening statutes and regulations in over half the states specify that newborn screening information is confidential, but these confidentiality provisions are often subject to exceptions. The most common exception allows disclosure of information for research purposes, provided that the child's identity is not revealed and researchers comply with applicable laws for the protection of humans in research activities. Other exceptions include use of information for law enforcement and for establishing paternity. Over half the states have statutes that govern the collection, use, or disclosure of genetic information, which may also apply to genetic information obtained from newborn screening [General Accounting Office, 2003].

Consent

In the United States as reported by the Government Accounting Office [2003], states usually do not require that parental consent be obtained before screening occurs. Many states allow exemptions from newborn screening for religious reasons.

In Canada none of the provinces require consent [Hanley, 2004]. In both countries consent is implied as it is considered a part of routine newborn care.

CRITERIA FOR SELECTION OF DISORDERS SUITABLE FOR SCREENING

The criteria used to determine which disorders to screen include both disease characteristics and testing technology. Disease characteristics include the severity of disease, the newborn incidence, a latent period between birth and clinical onset, the availability of appropriate treatment, and an understanding of the natural history of the disease both treated and untreated. As discussed previously, there is agreement on the need for the disease to be severe, but there is not unanimity on the remaining criteria.

Test Characteristics

Technology provides the capability to screen for specific disorders. Criteria for the suitability of a test include sensitivity, specificity, predictive value, and cost.

Sensitivity   probability of a positive test among patients with disease

Specificity   probability of a negative test among patients without disease

Positive predictive value   probability of disease among patients with a positive test

Negative predictive value   probability of no disease among patients with a negative test

The ideal screening test would have 100 percent sensitivity and 100 percent specificity with complete predictive value. In reality tests have false-positive and false-negative results. Most screening tests are structured with numerical thresholds set at a level such that all true positive tests are identified (high sensitivity), but there will also be a number of false positives (low specificity).

All positive test results need to be followed up with definitive testing and clinical assessment to establish whether a test result represents a true positive or a false positive. Minimizing the number of false-positive test results reduces the costs of follow-up.

For example, 240 μmol/L phenylalanine is frequently used as a threshold value in the Guthrie test. All specimens with a phenylalanine concentration greater than 240 μmol/L are followed up to determine whether the value is a true positive or a false positive. Follow-up can be with a repeat Guthrie test or quantitative phenylalanine assay. The number of false-positive results in the initial screening test is about 10 times the number of true positive results [Wang et al., 1997]. To put this in context, the upper limit of normal concentration of plasma phenylalanine is about 110 μmol/L. It is common for patients with untreated or inadequately treated PKU to have plamsa concentrations >1000 μmol/L. However, the threshold is set low with an acceptance of a high rate of false positives to minimize the risk of false-negative results.

This approach minimizes the risk of not identifying and treating an affected infant. However, repeat testing causes stress on all families whether false-positive or true positive results are obtained and consumes resources.

Waisbren et al., [2003] assessed the impact of false-positive test results in TMS-based screening programs. Mothers in the screened group experienced lower overall stress levels than mothers in clinically identified group. However, mothers of children in the false-positive group attained higher stress scores than mothers of children with normal stress results.

What are the sources of false-negative and false-positive results in PKU testing? Experience indicates that the greatest source of false-negative results occur on the basis of errors made in the specimen collection and resulting follow-up processes. Analytical false negatives may be at increased risk in infants that have been discharged early from hospital. It is recommended in most jurisdictions that specimens for PKU screening be collected after 24 hours of age. Screening programs for PKU were established at a time when newborns were kept in hospital for several days. In recent years the trend for earlier discharges from hospital has resulted in concerns that the false-negative rate may increase in infants who have not started to feed. False-positive results can be caused by parenteral feeding and immaturity.

Cost-Effectiveness

An estimate of the cost-effectiveness of screening for a particular disorder requires a systems approach to newborn screening incorporating all aspects of the program with consideration of both the costs of a test and the economic benefits of testing.

The costs of a test include specimen collection, laboratory testing, reporting, confirmatory testing of presumptive positive results, and treatment costs.

The costs of not screening for a disorder include the costs associated with symptomatic presentation that may often be tertiary acute care, long-term medical, social, and educational costs of managing and supporting individuals with lifelong disability, and the productivity losses of caregivers.

The mainstay for treatment of many patients with inborn errors of metabolism detected by newborn screening is based in medically indicated diets formulated to be replete in the metabolite or metabolite precursor that accumulates in the respective disease. Infants with PKU receive a synthetic formula that has very low levels of phenylalanine and progress as they grow older into a more adultlike low-phenylalanine diet. The standard of care is to stay on a low-phenylalanine diet for life. This diet costs about $10,000/year, which exceeds the resources of most patients.

Measures of the benefits of screening and presymptomatic management are health-related quality of life and the more encompassing quality-adjusted life-year [Pandor et al., 2004]. Several disorders have been identified that meet criteria for newborn screening. The most commonly screened disorders include PKU, congenital hypothyroidism, galactosemia, congenital adrenal hyperplasia, and sickle cell disease.

Waisbren et al. [2003] compared the outcomes of children identified in a TMS-based screening program to clinically ascertained children. The most frequently diagnosed disorder in the newborn screened group was MCAD deficiency. Within the first 6 months of life, children identified by screening had about half of the number of hospitalizations as those clinically identified. The children identified by screening required significantly fewer and less intensive medical services and developmental assessment scores were significantly higher.

The incorporation of tandem mass spectrometry into newborn screening programs has created a situation where the technology allows the addition of many disorders that may not meet established inclusion criteria, but the incremental cost at the detection level is negligible. However, there are increased follow-up costs.

Selection criteria that have been most difficult to satisfy have been the incidence of many inborn errors of metabolism, the natural history of the disorder, and the effectiveness of treatment. Frequently occurring inborn errors of metabolism such as PKU and MCAD have a newborn incidence of about 1 in 20,000 to 1 in 10,000 in many populations.

The newborn incidence of many rare inborn errors of metabolism may be in the range of 1/500,000 to 1/50,000. Published experiences in diseases rarely diagnosed is often only the occasional case report or at best the experience of a small series of patients. The evidence of the understanding of the natural history and the success of treatment of rare disorders is usually not robust and often based on expert opinion.

There is a circular argument in that although there is insufficient evidence to include many rare disorders in newborn screening programs, the most likely means of diagnosing sufficient numbers of patients to provide evidence is to ascertain patients through newborn screening programs. There is a responsibility for comprehensive follow-up and documentation of all patients with rare diseases.

For example, recent evidence obtained from newborn screening programs using tandem mass spectrometry raises the possibility that for many patients 3-methylcrotonyl-CoA carboxylase (OMIM #210200) deficiency is benign. Previous experience based on clinical ascertainment was that it was a serious disorder presenting with hypoglycemia, metabolic acidosis, hypotonia, and developmental delay.

Pandor et al. [2004] in a comprehensive assessment of the clinical effectiveness and cost-effectiveness of using tandem mass spectrometry for newborn screening in the United Kingdom concluded that PKU screening alone did not justify the introduction of tandem mass spectrometry. The inclusion of MCAD deficiency with PKU was sufficient to use tandem mass spectrometry. Tandem mass spectrometry is the only newborn screening technology able to diagnose MCAD deficiency and reduces by at about fivefold the number of false positives in PKU screening.

Venditti et al. [2003] assessed the cost-effectiveness of screening for MCAD deficiency by an established tandem mass spectrometry screening program and concluded that screening for MCAD deficiency reduces morbidity and mortality at an incremental cost that compares favorably with other health care interventions including PKU screening. The analysis indicated that universal newborn screening for MCAD can be achieved at an incremental cost of $5600 per quality-adjusted life-year saved.

Population characteristics may determine the newborn incidence of genetic disorders. Many disorders are panethnic while others have higher newborn incidences in certain racially, ethnically, or geographically determined populations. State and provincial jurisdictions should have the discretion to include disorders that may be relevant to local populations in the presence of national standards for newborn screening programs.

Wilcken et al. [2003] documented the impact of tandem mass spectrometry on the number of diagnoses of 31 inborn errors of metabolism in Australia by comparing the number of diagnoses made prior to the introduction of TMS and the number detected by TMS. The overall incidence of diagnosis approximately doubled to 1/6400. The increase in diagnosis was primarily due to MCAD deficiency.

Newborn Screening Technologies

Several technologies have been developed or adapted to newborn screening. No single approach is suitable for all disorders that meet the screening eligibility criteria. All methodologies use blood dot card specimens collected from heel pricks. The tests are developed to be robust tests that can be done in large volumes on blood dot card specimens that can require little special handling.

The bacteria growth inhibition assay developed by Guthrie for PKU has been adapted to other disorders of amino acid metabolism including maple syrup urine disease and homocystinuria. Phenylalanine has also been detected in newborn screening programs with a fluorometric assay.

Congenital hypothyroidism, congenital adrenal hyperplasia, and cystic fibrosis are typically screened with ELISA (enzyme-linked immunosorbent assay). There are specific enzyme tests for biotinidase and galactosemia. Sickle cell anemia is screened by isoelectric focusing electrophoresis.

The development of tandem mass spectrometry as a newborn screening test during the 1990s has enabled newborn screening programs to consider several disorders that were not detectable by other technologies.

Tandem mass spectrometry is a model for a multianalyte approach to newborn screening. It can simultaneously detect multiple metabolites enabling the diagnosis of about 30 inborn errors of metabolism. Tandem mass spectrometry has a short analytical time of about 3 minutes/specimen and is able to use blood dot specimens. High-throughput instruments have been developed that can reliably process over 400 specimens/day [Chace et al., 2003].

A mass spectrometer is essentially a mass detector. The mass can be the mass of an ion formed from the metabolite by electrospray ionization or it can be the mass of a fragment of the parent ion. The blood dot specimen may be analyzed after butyl ester derivitization or underivatized. The mass spectrometer identifies both the mass and number of ions. Interpretation of tandem mass spectrometer results is initially by computer algorithims.

Table 10.1   Errors of Metabolism Detected by Tandem Mass Spectrometry

Organic acidemias

3-OH 3-CH3 glutaric aciduria (HMG)
Beta-ketothiolase deficiency (BKT)
Glutaric aciduria I (GAI)
Glutaric aciduria II (GAII)
Isovaleric academia (IVA)
Methylcrotonyl-CoA carboxylase deficiency (3MCC)
Methylmalonic acidemia (Cbl A,B)
Methylmalonic acidemia (mutase deficiency)
Multiple CoA carboxylase deficiency
Propionic acidemia (PROP)

Urea cycle disorders

Arginase (ARG)
Argininosuccinic acidemia (ASA)
Citrullinemia (CIT)
Hyperammonemia, hyperornithinemia, homocitrullinuria (HHH)

Amino acid disorders

Homocystinuria (HCY)
Maple syrup urine disease (MSUD)
Phenylketonuria (PKU)
Tyrosinemia type I (TYR I)

Fatty acid oxidation disorders

2,4-Dienoyl-CoA reductase deficiency
Carnitine palmitoyl transferase II (CPT II)
Carnitine palmitoyl transferase I (CPT I)
Carnitine uptake defect (CUD)
Carnitine/acylcarnitine translocase deficiency (CAT)
Hydroxymethyl glutaryl-CoA lyase deficienct (HMGCoAlyase)
Long-chain hydroxy acyl-CoA dehydrogenase deficiency (LCHAD)
Medium-chain acyl-CoA dehydrogenase deficiency (MCAD)
Short-chain acyl-CoA dehydrognease deficiency (SCAD)
Trifunctional protein deficiency (TFP)
Very long chain acyl-CoA dehydrogenase deficiency (VLCAD)
Very long chain hydroxyacyl-CoA dehydrogenase (VLCHAD)

A list of 30 inborn errors of metabolism that have been detected by tandem mass spectrometry is provided in Table 10.1. This list is not complete in that there are some disorders not listed that are detectable, but the sensitivity of detection available is not clear. This may also apply to some disorders that are listed.

The age of the infant at the time of specimen collection is a factor in the ability to identify positive specimens. Specimens from infants in the first 2 days of life are likely to be more sensitive in the detection of fatty acid oxidation disorders. Conversely, specimens collected the first day of life may be less sensitive to disorders of amino acid metabolism that are exacerbated with feeding.

The specificity and sensitivity of tandem mass spectrometry screening varies with disorders and metabolites. The rate of false positives for PKU and MCAD is reported to be <0.05 percent; however, for tyrosine and methylmalonic acid the rate of false positives is much greater at 0.2 percent. The high false positive for tyrosine reflects the relatively large number of infants with liver disease or receiving parenteral nutrition [Chace et al., 2003].

CURRENT STATUS OF NEWBORN SCREENING

Most American states and Canadian provinces have advisory committees to newborn screening programs with committee membership being specialty medical care physicians, laboratory specialists, pediatricians, and/or other primary health care providers, health department staff who conduct follow-up activities, individuals with disorders or parents of children with disorders, and ethicists [General Accounting Office, 2003].

Starting in the mid-1990s tandem mass spectrometry was introduced in some jurisdictions but not others, resulting in a significant disparity in which disorders are included in newborn screening programs. The status of newborn screening programs in many jurisdictions is in a state of change as implementation decisions are made.

The General Accounting Office [2003] in a report to Congress on the status of newborn screening programs in the United States stated that the number of disorders included in state newborn screening programs ranged from 4 to 36 with most states screening for 8 or fewer disorders. Most states screened for PKU, congenital hypothyroidism, galactosemia, sickle cell disease, and congenital adrenal hyperplasia.

The National Newborn Screening and Genetics Resource Center (NNSGRC), http://genes-r-us.uthscsa.edu, is a cooperative agreement between the Maternal and Child Health Bureau (MCHB), Genetic Services Branch, and the University of Texas Health Science Center at San Antonio (UTHSCSA), Department of Pediatrics. The NNSGRC website maintains a current listing of screening programs. Currently (11/15/2004) 35 states have implemented or mandated newborn screening by mass spectrometry. In Canada, 5 of 10 provinces have implemented mass spectrometry [Hanley, 2004].

The March of Dimes Birth Defects foundation has an advocacy role for child health and the prevention of disability and recommends that every state screen every baby for at least the 30 disorders (listed in Table 10.1).

A similar trend toward the introduction of tandem mass spectrometry into newborn screening programs is occurring throughout the developed countries. In less developed countries newborn screening is either not a priority or not feasible. In South Africa, a developing country, a review of a pilot program screening for PKU in a European-derived community concluded that with an incidence of about 1/20,000 newborn "that newborn screening for PKU and other amino acidopathies is not cost-effective and justifiable, especially against the background of prevailing demographic conditions and more pressing health priorities in South Africa" [Hitzeroth et al., 1995].

FUTURE ISSUES

System Development

A major challenge facing most jurisdictions is how to most efficiently structure new-born screening programs. In the United States and Canada there is a need to adopt national standards but also to recognize that the incidence of disease can vary among regional populations. At a systems level newborns are being screened for not only genetic diseases but also for hearing loss and infectious diseases. These programs target the same population but often are administered independently with redundant activities.

Blood dot cards contain the complete genome of every newborn. This is potentially a resource to researchers and perhaps also to the long-term health of the newborn. There is a need for well-considered policies for retention of specimens and access to specimens that balance the potential long-term benefits with risks associated with confidentiality. There is a range in policies across the state screening programs with regard to how notification of results occurs, the purpose and duration of blood spot specimen storage, and the access provided to researcher [Mandl et al., 2002; General Accounting Office, 2003]. There is a similar variation among Canadian provinces. Confidentiality of information is important to protect access to health insurance, life insurance, employment, and to prevent social stigmatization.

Education

There are increasing needs for education to the public, families, and health care providers as to the benefits and limitations of expanded newborn screening. A negative screening test result does not mean the infant is not affected with the disorder. All tests have a false-negative rate. Physicians need to remember that PKU should be included in appropriate differential diagnosis even when there is evidence of a negative screening test result.

Newborn screening programs and physicians managing patients with rare diseases need to document and share information.

Inclusion Criteria

The general trend has been to gradually include more disorders in newborn screening programs. Decisions are sometimes the result of critical consideration of data and at other times in response to interest group lobbying. There are many competing perspectives as to which disorders to include in newborn screening programs. Do the traditional cost-effectiveness decision-making models of health care economists continue to apply or are the models too restrictive? It will take many years, if ever, for sufficient data to be collected to enable rare diseases to be assessed with the same rigor as PKU and MCAD deficiency. Should screening programs accept a lower level of evidence recognizing the burden of symptomatic disease on individuals and families?

Are screening programs able to accept the increase demand on resources and stress created by false-positive results?

Screening for New Types of Diseases

The technology to detect lysosomal storage diseases in newborn screening programs is under development using metabolite markers detected by tandem mass spectrometry [Meikle PJ et al., 2004]. The overall incidence of lysosomal storage diseases as a group in most populations is about 1/5000 births. Lysosomal storage diseases are typically progressive multisystem disorders for which treatment of the basic defects is approved for two disorders (Gaucher disease and Fabry disease) with several others in clinical or preclinical trials. The therapeutic approach has been enzyme replacement therapy, which is effective at preventing or ameliorating many aspects of Gaucher disease and possibly Fabry disease. Patients with other lysosomal storage diseases may benefit from bone marrow transplants and other medical interventions. The cost of enzyme replacement therapy is currently about $200,000/year for lifetime depending on the disorder and the size of the affected individual. The basic principles of newborn screening programs include the provision of treatment for which cost may prove to be cumulatively prohibitive.

At least 10 states have introduced newborn screening for cystic fibrosis. The incidence of cystic fibrosis is about 1/4000 births, and early diagnosis and management can prevent consequences of malnutrition and lung disease. The testing strategy often consists of a two-tiered approach using immunoreactive trypsin and DNA (deoxyribonucleic acid) testing. One of the challenges presented by cystic fibrosis testing is that large numbers of carriers are identified and require counseling and further testing to ensure carrier status [Comeau et al., 2004].

Newborn screening programs to date have considered only relatively rare single-gene disorders. There is some urgency to establish clear policies and criteria because in the future programs will be challenged by pressures to screen for genetic risk factors for common multifactorial diseases such as vascular disease, hypertension, and mental health diseases. The screening technologies may again make another leap forward to chip arrays to enable this testing. Society will need to be active participants in the decision-making processes.

CARRIER SCREENING

Lynn Holt and Bruce R. Korf
Department of Genetics, University of Alabama at Birmingham, Birmingham, Alabama

INTRODUCTION

Recessive disorders typically signal their presence in a family with the development of symptoms in a homozygous affected individual. The parents will be phenotypically normal carriers and face a one in four recurrence risk for each child. Usually there will not be other affected relatives, so the parents only become aware of their carrier status when an affected child is diagnosed. By this time they may already have had additional children, who are also at risk. Carrier testing for recessive disorders offers an option to avoid this scenario. Parents can become aware of their carrier status prior to pregnancy, allowing them to make choices regarding pregnancy, including the choice to avoid having an affected child. There is a long history of carrier screening for specific target populations, most notably through programs aimed at Tay-Sachs disease and hemoglobinopathies. The advent of DNA testing now raises the possibility of vastly increasing the menu of tests, increasing options for couples, but also raising complex questions regarding policy and medical decision making. We will consider some of these issues in this section.

CASE HISTORY

Cindy and Joseph are recently married and are thinking about starting a family. Cindy mentions this to her primary care physician, wondering whether there is anything for which the couple should be tested. Cindy is of Irish ancestry, and Joseph is of Ashkenazi Jewish background. The couple is referred to a genetic counselor. Family history reveals that Joseph has a sister with congenital deafness, but there are no other medical conditions that are known to run in the family. The counselor talks to the couple about various testing options, including a panel of tests offered to people of Ashkenazi Jewish ancestry and cystic fibrosis (CF) testing. Testing is done, and Joseph is found to be a CF carrier. Cindy does not carry a detectable cystic fibrosis mutation. The counselor explains their risk of having a child with cystic fibrosis. At the end of the visit, the couple also asks about whether they are at risk of having a child with hearing loss.

Issues Raised in Clinical Scenario

This is a couple of mixed ancestry, as is increasingly common, and therefore presents challenges in knowing what kind of screening to offer. Both are at risk of being cystic fibrosis carriers, and, indeed, one (Joseph) is found to be a carrier. Because of his Ashkenazi Jewish ancestry, the detection rate for CF carrier status is high ($\sim$97 percent). Cindy is of Irish ancestry, and the detection rate for her would be only $\sim$80 percent. Her risk of being a carrier is reduced from 1/29 to approximately 1/140. The risk to a pregnancy is then 1/560, but there is no capability of offering a prenatal test, since, if she is a carrier, her mutation is not known. The issue of congenital deafness raises similarly complex issues. The most common genetic form of congenital deafness is due to connexin 26 gene mutation. Joseph may be at risk of being a carrier based on his family history, and for Cindy the carrier risk for this can be 1 to 3 percent. It would be technically straightforward to offer carrier testing for the couple, but a major question is whether the nature of congenital deafness warrants carrier testing and possible prenatal diagnosis. There may be various views of the appropriateness of such testing, ranging from those who view deafness as a form of "normal" to those who might chose not to have a child who is going to be affected.

NEED TO KNOW

Development of a Screening Protocol

With the complete human genome sequence in hand, it is theoretically possible to test any individual for any genetic variant. We are a very long way from universal and comprehensive screening, however. Historically, screening protocols have focused on testing for a small number of severe disorders in populations at particularly high risk. Considerable effort has been made to tailor the programs to the needs of specific communities, often very effectively, sometimes not. Notable successes include the Tay-Sachs disease screening program in the Ashkenazi Jewish population [Kaback et al., 1993] and screening for thalassaemias in the Mediterranean and Asian populations [Cao et al., 2002]. In contrast, screening for sickle cell anemia carrier status in African Americans in the 1970s was a major failure [Modell, 1997]. In this section we will explore the criteria for a successful screening program and issues faced in implementation.

Goals of Screening

The goal of a screening program is to identify heterozygous carriers prior to their having an affected child in order to be able to offer them counseling regarding options for future pregnancy. Options include, but are not limited to, prenatal diagnosis. Other possible choices include use of a sperm or egg donor who is not a carrier, adoption, or not having a child. Those who choose prenatal diagnosis may elect termination of an affected pregnancy or may use prenatal testing hoping for reassurance or to plan for the care of an affected child. In any case, one would want to offer carrier screening for a disorder that is sufficiently severe to warrant prenatal testing. This is clearly the case for Tay-Sachs disease, which is lethal in childhood and is untreatable. Cystic fibrosis screening is being offered on a large scale, given its major effects on health in spite of treatments that now prolong life into the third or fourth decades. The argument for screening is much less clear as one gets to conditions that impose less of a medical burden, such as deafness.

Target Population

Disease frequency is another consideration in setting up a screening protocol. There should be an appreciable frequency of carriers in the population in order to justify launching a screening program. Carrier frequency often differs in different populations. This is due, in part, to the presence of founder effects and, in part, to a selective advantage to carriers in specific regions. Targeting a specific ethnic group for screening offers many advantages. The program can be tailored to unique cultural or educational backgrounds. Community facilities may be available to provide publicity and education. In some cares, there will be a limited repertoire of clinical phenotypes and genotypes in a distinct population, making it easier to provide testing and to offer counseling. There are, however, some pitfalls in targeted screening. To an increasing extent, individuals may be of mixed ancestry or may not know their ancestry. Also, for some groups there may be insufficient data to interpret the significance of a negative carrier screen if the spectrum of mutations in the population is not known.

Technical Considerations

Carrier screening may be done using a DNA test for gene mutation, by direct analysis of the gene product, or by enzyme assay. An example of protein analysis is hemoglobin electrophoresis. Enzyme assays may be done for metabolic disorders, such as Tay-Sachs disease. The specific test must be customized to the disorder and, to some extent, to the characteristics of the disorder in the target population.

Gene product or enzyme assays have the advantage that they directly test structure or function. Therefore, they are less prone to ambiguous results due to finding variants of unknown significance. The major pitfalls are overlaps between carrier and noncarrier ranges for enzyme activity and the existence of some alleles that result in deficient activity when tested with artificial substrates that are not clinically relevant. Another disadvantage is that some assays require biopsy of affected tissue, whereas DNA studies can be done using blood or any other tissue containing nucleated cells.

DNA testing can be highly sensitive and specific and can be performed on peripheral blood samples for carrier testing or amniotic fluid cells or fetal placenta for prenatal diagnosis. DNA testing is extremely robust for carrier testing in disorders with a limited repertoire of mutations, such as sickle cell anemia. In such cases a positive test proves carrier status and a negative test rules it out. Most genetic disorders, however, are associated with a variety of different mutation types. Mutation diversity may range from one or a few mutations to hundreds. It is not cost effective to do comprehensive DNA sequencing as part of routine carrier screening, and there is also risk of finding DNA variants of unknown significance with this approach. Most DNA-based carrier screens involve testing of a limited number of mutations of proven pathogenicity. Interpretation of a positive test is straightforward, but negative results must be interpreted with caution. Not finding a common mutation does not mean that there is *no* mutation. It is always possible that an individual carries a very rare abnormal allele that was not tested in the screen. The negative predictive value of the test may be known for some populations but not for others. The problem is illustrated by challenges in cystic fibrosis carrier screening. Over 90 percent of mutations are detectable in Caucasian carriers, but fewer than 30 percent are detectable in Asians. A Caucasian–Asian couple where the Caucasian partner is found to be a carrier but the Asian partner is not may still be at risk of having an affected child if the Asian parent carries a rare mutation not included on the screening panel.

Screening Approach

There are many potential points of contact for carrier screening. It is possible to screen newborns or children, but carrier screening requires informed consent and does not immediately benefit the child, so it is generally not done in children. In some cases, screening may be offered to individuals of childbearing age. This has been the case for Tay-Sachs disease screening in some communities. In one orthodox Jewish community where marriages are arranged, carrier screening is used to avoid pairing a carrier couple [Broide et al., 1993].

Various strategies have been used in preconception carrier screening. Both partners in a couple may be tested simultaneously, or one may be tested first, and the other tested only if the first partner was found to be a carrier. Cascade screening involves testing multiple family members of an affected individual. Counseling should be offered to relatives of probands, but this approach is unlikely to identify all at-risk individuals in a population.

Implementation of Carrier Screening

Implementation of a carrier screening program requires careful planning and overcoming multiple obstacles. Professional societies such as the American College of Medical Genetics can play a critical role by studying the question and issuing practice guidelines. A decision to recommend carrier screening has major medicolegal implications since failure to offer screening may expose a practitioner to liability if an affected child is born.

There are also demands for patient education. The most successful screening programs include efforts to provide community education. This can be critical to engaging participation by couples in screening. It is also important to help explain the implications of a positive screen. Finally, it is important to avoid misconceptions about carrier screening that can engender wariness about being tested or expose carriers to stigmatization. These problems occurred with sickle cell carrier screening in the United States in the 1970s and led to the discontinuation of that program.

A third important consideration is adequacy of human resources. Every new screen places on additional time burden on clinicians, including primary care physicians and obstetrician-gynecologists. Genetic counselors can play an important role in patient education and counseling, but their services may not be reimbursed by some insurance providers. Implementation of a screening program without addressing human resource needs can overwhelm the system and result in inability to deliver on a promise of screening.

Current Status of Carrier Screening

There are a number of established carrier screening programs. We will briefly review these here to illustrate fundamental principles of screening.

Cystic Fibrosis

Cystic fibrosis is an inherited condition that can result in pulmonary disease, pancreatic insufficiency resulting in malabsorption, and male infertility. Although many affected individuals have a significantly shortened life span, the phenotype can be very mild. CF is the most common autosomal-recessive lethal condition in the U.S. Caucasian population. It is estimated that 1 in 25 to 30 Caucasians is a carrier for CF, and the disease frequency is 1 in 2500 live births. CF carriers are asymptomatic, and 80 percent of couples that have an affected child have no family history of the condition [Balinsky and Zhu, 2004].

The *CFTR* gene responsible for CF was identified in 1989. Over 1000 disease-causing mutations have been identified in addition to numerous polymorphisms (Moskowitz et al., 2004). The wide variety of mutations likely explains the variable phenotype observed in CF. Clear genotype–phenotype correlations have not been established, however.

In 1997, the National Institutes of Health (NIH) recommended carrier screening for CF starting with relatives of an affected individual or couple planning to have a child. The NIH later worked in collaboration with the American College of Obstetricians and Gynecologists (ACOG) and the American College of Medical Genetics (ACMG) to develop implementation guidelines, educational materials, informed consent materials, and laboratory standards. The result of this collaboration was the 2001 statement Preconception and Prenatal Carrier Screening for Cystic Fibrosis: Clinical and Laboratory Guidelines, co-authored by ACOG and ABGC.

The ACOG and ABGC recommend offering CF carrier screening to the following populations:

1. Individuals with a family history of CF

2. Reproductive partners of individuals who have CF

3. Couples in whom one or both partners are Caucasian and are planning a pregnancy or seeking prenatal care

The recommendations further commented that screening should be offered to individuals at the greatest risk to have a child with CF and be made available to couples in racial or ethnic groups with a lower risk. (Preliminary discussions about the adoption of a similar recommendation have occurred in Canada, but to date there has been no clear agreement on an argument for or against population screening for CF carrier status.)

The recommended standard DNA screening panel includes 25 mutations that account for 80 percent of carriers in the European Caucasian population (Gilbert, 2001). One challenge in implementing carrier screening for CF is the fact that carrier detection rates vary based on the ethnic background of the individuals screened (Table 10.2). The issue is further complicated by extensive ethnic and racial mixing within the United States.

Screening should be offered to couples before they become pregnant in order for them to have the greatest number of reproductive options. However, most couples are offered screening once they become pregnant. Ideally, they should be offered screening during the first trimester in order for them to have time to consider prenatal diagnosis.

Table 10.2   Cystic Fibrosis Carrier Screening Detection Rate Based on Ethnic/Racial Group

Ethnic group	Detection rate (%)
Ashkenazi Jewish	97
European Caucasian	80
African American	69
Hispanic American	57

Source: Modified from Preconception and Prenatal Carrier Screening for Cystic Fibrosis: Clinical and Laboratory Guidelines, ABGC and ACOG [2001].

Cystic fibrosis carrier screening can be offered in a simultaneous or sequential fashion. In simultaneous screening, both parents are tested at the same time. This provides the most accurate risk to the couple for having an affected child because the carrier status of both parents is known. In sequential testing, one parent, usually the female, is screened first. If she is identified as a carrier, then her partner is subsequently screened. If she is not found to have a mutation, her partner is not offered testing. In this situation, the residual chance for the couple to have an affected child is low, but the risk is higher than if the carrier status of both parents were known.

When an at-risk couple is identified, there is a 25 percent chance that they will have a child with CF. If the couple is pregnant, prenatal testing should be offered to determine if the fetus is affected. Prenatal testing is possible via Chorionic Villus Sampling (CVS) or amniocentesis.

Many commercial and academic laboratories offer carrier screening for CF. The vast majority offer the recommended 25-mutation panel and a variety of additional mutations. For unaffected northern European Caucasian couples with no family history of CF, these screening panels are appropriate. Some laboratories offer testing for specific mutations for non-European Caucasian individuals. Specific common mutations in the African American, southern European (Italy and Spain), and Mexican American populations have been identified. Most of these mutations are not a part of the originally recommended panel. The carrier detection rate for low-risk populations will vary based on the type of mutation panel being used, so it is important to select a panel that includes mutations for the specific ethnic or racial group of the individual being screened.

For clinically affected individuals of reproductive age, genetic testing is required for accurate prenatal diagnosis and testing of at-risk family members. Many of today's adults with CF have not had genetic testing because it was not available at the time of their diagnosis. In most people with CF both mutated alleles can be identified. In some cases, though, a clinically affected individual will only have one identifiable mutation, and in rare cases no mutation will be identified (Stern, 1997). If both alleles are not identified with a screening panel, full gene sequencing is recommended. At a minimum, the partner of an individual with CF should be offered the best screening panel for his or her ethnic group. Some couples may consider full gene sequencing for the partner of the affected individual to provide the most accurate risk assessment.

Hemoglobinopathies

Hemoglobinopathies are a group of common autosomal-recessive conditions that affect the quality and quantity of hemoglobin. The two main categories are sickle cell disease and thalassemia. The ethnic origins of these conditions have been well studied. In general, African Americans and individuals of Asian, Middle Eastern, and Mediterranean descent are at greatest risk for a hemoglobinopathy (Table 10.3). All members of these ethnic groups and the partners of individuals with a known hemoglobinopathy should be offered carrier screening. Ideally, screening should be completed prior to pregnancy or early during the first trimester in order to consider prenatal diagnosis.

The genetics of sickle cell disease was discovered in the late 1940s and was followed by the first carrier screening for sickle cell trait in the early 1950s. However, there was no widespread implementation of sickle cell carrier screening despite the high frequency of the condition. In the 1970s and 1980s, a variety of U.S. federal government and community-based programs were initiated for sickle cell screening and education, yet there was never a national organized program to screen the African American population. These programs were successful in reducing the number of un-diagnosed affected school-age children [Whitten and Whitten-Shurney, 2001]. Similarly, there has not been a separate widespread, well-organized effort to screen at-risk populations for thalassemia.

Sickle cell disease is caused by a qualitative defect in the formation of the hemoglobin β chain. Common symptoms of the disease include chronic anemia, tissue ischemia, chronic pain, and vascular occlusion. Thalassemia is caused by decreased or absent production of normal hemoglobin. Affected individuals can have a wide variety of clinical presentations from asymptomatic to life-threatening anemia. Individuals can also have a combination of a sickle cell disease mutation and a thalassemia mutation.

Table 10.3   Hemoglobinopathy Carrier Frequency Based on Ethnic/Racial Group

Ethnicity	β-thal trait	α-thal trait	Sickle Cell trait	Hb C trait
Mediterranean	1/20–30	1/30–50 trans	1/30–50	Rare
African American	1/75	1/30 trans	1/12	1/50
Non-Hispanic Caribbean, West Indian	1/50–75	1/30 trans	1/12	1/30
West African	1/50	1/30 trans	1/6	1/20–30
Hispanic Caribbean	1/75	Variable	1/30	Rare
Hispanic Mexican, Central America	1/30–50	Variable	1/30–200	Rare
Asian	1/50	1/20 cis	Rare	Rare
Southeast Asian	1/30	>1/20 cis	Rare	Rare
Asian Subcontinent (India, Pakistan)	1/30–50	Variable	1/50–100	Rare
Middle Eastern	1/50	Variable	1/50–100	Rare

Source: Modified from Genetic Screening Pocket Facts, March of Dimes, 2001 (www.marchofdimes.com)

The most common hemoglobin mutation leading to sickle cell disease is a homozygous single amino acid substitution (valine to glutamate) at position 6 of the *HBB* gene responsible for β-globin (Hgb SS). Sickle cell carriers have one normal copy of the gene in addition to a mutated copy (Hgb AS) and are usually asymptomatic. Another common mutation, Hbg C, causes sickle cell disease when combined with the sickle cell trait. Sickle-β thal minor and sickle-β thal major are also observed. An additional 700 other mutations have been identified in the *HBB* gene.

Thalassemias fall into two major groups, α and β, based on which globin chain of hemoglobin is affected. In these conditions, there is decreased or absent production of the specific globin chain. Beta-thalassemia can present in two main forms: β-thalassemia minor and β-thalassemia major. Over 200 mutations have been described in β-thalassemia [Tuzmen and Schechter, 2001]. Individuals with β-thalassemia minor are carriers of a mutation in one copy of the gene responsible for β-globin synthesis. As a result, there is decreased production of β-globin. Most individuals with isolated β-thalassemia minor are asypmtomatic or may have mild anemia. Individuals with β-thalassemia major (Cooley anemia) are homozygous for mutations in the *HBB* gene and have absent β-globin synthesis. Affected individuals are transfusion dependent because of the severe anemia and require chelation therapy for iron overload.

Two genes, and therefore four alleles, are involved in α-globin synthesis. The number of alleles that are mutated as well as their distribution between the two genes can impact the clinical presentation of α-thalassemia. Many individuals with a mutation in only one of the four alleles are considered silent α-thalassemia carriers. Two mutations on the same chromosome represent *cis*-α-thal trait, while two mutations on opposite chromosomes constitute a trans-α-thal trait. Mutations in the *cis* configuration are more clinically significant than those in the *trans* configuration. In the most severe cases, all four genes will be mutated, leading to hydrops fetalis [Dumars et al., 1996].

Optimal screening for hemoglobinopathies includes a hemoglobin electrophoresis and a complete blood count (CBC) prior to pregnancy. However, most women are screened during pregnancy. The sickle cell prep or sickle dex is not an appropriate screening test for sickle cell trait status because it is unable to detect Hbg C and other less common mutations. Direct genetic testing for the hemoglobinopathies is not recommended because of the large number of mutant alleles.

A basic algorithm should be followed (Fig. 10.1) to guide identification of carrier status for hemoglobinopathies. In general, individuals with a Mean Corpuscular Volume (MCV) >80 fL and a normal hemoglobin electrophoresis do not need additional screening. If the MCV is <80 fL and/or the hemoglobin electrophoresis is abnormal, additional investigation is warranted. If an individual is found to be a sickle cell trait carrier or a thalassemia carrier, then his or her partner should be offered a CBC and hemoglobin electrophoresis. If both members of a couple are carriers, then there is a 25 percent risk for them to have a child with hemoglobinopathy. At that time, genetic counseling and DNA testing should be offered in order to facilitate prenatal diagnosis and screening for other at-risk family members. Prenatal diagnosis for the hemoglobinopathies is possible via CVS and amniocentesis.

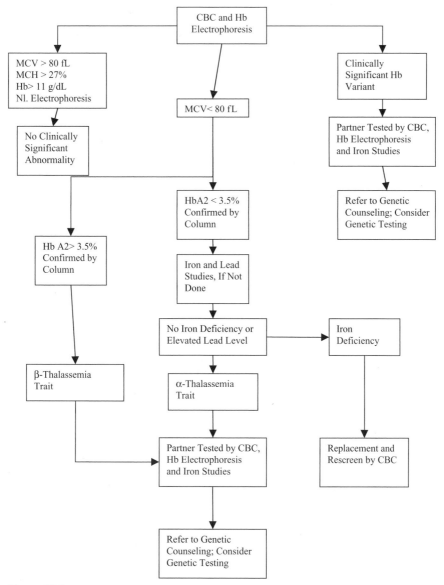

Figure 10.1 Hemoglobinopathy screening algorithm. (Modified from Genetic Screening Pocket Facts, March of Dimes,) 2001 (www.marchofdimes.com).

Common Diseases in the Ashkenazi Jewish Population

Organized carrier screening for a specific genetic disease was first successfully developed to screen Eastern European Ashkenazi Jewish (AJ) individuals for Tay-Sachs disease (TSD). Ninety-five percent of the Jewish population in the United States is

Table 10.4   Common Autosomal-Recessive Conditions in the Ashkenazi Jewish Population

Disorder	Incidence	Carrier frequency
Tay-Sachs disease	1/3,000	1/30
Canavan disease	1/6,400	1/40
Cystic fibrosis	1/2,500–3,000	1/29
Familial dysautonomia	1/3,600	1/32
Fanconi anemia group C	1/32,000	1/89
Niemann–Pick disease type A	1/32,000	1/90
Mucolipidosis IV	1/62,500	1/127
Bloom syndrome	1/40,000	1/100
Gaucher disease	1/900	1/15

Source: Modified from ACOG Committee Opinion, August 2004 [2004a].

AJ. In the 1970s, a biochemical carrier test was developed for TSD and large-scale population carrier screening resulted in the reduction of TSD in the United States by 90 percent. Advances in molecular genetics have identified AJ-specific mutations in TSD and other autosomal-recessive genetic conditions (ACOG committee opinion. Number 298, August 2004. Prenatal and preconceptional carrier screening for genetic diseases in individuals of Eastern European Jewish descent 2004).

The list of specific autosomal-recessive disorders with an increased carrier frequency in the AJ population is significant (Table 10.4). Carrier screening has been recommended for the most devastating of these disorders for which there is a reliable carrier test. In 2004, the American College of Obstetrics and Gynecology's Committee on Genetics published a committee opinion entitled, "Prenatal and Preconception Carrier Screening for Genetic Disease in Individuals of Eastern European Jewish Descent." Carrier screening for familial dysautonomia was added to the previous recommendation to screen for Tay-Sachs disease, Canavan disease, and cystic fibrosis. Some screening programs include Gaucher disease because of the high frequency of this disorder in the AJ population [Wallerstein et al., 2001].

Tay-Sachs disease is a progressive neurodegenerative disorder caused by deficient hexosimindase A (Hex A) enzyme production that often results in death in early childhood. The biochemical test for this condition is based on the ratio of Hex A to hexoaminidase B (Hex B). Biochemical carrier testing has been the standard of care for nonpregnant females and men. The test is not appropriate for pregnant females and females using birth control pills. Biochemical testing on leukocytes is possible for pregnant women and women using birth control pills, but the assay is technically more difficult. Three common mutations are responsible for 94 percent of the TSD mutations in the AJ population. Recent research suggests that DNA TSD carrier screening is a more appropriate carrier test because it can be offered to anyone, and it is cost effective without significantly sacrificing the sensitivity and specificity of TSD carrier screening. Prenatal biochemical and molecular testing are available for TSD [Bach et al., 2001].

Canavan disease is caused by a deficiency of the aspartoacylase enzyme, which leads to neurological decline and death in early childhood. DNA-based carrier screening of the most common AJ mutations can detect 98 percent of carriers. Biochemical and molecular prenatal diagnosis for Canavan disease is available.

Cystic fibrosis primarily affects the pulmonary, digestive, and male reproductive tracts. Genetic carrier screening for the most common AJ mutations can detect 98 percent of CF carriers. Additional CF screening recommendations are available for other ethnic groups. Molecular prenatal diagnosis is possible with informative parental carrier testing.

Familial dysautonomia is another neurological disorder that is present at birth and can cause abnormal feeding, pernicious vomiting, temperature and pain insensitivity, labile blood pressure, scoliosis, and absent tearing. DNA-based carrier screening for this condition has a 99 percent detection rate because one specific mutation accounts for 99 percent of mutations in the AJ population. Molecular prenatal diagnosis is available for this condition.

Gaucher disease is the most common autosomal-recessive condition in the AJ population. The incidence is 1 in 900 and the carrier frequency is 1 in 15. Commonly affected organs include the liver, spleen, and bone, causing hepatosplenomegaly, fractures, and limited joint movement. The condition has a variable age of onset from early childhood to late adulthood. Molecular carrier screening of the common AJ mutations can identify 95 percent of carriers. Genetic counseling regarding prenatal diagnosis for this condition is strongly recommended because of the variable expression of this condition.

Other conditions in which specific mutations have been identified in the AJ population include Bloom syndrome, Fanconi anemia group C, mucolipidosis IV, and Niemann–Pick disease type A. The carrier testing for these is 95 to 99 percent accurate in the AJ population, but the carrier frequency is significantly lower than in the conditions for which screening is currently recommended. Couples with a family history of any of these conditions or couples that express interest in carrier screening for any of these conditions should be provided genetic counseling about the disease incidence, carrier frequency, and available carrier and prenatal testing.

In general, genetic carrier screening is recommended for all AJ individuals, individuals with at least one family member of AJ descent, or a family history of a genetic disease known to have an elevated prevalence in this population. Carrier screening has become more challenging over time because of religious conversion, intermarriage outside of the AJ community, and the number of people who are unaware of their family's ethnic origin [D'Souza et al., 2000]. In couples in which only one member of the couple is AJ, accurate risk assessment can be particularly difficult because the mutation frequency outside the AJ population for many of these diseases is unknown and accurate carrier screening is not available.

If a couple is identified in which both members are a carrier of a specific disease, then there will be a 25 percent chance for them to have an affected child. The couple should be offered prenatal diagnosis for the condition. In a couple in which one member is of AJ descent and a known carrier and the other member of the couple is not AJ, prenatal diagnosis may be limited unless there is a biochemical test available.

Table 10.5 Fragile X Allele Sizes

Mutation category	Repeat size
Stable	6–50
Premutation	50–200
Full mutation	>200–230

Source: American College of Medical Genetics, Policy Statement: Fragile X Syndrome: Diagnostic and Carrier Testing [1994]

Other Potential Areas of Genetic Carrier Screening

Fragile X

Fragile X syndrome is the most common form of inherited metal retardation. It is estimated to be responsible for 40 percent of X-linked mental retardation. Estimates suggest that 1 in 4000 males and 1 in 4000 to 6000 females have fragile X syndrome. Clinical features of fragile X syndrome include variable mental retardation, physical features, and behavior problems. This condition is underdiagnosed because of the mild physical findings.

The *FMR1* gene contains a trinucleotide repeat, which expands resulting in a disruption of normal gene function. The genetics of fragile X are complex because of the presence of stable alleles, premutation alleles, and full expanded mutated alleles that can be modified by methylation (Table 10.5). Individuals with a premutation allele have no clinical symptoms but have up to a 50 percent risk to have a child with a fully expanded allele and fragile X syndrome. The general population carrier frequency of the premutation is not well known, but studies suggest 1 in 246 to 1 in 469 Caucasian females in the general population are carriers [Crawford et al., 2001]. General population screening for fragile X has been suggested because of the high carrier frequency.

The American College of Medical Genetics published a policy statement in 1994 entitled, Fragile X Syndrome: Diagnostic and Carrier Testing. Within that statement recommendations were made to offer prenatal screening to anyone with a family history of Fragile X syndrome or unexplained mental retardation. General population screening was not recommended unless as part of a research protocol because of the complex genetic issues surrounding the condition and need for appropriate genetic counseling. While widespread general population screening is not being offered in the United States, Finland and some centers in Israel now offer fragile X carrier screening to all pregnant women and subsequent prenatal diagnosis for at-risk individuals [Toledano-Alhadef et al., 2001].

Deafness

The role of genetics in congenital deafness has been a rapidly growing area of interest. Estimates suggest that 1 in 1000 births are affected with profound hearing loss. In 50 percent of cases a genetic etiology is responsible for the condition. Furthermore, 70 percent of the genetic cases are nonsyndromic forms of deafness.

Autosomal-recessive forms of nonsyndromic deafness are most common. Only 10 percent of children with nonsyndromic hearing loss are born to deaf parents, and 90 percent have normal hearing parents (Genetics Evaluation Guidelines for the Etiologic Diagnosis of Congenital Hearing Loss. Genetic Evaluation of Congenital Hearing Loss Expert Panel. ACMG statement 2002).

Mutations in *GJB2*, which is responsible for the production of connexin 26, are implicated in 50 percent of genetic nonsyndromic deafness. The carrier frequency of the most common mutation, *35delG*, has been identified in 1 to 3 percent of individuals of northern European descent and 3 percent Mediterranean populations. No one specific allele has been identified in the African American or AJ populations [Ryan et al., 2003].

Based on the high frequency of the carrier state in certain ethnic groups and a reliable carrier test, general population screening could be considered. However, it is important to bear in mind that the social, legal, and ethical issues of carrier screening for deafness are much different than for lethal genetic diseases that are currently screened on a population basis. Research suggests that deaf individuals are not interested in genetic testing, and members of the "deaf community" do not perceive deafness as a disability. Some deaf people have indicated that they would prefer to have a deaf child to a hearing child. Studies of hearing individuals reveal a greater interest in genetic screening for deafness, primarily for medical management and planning and not for pregnancy termination. Hearing parents of a deaf child have expressed interest in prenatal diagnosis of future pregnancies and possible termination of an affected fetus [Antoniadi et al., 2001]. At this time there is no policy statement regarding general population screening for nonsyndromic deafness.

CONCLUSION

Approaches to carrier screening have been in use for decades for specific disorders in specific populations. There programs have substantially reduced the frequency of targeted diseases in these populations. The sequencing of the human genome now makes it possible to significantly increase the scope of carrier testing. It will be important for the genetics and general medical communities to thoughtfully consider where and when such screening is appropriate, and how to effectively initiate screening programs.

Chapter 11

Susceptibility Testing

OVERVIEW

Prediction of the onset of a disease is an enticing concept because it raises the possibility of improved quality of life through primary prevention, increased surveillance, and/or early intervention. For an increasing number of common diseases, known genetic interactions are being investigated to see if genetic testing can identify individuals that will develop late-onset diseases (predictive testing) or are at increased risk for a disease (susceptibility testing). Susceptibility testing is an arena where commercial interests have pushed vigorously to present new test options. Concerns have been expressed that commercial susceptibility testing may be inappropriate in terms of the suitability of test indications and protocols, limitations of such test protocols, and whether the nature and scope of a particular DNA (deoxyribonucleic acid) molecular testing technology has received appropriate peer review for clinical validity. By way of example, a "negative" test result may not guarantee the absence of a mutation and only may refer to the specific panel of mutations tested. Even with complete sequencing, mutations may be missed because they are deletions or rearrangements or exist in the introns instead of the coding regions. For example, with regard to *BRCA1/2* susceptibility testing, contrary to possible patient and physician expectations, no current single technology can identify all mutations. Physicians need to be aware of the complexities of DNA molecular testing technologies in order to explain to patients their uses and respective limitations, and patients need to have a realistic perspective of the actual clinical value of such testing. This chapter includes a discussion about the debate with regard to the susceptibility testing of children.

GOALS AND APPLICATIONS OF SUSCEPTIBILITY TESTING

The goal of susceptibility testing is to use genetic tests to refine estimates of risk that a given individual will develop a disease at some time in the future. Susceptibility testing

Genetic Testing: Care, Consent, and Liability, by Neil F. Sharpe and Ronald F. Carter
© 2006 John Wiley & Sons, Inc.

does not identify a diagnosis, nor does it give a conclusive result; test results describe the presence or absence of genetic sequence variations that have been associated with an increased risk for a disease. Theoretically, susceptibility testing could be applied to any disease with a variable (later) age of onset and known to have a hereditary component involved in the etiology of the disease. In practice, to date susceptibility testing has primarily been used for:

- Hereditary breast/ovarian cancer syndromes associated with mutations of *BRCA1* and *BRCA2*
- Hereditary colorectal cancer syndromes, including
  - Familial adenomatous polyposis coli (*APC* gene)
  - Hereditary nonpolyposis colorectal cancer (*MSH2, MLH1, MSH6,* and other genes
- Selected adult-onset diseases, such as
  - Hereditary hemochromatosis (*HFE* gene)
  - Alpha-1-antitrypsin deficiency (*AAT*)

Other diseases under consideration for future susceptibility testing include common diseases of adult onset, including neurodegenerative conditions such as Alzheimer's disease (discussed in more detail in Chapter 9), cardiovascular disease, and diabetes.

The feasibility of susceptibility testing depends upon the:

- Heritability of the disease in question
- Complexity and predictability of interactions with nongenetic risk factors
- Number of different genes contributing to the pathogenesis of the disease
- Accuracy and complexity of laboratory test protocols
- Penetrance and expressivity of the mutations identified
- Options for disease surveillance and primary prevention
- Clinical value of surveillance and/or preclinical intervention

As a more complete understanding of etiologies, risk factors, and pathogenetic mechanisms involved in a given disease are elucidated over time, susceptibility testing will incorporate both genetic and nongenetic risk criteria for determining the risk of future disease. In a manner analogous to advances in prenatal screening, each additional independent variable should contribute to a more accurate estimate of risk.

BENEFITS AND RISKS OF SUSCEPTIBILITY TESTING

In the optimum, genetic tests applied to inform individuals of their risk for developing a disease in the future would permit:

- Targeted surveillance and early diagnosis for persons at high risk
- Reduction in surveillance for persons at low risk
- Reduction of risk by avoiding other risk-inducing behaviors

- Primary prevention through surgery, nutritional modification, and chemoprevention strategies
- Psychological benefit from alteration of self-perception, reduced anxiety, increased sense of control, and adoption of risk reduction strategies

However, risks and limitations also need to be identified:

- False reassurance from incorrectly low estimation of true risk, leading to failure to adopt risk-reducing strategies
- Adoption of irreversible, risk-inducing or expensive risk prevention strategies based upon incorrectly high estimations of risk (e.g., colonoscopies for colorectal cancer; bilateral prophylactic ovariectomy and mastectomy for hereditary breast/ovarian cancer; avoiding having children because of neurodegenerative disease)
- Psychological harm from knowledge of increased risk
- Feeling of powerlessness if surveillance and/or prevention strategies are ineffective
- Stigmatization and discrimination arising from recognition of increased risk

In particular, it is important to note that, by definition, susceptibility testing involves discussion of risks for a diagnosis; both the concept of risk and gaining an accurate perception of risk level, are difficult achievements for many people even after appropriate counseling.

Further, adoption of susceptibility testing requires an organized approach incorporating:

- Educational strategies for the public, patients at apparent risk, and providers
- Input and involvement of a wide spectrum of health care providers, due to the multidisciplinary nature of clinical management and the spectrum of clinical expression of diseases in question
- Standardized guidelines and protocols for counseling, testing, and clinical management, but individualized approaches to counseling patients and families
- Thorough examination of ethical, legal, and psychosocial implications of testing
- Resources for optimal clinical management of persons at increased risk

EXAMPLES OF SUSCEPTIBILITY TESTING

Hereditary Hemochromatosis

Hereditary hemochromatosis (HHE) is a disease of adult onset characterized by increased uptake and storage of iron from dietary sources. The basic defect is a failure by the body to recognize that enough iron has been absorbed from the gut; accordingly, iron is constantly taken up from the gut and delivered by the portal circulation to the liver and other tissues. Iron overload leads to multiorgan dysfunction and failure, including cirrhosis, diabetes, cardiomyopathy, arthritis, hypogonadotrophic hypogonadism, and cancer. In northern Europeans, the incidence of HHE is about 1 in 200, and the carrier rate is estimated to be as high as 1 in 10. The burden of morbidity and mortality associated with HHE is significant. HHE is asymptomatic in early stages, becomes symptomatic usually somewhere in midadult life, and is frequently clinically unsuspected until the late stages of disease. The major gene associated with HHE is *HFE*, a member of the *MHC* class I gene family. *HFE*-associated HHE has an autosomal-recessive pattern of inheritance, but there are strong environmental interactions, and there is considerable variability in genotype–phenotype correlations.

Age of onset and severity of *HFE*-associated HHE appear to vary with:

- Blood loss (and therefore utilization of iron stores), for example, menstruation
- Concurrent liver damage from other causes, for example, chronic alcohol poisoning, viral hepatitis, drug exposures
- *HFE* genotype
- Genotype of other proteins in the iron transport pathway

Clinical symptoms of HHE commonly include abdominal pain, lethargy, and weakness, progressing to weight loss, skin hyperpigmentation, arthritis, and congestive heart failure. Routine hematologic testing provides very useful screening tests for suspected HHE; abnormally high results for serum ferritin concentration and, in particular, fasting transferrin–iron saturation [2 or more tests with >60 percent (women) or >50 percent (men) saturation] are typical criteria for diagnostic confirmation of HHE by imaging [magnetic resonance imaging (MRI) of the liver], liver biopsy, and demonstration of increased iron stores by Prussian blue staining; and genetic testing for HFE mutations. Treatment consists of phlebotomy (iatrogenic loss of blood to stimulate red cell production and uptake of iron), at an initial rate to reduce stored iron to safe levels, followed by a maintenance rate to ensure iron toxicity does not develop.

Genetic testing commonly identifies the presence of a major mutation, *C282Y*, and a minor mutation, *H63D*. In approximate terms, about 85 percent of patients with HHE are homozygous for the major mutation; about 5 percent are compound heterozygotes; and, rarely, patients are homozygous for the minor mutation. A small proportion of patients are found to have large deletions and other types of mutations. The significant difficulty, however, is that correlations of clinical status show wide variation with genotype; prospective evaluation of individuals with mutation genotypes can show clinical states ranging from subclinical to severe with early onset. In essence, establishment of the genotype does not accurately predict whether a patient will or will not develop clinically significant disease [see Fletcher and Halliday, 2002; Kowdley et al., 2004; Genetests, 2004; Worwood, 2005]. Expression of the disease appears to be determined in part by the functionality of other proteins in the iron transport pathway as well, including mutations of other genes (e.g., *TfR2*) in addition to the environmental risk factors mentioned above. Thus, in practice, establishment of the genotype can be used to provide a revised estimate of risk for future disease, but careful comparisons of genotypes and phenotypes of various family members need to be made, and persons with a family history of the disease are monitored regardless of their genotype. In a similar concession to the lack of strong genotype–phenotype correlations, persons identified with major and/or minor mutations are routinely monitored for increased iron stores but need to be counseled that they may or may not develop the disease. Further details on genes in the iron transport pathway, genetic testing for *HFE*, and counseling and management guidelines for HHE can be obtained from Worwood [2005], the Genetests website [Kowdley et al., 2004], and other online reviews [e.g., Press RD; http://www.hosppract.com/genetics/9908mmc.htm].

Alpha-1-Antitrypsin Deficiency

Alpha-1-antitrypsin (AAT) is a disease that results in emphysema, pulmonary failure, and liver failure, but it varies in severity from subclinical to fatal in midadult life. AAT is an example of an underrecognized disease with detectable clinical symptoms, a well-defined, predictable hereditary component and known, modifiable environmental risk factors. AAT represents an optimum scenario for the successful utilization of susceptibility testing.

Alpha-1-antitrypsin is a protease inhibitor found in the plasma. One of its functions is to inhibit the activity of elastases released by neutrophils in the lung. If AAT is defective, the failure to inhibit elastase activity released by neutrophils during inflammatory responses results in destruction of pulmonary alveoli, emphysema, and chronic obstructive pulmonary disease in ∼90 percent of patients. The gene for AAT, also known as protease inhibitor (*PI*), has a normal allele (*M*) and 2 common mutant alleles, *S* and *Z*. The *Z* allele results in severe loss of enzyme activity, while the *S* allele causes partial loss. Carrier genotypes include *PiMS* and *PiMZ*, while enzyme deficiency is associated with *PiSS, PiSZ,* and *PiZZ* genotypes. The homozygous *PiZZ* genotype affects approximately 1 in 2500 northern Europeans, is associated with 10 to 15 percent enzyme activity, and accounts for most of the clinical burden of disease. Compound *SZ* heterozygotes have about 30 percent of normal enzyme activity and may also develop emphysema.

Homozygous *SS* genotype is associated with about 50 percent enzyme activity and a lack of clinical signs. While AAT is commonly considered a disease of white Europeans, recent data suggests that there is an underrecognized burden of disease in other populations as well, with at least 116 million carriers and 3.4 million affecteds in 58 countries recently surveyed [deSerres, 2003].

Of particular interest is the fact that environmental exposure to inhaled particulates can markedly exacerbate the rate of onset and severity of lung disease [reviewed by DeMeo and Silverman, 2004]. In particular, persons with a mutation genotype should not engage in smoking and certain occupations, such as mining. For example, the combination of smoking in Swedish men and women with a *PiZZ* genotype results in a substantially worse survival curve, with median ages of survival dropping from approximately 57 years down to about 35 years [Larsson, 1978]. OMIM (http://www.ncbi.nlm.nih.gov/entrez/dispomim.cgi?id=107400) and GeneTests (www.genetests.org) provide updated reviews of clinical and laboratory information for AAT.

HEREDITARY BREAST/OVARIAN CANCER SYNDROMES

About 1 in 20 breast cancers and 1 in 5 ovarian cancers are hereditary. There has been widespread uptake of susceptibility testing for hereditary breast and ovarian cancer associated with mutations of *BRCA1* and *BRCA2*. Together, these genes are estimated to account for approximately 90 percent of the incidence of hereditary breast and ovarian cancer. The environmental risk factors associated with breast cancer are primarily related to endogenous hormonal status and are relatively unmodifiable in North American culture. The penetrance of mutations of *BRCA1* and *BRCA2* in hereditary breast/ovarian cancer is incomplete and variable expression is described. It appears that the bias of ascertainment in initial studies of very high risk families led to rather high estimates of the penetrance of many mutations. Genotype–phenotype correlations are starting to be developed as a result of large multinational studies, but it is important to note that such correlations are variable; even a known, well-characterized mutation can vary in penetrance and expression in different families. Precise predictions of the clinical effect of a mutation in a given individual are usually not provided.

Various considerations relating to counseling for *BRCA1*- and *BRCA2*-related disease have been presented in detail elsewhere in this book. Guidelines for counseling, pedigree-based risk estimation, indications for testing, and provision of test results are available [e.g., OMA, 2001], and concise reviews for clinicians are available [e.g., Taylor 2001; Chang and Elledge, 2001; Srivastava et al., 2001; Frank and Critchfield, 2002; Narod and Foulkes, 2004] and GeneTests (www.genetests.org) is an updated resource. Pedigree information can be used to estimate the chance that a family has an inherited breast/ovarian cancer syndrome, the risk of a *BRCA1* or *BRCA2* mutation in a given individual, and the risk of developing breast cancer for a given individual [Couch et al., 1997; Frank et al., 1998; Gail, 1989, and online at http://brca.nci.nih.gov/brc/start.htm; Claus et al., 1994; Srivastava et al., 2001; MyriadTests, online at http://www.myriadtests.com/provider/mutprev.htm]. Table 11.1 presents a summary of some commonly used criteria applied to pedigree analysis for identification of families at increased risk.

It is worth emphasizing that the clinical significance of a described mutation of *BRCA1* or *BRCA2* needs to be carefully considered by the clinician. A useful starting resource is an online catalog of described mutations (The Breast Cancer Information Core registry, accessible at http://research.nhgri.nih.gov/bic/), which summarizes and categorizes the mutations recorded to date by the registry. A particular issue is the clinical significance of missense mutations (small sequence variations that result in the substitution of one amino acid for another in the protein). Many missense sequence variations would appear to have no clinical significance, while others may affect critical protein regions and therefore present some increase in risk for cancer. Details of associations between specific missense mutations and clinical outcomes require careful evaluation and will take years to assemble [for examples and details, see Mirkovic et al., 2004; Szabo et al., 2004; Abkevich et al., 2005; Phelan et al., 2005]. The laboratory report should also summarize the interpreted clinical significance of the result.

Patients demonstrated to have a clinically significant mutation, such as a frameshift or nonsense mutation, are provided with estimated lifetime risks for onset

Table 11.1 Example of Commonly Used Pedigree Criteria for Selection of Families with Hereditary Breast/Overian Cancer Syndromes[a]

Type of diagnosis	Who and how many affected	Age
Multiple cases of BC, and/or any case of OC	Close relatives (1st and 2nd degree); especially 2 or more generations	BC: especially if <50 OC: any age
BC	One or more individuals with very early onset	<35
BC and OC	Family member with both tumors	Any age
BC and/or OC	Ashkenazi Jewish	Any age
BC	Bilateral: i.e., two different primary tumors in same person	Any age, but especially if first diagnosis <50
Invasive serous OC	One or more individuals	Any age
BC	One of more males	Any age
Diagnosis of hereditary BC/OC syndrome: known mutation of BRCA1 or BRCA2	Identify individuals at risk by descent	
Other types of tumors suggestive of inherited cancer syndrome		

[a] Look for one or more of the following in families with breast and/or ovarian cancer to identify families that may benefit from counseling and/or genetic tesing. BC = breast cancer; OC = ovarian cancer.
Source: Compiled from a variety of sources, including the OMA Predictive Cancer Genetics Steering Committee, Ontario [2001] (J. Allanson, Chair): [OMA 2001 accessible at http://www.oma.org/pcomm/omr/nov/01 genetics.htm].

of breast, ovarian, and other associated cancers. For carriers of *BRCA1* and *BRCA2* mutations, lifetime risks are estimated to range between 60 and 85 percent for female breast cancer, 10 percent (*BRCA2*) to 85 percent (*BRCA1*) for ovarian cancer, 8 percent for male or female colon cancer, and 8 percent for prostate cancer [e.g., Burke et al., 1997b; Elit, 2001; Olivier et al., 2004; Cannistra, 2004]. Options for clinical management for patients at increased risk include increased surveillance, chemoprevention with tamoxifen and related drugs, and prophylactic excisional surgery [see Burke et al., 1997b; Cannistra, 2004]. Recent studies have suggested that MRI is substantially more effective than mammography or clinical breast examination for early detection of tumors in mutation carriers [Kriege et al., 2004]; MRI detected ~80 percent of invasive lesions compared to 33 percent for mammography and 18 percent for clinical breast examinations. In contrast, there is no clear evidence that screening options for ovarian cancer (transvaginal ultrasonography and serum CA-125) result in improved survival [reviewed by Cannistra, 2004]. Despite the obvious concerns about the risks and benefits of such options as chemoprevention and prophylactic bilateral

Table 11.2 Example of Criteria for Selection of Families with Hereditary Nonpolyposis Colorectal Carcinoma (HNPCC) or Familial Adenomatous Polyposis (FAP)[a]

Type of diagnosis	Who and how many affected	Age
CRC	One or more individuals with very early age of onset	<35
Multiple cases of cancers in HNPCC spectrum, including at least one CRC or endometrial cancer	Close relatives (1st or 2nd degree); especially if 2 or more generations and 3 or more diagnoses of CRC	Especially if at least one diagnosis <50 years
HNPCC spectrum	Individual with more than one primary tumor	
Individual with adenoma and/or CRC, positive for high microsatellite instability (MSI)		
Diagnosis of HNPCC or FAP: known mutation		
Indivdual(s) with 10 or more adenomatous polyps (e.g., FAP or attenuates) FAP)	Close relatives; especially if early age of onset	Onset in teens (FAP) or later (attenuated FAP)

[a] Look for one or more of the following to identify families that may benefit from counseling and/or genetic testing. The spectrum of cancers associated with HNPCC includes: colorectal (CRC), gastric, small bowel, hepatobiliary, pancreatic, endometrial, ovarian, ureter, renal pelvis, brain, sebaceous adenoma, or carcinoma. FAP is much less common than HNPCC and specifically associated with a history of multiple polyps (>10 to hundreds).
Source: Compiled from a variety of sources, including the Predictive Cancer Genetics Steering Committee, Ontario [2001] (J. Allanson, Chair): [OMA 2001 accessible at http://www.oma.org/pcomm/omr/nov/01genetics.htm].

For HNPCC specifically, the classic Amsterdam criteria [Vasen, 1991] require that the pedigree has each of three criteria:

- Three relatives with CRC; one a first-degree relative of the other two
- At least two generations affected
- At least CRC diagnosis before age 50.

In Amsterdam II criteria, the three or more relatives can have any type of tumors in the HNPCC spectrum (e.g., CRC, endometrial, small bowel, ureter, renal pelvis).

Modified Bethesda criteria for testing CRC tumors for HNPCC-related microsatellite instability are described by Umar et.al. [2004a]: [http://jncicancerspectrum. oupjournals, org/cgi/content/abstract/jnci; 96/4/261]. These criteria are highly sensitive but moderately specific. Tumors should be tested if:

- Diagnosis of CRC in patient <50 years old
- Presence of synchronous, metachronous CRC, or other HNPCC-spectrum tumors, at any age
- CRC with MSI-high histology in patient <60 years old
- CRC in one or more first-degree relatives with an HNPCC spectrum tumor, with at leat one diagnosis at <50 years old
- CRC in two or more first-degree relatives with HNPCC-related tumors, at any age

mastectomy and/or oophorectomy, recent clinical studies have shown >90 percent reductions in risk of a diagnosis of breast cancer, and >50 percent reduction in risk for ovarian cancer [Hartmann et al., 1999, 2001, 2004; Eisen et al., 2000; Kauff et al., 2002; Rebbeck et al., 1999, 2002, 2004; Rebbeck, 2000; Eisen and Weber, 2003; Narod and Foulkes, 2004]. Estimated gains of years of life expectancy can be demonstrated for chemoprevention, mastectomy, and oophorectomy, varying in degree based upon the strategy taken, age at adoption of a given strategy, and the predicted penetrance of mutation [Schrag et al., 2000]. For this disease, despite the limitations of the risk refinement provided by testing, the concerns associated with possible discrimination in employment and insurance (see Chapter 7, Informed Consent), the limited range of options for screening and prevention, and the risks for harm arising from surgery, it is apparent that the possible benefits of counseling, testing, and adoption of prevention strategies have encouraged a considerable uptake of care by patients at increased risk.

HEREDITARY COLORECTAL CANCER SYNDROMES

Approximately 5 to 10 percent of colorectal cancer can be attributed to hereditary cancer syndromes, including:

- HNPCC (hereditary nonpolyposis colon cancer)
- FAP (familial adenomatous polyposis)
- Turcot syndrome
- Cowden syndrome
- Familial juvenile polyposis
- Peutz–Jeghers syndrome

Hereditary Nonpolyposis Colon Cancer

Of these syndromes, HNPCC is by far the most common and is thought to account for about 1 in 20 colorectal cancers. HNPCC is associated with mutations of multiple genes: *MLH1, MSH2,* and *MSH6* account for most mutations, although *PMS1* and *PMS2* are also involved. These genes code for proteins that are known to have multiple functions including a critical role in the repair of small mismatches in DNA after synthesis. The pathogenetic mechanisms and clinical characteristics of

HNPCC-associated colorectal cancers have been reviewed recently [Robbins and Itzkowitz, 2002; Charames and Bapat, 2003; Brezden-Masley et al., 2003; Narayan and Roy, 2003; de la Chapelle 2004; Rowley, 2004].

Recognition of hereditary colorectal cancer by pedigree criteria is not an exact process (Table 11.2). About 30 percent of colorectal cancers are familial, but not hereditary, due to the influence of commonly shared environmental risk factors such as diet. The most stringent clinical criteria developed for clinical diagnosis of HNPCC are known as the 1990 Amsterdam criteria [Vasen et al., 1991]. Less restrictive criteria were developed later, known as Amsterdam II, and modified Amsterdam criteria, because the original criteria were shown to exclude some families that could be demonstrated to have mutations [see Vasen, 2000; Schoen 2000; Trimbath and Giardiello, 2002; Chung and Rustgi, 2003; Umar et al., 2004a, 2004b]. Once clinical criteria are applied, further selection for patients with mutations can be made by screening for microsatellite instability; this step makes use of a genetic test applied to histological biopsies to detect evidence of the defects in mismatch DNA repair associated with HNPCC. It is important to distinguish between germline evidence of mismatch repair deficiencies, and acquired mismatch repair detectable only in tumor tissue. A commonly used summary of the original criteria and various modifications are presented by Chung and Rustgi [2003], with estimates of the risk for detection of a mutation included:

Criteria applied	Probability of *MLH1* or *MSH2* mutation (%)
Amsterdam or modified Amsterdam	39–45
Bethesda	25–34
High microsatellite instability (MSI-H)	38–73
Low microsatellite instability (MSI-L)	3–8, except if MSH6: 22

It is evident that some HNPCC-positive families will be missed by any of these criteria. Myriad Genetics [2004] provides an updated website with details of the prevalence of *MLH1* and *MSH2* mutations in patients selected according to family history criteria.

Utilizing one or more selection criteria (Table 11.2), families can be investigated by genetic testing and mutation analysis. Umar et al. [2004a, 2004b] and Kievit et al. [2004] provide a review of testing guidelines for HNPCC. Laboratory analysis for HNPCC is more complicated than for the hereditary breast and ovarian cancers because a wide spectrum of tumors can be implicated, more genes are commonly involved, and the mutation spectrum for each gene is quite variable [Charames and Bapat, 2003; Brezden-Masley, 2003]. As with the hereditary breast and ovarian cancers, testing usually proceeds with individuals in the pedigree who have been confirmed affected by a tumor type associated with HNPCC. Review of pathology records is essential because family members may not accurately recall the type of tumor a given relative developed. Initial steps can include application of immunohistochemistry to detect loss of protein expression of MSH2 and MLH1 in tumor tissues, and (if not already applied) microsatellite DNA testing for evidence of impaired DNA mismatch repair. If either the immunohistochemistry or the microsatellite testing is abnormal, direct mutation analysis for germline mutations of *MSH2, MLH2*, and/or *MSH6* is applied. The immunohistochemistry testing is particularly valuable because an abnormal result will directly implicate a particular gene (~100 percent specificity) and therefore save the cost of unnecessary mutation screening in other genes. However, it is important to note that both the microsatellite instability testing and the immunohistochemistry tests can fail to identify some families with mutations [Lindor, 2002; Kievit et al., 2004]. Immunohistochemistry will detect approximately 92 percent of *MSH2-* and *MLH1*-associated tumors; if the pedigree information strongly suggests HNPCC, testing may proceed to mutation analysis of *MLH1, MSH2,* or *MSH6* even if the results from these preliminary tests are normal. Presently, mutation testing is complicated by the possibility that large deletions or splice-site mutations will be missed (particularly in *MSH2*), and our lack of understanding of the clinical significance of missense mutations (particularly in *MLH1* and *MSH6*) [Umar et al., 2004b].

An important outcome of definitive identification of an HNPCC-associated mutation (e.g., a mutation of *MLH1, MSH2,* or *MSH6*) is that the pathogenetic mechanism is clearly identified to be hereditary rather than sporadic. This provides prognostic information for specific diagnoses [e.g., Clark et al., 2004] as well as refined estimates of risks of recurrence and additional diagnoses of cancer in the future. This information permits more informed follow-up counseling for other family members at risk, if desired, and for mutation carriers, the increased risks for multiple tumor types are identified and can be acted upon through increased surveillance and preventive measures. Germline mutations of *MSH2* or *MLH1* are associated with significant increases in lifetime risks for multiple cancers, including approximately:

- 80 percent colorectal cancer
- 50 percent endometrial cancer
- 13 percent gastric cancer
- 12 percent ovarian cancer

as well as smaller increases in risks for small bowel, bladder, brain, urinary, and other cancers. These risks are substantial, and the implications for patients require careful review during counseling.

As with hereditary breast and ovarian cancers, it is important to carefully review the results from laboratory tests to ensure that the clinical significance of the mutation is clearly identified, and to provide counseling that allows for the variability in penetrance and expression that has been demonstrated; the clinical significance of some mutations, particularly missense mutations, are not clear at the time of writing.

Screening strategies for early detection of tumors in patients at high risk include colonoscopy, gynecological examinations including transvaginal ultrasonography, urine cytology, and gastroduodenoscopy [Vasen, 2000; McLeod, 2001].

Genetic counseling and knowledge of mutation status have been shown to affect patient compliance with screening recommendations [Hadley et al., 2004].

Familial Adenomatous Polyposis

A relatively rare syndrome, FAP affects approximately 1 in 10,000 people and accounts for ~1 percent of colorectal cancers. The clinical symptoms for classical FAP include the presence of retinal freckles in children (up to 85 percent of patients) and development of numerous polyps in the rectum, colon, and (less frequently) upper intestine. The multitude of polyps are a distinguishing characteristic; they usually are detected by the midteen years and are associated with a 100 percent risk for diagnosis of colorectal cancer 10 to 15 years later. An attenuated form of FAP is recognized, which is associated with a later age of onset and fewer polyps [Knudsen et al., 2003]. Patients with FAP usually present with colorectal symptoms, extracolonic signs, or a positive family history [Hyer and Fell, 2001]. Surgical removal of tissue is required to prevent the formation of tumors once polyps are evident; various surgical options are available depending upon the extent of tissue involvement. The FAP registry provides general information about clinical signs and treatments for patients and clinicians [http://www.mtsinai.on.ca/familialgicancer/Diseases/FAP/default.htm; accessed December 21, 2004].

Familial adenomatous polyposis is caused by mutations of the *APC* gene, which is central to mechanisms of tumor formation in both sporadic and inherited cancers [reviewed by Fearnhead et al., 2001; Fodde, 2002]. An autosomal-dominant pattern of inheritance with 100 percent penetrance is usually evident. Over 800 different mutations of *APC* have been reported. Attenuated FAP has been associated with the *I1307K* and *E1317Q* missense mutations of FAP as well as mutations of *MYH* [reviewed by Knudsen et al., 2003]. Once a mutation is definitively identified in a family with FAP, options for clinical management can be reviewed [Hyer and Fell, 2001; Knudsen et al., 2003]. Family members at risk of inheriting the disease can be clinically screened for the presence of disease and tested for the presence of the *APC* mutation. Because onset of the disease can occur relatively early, clinical screening of teen-aged children by annual colonoscopies is often instituted. The potential harm arising from increased clinical surveillance is commonly accepted as a justification for genetic testing of children at risk. This is one of relatively few accepted indications for susceptibility testing of minors; the moral, ethical, and legal implications are discussed in greater detail below.

SUSCEPTIBILITY TESTING IN CHILDREN

Genetic testing in children has raised a number of concerns that have received extensive public debate. Much of the debate has concerned whether susceptibility testing in children should be done, and if so, how and under what guidelines. The following outlines some of the relevant consideration for genetic testing of children in general, and more specifically, susceptibility testing in children.

Issues Arising in the Genetic Testing of Children and Adolescents

Genetic testing of children and adolescents is commonplace and an accepted practice in medical care. However, there are circumstances where genetic testing of minors may invoke consideration of conflicting principles. For example, parents may request susceptibility testing to determine the risk that a child will develop FAP, or an adolescent may request testing to determine the risk of transmitting the genetic alteration associated with a disease such as fragile X syndrome.

There is an overriding concern to ensure that such testing is in the best interest of the child, not only in the immediate time frame of pediatric care but also for future health and quality of life should the child become an adult. Often, there are significant risks for harm due to interference with the child's emotional well-being, autonomy, informed choice of testing and medical management, relationship to family members, future reproductive choices, and even adoption status. Requests for testing may originate from the child, parents, or guardians. The motivation, authority, and obligation to pursue testing each require careful evaluation. Parents or their substitutes expect to act as decision makers for a child who is deemed to be legally incompetent, but exercising parental authority may not always reflect the actual state of competence of the minor nor respect the best interests of the minor. Health care providers need to recognize situations where careful consideration of the impact of testing on a child is required.

Values, Guiding Principles, and Recommendations

The complexity of medical, social, ethical, and legal principles interacting in this area of genetic counseling have been reviewed and guidance for health care professionals has been published by both American and Canadian professional bodies [ASHG/ACMG, 1995; CCMG, 2003] as well as international bodies such as the WHO [1998]. The following briefly summarizes prevailing principles and recommendations.

Benefit

The first principle applicable to the testing of children is that it must be defensible by demonstrating the potential for benefit to the child. In particular, the primary focus is the value of a genetic test for the immediate care of the child, rather than the implications for health later as an adult. In situations where diagnostic testing is applied to optimize pediatric medical management, for example, confirmation of thalassemia, fragile X syndrome, or Tay-Sachs disease, this immediacy of benefit is clear. Also, in situations where: (1) the child is healthy but requires treatment or invasive screening for early detection or prevention of diseases of pediatric or early adult onset, for example, colonoscopies for early detection of FAP, or (2) the child is subclinically affected and the risk of adverse outcomes can be altered, for example, diagnosis of Marfan syndrome in a student athlete, the benefit also is clear. However, in testing a minor for diseases of later onset (such as susceptibility or predictive testing), or in situations where treatment is of limited value to the child, the benefits are not as clear; for example, how would knowledge of the genetic status of a child at risk for Huntington disease help the child [Bloch and Hayden, 1990; Committee of IHA and WFN, 1990; Sharpe, 1993; Wertz et al., 1994; AMA, 2003; Bioethics Committee (CPS) 2003; WHO, 2003; but also see Duncan, 2004]? On the other hand, knowledge of *AAT* genotype might provide a strong reason for an adolescent to avoid taking up smoking.

In some situations, it may be that psychosocial benefit can be anticipated, particularly for adolescents approaching the decisions of adult life, and this may be an acceptable justification for pursuing testing in this situation once the expectations and potential outcomes of testing are thoroughly discussed with the family. Potential benefits to the child can be summarized as [ASHG, 1995]:

- Improved diagnosis, treatment, and/or prevention of a disease
- Increased surveillance for disease
- Avoidance of unnecessary treatment or surveillance
- Improved prediction of clinical course of disease
- Psychosocial benefit to the child: reduction of uncertainty, informed plans for future lifestyle, and family planning

In summary, as stated by the WHO [1998]:

Every genetic test should be offered in such a way that individuals and families are free to refuse or accept according to their wishes and moral beliefs. All testing should be preceded by adequate information about the purpose and possible outcomes of the test and potential choices that might arise. Children should only be tested when it is for the purpose of better medical care.

Harm

A second principle is the necessity to consider potential harm. If the benefits to the child are not clear, the possibility of causing harm becomes a much more influential consideration. Counselors need to specifically assess potential for harm and discuss these issues with the family. This involves assessment of a variety of considerations [ASHG, 1995; AMA, 2003; CCMG 2003; Bioethics Committee (CPS) 2003; WHO, 2003; but also see Duncan, 2004], for example:

- Competence of the child (emotional maturity, capacity to comprehend, age, legal status)
- Current and future health of the child
- Future autonomy and privacy as an adult
- Possible clinical course of the disease in question and the effectiveness of treatment
- Potential for adverse psychosocial impact, for example, alterations to psychological status, self-concept, or relationships with peers and family members, and conflicts with cultural influences
- Parent(s)' motivations for testing and expectations for the child
- Possible conflicts between the child's wishes and the parents' wishes, for example, possible benefits to other family members, protection of the child, advocacy for the child, moral or cultural concerns of the parents and family as opposed to those of the child

Decision Making

A third principle is that education and comprehensive counseling for the family as a whole represents the best approach to ensure that the most appropriate decision is made.

A fourth principle is that the counseling process should identify and recognize the authority of the parents to act in the best interests of the child. This assumption of parental authority is an extension of moral and legal principles of autonomy, the right to informed choice, duties of parents to provide medical care for their children, and the incapacity of a minor to make informed decisions on medical care.

A fifth principle is that in situations where the competence of the minor approaches that of an adult [Buchanan and Brock, 1986, 1989], the stated wishes of the minor should be determined and considered. It also should be noted that populations at increased risk have proactively involved minors in educating themselves and their peers about the benefits of testing for carrier status.

A sixth principle is that health care providers are not obligated to conduct testing on a minor, or disclose a test result to a minor, or sequester a test result from a minor, if there is reason to believe that it is not in the best interests of the child [e.g., Huntington disease: Bloch and Hayden, 1990; Committee of IHA and WFN, 1990; Sharpe, 1993; Wertz et al., 1994]. The provider does have an obligation to carefully assess the circumstances and timing of the request, and, in particular, referral and consultation with other health care professionals may be valuable in developing a framework for an informed choice. Assessment should include consideration of potential harms and benefits, the capacity of the child, and the interests of the child. Recognition and promotion of these requirements, for example, through the publication of statements by professional bodies, has been linked to a reduction in genetic testing of minors in recent years [e.g., CCMG, 2003].

In summary, susceptibility testing for children and adolescents has to be carefully considered, with particular attention paid to the maturity of the minor, possible benefits of testing, the timing of the onset of disease, the timing of effective interventions, and the possible sources of harm. In particular, in situations where delay will not affect timely medical care and prevention of disease, families should consider deferring testing until the child has the right to make the decision.

GENETIC SUSCEPTIBILITY TESTING

Anne Summers, North York General Hospital, Toronto, Ontario, Canada

CASE SCENARIO

Jane is a 30-year-old woman whose 27-year-old sister, Ellen, has recently been diagnosed with breast cancer. Their mother died at the age of 33 years with ovarian cancer, and their grandmother developed breast cancer at age 50. Jane is concerned about her own risk so she asks her family doctor for a referral to a genetics clinic.

At the genetics clinic, the counselor explains to Jane that her family would be eligible for testing. However, in order to do testing, they would have to start with Ellen. Jane is reluctant to ask her sister but as she and her husband are considering starting a family, Jane decides to raise the subject with Ellen. Ellen readily agrees to participate and is found to have a mutation in the *BRCA1* gene, which would be consistent with the family history of breast and ovarian cancer. Jane then is tested and is found not to have the mutation. She and her husband have a healthy daughter 14 months later.

Their 24-year-old sister, Barbara, learns of Ellen's result and requests testing. Prior to the testing, she has always been healthy, has had no surveillance, and has not thought much about either breast cancer or her health in general. She is counseled about the potential medical and psychosocial effects of the testing. She seems to understand and is eager to proceed. Testing shows that she carries the same mutation as Ellen. She is told that this mutation is associated with a 70 percent chance of developing breast cancer and a 40 percent chance of developing ovarian cancer in her lifetime (as compared to the population risk of 10 percent for breast and ~2 percent for ovarian). The options of mastectomy and oophorectomy are discussed with her as considerations when she has completed her family. She is unmarried and is totally overwhelmed by this discussion. She is referred to a family physician for a clinical breast examination and although the result is normal, she becomes extremely anxious about developing cancer and requests a second opinion. The second normal examination does not seem to relieve her anxiety and her family also reports that she spends hours searching the Internet for possible ways to prevent breast cancer.

INTRODUCTION

Genetic testing to determine susceptibility to a variety of common adult-onset disorders has already become part of our medical armamentarium and will extend to involve many disorders. This type of testing does not diagnose or even definitely predict disease; it tells the person tested the chance that he or she will develop the disease in question. At the time of writing, that chance is given in the absence of other factors such as other genes, health status or environmental factors. It is expected that, over time, the risk calculation will become more complicated and will include a variety of factors.

For a number of cancers, it is estimated that 5 to 10 percent of cases are related to an inherited predisposition, and, as we begin to identify minor genes, that proportion could increase. Other disorders, such as Alzheimer's disease, Parkinson's disease, diabetes, and heart disease have varying levels of genetic contribution. At first glance, this early warning system would seem to have a tremendous advantage for those destined to develop these "hereditary" diseases. This would allow physicians to identify those at highest risk, introduce targeted surveillance program that would detect the disease at its earliest stage, to implement preventative therapy or lifestyle changes where appropriate, and, in the eventuality that the disease develops, to start earlier treatment where possible with a higher chance of cure. This is already a fairly common approach in preventative medicine, for example, calculating a patient's risk for cardiac disease based on a number of factors such as weight, smoking, and diabetes and then introducing lifestyle changes and medications, as appropriate. However, there is a slight difference when the risk is programmed into the person's genes. When a person is obese, a smoker, a drinker, or a diabetic, he or she may be considered salvageable because he or she could lose weight, stop smoking or drinking, or take medications for high cholesterol or diabetes. Persons who carry a defective gene that predisposes them to a disease may be viewed differently with the, not necessarily correct, belief there is nothing that they can do about their genetic constitution. Such a person may be seen as a potential burden to society particularly in a publicly funded health care system; or, as a less than perfect mate or student or employee; or may be judged to be ineligible for insurance coverage because of a preexisting condition. Apart from the manner in which a carrier of a susceptibility gene is viewed by others, there is also the consideration that this may affect the way he or she views him or herself. As a result, while genetic susceptibility testing may hold out great promise, there are undoubtedly significant concerns.

POINTS FOR DISCUSSION

1. Planning and funding (depending on the health care system) of genetic susceptibility testing must take into account not only the testing but also pre- and posttest counselling and the downstream costs such as surveillance, treatment, and psychosocial support. Each test should be viewed from the point of view of a service.

2. From a medical point of view, susceptibility testing is going to require more involvement of nongenetics health care providers because of the potential patient volume. Therefore, education is going to need to be ramped up in all areas of health care.

3. Health care providers, patients, and the general public will need a better understanding of genetic risk, especially in the face of the increased complexity of risk calculations. There will also be a need for an understanding of the difference between risk prediction and diagnosis both on an individual and a societal level.

4. Ethical concerns such as informed consent, confidentiality, discrimination, freedom from coercion, genetic reductionism, genetic determinism, duty to warn, testing of minors, and an understanding of the difference between testing for disease and testing for traits must be included in any health care planning for genetic susceptibility testing.

5. In dealing with the more immediate health care, planning and ethical concerns around susceptibility testing, some difficult areas have been put on the back burner. Two of these include the upcoming technologies that will allow the concurrent testing of multiple genes and the regulation of companies selling susceptibility testing directly to the consumer. There needs

to be some thought about how to deal with these issues, which are already upon us.

6. Government must reflect the views of society, and, to facilitate legislation and regulation where necessary in the area of genetics, there is a need for public debate on genetic issues such as that hosted by the SACGT and UK Human Genetics Commission.

HEALTH CARE PLANNING AND GENETIC SUSCEPTIBILITY TESTING

In taking into account the negative potential of genetic susceptibility testing, planning and funding (depending on the health care system) must take into account not only the testing itself but the associated necessary services. This was articulated in the recommendations of the Ontario Provincial Advisory Committee on New Predictive Genetic Technologies: "Genetic testing should be provided as an integrated, multidisciplinary service, incorporating genetic assessment and counselling, quality testing, psychosocial support and follow-up services, including surveillance, prevention and treatment, as appropriate" [Ontario Provincial Advisory Committee on New Predictive Genetic Technologies, 2001]. This recommendation is predicated on a preceding recommendation that a test must be evaluated and found to be technically accurate, clinically effective, and useful to tested individuals before even being considered for adoption by the province.

WHO WILL PROVIDE GENETIC SUSCEPTIBILITY TESTING?

As more and more genes associated with common adult-onset disorders are discovered, it will become decreasingly possible for all gene carriers to be seen by a geneticist or genetic counselor. For example, in the province of Ontario, Canada, there are fewer than 30 clinical geneticists and approximately 100 genetic counselors serving a population of 12 million people. A study by Harris et al. [1999] in the United Kingdom showed that fewer than half of patients with a risk for significant genetic disorders were referred for genetic consultation, and Lerman and Shields [2004] in the United States reported that among a population only 7 percent of patients with an increased familial risk of developing cancer where referred by their oncologists to a geneticist. It is possible that in the latter group, which was restricted to cancer, the physicians felt that they could do the genetic counseling themselves as Freedman et al. [2003] found in a survey of 1251 American physicians, 84 percent of the oncologists surveyed considered themselves qualified to recommend genetic testing to their patients.

This may well be appropriate. Genetics specialists cannot expect nor be expected to be knowledgeable in all areas of medicine. Therefore, primary health care providers will have to take the responsibility for identifying patients at genetic risk, providing counseling where they feel comfortable, and referring appropriately where they do not. Specialists will have to provide counseling around genes specific to their specialty. The genetics community can help with the transition of genetics from the genetics clinics to primary and secondary health care providers' offices or to multidisciplinary teams. Geneticists will continue to provide back up for the more complex cases particularly those that cross specialties [Pagon, 2002].

This will not be an easy transition; it will require a marked increase in genetics education at all levels of the medical, nursing, and other health care providers' curricula including postgraduate education. Freedman et al. [2003] looked at 1251 American physicians' preparedness for taking over genetic counseling of their patients and found that only 29 percent felt adequately trained to provide genetic counseling. This inadequacy was illustrated in a much cited American study that showed that nearly a third of physicians surveyed misinterpreted the meaning of a genetic test results for *APC*, the gene associated with familial adenomatous polyposis [Giardiello et al., 1997]. In order to be able to transmit the complexity, providers of genetics information must be supplied with the appropriate tools.

UNDERSTANDING RISK

The type of result given in genetic susceptibility testing denotes risk, sometimes quantitated as a risk figure, sometimes not. This is somewhat different from the usual information given in a health care provider–patient relationship. While patients may be told if they do not improve their lifestyle, they will have a heart attack or a stroke, they are more used to hearing a diagnosis from their physician. They expect to hear statements like, "your cough and fever are due to pneumonia," "the X-ray shows that you have a kidney stone," or "the biopsy shows that you have prostate cancer," not information such as, "you carry a gene that gives you a 70 percent chance of developing breast cancer at some point in your life" or "your genetic testing shows that you may develop hemochromatosis (a potentially lethal disorder related to increased iron storage in the body) at some point in your life, we're not entirely sure what that risk is at present."

Patients are not told that they have the disease or even that they will eventually develop it but that they have a chance of getting it. Based on this uncertainty, which will vary from gene to gene and disorder to disorder, patients will have to make major decisions that may be lifesaving. For example, in the case above, Barbara, who is perfectly healthy, has already had to start to think about mastectomy based on a 70 percent risk for breast cancer and oophorectomy based on a 40 percent risk for ovarian cancer. What if those risks were 10 and 5 percent, respectively, would this make her decision easier or more difficult? That people do not understand risk well is evidenced by the fact that, in this author's experience, at-risk persons almost invariably liken a 50/50 chance of inheriting a gene, such as the one that causes Huntington disease, to the chance of being hit by a bus. If much of medical care is going to rely on delivery of risk-based results, not only health care providers need to understand risk—patients and the general public need to be well grounded in risk concepts as well.

RISK PREDICTION VS. DIAGNOSIS

The difference between carrying a gene for a particular disease and being diagnosed with that disease is a concept that often gets blurred. For example, in the case above, Barbara is 24 years old and is found to carry the *BRCA1* gene. This does not mean that she has breast cancer or ovarian cancer. There is a 30 percent chance that she will not develop cancer. She will be offered surveillance and even prophylactic surgery, but that does not make her ill. To the medical community providing testing and support, it is felt to be important that gene carriers are not seen as affected with the particular illness, as "living with cancer" or "battling cancer" but as healthy persons who are taking steps to prevent illness. However, this may be viewed differently by the gene carrier: "I think in the minds of most people there are only two groups—those with breast cancer and those without it. To my friends, because I didn't have a diagnosis of breast cancer, I was simply in the healthy people group. To me, however, I was one small step away from the breast cancer group" [Prouser, 2000]. There is fear that gene carriers, by virtue of their knowledge regarding their genetic constitution, will be turned into patients. This can happen at any level, it may be personal in the way the carrier views the information, it may be the health care system medicalizing a healthy individual, or it may be society viewing the healthy gene carrier as ill.

ETHICAL CONCERNS ASSOCIATED WITH GENETIC SUSCEPTIBILITY TESTING

Many of the ethical issues associated with genetic susceptibility testing are discussed elsewhere in this book, and I will not review them here. I will briefly reiterate that susceptibility testing should always be done in the context of informed consent with equal weight given to a person's right to know and their right not to know genetic information. The latter concept leads into the concerns around coercion where a person at risk can be forced or manipulated into testing by others with possibly benign intentions but not considering the best course for the person facing the decision about testing. Coercion can come from family members, friends, health care providers, researchers, and insurance companies, among others. While it may be impossible to overcome, it is important to recognize this as a motivation for testing because it may affect the tested person's reaction to the result.

GENETIC DISCRIMINATION

In the discussion of the negative aspects of susceptibility testing, I have alluded to discrimination at all levels. This has always been one of the major concerns about testing that, to some degree, predicts the future. Such foreknowledge may expose the gene carrier to discrimination. Some aspects of discrimination can be prevented by legislation, for example, it could be made illegal for insurance companies to use a person's genetic background to make decisions about eligibility. It is impossible to legislate against others such as a partner who does not want a spouse who carries a harmful gene that will be passed on to their future children. As a society, we may never be able to get beyond the interpersonal areas of discrimination. On the other hand, it is possible that the "same boat" scenario will become commonplace as more genes are identified, that is, you have a gene for breast cancer but I have a gene for heart disease, which is worse?

GENETIC DETERMINISM

The discussion around discrimination also raises the issue of genetic determinism, a predestination-like belief that one's fate is dictated by one's genes. This belief assumes that genes are not influenced by environment, taking the nature vs. nurture argument to the nature extreme. It does not take much imagination to see the danger in this point of view, if a person's fate is determined genetically, then there is not much he or she can do to change it. Apart from being dangerous, it is also somewhat silly in that many negative genes rely on environmental factors for their effects. Hegele [2001] reported on a gene in the Oji-Cree of Northern Ontario that predisposes to obesity-related type II diabetes, a disorder which is found in 40 percent of adults in this population. There is excellent documentation that this disorder was virtually unknown among the Oji-Cree 50 years ago. Hegele postulates that a change in diet and lifestyle allowed the emergence of this genetic predisposition. The notion that a person is flawed because of carrying a gene associated with a disease such as cancer, heart disease, or a neurological disorder is taking genetic determinism to another extreme—reducing "you are your genes," to "you are your gene."

GENETIC REDUCTIONISM

Genetic reductionism is the tendency to exaggerate the effect that genetic technology will have on health care outcomes. The danger associated with this viewpoint is that such technologies can be adopted in health care over more effective strategies. For example, going back to *BRCA* testing, if a health care jurisdiction were to use genetic screening for breast cancer instead of mammography screening programs, the detection rate would drop markedly because 90 percent of women with breast cancer do not appear to carry a breast cancer gene. The use of information about genes that will predispose to a variety of disorders is predicted to be far less effective than current strategies focusing on lifestyle choices [Baird, 2002].

GENETIC SUSCEPTIBILITY TESTING IN CHILDREN

Genetic susceptibility testing in children seems to be a fairly straightforward area in the abstract but can become very difficult when dealing with anxious parents. Generally, it is recommended that unless testing will make a difference to the care or prognosis of the child, genetic testing for an adult-onset disorder should wait until the child is old enough to make his or her own decision. According to the American

Society of Human Genetics and the American College of Medial Genetics [1995]: "Timely medical benefit to the child should be the primary justification for genetic testing in children and adolescents .... If the medical or psychosocial benefits of a genetic test will not accrue until adulthood, as in the case of carrier status or adult-onset diseases, genetic testing generally should be deferred." These guidelines must be considered in the context of harms and benefits to the child, the child's ability to understand and make decisions personally, and, although the welfare of the child must be paramount, the well-being of the family. This philosophy has been echoed by the Ontario Advisory Committee on New Predictive Genetic Technologies [2001] and the American Society of Clinical Oncology [2003]. The minor controversy in this policy is the priority of the welfare of the child over that of the family. This becomes a difficult issue in clinical practice when there is no clear benefit to testing a child, but the extreme anxiety of the parents could have a negative effect on that child in terms of overprotective behavior or overinvestigation for symptoms perceived to be related to the disorder in question.

TESTING FOR DISEASE VS. TRAITS

In the area of genetic testing, there is always a discomfort around the possibility of eugenics—the wish to attain genetic "normality," whether it be in an individual, a family, or a society. There have been numerous discussions of this topic and I will not review them here other than to mention how this may apply to genetic susceptibility testing. One of the most extreme versions in recent history was the mandatory testing of all African Americans for sickle cell hemoglobin, whether they had sickle cell anemia or not. This resulted in discrimination in areas of employment and insurance and racial stigmatization even though this gene can be found in other populations [Lerman and Shields, 2004]. As we begin to identify minor genes that may predispose to increased environmental vulnerability in relation to a particular disorder, we could definitely get into a discussion of where susceptibility testing ends and testing for undesirable traits begins. For example, it is known that there are genes that predispose to obesity, addiction, and alcoholism—all of which are factors in a variety of very common diseases such as coronary artery disease, stroke, diabetes, lung and other cancers, and liver disease. Will we start testing for such genes as part of a disease susceptibility profile or would it be better to put our efforts into public health efforts for better nutrition and antismoking, alcohol, and drug campaigns. Where is the line drawn between health factors and discrimination against fat people, smokers, drunks, and drug addicts?

TESTING FOR MULTIPLE GENES SIMULTANEOUSLY

The above issues are going to become increasingly complicated by technology that will allow testing of hundreds to thousands of genes simultaneously. This can be done in a logical fashion by which concurrent testing is performed for all the known genes for a particular disease such as colon cancer, coronary artery disease, or kidney disease or even for related diseases such as heart disease and stroke. With this method, the physician or health care provider can explain to the person being tested that many genes are being assessed for one disease or a few related disorders, that some genes are more powerful indicators of disease than others, and that we frequently find genetic changes that we cannot entirely explain. He or she will not have to discuss testing for multiple diseases, which is the problem with the other possibility—testing for potentially hundreds of disorders at once. A major problem with the multiple disease approach is that it would be practically difficult, if not impossible, to obtain a separate informed consent for each specific test. However, the test provider still will have the obligation to disclose and discuss the nature of the tests, their purposes, and the associated risks and limitations. At some point, the health care community and potential users of these "shotgun" tests will have to decide on how they are to be used including the importance of informed consent.

DIRECT TO CONSUMER TESTING

Another issue that worries many health care providers is the over-the-counter genetic testing approach. There are many private companies that provide genetic testing. Most of these companies are highly regulated and will only deal with samples referred to by physicians or genetic counselors. However, there is an element that provides direct to consumer testing, mostly accessible through the Internet both for known genes and for more questionable susceptibilities. Sometimes this testing is linked to the sale of therapies. There are dangers associated with this practice. When dealing with reputable laboratories testing for well-researched disorders, the health care provider can generally be assured that there are safeguards in place in terms of quality and reliability of testing as well as the interpretation of results. If there is any question, then it is incumbent upon the health care provider ordering the test to find out whether the laboratory is reliable. For the consumer who is purchasing such services directly, the quality of the testing may be an issue, but the validity and the interpretation and follow-up of results are potentially much larger problems. The Internet is not regulated, it is international, and it is very difficult for anyone to keep track of all the claims made by persons or companies selling products through it. Therefore, it is quite difficult for the uninitiated consumer to know if the claims of a testing company are valid or not. If they are not, then maybe the worst that happens is that the person pays for useless "genetic" information. However, if the test is valid in a particular clinical context and is well-performed, the consumer may, for example, find out that he or she carries a gene that predisposes to a particular disease without necessarily having this information placed in its proper clinical and genetic context. For example, our consumer may find that he has a gene that is associated with increased clotting with the added information that increased clotting can be associated with problems such as thrombosis, embolus to the lungs, and stroke. Those with more knowledge of these genes would be aware that different thrombophilia (clotting) genes have different significance and whether one carries one or two copies of a particular gene makes a huge difference, but, most significantly, the patient's past history and family history contribute in a major way to the action taken on finding this information (and in fact doing the testing in the first place). If our consumer frantically searches the Internet and finds out that the treatment for thrombophilia is anticlotting or anticoagulant medication and then goes to his doctor and demands therapy, it is possible that a less than knowledgeable physician could prescribe the medication. Anticoagulant medication slows clotting and so can cause bleeding when given inappropriately. This is a possibly exaggerated scenario, but I think it illustrates the fact that genetic information in the wrong hands is not harmless.

GOVERNMENT AND GENETIC SUSCEPTIBILITY TESTING

A democratic government should reflect the views of society, and, in order to facilitate legislation and regulation where necessary in the area of genetics, there is a need for public debate on genetic issues such as those reflected above. A public debate process has been used by the U.S. Secretary's Advisory Commission on Genetic Testing (SACGT) and its subsequent form, the Secretary's Advisory Committee on Genetics, Health as well as Society, and The Human Genetics Commission in the United Kingdom. These organizations have invited both expert and public opinion on a variety of issues such as oversight of genetic testing, genetic discrimination, gene patents and licensing, direct to consumer marketing of genetic testing, and have used this opinion as well as extensive literature review to form the basis for public reports that have/will lead to regulatory or legislative changes.

SUMMARY

In summary, genetic susceptibility testing may have tremendous possibilities for health care in the future. Nevertheless, there are many concerns with this kind of testing that, rather than diagnosing, predicts possible disease. This outlook may have been best expressed by a patient: "I am a carrier of a *BRCA2* gene mutation. My genetic status is now as much a part of my personal identity as are my age (47 years old), my religion (Jewish), and my educational training (a master's in community health). I have one sister who won the coin toss and did not inherit the mutation. My husband and I have two adult children, a son and a daughter, both of whom intend to learn their genetic status some day. Without a doubt, this genetic journey has been not only one of the greatest challenges of my life but also one of the loneliest" [Prouser, 2000].

Each jurisdiction is going to have to deal with the many issues surrounding genetic susceptibility testing, and it is encouraging to see that this work is already underway in a number of countries as evidenced by the work cited in this chapter. The availability of much of this work on the Internet makes it accessible throughout the world. The value of having so much of this work available to those countries that are lagging a little is that they can adopt and adjust rather than having to "reinvent the wheel." The final decisions on each of the issues discussed above may vary among jurisdictions depending upon the underlying values of the population and, possibly, the structure of the health care system. Hopefully, they will be informed by a wide perspective.

Chapter 12

Test Samples and Laboratory Protocols

OVERVIEW

Collection and submission of specimens for genetic testing involves several critical steps. Clinicians need to identify the appropriate tests and laboratories for providing the testing. Labs need to receive the right type of specimen, properly taken, stored, transported, and identified, along with an appropriate indication and patient informed consent. As well, standards of quality assurance and reporting are improving, and technical advances promise rapid changes in the types of tests that can be performed. Clinicians need to understand test reports that clearly identify the test methodology, test findings, and clinical significance of the results. Microarrays, DNA (deoxyribonucleic acid) chips, automated mutation detection systems, and fluorescence microscopy techniques are all leading to new challenges for clinicians. This chapter reviews how clinicians can best take advantage of these new techniques, and includes a discussion of the debate regarding the impact of gene patents on genetic testing.

SCENARIO 1

A woman is referred for routine prenatal screening due to her age (41). Amniocentesis is performed and the fetal karyotype is reported as normal. At birth, the baby has multiple anomalies and subsequently fails to thrive. A repeat karyotype is requested for high-resolution banding and a blood specimen is submitted. A small, unbalanced structural rearrangement is found that, upon review of the prenatal karyotype, could not be detected at a routine level of resolution in the prenatal specimen. The parents are karyotyped and the father is found to carry a familial chromosomal translocation that led to the fetal anomalies. In follow-up counseling the family history is reviewed in detail; the father recalls a relative who is institutionalized because of multiple developmental anomalies and another relative who died shortly after birth. The baby

Genetic Testing: Care, Consent, and Liability, by Neil F. Sharpe and Ronald F. Carter
© 2006 John Wiley & Sons, Inc.

dies at 1 year of age due to the severe developmental anomalies despite multiple attempts at corrective surgery and constant in-home care. The parents want to know why the chromosome abnormality was missed in the fetus.

Issues Raised in This Scenario

The first issue is why the laboratory did not detect a chromosomal anomaly that was later found in high-resolution karyotyping upon the birth of the affected infant. G-banding studies should indicate the level of resolution obtained, expressed as banding level. The higher the banding level (the structural detail of the chromosomes), the higher the resolution and the better the ability to detect small structural rearrangements. In this situation, the amniocentesis specimen was reported at an acceptable and routine level of resolution, but this was inadequate for the detection of the small but clinically very significant anomaly. The later blood study was reported at a higher level of resolution in response to the clinician's request, and the chromosomal imbalance was detected. At first glance, this answers the questions of the patients. However, there is more to this story, as suggested by the family history.

The major issue concerns the inadequate review of the family history. If the prenatal counseling process had identified the positive family history for abnormal births, parental karyotyping would have been done on an urgent basis. This would have revealed the father's abnormal karyotype. The lab would have been alerted to the high risk for presence of the translocation and could have accurately detected the abnormal fetal karyotype by either paying special attention to banding patterns or using fluorescent probes [fluorescence in situ hybridization (FISH) studies] for the affected regions. Thus, the family did not receive an accurate prenatal karyotype because the indications for testing were inadequately informed by the prenatal counseling. This situation would have been averted by simple questions to the parents at presentation for prenatal counseling: "Do you know of anyone in your family who has ever had trouble having children, or had miscarriages or a stillbirth?" "Has anyone had a baby that was abnormal or died shortly after birth?"

Parental karyotyping for familial translocations may provide opportunities for a more conclusive analysis: (1) If the parent carries a balanced chromosome translocation, then the parent will have two abnormal chromosomes and therefore the lab has "twice" as much opportunity to detect the anomaly compared to a fetus inheriting the translocation in an unbalanced fashion with only one abnormal chromosome, and (2) blood specimens usually can be processed to provide a higher level of resolution than prenatal specimens. (*See also Chapter 5 clinical scenario for family history perspective.*)

SCENARIO 2

Amniocentesis is requested after maternal serum screening identifies an increased risk for Down syndrome in a fetus. The karyotype result is normal male. At delivery, the baby appears to be a girl with Down syndrome. Both the obstetrician and the family want to know what went wrong with the amniocentesis. A follow-up blood specimen is submitted to a second laboratory for karyotyping, which confirms that the karyotype is normal male. A clinical examination and molecular testing determine that the baby has Smith–Lemli–Opitz syndrome, which explains the apparent gender reversal and multiple developmental anomalies [Smith et al., 1964; Kelley and Hennekam, 2000; Nowaczyk and Waye, 2001]. The family wants to know why the syndrome was not diagnosed prenatally.

Issues Raised in This Scenario

The first response of the obstetrician and the parents was to assume that the laboratory had made an error. Usually, in this situation the lab is asked to find out what went wrong. Laboratory staff will review all aspects of the documentation for that specimen on that day, all available retained laboratory materials (slides, pictures, records), and all other accessioned specimens for the day, and even that week, in order to determine if there was any evidence of an error. Laboratories have standard protocols for investigating these incidents and maintaining communication with the family and the clinicians during this phase. In the meantime, however, in situations like this, there is a need to consider all possible explanations, to discuss those possibilities with the family, and to follow up with further investigations.

It is reasonable to request confirmation of the karyotype by testing peripheral blood lymphocytes; this was done, and in reflection of the parents' distrust of the first lab, the specimens were referred to a second lab (which confirmed the prenatal karyotype as accurate). However, in this situation, an additional clue was the presence of anomalies, which raised the possibility of a nonchromosomal syndrome involving sex reversal. Referral to a clinical geneticist and subsequent testing did indeed confirm Smith–Lemli–Opitz syndrome and ultimately showed that no laboratory error was made. In this situation, the fetus was affected by a genetic disease that could not be detected by the routine prenatal screening protocol. Molecular testing for Smith–Lemli–Opitz syndrome would be indicated for future pregnancies but, of course, would not have been offered to a couple without any positive family history presenting for routine prenatal counseling.

TYPES OF GENETIC TESTS

Genetic tests are used to screen for or specifically diagnose people who have or are at risk for a genetic condition. Indications for genetic testing vary from population screening for risk of disease, to confirmation of a diagnosis, to prenatal prediction of a disorder. Genetic testing can be categorized by the type of indication or by the technology used to investigate the indication. In a broader sense, a common blood film and complete blood count can be used to identify a patient with thalassemia. However, genetic testing more commonly refers to specialized biochemical, cytogenetic, or molecular test methodologies that specifically identify and categorize inherited genetic disorders by causal mechanism. Similar cytogenetic and molecular technologies are commonly used in oncology in the diagnosis and management of many tumor types. For the purposes of this chapter, discussion will be limited to specialty genetic laboratory testing for inherited disorders and cancer genetics.

Traditionally, cytogenetic tests have depended upon morphological analysis of chromosomes arrested in metaphase and stained with various techniques, such as G-banding, that permit counting and structural assessment of individual chromosomes (*karyotyping*). This requires viable tissue that can be cultured and harvested. Although karyotypes are digitally captured and prepared with computer assistance, karyotyping still depends upon expert assessment of chromosome banding patterns involving skills of pattern recognition and judgment of normal compared to abnormal morphology. While karyotyping surveys the whole genome and can detect a wide variety of different types of anomalies, routine G-banding resolution is limited to hundreds at genes at its most sensitive level of detail, even at the highest feasible levels of technical analysis in routine service. New techniques provide increased sensitivity and specificity. Since 1990, an increasing number of staining technologies have been applied to cytogenetics that utilize the ability of fluorescent dye-labeled DNA probes to hybridize to chromosomes in slide preparations. This FISH technology can offer substantial advantages over G-banding for specific applications, including the ability to assess dead and fixed tissues, greatly increased resolution (from whole chromosomes to single genes), definitive identification of deletions, duplications and chromosome rearrangements, and the ability to monitor for very low proportions of residual disease in cancer specimens.

Biochemical genetic tests include a wide variety of specialty tests that are typically categorized by the indication for testing (disease or test panel) and test methodology (e.g., urine amino acid analysis). Testing is applied to the diagnosis of inborn errors of metabolism and, therefore, typically consists of a biochemical screening or diagnostic test that may require confirmation by molecular testing to identify a specific mutation. Test methodologies and interpretive protocols have changed rapidly in recent years with advanced chromatographic techniques, tandem mass spectrometry, rapid DNA sequencing, and new generations of automated core laboratory analyzers. Some tests are available only in a very limited number of service or research laboratories, and lab directors have specialty training in the diagnosis and management of biochemical disorders. Since many biochemical disorders can present as urgent neonatal or pediatric crises, biochemical geneticists often work with neonatal care teams to provide a rapid diagnostic response while attempting to stabilize the patient.

Molecular testing comprises a variety of different technologies that provide information about DNA sequence. It is important to clearly delineate between molecular testing for genetic disease and the use of molecular test methodologies for other applications such as bacteriology, virology, tissue typing, and tumor diagnostics.

Hereditary mutations, which are passed on from generation to generation through meiosis, need to be distinguished from acquired (somatic) mutations that are limited to specific tissues; somatic mutations are by definition mosaic and are the critical point of interest in tumor diagnostics. Whereas indirect testing for hereditary mutations by linkage analysis commonly was used in the past, the cloning and sequencing of most genes now permits direct mutation analysis. As with cytogenetics, no one technique can detect all types of molecular mutations.

While automated gene sequencing is commonly used and recognized as a diagnostic standard, it does not usually detect certain types of splice site mutations and large deletions and duplications. Therefore, almost all molecular tests offered by laboratories today have less than 100 percent accuracy, and careful interpretation of the test result for clinical significance and risk assessment may be necessary. Because of the large number of different genetic conditions, no one laboratory provides a complete analytical menu.

Referring professionals can use websites (e.g., Genetests, www.genetests.org/) to identify specific laboratories that offer reliable testing in support of diseases in question. Future developments undoubtedly will include a marked reduction in the technical cost of sequencing, increased analytical accuracy, and a broader spectrum of available laboratory services.

In diagnostic cancer genetics, molecular testing is moving into the field of molecular tumor profiling. In these applications, proprietary assays combining molecular tests for multiple genes are being combined into single-step analytical methodologies that are micro in scale, robotic in application, and computerized in data assessment. Examples include DNA "chips" and microarrays. These applications provide significant advantages in cancer diagnostics because they assess the molecular mechanisms of up to hundreds of different mutations or abnormal expression patterns of many different genes. Aberrant expression of selected panels of tumor suppressor genes and DNA repair genes can be tested quickly and thereby provide much improved prognostic separation of tumor subtypes [e.g., Hedenfalk et al., 2001; Bertucci et al., 2003].

Similar chip-based screening panels are envisioned for prenatal and neonatal screening but, as of the time of this writing, are not in widespread commercial use. In these applications, the capacity to design chips that rapidly assay for thousands of different mutations promises to provide highly sensitive screening tests for dozens of different diseases from a single small specimen of DNA [e.g., Zammatteo et al., 2002]. The expected breadth and accuracy of these new test methodologies raise new issues for patient consent, professional interpretation, and clinical management.

Currently, the cost of genetic tests usually results in selected application by specific criteria, rather than as screening tests to rule out multiple diagnostic considerations with low probabilities of a positive result.

NEED TO KNOW

Burden of Disease: Indications for Testing

Chromosomal Anomalies as Causes of Developmental Disorders
- Prenatal Diagnosis
- Neonatal and Pediatric Cytogenetics
- Adults

Molecular and Biochemical Genetic Diagnosis of Disease
- Using Molecular Tests
- Using Biochemical Genetic Tests

Watch Out For
- Ordering and Interpretation of Tests: Obligations for the Clinician
- Laboratory Performance: Obligations for Laboratories

    Reporting Responsibilities

    Laboratory Accreditation, Quality Assurance, and Quality Management Protocols

Additional Considerations
- Use of Research Testing in Clinical Management
- Use of Archived Specimens
- Consent to Store Specimens Compared to Consent to Test
- Limitations of Test Methodologies

Clinical Significance of Test Results

Chromosomal Anomalies as Causes of Developmental Disorders

Cytogenetic testing is applied to two major types of indications: detection of inherited (constitutional) chromosomal anomalies and acquired anomalies (cancer diagnostics). For constitutional anomalies, the classic series of texts by Gardner and Sutherland [2004] represent definitive resources for the clinician. Cancer cytogenetics is not discussed in detail here, but for the interested useful summaries have been published by Dewald and colleagues [e.g., Dewald, 2002; Adeyinka and Dewald, 2003], and a particularly accessible and helpful source is a website maintained by a collaboration of French cytogeneticists (www.infobiogen.fr/services/chromcancer/).

Clinical anomalies occur when a constitutional chromosomal abnormality causes gain or loss of gene content, interrupts the coding of a critical gene, or interferes with the function of a gene (e.g., abnormal genetic imprinting). Chromosomal anomalies are highly variable in type, content, origin, and clinical significance (Table 12.1). The population burden of cytogenetic disorders has been determined in large surveys, most of which were conducted some years ago and at lower levels of technical resolution than are currently available [e.g., Hook et al., 1983; Hook and Cross, 1987a, 1987b]. Data from these surveys are summarized in Table 12.2. It is estimated that chromosomal anomalies cause approximately 50 percent of all recognized miscarriages. Note that the proportion of individuals with cytogenetic abnormalities is far lower at birth (approximately 0.6 percent).

Several important concepts should be recognized:

- There is heavy selection against conceptions with chromosomal anomalies; the vast majority of chromosomal errors result in miscarriage or stillbirth.

- There is an inverse association between the severity of a chromosomal anomaly and viability: The most severe chromosomal imbalances are found only in lost

Table 12.1   General Categories of Common Chromosome Anomalies[a]

Type	Examples	Mechanisms of clinical significance
Numerical	Gain: trisomies, triploidy tetraploidy Loss: monosomy (rare); mosaicism possible but rare	Unbalanced karyotypes with large errors in gene dosage (gain tolerated more easily than loss)
Structural	Reciprocal and Robertsonian translocations; inversions; deletions; microdeletions; ring chromosomes; insertions; duplications and amplifications; marker chromosomes, fragments	If balanced: risk for unbalanced segregations at meiosis If unbalanced: errors in gene dosage
Uniparental disomy	Abnormal inheritance pattern, e.g., for a given chromosome, both inherited from one parent, none from the other parent	Clinical abnormalities can result if the chromosome has imprinted regions
Epigenetic	Abnormality caused by abnormal imprinting or other mechanism unrelated to chromosome structure or DNA sequence	Errors of genetic regulation: can have functional abnormality even with normal chromosome structure and DNA sequence
Unbalanced	Overall effect of abnormality is gain, loss, or interruption of chromosome material that contains coding genes	Genetic imbalance usually infers increased risk for anomalies
Balanced	Chromosome segments may be rearranged but no gain or loss of coding material; associated with and implies normal growth and development	Carriers of balanced chromosome rearrangements may be at increased risk for unbalanced conceptions (fetal anomalies)
Familial	Inherited from a parent	If inherited without further rearrangement, infers same effect (or lack of) as in parent
De Novo	Arising at meiosis as a new abnormality	Breakpoints in de novo structural rearrangements may affect gene function, therefore risk for abnormal outcome even if apparently balanced
Normal variant (heteromorphism)	Variation in chromosome structure or banding pattern that is documented in normal population	By definition, not clinically significant

[a]Chromosome abnormalities can be classified according to the basic mechanism of the abnormality, the net effect on genetic dosage (balanced or unbalanced), and the apparent origin (familial, de novo, or normal variant).

Table 12.2 Frequency of Chromosomal Anomalies

Pregnancy losses: 50%	Livebirths: 0.6%
50%: trisomies	One-third: sex chromosomes
20%: 45,X (Turner syndrome)	One-third: trisomy 21, 18, 13
20%: triploidy and tetraploidy	One-third: structural
	Two-thirds: balanced[a]
	Two-thirds: unbalanced[a]
4%: structural[a]	
Three-fourths: de novo	
One-fourth: familial	

[a] Risk of recurrence needs to be carefully evaluated.
Source: Adapted from Gardner and Sutherland [2004].

pregnancies, while the mildest anomalies may be unsuspected throughout life and documented only as incidental findings.

- A wide variety of different structural rearrangements can occur (e.g., translocations, inversions, ring chromosomes, deletions); clinical significance will vary with type of inheritance (de novo vs. familial), number of chromosome breaks, the break sites in the chromosomes (segregation pattern at meiosis), genetic content, and degree of chromosomal imbalance.
- The vast majority of chromosomal anomalies arise de novo as errors of chromosome replication and segregation during meiosis.
- There is no significant variation by ethnic background in incidence or type of chromosomal anomalies (excluding rare chromosomal breakage syndromes).
- It is important to identify familial chromosomal anomalies with high risks of recurrence.
- The greatest clinical concerns relate to chromosomal imbalances that are compatible with viability but cause developmental anomalies.

For clinicians wishing to find more details, Gardner and Sutherland [2004] provide a definitive and detailed discussion of the origins and significance of chromosome anomalies as well as widely used and comprehensive guidelines for clinical counseling.

Prenatal Diagnosis

Down syndrome was an initial focus for prenatal screening of chromosomal disorders. Recognition of the relationship between maternal age and increased risk for Down syndrome in pregnancy led to the development of screening programs for mothers at increased risk of having fetuses with chromosomal aneuploidies (reviewed in Chapter 8). Numerous additional screening criteria have been added, including ultrasonography for fetal anomalies, maternal serum biochemical assays, and combinations of maternal serum analysis with ultrasound screening protocols. Cytogenetic analysis is routinely performed on chorionic villus sampling (CVS) of cells, amniotic fluid cells, and fetal blood cells. The timing and options for these analyses are discussed in detail in Chapter 8.

Advances in prenatal screening techniques are oriented toward improving the accuracy of prenatal screening by increasing the rate of detection of chromosomal anomalies, providing screening at earlier stages of pregnancy to reduce the risks associated with pregnancy termination, and reducing the number of false-positive screen results. In total, the aim is to provide a least-invasive test with a high accuracy for the detection of fetal chromosomal anomalies as early as possible in the pregnancy. Improved methodologies will reduce the anxiety and risks associated with unnecessary diagnostic testing. A corollary is that, in time, most amniocentesis and CVS testing will be done for confirmation of a screening result, and a much higher proportion of tests will provide abnormal results.

Neonatal and Pediatric Cytogenetics

As summarized in Table 12.2, it is important to note that:

- At least 1 in 200 live-born babies have a chromosomal anomaly of which approximately one third will be:
  - Autosomal trisomies such as trisomy 21, 18, or 13
  - Sex chromosome anomalies such as XXY, XYY, and 45,X
  - Structural anomalies (and only one fourth of those will be familial).

- Prediction of clinical impact is based upon assessment of the degree of chromosomal imbalance and the specific chromosomal segments involved (genetic content).

- Gain of any given chromosome segment is more likely to be associated with viability than loss.

- Chromosomal anomalies are a recognized cause of a wide variety of developmental abnormalities:
  - Multiple or isolated physical anomalies
  - Growth retardation, abnormal maturation
  - Developmental delay
  - Abnormal behavior, including autistic spectrum, attention deficit disorder, and other disorders
  - Infertility, recurrent, miscarriages, premature menopause

- Chromosomal anomalies are most likely to be found in patients with combinations of physical and developmental anomalies.

Karyotyping infants or children usually is initiated after recognition of congenital anomalies or developmental abnormalities. Informally, indications for testing often are lumped into "harder" signs that are associated with a higher chance of finding an abnormal karyotype (e.g., multiple congenital anomalies, severe developmental delay, failure to thrive, abnormal facies, suspected chromosomal syndrome) and "softer" signs that are more infrequently are found to be associated with a chromosomal error (e.g., abnormal behavior, autism, attention deficit disorder, learning disabilities).

Microdeletions, and microduplications affecting specific chromosomal segments can be very specifically associated with defined syndromes, for example, William syndrome (deletion of elastin gene region, chromosome 8) and the CATCH22/VCF/DiGeorge syndrome (deletion of chromosome 22). Referral criteria combining minor dysmorphism and behavioral anomalies can select for children with very subtle structural anomalies near the end of chromosomes that are detectable only by subtelomeric fluorescent DNA probes [e.g., Knight and Flint, 2000]. Thus, depending upon the specificity and the combination of clinical abnormalities that are presented, the laboratory may need to combine routine karyotyping with additional fluorescence-based special stains to rule out specific syndromes. Microarrays and gene chips also promise to provide new sensitivity in the detection of cryptic chromosomal rearrangements, with some versions now being offered for clinical applications (e.g., the Baylor microarray: http://www.bcmgeneticlabs.org/tests/alltests.html). It will therefore be even more important in the future that the laboratory receive an accurate clinical description on the requisition.

Adults

Chromosome anomalies in teenaged and adult patients are usually associated with: (1) abnormal maturation, behavior, or development delay; (2) infertility; (3) recurrent miscarriages and/or positive family history that suggest increased risk for cytogenetic abnormalities in children; and (4) parents of a fetus or child with an abnormal karyotype. Thus, the usual indications for cytogenetic testing in adults include a personal and/or family history that is positive for:

- Infertility, delayed puberty, premature menopause
- Recurrent miscarriages, stillbirths, unexplained neonatal deaths
- Physical or developmental anomalies
- Chromosomal anomaly identified in prenatal testing

Abnormal maturation and development can present as continued or delayed investigations following pediatric onset, or they may represent new clinical problems arising after puberty. Sex chromosome anomalies such as Turner syndrome (45,X), Klinefelter syndrome (47,XXY), 47,XYY, and 47,XXX can be variable in phenotypic effect and sometimes are entirely unsuspected well into adult life. Approximately half of all individuals with Klinefelter syndrome only are identified when laboratory investigations are initiated because of infertility. In this type of clinical context, the approach to providing the test result and counseling needs to be very carefully considered. Adults with gender ambiguities caused by sex chromosome disorders (e.g., 46,XY females, 46,XX males, 46,XX/46,XY, and other types of sex chromosome mosaicism) also present situations where genetic counseling can be difficult to provide without triggering extremely complex emotional and psychological issues for the patient.

Karyotyping is an essential component of investigation for both infertility and recurrent pregnancy losses; various studies present a range in figures, but about 1 in 20 couples with 3 or more miscarriages will be found to have a parental chromosomal anomaly upon testing (for more detailed review, see Gardner and Sutherland [2004]).

Cytogenetic testing is urgently required for parents when structurally abnormal chromosome(s) are discovered at prenatal diagnosis. The clinical impact of a chromosomal rearrangement can vary from innocuous to lethal, and parental karyotyping is a key component of assessing risk for abnormal outcomes. Determination of the origin of the fetal anomaly (de novo or familial), the type of rearrangement, the chromosome segments involved, and the number of chromosome breakpoints are important.

The most important determination is whether there is a familial or de novo origin, and whether the anomaly appears balanced or not. Detailed literature reviews for the outcomes of similar cases are usually necessary because, with only a few exceptions, structural breakpoints are assumed to be unique to a given case or family. Once the clinical impact on the fetus has been considered, the important issue of risk of recurrence is considered.

An important issue is the risk of another live-born abnormal child. In familial cases, this can range from near 0 to >30 percent, depending upon the type of rearrangement. De novo rearrangements usually have a very low risk of recurrence. However, when considering a de novo vs. familial origin for a fetal anomaly, it should be noted that the average North American chance of falsely stated paternity, as suggested by molecular testing for medical indications, is approximately 10 percent; this figure prompts the thought that perhaps some "de novo" rearrangements are actually familial.

Molecular and Biochemical Genetic Diagnosis of Disease

Indications for molecular or biochemical testing include:

- Family history of a disease
- Clinical signs and/or symptoms of a disease
- Increased risk for a disease due to ethnic background, clinical presentation
- Positive screening test for disease

Genetic disease commonly is caused by small DNA mutations detectable only by molecular techniques. The last 20 years have seen spectacular advances in the ability to diagnose and confirm the molecular basis of hundreds of genetic diseases. For many diseases, it is possible to correlate the presence of disease with a specific, identifiable molecular mutation, which then makes it possible to provide rather precise counseling as correlations are made between genotypes and clinical outcomes.

For biochemical disorders, biochemical assays can specifically identify proteins, enzymes, and metabolites that indicate a diagnosis. However, there are literally millions of possible mutations and thousands of possible diseases, and no technical

Table 12.3   Brief summary of Common Types of Molecularly Detected Mutations. Molecular mechanisms of disease can also be classified according to the functional group of the protein affected, which provides useful correlations with clinical effect (see Thompson and Thompson's *Genetics in Medicine, 6th ed.*, 2001, Chapter 12, for a review).

Type	Examples	Effect on protein
Point mutations (nucleotide substitution)	missense	substitution of one amino acid for another
	nonsencse	premature stop codon: truncated and/or unstable protein
Deletions, duplications, insertions	addition, loss, or insertion of one or more basepairs	if reading frame preserved: change in amino acid sequence; if reading frame shifted("frameshift") (mutation): usually lead to protein truncation
RNA processing mutations (splice site mutations)	loss or creation of splice sites; alteration of 5' cap or 3' polyA tail	abnormal and/or unstable protein product
Large deletions, insertions, duplications	loss or duplication of exons, gene fusions(often repeat-) (sequence mediated), insertions of large repeat sequences	loss or gain of whole exons or genes: aberrant expression of normal or abnormal protein
Regulatory mutiations	interference with promoter and/or enhancer sequences	aberrant or not protein expression

Splice site and regulatory mutations probably account for about 10% of total mutation spectrum in clinical diseases.

methodology—at the time of this writing—permits a complete "scan" of the patient's genome at tolerable cost. There is considerable variation in the types of sequence variations that cause genetic disease (Table 12.3), and over 6 billion base pairs of DNA in the genome of each cell nucleus. Molecular and biochemical diagnostics are caught—at the time of this writing—between new-found theoretical knowledge of the entire genome and a relative lack of affordable, accurate, high-throughput test capacity for addressing the clinical demands that arise. This means that:

- Many tests are based upon complicated manual analytical protocols.
- Tests for rare syndromes or recently cloned disease genes often are available only through research laboratories; while many research laboratories provide excellent diagnostic support, patient care may be adversely affected by a requirement to participate in research projects as a condition of testing, limits on the types of indications accepted for testing, limits on the clinical utility of the test results, delays in the reporting time, and laboratory practices not subject to

prevailing quality in assurance and quality management standards for service laboratories.

- Almost all molecular analyses, including tests provided for common diseases such as cystic fibrosis and muscular dystrophy, are based upon methodologies that cannot detect the entire spectrum of known mutations for the genes in question.
- Where direct mutation analysis cannot be performed, an indirect test such as linkage analysis may be the only option; these test methodologies introduce a further degree of uncertainty in the clinical significance of the result.

It should be noted that newborn screening has broadened to include expanded disease panels with the adoption of tandem mass spectroscopy (TMS) technology; as a result, many jurisdictions are adding new diseases to newborn screening programs based on Guthrie blood specimens (Chapter 10), and molecular testing is performed to follow up positive results.

Despite the sheer number of different genetic tests available, the clinician needs to assess the test results and act appropriately on the clinical implications for the patient (see Chapter 14, Test Results: Communication and Counseling, for more details). The basic steps are:

- Selection of an accredited clinical service laboratory rather than a research lab
- Referral of appropriate specimens, properly transported and accompanied by completed requisitions
- Careful review of the lab report to determine whether the test applied answered the clinical question, and whether there is a residual component of uncertainty due to limitations of laboratory methods
- Communication of the results to the patient and discussion of the clinical significance
- Continued contact with the patient particularly in situations where future retesting is likely

Molecular and biochemical diagnostic testing often fails to identify the cause of the clinical history. This may be because:

- The patient has a different disease (wrong test applied).
- No direct test is available (biochemical analyte or gene not identified).
- The test methodology could not identify the mutation (correct test, false negative result).
- The test could not be applied to the most appropriate person (affected member of family deceased or refuses to be tested, specimen degradation in storage).
- The wrong person was tested (false paternity, inaccurate family information).

In addition, even with the correct identification of a mutation or biochemical defect, patients do not always demonstrate the expected clinical outcomes. The effects of incomplete penetrance, variable expressivity, and genetic heterogeneity introduce variables that modify the expression of signs and symptoms. For this reason, particularly for molecular diseases, it is now more acceptable to speak in terms of *variants* rather than *mutations*, and we must think in terms of likelihood, risk, or possibility, rather than viewing a test result as a guarantee of a particular outcome.

Using Molecular Tests

The clinician needs to be assured that the test methodology answers the clinical question and identifies exactly the types of residual risk that the patient has a mutation that has not been detected (discussed in detail in Chapter 13). Good laboratory reports identify the test methodology, the accuracy of the result, possible sources of inaccuracy, and provide references. Written communication confirming the test result and the genetic counseling provided is especially useful for patients, given the complexity of the information provided and the need to retain it for future health care.

When a gene-specific molecular test is not available, linkage analysis may provide clinically useful information that modifies the assessed risk of disease. Linkage analysis is an indirect approach to testing, wherein the laboratory attempts to document the inheritance of the disease by genetic markers known to be inherited in combination with the disease allele. Multiple markers that map close to or within the disease gene are used. This approach still is used for some genes even when direct analysis by sequencing is possible; for example, in large genes with a complicated spectrum of mutations (where complete sequencing is prohibitively expensive for screening), linkage analysis can provide an acceptably accurate result at greatly reduced cost. Another application is situations where multiple closely linked genes are associated with a common clinical outcome, such as X-linked mental retardation.

In summary, health care professionals:

- Can access molecular testing for many different genetic disorders, both common and rare.
- Can keep up to date with availability of new test methodologies and providers.
- Need to understand the limitations of test methodologies and interpretations.
- Need to advise the patient accurately about the clinical significance of test results.
- Need to use accredited laboratories if possible; if a specific test is only available from a research laboratory, the potential clinical usefulness of the result needs to be communicated at the outset (it should be noted that many ethics review committees in the United States do not permit the release of results from research-based testing to patients).

Using Biochemical Genetic Tests

Biochemical genetic testing almost always involves consultation with biochemical geneticists in the clinical management and/or laboratory diagnosis components of care. Clinical biochemical geneticists commonly need to be consulted on an urgent basis because of the presentation of a baby or child in apparent metabolic crisis, while laboratory biochemical geneticists are experts in biochemical and molecular diagnosis of inborn errors of metabolism.

While some biochemical disorders can be implicated by routine serum or urine analytical chemistry tests, usually additional, more esoteric testing is required for confirmation of a diagnosis. As with molecular diagnostics, there are many candidate disorders for investigation, and only a limited number of laboratories for rare disorders. Hundreds of enzymopathies have been described, including aminoacidopathies [hyperphenylalaninemias (e.g., phenylketonuria (PKU)), lysosomal storage diseases (e.g., Tay-Sachs disease), purine metabolism defects, and the mucopolysaccharidoses (e.g., Hunter syndrome, Hurler syndrome, Scheie syndrome, and Sanfilippo syndrome)]. Abnormalities of mitochondrial enzymes, enzyme cofactor metabolism, receptor proteins (e.g., hypercholesterolemia), and transport proteins (e.g., CFTR (CFTR is the protein for the CF gene) and cystic fibrosis) are also recognized. Clinicians and laboratories use analytical algorithms employing a variety of different test methodologies. A definitive review of diseases, test options, and methodologies is beyond the scope of this chapter; for further information, Scriver et al. [2001] provide a definitive reference.

Testing is oriented around biochemical pathways, such as aminoacidopathies, purine metabolism disorders, lysosomal storage diseases, receptor defects, and mutations affecting enzyme cofactor function. Many disorders result from enzyme defects that are identified by analysis of serum, urine, or tissue specimens; detection of excess enzyme substrate, lack of enzyme product, or loss of enzyme activity, are common approaches with molecular diagnostics increasingly used to confirm the presence of a mutation and provide definitive confirmation of the diagnosis. Enzyme deficiencies are almost always inherited with an autosomal-recessive pattern of inheritance, and a patient may exhibit a loss of more than one enzyme function because of functional interdependence of cofactors or organelles. The molecular specificity of a diagnosis is becoming an important consideration as an increasing number of metabolic disorders become treatable with new drugs incorporating highly specific mechanisms of action.

Biochemical testing tends to be requisitioned by test methodology because one test can be used to survey multiple components of a biochemical pathway. It is important for the clinician to be familiar with presenting signs and symptoms of these disorders, and to provide sufficient information to the laboratory to permit the appropriate selection of tests. The clinician needs to work closely with the laboratory staff in order to ensure that all diagnostic possibilities are appropriately considered and the most informative test protocols are applied. Biochemical genetic testing is often provided in an urgent context because of the severity and timing of onset of clinical symptoms.

Many biochemical disorders exhibit a pronounced founder effect for specific mutations, leading to significant variations in incidence in different ethnic populations. Examples include the lysosomal storage disorders such as Tay-Sachs disease (hexosaminidase A enzyme deficiency) and Canavan disease, both of which are found at much higher incidence in people of Ashkenazi Jewish descent (estimated carrier rates of ~1 in 27 and ~1 in 40, respectively). Thus, ethnic background is an important consideration in evaluation and testing of a patient.

WATCH OUT FOR

Ordering and Interpretation of Tests: Obligations for the Clinician

Genetic test reports can be challenging. A study reported that 43 percent of primary care physicians interviewed about genetic tests were unfamiliar with the information provided, especially test methodology and sensitivity. Of pediatric specialists 52 percent were unfamiliar with at least some of the information regarding the interpretation of test results and counseling patients about the nature and implications [Goos et al., 2004]. Risk assessment and test interpretation are discussed at length in Chapter 13.

A health care professional has a reasonable obligation [*Hall v. Hilbun*, 1985; Quillin et al., 2003] to:

- Be aware whether a genetic test has received appropriate peer review to ensure clinical validity and clinical utility [Collins, 1996; ASHG, ACMG, 1996; Welch and Burke, 1998; Yang et al., 2000; Caulfield, 2000; Khoury et al., 2000, 2004; Burgess, 2001; Haga et al., 2003, and please see Chapter 7, Informed Consent: Test Validity and Utility] (e.g., cystic fibrosis) [Richards et al., 2002].

- Be aware of clinical guidelines and accreditation standards for genetic laboratories to determine what type of specialized services are provided by a laboratory and whether the services and laboratory personnel are appropriate for the test in question [e.g., ACMG, 2004; CCMG, 2004a]. Have the standards of proficiency testing been evaluated and if so by whom?

- Be aware of the criteria for specific tests [e.g., *Creason v. Department of Health Services*, 1998], the associated risk of false-positive and false-negative results associated with a specific test and to advise the patient for the purposes of informed consent.

- Know whether the laboratory has appropriate standards for personnel and quality assurance to ensure the reliability of the testing and to safeguard the rights of patients with regard to confidentiality and informed consent [Andrews, 1997].

- Be aware of the laboratory's procedure for the processing and reporting of test results, including estimated time for processing, and the mode of and time for communication of results whether by telephone and/or by print report. Some laboratories may provide a fact sheet and information about the test results to help health professionals to familiarize themselves with the condition before speaking to patients [Abramsky et al., 2001; Goos et al., 2004].

- Have an independent system in place to track test reports to ensure that test results are returned in timely fashion.

- Be able to appropriately interpret, and appropriately communicate, test results. Referral for genetic counseling, and the specialized practice of genetic counselors, should be considered a core activity in the testing process.

- Follow up test results in a timely manner. This process should be explained to patients to enable them to know what to expect in terms of the processing of the tests, including the expected time for reporting of the test results.

- Have an effective method to report test results and to facilitate patient understanding. Although reporting by telephone may be considered normal protocol for some medical tests, as discussed in Chapter 14, a health care professional has an obligation to undertake reasonable procedure to facilitate communication and comprehension of genetic tests results. Communication of test information only by telephone and/or by letter [Tluczek et al., 1991; Phelps et al., 2004] and/or electronic means [Cepelewicz et al., 1998; Hodge et al., 1999] without access to face-to-face discussion [AMA Report on Scientific Affairs, 1999] can increase the risk of misunderstanding.

- Provide, or refer the patient for, timely, comprehensive genetic counseling for test results.

Further, with respect to private testing by commercial laboratories, the health care professional has a reasonable obligation to inquire:

- Have the testing laboratory and personnel been accredited by a recognized professional organization?

- Has a genetic test received appropriate peer review prior to introduction to routine practice [Prence, 1999] to ensure clinical validity and clinical utility [Collins, 1996; American Society of Human Genetics, 1996; Holtzman and Watson, Task Force, 1997; Welch and Burke, 1998; Yang et al., 2000; Caulfield, 2000; Khoury et al., 2000, 2004; Burgess, 2001; Haga et al., 2003]?

- What are the limitations of the testing technology for the specific test?

- What is the laboratory's rate for false positives and false negatives for the specific test?

- What is the nature of the mutation panel? What percentage of known mutations will not be detected? What percentage of clinically significant mutations will be detected? What mutation panel will be applied given the particular ethnic and racial background of the patient?

- What is the time for processing?

- What is the time for reporting results and what is the mode of communication?

- Will the company provide or refer the patient for genetic counseling?

When listed out, these obligations appear daunting. However, in practice, health care professionals who routinely requisition genetic tests quickly develop referral patterns based upon functional and reliable relationships with a selected choice of supporting laboratories.

Laboratory Performance: Obligations for Laboratories

Reporting Responsibilities

Laboratories engaged in genetic testing have a duty to the patient(s) to use reasonable care in the processes of testing as well as in the administration of tests to provide accurate information about the test results. "Incorrect and inaccurate" information was reported to parents in the case of a prenatal test for Tay-Sachs disease: The child was born with mental retardation, convulsions, loss of motor reactions, muscle atrophy, blindness, and gross physical deformity [*Curlender V. Bio-Science Laboratories*, 1980].

Laboratories also have an obligation to ensure timely reporting of results. In a recent case, a pathology laboratory failed to send a diagnosis of malignant melanoma to the requesting physician. The laboratory had no system to confirm that the physician had received the report. As a result, no action was taken. When finally removed, the tumor had evolved significantly including the lymph nodes [Carter and Winkelaar, 2001].

Another case concerned a patient who had suffered rapid weight loss in 17 days with fever, frequent urination, and thirstiness. A tentative diagnosis of diabetes was made, and that same day the physician ordered laboratory tests. Four days later, the physician received the laboratory report indicating severely elevated blood sugar readings. Although the physician immediately contacted the patient and advised that he had diabetes and should go directly to the hospital, the patient died the same day. Had the patient been hospitalized the day after the test, he would have survived. The laboratory claimed that it had reported the test results by telephone the day after the test. The physician denied knowing any result until the written report arrived. The court found that protocol at the laboratory was to telephone the results immediately and deliver an interim report as soon as one could be prepared, one day after the test. The laboratory had failed to follow its procedure and was held liable [Carter and Winkelaar, 2001].

Laboratory Accreditation, Quality Assurance, and Quality Management Protocols

Recent years have brought a renewed emphasis on laboratory licensing, accreditation, and operating standards. The goals are, in essence, to improve the accuracy of laboratory testing and to ensure that clinical service laboratories adhere to uniform operational standards. These goals are met through a variety of methods:

- Educational and training standards leading to professional certification of staff
- Mandatory continuing education
- Accreditation and quality management standards for training programs and for laboratories
- Practice guidelines issued by professional bodies
- Standardized performance testing programs

Quality in laboratory practices is not determined solely by the analytical performance of the laboratory; it includes *all* the steps of the laboratory process that contribute to the overall quality and performance [Torresani, 2003; McGovern et al., 2003b] and ensure that the patient and the referring health care professional appropriately understand the test information [Quillin et al., 2003]. Laboratories are subject to both quality management and quality assurance expectations for accreditation. Quality management programs can very specifically dictate the expectations for all laboratory protocols [e.g., the Ontario Laboratory Accreditation standards outline more than 600 assessment criteria (http://www.qmpls.org/ola/ola.html)].

Although it is recognized that genetic testing services require high standards of laboratory work [Powledge and Fletcher, 1979; Williams et al., 2003], surveys have indicated: (1) concerns with quality assurance scores that may reflect "suboptimal" laboratory practices [McGovern et al., 1999, 2003a, 2003b; Grody and Pyeritz, 1999; Dequeker et al., 2001; Quillin et al., 2003] and (2) concerns about appropriate standards for personnel and quality assurance that ensure the reliability of the testing and safeguard the rights of patients with regard to confidentiality and providing informed consent [Andrews, 1997a].

Accreditation standards vary from country to country. As discussed in more detail below (SA Adams et al.), American service laboratories adhere to standards and certifications determined by Clinical Laboratory Improvement Amendments (CLIA), College of American Pathologists (CAP) and professional bodies such as the American Board of Medical Genetics, the American College of Medical Genetics, and a variety of other bodies (see also below). Other countries have similar requirements for both professional and laboratory accreditation. For example, in Canada, laboratories participate in the QMP-LS program (Quality Management Program—Laboratory Services, www.qmpls.org/) and/or the CAP accreditation program. In the United Kingdom, laboratories participate in the NQAS (National Quality Assurance Schemes), and Australia has the Cytogenetic & Molecular Genetic Quality Assessment Committees functioning as part of the Human Genetics Society of Australasia (www.hgsa.au).

ADDITIONAL CONSIDERATIONS

Use of Research Testing in Clinical Management

Clinical service laboratories have to be staffed and operated according to standards that are not applicable to research laboratories. Therefore, it usually is preferable to use a clinical service lab, as opposed to a research lab, if a clinical lab provides the required test. Difficulties can arise, however, when a test is only available from a research lab (as in the case for many molecular and esoteric biochemical tests). Although many research laboratories operate to very high standards, there is an increased risk that results from a research laboratory may be delayed, never released, inaccurate (especially if the test is still under development), or reported without appropriate interpretation and communication. These possibilities need to be considered before submitting specimens.

Also, it is increasingly common in the United States for ethics review panels to prohibit the release of results based upon research testing of specimens. Paradoxically, in Canada it is commonly accepted that there is a moral obligation to provide patients with any results from a research project that may have clinical significance for a patient. Whether or not the results are in fact made available to the patient, at the outset of testing the most important consideration is certainly to ensure that the patient is offered a completely informed choice before participating in the proposed research.

Use of Archived Specimens

It is common for clinicians and patients to wish to apply new tests to old specimens. However, if critical family members have become lost to follow-up, a tumor has been successfully treated, or a pregnancy lost, then hopes for getting an answer may rest with specimens of tissue, DNA, or RNA (ribonucleic acid) remaining in storage in a laboratory somewhere. It is important to note that this approach can be fraught with difficulties:

- The specimen may have degraded in storage (e.g., fragmentation of DNA in paraffin-embedded histology specimens; desiccation of formalin-fixed tissues; improper storage conditions).
- The specimen may be inaccurately identified or lost.
- The test may require a type of tissue that is not available (e.g., cytogenetics may require viable tissue; RNA is stable only for 2 years at most in frozen storage).
- There may be insufficient specimen (e.g., cytological specimens, small biopsies).

Consent to Store Specimens Compared to Consent to Test

Patients may request storage of a specimen without testing; usually, this is because the person does not wish to have testing done but does wish to ensure that other family members can make use of genetic testing in future for their own purposes. Instructions in regard to testing or specimen storage need to be clearly identified to the laboratory, and in particular, it is important for the counselor to communicate to the patient how long the specimen can be stored in a stable, useful form for the testing intended. The laboratory might consider performing testing immediately but not reporting the results to the patient, but this raises issues with regard to a failure to report information that may be clinically significant for the patient. It is important for the patient to clearly identify whether testing or storage is requested and for the laboratory to comply with the patient's expressed instructions.

Limitations of Test Methodologies

Genetic test reports should clearly identify the limitations of the methodology applied. Common limitations include:

Cytogenetics

- Inadequate specimen quality or quantity, including tissue contamination (especially, contamination of fetal specimens by maternal cells, which will remain unrecognized if the fetus is female)
- Structural anomalies not detectable at the level of resolution achieved
- Mosaicism for an abnormal cell line that is not detected within the number of metaphases analyzed

Molecular and Biochemical Genetics

- Inadequate specimen quality or quantity, including tissue contamination (especially, contamination of fetal specimens by maternal cells, which may remain unrecognized unless specifically ruled out by use of molecular identity testing with a control maternal specimen)
- Exact description of gene(s) or specific regions (exons) tested
- Accuracy of test method (proportions and types of mutations detected and not detected)
- Limitations to interpretation of the result
- Dependence upon accurate family history, including possibility of false paternity if relevant to test interpretation
- Possibility of specimen labeling errors or laboratory errors

Clinical Significance of Test Results

All reports for genetic tests should clearly provide a component of interpretation for the clinical significance of the results. For example, an abnormal cytogenetic result from an amniocentesis might refer to the likelihood of a particular type of diagnosis or an increase in risk for developmental anomalies, while a molecular test result might specifically identify the status of a patient as a carrier (if a mutation is found) or residual risk that the patient is not a carrier for a mutation (if no mutation is found, as discussed in more detail in Chapter 13). The interpretation must be written at a level that is understandable to nonspecialist health care professionals, and it is common practice to identify situations where the result suggests a need for follow-up genetic counseling and management.

Process Issues

Laboratory Requisitions: Legal and Medical Importance

Laboratory staff view the test requisition form as a critical requirement for initiating testing. The requisition identifies the patient, the date, the indications for testing (and therefore the test methodology), and the referring physician, and, moreover, the requisition provides an identifiable and documented link between the patient, the specimen submitted, and the lab. National standards exist for the inclusion of information on the requisition (see below).

Changing Indications for Testing

Indications for requesting a test may change with time and it is important to follow changes in diagnostic standards. Laboratory staff can be very helpful in identifying acceptable indications for referring a specimen for testing, and many offer websites and resources with constantly updated information.

Implementation of Test Results in Clinical Management

The implications of a test result for decisions on clinical management may not be immediately obvious. Again, laboratory staff and the health care professional need to work together to ensure that the result is neither over- or underinterpreted in regard to potential clinical significance. While it is not possible for laboratories to routinely confirm that a clinician has in fact received important lab results and acted appropriately upon receipt of the information, nevertheless many laboratories do utilize genetic counselors who contact clinicians in regard to particular lab results to ensure that the importance of the test is recognized and appropriate counseling is provided.

It is common practice for laboratory staff to document the send-out of a result as part of a quality management protocol. It is important for the health care professional to review the contents of a report and act in a timely fashion when required. One common source of difficulty, particularly in prenatal care, is for clinical staff to file lab reports away for the patient's next scheduled visit without careful review; if the report identifies an abnormality that requires urgent further assessment, this practice may result in an unnecessary delay in care (and see Chapter 6, Referral and Diagnosis). For example, even years after the introduction of maternal serum screening panels, a significant proportion of screen-positive patients still learn of the abnormal result only at their next routine visit, which can result in reflex amniocentesis testing being performed in the 19th, 20th, or 21st weeks of pregnancy, an uncomfortably late point for deciding on intervention.

Obsolete Technical Standards and Repeat Testing

As analytical standards for diagnostic laboratories continue to improve, it follows that a negative result reported in years past may represent inadequate information in the light of new techniques. It is not practical for laboratories to initiate requests for repeat testing. It is important that health care professionals are familiar with changing laboratory capabilities and routinely revisit the adequacy of previous lab results. For example, a patient thought to have Prader-Willi syndrome in the early 1990s most likely would have received only a routine G-banding karyotyping investigation, whereas microdeletion FISH and molecular testing protocols would add a much higher diagnostic standard to the same clinical question today. Similarly, new test methodologies are used to improve the diagnostic accuracy of molecular test panels; linkage tests are replaced by direct mutation analyses; and direct mutation assays become more accurate and complete over time. Clinicians need to consider repeat testing on a regular basis for patients whose clinical diagnosis has escaped laboratory confirmation in the past or could benefit from refined technical assessment.

Health Care Systems and International Laboratory Practices

In the United States, genetic testing is provided by a few large commercial laboratories and a large number of smaller commercial and academic laboratories. Because the cost of almost all testing is recovered by billing insurance plans, health maintenance organizations (HMOs), or patients, laboratory services are competitive and can be operated on a for-profit or not-for-profit basis. For this reason, American laboratories are very responsive to clinical expectations for rapid reporting times and communication. In contrast, almost all genetic testing in Canada is performed by nonprofit laboratories operated by the public health care system and most often located in tertiary health care facilities. Because these labs are usually funded by fiscal allocations rather than fee-per-service, most Canadian laboratories are primarily constrained to operate at maximum cost efficiency; this results in reporting times that are more variable and often delayed (for nonurgent indications) in comparison to American services. Similar variations in operating parameters for service laboratories can be found in other countries. Clinicians have to work within prevailing patterns of practice.

COMPLEXITIES IN ORDERING AND INTERPRETING GENETIC TESTING

S. Annie Adams, Daniel L. Van Dyke, Rhett P. Ketterling, and Erik C. Thorland
Mayo Clinic, Cytogenetics Laboratory, Department of Laboratory Medicine and Pathology, Rochester, Minesota

INTRODUCTION

The collection and submission of specimens for genetic testing involves several critical steps. Laboratories require the correct specimen type to be sent in the proper storage and transport media and delivered in a timely manner. Specimens must be identified clearly and shipped with a detailed reason for referral, including clinically relevant information regarding the test being ordered. When appropriate, a patient informed consent form should also accompany the specimen. This section reviews how clinicians can best order, utilize, and interpret the laboratory results from genetic testing.

CONSIDERATIONS BEFORE ORDERING GENETIC TESTING

Genetic testing is increasingly regarded as an integral part of patient care and management in medicine today. However, ordering appropriate genetic testing can be a complex process that requires significant consideration of a variety of ethical, legal, and practical issues by both the ordering physician and the patient. As is true for any high-complexity test, laboratory testing for genetic disorders should only be ordered by individuals with the expertise to appropriately counsel the patient, interpret the test results, and recommend proper follow-up for both the patient and extended family. In some situations, this testing can be ordered appropriately by primary care physicians or nongenetic specialists. In more complex cases it may be prudent to refer patients for a genetics consultation prior to testing. Genetic testing is unique in that test results may have significant repercussions for family members of the patient and may raise more complex psychosocial issues than other test types typically entail. Due to these factors, appropriate pretest counseling and informed consent from the patient are critical before initiating testing.

AVAILABILITY OF GENETIC TESTING

The genes responsible for various clinical disorders are being discovered at an ever increasing pace. Therefore, the availability of genetic testing for these disorders is also increasing at a rapid rate. Over the course of the past 5 years, the number of diseases for which genetic testing is available has increased approximately 40 percent from 774 to 1087 [Genereviews at Genetests, 2004]. However, testing is not available for all genetic disorders. When testing is available, clinicians are frequently confronted with a variety of testing options, making it difficult to determine which, if any, genetic testing is appropriate (Fig. 12.1). Care must be taken to consider all options before ordering a genetic test or group of tests to ensure that testing is ordered in the most efficient manner. For example, testing may be part of a complex algorithm and involve more than one test in more than one laboratory. Most genetic tests are focused on highly specific genes or disorders, thus careful clinical evaluation is essential to define the precise disorder or set of clinical features to target testing. Although genetic testing may be available for a particular disorder, if a diagnosis is clear based on phenotype or other clinical data, laboratory testing may not be necessary or even appropriate. In addition to genetic testing for germline abnormalities, testing is also available for acquired abnormalities in neoplastic disorders. Extensive literature and websites are available that detail recurrent abnormalities in hematologic disorders and solid

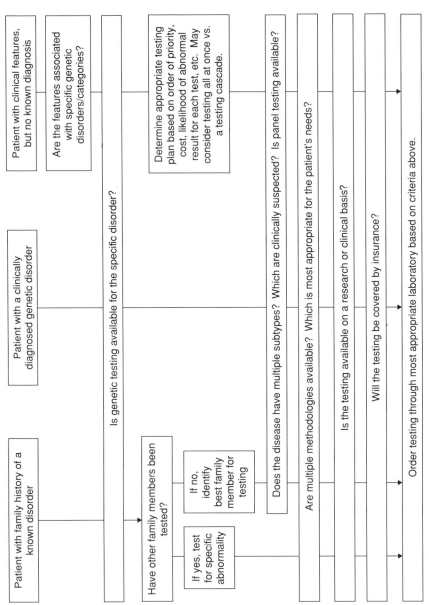

Figure 12.1 Choosing the appropriate genetic testing.

Table 12.4   Web-Based Resources for Requisitioning Genetic Tests

Name	Website	Purpose
OMIM [McKusick, 2002; OMIM, 2004]	http://www.ncbi.nlm.nih.gov/omim/	Clinical features, disorders, gene structure and function
GeneTests/GeneReviews	http://www.genetests.org/	Clinical features, disorders, differential diagnoses, laboratory testing directory
European Directory of DNA Diagnostic Laboratories (EDDNAL)	http://www.eddnal.com/	Laboratory testing directory
Atlas of Genetics and Cytogenetics in Oncology and Haematology	http://www.infobiogen.fr/ services/chromcancer/index.html	Recurrent rearrangements in neoplasms

tumors [*Atlas of Genetics and Cytogenetics in Oncology and Haematology*, 2004; Dewald, 2002]. Once a decision is made to initiate genetic testing, the availability of testing should be determined. Several Web-based resources are extremely useful to determine the availability of genetic testing (Table 12.4).

WHO SHOULD BE TESTED?

Genetic testing can be utilized for a variety of reasons. These include (1) confirmation of a clinical diagnosis, (2) rule out a particular genetic disorder as part of a differential diagnosis, (3) presymptomatic testing when there is a family history of a disorder, and (4) carrier testing due to family history of a disease or high carrier frequency in a specific population. Care must be taken to ascertain the appropriate person in the family in whom to initiate testing. When there is no family history, the affected individual will generally be the first in the family to seek medical attention and should be tested first, followed by testing of other family members if appropriate. However, when there is a positive family history, often an unaffected relative may request presymptomatic or carrier testing. In this situation, an affected individual should be tested to identify the familial mutation prior to testing other family members. The sensitivity of most genetic tests is not 100 percent, therefore, testing an asymptomatic individual without prior knowledge of the familial mutation may yield a relatively uninformative or misleading test result. In this situation, a negative test result is particularly problematic since it cannot be determined whether the individual truly does not carry the familial mutation or if the test was unable to detect the mutation, leading to a false-negative result. Misinterpretation of such results can lead to disastrous consequences and illustrates the importance of testing individuals within a family in a hierarchical fashion. For example, if an individual presents with a family history of severe hemophilia A and desires carrier testing, an affected family member should be tested first to identify the precise mutation in the family. This would allow carrier testing for that specific mutation in unaffected family members. An inversion mutation, detectable by Southern blot analysis, accounts for nearly half of all severe hemophilia A [Genereviews at Genetests, 2004]. Most other patients with hemophilia A have unique mutations that are detected by scanning or sequencing methods. Many laboratories only test for the inversion. In this example, if the patient were tested for the inversion without having previous family studies performed, a negative result would not rule out the possibility that the hemophilia in the family was caused by a less common mutation. However, if an affected family member were known to have the inversion, a negative test result would be fully informative for this patient.

TYPES OF GENETIC TESTING

Genetic diseases are associated with a wide variety of genetic alterations that range from mutation of a single DNA base pair to gains or losses of whole chromosomes. Thus, the specific biology of a genetic disorder and the spectrum of mutations associated with it govern the type of laboratory and specific testing methods employed to detect these alterations (Table 12.5). Molecular genetic testing typically examines

Table 12.5   Examples of Genetic Testing Methodologies and Their Utility

Category	Testing method	Utility
Molecular genetic	PCR based	Screening for single base pair changes or gene sequencing
	Southern blot	Analysis of deletion, duplication, dosage of large gene regions
Molecular cytogenetics	Fluorescence in situ hybridization (FISH)	Microscopic evaluation of specific chromosome regions using specific probes
Cytogenetics	Chromosome analysis	Microscopic evaluation of individual chromosomes
Biochemical genetics	High-pressure liquid chromatography (HPLC)	Separation, identification, and quantification of various compounds, such as amino acids and porphyrins
	Manual and electrophoretic methods	Quantitative analyses of enzymes and analytes such as biotinidase and galactose-1-phosphate; qualitative separation and detection of compounds and isoforms including acetylcholinesterase
	Gas chromatography/mass spectrometry (GC/MS)	Separation, identification, and quantification of compounds such as organic acids, acylglycines, and fatty acids
	Tandem mass spectrometry (MS/MS)	Rapid separation, identification, and quantification of compounds including acylcarnitines, carnitine, purines, and pyrimidines; commonly used for newborn screening

DNA for submicroscopic gene alterations ranging in size from a single base pair to several thousand base pairs of DNA. Cytogenetic testing examines chromosomes for larger scale rearrangements such as chromosomal deletions, duplications, inversions, or translocations, although the banding resolution can vary depending on sample quality, tissue type, and culture method. Traditional chromosome analysis can detect these types of alterations if they affect regions on the scale of millions of base pairs, whereas molecular cytogenetic techniques such as fluorescence in situ hybridization (FISH) may detect similar alterations on the scale of a few thousand base pairs. Biochemical genetic testing typically analyzes the function or activity of proteins rather than chromosomes or genes. This testing is performed via both qualitative and quantitative methods for detection of endogenous metabolites that abnormally accumulate in body fluids and tissues, allowing for the evaluation and detection of inborn errors of metabolism.

Although testing for many genetic disorders involves a single test using a single methodology in a single laboratory, testing for other disorders can be more complicated. Complex testing may involve multiple sample types, laboratories, and test methods that are incorporated into testing algorithms. For example, an algorithm for diagnosis of Prader-Willi syndrome might include methylation testing by Southern blot analysis to make the initial diagnosis, followed by FISH testing for deletions of 15q11.2-q12, microsatellite analysis of the patient and parents for uniparental disomy studies, and/or chromosome analysis to determine the genetic mechanism of disease for genetic counseling purposes [Monoghan et al., 2002]. In contrast to performing multiple tests for a single disorder, some disorders are genetically heterogeneous and have multiple clinically indistinguishable subtypes. In these cases, multiple tests may be necessary to identify abnormalities in several genes associated with the phenotype. In these situations, panels that include multiple gene tests may be an option. For example, at least 20 clinical subtypes of spinocerebellar ataxia (SCA) have been defined, of which at least 10 have clinical testing available [Genereviews at Genetests, 2004]. The clinical phenotypes of various SCA types are difficult to differentiate, so it is most efficient to test for multiple subtypes at one time. Thus, many laboratories have established panels that include several SCA-associated genes in one orderable test.

CHOOSING A TESTING LABORATORY

Since similar testing is available in multiple different laboratories, selection of an appropriate laboratory for testing can be difficult. Hundreds of genetic tests are available; thus, no single laboratory can offer comprehensive genetic testing. In addition, some laboratories hold exclusive rights to testing certain disorders in the United States. As a result, patent restrictions limit the test menu in many reference laboratories. Many aspects of genetic testing should be considered when selecting a laboratory, such as the reason for testing, methodology, mutation detection rate, and turnaround time. Different methodologies may be appropriate depending on the reason for testing. For example, it may be sufficient to test for a limited number of common mutations when performing carrier testing for cystic fibrosis (CF). However, complete gene sequencing may be an appropriate reflex test in a patient suspected of having CF if panel testing has not identified two mutations. In addition, methodology greatly influences mutation detection rates. For example, testing for deletions and duplications in Duchenne muscular dystrophy (DMD) will detect abnormalities in approximately 65 percent of patients with the disease, whereas adding additional mutation scanning or sequencing methods detect point mutations, which account for an additional 30 percent of disease-causing mutations [Genereviews at Genetests, 2004]. Testing methods also influence turnaround time. Polymerase chain reaction (PCR) techniques used for the analysis of a small number of mutations typically can be performed in a few days, versus Southern blotting techniques, which may take 1 and 2 weeks to perform. Interphase FISH studies are also routinely used for rapid detection of specific numeric abnormalities in prenatal and neonatal settings, as these studies may have a turnaround time as rapid as a day, whereas high-resolution chromosome studies take at least several days on blood specimens and over a week on prenatal specimens. Turnaround time may be an especially important variable for testing in a prenatal or neonatal setting when urgent medical or surgical intervention may be required.

Other questions should also be considered when choosing a laboratory. Insurance coverage for genetic testing is highly variable between individual policies, states, and countries. Therefore, prior investigation of a patient's insurance plan is prudent to determine whether testing will be covered. Certain genetic tests may only be available on a research basis, which may have a significant impact on turnaround time. In addition, while some laboratories may offer testing for a certain disorder, they may not include prenatal testing as an option. Therefore, if prenatal testing is a possibility for follow-up testing, a laboratory that performs prenatal testing should be chosen for the initial analysis. Lastly, the ethnicity of the patient should be taken into account. There may be wide variability in the mutation detection rate of certain disorders due to founder mutations within certain populations. For example, Tay-Sachs disease is most commonly found in individuals of Ashkenazi Jewish or French Canadian origin; thus, testing has been developed to specifically target these populations. However, the spectrum of mutations observed in patients of other ethnic backgrounds may be quite different from the common mutations in these two populations. Therefore, ethnic origin becomes a prime consideration in choosing the laboratory and methodology necessary to maximize patient benefit. Overall, careful planning and consideration of a number of factors prior to choosing a laboratory will ensure that appropriate and timely testing is performed.

UTILIZING NEW TECHNOLOGIES

Technical advances are driving change in the methodologies available to perform genetic testing, leading to gains in the speed, sensitivity, and throughput of genetic testing. When considering novel technologies, it can be difficult to determine their utility and justify cost as compared to more commonly utilized techniques. In addition, variables such as turnaround time, interpretation of new types of data, and potential follow-up difficulties should all play a role in the decision-making process. New techniques that offer more comprehensive genomewide testing, such as Comparative Genomic Hybridization (CGH) microarrays, are becoming available on a clinical basis. Advantages to this testing can depend on the patient being evaluated. For example, for patients in whom a specific clinical diagnosis is suspected, CGH microarray testing may be too comprehensive and expensive compared to more targeted testing by routine techniques. However, patients with nonspecific features and no suspected or established clinical diagnosis may benefit from a single global analysis rather than ordering several independent tests. These novel technologies also have potential pitfalls. CGH microarrays may not be adequate to provide evidence of the structural basis of rearrangements, so follow-up FISH and/or chromosome studies may be necessary. Global analysis of multiple loci may lead to unexpected or uninterpretable results. While this is generally not a problem with the CGH microarrays currently in clinical use, higher resolution genome arrays in the future have a much greater likelihood of uncovering novel, undescribed genomic alterations that may or may not be responsible for the clinical features in the patient. This type of alteration may require further characterization and by traditional techniques in both the patient and additional family members to differentiate between a polymorphic familial variant and a clinically significant mutation. Although global approaches to screening for genetic alterations may seem attractive, they should be utilized with caution due to limitations in the amount of information they can provide. CGH arrays are capable of detecting dosage differences in the genome but cannot detect balanced chromosome abnormalities, small gene-specific molecular changes, or metabolic disorders.

ORDERING PROCESS

Once a specific laboratory has been chosen for genetic testing, there are typically sample submission guidelines available that, if carefully adhered to, will ensure that the appropriate information and samples are forwarded to the testing facility and that testing will proceed efficiently with minimal delay. In most instances, the ordering process can be initiated through the referring physician's hospital laboratory. Many hospital laboratories have existing relationships with genetics laboratories. However, if testing is desired from a different laboratory, it is useful to contact the performing laboratory directly to order the testing on forms that are specific to that laboratory. Contact information and order forms are often available online and can be accessed through the online laboratory test directories (Table 12.4).

PROVIDING PHYSICIAN AND PATIENT INFORMATION

To ensure that appropriate, informative testing is performed, several key pieces of information need to be communicated to the performing laboratory through the order form. The American College of Medical Genetics (ACMG) has established guidelines for patient information that must accompany the specimen [American College of Medical Genetics, 2002]. It is critical for performing laboratories to receive the ordering physician's name and contact information in the event that they would need to request more clinical information, recommend additional testing, or discuss complex results. A detailed reason for referral, which should include suspected diagnosis, clinical features, and family history, must also be included. This enables the laboratory to determine why the test is being ordered so that the specimen can be appropriately processed and tested. For example, if a blood specimen is received for chromosome analysis on a 5-year-old with no reason for referral, the laboratory will not know whether to process the specimen to test for congenital chromosome abnormalities or for acquired abnormalities associated with hematologic disorders. A detailed clinical description of the patient is also helpful in guiding focused analysis. For example, if a chromosome analysis is ordered on a female patient with short stature and neck webbing, a cytogenetics laboratory may analyze additional cells to rule out sex chromosome mosaicism, which is common in Turner syndrome. However, if no clinical description were provided, the laboratory would process the specimen as a routine case. Family history is another critical component to the reason for referral. In familial disorders, such as inherited colon cancer predisposition syndromes, detailed pedigree information and prior test results of family members allows targeted testing for the appropriate syndrome, gene, and specific mutation causing the disease in the family. In summary, having complete information benefits the patient, makes laboratory testing more efficient, and minimizes unnecessary physician contact.

SAMPLE REQUIREMENTS

Genetic testing can be performed only if the appropriate sample is provided to the laboratory. Maintaining specimen integrity requires that several variables, such as temperature and transport time, be closely controlled (Table 12.6). In the United States, the College of American Pathologists (CAP) requires that a specimen collection manual be available either in print or electronically to clinicians and specimen collection facilities [CAP, 2004]. Various methods used by laboratories may limit the specimen types that can be used for testing. Thus, it is necessary to investigate the specimen requirements prior to collecting and shipping the specimen. For example, chromosome analysis is performed on cells in metaphase, so samples with viable, dividing cells are necessary. Many anticoagulants are available for blood and bone marrow collection, but the anticoagulation requirements are very specific for genetic tests and differ based upon the methodology used (Table 12.6). For example, blood collected in sodium heparin is desired for chromosome analysis but is problematic for most molecular assays because heparin may inhibit enzymatic reactions, such as those used in PCR. Testing requirements may also involve requesting specimens from additional family members. Specifically, molecular genetics laboratories often require that a maternal blood specimen accompany prenatal specimens to test for maternal cell contamination. Tests that analyze the inheritance pattern of polymorphic loci,

Table 12.6   Specimen Collection Information[a]

Testing	Specimen type	Collection media	Shipping temperature	Delivery requirements
Congenital chromosome studies	Whole blood	Sodium heparin	Ambient	48 h
Hematologic chromosome studies	Whole blood, bone marrow	Sodium heparin	Ambient	48 h
Prenatal molecular or cytogenetic studies	Amniotic fluid, chorionic villi	For villi: sterile tissue culture media	Refrigerated	Overnight
Chromosome studies on products of conception	Chorionic villi or fetal tissue biopsy	Hanks or other sterile tissue culture media	Refrigerated	Overnight
FISH or DNA testing for lymphoma or solid tumors	Paraffin-embedded tissue	N/A	N/A	N/A
DNA testing	Whole blood	EDTA or ACD-B	Ambient	48 h
RNA testing for solid tumors	Fresh frozen or paraffin-embedded tissue	N/A	N/A	N/A
RNA testing for hematologic disorders	Whole blood or bone marrow	EDTA	Refrigerated	Overnight
Biochemical genetic studies	Varies, i.e., urine, plasma, serum, skin biopsy	Varies by sample type	Varies by sample type	Varies by sample type

[a] May vary depending on laboratory; test list is not all-inclusive.

such as linkage studies or uniparental disomy (UPD) testing, may require specimens from parents or additional family members to appropriately track alleles through generations.

If specific specimen requirements are not researched prior to obtaining and sending the specimen for analysis, incorrect testing may result. For example, if testing for Morquio disease is desired, enzyme analysis is available for disease subtypes A and B. However, the specimen requirements differ for the two subtypes: Testing for type A requires a skin biopsy for fibroblast culture, whereas testing for type B can be performed on leukocytes. If Morquio type A is suspected but a laboratory receives a blood specimen with a reason for referral of Morquio syndrome, testing for type B might be performed in error.

REPORTING AND INTERPRETING RESULTS

Results of genetic testing can be reported in a variety of formats depending on the type of alteration detected. Nomenclature recommendations have been established for both molecular genetics and cytogenetics in an attempt to standardize reporting between laboratories. To assure consistent descriptions of molecular gene alterations, a system has been proposed to accurately and unambiguously describe mutations at the DNA level [Antonarakis 1998; HGVS, 2002]. In addition, the International System for Human Cytogenetic Nomenclature (ISCN) is a periodically updated set of guidelines established so that complex cytogenetic results can be described in a concise and reproducible manner [ISCN, 1995]. While the ISCN guidelines are well-developed for routine cytogenetics, the description of FISH results can be quite complex and are not yet adequately addressed by this nomenclature [Mascarello et al., 2003]. Results of biochemical genetic studies are typically reported as discreet values compared to a reference range. However, reference ranges and units can vary significantly between laboratories.

A result by itself may have several meanings depending on the context of the clinical indications. Therefore, a result should not be read without clarification in the context of an appropriate interpretation. The interpretation section of a genetic test report should include a clear description of the result, implications for the patient and/or family members, and test limitations. This section is essential to understanding the implications of the result.

Intuitively, a negative result should be the easiest to interpret. However, this type of result often requires the most complex interpretation. Negative results can be interpreted with high confidence in situations where a familial mutation has previously been identified. In these cases, the detection rate for a specific mutation is essentially 100 percent since the assay has previously been demonstrated to be capable of detecting the mutation. In contrast, due to technical limitations and incomplete understanding of the genetics of many disorders, the mutation detection rate is usually less than 100 percent. For this reason, a negative result often does not completely rule out the presence of a genetic abnormality. Patient ethnicity has a significant impact on mutation detection rates in some disorders and therefore affects the interpretation of a negative result. Because of the high cystic fibrosis (CF) carrier frequency in the general U.S. population, the American College of Obstetrics and Gynecology and the American College of Medical Genetics have published guidelines recommending that individuals considering reproduction be offered CF carrier testing for a panel of 25 common mutations [Grody et al., 2001; McKusick, 1998]. While most laboratories include the recommended 25 mutations, there is extensive variability regarding the number of mutations tested for between laboratories [Richards et al., 2002]. The majority of patients who have carrier testing will have a negative result, which reduces, but does not eliminate, their risk of being a carrier. The interpretation of negative results depends on the patient's ethnicity and involves statistical calculations that consider the ethnic group's carrier frequency and the mutation detection rate for the panel of mutations detected by the assay. Individuals of Ashkenazi Jewish background have a higher carrier frequency than individuals of many other ethnicities. Thus, using the recommended test panel, mutation detection rates in individuals of Ashkenazi Jewish ancestry versus someone of European Caucasian ancestry is approximately 97 percent versus approximately 80 percent. Although both groups have a population carrier frequency of 1/29 for all CF mutations, a negative result in an individual of Ashkenazi Jewish ancestry will reduce his or her risk to 1/930, whereas the same test for a patient of European Caucasian ancestry will reduce his or her risk to 1/140 due to the lower detection rate in the latter population [Grody et al., 2001].

When interpreting normal chromosome studies, a negative result must also be considered within the context of the test method. One must realize that the resolution of the chromosome analysis may not be sufficient to rule out subtle abnormalities or low-level mosaicism and certainly is not high enough to rule out other molecular genetic alterations. For example, Williams syndrome is caused by a recurrent 1.5-megabase deletion on chromosome 7 [OMIM, 2000]. This deletion is small enough that it cannot be detected by routine chromosome analysis. Therefore, FISH analysis must be performed to detect the deletion. Finally, when interpreting negative results, the result is highly specific to a particular gene or disorder. Testing does not exclude other genetic or nongenetic causes of a patient's clinical features.

Positive results are often more straightforward to interpret than negative results. If a specific disorder is suspected based on clinical features, a positive result can confirm the clinical diagnosis and provide information for prognosis, treatment, and counseling. Positive results after carrier screening are straightforward for the patient being tested, although this result may initiate additional steps such as testing the patient's partner, discussing prenatal testing, or following up with additional family members. However, interpretation of positive results can also be more complex. In patients with a family history of certain adult-onset disorders, a positive result may not necessarily predict with certainty whether the patient will develop the familial disorder or when symptoms will manifest. The predictive value of these tests is highly dependent on the penetrance of the clinical features. For example, patients with a family history of Huntington disease (HD) often pursue presymptomatic testing. In these individuals, a positive result is highly correlated with the future onset of symptoms [Genereviews at Genetests, 2004]. Thus, the predictive value of this test is high with regard to penetrance, but age of onset is more difficult to predict. In contrast, testing for hereditary hemochromatosis (HHC) is not necessarily predictive of disease. Approximately 1/10 Caucasians is a carrier of an *HFE* gene alteration, thus approximately 1/200 individuals carries two abnormal alleles [Pietrangelo, 2004a]. However, only a fraction of homozygotes have or will develop clinical symptoms of HHC, indicating that the penetrance of this genotype is very low. Thus, identification of two abnormal alleles in a patient should be interpreted in the context of clinical features.

Unfortunately, not all results are unambiguously negative or positive. Equivocal results are often encountered when a particular alteration has not been observed previously. As mutation scanning and sequencing technologies improve, data is produced at increasing rates. The analysis of this additional data inherently generates previously unidentified sequence changes. Novel nonsense and frameshift alterations are generally relatively easy to interpret; however, missense, splicing, silent, or intronic changes present unique difficulties [Antonarakis, 1998; HGVS, 2002]. In these cases, it is often difficult to predict whether the alteration has functional significance or is a polymorphism or rare variant. Several steps can be taken to clarify these results, but often these questions remain unsolved until additional data becomes available. In some cases, additional functional studies can be performed in an attempt to decipher whether the alteration affects protein function, although these studies are generally not available on a clinical basis. In addition, analysis of the evolutionary conservation of the altered amino acid may provide clues to the functional significance of an amino acid change. However, these analyses are not sufficient to predict significance with confidence. Testing can also be performed on both affected and unaffected family members to determine whether the alteration segregates with the disease.

Equivocal results are also a possible outcome of chromosome studies. For example, chromosome analysis may indicate that a patient has an apparently balanced translocation. Although this is an abnormal result, the interpretation depends on the clinical situation. In a child with developmental delays or multiple congenital anomalies, such a chromosome result raises additional questions. Balanced translocations are relatively common in the general population and typically do not cause clinical features. However, the resolution of chromosome analysis is not sufficient to determine whether such a translocation is truly balanced. This translocation could explain the child's features if a submicroscopic deletion or other rearrangement were present at the translocation breakpoints, or if the breakpoint disrupted an important gene or perturbed gene regulation in the region. Chromosome analysis of the parents can help to determine whether the translocation is familial and likely to be balanced. Identification of the translocation in a clinically normal parent makes a causal relationship between the translocation and clinical features in the child unlikely. This raises other issues for the family because a balanced translocation may confer an increased risk for infertility, miscarriages, and future children with unbalanced chromosome complements. Therefore, the family should be counseled about these risks and the extended family should be evaluated to determine who else may carry the translocation.

Chromosome analysis for hematologic disorders can cause equally challenging interpretive situations. For example, analysis of chromosomes in a child suspected to have an acute leukemia without other clinical information provided may reveal an extra copy of chromosome 21. This result can have multiple interpretations. Trisomy 21 can be seen as an abnormal acquired clone in various hematologic disorders. However, from a congenital perspective, trisomy 21 causes Down syndrome. Patients with Down syndrome are at an increased risk for developing hematologic malignancies, therefore when trisomy 21 is observed in a hematologic chromosome analysis, Down syndrome is not an unlikely explanation. Therefore, one must always consider constitutional explanations for abnormalities found when performing chromosome studies for hematologic disorders. In these situations, additional testing of a stimulated T-lymphocyte culture from a blood specimen may be required to differentiate between these two possibilities if the question cannot be resolved based on clinical features.

QUALITY SYSTEMS IN THE LABORATORY

A quality system is critical for promoting development, implementation, and improvement of laboratory processes. In 1988, the U.S. Congress passed the Clinical Laboratory Improvement Amendments (CLIA), which established quality standards for all laboratory testing [CLIA, 2003]. This amendment provides guidelines to ensure consistently accurate, reliable, and timely test results in laboratories across the country. CLIA created a government agency called Centers for Medicaid and Medicare Services (CMS), which regulates all clinical laboratory testing performed on humans in the United States [CLIA, 2004]. CMS allows the inspection of laboratories by approved accrediting organizations, such as the College of American Pathologists (CAP). CAP is a private not-for-profit accrediting organization that regulates and reviews laboratories through voluntary participation, professional peer review, education, and compliance with CLIA guidelines [CAP, 2004]. The Joint Commission on Accreditation of Healthcare Organizations (JCAHO) is an independent, not-for-profit organization that currently inspects and regulates quality systems within hospitals and laboratories [JCAHO, 2004]. Individual states may develop more stringent laboratory requirements that must be followed by any laboratory performing testing on samples from those states. In addition to state and federal regulations, the ACMG publishes standards and guidelines for clinical genetic testing[ACMG, 2002].

The CAP certification and review process ensures certain minimum criteria be met for a laboratory to be certified to perform clinical testing. These criteria include continual monitoring of quality indicators such as turnaround time, success rates, and abnormality rates. These regulations also provide guidelines for reporting and interpreting test results. Reports are required to include information identifying the performing laboratory, the patient, and the ordering physician; specimen type and collection information; clinical indications for the test; and specific result and interpretations depending on the type of testing [CAP, 2004]. Reporting of results must follow standard nomenclature recommendations [Antonarakis, 1998; HGVS, 2002; ISCN, 1995]. Interpretations must include a summary of results, correlation with previous studies, and clinical implications, such as penetrance, severity, and genotype/phenotype correlation [CAP, 2003a, 2003b]. In addition, associated risk estimates, testing limitations, and recommendations for genetic counseling must all be included as necessary [CAP, 2003a, 2003b].

POSTTEST RECOMMENDATIONS

The ordering physician is responsible for interpretation and follow-up on genetic test results in an appropriate manner. Newborn screening provides an example of the complexities encountered following receipt of genetic testing results. Tandem mass spectrometry (MS/MS) is a powerful multianalyte screening method commonly utilized for population-wide screening for many metabolic disorders using blood spots collected shortly after birth. Newborn screening using MS/MS is not diagnostic but identifies children who are more likely to have certain metabolic disorders. A positive screen result necessitates follow-up testing. For example, medium-chain-acyl-CoA dehydrogenase (MCAD) deficiency is a disorder of fatty acid oxidation that, if not detected by newborn screening, can result in metabolic crisis brought on by illness and may lead to sudden death if not appropriately treated [Genereviews at Genetests, 2004]. A positive screen for MCAD deficiency should result in immediate clinical intervention to prevent long periods of fasting until diagnostic biochemical genetic testing can be performed. Confirmation of MCAD deficiency by diagnostic testing should initiate referral to a genetics professional and discussion with the family regarding the implications for other family members. This illustrates the importance of proper follow up including clinical intervention, additional testing, referral to a genetics professional, and counseling the patient and other family members.

SUMMARY

In general, genetic testing is very complex and requires thoughtful consideration multiple variables both prior to and after testing. It is the physician's responsibility to submit suitable samples and all necessary clinical information to be sure that testing is performed appropriately. In addition, the physician is responsible for proper communication of results to the patient, which will likely be unique to each patient and/or family. The laboratory is responsible for performing quality testing and providing results and interpretations in a clear and unambiguous fashion. Communication between the ordering physician and the testing laboratory is essential to ensure maximum patient benefit throughout the testing process.

INFORMING GENOMIC PATENT POLICY

Timothy Caulfield,[1] *Lorraine Sheremeta,*[1] *E. Richard Gold,*[2] *Jon F. Merz,*[3] *and David Castle*[4]
[1]*Health Law Institute, Faculty of Law, University of Alberta, Alberta, Canada*
[2]*Centre for Intellectual Property Policy, McGill University, Montreal, Quebec, Canada*
[3]*University of Pennsylvania, Philadelphia, Pennsylvania*
[4]*Department of Philosophy, University of Guelph, Ontario, Canada.*

INTRODUCTION

For decades, a debate has raged about the ethical and legal appropriateness of patenting in the biotech sector. To a large extent, the debate has focused on the issues surrounding the patenting of human genes. The sheer number of gene patent applications filed [Cook-Deegan and McCormack, 2001; Cook-Deegan et al., 2003] along with mounting fears over the potential impact of already granted gene patents [Heller and Eisenberg, 1998] and evidence, albeit anecdotal, of monopolistic, anticompetitive behavior implicating patented genetic technologies have ignited a firestorm that has yet to be rationally addressed. Simultaneously, the economic role of biotech patents has been buttressed by the increasing focus of policy makers on innovation and the biotech sector. Intellectual property protection is widely heralded by industry, rightly or not, as a key factor in support of economic growth and advancement in the biotechnology sector. Patents are the de facto currency of the biotech industry—they imbue research findings with speculative value and provide incentive for private-sector investment in biotechnology development [Henderson et al., 1999].

In the context of human health, patents are expected to spur innovations to prevent, diagnose, and treat human disease. Patents provide inventors with a time-limited legal (but not necessarily economic) monopoly during which they have an opportunity to recoup research and development expenditures. Through this incentive, patents are intended to encourage innovation and the timely public disclosure of inventions. Having said this, the impact that patents have in facilitating the effective and timely delivery of quality health care is uncertain. Increasingly, concerns are raised that patents may impair research and unduly limit public access to innovative health technologies [Gold, 2000; Caulfield et al., 2003; Andrews, 2002; Resnik, 2003; Williams-Jones, 1999].

Unfortunately, discussions around the benefits and harms of the patent system tend to be fractious and based on anecdotal evidence with questionable relevance to the overall functioning of the system [OECD, 2002; ALRC, 2004a, 2004b]. The long recognized reality is that the patent system's effectiveness in attaining the broad social objectives that underlie its existence are inherently difficult to quantify [Kitch, 1977; Machlup, 1958].

In this section, we discuss a number of the most commonly articulated concerns associated with gene patents. We will show that the available evidence about the benefits and risks of patents on health care service delivery is equivocal. As a result, policy makers must be cautious in their attempts to rectify perceived problems—especially when the attempted remedies may have significant unintended consequences.

HUMAN GENE PATENT DEBATE

Although the legal question about the patentability of human gene sequences is largely settled, gene patents continue to attract a great deal of attention from the policy-making community [CBAC, 2002; Ontario, Ministry of Health, 2002; Nuffield Council, 2002; OECD, 2002; Cornish et al., 2003; ALRC, 2004a, 2004b]. Whereas in the past discussions about the intricate workings of the patent system were largely closed—involving economists, patent attorneys, patent examiners, and industrial elites—they now include citizens, consumers, nongovernmental organizations, nonpatent government policy makers, professionals, and academics from a variety of disciplines [Royal Society, 2003; WMA, 1992; Genewatch, 2001; Greenpeace, 2004]. In its 2002 report entitled "Genetic Inventions, Intellectual Property Rights and Licensing Practices" OECD member countries specifically sought to address the *public* concerns about systematic gene patenting [OECD, 2002]. The OECD notes that "[n]ew actors, including patient groups for particular diseases and doctors, have also joined the policy discussions and have helped to bring the rather esoteric subject of patents for genetic inventions to widespread public attention" [OECD, 2002,]. Furthermore, the OECD notes that the public's lack of trust in the patent system and its application to genetic inventions appears to be spreading to include the actions of scientists, doctors, universities, and government agencies. For example, those opposed to genetic patenting include numerous well-respected groups and organizations including Médecins sans frontiers, the Human Genome Organization, the American College of Medical Genetics, the Green parties in Europe, the European Parliament, and some Canadian provincial governments [OECD, 2002].

Public concerns about gene patents are reflective of a broader problem in science and technology development. Not until relatively recently was it recognized that the production of scientific and technological knowledge might also involve the co-production of social problems for which solutions lie outside the purview of the sciences [Ravetz, 1971]. While there have been calls for an "extended peer group" that would help to integrate social concerns into the production of scientific and technological knowledge [Funtowicz and Ravetz, 1993], and calls for greater social participation [Di Marchi and Ravetz 1999], technological progress has not yet been humbled by democratic engagement [Jasanoff, 2003]. Furthermore, what might constitute "meaningful public engagement" on the issue of gene patenting remains subject to debate.

The trend toward increased public discourse about the effects of patenting can be viewed as one facet of the democratization of science. News stories about the adverse effects of gene patents on health care delivery, the patenting of higher life forms, and the adverse impact of strong intellectual property protection on the provision of drugs to those in the developing world create anxiety in the mind of the public. For example, within industry it is recognized that patenting has become a public affairs problem. Dan Eramian, Vice President of Communications for the Biotechnology Industry Association (BIO) suggests that:

> *Part of the public affairs problem is that the average member of the public doesn't really know what a patent is, and even a fair number of health and economic policy experts don't comprehend their central role in biomedical innovation. Moreover, the very word patent leaves an impression of "greed" or "selfishness" with many members of the public, the media and even decision makers in our governments.*
> [Eramian, 2004]

At the extreme end of the antipatent spectrum some, for religious and moral reasons, abhor the idea that people can legally hold proprietary rights over human genes. "Commercialization" and "commodification" of human genes and biological materials are viewed as an affront to human dignity [Resnik, 2004]. Others view patents and the imposition of a strong intellectual property regime as an illegitimate tool of the corporatist elite to maintain power dominance in society [Shiva, 1997; ETC Group, 1999; Greenpeace, 2004]. In the global setting, this rationale translates to the view that intellectual property, and patents in particular, are created, used, and imposed by the industrialized world as a tool to exploit the developing world. Though these critical views have become more prominent, the prevailing view within the existing legal framework remains that patents on biological inventions ought to be permitted. Gene patents are like other patents in that they must meet the legal criteria of novelty, utility (or industrial application), and nonobviousness (or inventive step), and, in addition, gene patents must generally represent an appropriate trade-off between short-term private gain (a right to exclude others for up to 20 years) for long-term social progress (derived from continued innovation and economic development).

Given this backdrop, a multidisciplinary workshop was held in Banff, Alberta, Canada, to discuss a variety of issues associated with gene patenting.[1] This section was inspired by that meeting and provides a broad synthesis of the discussion. We conclude that the present understanding of the intellectual property regime is incomplete and that attempts to make it complete demand a multidisciplinary research agenda that can focus on the challenge of providing useful empirical data [Gold et al., 2004]. We should continue to ask and to attempt to answer whether *in sum*; the benefits conferred by the patent system outweigh any negative consequences on society. What, if any, are the societal harms that can be directly attributable to the patent system? Are these harms transient and self-correcting or enduring? To what extent ought patent policy reform be driven by this debate?

ROLE OF PATENTS IN STIMULATING INNOVATION

A patent system is appropriately judged on the basis of its effects on social welfare [Gold et al., 2004]. Through patents, social welfare may be enhanced by encouraging innovation, the dissemination of technically useful information, and by providing incentives for private and public investment in the commercialization of new products and processes that spur economic growth and advance other social goals [Merrill et al., 2004]. Proponents of the patent system consistently point to patents as an essential element of the innovation and commercialization process [Straus, 1998; Enriquez, 2001]. They claim that an erosion of patents rights will result in an inevitable loss of investment and innovation. From the industrial perspective it is argued that patents ensure private investment in the development of socially useful technologies that otherwise might not be developed [Doll, 1998; Bale, 1996; Kolker, 1997].

The need for private investment is a claim often heard in the context of drug development. For example, Mark McClellan, Commissioner of the U.S. Federal Drug Agency states that:

> By some estimates, it costs more than $800 million and typically takes well over a decade to develop a new drug product; and by all estimates, the cost of developing safe and effective new medical products has increased greatly, more than doubling over the past decade. In addition, the vast majority of the treatments that enter

---

[1] Participants of the workshop included Timothy Caulfield, Lorraine Sheremeta, Michelle Veeman (University of Alberta); Edna Einsiedel, Jennifer Medlock (University of Calgary); Grant Isaac, Chika Onwuekwe, Peter Phillips, Martin Phillipson, Ted Schrecker, Stuart Smyth (University of Saskatchewan); Richard Gold, Lara Khoury (McGill University); Scott Kieff (Washington University); Jon Merz (University of Pennsylvania); David Castle (University of Guelph); Yann Joly (University of Montreal); Donna Craig, Michael Jeffrey (Macquarie University); Bradley Bryan (University of Victoria); Ikechi Mgbeoji, Roxanne Mykitiuk (York University); Daryl Pullman (Memorial University); Marnie McCall (Canadian Biotechnology Advisory Committee); Jai Shah (Canadian Institutes of Health Research); and Valerie Howe (Justice Canada).

clinical testing don't succeed, and even among those products entering phase three trials, less than half result in an application for product approval to the [United States] FDA. [McClellan, 2003]

The strength of industry's pro-patent stance is well illustrated by the biotech industry's hostile reaction to the Supreme Court of Canada's decision in the Harvard mouse case where the court determined that the genetically modified mouse at issue in the case was not patentable in Canada [Harvard College v. Canada, 2002]. George Adams, CEO of Innovations Foundation, the University of Toronto office responsible for commercializing university research, saw the day of the decision as a dark day for the Canadian biotech industry. "If you can't patent it, you can't make a company out of it, you just have to dump it into the public domain and you can't get any investment"; fears were expressed that inventors would go somewhere else to protect and exploit their inventions [Abraham, 2002].

Despite such rhetoric, however, there is little reason to believe that the Canadian biotech industry has been significantly harmed by the Court's decision. Indeed, despite claims made by proponents of the patent system, debate remains about whether and how patents affect the rate of innovation in various sectors [Burk and Lemley, 2003]. Despite a burgeoning body of empirical research and theoretical analysis on the overall functioning of the patent system, our knowledge remains incomplete. The data are "beginning to describe the role patents play in important industrial sectors and to assess the effects of policy changes implemented" [Merrill et al., 2004]. For example, Arora and co-workers [2000] report, on the basis of interview data that patents have the most impact on the degree of research and development expenditures in the areas of pharmaceuticals, biotechnology, medical instruments, and computers [Arora et al., 2001].

The promotion of patents based on their perceived role in generating innovation and wealth for a society is a tempting position to adopt. Though not cash substitutes, patents are reasonable indicators of wealth flow once licensing is undertaken. Patents also can be reliable measures of networking and clustering within an innovation system and can act as proxy measures of research and development expenditures where real cost accounting of investments in innovation are difficult. For these reasons patents are often lauded for their efficiency producing, coordinating functions in innovation systems [Langenier and Moschini, 2002]. In fact, the economics are more complex because until a patent is filed, much upstream innovation is considered a sunk cost, which is not accountable as an asset [Nakamura, 1999].

Even if there are reasons for thinking that patents contribute to innovation, there is little evidence to support the notion that patents are *prima facie* necessary for innovation to occur [Walsh et al., 2004; Cohen et al., 2000; Baldwin et al., 2000]. There are also legitimate concerns that the current trend toward stronger patent protection may hinder rather than stimulate technological and economic progress [Mazzoleni and Nelson, 1998; Gold et al., 2004]. To the extent that research and development is sequential and builds upon previous discoveries, stronger patents and patents on foundational research tools may discourage the development of valuable, but potentially infringing, follow-on inventions [Merges and Nelson, 1990; Scotchmer, 1991; Gallini and Scotchmer, 2002]. As noted by Merrill et al., "both sides of [the] ledger are exceedingly complex" and empirical data cannot project the impact of eliminating patents altogether [Merrill et al., 2004].

Despite the misgivings expressed by some, without clear evidence that patents cause harm, the rational presumption must be that patents play a role in the stimulation of innovation, investment, and commercialization of new biotech products [OECD, 2002; Nuffield Council, 2002]. This conclusion harkens back to a statement made, in a 1958 report for the Subcommittee on Patents, Trademarks, and Copyrights as part of its larger study on the U.S. patent system, by Fritz Machlup, a political economist from Johns Hopkins University, wherein he notes that:

> *If one does not know whether a system "as a whole" (in contrast to certain features of it) is good or bad, the safest "policy conclusion" is to "muddle through"—either with it, if one has long lived with it, or without it, if one has lived without it. If we did not have a patent system, it would be irresponsible on the basis of our present knowledge of its economic consequences, to recommend instituting one. But since we have had a patent system for a long time, it would be irresponsible, on the basis of our present knowledge, to recommend abolishing it.* [Machlup, 1958]

Given that we clearly do not know whether the system as a whole is flawed—let alone how to conclusively determine if it is having an adverse impact on a given sector, such as biotechnology—we seem destined to continue to muddle through. Although academics have predicted the effect of gene patents on the research environment and in the provision of genetic testing services, the actual effects remain to be quantified. As such, policy makers should remain cautious of strident claims on either side of the policy debate.

IMPACT OF PATENTS ON THE RESEARCH AND CLINICAL ENVIRONMENT

One of the most frequently raised theoretical concerns about patents, particularly human gene patents, is that they unduly impair the research environment [Shapiro, 2001]. Because the biotech industry is rapidly developing and biotech products and services are developed from a cumulative, dynamic, and increasingly complex set of core technologies, there is a potential for patents to have a detrimental effect. In 1998, Michael Heller and Rebecca Eisenberg predicted a "tragedy of the anticommons" would develop as a result of gene patenting. They claimed that the scarce resource (genetic technology) would be prone to under use when multiple owners with overlapping claims would have a right to exclude others and no one would be left with an effective privilege of use [Heller and Eisenberg, 1998]. They specifically warned that:

> *Privatization must be more carefully deployed if it is to serve the public goals of biomedical research. Policy makers should seek to ensure coherent boundaries of upstream patents and to minimize restrictive licensing practices that interfere with downstream product development. Otherwise, more upstream rights may lead paradoxically to fewer useful products for improving human health.* [Heller and Eisenberg, 1998]

Specific concerns have been raised that researchers may avoid investigating a particular gene or gene region for fear of infringing an existing patent, thus stifling what may be useful research [Knoppers, 1999]. In addition, gene patents may increase the cost of research by compelling researchers to obtain multiple licenses from and pay royalties to multiple patent holders for the use of patented inventions. Increased transaction costs could result in "hold-up problems" that delay or prevent products and services from entering the marketplace [Shapiro, 2001]. Though not well understood or quantified, the potential effect of overlapping patents in the area of human genetics is mentioned in a number of policy documents calling for patent reform [CBAC, 2002; Nuffield Council, 2002; OECD, 2002; Ontario, 2002; ALRC, 2004a, 2004b].

However, at least one commentator has taken issue with the analysis put forward by Heller and Eisenberg (and others) and argues that many of the problems described would be worse if patents were not available. Patents, Kieff argues, increase output by increasing both input and, efficiency and, although not perfect, they are the best option. In part, this is because they act to ensure that biological research will be funded, for better or worse, through the private sector [Kieff, 2001]. From its analysis of the issues, the OECD concludes that:

> *The available evidence does not suggest a systematic breakdown in the licensing of genetic inventions. The few examples used to illustrate theoretical economic and legal concerns related to the potential for the over-fragmentation of patent rights, blocking patents, uncertainty due to dependency and abusive monopoly positions appear anecdotal and are not supported by existing economic studies.* [OECD, 2002]

Admittedly, there are studies that appear to implicate patents with a reduced level of collaboration and information sharing [Blumenthal et al., 1997]. This appears to be especially true as concerns medical geneticists [Rabino, 2001]. It has also been suggested that the patenting and the broader commercialization process may inappropriately skew the direction of research toward that which is potentially patentable and commercializable and away from basic research [Caulfield, 2003].

It is difficult to know if existing data actually reflects researcher behavior or broader systemic trends. For example, researchers may perceive that patents are having a particular impact when, in fact, there may be other, closely related, causes (e.g.,, the general push to commercialize research). In addition, public perception of researcher behavior may be influenced by a few, well-publicized, examples of monopolistic exploitation (e.g., the controversy surrounding the patenting of the *BRCA1* and *BRCA2* genes), rather than by standard practice. Indeed, researchers are often unaware that certain technologies they use are patented. In other circumstances, they may know that a patent exists but opt to run the risk and knowingly infringe recognizing that patent holders are unlikely to take legal action unless there is significant commercial gain and a real possibility to recoup profits from the infringer. In many cases, patents are frequently ignored by researchers and not enforced by owners against infringing researchers. What is described by Straus and others as an "informal research exemption" appears to play an important role in the interaction between researchers and patent holders [Straus, 2002; Walsh et al., 2004].

Of course, there are reasons to be apprehensive about the increasingly dominant role of the commercialization ethos in universities and research institutions. This ethos is now widely recognized as a hallmark of the high degree of coordination between research and development agendas of universities, governments, and industry [Etzkowitz and Leydesdorff, 2001]. One of the principal effects is that public funds are increasingly and inevitably used to promote private interests [Bar-Shalom and Cook-Deegan, 2002; Nelkin, 2002; Rai, 2002]. The U.S. case of *Madey v. Duke University* [Madey v. Duke University 2002] provides added cause for concern over this blurring between the private and the public [Saunders, 2003]. There is no doubt that the growing emphasis on patents has had some impact on the research environment, but the nature of that impact remains to be elucidated.

IMPACT OF GENE PATENTS ON CLINICAL GENETIC TESTING SERVICES

One of the most publicized concerns associated with gene patents is that they may have an adverse impact on the availability of genetic testing services. More broadly, there is concern that gene patents may add an unjust barrier to accessing a useful technology and create an inappropriate financial burden for health care systems. Indeed, numerous controversies have arisen that have caused policy makers to focus on the potential adverse implications of patenting in the context of health systems [Gold, 2003; Andrews, 2002; Caulfield and Gold, 2000]. Though in this discussion we focus on Myriad's exploitation of its *BRCA1* and *BRCA2* patents, other situations have raised concerns, for example:

- The EST patenting controversy in the United States sparked the National Research Council to hold a workshop to consider the impact of research tool patents on molecular biology [NRC, 1997].

- In 2002 Jon Merz and colleagues reported that 30 percent of survey respondents from genetic testing facilities had discontinued *HFE* testing or stopped test development because of patents. The authors concluded that gene patents had a measurable impact on *HFE* test delivery and availability in the United States. The authors also raised concerns about the effect of patents in delaying scientific publication, preventing the development of new tests for specific mutations, the effect of patents on the cost of testing (i.e., as a result of royalty stacking), an increased potential for laboratory errors, the impact of administrative bottlenecks, and time delay in the negotiation of license agreements [Merz et al., 2002].

- In an expanded second study, Mildred Cho and colleagues reported the results of a survey of 122 respondents from genetic testing laboratories in the United States [Cho et al., 2003]. Fifty-three percent of respondents reported that they had opted not to develop a test or provide a specific service because of an existing patent. The study revealed 12 specific tests subject to stoppage by at least one respondent laboratory. The authors concluded that "patents and licenses have a significant negative effect on the ability of clinical laboratories to perform already developed genetic tests" and that "the reported inhibition of clinical testing and research does not bode well for our ability to fully and efficiently use the results of the Human Genome Project and related work" [Cho et al., 2003].

In addition, specific concerns have been raised about patents in the context of testing for Canavan disease [Hahn, 2003; Anderlik and Rothstein, 2003]. Related concerns have also emerged in the context of patenting of human cells and cell lines, including human embryonic stem cells and therapeutic cloning techniques [Bar-Shalom and Cook-Deegan, 2002; Caulfield and Sheremeta, 2004a,b] and most recently in the context of nanotechnology research [Maebius and Rutt, 2004].

MYRIAD GENETICS

The practical impact of genetic patenting is well illustrated by the controversy generated by Myriad Genetics in Canada and Europe over its *BRCA1* and *BRCA2* patents [Gold, 2002]. The situation has been characterized as a harbinger of the "social and ethical implications of DNA patenting and the commercialization of genetic tests" [Williams-Jones, 2002].

Patents over the *BRCA1/2* genes and for the related testing methods are owned by Myriad Genetics. These two genes are associated with an increased risk of developing hereditary breast and ovarian cancer. Though there is much variation as to how diagnostic testing is evaluated and funded in Canada, historically, testing for the *BRCA1/2* genes has been provided by the provincial health ministries through the public health system. In the summer of 2001, Myriad Genetics took steps to enforce its patents over the *BRCA1/2* genes against publicly funded testing laboratories in Canada. Individual provincial health ministries received cease and desist letters directing them to stop performing the test procedures and to send all samples for testing to Myriad or its affiliated laboratories. If followed, this action would have resulted in a cost increase for breast and ovarian predisposition testing from approximately $1200 to $3850 per test. It would have also meant that provincial health authorities would have been unable to implement or to continue to test for breast and ovarian genetic predisposition using other testing methods not described in the Myriad patents. These effects led a number of Canadian provincial health ministers to declare that the public system could not afford the Myriad test [Mackie, 2001]. In large part, Canadian provinces have opted to ignore the Myriad patents and to continue testing despite the patents. Quebec is the only province to have reached a formal agreement with Myriad for sample testing.

In France, the Institute Curie, the Assistance Publique-Hopitaux de Paris, and the Institute Gustave Roussy filed a successful notice of joint opposition against Myriad Genetics. The opposition was supported by the French Hospital Federation, the French Ministries of Public Health and Research, and the European Parliament. The Belgian Society of Human Genetics and the genetics societies of Denmark, Germany, and the United Kingdom also filed notices of opposition [Wadman, 2001; Institute Curie, 2004].

Under this opposition, Myriad's patent for a "Method for Diagnosing a Predisposition for Breast and Ovarian Cancer" (EP 699754) was challenged on the basis that the invention was not inventive, there was no sufficient inventive step, and that the patent application failed to disclose a sufficient description of the invention. The appeal was granted by the European Patent Office in May 2004 for lack of inventiveness [Coghlan, 2004; Abbott, 2004]. An earlier decision of the EPO Opposition Division in February 2005 struck down another of Myriad's patents relating to *BRCA2* because the charity Cancer Research UK had filed a patent application on the gene first [Coghlan, 2004]. Although these decisions continue to make headlines and are seen by many as moral victories against industry writ large, it remains unclear how much weight the ethical concerns over gene patenting hold. Myriad faces two additional opposition hearings in 2005 relating to other of its granted patents [Abbott, 2004; Coghlan 2004].

The Myriad controversy has raised a number of policy questions, generated much press (popular and academic), and has led to the development of a number of advisory and policy documents on point [Nuffield Council, 2002; Ontario, 2002]. The controversy highlights the potential conflict between a government's desire to promote innovation—and, therefore, biotech patents—and timely access to affordable health care [Sheremeta et al., 2003; Caulfield, 2003]. It also raises the concern that gene patents might inhibit the development of less expensive and/or more effective ways of performing a test for a particular genetic condition [Merz et al., 2002]. A recent economic study by Christine Sevilla and colleagues on the cost-effectiveness of genetic testing strategies suggests that the broad scope of the *BRCA1* patent inhibits health care systems from choosing the most efficient testing strategy [Sevilla et al., 2003].

Clearly, we recognize that patents have the *potential* to impact health care delivery—either directly or indirectly—though the degree to which they will do so remains unclear [OECD, 2002; Nicol and Nielsen, 2003; ALRC, 2004a, 2004b]. Though these specific concerns appear to be valid, caution must be applied in presuming that the effects are, *in sum*, negative.

CONCLUSION

Although no substantial body of evidence relevant to this debate exists, preliminary data from the United States, Europe, and Australia suggests that gene patents, in combination with restrictive licensing practices, may pose a threat to the effective provision of genetic testing services. We are aware of no data, however, that shows patient access to genetic testing has been systematically impacted because of gene patents. The OECD [OECD, 2002] and the Australian Law Reform Commission [ALRC, 2004b] specifically note that there is a lack of empirical data demonstrating actual harm resulting from patents and that such data should be sought.

It is clear that industry favors the use of strong Intellectual Property (IP) protection and that industry relies on patents to attract long-term investment. But the traditional purpose of the patents is to encourage innovation *for the public good*. Without concrete evidence concerning the overall impact of patents on society, it is impossible for policy makers to make informed decisions about whether the public good is being served or whether there is a genuine need for patent reform. If reform is necessary, the scope of the reforms must be carefully considered.

Despite the equivocal nature of the evidence, many policy groups and government agencies have made formal recommendations about how the patent system should be reformed. These recommendations have come from respected, independent, policy groups [Nuffield Council, 2002] and from national and regional government advisory bodies [CBAC, 2002; Ontario, 2001; ALRC, 2004b; Department of Health and Human Services, 2004]. For the most part there is remarkable consistency with respect to the recommendations for reform, which typically include one or more of the following measures:

- Creating a statutory definition of *patentable subject matter* that includes or excludes certain biotechnological inventions
- Narrowing the scope of gene patents and restricting the interpretation of utility to near-term applications
- Adding an "ordre public" or morality clause to the Patent Act
- Adding a statutory opposition procedure that is similar to that which exists in Europe (and which has been successful thus far in striking down two of Myriad Genetics' breast cancer patents)
- Creating a narrow compulsory licensing regime that would facilitate access of key patented technologies to others
- Creating a specialized court to ensure that only judges with expertise in technology and patent law can hear intellectual property cases

Though there are a growing number of policy documents that recommend some degree of patent reform, one should not expect that reform will happen quickly. While all of the above suggested reforms are possible under existing trade agreements, including the TRIPS agreement, the growing integration of markets creates new challenges for policy makers. The push toward international harmonization of patent law seems to push against the types of reform that are often suggested. To the extent that reforms may disrupt or be perceived to disrupt a nation's competitive advantage in an important industrial sector, reform may proceed slowly and will require a great deal of cooperation between government departments [Caulfield, 2004].

Given the perceived importance of patents to the biotech industry and to the growing knowledge-based economy, governments will likely proceed with care. Time will tell whether Canadian Parliament will take steps to amend Canada's patent laws to reflect some or all of the various changes that have been recommended by critics of the present system. Nevertheless, given the high degree of consensus by those independent groups that have investigated the issues thus far, it would be unfortunate if the problems of coordination within governments were to lead to the failure to adequately seek additional evidence and to adopt recommendations that are appropriately tailored to address systemic problems.

Acknowledgment

The authors thank Genome Canada and Genome Prairie for their ongoing funding support.

Chapter 13

Risk Assessment

OVERVIEW

A physician has an obligation to undertake all reasonable procedures to evaluate test results. The physician should have a working knowledge of the manner in which risk information is calculated, the respective limitations of each model of analysis, and the potential error rates. Genetic test information can be complex and ambiguous; therefore, referral for genetic counseling generally is recommended.

INTRODUCTION

In genetics as in other medical disciplines, laboratory testing is a critical component of care. Laboratory analyses for genetics can be applied for a variety of indications beyond confirming or ruling out a specific diagnosis. In cytogenetics, for example, karyotyping can identify not only the fact that a familial chromosomal anomaly exists but also provide the critical information that informs estimates of risk for miscarriages and abnormal live births. Molecular genetic testing is commonly performed for a variety of applications in clinical genetics, but in practice most tests are requested in order to:

- Confirm a diagnosis, by
  - Identification of a causal genetic mutation
  - Identification of gene involved (e.g., by linkage analysis)
- Rule out a specific diagnosis
- Estimate a risk for future disease (susceptibility testing)

NEED TO KNOW

Molecular testing is often interpreted as a conclusive step that offers the refinement and specificity of DNA-based assessment. It is indeed tempting to assume that a molecular genetic test will provide "the" answer.

Genetic Testing: Care, Consent, and Liability, by Neil F. Sharpe and Ronald F. Carter
© 2006 John Wiley & Sons, Inc.

Molecular genetic tests can indeed be conclusive in specific scenarios, for example, the test:

- Identifies a causal mutation in a gene clearly known to be associated with the disease.
- Clearly identifies or rules out a familial mutation known to be associated with the disease.
- Methodology is highly accurate (i.e., near 100 percent) in ruling in or out the possibility of a causal mutation for a specific disease (e.g., testing for fragile X syndrome)

However, sometimes, in practice, our current technologies are limited in both obvious and subtle ways that may not be evident to the nongeneticist. For this reason, molecular diagnostic reports do not consist of just a statement of result (e.g., presence of absence of a specified mutation) but also provide the clinician and patient with an interpretation of the clinical significance of the result. The limitations of molecular testing can be significant, and they have to be specifically identified in the interpretive section of the result. Examples of common limitations include (in increasing levels of specificity):

- The genetic basis of the disease is unknown (no known test applicable).
- The disease is associated with one or more genes, some of which are unknown.
- The disease is associated with mutations of more than one known gene, but not all were tested.
- The disease is caused by mutations of one known gene, but not all mutations could be detected by the methodology used (as in the clinical scenario above).
- A mutation is correctly identified and known to be associated with disease, but the disease is subject to incomplete penetrance or variable expressivity, and the likelihood of being affected is not clear (e.g., situations of prenatal diagnostics or susceptibility testing).
- A variation in the sequence is identified, but it is not known whether it ever causes disease.
- An appropriate methodology was applied, but testing failed because of limitations with the specimen (e.g., degradation of DNA in a pathology specimen archived after death).
- An appropriate methodology was applied, but interpretation is inconclusive because of incomplete knowledge of the pedigree or false paternity or some other type of concern with specimen identification or availability.

Thus, two basic types of uncertainties often arise:

1. It is not clear that the test ruled in or out the possibility of a causal mutation.
2. A mutation was found, but the clinical significance of the mutation is not clear.

In regard to the latter, current terminology is moving away from the describing abnormalities of gene sequences as *mutations*. The preferred term is now *sequence variation*, given that the clinical significance of many "mutations" cannot be clearly shown. Further details on molecular terminology are available [Human Genome Variation Society website: http://www.hgvs.org/mutnomen/].

WATCH OUT FOR

How can a health care professional make appropriate use of testing? The key is to clearly identify the clinical significance of the results and understand any potential limitations. Laboratories report results in standardized formats to facilitate these objectives. If in doubt, calling the laboratory for clarification is very helpful.

When a report is received, the clinician needs to ensure that:

- It is documented as received and reviewed for further action in timely fashion *before* it is placed in the patient's file.

- The report is reviewed for correct identification of the patient and specimen, information on exactly what testing was provided, the result, and the interpretation of the result.

- The patient receives counseling that includes review of the clinical significance of the report and any limitations of the result or test methodology.

In practice, counseling the patient about the results of molecular testing can be complicated. Patients and physicians alike can be frustrated by inconclusive testing methodologies. Even some of the most common testing situations can be difficult:

- For example, prenatal assessment to rule out heterozygous carrier status for a *CFTR* mutation: if the test methodology applied is not 100 percent accurate, a residual risk remains and it must be calculated accurately using Bayesian analysis if the patient is to receive appropriate counseling. This process of risk assessment is described in detail with examples below. It is important to ensure that the report is neither over- nor underinterpreted for clinical significance.

- For example, identification of a sequence variation that is of unknown clinical significance. In susceptibility testing for hereditary breast/ovarian cancer, missense mutations (change of a single amino acid) may be identified that are known and recurrently reported, but to date have not been causally associated with a detectable increase in risk. The laboratory may use a variety of techniques to evaluate missense mutations for clinical significance (e.g., knowledge of protein structure and function; evolutionary conservation of the sequence in the region in the mutation; software programs that predict the effect of the variation on protein function; comparison to other known mutations affecting that codon or others in the region; and analysis of the whole family to demonstrate cosegregation of the missense mutation with disease). While these steps provide a more informed interpretation, only analysis of large extended families and very large clinical trials extending over many years will eventually provide the empirical data that illustrate how much these sequence variants actually mean to a family. Thus, in this situation, the test methodology reveals findings that cannot be interpreted unequivocally, and the patient is left with advice that in effect says "we found something, we do not think it means anything, but we will not know for sure for some years, so please keep in touch." Clinicians and patients may decide to continue with noninvasive and/or low-risk management strategies on the assumption that the sequence variant may at some point be shown to have a clinically significant association with risk.

RISK ASSESSMENT

Peter J. Bridge and Jillian S. Parboosingh Molecular Diagnostic Laboratory, Alberta Children's Hospital, Calgary, Alberta and Department of Medical Genetics, University of Calgary, Calgary, Alberta, Canada

INTRODUCTION

Although there are a few scenarios where a negative test result for a genetic disorder has a predictive accuracy of 100 percent, these represent a minority of such results now and, paradoxically, further genetic discoveries will ensure that they represent an even lesser fraction of cases in the future. The reason for this paradox is that the phenomenal progress of human genetic research, thanks to the Human Genome Project, has enabled us to start providing tests for complex diseases that have a genetic component combined with other risk factors. Prior to this, we could test only for diseases that were due entirely to the effects of mutations in a single gene. Thus, there will be a steady shift away from disorders where a single genetic test was all that was needed for accurate risk assessment. Case 1 will illustrate an example of a presymptomatic test for Huntington disease performed on a young man that indicates that he has a very low risk of developing the disease. In this case, because of the nature of the disease, the specificity of the test, and his family structure, the test results are as clear as we can possibly get. More often, we get a result that will reduce, but not entirely eliminate, a person's risk. Case 2 discusses the concept of residual risk following a negative screening test for cystic fibrosis mutations. Case 3 discusses a negative result for a major genetic contributor to familial cardiovascular disease that, although very useful in eliminating the extreme risk seen in other family members, does not really make any practical reduction in the approximately 50 percent population risk of developing cardiovascular disease.

Positive test results are subject to similar caveats. The finding of known disease-causing mutations for a disorder with full penetrance is usually considered diagnostic (or predictively diagnostic), but with the trend toward increased analysis of multifactorial conditions, again the genetic test will become just one line of evidence, albeit a very important one. Case 4 will illustrate a positive diagnosis of achondroplasia made by the detection of the *G380R* mutation in the *FGFR3* gene (see Appendix at end of chapter for mutation nomenclature) in a fetus with short limbs detected on ultrasound screening. Case 5 will illustrate an initially positive diagnosis (that later changed) of cystic fibrosis by finding two known severe mutations in the *CFTR* gene of an infant who was failing to thrive and who had recurrent respiratory infections. Case 6 will illustrate the detection of a *BRCA1* mutation in a woman with a strong positive family history of early-onset breast cancer. Case 7 will illustrate the use of genetic linkage analysis in an attempt to determine carrier status of the sister of a male affected with an X-linked recessive condition.

APPROACH TO RISK CALCULATION

Before we study individual cases, it is necessary to lay out the framework of how we approach almost all cases of risk calculation in medical genetics. The basic process is referred to as Bayesian analysis after an eighteenth-century mathematician, the Rev. T. Bayes. In this process, we consider the relative likelihood of two or more alternative hypotheses, given a set of known conditions or results. We set up a table with a column for each hypothesis and upon the rows place values for the following parameters: *prior probability* (the likelihood of each hypothesis before we consider any of the available evidence), *conditional probability(ies)* (the probability of observing the available evidence on the assumption that each hypothesis is true—there may be several such entries as long as each line of evidence is independent), and *joint probability* (the product of the prior probability and all of the conditional probabilities). Comparison of the magnitude of the joint probabilities tells us something about the relative likelihood of our original hypotheses. Finally, the *posterior probabilities* are normalized (each number divided by the total) so that the total of all columns (all hypotheses) equals unity. In medical genetics, the contrasting hypotheses usually take the form of the probability that someone has or has not inherited a disease-causing allele of a gene.

In the worked example (Fig. 13.1), the female in the third generation is the daughter of an obligate carrier of an X-linked recessive disorder. Her prior probability of being a carrier is therefore one-half, dependent only upon whether she inherited her mother's normal gene on one X chromosome or her mother's mutant gene on the other X chromosome. The conditional evidence that we can consider is that she

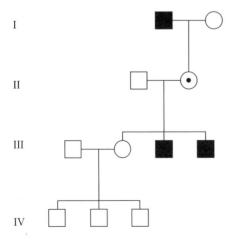

Figure 13.1 X-linked recessive pedigree. The female in the third generation, III-2, is the daughter of an obligate carrier of the X-linked recessive condition. At conception, her (prior) probability of being a carrier was 50 percent. The fact that she now has three normal sons, any of whom could potentially have been affected were she a carrier, reduces the (posterior) probability that she is a carrier.

has three healthy sons. The basic logic behind this Bayesian calculation is that if she were a carrier, there would be a one-half chance of passing on the mutant allele to each child or, conversely, a one-half chance of passing on the normal allele; but if she were not a carrier, it does not matter which allele she transmits since either would be normal. We therefore enter different numbers for the conditional probabilities for the two hypotheses: one-half for the probability of having a normal son if she were a carrier versus one for the probability of having a normal son if she were not a carrier. This is repeated for each of the three sons since each successive pregnancy acts as an independent test of the state of her genes.

	Carrier	Noncarrier
Prior probability	0.5	0.5
Conditional probabilities		
First son is normal	0.5	1
Second son is normal	0.5	1
Third son is normal	0.5	1
Joint probability	0.0625	0.5
Normalization	0.0625/(0.0625 + 0.5)	0.5/(0.0625 + 0.5)
Posterior probability	0.1111	0.8889

The joint probability of her being a carrier and having three normal sons is 0.0625, whereas the joint probability of her being a noncarrier and having three sons is 0.5. These two numbers do not add up to 1, but the process of normalization (dividing each number by the total of both) ensures that the posterior probabilities do add up to 1 (and can be directly converted to percentages). We could have performed this whole calculation using fractions instead of decimals if we had wished: The joint probability of her being a carrier would be $\frac{1}{16}$ and the joint probability of her being a noncarrier would be $\frac{1}{2}$, which equals $\frac{8}{16}$. In the normalization process, $\frac{1}{16}$ divided by the total of $\frac{9}{16}$ equals $\frac{1}{9}$, and $\frac{8}{16}$ divided by the total of $\frac{9}{16}$ equals $\frac{8}{9}$.

The power of Bayesian analysis is that the conditional probabilities may be of many different types and, as long as some basic rules are followed, these types may be mixed. In the example just given, the conditional probabilities were based upon reproductive history, in many other calculations, the conditional probabilities will be the probability of observing different DNA test results under each hypothesis or the probability of observing different biochemical assays.

CASE EXAMPLES

Case 1: Huntington Disease

Huntington disease is caused by an excessive number of repeats of a three-base sequence (CAG) in the *IT15* gene on chromosome 4. Ben and his affected mother, Alice, have both been tested for the trinucleotide repeat mutation that causes Huntington disease. The sizes of Alice's two alleles are 15 repeats and 45 repeats: 15 is very normal and 45 is definitely mutant. Thus, the molecular basis of Alice's Huntington disease is known. Since Ben's two alleles have 15 and 18 repeats, it is very obvious that he inherited Alice's normal allele (the one with 15 repeats). Ben's prior probability of having inherited a mutant allele from his mother was one-half, but we do not really need to set out a Bayesian table for this result since, having tested him, the conditional probability of his having a mutant allele is zero. This is the most accurate type of molecular genetic diagnosis that we can perform: The familial disease-causing mutation was identified in an affected parent, and it was demonstrated not to have been transmitted to the son. His risk of developing Huntington disease is zero (subject to the disclaimers given below that there was no sample mix-up and that the DNA tested actually came from Ben and Alice). Note also that incorrect paternity, if present, would not alter his risk: He has a normal maternal allele with 15 repeats and a normal paternal allele (regardless of its source) with 18 repeats.

Case 2: Cystic Fibrosis

Christine has a brother with cystic fibrosis (homozygous for the common *delF508* mutation) and upon being tested herself was found to be a heterozygous carrier of the *delF508* mutation. She recently married Dan and the couple are considering starting a family. Since Christine is a proven carrier of cystic fibrosis, it is very important to evaluate Dan's risk. He has no known family history of cystic fibrosis, but being Caucasian with family roots in Ireland, he is known to come from a population with an appreciable frequency of carriers. An appropriate prior probability that Dan is a carrier of cystic fibrosis would be 1 in 20 (equivalent to the carrier frequency for the Caucasian population). The conditional evidence that we might obtain would be the result of a DNA analysis that screens for the most common cystic fibrosis mutations. This screening test will identify approximately 90 percent of mutations in his ethnic background. Thus, if he has a negative result, the conditional probability of this result if he is a carrier is one-tenth (since nine-tenths of carriers will have a detectable mutation).

	Carrier	Noncarrier
Prior probability	1/20	19/20
Conditional probability		
Negative mutation screen	1/10	1
Joint probability	1/200	$19/20 = 190/200$
Posterior probability	1/191	190/191

Thus, Dan's residual risk of being a carrier following the negative mutation screen is 1 in 191 and a 1 in 764 chance of conceiving a child with cystic fibrosis. It is very important to realize that, despite Dan having undergone DNA analysis, his carrier risk is not zero: Only when the mutation screen detects 100 percent of mutations can a carrier risk be reduced to zero.

If Dan's mutation screen had instead detected a mutation, he and Christine would both be proven carriers of cystic fibrosis, and they would have a 1 in 4 chance of conceiving a child with cystic fibrosis. In this case, however, both parental mutations would be known to be detectable by the same mutation screen, and accurate prenatal diagnosis would be an option available to them.

Case 3: LDL Receptor and Heart Disease

Emil's mother died of a heart attack at age 35 and many of his maternal relatives also had very early heart attacks. Genetic analysis of some of his surviving maternal relatives revealed that a dominant mutation in the low-density lipoprotein (LDL) receptor was present in some family members, which causes familial hypercholesterolemia. Emil and his two sisters and their father were tested for this mutation, and it was established that neither Emil nor his father has this mutation but both of his sisters do. It can be inferred, therefore, that Emil's mother carried this dominant mutation and that this greatly increased her risk of cardiovascular disease (to approximately 100 percent without medical intervention). This mutation almost certainly was the cause of her early demise. If his father's family has no significant history of early heart disease, Emil's risk is reduced to that of the general population. The DNA test has removed the excess familial risk due to the known high-risk mutation, but makes no reduction in the general population risk of cardiovascular disease. Emil thus retains a lifetime risk of approximately 50 percent of cardiovascular disease. His sisters have greatly elevated risk of cardiovascular disease and need continuous medical supervision—paradoxically, they will be treated aggressively by cardiologists for this specific known risk factor (and likely benefit from general watchfulness over unknown risk factors), whereas Emil will probably just have to take his chances and may well have a poorer long-term prognosis than his sisters.

Case 4: *FGFR3*, Achondroplasia, and Thanatophoric Dwarfism

There has been a common public misconception over the last few decades that geneticists use prenatal testing to seek out and destroy mutant fetuses. While we do occasionally provide prenatal information to couples who do not wish to continue with a pregnancy in which their fetus has inherited a serious genetic disease, there are many cases where prenatal diagnosis can be used to improve the medical management of a pregnancy when termination never was the issue. One obvious example would be the prenatal diagnosis of severe hemophilia: Vaginal delivery is a difficult process for any fetus, but a male fetus with severe hemophilia needs special consideration to avoid intracranial bleeding (do not use forceps!). Fetuses shown to carry a mutation for the hereditary eye cancer, retinoblastoma, can potentially have their retinas monitored by high-resolution ultrasound and, if there is any sign of tumor formation, can be delivered a few weeks early so that laser treatment of the tumor can commence in time to preserve vision.

Rather more frequently, we are asked to perform prenatal testing to try to identify the cause of short limbs detected by ultrasound. Late in each pregnancy (e.g., 35 weeks) ultrasound is used to measure the size and maturity of each fetus. Short limbs can have many causes, but the differentiation between achondroplasia and thanatophoric dysplasia is important. These two conditions can look fairly similar on ultrasound, but achondroplasia is viable, whereas thanatophoric dysplasia is not (the lungs do not develop well and the baby usually fails to breathe). Both conditions have a large head-to-trunk ratio and are difficult to deliver. In one case, the parents can prepare for the delivery of a baby that will have achondroplastic dwarfism but otherwise be healthy, whereas in the other case the parents need careful counseling to bring them to the realization that the baby will almost certainly die within a few minutes of being delivered. The parents of a fetus with thanatophoric dysplasia may choose to induce labor early. These two conditions are caused by different mutations in the same gene (*FGFR3*) and mutation analysis will detect >99 percent of cases.

Case 5: Null Alleles

A newborn child in neonatal intensive care unit (NICU) is assessed for failure to thrive. There are many different causes of failure to thrive including cystic fibrosis. Treatment for the pancreatic insufficiency component of cystic fibrosis is available and successful. DNA can be extracted from a small blood sample obtained from the child and analyzed for the common mutations. Pressed to provide a quick report, the diagnostic lab releases a preliminary report indicating that the child is apparently homozygous for the severe *delF508* mutation, and the family is counseled that their child has cystic fibrosis and that any future pregnancy has a 1 in 4 risk of conceiving a child with cystic fibrosis. Based on these results, the pediatrician begins treatment for cystic fibrosis including prescribing pancreatic enzymes. The child, however, fails to respond to the treatment. A few days later, the final report is available and indicates that the child is not homozygous for the *delF508* mutation but carries one *delF508* allele and one *F508C* polymorphism. The *F508C* polymorphism is not detected by one of the commonly used commercial cystic fibrosis tests (i.e., it is a null allele) resulting in apparent homozygosity for the *delF508* allele in an individual with one copy of each. Reflex testing for this polymorphism is required when an unexpected apparently homozygous *delF508* is detected. If the parents had been tested at the same time, only one of the parents would have been a *delF508* carrier. This would have alerted the laboratory and no preliminary report would have been issued.

There are some genetic tests in which it is more important to obtain parental samples in order to provide accurate risk assessments. Congenital adrenal hypertrophy is an autosomal-recessive disorder caused by mutations in the *CYP21B* gene. Many of the mutations are due to gene conversion with the pseudogene (*CYP21A*), which differs from the *CYP21B* gene at only a few nucleotides distributed along the genes. These mutations cause the pseudogene to be inactive. A mutant allele can have one or

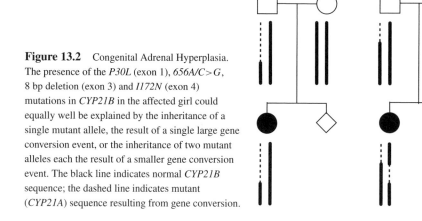

Figure 13.2   Congenital Adrenal Hyperplasia. The presence of the *P30L* (exon 1), *656A/C>G*, 8 bp deletion (exon 3) and *I172N* (exon 4) mutations in *CYP21B* in the affected girl could equally well be explained by the inheritance of a single mutant allele, the result of a single large gene conversion event, or the inheritance of two mutant alleles each the result of a smaller gene conversion event. The black line indicates normal *CYP21B* sequence; the dashed line indicates mutant (*CYP21A*) sequence resulting from gene conversion.

several of these mutations depending upon the length of the *CYP21B* gene involved in the conversion event. If we were to test for these mutations in an affected individual, we may detect one, two, three, or more mutations; yet we know the individual can only have two mutant *CYP21B* genes. Testing the parents who will each be carriers for mutant alleles ensures that we have identified two different mutant alleles and not just one allele containing multiple mutations resulting from a large gene conversion event (see Fig. 13.2).

Case 6: Founder Mutations and Breast Cancer Gene Testing

Dominique is 40 years of age and worried that she might develop breast cancer like her mother (first diagnosis at 42 years, second diagnosis at 45 years), sister (diagnosed at 38 years), and maternal grandmother (diagnosed at 48 years) and maternal great aunt (diagnosed at 80 years). All women have an approximately 1 in 12 lifetime risk of developing breast cancer, but in some families such as Joan's the risk is much higher due to the transmission of a disease-causing mutation in the family.

Approximately 5 to 10 percent of all breast cancer is clearly familial. In approximately 25 percent of these high-risk families, mutations in the *BRCA1* and *BRCA2* genes have been found. These genes are larger than average and mutations can occur at any position. Thus, unlike cystic fibrosis where a handful of mutations account for approximately 90 percent of mutations, each breast cancer family tends to have its own private mutation. This means that the entire sequence needs to be analyzed for all types of sequence changes. One exception to this is the presence of founder mutations in specific ethnic populations; these mutations were brought into the population by one of the founding members and transmitted to succeeding generations such that today they are overrepresented in the mutant gene pool. Dominique's family is French

Canadian and can be screened first for the small number of mutations that are known to cause the majority of breast cancer in the French Canadian population.

Although it is Dominique that would like to know her risk of developing breast cancer, testing must first start with a known family member with breast cancer. If we were to test Dominique first and did not identify a mutation, we would not be able to determine if this was because the cancer seen in her family was due to a mutation in some other gene (remember that 75 percent of breast cancer families do not have a mutation in *BRCA1* or *BRCA2*) or that she truly has no mutation. Identifying the most appropriate family member to test is critical and requires obtaining a good family history. Dominique's maternal great aunt is not the best person to test since her onset of breast cancer is much later than the others and thus may be sporadic and not due to the family mutation.

Dominique's mother would be the best person to test since she would be considered an obligate carrier having both an affected mother and an affected child; having had two independent diagnoses at an early age also makes her a good candidate. Mutation analysis for the *BRCA1* and *BRCA2* French Canadian mutations identified the *BRCA1 R1443X* mutation. Dominique can now be tested for this specific mutation. Testing shows that Dominique also carries this mutation and thus is at an increased risk of developing breast cancer. The clinical consequences of carrying a mutation in this case and all cancers is not as clear cut as for achondroplasia in which 100 percent of mutation carriers will develop the disease. Not all *BRCA1* mutation carriers develop breast cancer (the disease is not fully penetrant); by the age of 80 years, approximately 80 percent of carriers will develop breast cancer. Furthermore, by the age of 80 years, 65 percent of carriers will develop ovarian cancer, and there are increased risks for some additional cancers. The genetic counseling of breast cancer and other hereditary cancers is complicated.

In Dominique's family, the identified mutation is well known as are its effects on both the structure and function of the protein. However, more and more frequently, the sequence changes that are identified have not been previously identified, and their effect on protein or expression are not known. There is increased evidence in favor of the variant being disease causing if the variant is carried by all affected individuals within the family, is not found in a control normal population, or occurs in a nucleotide and/or amino acid that is invariant across different species (the fact that is nucleotide/amino acid has not changed over evolution provides evidence that this is a functionally important position). Ultimately, functional studies are necessary to determine pathogenicity. We frequently have no choice but to write on reports "the pathological consequence of this variant is unknown."

Case 7: Linkage Analysis and Alport Syndrome

Anne's brother Sam is in need of a kidney transplant and Anne wishes to know if she could be a donor. Sam has Alport syndrome as do two of his maternal uncles. Alport syndrome is a genetically heterogeneous disorder; autosomal-recessive (15 percent of cases) or autosomal-dominant (5 percent of cases) Alport syndrome is caused by

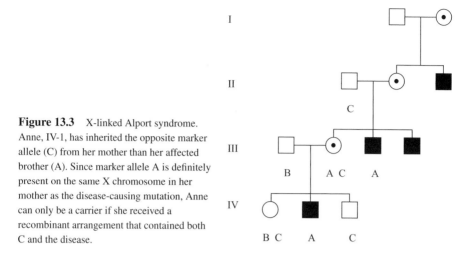

I

II

C

Figure 13.3   X-linked Alport syndrome.
Anne, IV-1, has inherited the opposite marker
allele (C) from her affected   III
brother (A). Since marker allele A is definitely
present on the same X chromosome in her
mother as the disease-causing mutation, Anne   IV
can only be a carrier if she received a
recombinant arrangement that contained both
C and the disease.

B

A C

A

B C

A

C

mutations in the *COL4A3* or *COL4A4* while X-linked Alport syndrome (80 percent of
cases) is caused by *COL4A5* gene. Direct mutation analysis is not available at the local
lab but indirect analysis by linkage can be performed. Given Anne's family history
(Fig. 13.3), it is clear that this family has the X-linked variety of Alport syndrome.
Linkage analysis can determine whether Anne has inherited the same region of the
X chromosome that contains the mutant *COL4A5* gene seen in her brother. Linkage
analysis involves tracking the region of the genome known to carry the defective
gene in family members. This is done by "labeling" the alleles in a family using
polymorphic markers. All affected individuals will inherit the same allele (or alleles
in an autosomal-recessive disorder). Unlike direct tests, linkage analysis requires
that DNA be available from other family members in order to establish accurately
which allele carries the disease-causing mutation. Anne's parents, affected brother,
unaffected brother, and one of her maternal uncles are willing to participate. A marker
located near the *COL4A5* gene is used in the analysis, and Anne is found to not share
this region of her X chromosome with her brother but is given a residual risk of
10 percent (i.e., there is a 10 percent chance that she does share the disease locus with
her affected brother). The arrangement of the genes in Anne's mother is known with
certainty (phase known) such that marker allele *C* is on the same X chromosome as
the normal Alport gene inherited from the grandfather and marker allele *A* is on the
same X chromosome as the mutant Alport gene inherited from the grandmother. If
recombination occurs between these two loci 10 percent of the time (and does not
occur 90 percent of the time), the probability that Anne would inherit the maternal *C*
allele and the disease (a recombinant genotype) is 10 percent and the probability that
she would inherit the maternal *C* allele and a normal Alport gene (a nonrecombinant
genotype) is 90 percent. Unlike direct tests where an individual either carries or does
not carry a mutation, indirect tests are not 100 percent accurate. This is because we
are not following the actual disease locus but a surrogate marker located near the

disease locus. Recombination between the disease locus and the surrogate marker results in the marker no longer being an accurate surrogate for the disease locus; the further the marker is from the disease locus the greater the chance of recombination occurring. The results of linkage tests are estimates with residual risks dependent upon the distance between the marker and the disease locus. Linkage analysis using markers within the disease locus where recombination is essentially zero are most accurate; however, for some very large genes, recombination may be significant within the gene itself (e.g., dystrophin). The use of two markers located on either side of the disease locus (flanking markers) greatly increases the accuracy of linkage analysis.

	Carrier	Noncarrier
Prior probability	0.5	0.5
Conditional probability		
Linkage result (maternal C allele)	0.1	0.9
Joint probability	0.05	0.45
Normalization	0.05/(0.05 + 0.45)	0.45/(0.05 + 0.45)
Posterior probability	0.1	0.9

LABORATORY TEST LIMITATIONS (DISCLAIMERS)

There are a large number of things that can cause misinterpretation of a molecular genetics assay: some common, many rare; some serious, others unimportant to the outcome. Let us consider some of the major limitations.

Critical Technical Disclaimer 1   All PCR (polymerase chain reaction) based tests are valid only if both alleles have been amplified equally. Failure to amplify an allele can be caused by rare variants that occur at the primer site; this results in an inability or reduced efficiency of the primer to bind to the template DNA. Similarly, if the primer binding site is deleted, amplification will not take place resulting in apparent homozygosity when only one allele is affected.

Critical Technical Disclaimer 2   Null alleles are defined as alleles that are invisible to the assay method used. A visible allele in combination with a null allele usually looks like two copies of the visible allele unless the assay is very carefully designed to be quantitative.

Critical Management Disclaimer   Correct specimen for the test. For example, tests for mitochondrial disorders may be performed on blood and, if positive, are valid; if the test result is negative, however, another tissue may need to be tested because not all mitochondrial mutations are expressed in blood cells. Any patients that have undergone a bone marrow transplant must have a tissue other than blood analyzed since the DNA in their blood now comes from the marrow donor and does not represent the constitutional genotype of the recipient.

Critical Clinical Disclaimer   Correct clinical diagnosis for the molecular test requested (particularly relevant when the test result is negative and absolutely essential whenever indirect testing is being performed).

SUMMARY

We have attempted to offer a variety of cases to illustrate how genetic risks are calculated by diagnostic laboratories. A great many more examples and significantly more theory can be found in Bridge [1997], Hodge [1998], and Young [1999]. We also have tried to make it clear that all analyses have limitations and have provided a brief description of the most common problems that we might encounter and that lead to the necessity of having disclaimers on diagnostic reports.

Final Word   If in any doubt, refer the patient to an appropriately qualified medical geneticist for risk assessment and genetic counseling.

APPENDIX: MUTATION NOMENCLATURE

There are many different ways to describe a mutation or sequence change: It can be described by its effect at the DNA, cDNA (DNA reverse transcribed from RNA) and protein level. Nomenclature must be accurate and unambiguous. This is important for the communication of mutations between diagnostic laboratories, allowing individual family members to be tested in different laboratories. A standardized method for naming mutations has recently been accepted by most laboratories (see references). There are three components to the naming of variations. The first identifies the reference sequence: a "g" prior to the variation indicates a genomic reference sequence; a "c" prior to a variation indicates a cDNA reference sequence; a protein reference sequence may be indicated with a "p" but is also recognizable by the presence of a letter first instead of a number. The second component identifies the position within the sequence that is altered. Nucleotides at both the genomic and cDNA level are numbered starting with the A of the start codon ATG (+1); nucleotides 5′ to this A will be negative numbers (note there is no position 0). At the protein level, the amino acids are numbered starting at the amino (N) terminal with the methionine start codon as codon 1. The third component indicates the sequence change that has occurred. The wild-type, normal, or reference nucleotide or amino acid is always given first, followed by an indication of the type of change (e.g., a substitution, deletion, or insertion) and the new nucleotide or amino acid. If these were all real examples, they would also indicate the gene in which the variation occurred (e.g., *BRCA1 R1443X*).

For example:

g. 67A>G	The nucleotide guanine (G) has replaced the 67th nucleotide adenine (A) of the genomic sequence.
c. 456_458delCTT	The CTT nucleotides found at position 456, 457, and 458 in the cDNA sequence have been deleted.
g. −677T>C	The nucleotide cytosine (C) has replaced the nucleotide thymine located 677 nucleotides 5′ of the start of the coding sequence, i.e., in the promoter region.
A67G	The amino acid glycine (G) has replaced alanine (A) at position 67 of the protein (may also be seen as ala67gly using the three-letter amino acid code).
delF508	The amino acid F (phenylalanine) at position 508 of the protein is deleted (note that this mutation does not begin with a number, indicating that the change is being described at the protein level).
R1443X	A stop codon represented by the letter X replaces the amino acid arginine (R) at position 1443 of the protein. This produces a 1442-amino-acid protein (missing all amino acids after the stop site).

Sequence changes can get quite complicated. For more information on mutation nomenclature refer to Antonorakis [1998] and den Dunnen and Antonorakis [2000, 2001].

Chapter 14

Test Results: Communication and Counseling

OVERVIEW

Medical tests play a crucial role in patient care. The information these tests provide affords physicians the increased knowledge upon which to expertly guide patients to make medical and life decisions.

Giving test results to a patient involves more than simply reading a laboratory report. It involves an interactive discussion centralized around the incorporation of the newly learned information into both the patient's health care plan and resulting perception of health [Alaszewski and Horlick-Jones, 2003]. Without the assimilation of this information, the test result, no matter how important, is useless. Genetic testing exemplifies this point.

Both genetic and nongenetic tests can have significant medical and social implications. Genetic testing information, like the medical diagnosis of serious infectious disease [e.g., human immune virus (HIV) and severe, acute respiratory syndrome (SARS)] or psychiatric disease (e.g., schizophrenia, bipolar), can have:

- Stigmatizing effects on the patient and the patient's relatives or significant other(s)
- Potential health risks to relatives or significant other(s)
- Limited effective health interventions
- Significant interpretive complexity [Green and Botkin, 2003]

Authors: Prepared by the following authors with a contribution from Neil Sharpe ("Follow Up and After Care")

Julianne M. O'Daniel, Educational & Training Programs, Institute for Genome Sciences & Policy, Duke University.

Allyn McConkie-Rosell, Division of Medical Genetics, Duke University Medical Center.

Genetic Testing: Care, Consent, and Liability, by Neil F. Sharpe and Ronald F. Carter
© 2006 John Wiley & Sons, Inc.

355

However, unlike schizophrenia or HIV, genetic testing also may predict or diagnose disease risks in individuals without any current measurable health concerns or exposures, including those not tested or even born, interpretation is intricately complicated relying on multiple factors, and it is centered around information felt by some in society to be part of their fundamental being—who they really are [Green and Botkin, 2003; Clayton, 2003].

Additionally, genetic testing may offer unique challenges because the interpretation is often incomplete and may be ambiguous [Van Zuuren et al., 1997; Geller et al., 1997; Penson et al., 2000]. Therefore, health care professionals need:

- To understand the application, nature, and limitations of genetic testing
- To know how to appropriately interpret and effectively communicate test results [Mansoura and Collins, 1998; Sandhaus et al., 2001]

Communication of test information is a two-way exchange of information, opinions, and questions. This dialog of explanation and empathy [Kessler in Capron, 1979; Sharpe, 1994a; Tudor and Dieppe, 1996; Edwards et al., 2002; Bennett, et al., 2003] is necessary if decision making is to reflect the attitudes, values, and beliefs of the patient and family, who will live with the outcomes of the information.

This chapter will explore the often difficult process of giving genetic test results. By focusing on the "three Cs," comprehension, communication, and counseling, practical steps will be described and illustrated through case examples, to incorporate the result-giving process into practice.

WEB RESOURCES

Most major medical centers have clinical genetics services. In addition, many of these will offer satellite clinics to the surrounding communities. If you need help locating clinical genetics professionals, online directories are available, such as:

Genetics Clinic Directory

GeneTests: www.geneclinics.org

Genetic Counselors

National Society of Genetic Counselors, Inc.: http://www.nsgc.org/ resourcelink.asp

Canadian Association of Genetic Counsellors: http://www.cagc-accg.ca/

Patient Education and Risk Communication

National Institutes of Health: National Library of Medicine: http://www.nlm .nih.gov/pubs/cbm/health_risk_communication.html

Genetics Home Reference: http://ghr.nlm.nih.gov/

Dolan DNA Learning Center, Cold Spring Harbor Laboratories: http://www .dnalc.org/

Support Groups

Disease and/or symptom-specific support groups can be incredibly helpful to patients and their families who often feel isolated and immobilized by a new diagnosis. Support groups can provide a sense of belonging and the security of knowing that others have "been in your shoes." Many support groups also provide educational materials through fact sheets and up-to-date newsletters as well as fund-raising opportunities to support ongoing disease research. Thus, these groups frequently can fulfill the patient and family's coping needs at many levels. There are several directories of support groups including:

> Alliance of Genetic Support Groups, Washington, D.C., 202-966-5557: http:// geneticalliance.org
>
> Family Village, Waisman Center, Madison, Wisconsin: http://www.familyvillage .wisc.edu/index.htmlx
>
> National Organization of Rare Disorders (NORD), Danbury, Connecticut, 800-999-6673 or 203-744-0100: http://rarediseases.org/search/rdblist.html

PRETEST FACTORS THAT INFLUENCE SUCCESSFUL GIVING OF GENETIC TEST RESULTS

The process of giving results begins with the initial visit. Therefore, prior to considering factors that influence effective communication of a genetic test result, it is important to first review the foundation built during the initial clinic visit. The initial clinic visit includes:

- Obtaining a detailed family and personal medical history
- Assessment of the patient's understanding of the inheritance of the disorder
- Discussion of the test sensitivity in general and within the family context

It is important to explore the meaning of genetic information to both the patient as an individual and as a member of a family, the family communication pattern, how this information has been dealt with in the past, and how it may be applied to future life and medical decisions. Also, the genetic disorder itself, and the morbidity and mortality associated with it, as well whether or not there is medical treatment for the disorder, will influence the appraisal of the disorder and the response to the test result [McConkie-Rosell and Sullivan, 1999].

Exploration of these issues with the patient helps develop an empathetic relationship between the health care provider and the patient. This relationship is key to effective communication; once results are given, the emotional response to the information may overshadow all else that transpires during that clinic visit. Successful communication of the test result is based upon this foundational relationship, tailored to the needs of the patient and their family. This solid foundation allows the clinician to communicate results in a meaningful way and anticipate essential questions. The initial visit ends with the development of a plan for how results will be given.

CASE EXAMPLES

Each patient is unique and no two genetic testing scenarios will ever be the same. There are, however, key points to consider based on the type of genetic testing performed. It is possible to divide genetic testing into three main "types":

1. Diagnostic testing
2. Increased risk testing
3. Carrier testing

Although there are numerous possible subdivisions within each group, the only distinction we wish to make is that diagnostic testing can be either symptomatic or nonsymptomatic. The following examples will be used to illustrate some of the key considerations to help guide your discussions about test results and their implications with your patients.

Example 1: Diagnostic Testing in a Symptomatic Patient

A 48-year-old man presents with a new diagnosis of colon cancer. He did not have many polyps. He had been screened based on a strong family history of nonpolyposis colorectal cancer: His mother was diagnosed at 52 years of age, his maternal uncle was diagnosed at 60 years of age, and his maternal first cousin was diagnosed at 45 years of age. He meets the clinical criteria [Vasen et al., 1999] for hereditary nonpolyposis colorectal cancer (HNPCC) or Lynch syndrome, an autosomal-dominant cancer syndrome, and is offered genetic testing for HNPCC.

Example 2: Diagnostic Testing in a Nonsymptomatic Patient

A 26-year-old woman presents with a diagnosis of anxiety disorder and a family history of Huntington disease (HD), an autosomal-dominant neurodegenerative disorder. She explains that her paternal grandfather was recently diagnosed at 72 years of age. She has become quite concerned about her risk to one day develop HD. She questioned her psychiatrist about possible early psychiatric symptoms and has also had a normal neurological exam by a neurologist who specializes in movement disorders. Currently, she has no symptoms of HD but based on her family history she is at 25 percent risk of one day developing them. She requests genetic testing of the HD gene.

Example 3: Increased Risk Testing

A 23-year-old male presents to the student health clinic for an annual check-up. His interim medical history is significant for recently participating in a study of the feasibility of population screening for hereditary hemochromatosis (HHC), an autosomal-recessive iron storage disorder. The study was offered through the School of Public Health where he is a student. As part of this study he had genetic testing of the HHC gene.

Example 4: Carrier Testing

A married couple both aged 31 present for a preconception visit. The woman is concerned about her family history of cystic fibrosis (CF) an autosomal-recessive disorder. Her older brother had been diagnosed with CF when he was 5 years old and later died when he was 19 years old. Based on that family history she has a 66 percent chance to be a carrier for CF. They understand that both of them must be carriers to have an affected child, so her husband would like to be tested, too. They are offered genetic testing for the CF gene.

COMPREHENSION OF GENETIC TESTS AND COMMUNICATING RISK

Several concepts related to genetic tests are essential to understand prior to communicating a test result to a patient. They are:

- Sensitivity and specificity
- Detection rates
- Population- and ethnicity-based data
- Clinical significance
- Test methodology
- Risk

Sensitivity and Specificity

Sensitivity refers to the frequency with which a test yields a positive result when the individual being tested is actually affected and/or has the gene mutation in question. *Specificity* refers to the frequency with which a test yields a negative result when the individual being tested is actually unaffected and/or does not have the gene mutation in question.

Positive predictive value is often confused with sensitivity. The positive predictive value will depend upon what you hope to "predict" and it can be tricky. It can refer to the likelihood that an individual with a positive molecular or clinical test result actually has the particular gene in question, is clinically affected, or will develop the disease in the future.

Detection Rates

Often described in genetic test results, the detection rate is almost synonymous with sensitivity. *Detection rate* is the frequency that a particular test will find a gene change within a certain, defined population. Often that population is defined as individuals who are actually affected with the genetic condition.

Sometimes, the detection rate will be subdivided to distinguish different patient populations by a specific symptom (i.e., individuals with or without blue sclerae in osteogenesis imperfecta testing) or by their ethnic origins (i.e., Ashkenazi Jewish ancestry in Gaucher disease testing). It is very important to know the population to which the detection rate is referring.

Test Methodology

There is a multitude of ways in which "genetic testing" as a whole can be performed. In fact, when a single genetic test is ordered, more often than not, several different methods will be employed to produce the end result. For example, if you were to order fragile X syndrome DNA (deoxyribonucleic acid) analysis, at least two different testing methods, specifically Southern blot and polymerase chain reaction (PCR), should be used to provide the most accurate result. It is not always necessary to understand the different test modalities, though it can be quite helpful especially when different information is provided by the different methods. In the case of fragile X syndrome, PCR provides information on the repeat number and Southern blot provides information on the methylation status [Bell et al., 1991]. Both are essential for clinical interpretation [McConkie-Rosell et al., 1993; Merenstein et al., 1996].

Take note of the method(s) used to provide the genetic test result. Different laboratories may use different testing methodologies, and, therefore, the detection rates, and perhaps even the end result interpretation may be different given the same test indication ordered.

Clinical Significance

Clinical significance refers to whether an abnormality found through genetic testing of a patient's sample is known to actually result in a measurable phenotype. In some instances the change, often referred to as a variant, has not been seen in other clinically affected individuals and/or its ability to affect the function of the gene in question is not fully understood. In these cases, the test result may classify the abnormal finding as a "variant of unknown significance." A functional gene change/variant is typically thought to have clinical significance because it is predicted or known to alter the function of the protein produced in some way.

In a symptomatic individual, be careful not to presume that any gene change found is clinically significant and thus responsible for his or her symptoms. If the result states that the significance of a genetic finding is unknown, you cannot be certain that the patient's symptoms are directly related to it. Additional studies of other affected family members can help prove or disprove the possible clinical significance of an otherwise variant of unknown significance.

Population- and Ethnicity-Based Data

Population-based data is often created through disease registries, where disease incidence numbers are compiled and compared to analyze prevalence trends among different demographic groups (i.e., age, ethnicity, sex) and with varying risk exposures (i.e., medications, tobacco products, diet). This data can produce enormously helpful, well-publicized risk estimates such as the 13.4 percent lifetime risk of breast

Table 14.1   Population-Based Risks

1 in 28	Risk of a child being born with some congenital abnormality[a]
1 in 125	Risk of a child being born with a structural heart defect[a]
45.59%	Lifetime risk of being diagnosed with cancer in a white American male[b]
31.76%	Lifetime risk of being diagnosed with cancer in a black American female[b]
12%	Risk of a child being born premature in the United States[a]
1 in 15	Risk of carrying one abnormal gene for Gaucher disease in Eastern European Jewish individuals[a]
1 in 75	Risk of carrying one abnormal gene for β-Thalassemia in African American individuals[a]
1 in 6	Risk of carrying one abnormal gene for sickle cell disease in West African individuals[a]

[a] March of Dimes [2004].
[b] Jemal et al. [2004].

cancer the average American woman faces [Jemal et al., 2004]. However, the patient is an individual, with his or her own specific environmental exposures, family and medical histories, and epidemiologic factors. The 13.4 percent risk of breast cancer does not take into account any of these additional, patient-specific factors. It is simply an estimate of the average risk to the American female to develop breast cancer at some point during her lifetime. Thus, population- and ethnicity-based data can provide baseline risk information from which your patient can move up (increased risk) or down (decreased risk) based on his or her specific factors including genetic test results. Be sure to examine the data to find the "best-fit" population- and ethnicity-based risk estimate for your patient. See Table 14.1 for examples of risk estimates within the general population.

Risk

Risk by definition refers to both the possibility of loss or peril as well as the thing that creates or suggests a hazard [Merriam-Webster, 2004]. In health care, it is often written and discussed in terms of the probability of a hazardous event, typically associated with morbidity or mortality, occurring.

Thus, a risk is defined by a specific health hazard:

- *Absolute risk* is the general population risk that an individual will develop a particular disease or condition over a defined time period—an example is the general lifetime risk of a woman for developing breast cancer.

- *Relative risk* is the ratio of risk for a disease comparing those who are exposed to a particular risk factor to those who are not [Calman, 1996].

When dealing with genetic disease, the risk can vary based on whose risk is being considered, the individual, his or her children, or other relatives. This can be quite confusing, especially to the patient. Health care professionals may presume that patients understand the risk information presented, however, this may not always be the case [Axworthy et al., 1996; Marteau and Michie, 1999a; Marteau et al., 1999b;

Bogardus et al., 1999; Lloyd, 2001], especially if health care professionals themselves are uncertain as to the nature and implications posed by test results [Sandhaus et al., 2001; McGovern et al., 2003b; Goos et al., 2004].

Take, for example, *risk of recurrence* for a simple, autosomal-recessive, single-gene disorder. The health hazard is the chance that there will be a recurrence of the disorder in the family. With this straightforward example, the recurrence of another affected individual only could occur through the birth of that individual. If patients are the parents of an affected child, then their recurrence risk is 25 percent based on Mendelian principles. If, however, the patient is the affected individual, then his or her recurrence risk will depend upon the genetic status of his or her reproductive partner. This genetic status is defined by yet more risks, specifically carrier risks (the risk to "carry" a single deleterious gene mutation in the heterozygous state).

A *carrier risk* may be estimated from the disease frequency in the general population or an ethnic-specific population, or it can be a combination of both population-based data as well as molecular data gained through genetic testing. Example 4 in this chapter examines this issue.

In not so straightforward examples, such as cancer syndromes, risk of recurrence takes on a new meaning and complexity. It becomes intertwined with risk of disease occurrence and recurrence. Such is the case with hereditary breast and ovarian cancer syndrome. The inheritance pattern of the hereditary gene mutation is autosomal dominant and thus has a 50 percent risk of recurrence to offspring of an individual known to have the mutation. If the mutation is, in fact, inherited, the risk of disease occurrence is up to 85 percent for breast cancer and 40 percent for ovarian cancer if the individual is female [Antoniou et al., 2003].

Yet another consideration is the state of being *at increased risk*. In order to discuss an increased risk status with a patient, it is imperative that the baseline, or a priori risk for the health hazard in question, be understood. The a priori risk is based on given information such as population-based data or Mendelian principles. Beginning with this baseline risk number, the increased risk state can be put into a broader context for comparison. A patient told that he is at a threefold increased risk of disease may perceive his risk to be much higher than if he is told that the population risk is 1 in 10,000 and thus his increased risk equates to 3 in 10,000.

Thus, risk is not a static number, but one that can change to reflect new information. Numerous breast cancer risk models have been developed to aid the incorporation of additional epidemiological and/or genetic data and the resulting risk adjustment [Antoniou et al., 2004; Claus et al., 1994; Gail et al., 1989]. Simple Bayesian statistical calculations are also commonly used to adjust a patient's risk, either higher or lower, from their a priori risk. Risk assessment is reviewed in depth in Chapter 13 of this text.

Key Points

- Risks may be perceived as higher when the same degree of risk is presented as odds compared to percentages [Kessler and Levine, 1987; Redelmeier et al., 1993] and when the disease or condition is perceived to be more threatening [Burke et al., 2000; Edwards et al., 2002].

- The use of terms such as "likely" or "rare" can imply significantly different meanings to individual patients [Budescu et al., 1988; Edwards et al., 2002; Paling, 2003].
- Risk may have different interpretations and will vary based on who is "at-risk"— self versus offspring.

INTERPRETATION OF THE TEST RESULT

Even given an appreciation of the concepts outlined above, it can be difficult to understand the information presented in the laboratory report. In order to make life and health decisions using the information gained from genetic testing, the result must be accurately understood and interpreted based on:

- The information in the laboratory report
- The health care provider's understanding of the information
- The patient's personalization of the information

First, the written laboratory report needs to be interpreted. The laboratory report should include the test outcome, methodology(s) used, detection rate, and predicted clinical significance of the result. The American College of Medical Genetics (ACMG) recommends [ACMG, 2004] that, in general, a clinical test report should include the following elements:

1. Repetition on the report of key interpretive information
2. Clear presentation of the result (with ranges, cutoffs as appropriate)
3. Interpretive statement that explains the result in the context of the test purpose (may include an estimate of risk)
4. Disclaimer or explanation of test limitations (e.g., analytic and clinical validity, nonpaternity)
5. Investigational test statement if appropriate
6. Information used for risk assessment calculations

Adherence with these recommendations is, however, voluntary, and the information in a test report from laboratory to laboratory may vary significantly and may not be complete. A study conducted by McGovern et al. [2003b] revealed that 28 percent of the genetic professionals surveyed felt the information contained within genetic test reports received from laboratories was inadequate to explain the result to the patient.

Second, the interpretation of the genetic test result is made by the physician who receives the report from the lab. If there is any concern regarding confusion or interpretation of a test result, the best remedy is to contact the testing laboratory. The laboratory should have staff whose role is to facilitate the entire testing process, from ordering the correct test, to sample collection and shipping, to accurate interpretation of the test result. In many cases, this individual is a genetic counselor, whose specialized training in genetic science and the ability to communicate that science in a comprehensible manner makes them ideally qualified. A study that surveyed the commercial uptake of genetic testing for familial adenomatous polyposis (FAP) revealed that in 31.6 percent of cases the physician misinterpreted the test result [Giardiello et al., 1997]. Ask to speak to the genetic counselor if available, or a laboratory director to clarify any questions you may have.

Lastly, the genetic testing information is interpreted by the patient. It has long been known that genetic risk is often reinterpreted into a binary form (either an event will happen or it will not happen) [Lippman-Hand and Fraser, 1979]. The patient/family will personalize and interpret the test result on several levels remembering generalities rather than specifics: good or bad news, high or low risk [Shiloh and Saxe, 1989], and decisions about health care will be made based on the outcome of this process.

Genetic risk interpretation is influenced by the lived experience of the disorder for which testing is being done. This life experience includes:

- Who in the family is affected
- Amount of exposure and relationship to those affected individuals
- The interpretive meaning of the disorder created by the family

Individuals whose lives have been affected by a genetic disorder often search for meaning and frequently have personal and family beliefs about the cause of the disorder that may or may not include the inheritance or genetic nature of the disorder [McAllister, 2003].

Patients and their families' bring with them their own lay construction of inheritance, what are genes and how traits are inherited, that influence understanding and incorporation of genetic test results [Shiloh and Berkenstadt, 1992]. Individuals appear to draw upon their personal theories of inheritance, developed from these lay constructions and personally salient aspects of the family history, in their interpretation and/or acceptance that a specific disorder is inherited [McAllister, 2003]. As part of exploring the family's interpretive meaning of the disorder, it is important to explore the family's perceptions of the cause of the illness as well as the view of the accuracy of genetic testing held by family members.

Personal experiences and values can play a significant role for both health care professionals as well as patients [Lloyd, 2001; Doust and del Mar, 2004] in the perception and interpretation of test information. Thus, interpretation by the laboratory, the health care provider, and the patient affects the process and success of providing test result information. Interpretation does not stand alone, however, but is intricately woven throughout the concepts of comprehension, communication, and counseling. Your interpretation of the test result relies heavily upon your comprehension of the disease, the patient, and the test itself. You must then communicate your interpretation of the test result and its significance to the patient. The patient thus begins his or her own interpretative process through comprehension of the complex information and the implied context in which it is communicated.

Through an empathetic, counseling relationship the test result information can be communicated in a manner that acknowledges the patient's life experiences and anticipates his emotional state. This approach facilitates accurate and meaningful inclusion of the new information. It is this interpretative meaning attributed to the genetic test result by the patient that is the basis upon which health and life decisions are made.

Key Points

Patient understanding and recall of genetic test information can be affected by:

- The health care professional's:

  - Choice of terminology, format, and context in which the information is presented
  - Choice of the timing, manner, and method of communication, including the health care professional's tone of voice and facial expressions

- The patient's:

  - Expectations of and motivations for seeking the information
  - Prior experiences including cultural and family beliefs and traditions
  - Level of education and understanding of the terminology used
  - Emotional and psychological responses
  - Personal experiences, values, and beliefs and prior coping strategies [Smith, 1998; Abramsky et al., 2001]

PRACTICAL STEPS TO COMMUNICATING GENETIC TEST RESULTS

In an ideal situation, the manner in which test results are to be communicated have been planned by the physician and patient prior to the test even being ordered. Such planning empowers the patient and family with a sense of control regarding the situation surrounding a test result over which they have no control. Pretest planning and its value to the communication process cannot be overstated. In communicating test results, there are several practical points to keep in mind.

Consider the Timing   If you receive a genetic test result on a Friday evening, do you inform the patient immediately or wait until Monday morning? Do you give it over the phone or schedule an appointment? The same formula may not work for all patients. In some instances reporting results by phone the day before a scheduled appointment will allow the patient/family some time to adjust to the news and thus be more able to actively engage in discussions and health care planning at the appointment. In other instances, reporting test results by phone can lead to increased anxiety and misunderstandings without immediately available clarification by their health care provider. If possible, plan this ahead of time with the patient based on anticipated receipt of the result, clinic policy, and patient and health care provider preference.

Consider the Setting   Is there a private room available? The information is likely to be technically and emotionally complex. Try to find a quiet area, free from distraction that will be available for more time than you anticipate needing.

Consider Your Language   Language here refers to both verbal and nonverbal means of communication. "The first communication to parents is important because it may affect how information is later interpreted or even sought" [Abramsky et al., 2001, p. 463]. There is no one correct way to present genetic testing information to your patient. Each patient is unique, and thus the presentation of information should be tailored to individual patients: their expectations, their learning style, and their questions. Present the information in a simple, straightforward manner. Do not assume that because a patient demonstrates conversational familiarity with genetic terminology that this is an indication of understanding. Lanie et al. [2004] found that individuals who were able to use genetic terminology often lack understanding about basic genetics.

The format in which information is presented can affect understanding among physicians and patients alike [Lippman-Hand and Fraser, 1979; Bryant et al., 1980; Slovic, 1987; Yamagishi, 1997; Phillips et al., 1999; Edwards et al., 2001].

- When physicians were asked to interpret statistical information provided by a cancer screening test, only 1 in 24 physicians were able to accurately solve the problem when the information was posed in a probability format, compared to 16 out of 24 physicians when presented in terms of natural frequency [Hoffrage et al., 2000; Greenhalgh et al., 2004].

- Numeric risk should be described as both percentages (25 percent) and fractions (1 in 4). Using "real-world" numbers (e.g., 1 in 5) and examples can also affect a patient's perception and comprehension of risk [Gigerenzer and Edwards, 2003] (see Table 14.2).

Additionally, presenting information from both a positive and negative risk perspective can provide a more realistic and supportive context [Phillips et al., 1999; Edwards et al., 2001; Paling, 2003]. Framing the result in a negative or positive manner can mean the difference between the patient perceiving the glass half full or half empty [Gigerenzer and Edwards, 2003]. For example, compare a 5 percent risk of complications paired with its corollary, a 95 percent chance of no complications. A treatment that *saves 8 lives out of 10* may be perceived to be better than one that *fails to save 2 in every 10*.

Perception of the resulting risk also may directly influence health care decisions. In their study of how women perceive screening results for maternal serum α-fetoprotein (AFP), Shiloh and her colleagues [2001] found that negatively framed test results raised more concerns about the fetus and increased the likelihood of amniocentesis than equivalent test results that were positively framed. Similarly, Michie et al. [2002a] found that even after negative testing for FAP, those who felt the test results were uncertain perceived themselves to be at greater risk, were more threatened by their negative test result, and indicated that they would continue with high-risk bowel screening.

Table 14.2 Conversion Between Commonly
Used Odds and Percentages

Odds	Percentage
1 in 2	50
1 in 3	33[a]
1 in 4	25
1 in 5	20
1 in 6	17[a]
1 in 7	14[a]
1 in 8	12[a]
1 in 9	11[a]
1 in 10	10
1 in 12	8[a]
1 in 15	7[a]
1 in 20	5
1 in 25	4
1 in 30	3[a]
1 in 40	2.5
1 in 50	2
1 in 100	1
1 in 200	0.5
1 in 500	0.2
1 in 1000	0.1
1 in 10,000	0.01

[a] Approximate value.

The use of graphs and pictures with written explanations provide added benefit [Kessler and Levine, 1987; Marteau, 1989; Shiloh and Sagi, 1989; Redelmeier et al., 1993; Grimes and Snively, 1999; Sandhus et al., 2001; Edwards et al., 2001, 2002; Woloshin et al., 2002]. Not only does it make use of a visual, concrete representation of the information, it can also be sent home with the patient for future reference. Additional tools include the use of metaphors (i.e., 50 percent risk as flipping a coin) and trying to put the risk number into a real-world context [Smith, 1998]. For example, when giving a carrier frequency of 1/20 for cystic fibrosis, it is easy for a patient to consider 40 friends and think that statistically at least 2 of them would also be carriers. Additionally, risk comparisons within the context of typical risks that patients encounter in everyday living can help to make the risk seem more understandable [Paling, 2003; Ghosh, 2004].

It is important to gage the patient and/or family's understanding of the information presented. This can be accomplished by asking them to describe in their own terms what they interpret their risk to be or how they plan to explain it to a close friend or relative who is not present in the clinic room. Address misunderstanding immediately. Medical and life decisions based on inaccurate interpretations of genetic test and risk information could be tremendously detrimental to all involved.

"Ultimately it is the counselee's perception of risk rather than actual risk that is meaningful" [Smith, 1998, p. 109].

Consider Patient/Family Reaction   The context in which risk information is presented and the resulting emotional responses can have a significant impact on perception and interpretation. [Vlek, 1987; Paling, 2003]. What were the patient's prior expectations? All genetic testing information is potentially life altering, and thus threatening to the patient's sense of "self."

- Patients often wish to discuss the psychological, socioeconomic, and familial concerns [Geller et al., 1998; Marteau and Croyle, 1998b; Menahem, 1998; McConkie-Rosell et al., 2001; Esplen et al., 2001; Speice et al., 2002; Parsons et. al., 2003]; however, the health care provider may be more focused on factual test results [Levinson and Roter, 1997; Michie et al., 1997a; Roter et al., 1997; Levinson et al., 2000; Carroll et al., 2000.; Phillips et al., 2000; Bensing et al., 2003].

- An empathetic approach [Bellett and Maloney, 1991; Zinn, 1993; Suchman et al., 1997] with a focus on the concerns of the patient and the family is associated with improved communication [Brody et al., 1989; Suchman et al., 1993; Ambady et al., 2002], understanding, reduction of patient dissatisfaction, and allegations of malpractice [Beckman et al., 1994; Levinson et al., 1997; Levinson, 1997; Bogardus et al., 1999; Blackston et al., 2002; Moore et al., 2000; Ambady et al., 2002].

- Patients may be uncertain as to the meaning and implications posed by terms such as *rare* and *carrier* and may wish to discuss choosing an appropriate time to discuss and disclose genetic information with their children [Fanos et al., 2001; McConkie-Rosell et al., 1999, 2002; Ciske et al., 2001; Tercyak et al., 2001b; McConkie-Rosell and Spiridigliozzi, 2004].

- Patients are influenced by not only what they are told, but who and how they are informed about genetic test results, suggesting that the health care provider's understanding, comfort with the type of testing, and clinical interpretation is transmitted to the patient [Michie et al., 2002b].

In stressful situations, it is important to adjust both the information provided as well as the manner in which it is communicated to the individual and families' coping style. Family coping styles for the purposes of disclosing genetic test results can be categorized into *monitors* (those who have a tendency to seek information and focus on heath threats) versus *blunters* (those who tend to avoid information) [Suls and Fletcher, 1985]. Providing detailed information tends to reduce distress in the information seekers while increasing distress in those who are avoiding information [Nordin et al., 2002]. Therefore, matching communication of test results to coping style of the patient and their family is important if the information is to be processed and used for medical and health decisions. As discussed in earlier chapters:

- The emotional and psychological responses of the patient may be more critical to risk perception than the actual facts (diagnosis, risk information, prognosis) [Lippman-Hand and Fraser, 1979; Wertz et al., 1986; Lerman et al., 1995; Hoskins et al., 1995; Phillips, 1999; Press et al., 2001; Horowitz et al., 2001; Lodder et al., 2001; Peshkin et al., 2001].

- Providing detailed information at the same time as a discussion of diagnostic information may be confusing and not be understood by the patient due to resulting stress and psychological responses [Lerman et al., 1995; Menahem, 1998; Hagerty et al., 2004].

- A negative test result may not reassure patients who continue to believe that they are at risk [Huggins et al., 1992; Lynch and Watson, 1992; Michie et al., 2003].

- Emotional distress has been associated with patient delay and a lack of compliance with surveillance and management options [Metcalfe et al., 2000; Kash et al., 2000].

- Emotional and psychological problems need to be recognized and appropriately addressed, with referral to a clinical psychologist or psychiatrist where appropriate [ABMG, 2004; ABGC, 2004; CCMG, 2004b].

Consider Your Own Reaction to the Information   Health care professionals may overestimate the burden and negative effects of a genetic disease or disorder [Blaymore et al., 1996; Kirschner et al., 2000; Abramsky et al., 2001]. Do not project your perception of the test result onto the patient. A positive result is not always a tragedy, and a negative result may not be good news as it can evoke feelings of guilt and not belonging, as well as remorse over a lost life lived in fear of the "family disease."

Most health care professionals feel a sense of moral duty to aid their patients and prevent distress if at all possible. This can lead to the premature urge to provide reassurance when giving results considered by the health care provider to be bad news [Buckman, 1992]. Although well intentioned, this act can interrupt the patient's assessment of the news. Allow the patient and family to adjust to the news even if that means allowing them to be upset. Providing additional information about the result and its health implication may not be comforting or helpful to the patient and their family. Ask open, empathetic questions. Do not be afraid to discuss emotions. Acknowledge the impact of the test result and be willing to take time to explore the meaning of the test result on the individual as well as the family.

Consider the Afterwards   Plan ahead for simple things, such as how the patient will get back home if he is under considerable psychological duress. Follow-up visits and written information [Baker et al., 2002; McConkie-Rosell et al., 1995; Michie et al., 1997a; Hallowell and Murton, 1998; Lobb et al., 2004] summarizing the test result(s) and its significance can help to facilitate understanding.

- Plan ahead and request that the patient bring a support person along.
- Provide written materials such as copies of the test result, explanatory brochures, graphs and pictures, contact information for local or national support and/or educational sources, and so forth.
- Make a plan with the patient for future contact through phone call and/or face-to-face appointment to review the test result, its implication, and any health plans that have been discussed.

The information presented is exceedingly complicated on many levels. Knowing that he or she has means to ask more questions and seek further clarification can provide some comfort and assurance to the patient who is attempting to adjust, cope, and make life and medical decisions regarding this information (see Table 14.3).

Table 14.3   Key Considerations When Communicating Genetic Test Results

Know the test result
    In context of the terminology and methodology
    In context of your patient
Know your patient
    From a medical standpoint
    From a genetic standpoint
    From a psychosocial standpoint
Know your own clinical limitations and when it is appropriate to involve a clinical genetics specialist
Present accurate information in an appropriate manner
    Recognize patient's level of cognitive/emotional functioning
Be prepared to present significant amounts of supplementary materials and information (i.e., disease fact sheets, explanatory graphs and pictures, support group contacts, prevention guidelines, etc.)
    Recognize when you should and should not give it to your patient
Be empathetic
    Acknowledge and adjust to the patient's coping style
Be silent
    Allow the patient time to adjust to the information
Be patient
    Be willing to repeat yourself
    Be willing to accept the blame if patients do not understand you
Plan ahead for future review of the information
    Return appointment for more in-depth discussion
    Allow for inclusion of other family members also affected by the result
Never assume a normal/negative result is good news or that an abnormal/positive result is bad news
Recognize that all genetic test information is potentially life altering
    Empower your patient to control how it will alter his or her life

CASE DISCUSSION

Now that several key considerations for giving and discussing genetic test results have been presented, we will apply them to our clinical examples introduced at the beginning of the chapter. The interactions with the patients in these examples have been italicized.

Example 1: Diagnostic Testing in a Symptomatic Patient

The patient's genetic test result is negative.

Know Your Test HNPCC is an autosomal-dominant cancer syndrome with which multiple genes have been associated. The patient's report states that it analyzed the *MLH1* and *MSH2* genes through sequencing and estimate the testing will detect approximately 90 percent of *detectable* mutations [Peltomaki, 2003]. No ethnic or population predilections are mentioned.

So what does this mean? The test method is rather good at picking up gene mutations in these two genes known to be associated with HNPCC. However, at most 50 percent of individuals who meet the revised clinical criteria will have a detectable mutation in either *MSH1* or *MLH2* [Wagner et al., 2003]. Therefore, although the detection rate seemed high (90 percent), it was misleading since the chance of actually finding a gene change in the patient was only about 50 percent.

What does this mean to the patient? Regardless of the genetic test result, his family still meets clinical criteria for a clinical diagnosis of HNPCC. With his diagnosis of colon cancer, the patient also is considered to be clinically diagnosed with HNPCC. Therefore, he is at risk to develop recurrences as well as additional cancers associated with HNPCC and should be followed according to the recommended preventive guidelines [Burke et al., 1997; Hadley et al., 2004].

Unfortunately, his genetic testing did not reveal the causative gene change for him. It is assumed that other affected members of his family would have the same undetectable gene change; genetic testing using the same test and method would not be useful in them. Instead, family history data must be used to provide risk estimates for currently unaffected, healthy relatives.

All first-degree relatives of affected individuals would have a 50 percent chance of also having inherited HNPCC. Thus, multiple other family members are at risk and should consider referral for genetic cancer risk assessment [Trepanier et al., 2004] and the development of cancer screening and surveillance plans with their physicians.

The negative test result is explained to the patient who expresses confusion regarding how he can have a genetic cancer and a normal genetic test result. He also states that he had always wondered if the cancer in his family was really due to the factory in the town where they grew up. This concern is acknowledged as there are many components that result in cancer, however, the type and pattern of cancer in his family members is consistent with the inherited colorectal cancer syndrome, HNPCC.

It is further explained that there was only about a 50 percent chance the genetic test would have found the specific gene change responsible for the colon cancer in him and his family. His clinical diagnosis of HNPCC is reviewed again as well as the preventative guidelines.

The inheritance pattern is discussed and the family history is reviewed where several "at-risk" members are identified, particularly the patient's two adult daughters. The patient expressed remorse over learning that his daughters were potentially facing an HNPCC-related cancer, too. He had looked up his family tree and not down; only considering his personal risk. The patient is strongly encouraged to discuss his diagnosis of HNPCC with his family members especially his two daughters who are at 50 percent risk to have inherited HNPCC just as he had. By informing them of this risk while still asymptomatic, he is giving them the ability to avoid some of the complications of HNPCC through screening and early detection of cancer.

The patient seems unsure of his ability to discuss the complexities of HNPCC with his daughters and other family members who live in a different state. This difficulty is acknowledged and the patient is reassured that he will not be left on his own to explain the complex information. The patient is engaged in brief "role play" of how he plans to explain the inheritance pattern and cancer risks of HNPCC to his family.

Lastly, he is given written materials regarding HNPCC, its inheritance, and the preventative guidelines as well as contact information for the clinic so that he and/or his family members can receive more information when future questions arise.

Key Points

1. A negative molecular test does not mean the cancer in the family is not genetic and thus not inherited. This is often a point of confusion for patients. A positive molecular test enables accurate diagnostic and/or increased risk testing (as is the case with most cancer syndromes) for other family members at risk of having inherited the condition in the family.

2. A negative molecular test does not override a positive clinical diagnosis.

3. The detection rate for finding "detectable mutations" can be misleading. You want to know the chance of finding a mutation in your patient, based on the patient's medical and family history.

4. Accurate family history data is essential to accurate recurrence risk information.

5. Recurrence risk information for your patient's family members requires intrafamilial communication. This brings up the issue of duty to warn which is addressed in Chapter 15 of this text.

6. Diagnostic testing for a symptomatic family member can allow accurate predictive risk testing in another family member.

Example 2: Diagnostic Testing in a Nonsymptomatic Patient

Know Your Test Huntington disease is an autosomal-dominant, neurodegenerative disorder caused by an expanded CAG trinucleotide repeat within the HD gene.

A CAG repeat number greater than or equal to 40 is considered diagnostic, though repeats greater than 36 are considered unstable and have been associated with symptoms in the individual [Rubinsztein et al., 1996; Brinkman et al., 1997]. HD is fully penetrant with a lifetime risk of 100 percent. The detection rate of the CAG repeat approaches 100 percent. Presymptomatic genetic testing for HD is available and can provide accurate predictive information for at-risk relatives of affected individuals. Although there are data providing age ranges for onset related to the size of the CAG repeat [Langbehn et al., 2004], caution should be taken in predicting the onset of the disease for an individual.

Know Your Own Clinical Limitations   The Huntington Disease Society of America (HDSA) has recommended guidelines for laboratories and health care providers who wish to offer this service [International Huntington Association and World Federation of Neurology, 1994]. This recommended presymptomatic testing protocol entails a multistep process including evaluations by medical geneticists, psychiatrists, and neurologists. These recommendations were made because presymptomatic testing for HD presents both complex social and medical concerns [Green and Botkin, 2003; O'Daniel and Wells, 2002].

Know the Patient   The patient does not have symptoms but presents in an at-risk state based upon her family history. Examining this aspect shows that she can only develop HD if two conditions are met: (1) her grandfather truly had HD and not another similar condition, and (2) she inherits the expanded HD repeat from her mother, the genetic link back to her affected grandfather. The first point requires that the HD diagnosis be confirmed in the grandfather. The second point requires that the mother have the expanded HD repeat, thus predicting disease onset in her. So who should be tested?

Confirming the grandfather's diagnosis is critically important to ensure that the appropriate test is offered to the patient and that the grandfather does not in fact have a different disorder with symptoms similar to HD. Confirmation of the diagnosis is best done through review of the medical records and/or initiation of molecular testing in the affected grandfather first prior to testing other relatives. To obtain this needed medical information, the guardian of the affected grandfather must provide written consent. This brings us back to the second point as the patient's mother is the grandfather's next of kin.

This testing has significant implications for multiple members of the family. These issues should be addressed with the patient prior to testing. If, however, the testing process has already begun, these issues can still be addressed before revealing the results to the patient. If the patient were tested and found to have the expanded HD repeat, it would reveal the genetic status of the untested mother.

The patient is contacted to review her testing in the context of needing to confirm the diagnosis in her grandfather and possibly revealing information about her mother. The patient states that her grandfather has had confirmatory HD testing, but that she had not considered her mother as the connecting link. She is still eager to learn her own risk of HD but recognizes the need to include her mother. The patient discusses her desire for HD testing with her mother who agrees to attend the pretest clinic visit to learn more.

When the mother came in for counseling, she was accompanied by her daughter (the original patient) as well as her three other adult children. The mother brought a copy of the DNA test on the grandfather confirming the diagnosis of HD.

During the clinic visit, the clinical symptoms, inheritance pattern, and testing protocol for HD are reviewed with the family. The mother expressed her concern and anguish over years of not knowing what was wrong with her father. For her, the diagnosis of HD was in many ways a relief. She finally knew what was wrong. The mother had not previously thought about her 50 percent risk for HD or how to go about getting tested. She states that she would like to know if she had inherited the HD gene because she feels it would make a difference in her retirement plans and where she would live. She also expressed concern for her children. The mother, although clearly upset by the diagnosis of HD, felt that the unknown of being at-risk was more difficult than learning of the certainty of HD. After much discussion, the mother, with the support of her children, made the decision to proceed with testing.

After completing the HD presymptomatic testing protocol [International Huntington Association and World Federation of Neurology, 1994], the mother came to her results appointment accompanied by her children. The mother is informed that she has inherited the expanded HD gene from her father. This means she will develop HD at some point in her lifetime. Her current neurological and psychiatric evaluations were normal, and thus at this time she does not have clinical symptoms of HD. The inheritance of HD as well as the difficulty in predicting the onset of symptoms is again reviewed with the mother and her children. A plan is developed with them for continued surveillance including periodic neurologic and psychiatric evaluations.

The risk to the daughter, your initial patient, and her siblings is now 50 percent based on the autosomal-dominant inheritance of HD.

The daughter and her siblings immediately state that they are interested in being tested. They are encouraged to take time to adjust to their mother's test result and subsequent increase in their own risk. They are also encouraged to consider their own unique family situations and how learning their own genetic status would impact their lives. Lastly they are encouraged to call to start the testing process for themselves when they are ready.

Key Points

1. The presenting patient may not always be the best person to test.
2. Confirm diagnosis and outcome of diagnostic testing in an affected relative.
3. Consider the implications of the test result on other relatives. It may indirectly disclose genetic status in a relative.

4. Respect family communication and coping styles. Not all family members may want testing, or their genetic status revealed. This can produce significant stress and tension on family relationships.

5. Explore the meaning of the diagnosis on an individual as well as a family basis.

6. Discuss implication and meaning of the genetic test prior to drawing blood.

Example 3: Increased Risk Testing

The patient's genetic testing revealed that he is homozygous for the *C282Y* mutation within the HHC gene.

Know Your Test   Hereditary hemochromotosis (HHC) is an autosomal-recessive disorder of iron metabolism that is characterized by iron overload in major body organs. Untreated, hemochromatosis can lead to multiorgan failure. However, the penetrance, or risk of developing clinical disease symptoms in the presence of a disease-causing geneotype is difficult if not impossible to predict [Pietrangelo, 2004a].

Penetrance of HHC is dependent on the particular gene change/mutation as well as other factors such as age, sex, alcohol use, and dietary consumption of iron [Pietrangelo, 2004a]. The two most common mutations are *C282Y* and *H63D*. Homozygotes for the *C282Y* mutation have the greatest risk of developing symptoms [Waalen et al., 2002]. Bulaj et al. [2000] has estimated the risk for disease-related morbidity to be 29 percent for homozygous men over the age of 40 and 11 percent for women over the age of 50 years. It is clear that the risk of developing symptoms increases with age. Thus, it is impossible to predict the future effects of being homozygous for the *C282Y* mutation for this young man. The most common finding in individuals with *C282Y* mutations is increased ferritin levels in the blood [Waalen et al., 2002]. Treatment to prevent the iron overload is periodic phlebotomy. Early diagnosis and treatment, prior to the onset of liver cirrhosis can prevent the long-term consequences of the iron overload [Niederau et al., 1996].

Know the Patient   As part of a feasibility study for population-based genetic screening of HHC the patient has learned that he is a homozygote for the *C282Y* mutation. Although he had received a pamphlet of information from the study, he does not understand what the test actually means to him in terms of his health. He has no current measurable signs of HHC and has not observed any of the symptoms from the pamphlet in his family members. He has become increasingly concerned and presents to clinic hoping to learn more.

Recognize the Impact   It is important to acknowledge the patient's concerns and attempt to give him a sense of control over the newly discovered information. This can be done by creating a preventative screening plan that will balance the benefits of early detection and reduced anxiety with the detriments of possible unnecessary treatment and potential societal or self-perceived stigma of being diagnosed with

HHC [Imperatore et al., 2003]. Additionally, because the test is "genetic" there may also be a perception that a disease process may also be inevitable.

The genetic test result is reviewed with the patient. He voices concerns of possible symptoms, stating that he is having some trouble sleeping and has vague abdominal pain that he thinks is coming from his liver. The test result is reviewed again, discussing that it is not considered diagnostic since he has no symptoms. The patient appears confused and reiterates his abdominal pain. A more in-depth discussion of HHC entails. The typical disease symptoms, prevalence and penetrance, age at onset, and means of preventative screening are reviewed. The patient had not realized what his prior "ferritin studies" meant and appears reassured that his abdominal pain is not likely related to HHC in light of his normal blood iron levels. A plan is developed for plasma ferritin studies at regular intervals in the future as well as preventative phlebotomy treatments if ever necessary.

The autosomal-recessive inheritance pattern is also reviewed as well as the high carrier rate of approximately 1 in 9. The commonality of the HHC mutations are emphasized to "normalize" the state of having the genetic changes since approximately 1 in 364 people have two HHC mutations just like he does (e.g., his college graduating class has 1800 graduates; suggesting that at least 4 other members of the graduating class also have two HHC mutations). Lastly, he is presented with additional information about HHC to share with his family including two siblings (who each have a 25 percent risk) and an appointment is made for the following year to repeat the ferritin levels.

Key Points

1. Genetic testing for "increased risk" of disease can be challenging and may require a review of the literature to determine the most accurate risk to give.

2. This information is very complex to the patient, too, and may require repeat explanations.

3. Such risk testing can create a perception of disease in a healthy individual. Attempt to frame the disease risk in terms of helpful preventative information that can be used to create health surveillance plans.

4. The diagnosis of a "genetic" disorder can have a stigmatizing effect. Providing information on gene and disease frequency using common comparisons can help the patient develop social references and thus aid in normalizing the information.

Example 4-: Carrier Testing

The CF carrier genetic testing is negative for both patients.

Know the Test   Only one gene is known to be associated with CF, the *CFTR* gene. It is a well-studied gene with over 1000 reported mutations. The American College of Medical Genetics (ACMG) issued a revised statement recommending the use of a

23-mutation panel for CF genetic testing when used as a population carrier screening tool [Watson et al., 2004b]. This standard panel is estimated to have a panethnic detection rate of approximately 83 percent. There are additional detection rates for specific ethnic groups.

The couple was offered the ACMG recommended 23-mutation panel. When family history information is collected, she reports her ethnicity to be English and German and he is Asian. She believes her affected brother did have genetic testing but does not know the results.

Know Your Patient   The detection rate in the non-Hispanic Caucasian population is approximately 87 percent and in the Asian American population it is approximately 49 percent. So what does this mean for the patients? In order to give accurate risk information based on the detection rates and negative molecular tests, it also is important to know the population-specific carrier frequencies. The carrier frequency for Asian Americans is approximately 1 in 90. The negative molecular test means that the male patient's carrier risk was reduced from the population carrier rate of 1/90 or ~1 to ~0.6 percent. Bayesian statistical calculations are used to incorporate the new genetic testing data to make the carrier risk adjustment.

The female patient has two factors to consider. First, she had an affected brother. The unaffected sibling of an individual with an autosomal-recessive condition has a two-thirds chance of being a carrier. Thus, although the carrier frequency of CF in the non-Hispanic Caucasian population is 1 in 25, her positive family history has greatly increased her carrier risk. If her affected brother's CF mutations were known, then she could simply be tested for those specific mutations and receive a yes/no answer to her carrier status. However, those test results are not available, so we must rely on the detection rate of the standard mutation panel to adjust her risk. In the female patient, her negative molecular test means her carrier risk was reduced using Bayesian statistical calculations from two-thirds or ~66 to ~17 percent.

The negative molecular test results were reviewed with the couple. Their original risks based on family history (66 percent) and ethnic background (1 percent) were explained. The new adjusted risks of 17 and 0.5 percent based on the testing and its ethnic-group-based detection rates were discussed. Finally, the risk for the couple to have an affected child based on their adjusted carrier risks and the autosomal-recessive inheritance pattern of CF was given to be roughly 1 in 5000.

The couple was encouraged to talk with the maternal family to see if the genetic results from her affected brother could be obtained as that information could significantly alter her carrier risk. The couple stated that was not an option and that her family would be very upset if they were even aware that she had undergone this preconception testing.

If the risk of an affected child had been "too high," the couple had planned to feign infertility and adopt because to them neither the birth of an affected child nor pregnancy termination were options. The couple stated they were very reassured by the 1 in 5000 risk. In light of this information, clinical testing for CF via sweat testing of any future children was discussed.

Key Points

1. It is important to know the patient's original carrier risk prior to testing in order to make accurate adjustments. This may include ethnic-based carrier frequencies and/or family histories.

2. Specific calculations using Bayesian statistics are often necessary when readjusting risk.

3. Medical records, especially genetic test results from other family members, can be invaluable to your patient's test result interpretation when they are available.

UTILIZING ADDITIONAL RESOURCES

Heath care professionals need to understand the potential difficulties created by often complex and often ambiguous test results, especially when helping patients to adjust to and manage sensitive genetic test information [Baird, 1993; Andrews, 1992; Sharpe, 1994b; Thompson and Thompson, 2001; also see Roter, 2004].

Due to the intricate complexities of genetic testing, health care providers should consider involving a genetic counselor or clinical geneticist in this aspect of their patient's care. The evidence of this can be found in the growing number of medical malpractice rulings in cases arising from genetic testing issues [Fortado, 2004].

The roles of the clinical medical geneticist and genetic counselor are to [Biesecker and Peters, 2001; National Human Genome Research Institute, 2004]:

- Assess the patient's family and medical history for risk of a genetic disorder.
- Weigh the medical, social, and ethical decisions surrounding genetic testing.
- Provide support and information to help a person make a decision about testing.
- Interpret the results of genetic tests and medical data.
- Communicate information to the patient/family and other health care professionals in an appropriate and comprehensible manner.
- Provide counseling or refer individuals and families to support services.
- Explain possible treatments or preventive measures.
- Discuss reproductive options.

In addition the physician geneticist provides physical examination, clinical diagnosis, and prescribed treatments when indicated.

FOLLOW-UP AND AFTERCARE

A physician's traditional duty of care extends to postoperative care [*Pike v. Honsinger*, 1898; *Monahan v. Weichert*, 1983; *Tacknyk v. Lake of Woods Clinic*, 1982]. The physician is required to provide necessary care as long as the case requires attention [*Ricks v. Budge*, 1937] or liability may result for abandonment [*Asher v.*

Gutierrez, 1976; *Tavkoli-Nouri v. Gunther*, 2000]. This obligation continues until such time when the physician–patient relationship comes to an end. Generally, this will occur if:

- It is terminated by the patient.
- It is terminated by mutual consent.
- It is terminated by the physician after reasonable notice has been provided to the patient [*Surgical Consultants, P.C. v. Ball*, 1989; *Matthies v. Mastromonaco*, 1998; *Tavakoli-Nouri v. Gunther*, 2000].
- The physician's services are no longer required [*Parknell v. Fitzporter*, 1923; *Cook v. Abbott*, 1980]. This may be condition upon the patient having been provided with reasonable notice [*Asher v. Gutierrez*, 1976; *Tavakoli-Nouri v. Gunther*, 2000; *Cook v. Abbott*, 1980].

A physician has an obligation to instruct a patient as to all appropriate precautions that must be carried out subsequent to treatment [*Bauer v. Friedland*, 1986; *Smith v. Rae*, 1919]. These can include the obligation:

- To provide the patient with an appropriate and *timely* follow-up especially for a newborn with a genetic disease to reduce morbidity and mortality [e.g., sickle cell disease, Miller et al. (1990); phenylketonuria (PKU), Koch (1999); *Humana of Kentucky, Inc. v. McKee*, 1992].
- To effectively communicate the nature of the medical condition or risk status and any continuing risks [Andrews, 1992; Gordon et al., 2003] such as the birth of a prospective child afflicted with a genetic disease or order.
- To warn of the necessity of taking proper precautions and, where appropriate, to monitor the patient's condition [*Crichton v. Hastings*, 1972; *Rosen v. Greifenberger*, 1999; *Cottrelle v. Gerrard*, 2003]. For example, optimal nutrition can maximize growth in children with sickle cell disease [Wethers 2000a, 2000b], and in the case of PKU, a low phenylalanine diet needs to be maintained years after birth.
- Health care professionals especially need to follow up on newborns with a genetic disease to reduce morbidity and mortality [e.g., sickle cell disease, Miller et al., 1990].

Referral to another physician will not necessarily terminate the original physician's duty of care to the patient. This may turn on the question of whether the patient is in continued need of the health care professional's care [*Longman v. Jasiek*, 1980; Mains, 1985; Andrews, 1991; *Lee v. O'Farrell*, 1988]. The duty to monitor in a continuing relationship can include the obligation to advise of any developments in management and treatment that would prove of benefit or detriment to the patient. There are American case law precedents [*Mink v. University of Chicago*, 1978; *Tresemer v. Barke*, 1976] that a duty of care will continue where the risk of future injury arises from the original physician–patient relationship. Please see Chapter 15 Recall for a discussion of this issue.

Psychological Aspects

The obligation to provide appropriate aftercare poses particular challenges given the psychological aspects of genetic testing and counseling. The emotional and psychological responses of the patient *may be more critical* to the interpretation and perception by the patient of genetic risk information than the actual facts (diagnosis, risk information, prognosis) [Lippman-Hand and Fraser, 1979; Wertz et al., 1986; Lerman et al., 1995; Hoskins et al., 1995; Phillips et al., 1999; Press et al., 2001; Horowitz et al., 2001; Lodder et al., 2001].

Although these standards continue to evolve, a health care professional who communicates a clinical diagnosis and/or provides information to clarify a patient's risk for developing a genetic disease or disorder may have a continuing duty of care with respect to the psychological and psychiatric responses of a patient as assessed on a case-by-case basis. The duty to provide appropriate aftercare may not be discharged by warning and discussing, with the patient and the referring physician, the possibility and nature of the psychological and psychiatric responses associated with a particular genetic disease or clinical scenario. In the same manner that a physician has a continuing duty of treatment until the patient completely recovers or terminates the relationship, the health care professional's duty of care may not terminate until such time as appropriate psychological and/or psychiatric counseling has been arranged [Sharpe, 1994b; *Hoeffner v. The Citadel*, 1993; ABMG, 2004; CCMG, 2004b]. For example, Huntington disease (HD) is a progressive, neurodegenerative, autosomal-dominant disease associated with psychiatric morbidity and a risk of suicide [Kessler and Bloch, 1989]. The application of rigorous and careful psychological assessment and counseling can help to alleviate such difficulties [Bloch et al., 1992, 1993; Wiggins et al., 1992; Tibben et al., 1992; Meiser and Dunn, 2000].

A health care professional should be aware of the potential emotional and psychological aspects of genetic testing, and be prepared to provide or to refer, on a case-by-case basis, the patient for appropriate counseling [Marteau and Croyle, 1998a; Halbert, 2004] especially in the case of a positive result for a potential life-threatening disease such as hereditary breast and ovarian cancer [Halbert, 2004]. American case law precedent indicates that a physician may have a duty to advise a patient about the attendant risks of not seeing a recommended specialist [*Moore v. Preventive Medicine Group*, 1986].

CONCLUSION

Informing a patient and often his or her family of a genetic test result involves comprehension, communication, and counseling. The process of informing a patient of genetic test results begins with the initial visit when detailed medical and family history is obtained and discussion of the perceived and actual benefits and limitations of genetic testing should take place. This helps to ensure that the best test to meet the expectations of the patient, family, and physician is used in a manner that is meaningful to the patient.

Giving genetic test results not only requires comprehension of the complex science but also how it relates to the patient as an individual and a member of their family. Physician comprehension necessitates accurate interpretation of the disease, the test result, and the resulting change in the patient's risk. Patient and family comprehension is more complex, requiring successful communication of this new information in an appropriate, understandable, and meaningful way.

Such effective communication of a genetic test result is much more than simplified explanations of complex risk information; it is also a counseling process:

> *If a patient cannot engage in the process of understanding the information disclosed, then informed choice cannot occur. Ensuring that information is provided in language the patient can understand, at a level of complexity that matches the patient's level of cognitive functioning, is necessary, but not sufficient. [Scheyett 2002, p. 381]*

This counseling process, begun before the test is ordered, explores the perceived meaning of the disorder, expectations of the test result, and coping and communication styles utilized and preferred by the patient. With this knowledge of the patient in mind, test results can truly be comprehended and, thus, incorporated into the health plan.

In October 2004 the Human Genome Project Consortium announced that the human genome sequence was complete to a 99.99 percent level of accuracy [International Human Genome Consortium, 2004]. This accomplishment will lead to the increasingly rapid discovery of new disease genes as well as genomic changes associated with increased or decreased risk of disease and/or response to therapeutics both of which are predicted to have great impact on multiple facets of health care [Collins et al., 2003; Bell, 2004; Bentley, 2004; Evans and Relling, 2004].

Our knowledge of the genomic determinants of susceptibility for common diseases such as asthma, heart disease, infectious disease, and diabetes as well as response to differing pharmaceuticals will continue to increase. This will ultimately result in the development of predictive genomic tests to give personalized risks of future health concerns. The availability of such testing information to the patient population, and the effect of physicians and patients acting upon this information to reduce morbidity and mortality is potentially enormous, as is the potential harm if it were to be misinterpreted or misused.

Numerous questions remain [Clayton, 2003; Green and Botkin 2003]. Will predictive genomic testing be incorporated into standard practice of care? Who will monitor the accuracy of the data behind these new tests? Several early genetic profile tests [Haga et al., 2003] have raised significant alarm. Should they be marketed directly to the consumer as pharmaceuticals frequently are? What will be the psychological impact of population-wide health labeling? Are special laws and policies needed to handle this new generation of predictive tests?

Health care practice will continue to evolve and adapt to the growing tide of genetic and genomic testing. Although the ultimate outcome of this revolution may be uncertain, the need for accurate and empathetic communication of the results of this new genre of genetic testing will be essential if it is to have any impact at all [Haga et al., 2003].

CANCER RISKS IN PERSPECTIVE: INFORMATION AND APPROACHES FOR CLINICIANS

Patricia T. Kelly 824 Cragmont Avenue Berkeley, California 94708

CASE HISTORIES

Andrea

Andrea a 35-year-old woman with a year-old child, had great concern about her breast cancer risk since her mother, grandmother, and aunt were all diagnosed with this disease in their 50s and 60s. "I'm hoping that since two of them were postmenopausal my risk is not that high," she said. "Still, I worry about it a lot and my husband gets upset with me sometimes because I can't stop thinking about it. We've decided that I should be tested to see if I am at increased risk so I will know once and for all what is ahead of me."

Andrea was surprised to learn that a substantial proportion of strongly inherited breast cancers occur postmenopausally, that women who carry a mutation in either the *BRCA1* or the *BRCA2* gene have a significantly increased risk of ovarian cancer, as well as breast cancer, and that not all who carry a *BRCA* mutation will develop cancer. She was even more surprised to learn that it would be useful to test her mother or surviving aunt before she, herself, was tested.

Andrea's mother was found to carry a *BRCA1* mutation, as was Andrea. In thinking about her cancer risks, Andrea was most concerned about the next 10 years because she and her husband wanted to have another child and hoped that new early detection modalities would become available within 10 years. She was also interested in information about the extent to which tamoxifen might help to reduce breast cancer risk once she completed her family.

Beverly

Beverly, a 53-year-old woman, learned she was a *BRCA* mutation carrier after taking hormone replacement therapy (HRT) for 3 years. When she tried to stop taking hormones, she experienced great difficulty in thinking clearly, sleeping, and controlling her temper. She wanted in-depth information about breast cancer risks associated with HRT use to help her decide how to proceed

Claire

Claire, a 52-year-old woman who carried a *BRCA1* mutation, was recently diagnosed with breast cancer. She had questions about her bilateral risk and whether she could safely have breast conservation and radiation treatment. She also had questions about ovarian cancer risk.

INTRODUCTION

Information about cancer risk is abundant and can be used to identify individuals whose risks may warrant consideration of procedures such as genetic testing for increased hereditary cancer risk, prophylactic mastectomy, prophylactic oophorectomy, and tamoxifen for prevention of breast cancer. However, the usefulness of cancer risk assessment is sometimes questioned since:

- Study results are sometimes inconsistent.
- The format in which risk is presented can obscure its clinical relevance.
- Current genetic testing can detect some, but not all, of the mutations increasing hereditary cancer risk.

As a result, physicians may be unsure about how and when to apply risk information or when to use genetic testing in their practices. Patients, who are seeking information and advice about how to best protect their health, may feel uneasy when they learn that genetic testing cannot always rule out an increased cancer risk. Some conclude that risk information is too confusing, that even with genetic testing too little is known, or that risk information is of questionable value since "the risk to any one person is either 100 percent or 0."

This section provides straightforward, commonsense information about cancer risks for clinicians to use in educating patients and in making treatment and follow-up decisions. The examples used are largely drawn from breast cancer studies and testing for mutations in the *BRCA1* and *BRCA2* genes, but the approaches apply to other diseases and mutations in other genes as well.

Clinically useful risks and risk formats are discussed for the following:

- Average woman's risk of breast and ovarian cancer
- Family history of cancer and genetic testing
- Cancer risks associated with *BRCA* mutations
- Tamoxifen and breast cancer prevention
- Bilateral breast cancer risk
- Hormone replacement therapy

As will be seen, when age and time are specified, when risks are presented as actual (also called absolute) risks, and when a distinction is made between diagnosis and prognosis, risk information is clinically useful and is essential in helping patients to make informed decisions. Moreover, when risk information is presented to patients in these ways, it is likely to be understood and used by them.

AVERAGE WOMAN'S RISK OF BREAST AND OVARIAN CANCER

Useful risk information has a time frame. For example, a 10 percent risk this year is different from a 10 percent risk spread over 10 years. With cancers of the breast, ovary, and many others, risk increases with age, so meaningful risk discussions specify both time and age. As Figure 14.1 shows, the average woman's breast cancer risk to age 80 is about 12 percent [Ries et al., 2004]. This means that 12 out of 100 women living to age 80 will develop breast cancer and 88 will not. (Here, and elsewhere, risks have been rounded to the nearest whole number.)

When the average woman reaches age 50 without a breast cancer diagnosis, she has left behind 2 percent of the risk. Just as she is not at risk of being in a car accident on a road she drove yesterday, she is not at risk of breast cancer in the years she lived without a diagnosis. Therefore, the average 50-year-old woman's risk of breast cancer *to age 80* is no longer 12 percent, but 12 minus 2 or 10 percent. Her risk to age 70 is 6 percent, and the risk from 70 to 80 is 4 percent.

An older woman has a higher risk in *any one year* than that does a younger woman. For example, the average 70-year-old woman has a higher breast cancer risk *in the next year* than the average 50-year-old. However, the 50-year-old has a higher risk of breast cancer *to age 80* than the 70-year-old because she has more years to travel before she reaches 80. Clearly, for risks to be meaningful, both age and the number of years need to be taken into account.

For ovarian cancer, the average woman's risk is 1.5 percent to age 80 [Reis et al., 2004]. The risk to age 50 is 0.3 percent. From 50 to 60 the risk is 0.3 percent, from 60 to 70 it is 0.4 percent, and from 70 to 80 it is 0.5 percent. Almost all of the average woman's ovarian cancer risk occurs after age 50.

The use of age-specific risks makes it possible to tailor risks to a woman's current age. For many women, information about risk in the next 10 years is very helpful in making treatment and follow-up decisions. For example, as seen in the case history for Andrea, some may want to start or complete a family. Others may feel comfortable postponing decisions about prophylactic oophorectomy or mastectomy for a few years

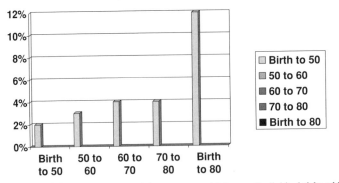

Figure 14.1 Average woman's breast cancer risk by age (Individual risks add up to more than 12% and risks from 60 to and 70 and 70 to 80 appear identical due to rounding up to next whole number).

Table 14.4 Deaths in 100,000 Women Aged 30 to 54 In the Next Ten Years

Age	Heart disease	Breast cancer	Colorectal cancer	Ovarian cancer	All causes
30–34	70	20	10	10	1,000
35–39	140	50	30	20	1,500
40–44	300	130	50	50	2,100
45–49	630	310	100	90	3,300
50–54	1,200	600	180	150	5,100

Schwartz et al. 1999.

to see if improvements in early detection and treatment modalities become available—if the risk year is not beyond their comfort level. It is important to be clear that the average woman's breast and ovarian cancer risks to age 80 do not accumulate over time, nor do they apply to the risk at any one particular time or in any one year, but are instead spread over 60 years.

DIAGNOSIS OR PROGNOSIS?

Individuals with concerns about cancer risk often fail to distinguish between risk of disease and prognosis following diagnosis [Kelly, 2000]. This is an essential distinction, as illustrated by breast and ovarian cancer risks. Breast cancer risks are higher than those for ovarian cancer. However, breast cancers are increasingly found at small sizes when they are likely to be node negative and associated with an excellent prognosis. For example, in one study node-negative breast cancers up to 1 cm in size, and even many up to 1.5 cm, were associated with 20-year survivals in excess of 90 percent, largely without chemotherapy and tamoxifen treatments [Tabar et al., 2000]. On the other hand, 75 percent of all ovarian cancers have spread by the time they are detected. The overall 5-year survival is 44 percent [Ries et al., 2004]. For these reasons, a lower risk of ovarian cancer may give rise to more concern than a higher breast cancer risk, since early ovarian cancer is so difficult to detect.

Some patients benefit from information to help them place their cancer risks in perspective. For example, as shown in Table 14.4, about 1000 out of 100,000 women age 55 to 59 will die of breast cancer in the next 10 years, while more than twice that number—2200—will die of coronary disease [Schwartz et al., 1999]. Over 8000 will die of all causes in this time. Most individuals with concerns about cancer, and especially breast cancer, are amazed to learn that other causes of death are so much greater than they had realized.

FAMILY HISTORY OF CANCER

Signs of Increased Hereditary Risk

About 10 percent of all breast, ovarian, and many other cancers are thought to be strongly hereditary. The other 90 percent are due to weaker hereditary causes, to environmental factors, and to interactions between hereditary and environmental factors. To assess the likelihood that cancers in a particular family are due to strong hereditary factors, a thorough family history is taken and scrutinized for features such as the following: [adapted from Thull and Vogel, 2004]:

- Breast or other cancer diagnosed before age 40
- Two or more breast or other cancers diagnosed before age 50
- Multiple primary tumors in an individual
- Ovarian cancer in individuals of Ashkenazi Jewish descent
- Male breast cancer
- Breast cancer diagnosed before age 50 and ovarian cancer
- Prostate cancer before age 55 and breast cancer before age 50
- Breast/ovarian cancer and melanin spots on lips or buccal mucosa
- Breast cancer and oral papillomatosis and/or facial trichilemmomas

The presence of features such as these in a family warrants referral for cancer risk assessment. However, not all families with one or more of the above criteria will have an increased hereditary cancer risk. And, families without these features may have cancers that are due to strong hereditary factors. For example, in families where several individuals were diagnosed with cancer after age 50, an increased hereditary risk may be present. Family size and other aspects, such as young age at death unrelated to cancer, make it possible for families with no or few cancer diagnoses to lack the above features but still contain individuals carrying a mutation that significantly increases cancer risk. An increased hereditary risk of breast and ovarian cancer can be passed through the paternal as well as the maternal side of the family, so no pattern suggesting increased hereditary risk may be apparent if there are few female relatives on the paternal side.

To some extent, analysis of family history to determine hereditary risk is analogous to making a diagnosis based on the feel of a breast lump. Just as a pathology report is required to provide definitive information about a breast lesion, so genetic testing is needed to definitively ascertain hereditary cancer. Unlike a pathology review, however, current genetic testing is not definitive for all families, since it cannot identify all mutations that increase hereditary cancer risk. For this reason, a negative test result does not always rule out the presence of a germline mutation that increases hereditary risk, as discussed below.

Origin of Hereditary and Nonhereditary Cancers

To help patients make informed decisions about genetic testing, it is useful for them to understand how hereditary and nonhereditary cancers arise. I generally start by pointing out that all cancers are due to changes in the genes, but few cancers are strongly hereditary. I then show pictures of the chromosomes, noting that each gene has its own place on its own chromosome. Because almost all cells in the body contain all the chromosomes, and hence all the genes needed for development and growth, suppressor genes are present to tell a breast cell, for example, not to make a kidney or a heart and to ensure that only breast cell products are made.

I point out that both *BRCA1*, which is located on chromosome 17, and *BRCA2*, located on chromosome 13, are strong suppressor genes. If an individual is born with a genetic change (mutation) in one copy of a *BRCA1* or *BRCA2* gene, that mutation is present in almost every cell of the body. All goes well until at some point the second copy of the gene, which was inherited from the other parent, also is damaged. At that point, with no suppression activity from either copy of the gene, a cell can begin unregulated growth, swing out of control, and become a cancer cell.

In discussing the origin of nonhereditary cancers, I find it useful to point out that in these instances the individual is born with no mutations in strong controlling genes. However, during a person's lifetime, mutations can collect in a cell due to every day living or exposure to carcinogens. At some point, when sufficient genetic changes accumulate in a cell, controls on cell division are released, enabling it to become a cancer cell. Nonhereditary cancers often occur at older ages than those that are hereditary, probably because it takes more time for the cell to accumulate the multiple genetic changes leading to a nonhereditary cancer than for the single change after conception that is thought to enable the development of strongly inherited cancers.

BRCA Mutations and Genetic Testing

Individuals may be more interested in considering hereditary genetic testing once they learn that it requires only a small blood sample. They sometimes resist even investigating testing if they are told or believe that they will be committed to having their breasts and ovaries removed if a mutation should be found. For this reason, follow-up decisions are more effectively made when patients have an opportunity to discuss their options in detail during the cancer risk assessment process. In this way, more are likely to agree to investigate testing and to be comfortable with their treatments and follow-up since they have participated in the decision-making process and more fully understand their risks [Kelly, 2000].

In explaining testing to patients, I find it useful to describe a gene as a building and to point out that just as a building has many rooms, so a gene has many chemicals. If even one of the thousands of chemicals in a gene is missing or changed, the product the gene makes will be a little "off," leading to increased cell growth and increased cancer risk. In the course of genetic testing, the laboratory will be "walking" through the *BRCA1* and the *BRCA2* buildings, looking for a damaged room (a mutation). If they find such damage in a gene, it will mean the individual has an increased risk of both breast and ovarian cancer.

If no mutation is found in the first individual in a family to be tested, the negative result could mean that:

- No mutation increasing cancer risk is present in the family, so the tested individual could not inherit one

- A mutation is present in the family, but the tested individual did not inherit it.
- A mutation is present in the *BRCA* or other genes, but it is not one that can be detected at present.

With a single individual tested, particularly one who has not been diagnosed with cancer, there is no way to distinguish between these three possibilities, so even with a negative finding, the individual's hereditary cancer risk might still be increased.

The finding of a mutation in an individual is evidence that testing can detect a mutation in that family. In these cases, close relatives can be definitively tested for that mutation. Those who are found to carry the same mutation have an increased hereditary cancer risk, while those who do not can be assured that their hereditary risk is not increased—at least from a mutation on that side of the family. Strongly inherited mutations increasing cancer risk are rare, so usually only a single one is present in a family. Cancer risk assessment can provide the specifics for each family.

Extra precautions are taken with individuals of Ashkenazi Jewish descent since about 2.5 percent of that population carries one of three *BRCA* mutations that increase cancer risk—two in the *BRCA1* gene and one in the *BRCA2* gene [Struewing et al., 1997]. Because of the (relatively) high frequency of these mutations, individuals of Ashkenazi descent are tested for all three mutations, even when a close relative is a known carrier of one of them.

Risks Associated with *BRCA* Mutations

Information about the risk of breast and ovarian cancer to women with a *BRCA* mutation is still evolving since testing is relatively new. In some families there appears to be an increased risk of only breast and ovarian cancer, while in others the risk of melanoma and other cancers such as pancreatic, gastric, cervical, endometrial, male breast cancer, and prostate cancer also seems to be increased [Brose et al., 2002; Eerola et al., 2001; Thompson and Easton, 2001]. To date, most reported risks are based on all known mutations within a gene, even though different mutations appear to be associated with different risks [Lubinski et al., 2004; Risch et al., 2001]. In addition, mutations and polymorphisms in other genes have been found to influence risks when a *BRCA* mutation is present [Kadouri et al., 2004]. Therefore, risks associated with *BRCA* mutations are likely to change as more families are studied and as risk profiles for specific mutations become available. The risks discussed below should therefore be viewed as approximations and are not definitive.

A recent study of *BRCA1* carriers found a breast cancer risk of 65 percent to age 70 [Antoniou et al., 2003]. As shown in Table 14.5, the risk to age 30 was less than 1 percent. From 30 to 40 the risk was 12 percent or about 1.2 percent a year. As a woman goes through each year, she leaves behind the risk for that year. For example, a 50-year-old woman has left behind the 12 and 26 percent risks associated with ages 30 to 40 and 40 to 50. She faces a 15 percent risk from 50 to 60 (1.5 percent a year). Her risk to age 70 is not 65 percent but 27 percent, or about 1.5 percent a year.

Table 14.5   Risk of Breast and Ovarian Cancer to *BRCA1* Mutation Carriers

Age	Breast cancer (%)	Ovarian cancer (%)
20–30	<1	—
30–40	12	2
40–50	26	11
50–60	15	9
60–70	12	17
20–70	65	39

Adapted from Antoniou et al. [2003].

In this study, the ovarian cancer risk to *BRCA1* carriers was 39 percent to age 70 (Table 14.5). No cases of ovarian cancer were found before age 30. From 30 to 40 the ovarian cancer risk was 2 percent, and from 40 to 60 it was 20 percent, or about 1 percent a year. By age 60 a woman had used up all but 17 percent of the risk of ovarian cancer and had a 1.7 percent risk a year up to age 70.

Breast and ovarian cancer risks associated with *BRCA2* mutations are shown in Table 14.6 [Antoniou et al., 2003]. To age 70 the risk of breast cancer was 45 percent. From 30 to 40 the risk was 6 percent or about 0.5 percent a year. From 40 to 50 it was 10 percent or about 1 percent a year and from 50 to 70 it was 29 percent, or about 1.5 percent a year. The ovarian cancer risk from birth to age 70 was 11 percent. All but 1 percent of the risk occurred after age 50. From age 50 on the risk of ovarian cancer was 10 percent, or about 0.5 percent a year.

The risks associated with the two *BRCA1* mutations most often found in individuals of Ashkenazi descent are shown in Table 14.7 [King et al., 2003]. As you can see, up to age 70 the breast cancer risk was 69 percent; for ovarian cancer the risk was 46 percent. Here also, as a woman leaves each age without a cancer diagnosis, she leaves behind the risk associated with it. Therefore, from 50 to 70 the breast cancer risk was not 69 percent but 48 percent, or about 2.5 percent a year. From 50 to 60 the risk was 19 percent or about 2 percent a year. Similarly, for ovarian cancer, a

Table 14.6   Risk of Breast and Ovarian Cancer to *BRCA2* Mutation Carriers

Age	Breast cancer (%)	Ovarian cancer (%)
20–30	<1	—
30–40	6	—
40–50	10	1
50–60	15	6
60–70	14	4
20–70	45	11

Adapted from Antoniou et al. [2003].

Table 14.7   Risks for *BRCA1* Mutations Found in Ashkenazi Jewish Population

Age	Breast cancer (%)	Ovarian cancer (%)
Up to 30	3	—
30–40	18	3
40–50	18	18
50–60	19	19
60–70	11	6
Age 30 to 70	69	46

Adapted from King et al. [2003].

40-year-old had a 37 percent risk to age 60, or about 2 percent a year. From 60 to 70 the risk was about 0.5 percent a year. Risks for the single *BRCA2* mutation most commonly found in those of Ashkenazi descent are shown in Table 14.8.

Note that much of the breast cancer risk and nearly all of the ovarian cancer risk associated with *BRCA* mutations occurs after age 50. Therefore, a woman can have several relatives diagnosed with cancer after age 50 and still be a mutation carrier. Furthermore, since not all of the mutations that increase breast and ovarian cancer risk can be detected at present, an individual who tests negative for a *BRCA* mutation may still be at significantly increased risk due to a *BRCA* or other gene mutation that cannot yet be identified.

BRCA Mutations and Breast Conservation

Three studies with 8 to 10 years of follow-up have found no significant difference in ipsilateral cancer risk to *BRCA* carriers and noncarriers treated with lumpectomy and radiation therapy [Pierce et al., 2000; Kirova et al., 2004; Metcalfe et al., 2004a]. One study of women diagnosed before age 42 did find a higher 10-year ipsilateral recurrence rate to *BRCA* mutation carriers treated with breast conservation [Haffty et al., 2002]. Another, of Ashkenazi women diagnosed before age 65 and treated with conservation therapy found a significantly higher ipsilateral recurrence rate in *BRCA*

Table 14.8   Risks for *BRCA2* Mutations Found in Ashkenazi Jewish Population

Age	Breast cancer (%)	Ovarian cancer (%)
Up to 30	—	—
30–40	17	2
40–50	17	—
50–60	•14	4
60–70	26	6
Age 30 to 70	74	12

Adapted from King et al. 2003.

carriers compared to noncarriers, but similar survival rates [El-Tamer et al., 2004]. Severe reactions following radiation therapy have not been found among women with *BRCA* mutations [Leong et al., 2000].

Decision Making

As discussed earlier, *BRCA* mutation carriers are generally more concerned about ovarian cancer than breast cancer risk, due to the difference in prognosis. If a woman's breasts are relatively easy to examine or to image on mammogram, or if she has access to other early detection modalities such as ultrasound and MRI (magnetic resonance imagins), a woman with a *BRCA* mutation may elect to be followed carefully, given the 90 percent and greater survival to 20 years with small node-negative disease [Tabar et al., 2000] and studies showing similar survival for those who choose breast conservation and mastectomy [El-Tamer et al., 2004; Gaffney et al., 1998; Turner et al., 1999; Pierce et al., 2000]. Women whose breasts are difficult to examine or who feel the quality of their lives would be diminished with worry about risk may choose to have prophylactic mastectomy. Many consider prophylactic oophorectomy if they are in their 40s and older, since ovarian cancer is difficult to detect before it has spread.

TAMOXIFEN AND BREAST CANCER PREVENTION

Women at increased risk of breast cancer, whether due to *BRCA* mutations or to other factors, may be advised to take tamoxifen to reduce breast cancer risk. A large 5.75-year study of women at increased risk of breast cancer reported a 49 percent reduction in risk to those who took tamoxifen for 5 years [Fisher et al., 1998]. However, breast cancers are thought to grow for 7 to 10 years before they are detected [Harris and Hellman, 1996], so all or most of the breast cancers found in this study were probably present, but undetected, at the beginning of the study.

More importantly, the absolute benefit represented by the 49 percent risk reduction is quite small—1.8 per 100 women. That is, in nearly 6 years, the group taking tamoxifen had 1.8 fewer breast cancers per 100 women than the placebo group. Another way of stating the 49 percent reduction is that 333 women would need to be treated with tamoxifen for 1 year to "prevent" one breast cancer. As this example demonstrates, the presentation of risk as a percent reduction can make even a small difference appear more substantial than it actually is.

BILATERAL BREAST CANCER RISK

Average Risk

Almost all cancers diagnosed in the contralateral breast are second primaries, not a spread from the first cancer. For women with breast cancer as a group, the bilateral risk is 0.5 percent to 1 percent a year [Rosen et al., 1989; Obedian et al., 2000].

As a woman goes through each year without a diagnosis in the opposite breast, that risk is left behind. The finding of bilateral disease is influenced by factors such as the proportion with a family history of breast cancer, the efficacy of detecting breast cancer, length of follow-up, and the amount of tissue examined [Ringberg et al., 1991]. Women with a family history of breast cancer have a higher risk of bilateral disease, as discussed below.

Tamoxifen

Tamoxifen is sometimes recommended as a way to decrease bilateral risk. However, the actual reduction in risk to those who take tamoxifen is quite small. For example, one study with an average 9-year follow-up compared contralateral risk in over 1300 women who took tamoxifen and over 1300 who did not [Rutqvist et al., 1995]. Compared to the placebo group, tamoxifen users had a statistically significant bilateral relative risk reduction of 0.6. Because the risk was reduced by almost 50 percent and was significant, many assumed that it was large. However, in absolute terms, the difference between the two groups was 1.9 percent. That is, at the end of 9 years, 1.9 fewer contralateral breast cancers were detected in 100 tamoxifen users compared to 100 women taking a placebo.

   Most other studies report similar small differences. In an overview of six studies with differing years of follow-up, those taking tamoxifen had absolute benefits of 0.2 to 1.8 percent [Curtis et al., 1996]. One study with a 15-year follow-up found a 2 percent difference between the tamoxifen and the control group to 15 years [Obedian et al., 2000]. Another study of over 700 women diagnosed premenopausally and followed for 9 years reported a 1.8 percent difference in contralateral risk between the tamoxifen and placebo group in the first 5 years, but not subsequently [Baum et al., 1992].

   The Early Breast Cancer Trialists' Collaborative meta-analysis on bilaterality [1998] found a 5-significant relative risk reduction of 0.53. This risk reduction was calculated to be a difference of 2.1 cancers per 100 women followed for 10 years. As these examples demonstrate, bilateral risk reduction due to tamoxifen appears to be far more beneficial when presented as a relative risk reduction than as an actual risk. Also the differences found refer to the group as a whole. Not all women in the tamoxifen group received equal benefit.

Family History of Breast Cancer

In a 20-year study, women who had at least two close relatives with breast cancer and who were themselves diagnosed with breast cancer before age 50 were found to have a 2 percent risk of bilateral disease a year [Anderson and Badzioch, 1985]. Those diagnosed after age 50 had a risk of 1 to 1.3 percent a year, depending on whether one or two generations were affected. If this study were done with today's improved early detection modalities, many of the cancers diagnosed in the first 5 years are likely to have been found at the time of the first breast cancer diagnosis, which would reduce the subsequent bilateral risk. Therefore, this study may overestimate the initial 5-year contralateral risk to women diagnosed today. It may underestimate the risk to modern women in subsequent years since breast cancer prognosis has greatly improved, enabling more women to live longer and so have more years to be at risk.

BRCA Mutation Carriers

An early study of families selected for very young ages at cancer diagnosis and multiple cases of breast and ovarian cancer found a 33 percent risk of contralateral disease to women with a *BRCA1* mutation who were diagnosed from 30 to 39— about 3 percent a year. The risk to those diagnosed from 40 to 49 was 17 percent, or 1.7 percent a year. Women diagnosed from 50 to 59 had a contralateral risk of 1 percent a year, and from 60 to 69 it was 0.5 percent a year [Easton et al., 1995]. Note that for women diagnosed after age 50, the contralateral risk is about the same as for the average woman with breast cancer.

In a study of 173 families with a *BRCA2* mutation, the contralateral risk to mutation carriers was also found to differ with age [Breast Cancer Linkage Consortium, 1999]. For *BRCA2* carriers diagnosed from 30 to 50 the bilateral risk was 37 percent, or 1.85 percent a year. Those diagnosed from 50 to 70 had a bilateral risk of 15 percent or 0.75 percent a year—about the same as that of the average breast cancer patient. Several studies reporting higher contralateral risks have small sample sizes [e.g., Verhoog et al., 1999] or are of women diagnosed before age 45 [e.g., Chabner et al., 1998; Ford et al., 1994].

A study of 336 *BRCA* carriers, in which about 80 percent were initially diagnosed before age 50, found a 5-year bilateral risk of 17 percent, or about 3 percent a year [Metcalfe et al., 2004a]. From 5 to 10 years the risk was 13 percent, also about 3 percent a year. In this study, the risk in the first 5 years could be inflated since cancers found as soon as 0.1 years after the initial diagnosis were included. In many studies contralateral cancers found within the first or even first several years are not counted as bilateral, but as synchronous.

About 80 percent of the women in this study were initially diagnosed before age 50, which could also account for the higher bilateral rate than found in some of the other studies of *BRCA* carriers. In this study, unlike others, women diagnosed after age 50 did not have a lower bilateral rate than those diagnosed at younger ages.

BRCA mutations were found in 25 percent of those with bilateral breast cancer tested by Myriad Genetics Laboratory [Parlanti et al., 2003]. Here also, bilateral frequency differed by age. Of those diagnosed with bilateral disease before age 50, 31 percent were *BRCA* mutation carriers, compared to 10 percent diagnosed from age 50 on. In a study of women of Ashkenazi Jewish descent with bilateral breast cancer, 30 percent were found to carry one of three *BRCA* mutations most frequently present in this ethnic group [Gershoni-Baruch et al., 1999]. Over 80 percent of these mutation carriers were diagnosed before age 42.

As a group, these studies suggest that with or without a *BRCA* mutation, women diagnosed before age 45 or 50 are more likely to develop a contralateral breast cancer than those diagnosed at older ages. Failure to consider age can lead to an overestimation or underestimation of risk, while consideration of age can aid in decision making about treatment to the contralateral breast.

In one study of women with a *BRCA* mutation, tamoxifen was reported to significantly reduce risk to the opposite breast, but this effect was confined to 2 to 5 years after the first diagnosis [Narod et al., 2000]. This short time frame makes it likely that all or most of the contralateral tumors were present at the time the study began, so the reduction in risk is unlikely to be due to tamoxifen.

Invasive Lobular Carcinoma

Women with invasive lobular breast cancer, which accounts for about 10 percent of all breast cancers, are generally said to have a significantly increased contralateral risk. However, in several studies invasive lobular carcinoma has been found to be associated with bilateral rates similar to those of invasive ductal carcinoma [Peiro et al., 2000; Sastre-Grau et al., 1996; Yeatman et al., 1997]. A large recent study found higher bilateral risk to women with invasive lobular than those diagnosed with other types of breast cancer [Arpino et al., 2004]. However, this finding was based on a subset of the study population, and information on factors that might affect bilateral risk such as age, family history, and length of follow-up was lacking. In at least two studies, tumors consisting of *both* lobular and nonlobular components were associated with a 20 percent risk of contralateral disease at 10 years [Sastre-Garau et al., 1996; Abner et al., 2000]. Moderate to marked amounts of lobular carcinoma in situ (LCIS) do not appear to lead to an increased risk of contralateral breast cancer [Abner et al., 2000].

HORMONE REPLACEMENT THERAPY

Family History of Breast Cancer

Hormone replacement therapy (HRT) is generally believed to be contraindicated for women with a family history of breast cancer. However, at least eight studies find that women with an affected mother, sister, or both who use HRT do not have a significantly higher risk of breast cancer than nonusers with similar histories, even with 5 or more years of use or with the use of both estrogen and a progestin [Brinton et al., 1986; Dupont et al., 1989, 1999; Palmer et al., 1991; Kaufman et al., 1991; Stanford et al., 1995; Sellers et al., 1997; Magnusson et al., 1998].

Women's Health Initiative Study

The Women's Health Initiative (WHI) study was widely reported as conclusively showing that HRT use at menopause significantly increases breast cancer risk [Writing

Group for the Women's Health Initiative Investigators, 2002]. However, a closer look reveals that the results of this study do not provide such evidence. Many or all of the small differences in risk between HRT users and nonusers may not be due to hormone use at all or may be the result of the particular hormones used in this study. A careful examination of this issue is important since *BRCA* carriers, who are at increased risk of ovarian cancer, generally investigate prophylactic oophorectomy and have questions about the effects of HRT on their breast cancer risk.

In the WHI study over 8000 women took daily Prempro, a conjugated equine estrogen plus medroxyprogesterone acetate [Writing Group for the Women's Health Initiative Investigators, 2002]. Other regimens that more closely approximate a woman's natural hormones and cycle may have different effects. For example, continuous but not sequential hormone use of 4 or more years was associated with a significant increase in breast cancer risk with continuous but not sequential hormone use [Olsson et al., 2003]. In another study, also, women with progestin use of less than 2 weeks a month had no increase in breast cancer risk [Porch et al., 2002].

Even with the use of a continuous regimen, the WHI study, as well as another large randomized prospective study, the Heart and Estrogen/Progestin Replacement Study (HERS) found no significant increase in breast cancer risk to hormone users [Hulley et al., 2002]. The estrogen-only arm of the WHI study also found no statistically significant increase in breast cancer risk to hormone users [WHI Steering Committee, 2004]. We now have three large randomized prospective studies as well as many retrospective studies that find no statistically significant increase in breast cancer risk to HRT users. Initial reports on WHI results stated that HRT users in this study had a statistically significant increase in breast cancer risk. However, it later became apparent that with correction for multiple statistical testing, no significant difference in breast cancer rate was present.

Correction for multiple significance testing is essential since with each statistical test there is a 5 percent chance that a false finding of significance will be obtained [Altman, 2000]. As the number of statistical tests increases, so does the likelihood of obtaining spurious statistical significance. For example, with only four statistical significance tests, there is a 20 percent chance that one of the differences will appear to be significant, but is actually not. In the WHI study, once correction was made for multiple testing, *no* statistically significant difference between the HRT and placebo group was present, except for venous thromboembolism.

Statistical significance is a way to compare disease rates in groups to see if they are sufficiently alike to be considered part of the same population. If they are sufficiently alike, the difference between them is not statistically significant. Statistical significance does not depend on size or importance, so a small difference can be statistically significant and a large one not.

When the findings of the WHI study are presented as actual or absolute risks, their importance can be more easily evaluated than when the hazard ratio, a comparison, is used. In this study, the absolute yearly difference between the Prempro and control groups was eight cases of breast cancer in 10,000 women or eight hundredths of one percent (0.08 percent). This actual difference is far less than the reported 26 percent increase in risk or hazard ratio of 1.26 seems to indicate—another example of the inflated importance a risk can assume when comparison formats are used instead of absolute risks.

Even this 0.08 percent yearly increase in breast cancer risk may not be due to Prempro use since:

- The increase in risk is so small it might well be due to other differences between the groups studied. For example, more women with a positive family history for breast cancer were present in the group assigned to take Prempro. Epidemiologic studies, such as the WHI, provide information about associations, not cause and effect. To evaluate the likelihood that a difference in disease rate is due to an agent such as Prempro, one looks for statistical significance in several studies (not present in this case or in many other studies), and a hazard ratio of three or greater (here the hazard ratio was 1.26, far lower than threefold) [Taubes, 1995]. When a hazard ratio is less than 3, the agent being studied is less likely to be causative than when it is 3 or greater [Taubes 1995; Kelly, 2000].

- Fully 42 percent of the women assigned to take Prempro stopped taking it during the study, but in calculating risk, those who discontinued Prempro use were counted as if they had taken it the entire time they were in the study.

- The mean age of women in the WHI study was 63. Most had never used hormones previously and were willing to enter a study in which they would be randomly assigned to take either Prempro or a placebo. They were therefore unlikely to have had severe menopausal symptoms. In this regard, they appear to differ physiologically from women whose menopausal symptoms are severe enough to lead them to take HRT at menopause. This study therefore does not address risk to women with troublesome menopausal symptoms or those whose use starts at menopause.

Some have wondered if Prempro users had a slightly higher rate of breast cancer in the WHI study because Prempro caused existing cancers to grow more rapidly and so to be detected sooner. This is unlikely since Prempro users and nonusers had breast cancers of similar grade and differed in size by only 2 mm [Chlebowski et al., 2003]. Also, in more than 15 studies, breast cancers found in women taking HRT do not differ in size, grade, or aggressiveness, compared to those in nonusers [Kelly, 2003]. In a study of mammographically detected breast cancers in women with equivalent screening, tumors in HRT users were more likely to be well differentiated than those of nonusers [Cheek et al., 2002]. HRT use therefore does NOT appear to increase most breast cancer growth rate.

Follow-up in the WHI study was 5.2 years, with all women followed at least 3.5 years. Breast cancers appear to take an average of 8 years to reach 1 cm [Harris and Hellman, 1996]. Therefore, all or most of the breast cancers detected in this study were probably present in an undetectable state at the time the study and HRT use began. The majority of the cancers detected in this study are therefore unlikely to be due to Prempro use. As this short review suggests, HRT use either does not increase breast cancer risk, or does so to such a small extent that it is difficult to measure.

CONCLUSION

Physicians, medical students, and the general public are more likely to understand and to be able to apply risk information in a clinically useful manner when it is presented in actual instead of relative risk terms [Chao et al., 2003; Naylor et al., 1992]. Even experienced medical investigators are swayed by the format used in presenting risk information [Elting et al., 1999]. In fact, when risks are presented only as a relative risk reduction, this "may lead a reader to believe that a treatment effect is larger than it really is" [Nuovo et al., 2002].

With the availability of a growing body of information about hereditary and nonhereditary risk factors for cancer, it is increasingly important to present risks in an understandable and clinically useful manner. Ways to do this include the use of:

- Age-specific risks
- Risks taking time and age into account
- Absolute risks

In evaluating risks derived from epidemiologic studies, it is important to be clear that they apply "to an aggregate of individuals, not to a specific individual" [Rockhill, 2001] and that these risks need to be put in context for them to be meaningful [Woloshin et al., 2002; Schwartz et al., 1999]. In addition, it is essential to be aware that a significantly increased risk may not be a large or important difference between groups, and that correction for multiple statistical testing is needed to reduce the likelihood of a spurious finding of significance.

As Hall et al. [1988] note, "Patients who received more information from their physicians were more satisfied and had higher compliance with medical regimens." Therefore, the time and skill that are required to help patients understand risks and their consequences are likely to be beneficial and even cost effective in the long run than are processes utilizing less comprehensive approaches.

Chapter 15

Confidentiality, Disclosure, and Recontact

OVERVIEW

A physician who treats a patient, impliedly agrees to keep in confidence all information divulged by the patient concerning the patient's physical and mental condition, as well as information provided by the patient in the course of examination or treatment [*MacDonald v. Clinger*, 1982; *Doe v. Roe*, 1977]. In the United States, this obligation is regulated by the Standards for Privacy of Individually Identifiable Health Information (Privacy Rule) [HIPAA Privacy; Hustead and Goldman, 2002; Frank-Stromberg, 2003; Cole and Fleisher, 2003], pursuant to the Health Insurance Portability and Accountability Act of 1996 [HIPAA, 1996]. The act provides for disclosure where the public interest is at risk [Sudell, 2001; Burnett, 1999], for example,

- A "serious and imminent threat to the health or safety of a person or the public"
- An imminent and serious threat to an identifiable third party and the physician has the capacity to prevent significant harm
- Infectious diseases
- Apprehension for law enforcement.

From a legal and ethical perspective, it long has been accepted in both medical custom and law that the health care professional is obligated to respect the patient's desire for medical confidentiality flowing out of the fiduciary nature of the physician–patient relationship [Hall et al., 1999; Rothstein, 1999; Anderlik and Lisko, 2000]. Although there are limited exceptions to this general rule [e.g., infectious diseases, *Simonsen v. Swenson*, 1920; *Tenuto v. Lederle Laboratories*, 1997], the assurance of confidentiality promotes a full and frank communication and discussion between the physician and the patient without fear of disclosure [*MacDonald v. Clinger, 1982*; Clayton, 1998; Offit et al., 2004].

Genetic Testing: Care, Consent, and Liability, by Neil F. Sharpe and Ronald F. Carter
© 2006 John Wiley & Sons, Inc.

The principle of medical confidentiality first appeared in the Hippocratic Oath [Berry, 1997]:

> *Whatever, in connection with my professional practice, or not in connection with it, I see or hear, in the life of men, which ought not to be spoken abroad, I will not divulge, as reckoning that all such should be kept secret.*

Beneficence (an act done for the benefit of others), nonmaleficience (the concept of doing no harm), respect for autonomy, justice, and the obligations of a physician with regard to privacy, truth telling, and to act in good faith are cornerstones for medical ethics and the moral principles of patient care [Beauchamp and Childress, 1994; Laurie, 2001].

> *Rights to privacy are valid claims against unauthorized access that have their basis in the right to authorize or decline access. These rights are justified by rights of autonomous choice . . . expressed in the principle of respect for autonomy.*
> [Beauchamp and Childress, 1994]

These principles are concerned with enabling and protecting the right of the patient to make an informed, autonomous choice. However, beneficence and a patient's right of autonomy may find conflict. What if a physician believes that a medical disclosure could prevent or reduce possible harm to others? Which principle is to be given priority and by whom? The sincerity of the physician in seeking to prevent harm and to provide benefit is not in question. However, the health care professional, the patient, and members of the patient's family, each may have sharply different perspective as to what constitutes *benefit* and *harm*. Will the physician effectively impose upon the patient and others her or his own values about what constitutes benefit and harm [Kottow, 1986; Sharpe, 1993]. Why are these deemed more valid, and given a higher priority, than those of the patient? What of the harm that could result from disclosure [Laurie, 2001]?

- A patient may not wish to disclose the test information out of concern for privacy, confidentiality, emotions, family considerations [e.g., McConkie-Rosell et al., 1995; Benkendorf et al., 1997; Metcalfe et al., 2000; Julian-Reynier et al., 2000; Dugan et al., 2003; Falk et al., 2003], and potential discrimination and stigmatization [Fanos et al., 1995; Liang, 1998; Metcalfe et al., 2000; ASCO, 2003; Dugan et al., 2003; Gordon et al., 2003; Kass et al., 2004].
- Disclosure may represent a violation of an individual's right of autonomy, of the moral "right not to know" [De Wert, 1992; Laurie, 2001; Garrison, 2003; Taylor, 2004—this study reports that with regard to Huntington disease, "predictive testing was regarded as a significant life decision with important implications for self and others, while the right 'not to know' genetic status was staunchly and unanimously defended"].

Will the physician be appropriately aware of the potential social, economic, and familial consequences of disclosure? May one argue with equal force that the patient alone can best know and weigh the social, familial, and emotional factors?

But to manipulate men, to propel them towards goals which you—the social reformer—see, but they may not, is to deny their human essence, to treat them as objects without wills of their own, and therefore to degrade them. This is why . . . to use them as means for my, not their own, independently conceived ends, even if it is for their own benefit is, in effect, to treat them as sub-human, to behave as if their ends are less sacred and ultimate than my own. [Berlin, 1969]

I recognize, of course, that the choice between violating a patient's private convictions and accepting her decision is hardly an easy one for members of a profession dedicated to aiding the injury and preserving life . . . If patient choice were subservient to conscientious medical judgment, the right of the patient to determine her own treatment, and the doctrine of informed consent, would be rendered meaningless . . . For this freedom to be meaningful, people must have the right to make choices that accord with their own values, regardless of how unwise or foolish those choices appear to others. [*Malette v. Shulman,* 1990; also see: Kottow, 2004]

GENETIC TEST RESULTS AND DISCLOSURE

Genetic testing and the capacity to identify individuals at risk for life-threatening genetic diseases years before the age of onset have triggered a fundamental reevaluation of these long held principles. Health care professionals may feel morally and ethically obligated to prevent harm and to promote the welfare of the other family members through disclosure of genetic test results to facilitate testing, surveillance, and preventive strategies [Suslak, et al., 1985; Hakimian, 2000; d'Agincourt-Canning, 2001; Avard and Knoppers, 2001]. By way of example:

- The patient has an affected *BRCA1* and/or *BRCA2* gene. Given the potential implications for family members, should the term "patient" be expanded to include the immediate family [DudokdeWit et al., 1997; Skene, 1998; Burnett, 1999; Parker and Lucassen, 2004]? Notifying other family members arguably could enable them to be aware of their potential risk status, to consider testing, and to undertake targeted surveillance and management programs [Fanos and Johnson, 1995; Julian-Reynier et al., 2000; Lehmann et al., 2000; Esplen et al., 2001; Hughes et al., 2002; Forrest et al., 2003; Koehly et al., 2003].

- The patient is at risk for Huntington disease (HD). A patient who has been identified as being at risk or has been diagnosed with Huntington disease may refuse to allow the geneticist to discuss the test results with family members or relatives out of shame, fear of social stigmatization, concern for insurance or employment discrimination, and familial rejection and breakup [Kessler and Bloch, 1989; Lynch et al., 1999] However, being identified as at risk for HD and the actual expression of the disease could have far-reaching implications not only for the patient but also the family, with significantly high levels of depression [Kessler, 1993; Quaid and Wesson, 1995]. Notifying other family members could extend testing to identify risk status and to provide psychological counseling.

In 1983, the President's Commission for the Study of Ethical Problems in Medicine and Biomedical and Behavioral Research stated that disclosure of genetic information could be justified in exceptional circumstances where:

- There is a high probability that harm would occur if the information was withheld.
- The disclosed information actually will be used to prevent harm.
- The harm that identifiable individuals would suffer would be serious.
- Only genetic information needed for diagnosis and/or treatment of the disease in question would be disclosed.
- Reasonable efforts to elicit voluntary consent to disclosure have failed [President's Commission, 1983; also see Committee for Assessing Genetic Risks, 1994; the Committee added the requirement that the "burden should be placed on the person who wishes to disclose to justify to the patient, to an ethics committee, and perhaps in court that the disclosure was necessary and met the committee's test"].

The Science Council of Canada, in 1991, adopted similar guidelines for physician disclosure to third parties:

- Reasonable efforts to elicit voluntary consent to disclosure have failed.
- There is a high probability both that harm will occur if the information is withheld and that the disclosed information will actually be used to avert harm.
- The harm that identifiable individuals would suffer would be serious.
- Appropriate precautions are taken to ensure that only genetic information needed for diagnosis and/or treatment of the disease in question is disclosed.

In 1998, the Social Issues Subcommittee on Familial Disclosure of the American Society of Human Genetics largely adopted the principles of the President's Commission in its policy statement [Knoppers et al., 1998]. Although the Subcommittee recognized as a general rule that genetic information, like medical information, should be protected by the ethical and legal principles of confidentiality within in the patient–physician relationship, it stated that disclosure could be permitted to family members "where there is a substantially higher risk of suffering from a serious and otherwise undetected genetic disorder and where prevention or treatment is available," in those circumstances where:

- Attempts to encourage disclosure on the part of the patient have failed.
- The harm is highly likely to occur and is serious, imminent, and foreseeable; the at-risk relative(s) is identifiable.
- The disease is preventable and treatable.
- Medically accepted standards indicate that early monitoring will reduce the genetic risk.

Although the Committee stated that "harm" should not be defined solely from the medical perspective of treatment and cure, disclosure only would be justified where the harm from failing to disclose should outweigh the harm from disclosure.

Given that effective therapeutic treatments for life-threatening genetic conditions such as Huntington disease and hereditary breast and ovarian cancer are only just emerging, the underlying rationale for disclosure, effective intervention, is open to question [ASHG, 1998; Dugan et al., 2003, Carroll, 2001; Falk et al., 2003; Offit et al., 2004]. The Canadian College of Medical Geneticists [2004] has acknowledged the difficulties. Its training standards for clinical genetics state that the trainee should "be aware of patient confidentiality and the difficulties it poses in instances where relatives are at risk for serious and potentially preventable diseases."

In June, 2003, the American Medical Association published policy guidelines on the disclosure of familial risk information in genetic testing to family members and relatives (AMA, 2003). The statement recognized a physician's professional duty to "the confidentiality of their patients' information, including genetic information," and stated that:

> *Physicians should discuss with them whether to invite family members to participate in the testing process. Physicians also should identify circumstances under which they would expect patients to notify biological relatives of the availability of information related to risk of disease. In this regard, physicians should make themselves available to assist patients in communicating with relatives to discuss opportunities for counseling and testing, as appropriate.*

The American Society of Clinical Oncology [2003] adopted a similar position with regard to susceptibility testing for cancer, recommending that:

> *Providers make concerted efforts to protect the confidentiality of genetic information. However, they should remind patients of the importance of communicating test results to family members, as part of pretest counseling and informed consent discussions. ASCO believes that the cancer care provider's obligations (if any) to at-risk relatives are best fulfilled by communication of familial risk to the person undergoing testing, emphasizing the importance of sharing this information with family members so that they may also benefit.*

Studies indicate that although geneticists and counselors, in principle, may support disclosure in specific scenarios to family members, in actual practice they generally will respect a patient's right of confidentiality [Benkendorf et al., 1997; Dugan et al., 2003; Falk et al., 2003] and leave the decision to disclose test results to the patient [Lermann et al., 1997; Lehmann et al., 2000; Hughes et al., 2002: Plantinga et al., 2003; also see Kapp, 2000]. These policy statements and surveys appear to largely reflect the position adopted, at the time of this writing, by a number of American courts of law.

LEGAL PERSPECTIVES

Duty to Warn

Questions about a health care professional's obligation to disclose genetic information to family members have been considered by courts of law. American courts of law traditionally have recognized that a physician can have a reasonable duty to protect third parties against "dangers emanating from the patient's illness" [*Simonsen v. Swenson*, 1920]. Examples include the duty to prevent the spread of contagious diseases [*Troxel v. A.I. Dupont Institute*, 1996; *Reisner v. Regents of University of California*, 1995; *Tenuto v. Lederle Laboratories*, 1997], to notify the appropriate authority if a patient is unfit to drive an automobile [*Freese v. Lemmon*, 1973; *Joy v. Eastern Maine Medical Center*, 1987; *Joyal v. Starreveld*, 1996], and to report child abuse [e.g., *Hope v. Landau*, 1986].

The landmark 1976 decision of *Tarasoff v. Regents of University of California* substantially expanded the scope of the duty to protect third parties. In this case the parents of a young woman who was murdered sued the therapist, alleging that prior to the killing, the boy friend had confided to the therapist his intention to kill the woman. The court held that where the psychotherapist determines, or should have determined, that a patient poses a serious risk of physical violence to a third party, the therapist has a duty to use reasonable care to warn the intended victim.

It has been argued that this obligation to warn third parties could lend itself "to the development of more encompassing duties to warn patients—both past and present— who have been under the physician's care" [Berg and Hirsh, 1980]. However, a review of later American and Canadian case law indicates that, although judicial interpretation and application of what became known as the "*Tarasoff* doctrine" has varied *significantly*, the obligation to warn third parties generally has been restricted to acts of potential physical violence [*Leedy v. Hartnett*, 1982; *Lipari v. Sears-Roebuck*, 1980; *Hedlund v. Superior Ct.*, 1983; but see *Peck v. Counseling Services*, 1985] or to warn the community of a potentially dangerous person [for Canada, see *Tanner v. Noys*, 1980]. Given the manner in which both American and Canadian courts of law generally have interpreted and applied the *Tarasoff* doctrine over the past 25 years, it is reasonable to conclude that the argument that this decision would lead to the development of more "encompassing duties to warn patients—both past and present" [Berg and Hirsch, 1980] has not materialized—at least based upon the *Tarasoff* doctrine alone.

Duty to Warn: Genetic Information

Three cases have generated significant attention and debate with regard to the issue of a duty to warn family members and relatives regarding genetic risk information. In *Pate v. Threkel*, a 1995 Florida case, the daughter of a mother who had been diagnosed with hereditary thyroid cancer sued the mother's physician for failure to warn the daughter that she might be at risk. The daughter was diagnosed with advanced stage thyroid cancer 3 years after her mother's. The daughter alleged that if she had been warned, prophylactic removal of the thyroid gland could have been undertaken before cancer was detected. The trial courts rejected the plaintiff's claim that a physician's duty to inform others of a patient's contagious disease should be extended to include the child of a patient who suffers from an inheritable disease [Clayton, 1998: Editorial Note: The author distinguishes between infectious disease cases where the individual only becomes sick because of exposure to the infected person compared to genetic risk information where the third party already may be at risk to develop the genetic disease].

The appellate court, however, reversed this finding. The court cited the specific wording of a Florida statute that defines the legal duty owed by a health care provider in a medical malpractice case section (766.102, Florida Statutes: 1989):

The prevailing professional standard of care for a given health care provider shall be that level of care, skill, and treatment which, in light of all relevant surrounding circumstances, is recognized as acceptable and appropriate by reasonably prudent similar health care providers. [s. 768.50(2)(b)]

The appellate court recognized that in applying the statue, a physician had a duty to warn a patient "of the genetically transferable nature of the condition for which the physician was treating the patient," and that when the prevailing standard of care is created for the benefit of identifiable third parties, the physician's duty to warn also extended to those third parties, in this case, the children of the patient.

The court, however, limited the duty to warn to the patient [Burnett, 1999]:

If there is a duty to warn, to whom must the physician convey the warning? Our holding should not be read to require the physician to warn the patient's children of the disease. In most instances the physician is prohibited from disclosing the patient's medical condition to others except with the patient's permission. Moreover, the patient ordinarily can be expected to pass on the warning. To require the physician to seek out and warn various members of the patient's family would often be difficult or impractical and would place too heavy a burden upon the physician. Thus, we emphasize that in any circumstances in which the physician has a duty to warn of a genetically transferable disease, that duty will be satisfied by warning the patient.

The *Pate* decision was followed by the 1996 decision of *Safer v. Estate of Pack* by the New Jersey Superior Court. A daughter sued the estate of the late family physician, Dr. Pack who, 30 years earlier, had treated her father for a hereditary condition, hereditary familial adenomatous polyposis, as well as her mother for a vaginal ulcer. The father had been diagnosed with retroperitoneal cancer; Dr. Pack had performed a total colectomy and an ileosigmoidectomy for multiple polyposis of the colon. However, despite subsequent interventions, the carcinoma of the colon metastasized to the liver, and the father died in 1963 at 45 years of age. In 1990, the daughter, who was 36 years of age, was diagnosed with a cancerous blockage of the colon and multiple polyposis. After examination of her father's medical records, the daughter alleged professional negligence on the part of the late Dr. Pack for his failure to warn of the risk to the daughter Donna Safer's health, and more specifically that:

- Multiple polyposis is a hereditary condition.
- The hereditary nature of the disease was known at the time Dr. Pack was treating the father.
- The then prevailing medical standard of care required the physician to warn those at risk to enable an early examination, monitoring, detection, and treatment.

The trial judge rejected the claim, following the decision of *Pate v. Threlkel* [1994], finding that no patient–physician relationship existed with the daughter and that genetically transmissible diseases differ from contagious or infectious diseases because "the harm is already present within the non-patient child, as opposed to being introduced, by a patient who was not warned to stay away. The patient is taking no action in which to cause the child harm."

The appellate court overturned the trial judge's decision, finding that the prevailing standard of care was for the benefit of identifiable third parties, the physician knew of those third parties, and that children fell within a zone of foreseeable risk. The court stated:

> We see no impediment, legal or otherwise, to recognizing a physician's duty to warn those known to be at risk of avoidable harm from a genetically transmissible condition. In terms of foreseeability especially, there is no essential difference between the type of genetic threat at issue here and the menace of infection, contagion or a threat of physical harm.... The individual or group at risk is easily identified, and substantial future harm may be averted or minimized by a timely and effective warning.
>
> Although an overly broad and general application of the physician's duty to warn might lead to confusion, conflict or unfairness in many types of circumstances, we are confident that the duty to warn of avertible risk from genetic causes, by definition a matter of familial concern, is sufficiently narrow to serve the interests of Justice.

Courts of law appear more willing to extend a duty to disclose to third parties where the third parties are the parents of a minor child patient who received negligent care in the field of genetics with resulting serious harm, including the birth of an afflicted child. By way of example, in the decision of *Schroeder v. Perkel* [1981], a doctor failed to diagnose a child born with cystic fibrosis (CF). This led to the birth of a second child with CF. The court held that liability could extend to the patient's family where it was foreseeable that the parents would rely on the diagnosis (however, the trial court in the case of *Pate v. Threlkel* declined to follow the *Schroeder* decision, but see Burnett [1999] the author argues that courts have "begun to de-emphasize the privity requirement" in malpractice actions).

In the 2004, the decision of *Molloy v. Meier, Backus*, by the Minnesota Supreme Court, concerned a 3-year-old child who the physician noticed was developmentally delayed. Further testing did not reveal the source of difficulties. In discussions with the child's mother, the physician learned that the mother's half-brother was mentally challenged. The physician ordered genetic testing for "? Chromosomes + fragile X," with the intent to order chromosomal testing and testing for fragile X syndrome.

- The laboratory subsequently reported that the chromosome testing results were negative, that is, normal. The physician telephoned the parents, told them that the test results were "normal" but failed to mention that fragile X testing had not been performed.
- The mother assumed that the negative test results include a negative result for fragile X.
- The mother also had been referred to another a specialist (neurology) for consultation. This physician told the mother that the child's difficulties were not genetic in origin and the risk that another child would be born with similar difficulties "was extremely remote." The physician made this assessment before the test results were known.
- Several years later, the child was referred to another physician "G" who knew of the half-brother's condition but did not order testing for fragile X assuming that it already had been done 3 years before. This physician did not meet with the parents.

The mother subsequently gave birth to another child who showed similar developmental difficulties and was diagnosed with fragile X. The mother commenced a lawsuit alleging that she and her husband would not have conceived another child if they had known of the diagnosis and that the physicians were negligent by:

- Failing to order fragile X testing
- Failing to properly read those lab tests that were performed
- Mistakenly reporting that S.F. had been tested for fragile X
- Failing to provide counseling regarding the risk of passing an inheritable genetic abnormality to future children

The Minnesota Court of Appeals held that the physicians were liable. In its decision, the court stated:

> *Our decision today is informed by the practical reality of the field of genetic testing and counseling; genetic testing and diagnosis does not affect only the patient. Both the patient and her family can benefit from accurate testing and diagnosis. And conversely, both the patient and her family can be harmed by negligent testing and diagnosis. Molloy's [the plaintiff] experts indicate that a physician would have a duty to inform the parents of a child diagnosed with Fragile X disorder. The appellants admit that their practice is to inform parents in such a case. The standard of care thus acknowledges that families rely on physicians to communicate a diagnosis of the genetic disorder to the patient's family. It is*

foreseeable that a negligent diagnosis of Fragile X will cause harm not only to the patient, but to the family of the patient as well. This is particularly true regarding parents who have consulted the physicians concerning the patient's condition and have been advised of the need for genetic testing.

We therefore hold that a physician's duty regarding genetic testing and diagnosis extends beyond the patient to biological parents who foreseeably may be harmed by a breach of that duty.

Although these three cases appear to be contradictory, two of the decisions were based, in part, on the prevailing standard of care, and in the *Safer* decision, a medical intervention could have prevented significant harm through early detection and effective treatment. The *Safer* decision, however, must be questioned for its failure to adequately address the issue of how to protect confidentiality and the patient's right of autonomy [Liang 1998].

RISKS AND HARM

Genetic risk information poses special challenges and limitations. By way of example: What constitutes a serious or imminent harm? With regard to hereditary breast and ovarian cancer, does the 50 percent risk of having inherited a mutation within the *BRCA1* or *BRCA2* genes qualify [Offit et al., 2004]? How does this compare to the lifetime range of risk for actually expressing the disease? How are these risks balanced against the difficulties in predicting, among others, penetrance and age of onset?

What will be the outcome of the disclosure? What are the limitations of the specific test(s) and the testing technologies, if any, that could be provided to the family member or relative? What is the possibility of an inconclusive or ambiguous result? What proven therapies and preventive strategies are available, if any; what if the treatment will be painful and ultimately could prove ineffective [Takala and Gylling, 2000]? By way of example, with regard to Huntington disease [Garrison, 2003], the age of onset, the nature and degree of severity of expression of symptoms [Maat-Kievit et al., 2001], and the potential emotional and psychological impact on the patient and the family, all may vary significantly, making it difficult to postulate as a general rule that a family member or relative will exhibit certain symptoms and trigger specific emotional and psychological responses with any reliable correlation to age and gender [Farrer and Conneally, 1987; Gomez-Tortosa et al., 1998; Louis et al., 2000]. Perhaps more importantly, will the family member or relative want to know of the risk status [Laurie, 2001; Taylor, 2004] given the lack of effective treatment and cure and the risk of potential psychological harm, familial unrest, and insurance and employment discrimination [Takala and Gylling, 2000]?

The issue of confidentiality and the duty to warn will continue to evolve. Although disclosure may prove of medical value to a family relative who may be at risk for a life-threatening condition thereby enabling testing and targeted surveillance, disclosure brings with it the potential for loss of privacy, insurance and employment discrimination, social stigmatization, and psychological consequences [Lerman et al., 1998a]. One may seek to justify disclosure in order to enhance personal autonomy in decision making and prevent harm to the recipient, however, the disclosure may itself cause harm and infringe on an individual's right of autonomy, including the right not to know. In these scenarios, either course of action may result in harm, and neither the health care professional, nor the patient, nor the patient's family and relatives may understand, nor accept, what the other perceives to be the most "beneficial," "harmful," "rational," or "irrational" decision. For example, how does one balance the harm between a child being born with a severe genetic disease or not being born at all [Garrison, 2003]? It is because of these complex and troubling issues that the value of nondirectiveness has been advocated in terms of patient decision making.

If disclosure were to be adopted as standard policy for genetic risk information in general, such a policy arguably would not be undertaken in response to the individual patient's needs, beliefs, values, and right of personal autonomy but rather upon preconceived assumptions and values that may prove invalid on a case-by-case basis. Until such time that effective management and treatment options are developed, including proven preventive strategies such as individually targeted pharmacogenetic drugs and therapies, the rationale for disclosure to family and relatives remains open to question.

CONCLUSION

Given the respective statements of professional organizations and courts of law, at the time of this writing, the health care professional's duty arguably is most appropriately fulfilled through discussion with the patient about the implications of the test result for other family members and relatives including the attendant benefits and risks of disclosure.

WEB RESOURCES

American Medical Association [2003] Disclosure of Familial Risk in Genetic Testing: http://www.ama-assn.org/ama/pub/category/11963.html

American Society of Clinical Oncology [2003]: http://www.asco.org/ac/1,1003, _12-002228-00_18-0011991-00_19-0011993-00_20-001,00.asp

HIPAA Privacy: Summary of the HIPAA Privacy Rule: http://www.hhs.gov/ocr/ privacysummary.pdf. Federal regulations for the protections of patients' medical records (Privacy Rule or HIPAA—Health Insurance Portability and Accountability Act—regulations) on April 14, 2001. Additional changes were made to the Privacy Rule on August 14, 2002. All covered entities (health care plans, providers, and clearinghouses) were to be in compliance with the regulation by April 14, 2003.

HIPAA [1996] Health Insurance Portability and Accountability Act of 1996: http://www.aspe.hhs.gov/admnsimp/pl104191.htm

Knoppers et al. (1996): http://genetics.faseb.org/genetics/ashg/policy/pol-29 .htm.

EMERGING DUTIES RE: PROFESSIONAL DISCLOSURE

Béatrice Godard, Université de Montréal, Programmes de Bioéthique, Montréal, Canada

Bartha-Maria Knoppers, Université de Montréal, Faculté de Droit, Montréal, Canada

INTRODUCTION

The proliferation of tests to identify individuals with an increased risk of genetic disease has led to a reevaluation of the duties of medical geneticists, genetic counselors, and other health care professionals. Considering the speed at which genetic technologies are moving from bench to bedside, the questions of (1) whether researchers should communicate research results, (2) whether physicians should recontact patients concerning new information that might be useful to them, and (3) whether physicians should inform family members of relevant genetic information are becoming increasingly important. These questions correspond to the current changes taking place in the spheres of knowledge, social and cultural values and norms, and individual life experience.

Currently, changes in professional attitudes are reflected in the increased level of communication and involvement in relationships with patients and research participants. Fully informed communication and the presentation of choices and alternatives is now heralded as indispensable to ethical care. Patients are seen as full partners in the process of informed and involved decision making. Debate surrounding the principles of confidentiality and beneficence have led to a reexamination of the Hippocratic Oath and the sacrosanct principle of confidentiality. Yet, there are still few therapies in the field of human genetics. It seems that genetic information is the "treatment"! Once treatment does become available, however, a whole range of new questions arise, from access to care, the choice and costs of treatment, and the facilities available. In short, the sophistication and broadening scope of genetic technologies along with new knowledge about existing problems or techniques, have transformed the way in which we view the physician–patient relationship and accompanying social and ethical issues in genetic research and treatment. New knowledge may provide information and choices but not solutions. This information does have consequences for prevention, reproduction, and health promotion. The question is: Ought we to disclose? In previous practice the wishes of the patient were paramount, but with the rapidity of genetic discoveries, research results might change over time, or the shifting of the significance of genetic diagnosis might prompt the recontact of a patient, or, finally, relatives wish to receive information for their own benefit. This could mean telling the patient or even a third party against the express wishes of the patient. Shifts in social values and attitudes have generated evolving public views of the role and authority of health professionals and created expectations as concerns the power of information technology to track over individual and families over time [Calman, 2004, p. 368].

These issues, however, do not require an outright renunciation of contemporary ethical values but rather point to the need to respond to new developments in relation to existing ethical issues. An ethical framework for the practice of genetics may thus introduce new professional responsibilities in the existing duty of care that includes the duty to inform, to provide appropriate care, to follow-up, to refer, and to refrain from abandoning a patient [Hunter et al., 2001, p. 270]. To this, should we add the duty to communicate results in the research setting, the duty to recontact patients, and the duty to warn relatives?

When it comes to potentially preventable or treatable hereditary diseases, it is important for individuals to be informed of their own genetic risk. Obviously, this provides information on family members as well. In both situations, professionals are faced with conflicting ethical obligations. In the former, they must evaluate whether the decision to inform a patient of her or his genetic risk outweighs all the possible harm that may result from disclosure; in the latter, they must weigh their duty to preserve their patient's privacy and autonomy against their obligation to prevent harm [Dugan et al., 2003; Falk et al., 2003].

This chapter addresses emerging professional responsibilities in genetics. It stresses the possible ethical duties of genetic services providers as well as the rights of patients and relatives. In particular, we examine responsibilities as they relate to the communication of research results, the recontact of patients, and the possible disclosure of genetic information by physicians to at-risk family members. These emerging responsibilities aim to promote health and to prevent medical harm by providing patients with the latest information on either patterns in research data or individual results when available and validated or on diagnostic or therapeutic interventions that may be of interest to family members as well. More importantly, this ethical framework for genetics strives to promote a "more human vision of care," based on research participants' and patients' informational, emotional, and psychological needs [Sharpe, 1999, p. 1202]. Ultimately, we proposes have an ethical framework to guide researchers and health professionals involved in genetic risk notification to better define their emerging responsibilities.

COMMUNICATION OF RESULTS IN RESEARCH SETTINGS

The elaboration of an ethical framework must first and foremost address issues surrounding the communication of research results. To clarify the discussion on the communication of results in research settings, there is a need to distinguish between the established ethical obligation to communicate general basic research results from the eventual return of specific findings that may be relevant to individuals (see infra: Duty to Recontact) [Godard and Simard, 2003; Knoppers and Joly, 2004; Letendre and Godard, 2004]. By their very nature, most research results are usually in the aggregate form and not at all "individual." As stated in 2003 by WHO in its report on genetic databases, in most cases genetic research data will remain of abstract significance [WHO, 2003]. Indeed, there has been insufficient discussion on the communication of research results outside of the context of clinical trials. Nevertheless, a suggested avenue is to communicate results in basic research if and when they have been scientifically and clinically validated and when they have implications for the health of the participant. The Quebec Network of Applied Genetic Medicine requires that scientifically valid results be communicated to the participant by the treating physician when they have "significant implications for the health of the participant and [when] prevention or treatment is available ... unless the participant has chosen not to receive any results" [RMGA, 2000, IV.3]. The Consortium on Pharmacogenetics offered similar advice, directing that researchers should "offer the research subject the option of disclosure of research information when its reliability has been established and when the disclosure is of potential benefit to the subject" [Consortium on Pharmacogenetics, 2002].

Turning to results in clinical trials, international guidelines for good clinical practice stipulate that investigators as well as institutions involved in trials "should ensure that adequate medical care is provided to a subject for any adverse events, including clinically significant laboratory values, related to the trial" [ICH, 2001]. The communication of research results in clinical trials may occur when test results are relevant to the patient's health: The patient has the right to receive information (recorded or not) and to be fully informed about her or his health status including the medical facts about her or his condition [World Health Organization (WHO), 1997; World Medical Association (WMA), 1995].

Ethical considerations require health professionals to evaluate the specificity of research results, differentiating between the obligations of a researcher who usually only has general results and those of the physician vis-à-vis their patients involved in clinical trials, as we will see below. Failure to provide information about study results may be one of the many factors that adversely affect accrual [Hébert et al., 2001]. Providing results to study participants or patients might maximize the good, that is, respect for the person, quality of life, although no empirical data support this premise. The communication of research results might lead to better patient–physician communication, which would in turn lead to greater satisfaction with medical care.

DUTY TO RECONTACT IN CLINICAL SETTINGS

In addition to the communication of research results, we must examine the duty that professionals have to recontact patients when confronted with new genetic information, more precise tests or new therapies. A physician may consider that recontact is needed following the discovery of preventive measures or treatment for a specific condition or when the patient requires immediate and continuing treatment or management. This degree of incertitude surrounding genetic risk notification warrants guidelines for determining the circumstances in which recontact efforts should be undertaken. It is important to remember that novel ethical questions come into play, as it is doubtful that consent to be recontacted could be autonomous since patients do not know in advance what kind of information might be disclosed. While recontact may be attempted if treatment or prevention is available, professionals should consider the emotional impact of communicating new sensitive information. They must above all ensure that the patient's privacy and familial interactions are protected [Fitzpatrick et al., 1999; Hunter et al., 2001]. The Canadian College of Medical Geneticists stresses that the disclosure process in testing "should avoid needlessly informing individuals who do not wish to learn their genotype or informing one family member of another family member's genotype" [CCMG, 1991, p. 3].

Most professional bodies, however, have no policies requiring physicians to disclose new relevant information, except in the case of adverse events or medical incidents. This may be changing: Ethical guidelines are directed at empowering patients to assume ownership of their ongoing care and to use new findings and therapies to improve their quality of life (Partridge and Winer (2002)). Presently, the duty to recontact is not addressed explicitly in the Quebec Code of Ethics for physicians, although it requires medical follow-up or referral as dictated by the patient's condition (2002, article 32). The American College of Medical Geneticists [1999b], on the other hand, has issued a policy statement on the duty to recontact:

> It is the medical geneticist's responsibility to provide clinical updates to those patients to whom they provide an on-going service. However, since this represents the smaller percentage of the caseload, it should be incumbent upon the primary care physician to alert his/her patient to the need for a recontact as necessary. The patient should also be included in the process by being adequately counseled to contact the primary care physician or genetics unit as relevant changes in their lives occur.

Clearly, the duty to recontact may be crucial because failing to disclose relevant updated information undermines public trust in medicine. Failure to recontact not only involves deception but hints at a preservation of professional interests over the well-being of patients. It may cause harm if patients are injured further by the failure to disclose; moreover, it may also undermine efforts to improve the involvement of patients in their care. And because it impinges on a patient's right to know, failure to disclose relevant genetic information could be deemed a breach of professional ethics. Moreover, it has been shown that genetic service providers assign a higher degree of responsibility for maintaining contact to patients (3.9 on a scale of 1 to 5) than to themselves (3.4) [Fitzpatrick et al., 1999].

In practice, more than half of the respondents had recontacted a past patient about recent clinical findings, but they perceived significant barriers for doing so systematically and viewed the implementation of a formal duty to recontact as requiring a major investment of resources. Only 13 percent indicated that they had a formal system in place for doing so [Fitzpatrick et al., 1999]. From the point of view of a patient, Huggins' study [1996] concluded that the responsibility for staying in contact should be shared between health professionals and patients. Very few respondents indicated that they were responsible for staying informed. Of responses 25 percent attributed the responsibility to health care providers exclusively; 66 percent of the responses indicated that patients share this responsibility with providers. Patients did not assign the same duties to the health professionals. Respondents consistently attributed the greatest level of duty to clinical geneticists and genetic counselors, and relatively less duty to pediatricians, family doctors, and other physician consultants [Fitzpatrick et al., 1999].

While there are benefits to recontacting and to communicating new clinical data, there remain considerable disadvantages. Patients might not want to be recontacted or receive new information. Clinicians need to anticipate negative reactions and be prepared to provide the necessary support. Disclosure necessitates the use of substantial resources: Participants may want more or less information and it could become quite time consuming. How should professionals recontact for new tests or treatments? Both the processes for the communication of research results and for recontact should be built into the informed consent process and, if acceptable, reoffered to the individual at the recontact. And by engaging the individual in a comprehensive discussion of what disclosure entails (and always estimating the perceived benefits conservatively), physicians must ensure a balance between harms and benefits. Provisions for anticipatory follow-up have to be made explicit. Patient advocacy groups, which lobby for the protection of patient rights, may prove very helpful in this process.

DUTY TO WARN FAMILY MEMBERS

The personal and familial nature of genetic information further complicates confidentiality and privacy issues. It may be important for an individual to be informed of a family member's genetic test results and thus of his or her own risk when it comes to potentially preventable or treatable conditions. Hence, the duty to warn family members could arise when the warning is necessary to avert serious harm for treatable or preventable conditions. A Code of Ethics being prepared by the Canadian Medical Association [2003] maintains:

> *Disclose your patients' health information to third parties only with their consent, or as required by law, or when the maintenance of confidentiality would result in a significant risk of substantial harm to others or to the patient if the patient is incompetent; in such cases, take all reasonable steps to inform the patient that confidentiality will be breached.*

Thus, health professionals must determine whether the family member is at a high risk of serious harm; they must also decide if, in the event where a patient has repeatedly refused to disclose the information, a breach of confidentiality is necessary to prevent or minimize this harm. How is a physician to resolve the competing ethical mandates of autonomy and beneficence and nonmalificence? This requires an evaluation and comparison of the actual risk of a genetic disease, the efficacy of potential preventive interventions, and emerging legal considerations and potential liabilities. The physician should strive for a balance between the right to confidentiality, the right to know, and the right not to know, including that of others who are not even his or her patients!

Examining the views held by medical geneticists, genetic counselors, and patients enables a better understanding of the duty to warn. A study conducted by Falk et al. [2003] reveals that although 69 percent of medical geneticists believe they bear responsibility to warn their patients' relatives when found to be at risk for genetic disease and 25 percent who faced the dilemma of a patient refusing to notify their at-risk relatives seriously considered disclosure without patient consent, only four respondents proceeded to warn at-risk relatives of their status. Similarly, 46 percent of genetic counselors have had a patient refuse to notify an at-risk relative: 21 percent seriously considered warning at-risk relatives without patient consent but only one genetic counselor did go on to disclose [Dugan et al., 2003]. Concerning relatives, a study [Porteous et al., 2003] showed that 78 percent of participants in genetic testing for hereditary colon cancer thought it acceptable to have the genetic information brought to their attention, the remaining 22 percent did not mind being approached with the information and 91 percent decided to take the genetic test.

In parallel to these studies, we are seeing more detailed advice concerning the duty to warn relatives, urging physicians to assist patients in their communication with relatives. Physicians, according to the American Medical Association Code of Medical Ethics [2004]: "should make themselves available to assist patients in communicating with relatives to discuss opportunities for counseling and testing, as appropriate." This view is reaffirmed by the American Society of Clinical Oncology [2003] in its policy update on genetic testing for cancer susceptibility: "the cancer care provider's obligations (if any) to at-risk relatives are best fulfilled by communication of familial risk to the person undergoing testing, emphasizing the importance of sharing this information with family members so that they may also benefit." Advocacy groups, meanwhile, recommend asking the patient's consent in advance of genetic testing for permission to disclose genetic information to relatives [Genetic Interest Group, 1998]. Under Danish law, patients may be asked to disclose the identities of family members to a physician to ensure diffusion of genetic risk information within affected families. In Quebec, communication to third parties is prohibited to ensure confidentiality of medical records [L.S.S.S.S., C.p.C article 19], except after the death of a patient, when a spouse, close relative [23 al.2] or blood related individual [23 al.3] is affected. The Quebec Code of Ethics of Physicians permits disclosure by physicians "when the patient or the law authorizes him to do so, or when there are compelling and just grounds related to the health or safety of the patient or of others" [2002, article 20.5].

DISCUSSION

In the joint account model elaborated by Parker and Lucassen [2004], it is assumed that genetic information should be available to all account holders (health professionals, relatives) unless there are good reasons to do otherwise. The justification in favor of a joint account adheres to the ethical principles of justice and reciprocity; there are also benefits to be gained by sharing genetic information. Because geneticists work with families, the chosen approach appears to be consistent with the very nature of practice in genetics. However, more often, a case-by-case approach is chosen. This approach enables professionals to intervene when the disease is fatal if diagnosed late but treatable if diagnosed early. On the other hand, in cases in which the disease is untreatable, the duty to warn would be tenuous, except for reproductive decisions. Some diseases would fit in between, such as hereditary breast cancer, which is treatable by aggressive surgery.

Ultimately, emerging responsibilities in genetic risk notification require an over expandary duty of care for professionals. Duties to communicate research results and to recontact with clinical findings affect the concept of an "ongoing duty of care" in order to include patients' informational needs. The duty to warn, moreover, is encouraged by human genome research, which creates a shift from "individual therapy" to one that is "family-and-future-generation oriented." The importance of this issue seems unlikely to diminish in the future. As public awareness grows and new technologies develop, the pressure for greater genetics services and information is inevitable. Individuals at risk push for research that is more process oriented and patient oriented. Consequently, requirements of informed consent and confidentiality may be modified relative to other forms of medical care. These issues, along with that of nonmaleficence, the duty to not harm and yet also to prevent harm are omnipresent. The current sociocultural context and pressures of the "promise" of genetic information cannot be ignored.

A fair process in health care ethics remains a priority. This approach is more likely to be successful and useful than agreement on a set of principles. A fair process would be one in which there was transparency about the grounds of the decisions to return results, recontact, or warn based on prior discussion. Appealing to rationales that all can accept as relevant to meet health needs fairly, it would allow procedures for revising decisions in the light of challenges to them.

If we adopt this extensive duty of care, are we moving toward an effective ethical management of day-to-day genetic testing? There remains no systematic assessment of effectiveness of different models of ethical management. We must therefore consider other questions. What kinds of ethics support would genetic services providers find useful? What is the most effective model of ethics support in human and clinical genetics? It is time to build this model before the courts decide. At a minimum, we need to provide guidance on what could be seen to be a new form of "professionalism."

Acknowledgment

This section forms part of INHERIT-BRCAs research program financed by the Canadian Institutes of Health and Research.

ETHICAL, LEGAL, AND PRACTICAL CONCERNS ABOUT RECONTACTING PATIENTS TO INFORM THEM OF NEW INFORMATION: THE CASE IN MEDICAL GENETICS

Alasdair Hunter, Ottawa, Ontario Canada K1N 5S4
Neil F. Sharpe, Genetic Testing Research Groups, Hamilton Ontario L8R 3B7
Michelle Mullen, University of Ottawa, Ottawa Ontario, Children's Hospital of Eastern Ontario, Ontario, Canada K1H 8L1
W.S. Meschino, North York General Hospital, Department of Genetics, Toronto, Ontario Canada M2K 1E1

OVERVIEW

Medical genetics is in the midst of an exponential growth in knowledge and the technical ability to apply new information to genetic testing, patient diagnosis, and genetic counseling of families. The question that has become paramount is whether the geneticist, the laboratory, or the family physician have a *duty to recontact* these patients and to inform them of research and service changes that may be relevant to their situation. Or, is it to be the patient who ultimately bears responsibility for maintaining this contact?

In 1999, the Social Ethics and Legal Committee of the American College of Medical Genetics (ACMG 1999b) published a policy statement that stated:

> *Rapid and continued advances in human genetics have resulted in new opportunities for the early detection of disease, presymptomatic diagnosis, and the possibility for new interventions and therapies. The vast majority of physicians would agree that optimal medical care calls for informing patients of these new developments. The difficult issue is determining which physician or other health professional should be responsible for informing the patient of advances which become available after the last scheduled consultation.*

The committee placed the duty to do so at the feet of patients and primary care physicians:

> *After an initial genetics consultation, the referring physician and, as appropriate, the patient and the designated primary care provider should receive a written summary of the consultation that includes the recommendation to contact the genetics unit for new advances. It is the medical geneticist's responsibility to provide clinical updates to those patients to whom they provide an on-going service. However, since this represents the smaller percentage of the caseload, it should be incumbent upon the primary care physician to alert his/her patient to the need for a recontact as necessary. The patient should also be included in the process by being adequately counseled to contact the primary care physician or genetics unit as relevant changes in their lives occur.*

However, the committee was not unanimous on whether patients should receive a written summary of the counseling and advice as to reasons to recontact the genetics program [Hirschhorn et al., 1999].

Clinical situations where this scenario could arise include:

- A diagnosis has been suspected but not made, and a new diagnostic test becomes available.
- A more accurate diagnostic and/or prognostic test has been developed.
- New information may alter the prognosis or recurrence risk estimates [Fitzpatrick et al., 1999].

Recontact is of particular concern because clients may not act on information, for example, with respect to reproduction, for years or even a generation after the genetic consultation and because information that is now incomplete or even incorrect may be relevant to the extended family. The question that has become paramount is whether the medical geneticists, and/or their laboratories, and/or primary care physicians have a *duty to recontact* these patients and to inform them of research and service changes that may be relevant to their situation.

Most geneticists appear to accept that recontact is ethically desirable [Fitzpatrick et al., 1999; Hirschhorn et al., 1999, Dean et al., 2000], but debate has centered mainly around whether this obligation is practically feasible, and if so, whether it is the geneticist, the family physician, or the patient who ultimately bears responsibility for maintaining this contact.

A legal argument had been raised [Berg and Hirsch, 1980; Pelias, 1991; Andrews, 1991] that suggested that prior case law leaves physicians at legal risk should they fail to recontact previous patients and to provide them with news of advances. However, in 2001, it was argued that a duty to recontact former patients generally was not supported by case law and would represent an inequitable and onerous clinical burden [Hunter et al., 2001]. *The following review is a synopsis of that discussion* [Hunter et al., 2001].

Throughout, it is assumed that the question is whether or not to recontact patients who have been seen in consultation for one or more visits. It is accepted that those who are in an *ongoing* care or a follow-up relationship must be kept informed, unless the relationship has come to an end.

Discussion is restricted to new information derived from a technological advance and not to the later discovery of errors in the application of knowledge or a technical procedure that was current at the time of the initial consultation. Ethics and the law are clear that patients have the right to be informed when a mistake has been made, and this duty flows from the original relationship.

WHEN IS RECONTACT APPROPRIATE?

Patients are undergoing susceptibility testing for mutations in the breast cancer genes *BRCA1* and *BRCA2*. At the time of writing, no single technology can detect all known mutations. In the future, automated methods may detect nearly 100 percent of mutations. Should one then offer to retest a woman who, despite an initial "negative" test result, but because of her family history, is now considering whether to undergo bilateral oophorectomies and mastectomies? Similarly, if by the time such testing becomes available, it has been shown that increased surveillance of *BRCA* mutation carriers has no impact on morbidity and mortality, do health care professionals still need to recontact patients and offer retesting?

If the potential impact of new information must be considered in assessing the need to recontact a family, geneticists and physicians may be left open to the criticism of being paternalistic. Yet, it must be considered. Notwithstanding the individuality of patients with respect to their ambitions, needs, and tenets, the potential relevance of new information is not static. Rather it changes with time and with the life history of a particular patient or family. A prenatal test for spinal muscular atrophy may be highly relevant to a 30-year-old woman contemplating having children but completely irrelevant 10 years later once her family is complete. It will be far more compelling for a patient to be told that there is a new test that will detect 85 percent of mutations when none was available before than to inform a patient that a test that was 85 percent effective is now 95 percent effective.

SERVICE IMPLICATIONS

The potential obligation to recontact former patients has raised questions about the practical ability of physicians and hospitals to appropriately respond to patient needs. A shortage of trained personnel, appropriate time, and the complexity of the resources required to institute a policy of patient recall should not be underestimated. How will health care professionals balance existing resources against not only current patient needs but also those of patients who, after they were recontacted, have requested services? If new information involves application of a new or more accurate technology, it is important to consider whether the patient, the patient's insurance company, or a government health plan, in whole or in part, will pay for the new service. Therefore, any decision about whether to recontact former patients may practically have to take into account whether adequate and appropriate resources are available to reasonably meet the patient's needs, either directly or by referral [Morreim, 1997].

DOES THE PATIENT WANT TO KNOW?

The principle of *respect for persons* includes respect for the autonomy of capable persons and the protection of vulnerable persons: *Beneficence* (doing good), *nonmaleficence* (doing no harm), and *justice* (minimally, treating like cases alike) [Beauchamp and Childress, 1994]. A *prima facie* interpretation of the respect for persons principle suggests that recontact of patients to inform them of advances in genetic science affecting their health may be indicated, promoting the truism that information promotes autonomy. Autonomy presumably also includes the "right" to decide whether or not to seek information: that is, the right to remain uninformed.

When a patient is referred for a genetic consultation, he or she has a right to refuse the appointment and, for a variety of reasons, about 10 percent do not to appear for a scheduled appointment [Humphreys et al., 2000]. Nor can it be assumed that because a patient attended the initial visit that he or she would automatically wish to be contacted with new information:

- Patients may have made a decision based on the information current then and have no interest in reopening the question.
- Patients may experience anxiety and psychological unrest surrounding an issue previously laid to rest.
- A couple may have forgone reproduction or have undergone prenatal diagnosis and terminated an *affected* fetus and may come to suffer greatly in the new knowledge that it was for a nonexistent risk.
- A patient's prior genetic history and counseling may not have been revealed to a new partner, raising issues of privacy and confidentiality.

While the potential benefits of new information to families may be self-evident, it seems very likely that a standard policy of recontact by geneticists would help some families while harming others or help certain members of a given family while harming others. The proportion in these groups might be expected to vary by disease and with the nature and timing of new information. For example, there may be more justification for recontact when there is significant new information about a gene causing a life-threatening disease than for one that results in a 5 percent increased risk for Alzheimer's disease. Clearly, inclusive discourse would be required to bring about some consensus as to *thresholds for recontact*. At this time, there are few empirical studies that help inform analysis of the desirability of a recontact policy.

Nonmaleficence remains an important ethical consideration, and the uninvited recontacting of patients includes significant risk of causing harm. We have already mentioned the potential resurgence of anxiety surrounding an issue previously laid to rest, and there are other very concrete issues. A couple may have forgone reproduction, or have undergone prenatal diagnosis and terminated an affected fetus and come to suffer greatly in the new knowledge that it was for a nonexistent risk. The prior genetic history and counseling may not have been revealed to a new partner, raising issues of privacy and confidentiality. The commonly suggested approach of stating routinely that the department will contact the patient (or vice versa) in the case of new information may have an unnecessary damaging effect on the patient's confidence and reassurance concerning the information provided during the initial visit.

A documented account of patient's preference, perhaps with written consent, at the time of the initial consultation could address this question, although this does not take account of the matter that the patient cannot foresee the future and the possible impact that new information could have, or that his or her perspective may change with time.

Neither the nature of new information or the course of a patient's life can be predicted. Accordingly, it is unknown whether informing a patient, at an unknown time in the future will be of benefit. The issue of whether the patient wishes to be kept informed of new developments should be clearly and extensively discussed *and documented* with each patient. The patient always should have the right to choose [De Wert, 1992; Weaver, 1997] whether or not he or she wishes to receive additional updated information, especially given the potential psychological and social impact [Genetic Information and Health Insurance, 1993; Birmingham, 1997; Weaver, 1997; Lemmens and Bahamin, 1998; Hall and Rich, 2000b] on the patient and the family, and the unknown scope and nature of the "future information."

The range of issues suggests that recontact policy is not an unqualified good but a complex issue that we have only begun to discuss.

PATIENT OBLIGATIONS

If the duty to recontact were adopted as a standard of care, the fundamental values and objectives of the genetic counseling process [Ad Hoc Committee, 1975] would need to be taken into account. These principles recognize that the patient will play an important and integral role in the therapeutic process. Therefore, with respect to the duty to recontact, the patient should accept a reasonable degree of responsibility. These duties [Meyer, 1992] would include the obligations to contact the physician at previously agreed to periods of time for new information, to make a reasonable effort to understand the nature and implications of the new information, to make reasonable use of resources available to patients to keep informed of developments [*Jolly v. Eli Lilly Co.*, 1988], and to request an appointment for clarification or if counseling is required.

METHODS OF CONTACT

Health care professionals would be required to develop a method to decide what constitutes sufficient medical progress to justify recontacting a family. There are different potential approaches.

- An institution or practice might simply decide to review routinely every chart on a periodic basis. The caseload could be reduced by excluding routine prenatal diagnosis patients and by having genetic counselors mark only those cases relevant for future review. However, this selective approach might have legal consequences if a patient learns that he or she was not informed about some new piece of information. This approach would require that files, presumably selected from a computer-generated list, be pulled, perhaps from off-site storage, for review by a genetic counselor and/or physician. Decisions about recontact would be made on a case-by-case basis and would be affected in part by some of the matters we have discussed already. A review of the literature may be required before such decisions could be made, a potentially-time consuming process.

- A second approach would be to make a decision regarding recontact on a disease-specific basis as progress occurs. Here the geneticist and/or physician assumes responsibility for staying current with the literature on all diseases for which patients have been seen. Again there must be criteria to decide when recontact is appropriate, and it is unlikely that a single set of criteria will suffice for all diseases or for all patients with a particular disease. The disease-specific approach requires that geneticists and physicians have a failsafe way of recording diagnoses in order to assure that patients are not overlooked. It is very unlikely that most medical centers have a high degree of consistency in diagnostic nomenclature data entry because diagnoses are often entered into the computer by clerical staff from the notes of the physicians and genetic counselors.

Regardless of which approach to recontact is taken, centers will be faced with the daunting fact that annually up to 10 percent of North Americans move residence and that they are very unlikely to include the genetics department in their change of address notifications. How far is far enough in an effort to recontact a patient? A call and/or letter to the referring doctor; to the patient's home and/or business, to a former place of employment? Is recontact the responsibility of the individual counselor/physician or of the department? The latter point is of particular importance if the geneticist or physician leaves the center.

Every single step, question, and each potential decision discussed has resource implications for a medical facility. The time required per patient is difficult to predict and would vary with the condition. It is likely to vary inversely with the frequency of the condition, but an hour per patient is probably not unrealistic, and in some cases this would be followed by a clinic visit.

The advent of the birth-to-death electronic medical record would simplify some of the technical issues, but many resource problems would remain, and such a record may be problematic from legal, ethical, and privacy perspectives. It has been suggested [Fitzpatrick et al., 1999] that the use of innovative technologies such as the Internet may provide a practical solution. Commentators, however, have questioned whether:

- An effective physician–patient relationship is less likely through the use of the telephone, letters, websites [Jadad and Gagliardi, 1998; Kim et al., 1999; Jadad, 1999; Jadad et al., 2000], or e-mail [Bovi and Council on Ethical and Judicial Affairs, 2003; Houston et al., 2003; Katz et al., 2003; Patt et al., 2003] than with a face-to-face relationship, and the risk of a lawsuit is potentially greater [Cepelewicz et al., 1998; Hodge et al., 1999].

- With the growing use of electronic means to communicate and store health information [Ferguson, 1998], the same standards that apply for informed consent [Rosoff, 1999; Spielberg, 1999; Blum, 1999] and confidentiality of medical records [Kuszler, 1999; Spielberg, 1999; Terry, 1999; Gaster et al., 2003; HIPAA, 2001] equally should apply to the use of such technologies.

Medical malpractice encompasses errors of omission and commission by health care professionals that fall below the normal, or appropriate, standard of care. However, courts of law have recognized that a standard of care is not an absolute standard, but what can be considered reasonable given the facts and exigencies of a medical situation [Sharpe, 1994b, 1996]. To illustrate, the standard of care generally applicable to a physician may be adjusted to take into account the resources available to the same physician if he or she provides treatment in an emergency clinic. It is uncertain whether courts would make similar allowance when a physician tries to respond to the inflow of "recontacted" patients.

LEGAL ASPECTS

Continuing Duty of Care

Various codes and principles of ethics of organizations such as the American Medical Association and the Canadian Medical Association and courts of law [*Longman v. Jasiek*, 1980; Sharpe, 1994b, 1996] have long recognized that a physician has a posttreatment duty of care to a patient, including the duties to monitor a patient's condition, to provide appropriate aftercare, to refer, and to not abandon the patient. Generally speaking, this posttreatment relationship will continue until [Sharpe, 1994b, 1996; *Blanchette v. Barrett*, 1994]: (1) it is terminated by the patient; (2) it is terminated by mutual consent; (3) it is terminated by the physician after reasonable notice has been provided to the patient; or (4) the physician's services are no longer required (this may be conditional upon the patient having been provided with reasonable notice).

Referral to another physician may not necessarily terminate the duty of care if the patient is in continued need of the physician's expert care [*Longman v. Jasiek*, 1980; Mains, 1985; Sharpe, 1994b, 1996]. The *Longman v. Jasiek* [1980] decision and discussions made about a geneticist's duty of continuing care in Huntington disease [Sharpe, 1994b] illustrate situations where the patient may require *immediate and continuing* treatment and management. This continuing duty of care arguably could include the obligation to advise of any developments in management and treatment that would prove of benefit or detriment to the patient [Sharpe, 1994b]. What each of these situations has in common, however, is that the patient is in *continued* need of a physician's expert care [Sharpe, 1994b].

Do Legal Precedents Support Recall?

American and Canadian courts of law have recognized that where drugs and medical devices have been prescribed, a physician may have a *continuing* duty to disclose subsequently discovered risks to patients and former patients [Andrews, 1991; Pelias, 1991; Berg and Hirsch, 1980, 1983].

It has been argued [Andrews, 1991; Pelias, 1991; Berg and Hirsch, 1980], given these legal precedents, that courts are "likely" [Andrews, 1991] to create a *new* duty to disclose subsequently discovered risks of treatment to *former* patients. Also referred to as a "duty to recall" [Berg and Hirsch, 1980] or a "duty to recontact" [Fitzpatrick et al., 1999], it is further argued that this continuing duty of disclosure could be: (a) expanded to include "complete disclosure of new information" and therefore extended to geneticists who provide patients only with information [Berg and Hirsch, 1980; Pelias, 1991; Andrews, 1991]; and (b) extended for an indefinite period of time [Pelias, 1991].

The use of *Mink v. University of Chicago* [1978] as a legal precedent to support the duty to recontact is problematic. Although the court did find that the obligation to warn was a continuing one flowing out of the duty to secure an informed consent, at the same time, this precedent arguably is restricted to novel and experimental medical procedures and, therefore, the application of a significantly broader standard of disclosure [Sharpe, 1994a]. More importantly, since the 1978 decision of *Mink v. University of Chicago*, given that the standards applied by courts to evaluate a physician's duty of disclosure for informed consent have proven unduly restrictive [Schuck, 1994; Klimchuk and Black, 1997], it is unlikely that a court will find a physician has a continuing duty to recontact and warn flowing solely from the obligation to secure an informed consent.

Duty to Protect Third Parties

Although courts of law generally apply restrictive standards of disclosure for informed consent, in other areas, the nature and scope of a physician's duty to warn and disclose has been substantially expanded. American courts of law traditionally have recognized that a physician can have a reasonable duty to protect third parties against "dangers emanating from the patient's illness" [*Simonsen v. Swenson*, 1920]. Examples include the duty to prevent the spread of contagious diseases [*Troxel v. A.I. Dupont Institute*,

1996; *Reisner v. Regents of University of California*, 1995] or to notify the appropriate authority if a patient is unfit to drive an automobile [*Freese v. Lemmon*, 1973; *Joy v. Eastern Maine Medical Center*, 1987; *Joyal v. Starreveld*, 1996] or where a patient poses a serious risk of physical violence to a third party, a therapist has a duty to use reasonable care to warn the intended victim.

However, a review of later American and Canadian case law indicates that, although judicial interpretation and application of what became known as the "*Tarasoff* doctrine" has varied significantly, the obligation to warn third parties generally has been restricted to acts of potential physical violence [*Leedy v. Hartnett*, 1981; *Lipari v. Sears-Roebuck*, 1980; *Hedlund v. Superior Ct.*, 1983] or to warn the community of a potentially dangerous person [*Tanner v. Noys*, 1980].

As discussed, the argument that the *Tarasoff* decision would lead to the development of more "encompassing duties to warn patients—both past and present" [Berg and Hirsch, 1980] has not materialized—at least based upon the *Tarasoff* doctrine alone.

Drugs and Devices

American [Nadel, 1982] and Canadian courts [*Vasdani v. Sehmi*, 1993] have recognized that physicians can have a duty to warn patients of subsequently discovered adverse risks of drugs and medical devices [*Tresemer v. Barke*, 1978; Andrews, 1991; Pelias, 1991]. However, this duty to warn generally is restricted to reasonable care for a reasonable period of time—assessed on a case-by-case basis—after performance of the original service [*Blanchette v. Barrett*, 1994]. Courts have stated [*Snow v. AH Robins Company Inc.*, 1985; *Getty v. Hoffman-La Roche Inc.*, 1987; *Havens v. Ritchey*, 1991] that the statute of limitations begins to run when the plaintiff-patient knows, or should know by the exercise of reasonable diligence [*Neudeck v. Vestal*, 1931; *Jolly v. Eli Lilly Co.*, 1988], the cause of the injury. Therefore, it is reasonable to conclude that a physician's duty to recontact and warn will not extend for an indefinite period of time. This continuing duty to warn generally has been restricted to those situations where the risk of *physical* injury was created by the application of a drug or device [*Reyes v. Anka Research*, 1981]. What then of the situation where the physician provides patients only with information rather than drugs or medical devices?

INFORMATION-ONLY SERVICES

The New Jersey Superior Court decision of *Procanik by Procanik v. Cillo* [Law Division Court, 1985; Appellate Court, 1988] considered this type of situation and provides insight into the manner in which a court of law may interpret the proposed duty to recontact. This case concerned the allegation that the defendant, an attorney who was a specialist in medical malpractice claims—an action for wrongful birth, where the child was born with rubella syndrome—had not fulfilled a "posttermination duty" to the client, by reason of his failure to contact the plaintiff-clients about an important judicial decision. The Appellate Division of the Superior Court of New Jersey [*Procanik by Procanik v. Cillo*, 1988] held that the attorney had fulfilled his original obligation to the client. The court stated that because the attorney only could express a reasonable opinion based upon the facts at hand and was not obliged to anticipate changes in current legal precedent, no posttermination duty to recontact was created. The court also stated that the reporting letter of the attorney made a full and reasonable disclosure because it correctly outlined the procedures that would have to be followed in order to overturn existing legal precedents that at the time effectively barred causes of action for wrongful birth.

It is reasonable to conclude that if a geneticist and/or physician, for example:

1. Has conducted a DNA diagnostic test in compliance with the generally accepted standard of care;
2. Administered the test in a manner that is appropriate for the particular patient;
3. Has disclosed, pursuant to the doctrine of informed consent, all material information including the test's risks and limitations; and,
4. Has fully disclosed the test results and their implications [Sharpe, 1994a; National Society of Genetic Counselors, 1997] including any associated limitations or reservations, this arguably reasonably would fulfill—following traditional medical standards as well as the principles enunciated in *Procanik v. Cillo* [Law Division Court, 1985; Appellate Court, 1988]—the purpose for which the geneticist has been retained [H. (R.) v. Hunter et al., 1996]. Therefore, assuming that the physician clearly terminates the relationship, a continuing duty to recontact arguably would not be created.

SUMMARY

Although American courts have speculated about the creation of a *new* duty to disclose subsequently discovered risks to former patients [*Pate v. Threlkel*, 1995; *Gorab v. Zook*, 1997; *Blaz v. Michael Reese Hospital Foundation*, 1999], others have questioned the practical ability of physicians to fulfill such a duty [*Hoemke v. New York Blood Center*, 1989; *Havens v. Ritchey*, 1991]. As a general principle, Courts of law have recognized that a physician's obligation to a patient requires that the physician exercise a reasonable degree of care toward the patient, and that the nature and scope of a physician's obligations will be adjusted to the particular facts and exigencies of a medical situation.

The proposed duty to recontact *former* patients who have been provided with information-only services does not find strong support, at the time of this writing, by existing American or Canadian legal precedents for (a) informed consent, (b) a duty to warn and protect third parties, or (c) a duty to warn of subsequently discovered risks except in the case of the application of drugs and medical devices. With respect to those situations where the physician provides only information, in keeping with a physician's traditional duties of posttreatment care, a *continuing* duty to inform and to recontact patients may be created, but arguably only in those situations where the physician has not reasonably fulfilled the purpose for which he or she was originally retained.

Appendix 1

Reproduced with permission from: Weisbrot D (204) The Human Genome: Lessons for Life, Love and the Law. J. Law Med.; 11(4) 428–435.

New Genetics and the Protection of Information

INTRODUCTION

In 2003, the Australian Law Reform Commission (ALRC) and the Australian Health Ethics Committee (of the National Health and Medical Research Council) completed a major inquiry into the protection of human genetic information, focusing on privacy protection, protection against unlawful discrimination based on genetic status, and the establishment and maintenance of high ethical standards. The joint inquiry considered these matters across a wide range of contexts, including such diverse areas as medical research, clinical genetic services, genetic research databases, employment, insurance, immigration, sport, parentage testing, and law enforcement. This appendix discusses some of the major themes that emerged in preparing for future challenges in genetics.

David Weisbrot

Australian Law Reform Commission, Sydney, Australia

INQUIRY INTO THE PROTECTION OF HUMAN GENETIC INFORMATION

Objectives

The objectives of the inquiry were to:

- Protect privacy
- Protect against unfair discrimination
- Ensure the highest ethical standards in research and practice

while promoting

- Innovations in genetic research and practice that serve humanitarian ends
- Reassurance to the community that such innovations will be subject to proper ethical scrutiny and legal (and other) controls

These objectives brought into consideration a wide array of concerns centering on legal and ethical issues, social dimensions, medical practices, scientific advances, and opportunities for consultation. Expert advice was solicited from the areas of genetic and molecular biological research; medicine, clinical genetics, and genetic counseling; community health; indigenous health; health administration and community education; insurance and actuarial practice; regulatory practice; and; labor, privacy, and antidiscrimination law. A separate group worked with experts on forensic medicine, DNA (deoxyribonucleic acid) profiling, policing, and trial practice. Examination of ethical issues was not restricted to just those aspects arising in the practice of medicine and the conduct of research; the ethical dimension was invoked across the full range of human interactions and institutions. The methodology and findings of the inquiry have been reported in detail in a report entitled *Essentially Yours* [Australian Law Reform Commission (ALRC), 2003], which constitutes the most comprehensive assessment of these issues to date.

Background of Current Regulatory Frameworks and Issues

As in many other countries, current methods of regulation and conflict resolution in this field in Australia involve a patchwork of:

- Federal and regional laws
- Official guidelines
- Personal and professional ethics
- Institutional restraints
- Peer review and pressure
- Oversight by public funding authorities and professional associations
- Supervision by public regulatory and complaints-handling authorities
- Occasional media scrutiny and exposés
- Private interests
- Market pressures

This complexity adds substantially to the difficulties in describing, much less reforming, law and practice in this field. The inquiry's brief was to scrutinize the existing regimes and then tailor them—where necessary and to the extent possible—to the particular needs and demands of genetic testing and information. Successfully fulfilling this brief involved not only providing adequate protections against the unlawful use of genetic information but also putting into place measures and strategies aimed at ensuring a higher-order goal: When such information may be used lawfully, it will be used properly, fairly, and intelligently.

Many of the public submissions to the inquiry adopted the language of rights. However, achieving justice in this complex area is not susceptible to a simple vindication of individual rights. Careful consideration of the legal and policy issues thrown up by the use of genetic samples and information requires a wide range of interests to be balanced. Although relatively easy to articulate in the abstract, achieving the proper balance is difficult in practice since various interests will compete and clash across the spectrum of activity.

For example, human genetic information has a powerful familial dimension. The inquiry noted that human genetic information calls from before the cradle and lasts well beyond the grave—an individual's genetic information usually will reveal information about, and have implications for, her or his parents, grandparents, siblings, children, and generations to come. (Or it may reveal that the person is not biologically related to her or his social relatives—another sensitive matter that may or may not have been known or openly disclosed.) Thus, there may be circumstances in which an individual's presumptive right to privacy, and to the confidentiality of the doctor–patient relationship, may be called into question by the competing needs of genetic relatives.

Similarly, a balance must be struck in a number of other areas in such a way as to recognize and accommodate broad societal interests that transcend individual ones, for example, in the compulsory acquisition of DNA samples by law enforcement authorities, the ability of researchers to gain a waiver of individual consent requirements, the imposition of restrictions on employers from requiring genetic testing and information from their employees, or in limiting the ability of a person to initiate parentage testing without the knowledge and consent of the child and the other parent.

In an earlier era, the centerpiece of any significant law reform effort typically was the recommendation of a major new piece of legislation. However, in a more complex environment in which authority is much more diffused, modern law reform efforts are likely to involve a sophisticated mix of strategies and approaches, including legislation and regulations, official standards and codes of practice (in Australia, such as those promulgated by the National Health and Medical Research Council (NHMRC) and the Office of the Federal Privacy Commissioner), industry codes and best practice standards (such as the policy on genetic testing developed by the peak life insurance body, the Investment and Financial Services Association), education and training programs (ranging from basic community education through to continuing and specialist medical education), better coordination of governmental and intergovernmental programs, and so on. Thus, the final report addressed 144 recommendations to 31 different parties, not merely to the Australian government, and therefore implementation of the recommendations and subsequent monitoring will require coordination of diverse regulatory components.

Key Recommendations Relating to Medical Genetics

- A standing Human Genetics Commission of Australia (HGCA) should be established to provide high-level technical and strategic advice to Australian governments, industry, and the community about current and emerging issues in human genetics, as well as providing a consultative mechanism for the development of policy statements and national standards and guidelines in this area.

- Discrimination laws should be amended expressly to prohibit unlawful discrimination based on a person's real or perceived genetic status.

- Privacy laws should be harmonized and tailored to address the particular challenges of human genetic information; among other things, this will require something of a radical departure from the focus on "data" and "information" in existing laws to extend privacy protection to genetic samples as well as genetic information.

- The powerful familial dimension of genetic information requires formal acknowledgment; for example, doctors should be authorized to disclose personal genetic information to a genetic relative in circumstances where such disclosure would be necessary to lessen or prevent a serious threat to an individual's life, health, or safety.

- Protection of the integrity of the individual warrants the creation of a new criminal offense to prohibit an individual or a corporation from submitting another person's sample for genetic testing, or conducting such testing, knowing (or recklessly indifferent to the fact) that this is done without the consent of the person concerned or other lawful authority.

- A series of requirements for strengthening the ethical oversight of genetic research, including better support of human research ethics committees, providing guidances to researchers and participants about ethical best practice, improved governance of human genetic research databases, and stricter reporting requirements.

- Addition of an ethical dimension to the accreditation standards of the National Association of Testing Authorities, Australia (NATA).

- Restriction of genetic testing for health and medical purposes to laboratories accredited for the purpose, and stricter regulation of genetic testing devices.

- Adoption of strategies for assessing and responding to the need for increased and adequately resourced genetic counseling services.

- Employers should not gather and use genetic information except in rare circumstances, for example, when this is necessary to protect the health and safety of workers or third parties, and the action complies with stringent standards.

- A range of safeguards and improved policies and practices should be applied to the insurance industry's use of genetic information (including family history) for underwriting purposes; this will be aimed at ensuring that genetic information must be used in a scientifically reliable and actuarially sound manner, and reasons must be provided for unfavorable underwriting decisions; industry complaints-handling processes must be improved and extended to cover review of underwriting decisions based on genetic information; and industry education and training in this area must be significantly enhanced.

- DNA parentage testing should be conducted only with the consent of each person sampled or pursuant to a court order. In the case of a child who is unable to make an informed decision, testing should go ahead only with the consent of both parents or pursuant to a court order. In those cases in which agreement cannot be reached—for example, because a mature child or a person with parental responsibility withholds consent or is unavailable—a court may authorize testing, after taking the child's interests into account. In order to ensure high ethical standards and technical competence, DNA parentage testing should be conducted only by laboratories operating in accordance with specific accreditation standards in this area. Information about the availability of genetic counseling should be provided to the parties.

General Findings of the Inquiry

People Are Interested

People in Australia are very interested in learning about the "New Genetics" and the implications of this for themselves and their families. The inquiry organized 15 public forums in all of the capital cities and major regional centers, over 200 meetings and consultations with key stakeholders and community organizations (e.g., research laboratories, doctors, health officials, genetic support groups, counselors, the insurance industry, employer groups, trade unions, professional associations, academics, students, privacy groups, human rights lawyers, etc.) and received more than 300 written submissions. In just over 2 years, the inquiry engaged directly with many thousands of Australians (and this is a conservative figure). This included the expected professionals and "experts" but also many others from the general community and from affected communities. Many more Australians followed the development of the inquiry through the media, or by requesting literature from the ALRC, or by consulting our website, which has recorded exceptionally heavy usage for the documents associated with the inquiry.

Usage levels appear to correlate significantly with media coverage, but the public is often badly let down by the media in this area. With some notable exceptions, reports about developments in genetics too often have a "dumbed-down," "gee whiz" flavor that avoids the complexity and the need for balance in policy formation. There also appears to be far too little in the way of corporate memory. For example, stories appearing soon after the release of *Essentially Yours*, and dealing with some of the same issues, failed even to mention that Australia had just gone through a major consultative process and policy development exercise. Media coverage is much more likely to focus on individual cases or to report on parallel developments in other countries, lifted straight off the wire services.

Public Ambivalence . . . and Anxiety, But Not Cynicism

Second, the inquiry's experience in dealing with the Australian public confirms the local and overseas literature with respect to social attitudes to the rise of the New Genetics. Put briefly, there are strong but conflicting feelings in the community about biotechnology and its regulation.

On the one hand, there is considerable optimism about potential for genetic research to produce important medical breakthroughs in the diagnosis, treatment, and prevention of some terrible debilitating diseases, such as diabetes, Alzheimer's and Parkinson's, as well as leading to the development of whole new fields of medicine, such as gene therapy, regenerative medicine, and pharmacogenomics. While Australia has only 0.3 percent of the world's population, it produces 2.7 percent of the world's medical research, and we are ranked among the top six biotechnology countries in the world. Federal and state governments in Australia have recognized these achievements and have been especially supportive. For example, biotechnology research has been assigned a high priority by the Australian government in the allocation of competitive research grant funds, and the Queensland government has made a substantial investment in biotech infrastructure and research in that state.

At the same time, there is an (understandable) underlying anxiety in the Australian community about the pace of change—concerns about the loss of control, fears about the beginnings of "genetic determinism" or perhaps even eugenics, and doubts about the ability of public authorities to regulate this area effectively in the public interest. Judging from the public meetings, it appears that these concerns are based, to some extent, on a "genetic muddle," a generalized anxiety that wraps up doubts, misunderstandings, and appropriate concerns about genes, genetics, genetic testing, genetic engineering, genetically modified foods, stem cell experimentation, human cloning, and xenotransplantation (not to mention Chernobyl, mad cow disease, foot and mouth disease, the Ebola virus, and all things deadly).

Another commonly expressed concern was about access and equity—the fear that yet another major modern technology with the potential to make life better might, in practice, tend to drive up the costs of health care and increase the divide between the "haves" and the "have nots." These issues also featured in the ALRC's recent reference on the intellectual property aspects of genetic materials and technologies, which culminated in the 2004 report Genes and Ingenuity [ALRC, 2003]. That project explored the balance between encouraging investment and innovation in biotechnology and ensuring that further research and the delivery of cost-effective clinical genetic services are not compromised.

At every public meeting, the same concern was expressed, in almost identical terms: "We can see the value of medical research into genetics, and we generally would be happy to participate by giving information, blood or tissue to facilitate this research. However, we are not comfortable with the heavy degree of commercialisation of this research, and we definitely do not want our altruism to lead to billion dollar profits for multinational pharmaceutical companies" (usually expressed as "American pharmaceutical companies!)". A number of people at the public forums and consultations commented that while developing "smart drugs" based on modern pharmacogenomics is ostensibly a good thing, the resulting markets for individualized and customized drugs would be smaller and more fragmented, resulting in more effective drug therapies but at much higher prices. Concerns were expressed, by indigenous people among others, that this would shape the research programs for drug companies, prompting them to focus more on "white, middle-class, lifestyle diseases" than on diseases associated with poverty or those that primarily affect indigenous Australians and people in the Third World (e.g., malaria).

Another thing that clearly emerged at the public forums is the primal fear among members of the community about their genetic material being sent overseas (again, often expressed as being "sent to the United States"). At almost every event, someone in the audience expressed concern about volunteering for an experiment at an Australian university research laboratory or teaching hospital, then finding that the researchers had "spun off" into a private biotech company that merged with or was taken over by American interests—and "the next thing you know, your DNA is in California."

Beyond the implications for the health and well-being of individuals, the new genetics gives rise to questions as fundamental as what it means to be "human." It is noteworthy that the recent book by American political scientist Francis Fukuyama, *Our Posthuman Future: Consequences of the Biotechnology Revolution*, contains a plea for strong government regulation of the biotechnology sector—this from an ardent conservative advocate for small government, limited regulation, and free markets!

However, for all of that—and quite contrary to the European situation [ALRC, 2003]—it was clearly evident in the meetings, consultations, and submissions that Australians have not lost faith in the possibility of effective regulation of biotechnology in the public interest. In part, this is as a result of good management to date; in part, it is a result of good fortune, insofar as Australia (unlike Europe) has not suffered any public health crises or major scandals in this area that have sapped public confidence. This leaves open a window of opportunity for Australian governments and policy makers [ALRC, 2003].

Rapid Pace of Change in Genetic Science and Technology

Third, it is striking how rapidly genetic science and technology has developed and is continuing to develop, with "generational change" in the knowledge base occurring every few years. The New Genetics is no longer the stuff of science fiction. Consider the developments that have occurred in just the past few years alone:

The Human Genome Project has been completed, before schedule and on budget [ALRC, 2003]. Significant progress already has been made on the next phase, the Haplotype Mapping (HapMap) Project, also being undertaken by an National Human Genome Research Institute (NHGRI)-led international public consortium [ALRC, 2003], which is committed to making all of its results available on a public access database, subject to a "clickwrap" licence to avoid "parasitic" use of this material, according to Dr. Collins [ALRC, 2003].

Work on "comparative genomics" is providing increasingly greater insights, as scientists map a range of species for comparison: the human being, the mouse, yeast, roundworm, *Arabidopsis thaliana* (mouse-ear cress), fruit fly, bacteria *Escherichia coli* and *Haemophilus influenzae*, and so on. It is already part of popular culture that other primates are 97 percent or more genetically identical to humans, and chimpanzees in particular are about 99 percent identical.

However, even many "lower-order" animals and plants have a significant enough similarity to humans to be valuable in studying gene function. For example, mice and humans share about 70 to 90 percent of their DNA (an average of 85 percent), which is why researchers use "knock-out mice" (in which certain genes are "knocked out") in experiments aimed ultimately at understanding the dynamics of human genetics.

We still have much to learn, however. A fascinating talk delivered at the HUGO Conference in Shanghai in April 2002 by Dr. Eric Lander of MIT's Whitehead Institute pointed to both the potential for comparative genomics and the complexities [ALRC, 2003]. Researchers noted that certain genetic sequences may be found across a large number of species and over a long period of time, from basic organisms first appearing billions of years ago to much more complex and contemporary species, including humans. Assuming that nature must value highly such "highly conserved" sequences, it is logical to assume that these code for proteins with important functions. Researchers then knocked out these genetic sequences in laboratory animals, and found . . . no discernable result whatsoever! Along similar lines, the emerging field of "epigenetics" suggests there is also a great deal more to learn about direct environmental influences on the heritable genome, not merely the interplay between genes and the environment in individuals [ALRC, 2003].

Advances also are being made in related new fields, such as proteomics (focusing on the proteins expressed by genes, which is more important to health than the genes themselves), pharmacogenomics (as discussed above), and bioinformatics, which is an attempt to harness supercomputing power to explore the genome. As Professor John Mattick, Director of the Institute for Molecular Biosciences at the University of Queensland and one of Australia's leading geneticists, has pointed out, this makes sense not only because high-speed computers are useful tools in sequencing and correlating but also because the genome itself is so mathematical in nature. And apart from health and medical applications, very rapid progress has been made in the use of genetic science and technology by law enforcement authorities.

It is nevertheless worth sounding a cautionary note at this point because, for all of this amazing progress, we also need to maintain some perspective. Although we are getting close to some exciting breakthroughs in advanced medicine, for much of the Third World—at least the "two-thirds world" in terms of population—there is still no access to fresh drinking water or to basic health care or affordable drugs (although the latter is looking more promising now that the World Trade Organization finally may reach agreement on the production and distribution of cheaper generic drugs). And in Australia, with our world-class health care system and well-managed growth economy, the state of indigenous health is still a matter of national disgrace [ALRC, 2003].

Even in the affluent Western world, there are significant and growing health problems, although these are primarily the by-products of overconsumption. We read about epidemic levels of adult-onset diabetes and juvenile obesity. The *Essentially Yours* report argued for much more community and professional education about genetics, on the basis that an understanding of genetic health information can be empowering [ALRC, 2003]. However, even without knowing much about the secrets of the human genome, our community already knows well the secrets to a long and healthy life: Eat a good balanced diet, with lots of fresh fruits and vegetables, and get plenty of exercise. When it comes to human nature, however, the double cheeseburger always seems to trump the double helix.

Wide Breadth of Actual/Potential Applications

The fourth striking feature of the inquiry was the wide breadth of actual and potential applications for the use of human genetic information. The advent of the New Genetics represents one of those fundamental paradigm shifts that occur rarely in human history, forcing us to reappraise almost every area of social interaction and legal regulation. It is likely to be similar in kind to the impact of that other contemporary paradigm shift, the information technology and communications (IT&C) revolution, which has the Internet as its leading vector.

As documented at length in *Essentially Yours*, the impact of the New Genetics now extends well beyond science [ALRC, 2003], medicine, and law enforcement [ALRC, 2003], to such disparate areas as:

- Insurance (allegations of genetic discrimination by insurers was one of the main factors leading to the establishment of the inquiry) [ALRC, 2003]
- Immigration [ALRC, 2003]
- The management of tissue banks, genetic registers, and databases [ALRC, 2003]
- Family relations (through DNA parentage and kinship testing) [ALRC, 2003]
- The construction of racial, ethnic, and cultural (including aboriginal) identity [ALRC, 2003]
- Employment [ALRC, 2003], which may become, in the next few years, the major battleground with respect to the socially (and then legally) permissible uses of human genetic information

Even in sports [ALRC, 2003], advances in genetic science already have begun to raise questions about the value we place on "the fair go," on aspiration and hard work, and the triumph of the human will. For example, the Australian Institute of Sport (AIS) runs a talent identification program (TIP) to select promising young athletes for its many sporting programs. Would it be proper, and socially acceptable, to incorporate genetic testing and information into the TIP? And if so, what weight should such information be given? Would we tell promising young athletes that they will not be supported unless they have the genetic markers for the proteins found in the elite performers in their sport?

Our state of scientific knowledge is already such that we can identify some genetic markers that are associated with high performance in certain sports, and there are others that are associated with diseases and conditions relevant to certain sports. For example, Australian researchers have identified a protein (ACTN3) that appears to produce "fast twitch" muscle fiber. It is also known that the ACE protein enhances heart function—and the combination of ACTN3 and ACE is found in 95 percent of the world's elite sprinters and power athletes [ALRC, 2003]. (Conversely, a deficiency of ACTN3 appears to aid success in the marathon and other high-endurance sports.) A private genetic testing lab in Australia is already marketing this test for $110, aimed at both anxious parents and elite training centers and sports clubs [ALRC, 2003].

As another example, the Professional Boxing and Combat Sports Board of Victoria has floated the idea that it might require anyone seeking a professional boxing license under the Professional Boxing Control Act 1985 (Vic) to undergo a genetic test for *ApoE*4, but it has not followed up on this [ALRC, 2003]. Research suggests that this gene, which is connected with late-onset familial and sporadic Alzheimer's disease, may also be associated with an increased risk of chronic traumatic encephalopathy (CTE), or "punch drunk" syndrome, in boxers. It has been suggested that a milder form of this condition can occur in players of rugby, soccer and other sports associated with repetitive blows to the head [ALRC, 2003]. The AIS also runs a boxing program. Should it require its young athletes to undergo a genetic test? What if the public liability insurer for the Victorian boxing authority or the AIS begins to insist upon this as a condition for providing coverage? It is already the case that the AIS employs genetic screening for some athletes, notably in the basketball and volleyball programs, for Marfan's syndrome (which can result in a rupturing of the aorta and sudden death). The inquiry recommended that the AIS establish policies and guidelines governing the use of genetic testing and information in its programs [ALRC, 2003]. To its credit, the authorities at the AIS have taken this to heart and established a Reference Group to develop draft policies and guidelines to be taken to the Australian Sports Commission for formal adoption [ALRC, 2003].

Action Plan for Implementing Sound Policy

As mentioned, *Essentially Yours* contained 144 recommendations for reform, many of them with multiple strands, addressed at a wide range of actors. The "10-point action plan" below is not intended as a summary or prioritized list of the many recommendations in *Essentially Yours*; rather, this is an attempt to identify the major philosophical underpinnings for those recommendations.

First, all public policy developed in this area must incorporate a strong ethical dimension. Second, we must place a very high premium on the dignity of the individual. Thus, we should:

- Take the concept of "informed consent" seriously (including when it is used in the context of criminal law and procedure).
- Create a new criminal offense of submitting another person's genetic material for testing without his or her consent or other lawful authority (e.g., statutory authority or a court order) [ALRC, 2003].
- Take a firmly interventionist approach to the use of genetic testing and information in the employment area (we need to decide what sort of society we want: the one depicted in *Brave New World* or *GATTACA* or one in which Australian people are free to pursue their full human potential) [ALRC, 2003].
- Also take this seriously in the area of parentage testing, and resist the attempts of fathers' rights groups to promote the validity of unilateral, nonconsensual testing of children [ALRC, 2003], while recognizing their grave reservations

about the processes of the Family Court of Australia, particularly given the ALRC's very strong criticism of the management culture and procedures of the Family Court in the Managing Justice report [ALRC, 2003].

Third, we should be vigilant in not accepting notions of genetic essentialism or genetic determinism, or incorporating these into policy (even unwittingly) [ALRC, 2003]. People are worth more than the sum total of their genetic sequence. People are not "gene machines." This is easy to articulate now, but it may become increasingly hard to remember as "behavioural genetics" develops, particularly in the hothouse atmosphere of law-and-order campaigns conducted during elections or by crusading media organizations. For example, in future, how will we calculate the legal—and moral—effects of a scientifically established predisposition to drug and alcohol use? Or a predisposition to "risk-taking" behaviors? Or a predisposition to violence? Will our traditional common law emphasis on free will and *mens rea* preclude or diminish criminal responsibility for someone found to be acting under such a genetic predisposition? Or conversely, will we get calls for compulsory genetic testing to identify and deal with people predisposed to criminality, before they commit an offence, or before they reoffend—and will the presence of such genetic markers result in isolation, tracking, or treatment [ALRC, 2003].

Fourth, we should not develop policy and practice on the basis of genetic exceptionalism. A threshold question for the inquiry was whether we should embrace notions of "genetic exceptionalism," that is: "The idea that genetic information is so fundamentally different from, and more powerful than, all other forms of personal information that it requires different and higher levels of legal protection" [ALRC, 2003 p. 132]. The initial public policy responses to the New Genetics largely followed this approach, most clearly represented by the work of Professors Annas, Glantz, and Roche of the Boston University School of Public Health, who produced the influential Model Genetic Privacy and Non-Discrimination Bill—which was introduced into the federal Parliament by Senator Natasha Stott Despoja in 1998, a first attempt to debate these issues in Australia [ALRC, 2004]. In the words of Professors Annas, Glantz, and Roche: "Genetic information is uniquely powerful and uniquely personal, and thus merits unique privacy protection" [ALRC, 2003]. This approach is predicated on the basis that one's DNA amounts to "a coded probabilistic future diary [which] describes an important part of a person's unique future" [ALRC, 2003].

Now that the "shock of the new" has passed, however, the inquiry firmly concluded that an "inclusivist" approach was much to be preferred in which we refrained from making artificial and unproductive distinctions between "genetic" and "nongenetic" information and adapted existing laws and practices to meet the special features and challenges of genetic information, rather than creating new, specialist regimes. For example, we can build upon what we as a community have learned in recent years from dealing with the challenges of HIV–AIDS (human immune virus–acquired immunodeficiency syndrome), in terms of privacy and nondiscrimination/nonstigmatization, as well as in terms of community education, pre- and posttest counseling, sound laboratory practices, and effective public health administration.

Similarly, we can adapt existing privacy laws and safeguards (e.g., to cover genetic samples, and to recognize the familial dimension of genetic information—see below), as well as antidiscrimination laws and watchdog bodies, rather than establish new ones expressly to cover disputes arising out of one's real or perceived genetic status.

The inquiry's lack of attraction to genetic exceptionalist approaches also influenced the recommendations it ultimately made with respect to insurance underwriting law and practice. The inquiry was not persuaded that there was a case for interfering with long-standing legal requirement that parties to an insurance contract make full disclosure of all material facts—which certainly would include known, significant risks of genetic disorder, whether this knowledge comes from family medical history or a genetic test. Many of these arguments ran along the lines that persons with a genetic disorder were being punished unfairly, in insurance terms, for something that was "not their fault." If these arguments are accepted, then would someone with a genetic-linked cancer be privileged for insurance purposes over someone with (so far as we are presently aware) a non-genetic-linked cancer? And would a person with a genetic-linked disorder be privileged over, say, someone who had been severely injured in a car accident or an industrial accident? Or someone who had suffered brain injury or another serious disability, having been the innocent victim of a criminal assault? Or someone who had developed depression because of stressful family (or world) circumstances, rather than (identifiable) genetic factors?

While remaining sympathetic to the human dimension, the inquiry considered that such arguments muddled up the concepts of fault and risk, the latter of which is central to underwriting. The ALRC has no particular philosophical problem with no-fault schemes or community-rated underwriting; however, it also must be understood that such schemes cannot operate in the private marketplace without a heavy public subsidy [ALRC, 2003]. The inquiry did have a principled objection to recommending the imposition of a community-rating system, in effect, on the private, traditionally risk-rated, life insurance market only in respect of persons shielding adverse genetic health information.

However, the inquiry did recommend a range of significant consumer-oriented reforms, for example, that insurers must:

- Develop and publish sound policies for handling family medical history information (vastly more prevalent than genetic test information)—and indeed, the industry recently produced a good draft policy that is currently the subject of consultation.

- Refine their actuarial processes and utilization of cutting-edge scientific information, if they are to justify the exemption currently granted under antidiscrimination laws.

- Improve internal industry education and processes in relation to dealing with predictive genetic information—as the industry is doing.

- Provide reasons in writing for any adverse underwriting decision (which is included in the draft policy mentioned above).

- Establish an independent, effective new mechanism to review such decisions (whereas, at present, dissatisfied consumers basically must go to the courts for redress) [ALRC, 2003].

Fifth, what are those special features of genetic information that will require some acknowledgment in practice? Genetic information is:

- Ubiquitous—unlike most other forms of sensitive personal information, we are constantly leaving behind genetic material (saliva, blood, etc.) that can be tested and analyzed.

- Very stable—as evidenced by the ability to test DNA from dinosaurs and from terrorist disasters such as the Bali bombing and the destruction of the World Trade Center.

- Very personal and sensitive—but this is also true of HIV status, sexually trans-mitted diseases (STDs), depression, schizophrenia, cancer, and so on.

- Uniquely individual, on the one hand, since each person's 3.2 billion base-pair genetic sequence will be different, but has a very strong familial dimension, on the other; indeed, we share 99.9 percent of our genome with all other human beings, and an even higher percentage with members of our families and communities.

- Predictive in certain circumstances, which will only increase with our growing understanding.

Sixth, good public policy in the area of genetic information will involve a careful balancing of interests and, as noted above, the language of absolute "rights" will not be useful since interests inevitably will compete, conflict, and collide across the whole spectrum. One matter identified by the inquiry that requires urgent public debate, and perhaps fundamental rethinking, is the powerful preference in Western societies (es-pecially in the English-speaking world) to focus entirely on the individual and the primacy of the doctor–patient relationship. Yet, as already mentioned, genetic infor-mation is by definition shared information—with family members, with communities, and with all other humans.

The inquiry heard often from doctors and, especially, familial cancer registries that they "live in dread" of receiving a telephone call that might go something like this:

> You treated my sister, whose test showed a predisposition to a cancer that runs in families (such as BRCA1, or colon cancer, or FAP), but we are estranged and she never told me about it. But you are a health professional, and all you had to do was make one phone call and I would have sought my own medical advice—and I probably would not be in the terminal stages of cancer now.

Not surprisingly, health professionals expressed concern about both the legal ramifi-cations of this scenario as well as the ethical and moral dimensions.

The late Professor Dorothy Wertz of the University of Massachusetts conducted a number of fascinating cross-cultural empirical surveys that revealed a marked divergence in approach among differing cultures [ALRC, 2003]. For example, in response to questions about whether it would be proper to reveal to genetic relatives the fact that a patient tested positive for Huntington disease (HD) or for a familial cancer mutation, health professionals in Northern Europe, Western Europe and most especially the English-speaking countries (including Australia, the United States, the United Kingdom, Canada, and New Zealand) placed their focus squarely on the individual doctor–patient relationship and were reluctant to breach this confidence, whatever the consequences for other family members. By way of contrast, health professionals in African, Asian, Latin American, Middle Eastern, Eastern European, and Southern European, societies were much more likely to value familial and communal interests over individual autonomy [ALRC, 2003].

Along the same lines, it is important for us now to have a public debate in Australia about how we wish to proceed with work on population genetics, especially in relation to the use of collected genetic material (such as neonatal blood spots cards, also known as "Guthrie cards") [ALRC, 2003]. The use of such material—in the case of Guthrie cards, virtually a complete national collection for the last 40 years, even if unsystematized—for epidemiological purposes would have beneficial society-wide effects in terms of research, planning, and public health administration. But, of course, this would mean some trade-offs against privacy protection for individuals. Would that be a price we are happy to pay?

Such a national discussion also could address our cultural preoccupation with biological parenting (and the establishment of paternity) at the expense of social parenting. And while we are on the subject, would it not be wonderful if we could use these national debates, assuming we can get them started, to promote the genetic basis for tolerance? One of the major findings of the Human Genome Project is that there is no genetic foundation for "race" [ALRC, 2003]. It is now well accepted among medical scientists, anthropologists, and social scientists that "race" and "ethnicity" are social, cultural, and political constructs rather than matters of scientific "fact." In 1997, the American Anthropological Association (AAA) recommended that the U.S. government no longer use the term "race" on census forms or other official data collection documents because the term has "no scientific justification in human biology." The AAA noted:

> *Ultimately, the effective elimination of discrimination will require an end to such categorisation, and a transition toward social and cultural categories that will prove more scientifically useful and personally resonant for the public than are categories of "race" [ALRC, 2003]. p. 922*

Genetic research increasingly appears to verify the anthropological/ archaeological "out of Africa" theory of human origins. There is a very low degree of genetic variation among humans (only 1 Single Nucleotide Polymorphism (SNP) per 1300 bases, which is much lower than other complex species) tending to confirm that we have all emerged from a small "starter population." What this means is that every Palestinian and every Israeli, every Hindu and every Muslim in India, every Catholic and every Protestant in Northern Ireland, every Bosnian and every Serb, and every Hutu and every Tutsi in Rwanda is 99.9 percent genetically identical. What a tragedy that the "genius" of the human race always seems able to find the murderous potential in that 0.1 percent.

Seventh, we need to develop flexible approaches that can respond to the rapid changes in science and technology, and a broad mix of regulatory strategies tailored to the needs and circumstances of each context—not "one size fits all" solutions or "big law." The strategies useful for ensuring more effective oversight of medical research are not likely to be the same as those that will work in overseeing the work of law enforcement officials or ensuring best practice in our hospital laboratories.

Thus, as mentioned, the inquiry pushed for strong intervention in the employment context, to avoid the creation of a "genetic underclass" in Australia of people fit and willing to work, but with some predisposition that employers may use to rule them out [ALRC, 2003]. However, in the area of risk-rated private insurance (e.g., life insurance and critical illness insurance, but not private health insurance, which is "community rated" in Australia), the inquiry did not favor intrusive government intervention, preferring instead to call upon the industry and rely on the market to provide more effective regulation, greater adherence to scientific and medical advances (and their actuarial implications), more transparency, and greater responsiveness to legitimate consumer interests and concerns [ALRC, 2003].

Eighth, we have a clear need for much more community and professional education about how to deal with the New Genetics. There are some sobering surveys available for those who may doubt this need. For example, a Eurobarometer survey conducted in European Union countries in March 2003 put forth the following proposition: "Ordinary tomatoes do not contain genes, while genetically modified tomatoes do." Over 60 percent of respondents agreed with this [ALRC, 2003]. The same question put in the United States attracted support from 43 percent of respondents [ALRC, 2003].

The inquiry also heard consistent complaints from members of affected communities (patients, genetic support groups, etc.) about most doctors' lack of knowledge of genetics. This included not only general practitioners but also (or perhaps especially) leading specialists (in fields other than medical genetics) [ALRC, 2003].

The literature about how to communicate risk to patients contains many fascinating studies about how members of the community understand and apply concepts of risk and probability. The short answer is, they do not. Even the well educated do very poorly, yet this skill is essential for coming to grips with the predictive power of genetics. Some of this literature emerges from the health and medical context, while some of it comes from researchers in logic and mathematics. However, the conclusions are mutually reinforcing insofar as most people are unable to distinguish between good risks and bad risks in the surgery or at the casino [ALRC, 2003].

This points very strongly to the need for increasing the availability of qualified genetic counselors in Australia [ALRC, 2003]. The inquiry heard story after story, from individuals and families, about learning through a diagnosis, test, or family history that they were affected by a genetic disorder. In the initial stages, this rarely involved much accompanying information since so many doctors do not possess the necessary level of knowledge about clinical genetics, nor do they have the communications skills to explain to laypeople concepts such as penetrance, predisposition, and probability.

In the normal way of the modern world, people would go home, type the name of the genetic condition into "Google"—and then scare themselves witless. It is only after they had been able to meet with a genetic counselor that affected individuals and their families got some real sense of their position, prospects, and options. Often they are put in touch with genetic support groups, which also play an extremely important role in providing support, information, practical advice, advocacy services, and a sense of community. In the case of rare genetic conditions, virtual communities emerge and engage through the Internet. The inquiry was told very often about how reassuring it was for affected families to meet others experiencing the same feelings and coping with the same problems. The present author has said on a number of occasions that if a "hero" emerged during the course of this inquiry, it would have to be the genetic counselor.

Apart from those people forced to deal with serious genetic health issues, it will be necessary to empower health consumers more generally since "lifestyle" genetic tests and products are already beginning to be marketed, and this only will increase. For example, the United Kingdom company Sciona directly markets a "do-it-yourself" genetic testing kit for 120 British pounds, which includes a buccal swab used to collect cheek cells and then to be posted to the laboratory for analysis. The test, sold as "You and Your Genes," is said to look for particular SNPs on nine genes that may have some health consequences. Much of the marketing is now being conducted via the Internet. The Body Shop stores initially carried the product but apparently no longer do, and a number of major pharmacy chains have rejected it. The chief executive officer of Sciona, Dr. Chris Martin, has defended the direct marketing of these tests in the following terms:

There is already a lot of information out there recommending, for instance, a diet high in fruit, broccoli, and grains and low in char-grilled red meat, smoked, and preserved foods and alcohol. Consumers find this advice daunting, as they are not sure to what extent it pertains to them as individuals [ALRC, 2003].

It strains credibility to suggest that people who have difficulty processing simple health messages such as "eat more salads" and "avoid fatty foods" will benefit significantly from the highly complex and contingent information that would be derived from analysis of their DNA. This trade is probably not of sufficient importance or harm as to necessitate heavy-handed regulation: Such "curiosity" testing simply may be the genetic age's equivalent of "mood rings," requiring only some public education that instils a healthy sense of "buyer beware."

Ultimately, we should be optimistic that it is possible to mount an effective, comprehensive, public education campaign about the New Genetics. As a career academic, the present author has a strong stake in maintaining the belief that people are educable. Experience with the inquiry over the past 2 years also suggests that the public will be engaged by developments in genetic science and technology, and that there are interesting and effective ways of disseminating this information widely [ALRC, 2003].

Ninth, we need to start planning now for the imminent time when "all medicine will be genetic medicine." The advances in gene chip technology and bioinformatics will lead to a rapid expansion in the number and scope of genetic tests available and a substantial decrease in unit cost. This, in turn, will dramatically increase the pressure for access to and use of genetic tests (and not only in the medical/health context). To provide an example of the sort of advances that are being made in this regard, the Australian media recently reported the story of a University of New South Wales Ph.D. student's success in making correlations among 15,000 mouse genes. This took the "Barossa" supercomputer 32 hours; a standard desktop computer would take 5700 years to complete the same calculations [ALRC, 2003].

A recent Canadian report estimated that 60 percent of all Canadians will experience a disease with some form of genetic component during their lifetime, usually a complex, multifactorial but common disease, such as diabetes or hypertension [ALRC, 2003]. If you factor in family, friends, employers, and so on, we are already talking about an impact of close to 100 percent, and there is no reason to think that Australia is any different in this respect. At present, only three genetic tests are listed on the Medical Benefits Schedule, but in only 5 years or so, we will have the reality of the "$1000 genome"; that is, most Australians soon will be able to afford to have their genome fully sequenced and recorded on a CD.

If we wish to take full advantage of this science and technology, it is critical that our health systems immediately begin strategic planning to:

- Address cost issues and training needs.
- Prepare family doctors to be the key "gatekeepers" for genetic testing.
- Develop an integrated genetics education program running through from university medical schools to professional bodies and other institutions.
- Provide (much) more resources for genetic counseling and genetic support groups.
- Manage more effectively the protection and use of genetic databases.

Tenth, all of these initiatives need to be pulled together by a Human Genetics Commission of Australia (HGCA), as recommended in *Essentially Yours* [ALRC, 2003], and as already exists in the United Kingdom, and less formally in Canada (the Ontario Genetics Advisory Committee) and the United States (the Secretary's Advisory Committee on Genetics, Health and Society). The HGCA is envisaged as a broad-based body capable of providing cutting-edge advice to governments, industry, and the general community about the scientific and technological advances in human genetics (including those "over the horizon") as well as about the ethical, legal, and social implication of these advances.

The HGCA also would have important roles to play in:

- Coordinating community and professional education at the national level
- Setting standards for insurers and employers about the permissible uses of genetic tests and information
- Advising regulators (such as the Therapeutic Goods Administration) on best practice standards
- Monitoring international developments

- Liaising with other relevant bodies, such as the NHMRC, the TGA, the Office of the Gene Technology Regulator (OGTR), which handles plant and animal genetics, and the Australian Health Ministers Advisory Committee (AHMAC), which comprises the federal, state, and territory health ministers) to help ensure that Australia maintains clear, consistent, and high national standards in this area

CONCLUSIONS

It may well be human nature—perhaps it is even genetic—to leave everything to the last minute and to react to emerging events rather than to plan ahead. However, good management requires that we anticipate and prevent problems, and we currently have an excellent opportunity to make policy based on sound principle rather than on crisis management. The area of genetic research, testing, and information is so sensitive that it is critical we get this right—and do so now—to avoid the crisis of confidence and the public backlash that inevitably would follow from poor or unethical practices, as it has elsewhere.

What the ALRC/AHEC has proposed is not radical. In 1997, the then governor of Texas, George W. Bush, praised and signed into law a bill prohibiting genetic discrimination in the workplace and in group health insurance plans: this in the spiritual epicenter of the free market. President Bush now supports a similar law that recently passed through the U.S. Senate with all-party support.

These initiatives are broadly similar to what we have recommended for Australia (although our comprehensive public health care system and community-rated private health insurance mean that the landscape is different). However, the reforms recommended in *Essentially Yours* are much more comprehensive, are more finely tuned, and extend well beyond legislation to involve many more stakeholders than government(s). According to the Human Genome Program's Dr. Francis Collins, *Essentially Yours* is "a truly phenomenal job, placing Australia ahead of what the rest of the world is doing" [ALRC, 2003]. In Australia, the community is already engaged, and some key sectors (research, insurance, law enforcement) are operating at the cutting edge. If our political leaders can grasp the opportunity, Australia also can be an international pacesetter in the protection and intelligent use of human genetic information.

Appendix 2

Web Resources

All websites listed were accessed December 15, 2004.

Cancer Genetics

Trepanier A, Ahrens M, McKinnon W, Peters J, Stopfer J, Grumet SC, Manley S, Culver JO, Acton R, Larsen-Haidle J, Correia LA, Bennett R, Pettersen B, Ferlita TD, Costalas JW, Hunt K, Donlon S, Skrzynia C, Farrell C, Callif-Daley F, Vockley CW (2004) Genetic cancer risk assessment and counseling: Recommendations of the National Society of Genetic Counselors. *J Genet Counsel* **13**(2):83–114: http://www.guideline.gov/summary/summary.aspx?view_id=1&doc_id=5274

National Cancer Institute: Elements of Cancer Genetics Risk Assessment and Counseling: http://www.cancer.gov/cancertopics/pdq/genetics/risk-assessment-and-counseling/HealthProfessional/page4

U.S. Cancer Genetics Services Directory: http://www.cancer.gov/search/genetics_services/

Atlas of Genetics and Cytogenetics in Oncology and Hematology: www.infobiogen.fr/services/chromcancer

Continuing Medical Education: Genetics

American Medical Association. Genetics in Clinical Practice: A Team Approach, an Interactive Medical Laboratory Virtual Clinic. This CD-ROM is now available at no cost to physicians and other health care professionals that have an interest in clinical genetics: http://www.ama-assn.org/ama/pub/article/1615-8311.html

National Coalition for Health Professional Education in Genetics: http://www.nchpeg.org

Genetic Testing: Care, Consent, and Liability, by Neil F. Sharpe and Ronald F. Carter
© 2006 John Wiley & Sons, Inc.

National Center for Biotechnology Information: Genes and disease: A collection of articles that discuss genes and the diseases that they cause: http://www.ncbi.nlm.nih.gov/books/bv.fcgi?rid=gnd

National Institutes of Health: National Library of Medicine: http://www.nlm.nih.gov/pubs/cbm/health_risk_communication.html

Dictionary of Genetic Terms

Human Genome Project: http://www.ornl.gov/sci/techresources/Human_Genome/publicat/primer2001/glossary.shtml

Ethics Resources

Genethics: http://genethics.ca/

Genethics: Murdoch Childrens Research Institute: http://www.genecrc.org/site/ge/index_ge.htm

U.S. Department of Energy: Legal, Ethical, and Social Issues: http://www.ornl.gov/sci/techresources/Human_Genome/elsi/elsi.shtml

Family History Resources—Professional

American Medical Association: http://www.ama-assn.org/ama/pub/category/2380.html

Centers for Disease Control: http://www.cdc.gov/genomics/public/famhistMain.htm

Centers for Disease Control: Public Health Perspective: Family History Tools: http://www.cdc.gov/genomics/info/perspectives/famhistr.htm

National Coalition for Health Professional Education in Genetics (2004) The Genetic Family History in Practice (pdf only): http://www.nchpeg.org/newsletter/inpracticesum04.pdf

National Cancer Institute: Elements of Cancer Risk Assessment and Counseling: Taking a Family History: http://www.cancer.gov/templates/doc.aspx?viewid=c0fc1ac3-607b-44a5-9d24-39b0a2a4703c&version=HealthProfessional&sectionID=1&#Section_18

Family History Resources—Public

American Medical Association: http://www.ama-assn.org/ama/pub/category/2380.html

Centers for Disease Control: http://www.cdc.gov/genomics/public/famhistMain.htm

March of Dimes: http://www.marchofdimes.com/pnhec/4439_1109.asp

U.S. Department of Health and Human Services: Family History Intiative: http://www.genome.gov/12513847

For a practical family history tool, please see: www.hhs.gov/familyhistory

Genetic Counseling Organizations

American Board of Genetic Counseling: http://genetics.faseb.org/genetics/abgc_ diplomates.html

National Society of Genetic Counselors: http://www.nsgc.org/resourcelink.asp

How to Find a Genetic Counselor (U.S.): http://www.genetichealth.com/ Resources_How_to_Find_a_Genetic_Counselor.shtml

Canadian Association of Genetic Counsellors: http://www.cagc-accg.ca/

Genetic Research and Resources

Centers for Disease Control: Human Genome Epidemiology Network: http:// www.cdc.gov/genomics/hugenet/default.htm

National Center for Genome Resources (NCGR): http://www.ncgr.org/

National Human Genome Research Institute: http://www.genome.gov/

OMIM, Online Mendelian Inheritance in Man: A catalog of human genes and genetic disorders: http://www.ncbi.nlm.nih.gov/entrez/query.fcgi?db=OMIM

Genetic Testing Resources

GeneTests: a publicly funded medical genetics information resource developed for physicians, other health care providers, and researchers, available at no cost to all interested persons: http://www.genetests.org/

Association of Public Health Laboratories: Newborn Screening and Genetics: http://www.aphl.org/Newborn_Screening_Genetics/index.cfm

Centers for Disease Control: Genetic Testing: http://www.cdc.gov/genomics/ gTesting.htm

Centers for Disease Control: Regional and State Genetics Directory: http://www .cdc.gov/genomics/links/regional.htm

Directory of Medical Cytogenetic Laboratories in Canada: http://www.hrsrh.on .ca/genetics/canlabs.htm

Genetics and Public Policy Center: http://www.dnapolicy.org/genetics/testing .jhtml;$sessionid$ETC3MSIAAAYYUCQBAT3RVQQ

Glossary of Terms

University of Kansas Medical Center: http://www.kumc.edu/gec/glossnew.html

National Human Genome Research Institute: http://www.ornl.gov/sci/ techresources/Human_Genome/glossary/

National Human Genome Research Institute: Glossary in Spanish: http://www .genome.gov/sglossary.cfm

"Talking" Glossary of Genetic Terms: http://www.genome.gov/10002096

Insurance

Georgetown University Health Policy Institute: http://www.healthinsuranceinfo. net/

Laboratories

Association of Public Health Laboratories: Newborn Screening and Genetics: http://www.aphl.org/Newborn_Screening_Genetics/index.cfm

GeneTests: A publicly funded medical genetics information resource developed for physicians, other health care providers, and researchers, available at no cost to all interested persons: http://www.genetests.org/

Canadian College of Medical Geneticists: Cytogenetics: http://www.hrsrh.on.ca/ genetics/CanCyt/

Directory of Medical Cytogenetic Laboratories in Canada: http://www.hrsrh.on .ca/genetics/canlabs.htm

Atlas of Genetics and Cytogenetics in Oncology and Hematology: www .infobiogen.fr/services/chromcancer

Language

Genetic Information Websites in Spanish for Public Education: http://www.ornl.gov/ sci/techresources/Human_Genome/education/spanish.shtml

Law and Legislatures

Council for Responsible Genetics (CRG): Genetics and the Law: http://www .genelaw.info/

Genetics and Public Policy Center: http://www.dnapolicy.org/policy/legalIssues .jhtml

Genetic Laws and Legislative Activity (U.S.): http://www.ncsl.org/programs/ health/genetics/charts.htm

National Cancer Institute: Elements of Cancer Risk Assessment and Counseling: Informed Consent: http://www.cancer.gov/templates/doc.aspx?viewid =c0fc1ac3-607b-44a5-9d24-39b0a2a4703c&version=HealthProfessional& sectionID=1&#Section_89

National Conference of State Legislatures (U.S.): Newborn Genetic and Metabolic Screening: http://www.ncsl.org/programs/health/genetics/nbs.htm

Policy Statements

American College of Medical Genetics—Policy Statements: http://www.acmg .net/resources/policy-list.asp

Prenatal/Neonatal—Professional Resources

American Academy of Pediatrics—Section on Birth Defects: Policy Statement and Information on Specific Disorders: http://www.aap.org/healthtopics/birthdefects.cfm

American College of Medical Genetics Foundation: Evaluation of the Newborn with Single or Multiple Congenital Anomalies: A Clinical Guideline: http://www.health.state.ny.us/nysdoh/dpprd/index.htm

American College of Obstetricians and Gynecologists: Committee opinion on first-trimester prenatal screening methods for chromosome abnormalities, including nuchal fold translucency screening. To read more about first trimester screening and the policy issues surrounding it, visit the Center's website: http://www.dnapolicy.org/FirstTrimesterScreening

Association of Public Health Laboratories: Newborn Screening and Genetics: http://www.aphl.org/Newborn_Screening_Genetics/index.cfm

Catalogue of Rare Genetic Diseases in Children: http://www.med.nyu.edu/rgdc/disease.htm

Frequency of Inherited Disorders Database: http://archive.uwcm.ac.uk/uwcm/mg/fidd/

Genetics and Public Policy Center: Prenatal Genetic Testing: http://www.dnapolicy.org/genetics/prenatal.jhtml

March of Dimes Fact Sheets: These cover a wide range of prenatal and genetic topics in English and Spanish: http://www.marchofdimes.com/professionals/681_1116.asp

March of Dimes Fact Sheets: Newborn Screening: http://www.marchofdimes.com/professionals/580.asp

National Newborn Screening and Genetics Resource Center: http://genes-r-us.uthscsa.edu/

National Conference of State Legislatures (U.S.): Newborn Genetic and Metabolic Screening: http://www.ncsl.org/programs/health/genetics/nbs.htm

Rare genetic diseases in children: http://www.med.nyu.edu/rgdc/disease.htm

Prenatal/Neonatal—Public Resources

Birth Defect Research for Children provides parents and expectant parents free fact sheets about the most common categories of birth defects, the National Birth Defect Registry, and research resources: http://www.birthdefects.org/

Canadian Organization for Rare Disorders: http://www.cord.ca/

Centers for Disease Control: Fact Sheets (e.g., carrier testing, prenatal diagnosis, etc., including information about specific genetic diseases and disorders): http://www.cdc.gov/genomics/public/facts.htm

Cystic Fibrosis Foundation (U.S.): Directory of care centers and chapters: http://www.cff.org/chapters_and_care_centers/

Genetic and Rare Diseases Information Center: http://www.genome.gov/10000409

March of Dimes Pregnancy & Newborn Health Education Center: http:// www.marchofdimes.com/pnhec/pnhec.asp

Medline Plus: Genetic Testing: http://www.nlm.nih.gov/medlineplus/genetic testing.html

National Newborn Screening and Genetics Resource Center: http://genes-r-us.uthscsa.edu/

National Organization for Rare Disorders: http://www.rarediseases.org/

National Conference of State Legislatures (U.S.): Newborn Genetic and Metabolic Screening: http://www.ncsl.org/programs/health/genetics/nbs.htm

Rare genetic diseases in children: http://www.med.nyu.edu/rgdc/disease.htm

Professional Organizations

American Board of Genetic Counselors: http://genetics.faseb.org/genetics/abgc_diplomates.html

American Board of Genetic Counseling: http://www.abgc.net/

American Board of Medical Genetics: http://www.abmg.org/

American College of Medical Genetics: http://www.acmg.net/

American Society of Human Genetics: http://genetics.faseb.org/genetics/ashg/ashgmenu.htm

Association of Genetic Technologists: http://www.agt-info.org/

Australasia: The Human Genetics Societies of Australasia: http://www.hgsa.com.au/

Canadian College of Medical Genetics: http://ccmg.medical.org/

Council of Medical Genetics Organizations: http://genetics.faseb.org/genetics/ashg/comgo.htm

European Society of Human Genetics: http://www.eshg.org/

Genetics Society of America: http://www.genetics-gsa.org/

International Federation of Human Genetics Societies: http://www.ifhgs.org/

Latin American Network of Human Genetics Societies: http://www.relagh.ufrgs.br/

International Society for Nurses in Genetics: http://www.isong.org/

National Society of Genetic Counselors: http://www.nsgc.org/

National Coalition for Health Professional Education in Genetics: http://www.nchpeg.org/

National Society of Genetic Counselors: http://www.nsgc.org/resourcelink.asp

International Federation of Human Genetics Societies: http://genetics.faseb.org/genetics/ifhgs/

Coalition of State Genetics Coordinators (CSGC): http://www. stategeneticsco-ordinators.org/

Psychological

American Psychological Association: http://www.apa.org/science/genetics/homepage.html

National Cancer Institute: Elements of Cancer Risk Assessment and Counseling: Psychological Impact of Genetic Information/Test Results on the Family: http://www.cancer.gov/templates/doc.aspx?viewid=c0fc1ac3-607b-44a5-9d24-39b0a2a4703c&version=HealthProfessional&sectionID=1&#Section_108

Patient and Public Education

Medline Plus: Genetic Testing: http://www.nlm.nih.gov/medlineplus/genetictesting.html

Genetic Disorders: Human Genome Project Information: http://www.ornl.gov/sci/techresources/Human_Genome/medicine/assist.shtml

Genetic Information Websites in *Spanish* for Public Education: http://www.ornl.gov/sci/techresources/Human_Genome/education/spanish.shtml

Human Genome Project: Exploring Our Molecular Selves: Online Multimedia Education Kit: http://www.genome.gov/Pages/EducationKit/

Medicine and the New Genetics: Human Genome Project Information: http://www.ornl.gov/sci/techresources/Human_Genome/medicine/medicine.shtml

March of Dimes Pregnancy & Newborn Health Education Center: http://www.marchofdimes.com/pnhec/pnhec.asp

Public Genetics Education: *Gene Almanac* (produced by Dolan Learning Center, Cold Springs Harbor, New York): http://www.dnalc.org/

The 21st Century Community Schoolhouse: Websites for Genetic Disorders: http://www.communityschoolhouse.org/websites.geneticdisorders.htm

Understanding Genetic Testing: http://www.pitt.edu/~super1/lecture/lec2631/001.htm

U.S. Department of Energy Office of Science: http://www.doegenomes.org/

Public Health

Genomics for Public Health Practitioners is a 45-minute introductory presentation on genomics and public health. Resource is intended for public health practitioners who have minimal experience in the area of genomics as it pertains to public health: http://www.cdc.gov/genomics/training/GPHP/default.htm

Centers for Disease Control: Public Health Perspective: Family History Tools: http://www.cdc.gov/genomics/info/perspectives/famhistr.htm

Secretary's Advisory Committee on Genetic Testing (U.S.): http://www4.od.nih.gov/oba/sacghs/sacghslinks.html

Policy Issues

American College of Medical Genetics Policy Statements: http://genetics.faseb.org/genetics/acmg/pol-menu.htm

American Society of Human Genetics Policy Papers and Reports: http://genetics.faseb.org/genetics/ashg/policy/pol-00.htm

Public Issues

The Foundation for Genetic Education and Counseling (FGEC) is a nonprofit organization dedicated to raising the understanding of human genetics and genetic medicine among the general public and their health professionals: http://www.fgec.org/

Foundation for Genetic Medicine: To help create and strengthen a supportive environment for the ethical development of genetic medicine to improve human health: http://www. geneticmedicine.org/

Genetics and Public Policy Center: http://www.dnapolicy.org/

Risk Communication

National Institutes of Health: National Library of Medicine: http://www.nlm.nih.gov/pubs/cbm/health_risk_communication.html

Support Organizations

American Directory of Genetic Support Groups: http://www.kumc.edu/gec/support/

Birth Defect Research for Children: Provides parents and expectant parents free fact sheets about the most common categories of birth defects, the National Birth Defect Registry, and research resources: http://www.birthdefects.org/

Canadian Directory of Genetic Support Groups: http://www.lhsc.on.ca/programs/medgenet/

Canadian Organization for Rare Disorders: http://www.cord.ca/

Cystic Fibrosis Foundation (U.S.): Directory of care centers and chapters: http://www.cff.org/chapters_and_care_centers/

Genetic Alliance (U.S.): http://www.geneticalliance.org/

March of Dimes: http://www.marchofdimes.com/

National Newborn Screening and Genetics Resource Center: http://genes-r-us.uthscsa.edu/

National Organization for Rare Disorders: http://www.rarediseases.org/

Disclaimer

The authors of this text did not develop, review, participate, or control the content at the above sites; therefore, these links do not constitute an endorsement by the authors of the content. Although every effort is made to ensure all links are active, any of the above organizations may unlink without prior notice. Please excuse any inconvenience.

References

Aalfs CM, Smets EM, de Haes HC, Leschot NJ (2003) Referral for genetic counselling during pregnancy: Limited alertness and awareness about genetic risk factors among GPs. *Fam Pract* **20**(2):135–141.

AAMC (1999) Contemporary issues in medicine: Communication in medicine. Association of American Medical Colleges. October: http://www.aamc.org/meded/msop/msop3.pdfm.

Abbot A (2004) Clinicians win fight to overturn patent for breast-cancer gene. *Nature* **429**: 329.

Abkevich V, Zharkikh A, Deffenbaugh AM, Frank D, Chen Y, Shattuck D, Skolnick MH, Gutin A, Tavtigian SV (2005) Analysis of missense variation in human BRCA1 in the context of interspecific sequence variation. *J Med Genet* **41**:492–507.

Abner AL, Connolly JL, Recht A, Bornstein B, Nixon A, Hetelekidis S, Silver B, Harris JR, Schnitt SJ (2000) The relation between the presence and extent of lobular carcinoma in situ and the risk of local recurrence for patients with infiltrating carcinoma of the breast treated with conservative surgery and radiation therapy. *Cancer* **88**:1072–1077.

Abraham C (2002) Mouse ruling may stall research, 6 December 2002, *Globe & Mail*, A1.

Abramsky L, Fletcher O (2002) Interpreting information: What is said, what is heard— a questionnaire study of health professionals and members of the public. *Prenat Diagn* **22**(13):1188–1194. Cytogenetic laboratories began faxing an information leaflet to the health professional. This helped health professionals to familiarize themselves with the condition and information before speaking to patients.

Abramsky L, Hall S, Levitan J, Marteau TM (2001) What parents are told after prenatal diagnosis of a sex chromosome abnormality: Interview and questionnaire study. *BMJ* **322**(2784):463–466.

ACCE (Analytic validity, Clinical validity, Clinical utility and Ethical, legal and social implications) Review of CF Prenatal. Draft Report. Version. 2002.6: http://www.cdc.gov/genomics/info/reports/research/FBR/ELSIJune2002.pdf; http://www.fbr.org/research/acce-cdc/acce_doc.html.

Access to Abortion Services Act, S.B.C. 1995, c.35.

Acheson LS, Wiesner GL, Zyzanski SJ, Goodwin MA, Stange KC (2000) Family history-taking in community family practice: Implications for genetic screening. *Genet Med* **2**(3):180–185. Methods: Research nurses directly observed 4454 patient visits to 138 family physicians

Genetic Testing: Care, Consent, and Liability, by Neil F. Sharpe and Ronald F. Carter
© 2006 John Wiley & Sons, Inc.

and reviewed office medical records. Results: Family history was discussed during 51% of visits by new patients and 22% of visits by established patients. Physicians' rates of family history taking varied from 0 to 81% of visits. Family history was more often discussed at well care rather than illness visits. The average duration of family history discussions was 2.5 minutes.

Achiron R, Lipitz S, Gabbay U, Yagel S (1997) Prenatal ultrasonographic diagnosis of fetal heart echogenic foci: No correlation with Down syndrome. *Obstet Gynecol* **89**(6):945–948.

ACMG (1995) Genetic testing in children and adolescents, points to consider: Ethical legal and psychosocial implications of (ACMG/ASHG). *Am J Hum Genet* **57**:1233–1241.

ACOG (2004) Medical liability survey reaffirms more ob-gyns are quitting obstetrics. American College of Obstetricians and Gynecologists. July 16, 2004: "The national survey of ACOG ob-gyn members confirms that the medical liability insurance crisis has worsened in recent years, with one in seven ACOG Fellows reporting that they had stopped practicing obstetrics because of the high risk of liability claims. Ob-gyns have an average of 2.6 claims filed against them during their career." http://www.acog.org/from_home/publications/press_releases/nr07-16-04.cfm.

ACOG (1996) American College of Obstetricians and Gynecologists. Maternal serum screening. *Int J Gynecol Obstet* **55**:299–308.

Acton RT, Go R, Roseman J (1989) Strategies for the utilization of genetic information to target preventive interventions. Proceedings of the 25th Annual Meeting of the Society of Prospective Medicine, Indianapolis, Indiana, pp. 88–100.

Adair v. Weinberg (1995) 79 Wash. App. 197, 901 P. 2d 340.

Adams KE (2003) Ethical considerations of applications of preimplantation genetic diagnosis in the United States. *Med Law* **22**:489–494.

Adamson TE, Tschann JM, Gullion DS, Oppenberg AA (1989) Physician communication skills and malpractice claims. A complex relationship. *West J Med* **150**(3):356–360.

Addington AM, Gornick M, Sporn AL, Gogtay N, Greenstein D, Lenane M, Gochman P, Baker N, Balkissoon R, Vakkalanka RK, Weinberger DR, Straub RE, Rapoport JL (2004) Polymorphisms in the 13q33.2 gene G72/G30 are associated with childhood-onset schizophrenia and psychosis not otherwise specified. *Biol Psychiatry* **55**(10):976–980.

Adeyinka A, Dewald GW (2003) Cytogenetics of chronic myeloproliferative disorders and related myelodysplastic syndrome. *Hamtol Oncol Clin North Am* **17**:1129–1149.

Ad Hoc Committee on Genetic Counseling (1975) Genetic counseling. *Am J Hum Genet* **27**:240–242.

Aiken v. Clary (1965) 396 SW 2d 668 (Mo.).

Alan Guttmacher Institute (2004a) *State Policies in Brief: Restrictions on Postviability Abortions.* New York: Alan Guttmacher Institute.

Alan Guttmacher Institute (2004b) *Mandatory Counseling and Waiting Periods for Abortion.* New York: Alan Guttmacher Institute.

Alaszewski A, Horlick-Jones T (2003) How can doctors communicate information about risk more effectively? *BMJ* **327**:728–731.

Albertson GA, Lin CT, Kutner J, Schilling LM, Anderson SN, Anderson RJ (2000) Recognition of patient referral desires in an academic managed care plan frequency, determinants, and outcomes. *J Gen Intern Med* **15**(4):242–247.

Alexander M (1990) Informed consent, psychological stress, and noncompliance. *Humane Med* **6**(2):113–119, at 119. The author argues psychological stress has "considerable influence on informed consent," and "may interfere with the patient's ability to give true, voluntary informed consent."

Almqvist E, Adam S, Bloch M, Fuller A, Welch P, Eisenberg D, Whelan D, Macgregor D, Meschino W, Hayden MR (1997) Risk reversals in predictive testing for Huntington's disease. *Am J Hum Genet* **61**:945–952.

Almqvist EW, Bloch M, Brinkman R, Crawford D, Hayden MR (1999) On behalf of an international Huntington disease collaborative group. A worldwide assessment of the frequency of suicide, suicide attempts, or psychiatric hospitalization after predictive testing for Huntington disease. *Am J Hum Genet* **64**:1293–1304.

Alper J (2002) Genetic complexity in human disease and behavior. In: Alper JS, Ard C, Asch A, Beckwith J, Conrad P, Geller LN, eds. *The Double-Edged Helix: Social Implications of Genetics in a Diverse Society.* Baltimore: Johns Hopkins University Press, pp. 17–38.

Alper JS, Beckwith J (1998) Genetic vs. nongenetic medical tests: Some implications for antidiscrimination legislation. *Sci Eng Ethics* **4**:141–150.

Alper J, Geller LN, Barash CI, Billings PR, Laden V, Natowicz M (1994) Genetic discrimination and screening for hemochromatosis. *J Public Health Policy* **15**:345–358.

Altman D (2000) A review of experimentwise Type I error; Implications for univariate post hoc and for mulivariate testing. Southwest Educational Research Association, Dallas, January 29.

Altmuller J, Palmer LJ, Fischer G, Scherb H, Wjst M (2001) Genomewide scans of complex human diseases: True linkage is hard to find, *Am J Hum Genet* **69**:936–950.

ALRC (2004a,b) Please see Australian Law Reform Commission.

Alvarado M (1999) Personal communication. Genetic Counselor, Kaiser Permanente Medical Group. Woodland Hills, CA.

AMA (1998) E-2.139 Multiplex Genetic Testing. June, 1998. http://www.ama-assn.org/ama/pub/category/8440.html.

AMA Report on Scientific Affairs (1999) Report 5 of the Council on Scientific Affairs (A-99) Full Text. http://www.ama-assn.org/ama/pub/article/2036-4077.html.

AMA, Genetic Testing of Children (2003) E-2.138 Genetic Testing of Children http://www.ama-assn.org/ama/pub/category/8439.html.

Ambady N, Laplante D, Nguyen T, Rosenthal R, Chaumeton N, Levinson W (2002) Surgeons' tone of voice: A clue to malpractice history. *Surgery* **132**(1):5–9.

American Academy of Pediatrics (2000) Newborn Screening Task Force, Serving the family from birth to the medical home—Newborn Screening: A blueprint for the future. *Pediatrics* **106**(Suppl):383–427.

American Academy of Pediatrics, Committee on Bioethics (2001) Ethical issues with genetic testing. *Pediatrics* **107**:1451–1455. http://aappolicy.aappublications.org/cgi/content/full/pediatrics%3b107/6/1451.

American Board of Genetic Counseling (2004) 2005 Certification Examination ProgrAmer http://www.abgc.net/genetics/abgc/abgc-cert/2005/step-01.htm.

American Board of Medical Genetics (1998) Bulletin of Information—Description of Examinations. National Board of Medical Examiners.

American Board of Medical Genetics (2001) Genetics and managed care: Policy statement of the American College of Medical Genetics. *Genet Med* **3**(6):430–435. http://www.acmg.net/resources/policies/pol-019.pdf.

American Board of Medical Genetics (2004) Certification Standards: http://www.boardcertifieddocs.com/bcd/supplement/mdgen.pdf.

American Cancer Society (2003) *Cancer Facts & Figures.* Atlanta: American Cancer Society.

American College of Medical Genetics (ACMG) (1996) Statement on guidance for genetic counseling in advanced paternal age. *ACMG Newslett* **6**:13.

American College of Medical Genetics Clinical Practice Committee (1997) Principles of Screening: Report of The Subcommittee on Screening, http://www.acmg.net/resources/policies/pol-026.asp.

American College of Medical Genetics (1999a) Genetic Susceptibility to Breast and Ovarian Cancer, Assessment, Counseling, and Testing Guidelines. http://www.health.state.ny.us/nysdoh/cancer/obcancer/pp27-35.htm.

American College of Medical Genetics (1999b) Duty to recontact. *Genet Med* **1**(4):171.

American College of Medical Genetics (2000) Genetics evaluation guidelines for the etiologic diagnosis of congenital hearing loss. Genetic Evaluation of Congenital Hearing Loss Expert Panel. ACMG statement. *Genet Med* **4**(3):162–171.

American College of Medical Genetics (2004) Standards and Guidelines for Clinical Genetics Laboratories. http://www.acmg.net/resources/s-g/s-g-yes-no.asp. Also see GeneTests.Org, Laboratory Directory: http://www.genetests.org/servlet/access?id=8888891&key=S0NWW 52w0fvZU&fcn=y&fw=vEHp&filename=/labsearch/searchdztest.html; accessed December 19, 2004.

American College of Medical Genetics Subcommittee on Cystic Fibrosis Screening (2001) Laboratory standards and guidelines for population-based cystic fibrosis carrier screening. *Genet Med* **3**(2):149–154. http://www.acmg.net/resources/policies/pol-005.asp.

American College of Obstetricians and Gynecologists (2001a) Ob-gyns offering large-scale cystic fibrosis screening. http://www.acog.org/from_home/publications/press_releases/nr12-12-01-2.cfm.

American College of Obstetricians and Gynecologists Committee on Practice (2001) ACOG Practice Bulleting. Clinical management guidelines for obstetricians and gynecologists. Prenatal diagnosis of fetal chromosomal abnormalities. *Obstet Gynecol* **97**(suppl):1–12.

American College of Obstetricians and Gynecologists (2004a) ACOG Committee Opinion. Number 298, August 2004. Prenatal and preconceptional carrier screening for genetic diseases in individuals of Eastern European Jewish descent. *Obstet Gynecol* **104**(2):425–428.

American College of Obstetricians and Gynecologists (ACOG) (2004b) Code of Professional Ethics. American College of Obstetricians and Gynecologists, http://www.acog.org/from_home/acogcode.pdf.

American Heart Association (2003) *Heart and Stroke Statistics*. Dallas, Texas: American Heart Association. http://www.americanheart.org/presenter.jhtml?identifier=1200026.

American Jurisprudence (2002) *Physicians, Surgeons and Other Healers*, 2nd ed. St Paul, MN: West Press, Vol. 61, Sections 211–212, 213–214.

American Medical Association (2003) E-2.131 Disclosure of familial risk in genetic testing. AMA Code of Medical Ethics. http://www.ama-assn.org/ama/pub/category/11963.html.

American Medical Association (2004) The importance of gathering a family history. http://www.ama-assn.org/ama/pub/category/2380.html.

American Medical Association (AMA) (2003) Code of Medical Ethics. American Medical Association.

American Society for Human Genetics/American College of Medical Genetics (1995). Points to consider: Ethical, legal and psychosocial implications of genetic testing in children and adolescents. *Am J Hum Genet* **57**:1233–1241.

American Society for Reproductive Medicine (ASRM) Ethics Committee (1999) Sex selection and preimplantation genetic diagnosis. *Fertil Steril* **72**(4):595–598.

American Society for Reproductive Medicine (ASRM) Ethics Committee (2001) Preconception gender selection for nonmedical reasons. *Fertil Steril* **75**(5):861–864.

American Society of Clinical Oncology (1996) Genetic testing for cancer susceptibility. *J Clin Oncol* **14**:1730–1736.

American Society of Clinical Oncology (2003) Policy statement update: Genetic testing for cancer susceptibility: Confidentiality and communication of familial risk. *J Clin*

Oncol **21**(12):2397–2406. Epub 2003 Apr 11: American Society of Clinical Oncologists Genetic Testing for Cancer Susceptibility: Confidentiality and Communication of Familial Risk. http://www.asco.org/ac/1,1003,_12-002228-00_18-0011991-00_19-0011993-00_20-001,00.asp.

American Society of Human Genetics (1996) Statement on informed consent for genetic research. *Am J Hum Genet* **59**:471.

American Society of Human Genetics (1998) ASHG statement. Professional disclosure of familial genetic information. The American Society of Human Genetics Social Issues Subcommittee on Familial Disclosure. *Am J Hum Genet* **62**(2):474–483. http://genetics.faseb.org/genetics/ashg/policy/pol-29.htm.

American Society of Human Genetics (2004) Your family history. American Society of Human Genetics. http://genetics.faseb.org/genetics/ashg/educ/007.shtml.

American Society of Human Genetics (ASHG) and American College of Medical Genetics (ACMG) Test and Technology Transfer Committee (1996) Diagnostic testing for Prader-Willi and Angelman syndromes: Report of the ASHG/ACMG Test and Technology Transfer Committee. *Am J Hum Genet* **58**:1085–1088.

American Society of Human Genetics (ASHG) Social Issues Committee and The American College of Medical Genetics (ACMG) Social, Ethical, and Legal Issues Committee (2000) Genetic testing in adoption. *Am J Hum Genet* **66**:761–767.

American Society of Human Genetics Ad Hoc Committee on Genetic Counseling (1975) Genetic counseling. *Am J Hum Genet* **27**:240–242.

American Society of Human Genetics Social Issues Subcommittee on Familial Disclosure (1998) Professional disclosure of familial genetic information. *Am J Hum Genet* **62**:474–483.

Amir RE, Van den Veyver IB, Wan M, Tran CQ, Francke U, Zoghbi HY (1999) Rett syndrome is caused by mutations in X-linked MECP2, encoding methyl-CpG-binding protein 2. *Nat Genet* **23**(2):185–188.

Amos CI, Shaw GL, Tucker MA, Hartge P (1992) Age at onset for familial epithelial ovarian cancer. *JAMA* **268**:1896–1899.

Anderlik MR, Lisko EA (2000) Medicolegal and ethical issues in genetic cancer syndromes. *Semin Surg Oncol* **18**(4):339–346.

Anderlik MR, Rothstein MA (2003) Canavan decision favours researchers over families. *J Law Med Ethics* **31**:450–453.

Andermann E (1982) Multifactorial inheritance of generalized and focal epilepsy. In: Anderson VE, Hauser WA, Penry JK, eds. *Genetic Basis of the Epilepsies*. New York: Raven, pp. 355–374.

Anderson DE, Badzioch MD (1985) Bilaterality in familial breast cancer patients. *Cancer* **56**:2092–2098.

Anderson G (1999) Nondirectiveness in prenatal genetics: Patients read between the lines. *Nurs Ethics* **6**(2):126–136.

Andre J, Fleck LM, Tomlinson T (2000) On being genetically "irresponsible." *Kennedy Inst Ethics J* **10**(2):129–146.

Andrews LB (1985) The rationale behind the informed consent doctrine. *J Med Pract Manage* **1**(1):59–65.

Andrews LB (1989) Newborn screening for sickle cell disease and other hemoglobinopathies: Overview of legal issues. *Pediatrics* **83**(5 part 2):886–890.

Andrews LB (1991) Legal aspects of genetic information. *Yale J Biol Med* **64**(1):29–40.

Andrews LB (1992) Torts and the double helix: Malpractice liability for failure to warn of genetic risks. *Houst Law Rev* **29**(1):149–184.

Andrews LB (2002) The gene patent dilemma: Balancing incentives with health needs. *Houst J Health Law Policy* **2**:65–106.

Andrews, LB (1997) Past as Prologue: Sobering Thoughts on Genetic Enthusiasm. 27 Seton Hall L. Rev. 893, at 901: The author reviews incidents where physicians either have surreptitiously tested pregnant women's blood for carrier status for genetic diseases or have tried to coerce the patient to undergo carrier screening. At 917, the author argues that an information brochure given to patients by Oncor Med with regard to commercial testing for susceptibility to hereditary breast and ovarian cancer, "overestimated the chances of women getting cancer, perhaps frightening people into testing." Could a health care professional be held accountable for an act of negligence committed by a laboratory employed by the health care professional. The concept of the "captain of the ship" is frequently referred to by courts of law. An example is the responsibility of a surgeon for the actions of other professional personnel during an operation. If the patient is injured as the result of a negligent act during the operation, the surgeon may be found liable for failure to exercise reasonable care in the supervision of the other personnel. The issue generally will turn on two questions of fact. Was the negligent party under the "actual control" of the surgeon [*Szabo v. Bryn Mawr Hospital*, 1994]? For example, in some jurisdictions, courts have found that surgeons were not liable where the negligent personnel were employees of the hospital, the negligent acts were not reasonably discoverable by the surgeon, and the surgeon did not have the right to control the personnel at the time of the negligent act [e.g., *Harris v. Miller*, 438 S.E.2d, 731, NC 1994]. In the case of *Taniv v. Taub* [1998], a physician who was responsible for supervision of the office staff, implementing office policy, transmitting radiology reports, and was a principal of the professional services corporation, was held liable when a patient's X-ray was not sent to the treating physician, resulting in the failure to diagnose a patient's cancer.

Andrews LB, Fullarton JE, Holtzman NA, Motulsky AG, eds. Committee on Assessing Genetic Risk: (1994) *Assessing genetic risk: implications for health and social policy*, Institutes of Medicine. National Academy Press, Washington, DC: 1994.

Andrews LB, Zuiker ES (2003) Ethical, Legal, and Social Issues in Genetic Testing for Complex Genetic Diseases, 37 Val. U. L. Rev. 793–829. Should a distinction be drawn between genetic testing for single-gene disorders and testing for complex, multifactorial genetic diseases? Can, or should, the doctrine of informed consent be waived where genetic tests meet criteria for clinical utility and validity? Are such practices ethically, medically, and legally defensible? Although over 1000 genetic tests are now available—www.genetests.org—much of the research pertaining to what constitutes an appropriate model for informed consent for genetic testing, has been derived primarily from experiences with susceptibility testing for single gene disorders like Huntington disease. In contrast, complex, multifactorial genetic diseases are determined by a combination of multiple genetic and environmental factors and the interaction between them [Thompson and Thompson, 2001; Becker, 2000]. Uncertainties about the nature and effect of this interaction significantly hamper the ability to identify the role of heredity and to make accurate predictions of risk compared to single-gene disorders. Accordingly, genetic testing for multifactorial genetic diseases is less precise and less definitive compared to single-gene diseases. This raises the issue of whether the model of informed consent for single-gene disorders should be applicable for multifactorial testing. The authors note that although the nature of a genetic disease or disorder may differ, the identification of an individual whose genotype is classified as being "at risk" for a genetic condition can have "profound implications for individual self-esteem and self-perception" including emotional and psychological distress, as well as familial, social, and employment relationships: "Knowing the presence of one 'defective' gene can lead a person into severe depression, but the knowledge of several 'defective' genes without a clear sense of their meaning and implication for future diseases has the potential for devastating results. Genetic knowledge, despite its nuances and inaccuracies, can alter people's idea of self-efficacy, esteem, personal locus of control, and even risk-taking behaviours" see also: [Holtzman and Watson, 1997]. This report drew a distinction between genetic tests for infants that would not require parental informed consent (tests that met criteria for clinical validity and utility in terms of being in the "best interests" of the infant—however, parents should receive "sufficient information to understand the reasons for the screening") and those tests that did not meet such criteria and therefore would require parental consent. The U.S. 1999 Task Force on New Born Screening [Newborn Screening Task Force, 1999, at pages 395, 410] endorsed the concept of "informed refusal" where parents have the right to be informed of genetic screening tests and to refuse screening. See Wildeman and Downie [2001]. What of scenarios where more genetic information is generated than expected? The authors Andrews & Zwiker, cite the example where during the testing process information is gained that indicates the patient is at risk for another genetic disease other than the one for which the patient had granted informed consent. Is there any obligation to disclose this additional information or only the information for which informed consent has been granted? In this scenario, the health care professional and the patient may have sharply distinctive opinions as to what constitutes "benefit" and "harm". The health care provider may wish to disclose information for which an informed consent has not been obtained in order to prevent "medical harm" and to provide appropriate management -if any- and counseling. The patient, however, may not wish to learn of this information where effective management and therapies are not available, such as in the case of untreatable late-onset disorders, and due to the risk of resulting "harm" in the form of potential stigmatization, familial unrest, psychological unrest, and insurance and/or employment discrimination. For a further discussion, please see Chapter 7, "Informed Consent" and Chapter 15, "Confidentiality".

Ang P, Garber JE (2001) Genetic susceptibility for breast cancer—Risk assessment and counseling. *Semin Oncol* **28**(4):419–433.

Annas GJ (1993) Privacy rules for DNA databanks. Protecting coded "future diaries," *JAMA* **270**:2346–2350.

Annas GJ, Elias S (1992). Social research priorities for the Human Genome Project. In: Annas GJ, Elias S, eds. *Gene Mapping Using Law & Ethics as Guides*. New York: Oxford University Press, pp. 272–273.

Annas G, Glantz L, Roche P (1995) Drafting the Genetic Privacy Act: Science, Policy and Practical Considerations. *J Law, Medicine and Ethics* **23**:360–366, at 365.

Antonarakis SE (1998) Recommendations for a nomenclature system for human gene mutations. *Hum Mut* **11**:1–3.

Antoniadi T, Pampanos A, Petersen MB (2001) Prenatal diagnosis of prelingual deafness: Carrier testing and prenatal diagnosis of the common GJB2 35delG mutation. *Prenat Diagn* **21**(1):10–13.

Antoniou A, Pharoah P, Narod S, Risch HA, Eyfjord JE, Hopper JL, Loman N, Olsson H, Johannsson O, Borg A, Pasini B, Radice P, Manoukian S, Eccles DM, Tang N, Olah E, Anton-Culver H, Warner E, Lubinski J, Gronwald J, Gorski B, Tulinius H, Thorlacius S, Eerola H, Nevanlinna H, Syrjakoski K, Kallioniemi O-P, Thompson D, Evans C, Peto J, Lalloo F, Evans DG, Easton DF (2003) Average risks of breast and ovarian cancer associated with BRCA1 or BRCA2 mutations detected in case series unselected for family history: A combined analysis of 22 studies. *Am J Hum Genet* **72**(5):1117–1130.

Antoniou A, Pharoah P, Smith P, Easton D (2004) The BOADICEA model of genetic susceptibility to breast and ovarian cancer. *Br J Cancer* **91**(8):1580–1590.

Apfel RJ, Fisher SM (1984) *To Do No Harm: DES and the Dilemmas of Modern Medicine.* New Haven, CT: Yale University Press.

Appleton S, Fry A, Rees G, Rush R, Cull A (2000) Psychosocial effects of living with an increased risk of breast cancer: An exploratory study using telephone focus groups. *Psycho-Onc* **9**:511–521.

Arato v. Avedon (1993) 23 Cal Rptr 131, 858 P.2d 598. The court held that there was no duty to inform patient of nonmedical information. Physicians were not held in breach of the duty of care to the patient when they failed to disclose the statistical life expectancy when recommending a course of chemotherapy and radiation treatment.

Arena v. Gingrich (1987) 84 Or. App. 25, 733 P. 2d. 75.

Armstrong K, Stopfer J, Calzone K, Fitzgerald G, Coyne J, Weber B (2002) What does my doctor think? Preferences for knowing the doctor's opinion among women considering clinical testing for BRCA1/2 mutations. *Genet Test* **6**(2):115–118; Comment in *Genet Test* **6**(2):71–74.

Armstrong K, Weber B, FitzGerald G, Hershey JC, Pauly MV, Lemaire J, Subramanian K, Asch DA (2003) Life insurance and breast cancer risk assessment: Adverse selection, genetic testing decisions, and discrimination. *Am J Med Genet* **120A**(3):359–364. Compare to Geer et al. [2001]. A majority of those who refused to proceed with cancer genetic counseling cited concern for insurability for the patient and the family.

Arndt v. Smith (1997) 2 S.C.R. 539.

Arnoff SL, Bennett PH, Gorden P, Rushforth N, Miller M (1977) Unexplained hyperinsulinemia in normal and "prediabetic" Pima Indians compared with normal Caucasians. *Diabetes* **26**:827–840.

Arora A, Gambardella A, Pammolli F, Riccaboni M (2000) The nature and extent of the market for technology in biopharmaceuticals; online: EPRIS <http://www.unisi.it/ricerca/prog/epris/3.htm>.

Arpino G, Bardou VJ, Clark GM, Elledge RM (2004) Infiltrating lobular carcinoma of the breast: Tumor characteristics and clinical outcome. *Breast Cancer Res* **6**:R149–R156.

Ashcroft R (2003) Back to the future: Response to: Extending preimplantation genetic diagnosis: Medical and non-medical uses. *J Med Ethics* **29**(4):217–219.

Asher v. Gutierrez (1976) 533 F. 2d 1235 (D.C. Cir).

ASHG (2004) Your family history. American Society of Human Genetics. http://genetics.faseb.org/genetics/ashg/educ/007.shtml.

ASHG/ACMG Report (1995) Points to consider: Ethical, legal, and psychosocial implications of genetic testing in children and adolescents. *Am J Hum Genet* **57**:1233–1241. http://genetics.faseb.org/genetics/ashg/policy/pol-13.htm; http://www.acmg.net/resources/policies/pol-018.asp.

ASHG/ACMG (1996) Diagnostic testing for Prader-Willi and Angleman syndromes: Report of the ASHG/ACMG Test and Technology Transfer Committee. *Am J Hum Genet* **58**:1085–1088.

ASRM (1999) Sex selection and preimplementation genetic diagnosis. American Society of Reproductive Medicine. *Fertil Steril* **72**(4):595–598.

ASRM Ethics Committee (1999) American Society for Reproductive Medicine-Ethics committee reports and statements (includings updates) http://www.asrm.org/Media/Ethics/ethicsmain.html.

Atlas of Genetics and Cytogenetics in Oncology and Haematology (2004) World Wide Web URL: http://www.infobiogen.fr/services/chromcancer/.

Audrain J, Lerman C, Rimer B, Cella D, Steffens R, Gomez-Caminero A (1995) Awareness of heightened breast cancer risk among first-degree relatives of recently diagnosed breast cancer patients. *Cancer Epidemiol Biomarkers* **4**:561–565.

Audrain J, Schwart MD, Lerman C, Hughes C, Peshkin BN, Biesecker B (1998) Psychological distress in women seeking genetic counseling for breast-ovarian cancer risk: The contributions of personality and appraisal. *Ann Behav Med* **19**:370–377.

Auranen M, Vanhala R, Varilo T, Ayers K, Kempas E, Ylisaukko-Oja T, Sinsheimer JS, Peltonen L, Jarvela I (2002) A genomewide screen for autism-spectrum disorders: Evidence for a major susceptibility locus on chromosome 3q25-27. *Am J Hum Genet* **71**(4):777–790.

Australia Law Reform Commission (2004a) Gene patenting and human health. Discussion Paper 68, online: ALRC, <http://www.austlii.edu.au/au/other/alrc/publications/dp/68/>.

Australia Law Reform Commission (2004b) Genes and ingenuity: Gene patenting and human health, Report 99, online: ALRC, <http://www.austlii.edu.au/au/other/alrc/publications/reports/99/>.

Australian Law Reform Commission (ALRC) and Australian Health Ethics Committee (2003) Essentially Yours: The Protection of Human Genetic Information in Australia. 96, ALRC.

Avard D, Knoppers BM (2001) Screening and children—Policy issues for the new millennium. ISUMA, 2(3); http://www.isuma.net/v02n03/avard/avard_e.shtml. The authors argue in the case of screening for PKU, if parents refuse to proceed with testing, the refusal should be documented in writing and honored. However, in some circumstances such as when a severe disorder such as PKU can be prevented, health professionals have a duty to the newborn to persuade the parents to consent and to override parental refusal to the child being screened. Also, see references 36 and 37 therein.

Axworthy D, Brock DJ, Bobrow M, Marteau TM (1996) Psychological impact of population-based carrier testing for cystic fibrosis: 3-year follow-up. UK Cystic Fibrosis Follow-Up Study Group. *Lancet* **347**(9013):1443–1446.

Azzolino v. Dingfelder (1985) 337 S.E. 2d 528 (N.C.).

Babul R, Shelin A, Kremer B, Dufrasne S, Wiggins S, Huggins M, Theilmann J, Block M, Hayden MR (1993) for the Canadian Collaborative Group on Predictive Testing for Huntington Disease. Attitudes toward direct predictive testing for the Huntington disease gene: Relevance for other adult-onset disorders. *JAMA* **270**:2321–2325.

Bach G et al. (2001) Tay-Sachs screening in the Jewish Ashkenazi population: DNA testing is the preferred procedure. *Am J Med Genet* **99**(1):70–75.

Bader v. Johnson (2000) 732 N.E.2d 1212.

Badger v. Surkan (1970) 16 D.L.R. (3d) 146 at 153 (Sask. Q.B.), aff'd [1973] 1 W.W.R. 302 (C.A.).

Badner JA, Gershon ES (2002) Meta-analysis of whole-genome linkage scans of bipolar disorder and schizophrenia. *Mol. Psychiatry* **7**(4):405–411.

Bahado-Singh RO, Deren O, Tan A, D'Ancona RL, Hunter D, Copel JA, Mahoney MJ, (1996) Ultrasonographically adjusted midtrimester risk of trisomy 21 and significant chromosomal defects in advanced maternal age. *Am J Obstet Gynecol* **175**(6):1563–1568.

Bailey A, Le Couteur A, Gottesman I, Bolton P, Simonoff E, Yuzda E, Rutter M (1995) Autism as a strongly genetic disorder: Evidence from a British twin study. *Psychol Med* **25**(1):63–77.

Baird P (1993) *Proceed with Care: Final Report of The Royal Commission on New Reproductive Technologies*, Vol. 2, Ottawa: Minister of Government Services, pp. 788–789. As to the issue of prenatal diagnosis, the Commission reported (at 766–767) that a "disturbing proportion of referring physicians who do not accept the principle that patients should make their own informed choice about whether to have PND and when to have an abortion after the diagnosis of a fetal disorder . . . This is of great concern to the Commissioners, respect for the pregnant woman's autonomy requires that it be her values and priorities, not the doctor's, that determine her decision to accept or decline PND testing."

Baird PA (2002) Identification of genetic susceptibility to common diseases. The case for regulation. *Perspect Biol Med* **45**(4):516–528.

Baker DL (1998) Interviewing techniques. In: Baker DL, Eash T, Schuette JL, Uhlmann WR eds. *A Guide to Genetic Counseling.* New York: Wiley-Liss, pp. 55–74.

Baker MT, Taub HA (1983) Readability of informed consent forms for research in a Veterans Administration medical center. *JAMA* **250**(19):2646–2648.

Baker DL, Eash T, Schuett JL, Uhlmann WR (2002) Guidelines for writing letters to patients. *J Genet Counseling* **11**(5):399–418.

Baldwin J, Hanel P, Sabourin D (2000) Determinants of innovative activity in Canadian manufacturing firms: The role of intellectual property rights, Statistics Canada Working Paper No. 122. Online: Statistics Canada <http://www.statcan.ca/cgi-bin/downpub/listpub.cgi?catno=11F0019MIE2000122>.

Bale HE Jr (1996) Patent protection and pharmaceutical innovation. *Int J Law Politics* **29**:95–107.

Balinsky W, Zhu CW (2004) Pediatric cystic fibrosis: Evaluating costs and genetic testing. *J Pediatr Health Care* **18**(1):30–34.

Balint J, Shelton W (1996) Regaining the intiative: Forging a new model of the patient-physician relationship. *JAMA* **275**(11):887–891.

Barash CI (2000) Genetic discrimination and screening for hemochromatosis: Then and now. *Genet Test* **4**(2):213–218.

Barber M, Whitehouse PJ (2002) Susceptibility testing for Alzheimer's disease: race for the future. *Lancet Neurol* **1**(1):10.

Bardessono v. Michels (1970) 478 P.2d 480, 484 (Cal.).

Barnes EA, Hanson J, Neumann CM, Nekolaichuk CL, Bruera E (2000) Communication between primary care physicians and radiation oncologists regarding patients with cancer treated with palliative radiotherapy. *J Clin Oncol* **18**(15):2902–2907.

Baron JA, Beach M, Mandel JS, van Stolk RU, Haile RW, Sandler RS, Rothstein R, Summers RW, Snover DC, Beck GJ, Bond JH, Greenberg ER, for the Calcium Polyp Prevention Study Group (1999) Calcium supplements for the prevention of colorectal adenomas. *N Engl J Med* **340**:101–107.

Bar-Shalom A, Cook-Deegan R (2002) Patents and innovation in cancer therapeutics: Lessons from Cellpro. *Milbank Quart* **80**(4):637–676.

Baron JA, Cole BF, Sandler RS, Haile RW, Ahnen D, Bresalier R, McKeown-Eyssen G, Summers RW, Rothstein R, Burke CA, Snover DC, Church TR, Allen JI, Beach M, Beck GJ, Bond JH, Byers T, Greenberg ER, Mandel JS, Marcon N, Mott LA, Pearson L, Saibil F, van Stolk RU (2003) A randomized trial of aspirin to prevent colorectal adenomas. *N Engl J Med* **348**:891–899.

Barrett S, Hall H (2003) Dubious genetic testing. http://www.quackwatch.org/01Quackery RelatedTopics/Tests/genomics.html.

Barrett S, Beck JC, Bernier R, Bisson E, Braun TA, Casavant TL, Childress D, Folstein SE, Garcia M, Gardiner MB, Gilman S, Haines JL, Hopkins K, Landa R, Meyer NH, Mullane JA, Nishimura DY, Palmer P, Piven J, Purdy J, Santangelo SL, Searby C, Sheffield V, Singleton J, Slager S, et al (1999) An autosomal genomic screen for autism. Collaborative linkage study of autism. *Am J Med Genet* **88**(6):609–615.

Bartels DM, LeRoy BS, Caplan AL (1993) *Prescribing our Future, Ethical Challenges in Genetic Counseling.* Walter de Gruyter. New York.

Bartels DM, LeRoy BS, McCarthy P, Caplan AL (1997) Nondirectiveness in genetic counseling: A survey of practitioners. *Am J Med Genet* **72**(2):172–179.

Bassett M, Dunn C, Battese K, Peek M (2001) Acceptance of neonatal genetic screening for hereditary hemochromatosis by informed parents. *Genet Test* **5**(4):317–320. Also see Delatycki MB, Powell LW, Allen KJ (2004) Hereditary hemochromatosis genetic testing of at-risk children: What is the appropriate age? *Genet Test* **8**(2):98–103.

Bassett SS, Havstad SL, Chase GA (2004) The role of test accuracy in predicting acceptance of genetic susceptibility testing for Alzheimer's disease. *Genet Test* **8**(2):120–126.

Baty BJ, Kinney AY, Ellis SM (2003) Developing culturally sensitive cancer genetics communication aids for African Americans. *Am J Med Genet* **118A**(2):146–155.

Bauer v. Friedland (1986) 394 N.W. 2d 549 (Minn. App.).

Baulac S, Huberfeld G, Gourfinkel-An I, Mitropoulou G, Beranger A, Prud'homme JF, Baulac M, Brice A, Bruzzone R, LeGuern E (2001) First genetic evidence of GABAA receptor dysfunction in epilepsy: A mutation in the gamma2 subunit gene. *Nat Genet* **28**:46–48.

Baum M, Houghton J, Riley D (1992) Results of the cancer research campaign adjuvant trial for preoperative cyclophosphamide and long-term tamoxifen in early breast cancer reported at the tenth year of follow up. *Acta Oncol* **31**:251–257.

Baylis F, Downie J, Freedman B, Hoffmaster B, Sherwin S. (1995) Theory and method in health care ethics. In: *Health Care Ethics in Canada*. Toronto: Harcourt Brace, pp. 4–8.

Beauchamp TL, Childress JF (1994) *Principles of Biomedical Ethics*, 4th ed. New York: Oxford University Press, pp. 31–43.

Beaudet AL (1990) Invited editorial: Carrier screening for cystic fibrosis. *Am J Hum Genet* **47**:603–605.

Beck AT, Beamesderfer A (1978) Assessment of depression: The depression inventory. *Mod Probl Pharmacopsychiatry* **7**:51–69.

Beckcom v. U.S. (1984) 584 F.Supp. 1471 (D.C.N.Y.). In the Canadian decision of *Champigny v. Ste-Marie* (1994) 19 C.C.L.T. (2d) 307 (Que. S.C.) the court confirmed that a physician has a duty to pay attention to a patient's complaints.

Becker AB (2000) Is primary prevention of asthma possible? *Pediatr Pulmonol* **30**:63–72.

Becker v. Schwartz (1978) 46 N.Y.2d 401, 413 N.Y.S.2d 895, 386 N.E.2d 807. Also see *Phillips v. United States* (1983) 575 F. Supp. 1309. (D.S.C. 1983). Parents entitled to damages where doctor failed to advise and counsel them of risk of Down syndrome, and the child was born with Down syndrome.

Beckman HB, Frankel R (1984) The effect of physician behavior on the collection of data. *Ann Intern Med* **101**:692–696.

Beckman HB, Markakis KM, Suchman AL, Frankel RM (1994) The doctor-patient relationship and malpractice. Lessons from plaintiff depositions. *Arch Intern Med* **154**(12):1365–1370.

Beckwith J (1991) The Human Genome initiative: Genetics' lightning rod. *Am J Law Med* **17**:1–14.

Beckwith J (2005) Whither human behavioral genetics? In: Parens E, Chapman A, Press N, eds. *Wrestling with Behavioral Genetics: Implications for Understanding Selves and Society*, Baltimore: Johns Hopkins University Press.

Begley S (1999) Screening for genes; matching medications to your genetic heritage. *Newsweek* Feb. 8, p. 66.

Bekker H, Modell M, Denniss G, Silver A, Mathew C, Bobrow M, Marteau T (1993) Uptake of cystic fibrosis carrier testing in primary care: Supply push or demand pull? *BMJ* **306**:1584–1586.

Belkin L (1999) What the Jumans didn't know about Michael. *New York Times Magazine*, March 14, pp. 42–49.

Bell M, Hirst M, Nakahori Y, Mackinnon R, Roche A, Flint T, Jacobs P, Tommerup N, Tranebjaerg L, Froster-Iskenius U, Kerr B, Turner G, Lindenbaum R, Winter R, Penbrey M, Thibodeau S, Davies K (1991) Physical mapping across the fragile X: Hypermethylation and clinical expression of the fragile X syndromes. *Cell* **64**:861–866.

Bellet PS, Maloney MJ (1991) The importance of empathy as an interviewing skill in medicine. *JAMA* **266**(13):1831–1832.

Belloni E, Martucciello G, Verderio D, Ponti E, Seri M, Jasomi V, Torre M, Ferrari M, Tsui LC, Scherer SW (2000) Involvement of the HLXB9 homeobox gene in Currarino syndrome. *Am J Hum Genet* **66**(1):312–319.

Benacerraf BR, Frigoletto FD Jr (1987) Soft tissue nuchal fold in the second trimester fetus: Standards for normal measurements compared with those with Down syndrome. *Am J Obstet Gynecol* **157**(5):1146–1149.

Benacerraf BR, Gelman R, Frigoletto FD Jr (1987) Sonographic identification of second-trimester fetuses with Down syndrome. *N Engl J Med* **317**:1371–1376.

Benacerraf BR, Mandell J, Estroff JA, Harlow BL, Frigoletto FD Jr (1990) Fetal pyelectasis: A possible association with Down syndrome. *Obstet Gynecol* **76**(1):58–60.

Benassayag C, Rigourd V, Mignot TM, Hassid J, Leroy MJ, Robert B, Civel C, Grange G, Dallot E, Tanguy J, Nunez EA, Ferre F (1999) Does high polyunsaturated free fatty acid level at the feto-maternal interface alter steroid hormone message during pregnancy? *Prostaglandins Leukot Essent Fatty Acids* **60**:393–399.

Bender PL (1998) Genetic family history assessment. *AACN Clin Issues* **9**(4):483–490; quiz 615–617.

Benkendorf JL, Reutenauer JE, Hughes CA, Eads N, Willison J, Powers M, Lerman C (1997) Patients' attitudes about autonomy and confidentiality in genetic testing for breast-ovarian cancer susceptibility. *Am J Med Gen* **73**:296–303. Of women 95% of the 238 women Surveyed in the survey thought that they should be able to get testing notwithstanding a physician's opinion to the contrary.

Benn PA, Leo MV, Rodis JF, Beazoglou T, Collins R, Horne D (1999) Maternal serum screening for fetal trisomy 18: A comparison of fixed cutoff and patient-specific risk protocols. *Obstet Gynecol* **93**:707–711.

Benn PA, Egan JFX, Fang M, Smith-Bindman R (2004) Changes in the utilization of prenatal diagnosis. *Obstet Gynecol* **103**(6):1255–1260.

Bennett B (2001) Prenatal diagnosis, genetics and reproductive decision-making. *J Law Med* **9**:28–40.

Bennett RL, Steinhaus KA, Uhrich SB, O'Sullivan CK, Resta RG, Lochner-Doyle D, Markel DS, Vincent V, Hamanishi J (1995) Pedigree Standardization Task Force of the National Society of Genetic Counselors. Recommendation for standardized human pedigree nomenclature. *Am J Hum Genet* **56**:7445–7452.

Bennett RL, Hampel HL, Mandell JB, Marks JH (2003) Genetic counselors: Translating genomic science into clinical practice. *J Clin Invest* **112**(9):1274–1279.

Bensen JT, Liese AD, Rushing JT, Province M, Folsom AR, Rich SS, Higgins M (1999) Accuracy of proband reported family history: The NHLBI Family Heart Study (FHS). *Genet Epidemiol* **17**:141–150.

Bensing JM, Roter DL, Hulsman RL (2003) Communication patterns of primary care physicians in the United States and the Netherlands. *J Gen Intern Med* **18**(5):335–342.

Benson v. Massachusetts General Hospital (2000) 49 Mass. App. Ct. 530, 731 N. E. 2d 85. In *Benson v. Massachusetts General Hospital*, the court applied a relatively restriative standard holding that the nature of the information to be disclosed for the purpose of informed consent is that possessed by an average qualified physician, or in the case of a specialty, by an average qualified physician practicing that specialty.

Bentley D (2004) Genomes for medicine. *Nature* **429**:440–445.

Berg D, Hirsch HL (1980) Duty to recall. *S Med J* **73**:1041–1043.

Berg D, Hirsch HL (1983). Physician's duty to recall revisited. *Legal Aspects Med Practice* **11**:5.

Berger K (2003) Informed consent: Information or knowledge? *Med Law* **22**:743–750.

Bergstrand CG, Czar B (1956) Demonstration of a new protein fraction in serum from the human fetus. *Scan J Clin Lab Invest* **174**:8.

Berkovic SF, Scheffer IE (2001) Genetics of the epilepsies. *Epilepsia* **42**(Suppl 5):16–23.

Berkovic SF, Howell RA, Hay DA, Hopper JL (1998) Epilepsies in twins: Genetics of the major epilepsy syndromes. *Ann Neurol* **43**(4):435–445.

Berlin I (1969) Two Concepts of Liberty. In: *Four Essays on Liberty.* Oxford University Press, p. 118. Reproduced by permission of Curtis Brown on behalf of Isaiah Berlin Literary Trust copyright Isaiah Berlin 1958, 1969, 1997.

Bernhardt BA (1997) Empirical evidence that genetic counseling is directive: Where do we go from here? *Am J Hum Genet* **60**:17–20.

Bernhardt BA, Geller G, Doksum T, Larson SM, Roter D, Holtzman NA (1998) Prenatal genetic testing: Content of discussions between obstetric providers and pregnant women. *Obstet Gynecol* **91**(5 Pt 1):648–655.

Bernhardt BA, Biesecker BB, Mastromarino CL (2000) Goals, benefits, and outcomes of genetic counseling: Client and genetic counselor assessment. *Ar J Med Genet* **94**:189–197.

Berry R (1997) The genetic revolution and the physician's duty of confidentiality. The role of the old Hippocratic virtues in the regulation of the new genetic intimacy. *J Legal Med* **18**:401–441.

Berry DA, Iversen ES Jr, Gudbjartsson DF, Hiller EH, Garber JE, Peshkin BN, Lerman C, Watson P, Lynch HT, Hilsenbeck SG, Rubinstein WS, Hughes KS, Parmigiani G (2002) BRCAPRO validation, sensitivity of genetic testing of BRCA1/BRCA2, and prevalence of other breast cancer susceptibility genes. *J Clin Oncol* **20**:2701–2712.

Bersinger NA, Brizot ML, Johnson A, Snijders RJ, Abbott J, Schneider H, Nicolaides KH (1994) First trimester maternal serum pregnancy-associated plasma protein A and pregnancy-specific beta 1-glycoprotein in fetal trisomies. *Br J Obstet Gynaecol* **101**:970–974.

Bertakis KD, Roter D, Putnam SM (1991) The relationship of physician medical interview style to patient satisfaction. *J Fam Pract* **32**(2):175–181.

Bertoli-Avella AM, Oostra BA, Heutink P (2004) Chasing genes in Alzheimer's and Parkinson's disease. *Hum Genet* **114**:413–438.

Bertram L, Tanzi RE (2004) The current status of Alzheimer's disease genetics: What do we tell the patients? *Pharmacol Res* **50**(4):385–396.

Bertucci F, Viens P, Hingamp P, Nassser V, Houlgatte R, Birnbaum D (2003) Breast cancer revisited using DNA array-based gene expression profiling. *Int J Cancer* **103**:565–571.

Bettelheim D, Deutinger J, Bernaschek G (1999) The value of echogenic foci ("golfballs") in the fetal heart as a marker of chromosomal abnormalities. *Ultrasound Obstet Gynecol* **14**(2):98–100.

Bickel H (1996) The first treatment of phenylketonuria. *Eur J Pediatr* **155**(Suppl 1):S2–3.

Biesecker BB (2001) Goals of genetic counseling. *Clin Genet* **60**(5):323–330.

Biesecker B, Peters K (2001) Process studies in genetic counseling: Peering into the black box. *Am J Med Genet* **106**(3):191–198.

Biesecker B, Ishibe N, Hadley DW, Giambarresi TR, Kase RG, Lerman C, Struewing JP (2000) Psychosocial factors predicting BRCA1/BRCA2 testing decisions in members of hereditary breast and ovarian cancer families. *Am J Med Genet* **93**:257–263.

Bill C-6 (2004) Assisted Human Reproduction Act, 3rd Sess., 37th Parl., 2004, cl. 5(1)(e) (as passed by the House of Commons 29 March 2004).

Bill C-43, *An Act Respecting Abortion*, 2d sess., 34th Parliament, 1989–90, (defeated in Senate, 31 January 1991).

Billings PR, Kohn MA, de Cuevas M, Beckwith J, Alper JS, Natowicz MR (1992) Discrimination as a consequence of genetic testing. *Am J Hum Genet* **50**(3):476–482.

Binedell J, Soldan JR, Haper PS (1998) Predictive testing for Huntington's disease: Predictors of uptake in South Wales. *Clin Genet* **54**:477–488.

Bioethics Committee (CPS) (2003) Guidelines for genetic testing of healthy children. Bioethics Committee, *Canadian Paediatric Society Paediatrics & Child Health* **8**(1):42–45. http://www.cps.ca/english/statements/B/b03-01.htm.

Birminghan K (1997) Insurers admit genetic discrimination. *Nature Med* **3**:710.

Bish A, Sutton S, Jacobs C, Levene S, Ramirez A, Hodgson S (2002) Changes in psychological distress after cancer genetic counselling: A comparison of affected and unaffected women. *Br J Cancer* **86**(1):43–50.

Bjornholt JV, Erikssen G, Liestol K, Jervell J, Thaulow E, Erikssen J (2000) Type 2 diabetes and maternal family history: An impact beyond slow glucose removal rate and fasting hyperglycemia in low-risk individuals? Results from 22.5 years of follow-up of healthy nondiabetic men. *Diabetes Care* **23**:1255–1259.

Black WC, Nease Jr RF, Tosteson ANA (1995) Perceptions of breast cancer risk and screening effectiveness in women younger than 50 years of age. *J Nat'l Ca Inst* **87**:720–731.

Blackston JW, Bouldin MJ, Brown CA, Duddleston DN, Hicks GS, Holman HE (2002) Malpractice risk prevention for primary care physicians. *Am J Med Sci* **324**(4):212–219.

Blake v. Cruz, 698 P.2d 315, 319 (Idaho 1984).

Blanchette v. Barrett (1994) 640 A. 2d. 74 (Conn.).

Blaymore Bier JA, Liebling JA, Morales Y, Carlucci M (1996) Parents' and pediatricians' views of individuals with meningomyelocele. *Clin Pediatr (Phila)* **35**(3):113–117.

Blaz v. Michael Reese Hospital Foundation (1999) 74 F. Supp. 2d 803 (N.D.Ill.). The court ruled that a physician in charge of the hospital's research program for radiation treatment had a duty to warn a former patient, who had undergone radiation treatments as a child and had developed tumors, of a strong association between the treatments and the tumors, due to a special relationship although he was not the patient's original attending physician. This decision is distinguishable in that it applied Illinois law.

Blenkinsopp A, Bashford J, Dickinson D (1998) Partnership with patients. Health professionals need to identify how much information patients want. *BMJ* **317**(7155):413–414.

Bloch M, Hayden MR (1990) Opinion: Predictive testing for Huntington disease in childhood: Challenges and implications. *Am J Hum Genet* **46**:1–4.

Bloch M, Fahy M, Fox S, Hayden MR (1989). Predictive testing for Huntington's disease II: Demographic characteristics, life-style patterns, attitudes and psychosocial assessments of the first fifty-one test candidates. *Am J Med Genet* **32**:217–224.

Bloch M, Adam S, Wiggins S, Huggins M, Hayden M (1992) Predictive testing for Huntington diseases in Canada: The experience of those receiving an increased risk. *Am J Med Genet* **42**:449–507. Also see: Frets and Niermeijer, 1990; Jedlicka-Kohler et al. 1994.

Bloch M, Adam S, Fuller A, Kremer B, Welch JP, Wiggins S, Whyte P, Huggins M, Theilmann J, Hayden MR (1993) Diagnosis of Huntington disease: A model for the stages of psychological response based on experience of a predictive testing program. *Am J Med Genet* **47**(3):368–374.

Blum JD (1999) Telemedicine poses new challenges for the law. *Health Law Canada* **20**:115–126.

Blumenfeld Z, Siegler E, Bronshtein M (1993) The early diagnosis of neural tube defects. *Prenat Diagn* **13**(9):863–871.

Blumenthal D, Campbell EG, Anderson MS, Causino N, Louis KS (1997) Withholding research results in academic life science: Evidence from a national survey of faculty. *JAMA* **277**:1224.

Boardman LA (2002) Heritable colorectal cancer syndromes: Recognition and preventive management. *Gastroenterol Clin North Am* **31**:1107–1131.

Bogan v. Altman, McQuire and Pigg (2003) PSC. 2001. Ky. App. WL 201848; consolidated with the case of *Grubbs v. Barbourville Family Clinic* (2003) Ky. S. Ct. In the *Grubbs* case, the plaintiff alleged that the physicians failed to accurately interpret and/or to report the test results from an ultrasound that indicated spina bifida. In the *Bogan* case, the plaintiffs alleged that the defendant physician failed to inform them that prenatal genetic tests were available and failed to accurately interpret an ultrasound. The child was born without any eyes an underdeveloped brain, a cleft palate, and could not speak. The court in the *Bogan* case held that the alleged failure to provide the parents with "information necessary to make a decision regarding the continuation of pregnancy stated a viable cause of action for medical negligence." After these cases were consolidated, the Supreme Court of Kentucky declined to recognize a cause of action for wrongful birth or wrongful life.

Bogardus ST, Holmboe E, Jekel, J (1999) Perils, pitfalls, and possibilities in talking about medical risk. *JAMA* **281**:1037–1041.

Bogart MH, Pandian MR, Jones OW (1987) Abnormal maternal serum chorionic gonadotropin levels in pregnancies with fetal chromosome abnormalities. *Prenat Diagn* **7**:623–630.

Bonacruz Kazzi G, Cooper C (2003) Barriers to the use of interpreters in emergency room paediatric consultations. *J Paediatr Child Health* **39**(4):259–263.

Botkin J (2003) Prenatal diagnosis and the selection of children. *Florida State University Law Rev* **30**:265–293.

Bottorff JL, Ratner PA, Johnson JL, Lovato CY, Joab SA (1998) Communicating cancer risk information: The challenges of uncertainty. *Pt Educ Counsel* **33**:67–81.

Bottrell MM, Alpert H, Fischbach RL, Emanuel LL (2000) Hospital informed consent for procedure forms: Facilitating quality patient-physician interaction. *Arch Surg* **135**(1):26–33.

Bovbjerg R (1975) The medical malpractice standard of care: HMOs and customary practice. *Duke Law J* 1375, at pp. 1384–1407.

Bovi AM (2003) Council on Ethical and Judicial Affairs of the American Medical Association. Ethical guidelines for use of electronic mail between patients and physicians. *Am J Bioeth* **3**(3):W-IF2.

Bowen D, McTiernan A, Burke W, Powers D, Pruski J, Durfy S, Gralow J, Malone K (1999) Participation in breast cancer risk counseling among women with a family history. *Ca Epi, Bio Prev* **8**:581–585.

Bower MA, Veach PM, Bartels DM, LeRoy BS (2002) A survey of genetic counselor's strategies for addressing ethical and professional challenges in practice. *J Genet Couns* **11**(3):163–186.

Bowling A, Ebrahim S (2001) Measuring patients' preferences for treatment and perception of risk. *Qual Health Care* **10**(Suppl. I):i2–8.

Boyd PA, Wellesley DG, DeWalle HEK, Tenconi R, Garcia-Minaur S, Zandwijken GRJ, Stoll C, Clementi M (2000) Evaluation of the prenatal diagnosis of neural tube defects by fetal ultrasonographic examination in different centres across Europe. *J Med Screen* **7**:169–174.

Boyle RJ, de Crespigny L, Savalescu J (2003) An ethical approach to giving couples information about their fetus. *Hum Reprod* **18**(11):2253–2256.

Braddock CH 3rd, Fihn SD, Levinson W, Jonsen AR, Pearlman RA (1997) How doctors and patients discuss routine clinical decisions. Informed decision making in the outpatient setting. *J Gen Intern Med* **12**:339–345.

Braddock CH 3rd, Edwards KA, Hasenberg NM, Laidley TL, Levinson W. (1999) Informed decision making in outpatient practice: Time to get back to basics. *JAMA* **282**(24):2313–2320.

Bradford v. O'Neill (1996) 688 so. 2d. 33 (La. Ct. App. 4th Cir.), writ denied, 693 So. 2d 769 (La. 1997).

Bradley LA, Palomaki GE, McDowell GA (2004) ACMG technical standards and guidelines: Prenatal screening for open neural tube defects. Available at: <http://www.acmg.net/Pages/ACMG_Activities/stds-2002/ONTD.htm>. Accessed October 26, 2004.

Brain K, Gray J, Norman P, France E, Anglim C, Barton G, Parsons E, Clarke A, Sweetland H, Tischkowitz M, Myring J, Stansfield K, Webster D, Gower-Thomas K, Daoud R, Gateley C, Monypenny I, Singhal H, Branston L, Sampson J, Roberts E, Newcombe R, Cohen D, Rogers C, Mansel R, Harper P (2000) Randomized trial of a specialist genetic assessment service for familial breast cancer. *J Natl Cancer Inst* **92**(16):1345–1351.

Braithwaite D, Emery J, Walter F, Prevost A, Sutton S (2004) Psychological impact of genetic counseling for familial cancer: A systematic review and meta-analysis, *J Natl Cancer Inst* **96**:122–133.

Brambati B, Tului L, Bonacchi I, Shrimanker K, Suzuki Y, Grudzinskas JG (1994) Serum PAPP-A, free á-hCG are first trimester screening markers for Down syndrome. *Prenat Diagn* **14**:1043–1047.

Bramwell R, Carter D (2001) An exploration of midwives' and obstetricians' knowledge of genetic screening in pregnancy and their perception of appropriate counselling. *Midwifery* **17**(2):133–141.

Branca v. Miro (2001) 0735/2001 Rockland Co. N.Y. Sup. Ct.; and see Fortado L [2004]. Genetic testing maps new legal turf; Doctors' liability grows as tests are more widely used. National Law Jour.; 26 (50):1.

Breast Cancer Linkage Consortium (1999) Cancer risks in BRCA2 mutation carriers. *J Natl Cancer Inst* **91**:1310–1316.

Breyne v. Potter (2002) 258 Ga. App. 278, 574 SE.2d 916.

Brezden-Masley C, Aronson MD, Bapat B, Pollett A, Gryfe R, Redston M, Gallinger S (2003) Hereditary nonpolyposis colorectal cancer—molecular basis. *Surgery* **134**:29–33.

Briellmann RS, Torn-Broers Y, Busuttil BE, Major BJ, Kalnins RM, Olsen M, Jackson GD, Frauman AG, Berkovic SF (2000) APOE epsilon4 genotype is associated with an earlier onset of chronic temporal lobe epilepsy. *Neurology* **55**:435–437.

Briguglio J, Cardella JF, Fox PS, Hopper KD, TenHave TR (1995) Development of a model angiography informed consent form based on a multiinstitutional survey of current forms. *J Vasc Interv Radiol* **6**:971–978.

Brinkman R, Mexei M, Theilmann J, Almqvist E, Hayden M (1997) The likelihood of being affected with Huntington disease by a particular age, for a specific CAG size. *Am J Hum Genet* **60**:1202–1210.

Brinton LA, Hoover R, Fraumeni JF (1986) Menopausal oestrogens and breast cancer risk: An expanded case-control study. *Br J Cancer* **54**:825–832.

Brizot ML, Snijders RJ, Bersinger NA, Kuhn P, Nicolaides KH (1994) Maternal serum pregnancy-associated plasma protein A and fetal nuchal translucency thickness for the prediction of fetal trisomies in early pregnancy. *Obstet Gynecol* **84**:918–922.

Brizot ML, Snijders RJ, Butler J, Bersinger NA, Nicolaides KH (1995) Maternal serum hCG and fetal nuchal translucency thickness for the prediction of fetal trisomies in the first trimester of pregnancy. *Br J Obstet Gynaecol* **102**:127–132.

Broadstock M, Michie S (2000) Processes of patient decision making: Theoretical and methodological issues. *Psych Health* **15**(2):191–204.

Broadstock M, Michie S, Marteau T (2000) Psychological consequences of predictive genetic testing: A systematic review. *Eur J Hum Genet* **8**:731–738.

Brock DJH, Sutcliffe RG, (1972) Alpha-fetoprotein in the antenatal diagnosis of anencephaly and spina bifida. *Lancet* **2**:197–199.

Brock DW, Wartman SA (1990) When competent patients make irrational choices. *New Eng J Med* **322**:1595–1599, at 1595. "Even the irrational choices of a competent patient must be respected if the patient cannot be persuaded to change them." But see *Turner v. Children's Hosp, Inc*. (1991) 602 N.E.2d 423, 431 (Ohio Ct. App. 1991): "This doctrine has not emerged to educate the patient generally on medical matters." Also see *McGeshick v. Choucair* (1993) 9 F.3d 1229, 1233–35 (7th Cir. 1993). However, neither of these cases considered this issue within the context of genetic testing.

Brock DHJ, Bolton AE, Monaghan JM (1973) Prenatal diagnosis of anencephaly through maternal serum-alphafetoprotein measurement. *Lancet* **2**:923–924.

Brody DS, Miller SM, Lerman CE, Smith DG, Lazaro CG, Blum MJ (1989) The relationship between patients' satisfaction with their physicians and perceptions about interventions they desired and received. *Med Care* **27**(11):1027–1035.

Broide E, Zeigler M, Eckstein J. Bach G (1993) Screening for carriers of Tay-Sachs disease in the ultraorthodox Ashkenazi Jewish community in Israel. *Am J Med Genet* **47**(2):213–215.

Bromley B, Doubilet P, Frigoletto F, Krauss C, Estroff J, Benacerraf B (1994) Is fetal hyperechoic bowel on second trimester sonogram an indication for amniocentesis? *Obstet Gyneacol* **83**:647–651.

Bromley B, Lieberman E, Shipp TD, Richardson M, Benacerraf BR (1998) Significance of an echogenic intracardiac focus in fetuses at high and low risk for aneuploidy. *J Ultrasound Med* **17**(2):127–131.

Bronshtein M, Rottem S, Yoffe N, Blumenfeld Z (1989) First-trimester and early second-trimester diagnosis of nuchal cystic hygroma by transvaginal sonography: Diverse prognosis of the septated from the nonseptated lesion. *Am J Obstet Gynecol* **161**:78–82.

Brookes A (2001) Women's voices: Prenatal diagnosis and care for the disabled. *Health Care Anal* **9**:133–150.

Brose MS, Rebbeck TR, Calzone KA, Stopfer JE, Nathanson KL, Weber BL (2002) Cancer risk estimates for BRCA1 mutation carriers identified in a risk evaluation program. *J Natl Cancer Inst* **94**:1365–1372.

Brostoff S (1992) CEOs: Defend genetic test use in underwriting. *Natl Underwriter* April 27:24.

Brown v. Dahl (1985) 41 Wn. App. 565, 580–83, 705 P.2d 781.

Browner CH, Preloran HM (2000a) Interpreting low-income Latinas' amniocentesis refusals. *Hispanic J Behav Sci* **22**(3):346–368.

Browner CH, Preloran HM (2000b) Para sacarse la espina (to get rid of the doubt): Mexican immigrant women's amniocentesis decisions. In: Saetnan A, Nelly Oudshoorn N, Kirejczyk M, eds. *Bodies of Technology: Women's Involvement with Reproductive Medicine*. Columbus, OH: Ohio State University, pp. 369–383.

Browner CH, Preloran HM, Cox SJ (1999) Ethnicity, bioethics, and prenatal diagnosis: The amniocentesis decisions of Mexican-origin women and their partners. *Am J Public Health* **89**(11):1658–1666.

Browner CH, Preloran HM, Casado MC, Bass HN, Walker AP (2003) Genetic counseling gone awry: Miscommunication between prenatal genetic service providers and Mexican-origin clients. *Soc Sci Med* **56**:1933–1946.

Browner CH, Preloran HM (2004) Expectations emotions, and medical decision making: a case study on the use of amniocentesis. *Transcult Psychiatry* **41**(4):427–44.

Brunner HG, Nelen M, Breakefield XO, Ropers HH, van Oost BA (1993) Abnormal behavior associated with a point mutation in the structural gene for monoamine oxidase A. *Science* **262**(5133):578–580.

Bryant GD, Norman GR (1980) Expressions of probability: Words and numbers. *N Engl J Med* **302**:411.

Buchanan A, Brock DW (1986) Deciding for others. Milbank Q.;64 (Suppl. 2):17–94.

Buck PS (1950) The Child Who Never Grew. Woodbine House, Bethesda, Maryland.

Buckman R (1992) *How to Break Bad News: A Guide for Health Care Professionals.* Baltimore: Johns Hopkins University Press.

Budescu DV, Weinberg S, Wallsten TS (1988) Decisions based on numerically and verbally expressed uncertainties. *J Exp Psychol Gen* **14**:281–294.

Bulaj Z, Ajioka RS, LaSalle B, Jorde L, Griffen L, Edwards C, Kushner J (2000) Disease-related conditions in relatives of patients with hemochromatosis 23. *New Engl J Med* **343**(21):1529–1535.

Bunting v. Jamieson (1999) WY. 105, 984 P.2d 467. A couple, the Buntings, alleged that the physician's failure to timely refer their son to a genetic specialist delayed a diagnosis of Hurler syndrome and significantly compromised the child's recovery ability. The plaintiffs alleged that had the child, Raef, received a transplant following an accurate diagnosis at approximately 12 months, rather than allowing the disease to progress until he received his transplant at 31 months old, Raef would have higher intelligence and fewer physical problems in the future. The defendant physician Dr. Jamieson challenged this evidence, claiming that it was inconsistent with medical literature and that the research is incomplete and ongoing. http://www.oscn.net/applications/oscn/DeliverDocument.asp?citeID=123916.

Buono RJ, Lohoff FW, Sander T, Sperling MR, O'Connor MJ, Dlugos DJ, Ryan SG, Golden GT, Zhao H, Scattergood TM, Berrettini WH, Ferraro TN (2004) Association between variation in the human KCNJ10 potassium ion channel gene and seizure susceptibility. *Epilepsy Res* **58**:175–183.

Burd L, Wilson H (2004) Fetal, infant, and child mortality in a context of alcohol use. *Am J Med Genet* **127C**(1):51–58. Also see Burd L, Klug M, Martsolf J (2004) Increased sibling mortality in children with fetal alcohol syndrome. *Addict Biol* **9**(2):179–186; discussion 187–188.

Burgermeister J (2003) Teacher was refused job because relatives have Huntington's disease. *BMJ* **327**:827.

Burgess MM (2001) Beyond consent: Ethical and social issues in genetic testing. *Nat Rev Genet* **2**:147–151.

Burgess MM, Laberge CM, Knoppers BM (1998) Bioethics for clinicians: Ethics and genetics in medicine. *CMAJ* **158**:1309–1313.

Burk DL, Lemley MA (2003) Policy levers in patent law. *Virginia L Rev* **89**(7):1575–1696.

Burke W (2004) Genetic testing in primary care. *Annu Rev Genomics Hum Genet* **5**:1–14.

Burke W, Austin MA (2002) Genetic risk in context: Calculating the penetrance of BRCA1 and BRCA2 mutations, *J Nat Cancer Inst* **94**:1185–1187.

Burke W, Daly M, Garber J, Botkin J, Kahn MJ, Lynch P, McTiernan A, Offit K, Perlman J, Petersen G, Thomson E, Varricchio C (1997) Recommendations for follow-up care of individuals with an inherited predisposition to cancer. II. BRCA1 and BRCA2. *JAMA* **277**(12):997–1003. Comment in: *JAMA* (1997) **278**(4):289–290; discussion 290; comment in: *JAMA* (1997) **278**(4):290.

Burke W, Culver JO, Bowen D, Lowry D, Durfy S, McTiernan A, Andersen MR (2000) Genetic counseling for women with an intermediate family history of breast cancer. *Am J Med Genet* **28;90**(5):361–368.

Burke W, Pinsky LE, Press NA (2001) Categorizing genetic tests to identify their ethical, legal, and social implications. *Am J Med Genet* **106**(3):233–240.

Burke W, Acheson L, Botkin J, Bridges K, Davis A, Evans J, Frias J, Hanson J, Kahn N, Kahn R, Lanier D, Pinsky LE, Press N, Lloyd-Puryear MA, Rich E, Stevens N, Thomson E, Wartman S, Wilson M (2002) Genetics in primary care: A USA faculty development initiative. *Community Genet* **5**(2):138–146.

Burnett JW (1999) A physician's duty to warn a patient's relatives of a patient's genetically inheritable disease. *Houst Law Rev* **36**(2):559–582.

Burris S, Gostin LO, Tress D (2001) Public health surveillance of genetic information: Ethical and legal responses to social risk. In: Khoury MJ, Burke W, Thomson EJ, eds. *Genetics and Public Health in the 21st Century. Using Genetic Information to Improve Health and Prevent Disease.* Oxford: Oxford University Press. http://www.cdc.gov/genomics/oldWeb01_16_04/info/books/21stcent5.htmthe.

Buxbaum JD, Silverman JM, Smith CJ, Greenberg DA, Kilifarski M, Reichert J, Cook EH Jr, Fang Y, Song CY, Vitale R (2002) Association between a GABRB3 polymorphism and autism. *Mol Psychiatry* **7**:311–316.

Calderon-Margalit R, Paltiel O (2004) Prevention of breast cancer in women who carry BRCA1 or BRCA2 mutations: A critical review of the literature. *Int J Cancer* **112**:357–364.

Caldwell S (2000) Personal communication. Genetic Counselor, Kaiser Permanente Medical Group. Panorama City, CA.

California Genetic Testing Dispute (no date) A series of decisions by the Independent Medical Review, of the California Department of Managed Health Care was reported to have denied insurance coverage for testing. http://www.genelaw.info/pages/caseresults.asp?area=5. Also see Klanica, K (2004) Genetic information and health insurance: A report on the law in the United States. http://www.genelaw.info/reports/health.pdf.

Calman KC (1996) Cancer: Science and society and the communication of risk. *BMJ* **313**:799–802. And see Berry DC, Raynor DK, Knapp P, Bersellini E (2003) Patients' understanding of risk associated with medication use: Impact of European Commission guidelines and other risk scales. *Drug Safety* **26**:1–11.

Calman KC (2004) Evolutionary ethics: Can values change. *J Med Ethics* **30**:366–370.

Campbell A, Glass KC (2000) The legal status of clinical and ethics policies, codes, and guidelines in medical practice and research. *McGill Law J* **46**:473. http://www.journal.law.mcgill.ca/abs/462glass.htm?french=1.

Canadian abortion rights action league, protecting abortion rights in Canada, Canadian abortion rights action league, Ottawa, 2003.

Canadian Biotechnology Advisory Committee, CBAC (2002) Patenting of Higher Life Forms and Related Issues. Online: CBAC <http://cbac-cccb.ic.gc.ca/epic/internet/incbac-cccb.nsf/en/ah00188e.html>.

Canadian College of Medical Geneticists (1991) Policy Statement concerning DNA banking and molecular genetic diagnosis.

Canadian College of Medical Geneticists (1998) Training Guidelines for Genetic Centres and Candidates for Fellowship.

Canadian College of Medical Geneticists (2004a) Information on individual laboratories. http://www.ucalgary.ca/UofC/faculties/medicine/medgenetics/CCMG/labs.htm. Accessed December 19, 2004.

Canadian College of Medical Geneticists (2004b) Training Guidelines: http://ccmg.medical.org/pdf/training/train_clinical.pdf.

Canesi v. Wilson (1999) 158 N.J. 490, 730 A.2d.805. The court mandated that an informed consent charge should be given in every wrongful birth case.

Canick JA, Knight GJ, Palomaki GE, Haddow JE, Cuckle HS, Wald NJ (1988) Low second trimester maternal serum unconjugated oestriol in pregnancies with Down syndrome. *Br J Obstet Gynaecol* **95**:330–333.

Canick JA, Panizza DS, Palomaki GE (1990) Prenatal screening for Down syndrome using AFP, uE3 and hCG: Effect of maternal race, insulin-dependent diabetes and twin pregnancy. *Am J Hum Genet* **47**:a270.

Cannistra SA (2004) Cancer of the ovary. *N Engl J Med* **351**:2519–2529.

Canterbury v. Spence (1972a) 464 F. 2d 772, 781 (D.C. Cir. App.), cert. denied, 409 U.S. 1064, 93 S Ct. 560: "[I]t is the prerogative of the patient, not the physician, to determine for himself the direction in which his interests seem to lie." Three distinct standards to evaluate the duty of disclosure have evolved in the United States. The traditional "professional standard" inquires what a reasonable medical practitioner would disclose in the same or similar circumstances— *Largey v. Rothman* 540 A. 2d. 504 (N.J. 1988). The "reasonable patient" standard applied in *Cantebury v. Spence* is founded upon the principle that "physicians communications to the patient . . . . must be measured by the patient's need and that need is material to the decision" (464 F 2d. 772, at 786). A small number of American jurisdictions have adopted a subjective standard—*Arena v. Gingrich* 84 Or. App 25, 733 P. 2d. 75 (1987). For a review of the various judicial standards for disclosure and causation see: *Ketchup v. Howard* (2000) 247 Ga. App. 54, 543 S.E.2d 371; Schuck PH (1994) Rethinking informed consent. *Yale LJ* **103**:899; Merz JF (1993) On a decision-making paradigm of medical informed consent. *J Legal Med* **14**:231, 243–264. A minority of jurisdictions limit consent to surgical procedures, injections, and diagnostic tests; e.g., *Morgan v. MacPhail* (1997) 550 Pa. 202, 704 A. 2d.617, 1997—compared to the majority of jurisdictions that hold a physician's duty to obtain an informed consent applies whether a treatment is invasive or noninvasive: e.g., *Matthias v. Mastromonaco* (1999) 160 N.J. 26, 40, 733 A. 2d. 456 [N.J. 1999].

Cao A, Rosatelli MC, Monni G, Galanello R. (2002) Screening for thalassemia: A model of success. *Obstet Gynecol Clin North Am* **29** (2):305–328,vi–vii.

Cappelli M, Surh L, Humphreys L, Verma S, Logan D, Hunter A, Allanson J (1999) Psychological and social determinants of women's decisions to undergo genetic counseling and testing for breast cancer. *Clin Gen* **55**:419–430.

Cappelli M, Surh L, Walker M, Korneluk Y, Humphrey L, Verma S, Hunter A, Allanson J, Logan D (2001) Psychological and social predictors of decisions about genetic testing for breast cancer in high-risk women. *Psychol Hlth Med J* **6**:323–335.

Capron A (1979) Tort liability in genetic counseling. *Colum Law Rev* **79**(4):618–684, at page 623 fn. 17. During the 1970s, electronic fetal monitoring was rapidly adopted to monitor patients during labor, notwithstanding questions about its safety and efficacy. Also see Gilfix M (1984) Electronic fetal monitoring: Physician liability and informed consent. *Am J Law Med* **10**(1):31–90. The author reviews the difficulty physicians faced, with regard to EFM, in trying to balance standards concerning "customary practice" against a physician's obligations to apply the "best judgment" and "to keep abreast" standards of care.

Carraccio C, Wolfsthal SD, Englander R, Ferntz K, Martin C (2002) Shifting paradiagms: from Flexner to competencies. Academic Medicine **77**:361–7.

Carroll AM (2001) Closing the Gaps in Genetics Legislation and Policy: A Report by the New York State Task Force on Life and the Law. *Genet Test* **5**(4):275–280.

Carroll JC, Brown JB, Reid AJ, Pugh P (2000) Women's experience of maternal serum screening. *Can Fam Physician* **46**:614–620.

Carter RB (1983) *Descartes' Medical Philosophy—The Organic Solution to the Mind-Body Problem*. Baltimore: Johns Hopkins University Press, at 7.

Carter R, Winkelaar P (2001) Test results, patients and doctors: Legal status. *Alberta Doctors' Digest* July/August:2001.

Caspi A, McClay J, Moffitt TE, Mill J, Martin J, Craig IW, Taylor A, Poulton R (2002) Role of genotype in the cycle of violence in maltreated children. *Science* 297(5582):851–854.

Castermans D, Wilquet V, Steyaert J, Van de Ven W, Fryns JP, Devriendt K (2004) Chromosomal anomalies in individuals with autism: A strategy towards the identification of genes involved in autism. *Autism* 8(2):141–161.

Caulfield T (2000) Underwhelmed: Hyperbole, regulatory policy, and the genetic revolution. *McGill LJ* 45:437.

Caulfield T (2001) The informed gatekeeper? A commentary on genetic tests, marketing pressure, and the role of primary care physicians. *Health Law Rev* 9(3):14–18.

Caulfield TA (2003) Sustainability and the balancing of the health care and innovation agendas: The commercialization of genetic research. *Sask L Rev* 66:629–645.

Caulfield T (2004) Gene patents, human clones and biotechnology policy: The challenge created by globalization. *Alberta L Rev* 3:713–724.

Caulfield T, Gold R (2000) Genetic testing, ethical concerns and the role of patent law. *Clin Genet* 57:370–375.

Caulfield T, Sheremeta L (2004a) Biotechnology patents and embryonic stem cell research: Emerging issues (Part I). *J Int Biotech Law* 1:98.

Caulfield T, Sheremeta L (2004b) Biotechnology patents and embryonic stem cell research: Emerging issues (Part II). *J Int Biotech Law* 1:142.

Caulfield TA, Knoppers BM, Gold ER, Sheremeta L, Bridge P (2003) Genetic technologies, healthcare policy and the patent bargain. *Clin Genet* 63:15–18.

CCMG (2003) Guidelines for genetic testing of healthy children: Bioethics Committee, Canadian Paediatric Society (CPS): Approved by the Board of Directors of CCMG (Canadian College of Medical Geneticists). Paediatrics & Child Health; 8(1):42–45. http://www.cps .ca/english/statements/B/b03-01.htm.

CDC (2003) Centers for Disease Control: Genomics and the Population Health. The Family History Public Health Initiative. http://www.cdc.gov/genomics/activities/ogdp/2003/chap06. htm.

Cepelewicz BR, Dunn LD, Fetch DM, Levin LJ, Nelson BC, Rothschild IS, Rust ME, Seremski MJ (1998) Recent developments in medicine and law. *Tort Insur Law J* 33:583–603.

Cerhan JR, Parker AS, Putnam SD, Chiu BC, Lynch CF, Cohen MB, Torner JC, Cantor KP (1999) Family history and prostate cancer risk in a population-based cohort of Iowa men. *Cancer Epidemiol Biomarkers Prev* 8:53–60.

Chabner E, Nixon A, Geloman R (1998) Family history and treatment outcome in young women after breast-conserving surgery and radiation therapy for early-stage breast cancer. *J Clin Oncol* 16:2045–2051.

Chace DH, Kalas TA, Naylor EW (2003) Use of Tandem Mass Spectrometry for Multianalyte Screening of Dried Blood Specimens from Newborns. *Clinical Chemistry* 49:1797–1817.

Chakrabarti S, Fombonne E (2001) Pervasive developmental disorders in preschool children. *JAMA* 285(24):3093–3099.

Chang J, Elledge RM (2001) Clinical management of women with genomic BRCA1 and BRCA2 mutations. *Breast Cancer Res Treatment* 69:101–113.

Chao C, Studts JL, Abell T, Hadley T, Roetzer L, Dineen S, Lorenz D, YoussefAgha A, McMasters KM (2003) Adjuvant chemotherapy for breast cancer: How presentation of recurrence risk influences decision making. *J Clin Oncol* 21:4299–4305.

Charames GS, Bapat B (2003) Genomic instability and cancer. *Curr Mol Med* 3:589–596.

Charles C, Gafni A, Whelan T (1997) Shared decision-making in the medical encounter: What does it mean? (or it takes at least two to tango). *Soc Sci Med* **44**:681–692.

Charles C, Gafni A, Whelan T (1999) Decision-making in the physician-patient encounter: Revisiting the shared treatment decision-making model. *Soc Sci Med* **49**(5):651–661.

Charlier C, Singh NA, Ryan SG, Lewis TB, Reus BE, Leach RJ, Leppert M (1998) A pore mutation in a novel KQT-like potassium channel gene in an idiopathic epilepsy family. *Nat Genet* **18**: 53–55.

Chartier-Harlin MC, Crawford F, Houlden H, Warren A, Hughes D, Fidani L, Goate A, Rossor M, Roques P, Hardy J, et al (1991) Early-onset Alzheimer's disease caused by mutations at codon 717 of the beta-amyloid precursor protein gene. *Nature* **353**:844–846.

Chartier-Harlin MC, Parfitt M, Legrain S, Perez-Tur J, Brousseau T, Evans A, Berr C, Vidal O, Roques P, Gourlet V, et al (1994) Apolipoprotein E, epsilon 4 allele as a major risk factor for sporadic early and late-onset forms of Alzheimer's disease: Analysis of the 19q13.2 chromosome region. *Hum Mol Genet* **3**:569–574.

Chartrand LN, Forbes-Chilibeck EM (2003) The sentencing of offenders with fetal alcohol syndrome. *Health Law J* **11**:35–70.

Cheek J, Lacy J, Toth-Fejel S (2002) The impact of hormone replacement therapy on the detection and stage of breast cancer. *Arch Surg* **137**:1015–1019.

Chen EA, Schiffman JF (2000) Attitudes towards genetic counseling and prenatal diagnosis among of group of individuals with physical disabilities. *J Genet Counseling* **9**(2):137–152.

Chen YS, Akula N, Detera-Wadleigh SD, Schulze TG, Thomas J, Potash JB, DePaulo JR, McInnis MG, Cox NJ, McMahon FJ (2004) Findings in an independent sample support an association between bipolar affective disorder and the G72/G30 locus on chromosome 13q33. *Mol Psychiatry* **9**(1):87–92.

Cherewayko v. Grafton (1993) 3 WWR 604; 16 C.C.L.T. 115 at 132. (Man. Q.B.): "The defendant acknowledged he gave them no written advice other than the pamphlet that was handed to the plaintiff prior to the first operation some nine months before . . . In my view, any benefit from the pamphlet had long since expired insofar as warning the plaintiff . . . both the plaintiff and her husband testified that they were in a state of stress during the March 16 interview, with the plaintiff unable to open her mouth without crying. She said a lot of things were very vague, and she described herself as uptight with the result. She felt she was 'a little foggy on that day'. In these circumstances, it is quite probable the plaintiff and her husband missed the comments of the defendant."

Cherry v. Borsman (1990) 5 C.C.L.T. (2d) 243 (B.C.S.C.), (1991), 75 D.L.R.(4th) 668, 16 BCAC, 93; 28 WAC 93.

Cheung v. Cunningham (1988) 520 A. 2d. 832, 832–835 (N.J. Super) A.D. (1987), rev;d., 111 N.J. 573, 546 A. 2d. 501 (1988).

Childbirth by Choice (2003) Abortion in Canada today: The Situation Province-by-Province, Childbirth by Choice, Toronto.

Chitayat D, Toi A, Babul R, Levin A, Michaud J, Summers A, Rutka J, Blaser S, Becker LE. (1995) Prenatal diagnosis of retinal nonattachment in the Walker-Warburg syndrome. *Am J Med Genet* **56**(4):351–358.

Chitty LS, Altman DG (2003) Charts of fetal size: Kidney and renal pelvis measurements. *Prenat Diagn* **23**(11):891–897.

Chitty LS, Pandya PP (1997) Ultrasound screening for fetal abnormalities in the first trimester. *Prenat Diagn* **17**:1269–1281.

Chlebowski RT, Hendrix SL, Langer RD, Stefanick ML, Gass M, Lane D, Rodabough RJ, Gilligan MA, Cyr MG, Thomson CA, Khandekar J, Petrovitch H, McTiernan A, WHI Investigators (2003) Influence of estrogen plus progestin on breast cancer and mammography in healthy postmenopausal women: The Women's Health Initiative Randomized Trial. *JAMA* **289**:3243–3253.

Cho MK, Arruda M, Holtzman NA (1997) Educational material about genetic tests: Does it provide key information for patients and practitioners? *Am J Med Genet* **73**(3):314–320.

Cho MK, Sankar P, Wolpe PR, Godmilow L (1999) Commercialization of BRCA1/2 testing: Practitioner awareness and use of new genetic test. *Am J Med Genet* **83**:157–163.

Cho MK, Illangasekare S, Weaver MA, Leonard DGB, Merz JF (2003) Effects of patents and licenses on the provision of clinical genetic testing services. *J Mol Diagnostics* **5**:3–8.

Chodirker BN, Cadrin C, Davies GAL, Summers AM, Wilson RD, Winsor EJT, Young D (2001) Canadian guidelines for prenatal diagnosis: Genetic indications for prenatal diagnosis. *J Obstet Gynaecol Can* **23**:525–531.

Chou IC, Lee CC, Huang CC, Wu JY, Tsai JJ, Tsai CH, Tsai FJ (2003a) Association of the neuronal nicotinic acetylcholine receptor subunit alpha4 polymorphisms with febrile convulsions. *Epilepsia* **44**(8):1089–1093.

Chou IC, Peng CT, Huang CC, Tsai JJ, Tsai FJ, Tsai CH (2003b) Association analysis of GABRG2 polymorphisms with febrile seizures. *Pedatr Res* **54**:1–4.

Chumakov I, Blumenfeld M, Guerassimenko O, Cavarec L, Palicio M, Abderrahim H, Bougueleret L, Barry C, Tanaka H, La Rosa P, Puech A, Tahri N, Cohen-Akenine A, Delabrosse S, Lissarrague S, Picard FP, Maurice K, Essioux L, Millasseau P, Grel P, Debailleul V, Simon AM, Caterina D, Dufaure I, Malekzadeh K, Belova M, Luan JJ, Bouillot M, Sambucy JL, Primas G, Saumier M, Boubkiri N, Martin-Saumier S, Nasroune M, Peixoto H, Delaye A, Pinchot V, Bastucci M, Guillou S, Chevillon M, Sainz-Fuertes R, Meguenni S, Aurich-Costa J, Cherif D, Gimalac A, Van Duijn C, Gauvreau D, Ouellette G, Fortier I, Raelson J, Sherbatich T, Riazanskaia N, Rogaev E, Raeymaekers P, Aerssens J, Konings F, Luyten W, Macciardi F, Sham PC, Straub RE, Weinberger DR, Cohen N, Cohen D, Ouelette G, Realson J (2002) Genetic and physiological data implicating the new human gene G72 and the gene for D-amino acid oxidase in schizophrenia. *Proc Natl Acad Sci USA* **99**(21):13675–13680, Erratum in same journal (2002) **99**(26):17221.

Chung DC, Rustgi AK (2003) The hereditary nonpolyposis colorectal cancer syndrome: Genetics and clinical implications. *Ann Intern Med* **138**:560–570.

Christen Y, Mallet J (2003) Introduction: Neurogenomining. In: Mallet J, Christen Y, eds, *Neurosciences at the Postgenomic Era*. Berlin:Springer.

Christensen v. Munsen (1994) 123 Wash 2d 234, 867 P 2d 626, 30 ALR 5th 822.

Christiansen M, Oxvig C, Wagner JM, Qin QP, Nguyen TH, Overgaard MT, Larsen SO, Sottrup-Jensen L, Gleich GJ, Norgaard-Pedersen B (1999) The proform of eosinophil major basic protein: A new maternal serum marker for Down syndrome. *Prenat Diagn* **19**:905–910.

Chudleigh T (2001) Mild pyelectasis. *Prenat Diagn* **21**:936–941.

Chudleigh PM, Chitty LS, Pembrey M, Campbell S (2001) The association of aneuploidy and mild fetal pyelectasis in an unselected population: The result of a multicenter study. *Ultrasound Obstet Gynecol* **17**:197–202.

Ciarlariello v. Schacter (1993) 2 SCR 119, at 140: "It was incumbent on the doctor to make sure that he was understood, particularly where it appears that the patient had some difficulty with the language spoken by the doctor. Indeed, it is appropriate that the burden should be placed on the doctor to show that the patient comprehended the explanation and the instructions given." Other Canadian courts have held that a physician can be obligated to ensure that communication accurs at a suitable time and place—*Coughlin v Kuntz* (1987) 17 B.C.L.R. (2d) 365 at 399 (B.C.S.C.); aff"d. 1989. 42 B.C.L.R. (2d) 108 (B.C.C.A.)—and in a manner appropriate to the communication needs of the patient—*Cherewayko v Grafton* (1993) 16 C.C.L.T. 115.

Cicero S, Curcio P, Papageorghiou A, Sonek J, Nicolaides K (2001) Absence of nasal bone in fetuses with trisomy 21 at 11–14 weeks of gestation: An observational study. *Lancet* **358**:1665–1667.

Cicero S, Bindra R, Rembouskos G, Spencer K, Nicolaides KH (2003) Integrated ultrasound and biochemical screening for trisomy 21 using fetal nuchal translucency, absent fetal nasal bone, free á-hCG and PAPP-A at 11 to 14 weeks. *Prenat Diagn* **23**:306–310.

Ciceron v. Jamaica Hospital (1999) 264 A.D. 2d 497, 694 N.Y.S. 2d 459 (2nd Dept.).

Ciruzzi M, Pramparo P, Rozlosnik J, Zylberstjn H, Delmonte H. Haquim M, Abecasis B, de la Cruz Ojeda J, Mele E, La Vecchia C, Schargrodsky H (1997) Frequency of family history of acute myocardial infarction in patients with acute myocardial infarction. Argentine FRICAS (Factores de Riesgo Coronario en America del Sur) Investigators. *Am J Cardiol* **80**:122–127.

Ciske DJ, Haavisto A, Laxova A, Rock LZ, Farrell PM (2001) Genetic counseling and neonatal screening for cystic fibrosis: An assessment of the communication process. *Pediatrics* **107**(4):699–705. But see Jarvinen O, Hietala M, Aalto AM, Arvio M, Uutela A, Aula P, Kaariainen H (2000) A retrospective study of long-term psychosocial consequences and satisfaction after carrier testing in childhood in an autosomal recessive disease: Aspartylglucosaminuria. *Clin Genet* **58**(6):447–454. The authors report that 23 out of 25 patients knew and understood their test results correctly. It should be noted that this study surveyed Finish patients and concerned a different testing process.

Claes L, Del-Favero J, Ceulemans B, Lagae L, Van Broeckhoven C, De Jonghe P (2001) De novo mutations in the sodium-channel gene SCN1A cause severe myoclonic epilepsy of infancy. *Am J Hum Genet* **68**:1327–1332.

Claes E, Evers-Kiebooms G, Boogaerts A, Decruyenaere M, Denayer L, Legius E (2004) Diagnostic genetic testing for hereditary breast and ovarian cancer in cancer patients: Women's looking back on the pre-test period and a psychological evaluation. *Genet Test* **8**(1):13–21.

Clarke A (1991) Is non-directional genetic counselling possible? *Lancet* **338**:998–1001.

Clarke A (1997) The process of genetic counselling. In: Harper P, Clarke A, eds., *Genetics, Society and Clinical Practice*. Oxford: BIOS Scientific Publishers, pp. 179–200.

Clark AJ, Barnetson R, Farrington SM, Dunlop MG (2004) Prognosis in DNA mismatch repair deficient colorectal cancer: Are all MSI tumours equivalent? *Fam Cancer* **3**:85–91.

Clarkeburn H (2000) Parental duties and untreatable genetic conditions. *J Med Ethics* **26**(5):400–403.

Claus EB, Risch N, Thompson WD (1993) The calculation of breast cancer risk for women with a first degree family history of ovarian cancer. *Breast Cancer Res Treat* **28**:115–120.

Claus EB, Risch N, Thompson WD (1994) Autosomal dominant inheritance of early-onset breast cancer: Implications for risk prediction. *Cancer* **73**:643–651.

Clayton EW (1998) What should the law say about disclosure of genetic information to relatives. *J Health Law Policy* **1**:373–390.

Clayton EW (2003) Genomic medicine: Ethical, legal, and social implications of genomic medicine. *New Engl J Med* **349**(6):562–569.

CLIA (Clinical Laboratory Improvement Amendments) (2003) U.S. Food and Drug Administration, Department of Health and Human Services, Center for Devices and Radiological Health, Rockville, MD. World Wide Web URL: http://www.fda.cov/cdrh/clia/.

CLIA Program: Clinical Laboratory Improvement Amendments (2004) Centers for Medicare & Medicaid Services, Baltimore, MD. World Wide Web URL: http://www.cms.hhs.gov/.

Clinton HR (2004) Now can we talk about health care? *NY Times Mag* April, **18**:28.

Cobbs v. Grant (1972) 502 P.2d 1, 12 (Cal.).

Code of Ethics of Physicians (2002) Professional Code, R.S.Q., c. C-26, s. 87. Quebec.

Codori A (1997) Psychological opportunities and hazards in predictive genetic testing for cancer risk. *Gastroenterol Clin North Am* **26**:19–39.

Codori AM, Slavney PR, Young C, Mighoretti DL, Brandt J (1997) Predictors of psychological adjustment to genetic testing for Huntington's disease. *Hlth Psychol* **16**:36–50.

Codori A, Petersen G, Miglioretti D, Larkin E, Bushey M, Young C, Brensinger J, Johnson K, Bacon J, Booker S (1999) Attitudes toward colon cancer gene testing: Factors predicting test uptake. *Ca Epi, Bio Prev* **8**:345–351.

Coghlan A (2004) Europe revokes controversial gene patent. *New Scientist*, May 19; online: New Scientist.com; <www.newscientist.com/news/news.jsp?id=ns99995016>.

Cohen LH, Fine BA, Pergament E (1998) An assessment of ethnocultural beliefs regarding the causes of birth defects and genetic disorders. *J Genet Counseling* **7**(1):15–29.

Cohen W, Nelson R, Walsh J (2000) Patenting their intellectual assets: Appropriability conditions and why US manufacturing firms patent or not. NBER paper 7522, Boston; online: SSRN <http://www.nber.org/papers/W7552>.

Cole LJ, Fleisher LD (2003) Update on HIPAA privacy: Are you ready? *Genet Med* **5**(3):183–186. Also see Fleisher LD, Cole LJ (2001) Health Insurance Portability and Accountability Act is here: What price privacy? *Genet Med* **3**(4):286–289.

Collaborative Group on Hormonal Factors in Breast Cancer (2001) Familial breast cancer: Collaborative reanalysis of individual data from 52 epidemiological studies including 58, 209 women with breast cancer and 101, 986 women without the disease. *Lancet* **358**:1389–1399.

Collaborative Linkage Study of Autism (2001) An autosomal genomic screen for autism. *Am J Med Genet* **105**(8):609–615.

College of American Pathologists (CAP) (1996–2003) World Wide Web URL: http://www.cap .org/.

College of American Pathologists (2003a) Laboratory Accreditation Program: Cytogenetics Checklist. Commission on Laboratory Accreditation, College of American Pathologists, Northfield, IL, July.

College of American Pathologists (2003b) Laboratory Accreditation Program: Molecular Pathology Checklist. Commission on Laboratory Accreditation, College of American Pathologists, Northfield, IL, December.

College of American Pathologists (2004) Laboratory Accreditation Program: Laboratory General Checklist. Commission on Laboratory Accreditation, College of American Pathologists, Northfield, IL, September 30.

Collins FS (1996) BRCA1-lots of mutations, lots of dilemmas. *N Engl J Med* **334**(3):186–188.

Collins FS (2003) A Brief Primer on Genetic Testing. Available from http://www.genome .gov/10506784.

Collins FS, McKusick VA (2001) Implications of the Human Genome Project for medical science. *JAMA* **285**:540–544.

Collins V, Halliday J, Warren R, Williamson R (2000) Cancer worries, risk perceptions and associations with interest in DNA testing and clinic satisfaction in a familial colorectal cancer clinic. *Clin Genet* **58**(6):460–468.

Collins F, Green E, Guttmacher A, Guyer M (2003) A vision for the future of genomics research: A blueprint for the genomic era. *Nature* **422**:835–847.

Colp v. Ringrose (1976) 3 L. Med. Q. 72 (Alta. T.D.). Also see *Doiron v. Orr* (1978) 86 D.L.R. (3d) 719 (Ont. H.C.); *Cataford v. Moreau* (1978) 114 D.L.R. (3d) 585 (Que. S.C.).

Combi R, Dalpra L, Malcovati M, Oldani A (2004) Evidence for a fourth locus for autosomal dominant nocturnal frontal lobe epilepsy. *Brain Res Bull* **63**(5):353–359.

Comeau AM, Parad RB, Dorkin HL, Dovey M, Gerstle R, Haver K, Lapey A, O'Sullivan BP, Waltz DA, Zwerdling RG, Eaton RB (2004) Population-based newborn screening for genetic disorders when multiple mutation DNA testing is incorporated: A cystic fibrosis newborn screening model demonstrating increased sensitivity but more carrier detections. *Pediatrics* **113**(6):1573–1581.

Commission on Classification and Terminology of the International League Against Epilepsy (1989) Proposal for revised classification of epilepsies and epileptic syndromes. *Epilepsia* **30**:389–399.

Committee of the International Huntington's Association (IHA) and the World Federation of Neurology (WFN) (1990) Ethical issues policy statement on Huntington disease's molecular genetics predictive test. *J Med Genet* **27**:34–38.

Committee on Assessing Genetic Risk (1994) Andrews LB, Fullarton JE, Holtzman NA, Motulsky AG, eds. *Assessing Genetic Risk: Implications for Health and Social Policy.* Washington, DC: Institutes of Medicine. National Academy Press. At page 50 concerning the impact of the medical and legal obligations to the duty to inform a patient about the existence of genetic risks, to refer for appropriate testing, and to provide, or to refer for, genetic counseling. At page 148 with regard to nondirective counseling. At page 190 concerning the situation where patients have difficulty with information and wish to leave it to the health care professional to make decisions.

Consortium on Pharmacogenetics (2002) (Buchanan A. McPeherson E, Brudy BA, Califano A, Kahn J, et al.) Pharmacogenetics: ethical, legal and regulatory issues in research and clinical practice.

Cook EH Jr (2000) Genetics of psychiatric disorders: Where have we been and where are we going? *Am J Psychiatry* **157**(7):1039–1040.

Cook v. Abbott (1980) 112 D.L.R. (3d) 234, 11 C.C.L.T. 217; aff'd (1980) 13 C.C.L.T. 264, 113 D.L.R. (3d) 88 (N.S.C.A.).

Cook EH Jr, Lindgren V, Leventhal BL, Courchesne R, Lincoln A, Shulman C, Lord C, Courchesne E (1997) Autism or atypical autism in maternally but not paternally derived proximal 15q duplication. *Am J Hum Genet* **60**(4):928–934.

Cook EH Jr, Courchesne RY, Cox NJ, Lord C, Gonen D, Guter SJ, Lincoln A, Nix K, Haas R, Leventhal BL, Courchesne E (1998) Linkage-disequilibrium mapping of autistic disorder, with 15q11–13 markers. *Am J Hum Genet* **62**(5):1077–1083.

Cook RJ, Dickens BM Fathalla MF (2001) *Reproductive Health and Human Rights: Integrating Medicine, Ethics, and Law*, Oxford:Clarendon Press.

Cook-Deegan RM, McCormack SJ (2001) Intellectual property: Patents, secrecy and DNA. *Science* **293**:217.

Cook-Deegan RM, Walters L, Pressman L, Pau D, McCormack S, Gatchalian J, Burgess R (2003) Preliminary data on U.S. DNA-based patents and plans for a survey of licensing practices. In: Knoppers BM, ed. *Populations and Genetics: Legal and Socio-Ethical Perspectives*, Boston:Martinus Nijhoff, pp. 453–470.

Corder EH, Saunders AM, Strittmatter WJ, Schmechel DE, Gaskell PC, Small GW, Roses AD, Haines JL, Pericak-Vance MA (1993) Gene dose of apolipoprotein E type epsilon 4 allele and the risk of Alzheimer's disease in late-onset families. *Science* **261**:921–923.

Corder EH, Saunders AM, Risch NJ, Strittmatter WJ, Schmechel DE, Gaskell PC Jr, Rimmler JB, Locke PA, Conneally PM, Schmader KE, et al (1994) Protective effect of apolipoprotein E type 2 allele for late onset Alzheimer disease. *Nat Genet* **7**:180–184.

Corey G, Forey MS, Callanan P, eds (1988) *Issues and Ethics in Helping Professions*, 3rd ed., Belmont CA: Cole Press, pp. 237–267.

Cornish WR, Llewelyn M, Adcock M (2003) Intellectual Property Rights (IPRs) and Genetics, July; online: Public Health Genetics Unit, <http://www.phgu.org.uk/about_phgu/intellect_prop_rights.html>.

Cosmides L, Tooby J (1996) Are humans good intuitive statisticians after all? Rethinking some conclusions from the literature on judgment under uncertainty. *Cognition* **58**:1–73.

Cossette P, Liu L, Brisebois K, Dong H, Lortie A, Vanasse M, Saint-Hilaire JM, Carmant L, Verner A, Lu WY, Wang YT, Rouleau GA (2002) Mutation of GABRA1 in an autosomal dominant form of juvenile myoclonic epilepsy. *Nat Genet* **31**:184–189.

Costa v. Boyd (2003) No. 36,584-CA (La.App. Cir.2 01/31/2003): Physician must perform necessary tests or refuse to treat unless it is an emergency. http://biotech.law.lsu.edu/cases/la/medmal/Costa_v_Boyd.htm.

Cottrelle v. Gerrard (2003) 178 O.A.C. 142, 233 D.L.R. (4th) 45, 20 C.C.L.T. (3d) 1, 67 O.R. (3d) 737.

Couch FJ, DeShano ML, Blackwood MA, Calzone K, Stopfer J, Campeau L, Ganguly A, Rebbeck T, Weber BL (1997) BRCA1 mutations in women attending clinics that evaluate the risk of breast cancer. *N Engl J Med* **336**:1409–15.

Coughlin v. Kuntz (1986) 42 C.C.L.T. 142 (B.C.S.C.), aff'd (1990), 2 C.C.L.T. (2d) 42 (B.C.C.A.). The court stated the physician must provide adequate time for the explanation and to respond to the patient's questions.

Coulter A (1997) Partnerships with patients: The pros and cons of shared clinical decision-making. *J Health Serv Res Policy* **2**(2):112–121.

Coulter A, Entwistle V, Gilbert D (1999) Sharing decisions with the patient: Is the information good enough? BMJ **318**:318–322.

Council of Europe (1997) Convention for the protection of human rights and dignity of the human being with regard to the application of biology and medicine: Convention on Human Rights and Biomedicine. Available at <http://conventions.coe.int/treaty/en/Reports/Html/164.htm>. Accessed August 30, 2004.

Council on Ethical and Judicial Affairs, American Medical Association (1994) Ethical issues related to prenatal genetic testing, *Arch Family Med* **3**:633–642.

Cox B (1992) Genetic tests become next underwriting frontier, *Natl Underwriter* July 27:3,6.

Coyne J, Benazon N, Gaba C, Calzone K, Weber B (2000) Distress and psychiatric morbidity among women from high-risk breast and ovarian cancer families. *J Consult Clin Psychol* **68**(5):864–874.

Coyne J, Kruus L, Raciioppo M, Calzone K, Armstrong K (2003) What do ratings of cancer-specific distress mean among women at high risk of breast and ovarian cancer? *Am J Med Gen* **116A**:222–228.

Craddock N, Jones I (1999) Genetics of bipolar disorder. *J Med Genet* **36**(8):585–594.

Crandall BF, Lebherz TB, Schroth PC, Matsumoto M (1983) Alpha-fetoprotein concentrations in maternal serum: Relation to race and body weight. *Clin Chem* **29**:531–533.

Crawford DC, Acuna JM, Sherman SL (2001) FMR1 and the fragile X syndrome: Human genome epidemiology review. *Genet Med* **3**(5):359–371.

Creason v. Department of Health Services (1998) 957 P.2d 1323. At trial, evidence indicated that there were two components of a congenital hypothyroidism test: thyroid stimulation hormone levels and thyroxin 4 (T4) factor. The plaintiff argued that a test result was positive when both levels were low. The defendant state argued that positive test results were reported only when the thyroid stimulation hormone was high and the T4 factor was low. The action was dismissed. The court ruled that the state had only a discretionary, rather than a mandatory, duty under the relevant legislation to formulate appropriate standards for testing and reporting test results.

Crichton v. Hastings (1972) O.R. 859, 29 D.L.R. (3d) 692; aff'g [1972] 3 O.R. 860, 29 D.L.R. (3d) 693 (C.A.).

Crits v. Sylvester (1956) O.R. 132, 1 D.L.R. (2d) 502; reversing in part, [1955] O.R. 332; 3 D.L.R. 181; aff'd [1956] S.C.R. 991.

Crossley JA, Aitken DA, Cameron AD, McBride E, Connor JM (2002) Combined ultrasound and biochemical screening for Down's syndrome in the first trimester: A Scottish multicentre study. *BJOG* **109**:667–676.

Croyle RT, Lerman C (1999) Risk communication in genetic testing for cancer susceptibility. *J Nat'l Ca Inst Monographs* **25**:59–66.

Croyle RT, Smith K, Botkin JR, Baty B, Nash J (1997) Psychological responses to BRCA1 mutation testing. *Health Psychol* **16**:63–72.

Crundwell v. Becker (1998) 981 S.W. 2d 880 (Tex. App. Houston 1st Dist. reh'g overruled, (Jan. 11, 1999), review denied (March 25, 1999). The court held that where a patient signs a written consent to surgery that lists the risk of surgery and the signing of the form is witnessed by a credible person, the "presumption" is that the physician has obtained the patient's informed consent.

Csaba A, Papp Z (2003) Ethical dimensions of genetic counselling. *Clin Perinatol* **30**:81–93.

Cubells JF, Amin F, Asherson P, Banaschewski T, Brandt H, Berrettini W, Bierut L, Black D, Cloninger CR, Coryell W, Crowe R, Crowley T, Devlin B, El-Mallakh RS, Faraone SV, Fichter MM, Freedman R, Gelernter J, Gershon E, Geschwind D, Kaplan A, Kreek MJ, Lachman H, Levinson D, Li MD, Madden P, Murphy DL, Nimgaonkar V, Nurnberger JI Jr, Pollice C, Pomerleau OF, Rice J, Silverman JM, Smith C, Sprock J, Sullivan B, Sutcliffe J, Swan GE, Tischfield JA, Tsuang M, Vanyukov M, Yu L (2004) Unpublished communication.

Cuckle H (1999) Down syndrome fetal loss rate in early pregnancy. *Prenat Diagn* **19**:1177–1179.

Cuckle H, Arbuzova S (2002) The efficiency and clinical practicality of multi-modality screening strategies. In: Macek M, Bianchi D, Cuckle H, eds. *Early Prenatal Diagnosis, Fetal Cells and DNA in the Mother: Proceedings of the 12th Fetal Cell Workshop*. Prague:Charles University Press.

Cuckle HS, van Lith JM (1999) Appropriate biochemical parameters in first-trimester screening for Down syndrome. *Prenat Diagn* **19**:505–512.

Cuckle HS, Wald NJ, Lindenbaum RH, Jonasson J (1985) Amniotic fluid AFP levels and Down syndrome. *Lancet* **1**:290–291.

Cuckle HS, Wald NJ, Thompson SG (1987) Estimating a woman's risk of having a pregnancy associated with Down's syndrome using her age and serum alpha-fetoprotein level. *Br J Obstet Gynaecol* **94**:387–402.

Cuckle HS, Holding S, Jones R, Groome NP, Wallace EM (1996) Combining inhibin A with existing second-trimester markers in maternal serum screening for Down's syndrome. *Prenat Diagn* **16**:1095–1100.

Cullen MT, Gabrielli S, Green JJ, Rizzo N, Mahoney MJ, Salafia C, Bovicelli L, Hobbins JC (1990) Diagnosis and significance of cystic hygroma in the first trimester. *Prenat Diagn* **10**:643–651.

Cunningham v. Helping Hands, Inc. (2001) 346 S.C. 253, 550 S.E.2d 872 (Ct. App.) certiorari. Opinion No. 25575; affirmed, as modified (2003).

Curlender v. Bio-Science Laboratories (1980) 165 Cal.Rptr. 477 106 call App. 811 (Ct. App.). The defendant laboratory was hired by plaintiff parents to conduct genetic test for Tay Sachs disease. False-negative results were given, and the plaintiff child was born with Tay Sachs. The court allowed the claim for wrongful life on the basis that the child had a right to recover damages for pain and suffering endured during the span of the child's life as opposed to the life span if the child had been born without Tay Sachs disease.

Curtis RE, Boice JD, Shriner DA, Hankey BF, Fraumeni JF Jr (1996) Second cancers after adjuvant tamoxifen therapy for breast cancer. *J Nat Cancer Inst* **88**:832–833.

Cutillo D, Ramsey D, Cohen A, McCann CL, Hedrick J, Forley C, Nogel HL (2002) Risk factors identified through genetic counseling: The value of a comprehensive risk assessment. *Annu Clin Genet* Meeting Program and Abst **50**:

d'Agincourt-Canning L (2001) Experiences of genetic risk: Disclosure and the gendering of responsibility. *Bioethics* **15**(3):231–247.

Dale v. Munthali (1978) 21 O.R. (2d) 554, 90 D.L.R. (3d) 763, 2 L. Med. Q. 231 (Ont. C.A.). The patient was suffering from the sudden onset of extreme headaches. The physician was found negligent in failing to suspect meningitis and in not questioning sufficiently into the previous high fever that had been reported. The court held that the physician should have called for further tests that were available. In *Meyer v. Gordon* (1981) 17 C.C.L.T. 1, an infant was delivered very rapidly. Her lungs became congested and the flow of oxygen to her body impeded, resulting in brain damage. The hospital was found negligent because its staff failed to take the routine history from the mother that would have revealed that her previous delivery had also been a rapid one. Had they done so, the mother could have been kept under closer observation and the staff would have been able to relieve the child's distress.

Daley S (2002) France bans damages for "wrongful births." *New York Times*, January 11.

Dallaire L, Lortie G (1993) Parental reaction and adaptability to the prenatal diagnosis of genetic disease leading to pregnancy termination. In: *Research Volumes of the Royal Commission on New Reproductive Technologies*, Ottawa, Minister of Government Services 1993.

Darling v. Charleston Community Memorial Hospital (1965) 33 Ill.2d 326, 211 N.E.2d 253, 14 A.L.R.3d 860 (Ill. Sep 29). http://biotech.law.lsu.edu/cases/Medmal/darling.htm.

Davies JC (2004) Modifier genes in cystic fibrosis. *Pediatr Pulmonol Suppl* **26**:86–87.

Davis DS (1997) Genetic dilemmas and the child's right to an open future. *Hastings Center Report* **2**:7–15.

Davis DS (1998) Discovery of children's carrier status for recessive genetic disease: Some ethical issues. *Genet Test* **2**(4):323–327.

Davis v. Patel (2001) 287 AD2d 479, 480 (2d Dept.).

Davis RO, Cosper P, Huddleston JF, Bradley EL, Finley SC, Finley WH, Milunsky A (1985) Decreased levels of amniotic fluid alpha-fetoprotein associated with Down syndrome. *Am J Obstet Gynecol* **153**:541–544.

Dawes PJ, O'Keefe DL, Adcock S (1993) Informed consent: Using a structured interview changes patients' attitudes towards informed consent. *J Laryngol Otol* **107**(9):775–779.

Daya S (1993) Accuracy of gestational age estimation by means of fetal crown-rump length measurement. *Am J Obstet Gynecol* **168**:903–908.

Dean JCS, FitzPatrick DR, Farndon PA, Kingston H, Cusine D (2000). Genetic Registers in Clinical Practice—A survey of UK clinical geneticists. *J Med Genet* **37**:636–640.

De Biasio P, Venturini PL (2002) Absence of nasal bone and detection of trisomy 21. *Lancet* **359**:1344.

Decruyenaere M, Evers-Kiebooms G, Boogaerts A, Cloostermans T, Cassiman J, Demyttenaere K, Dom R, Fryns J, Van den Berghe H (1997) Non-participation in predictive testing for Huntington's disease: Individual decision-making, personality and avoidant behaviour in the family. *Euro J Hum Gen* **5**:351–363.

Decruyenaere M, Evers-Kiebooms G, Denayer L, Welkenhuysen M, Claes E, Legius E, Demyttenaere K (2000) Predictive testing for hereditary breast and ovarian cancer: A psychological framework for pre-test counseling. *Euro J Hum Gen* **8**:130–136.

De Dinechin O, Harris R, Kettner M, Koch L, Zwierlein E (1993) Workshop of the Commission of the European Communities on Ethics of Human Genome analysis: Survey of the European discussion. *J Med Genet* **30**:257–260, at 259. The paper described nondirective counseling as the "gold standard" of genetic counseling.

de la Chappelle A (2004) Genetic predisposition to colorectal cancer. *Nature Rev Cancer* 4:769–780.

Delaporte C (1996) Ways of announcing a late-onset, heritable, disabling disease and their psychological consequences. *Genet Couns* 7:289–296.

DeMeo DL, Silverman EK (2004) Alpha1-antitrypsin deficiency. 2: Genetic aspects of alpha(1)-antitrypsin deficiency: Phenotypes and genetic modifiers of emphysema risk. *Thorax* **59**:259–264.

den Dunnen JT, Antonarakis SE (2000) Mutation nomenclature extensions and suggestions to describe complex mutations: A discussion. *Hum Mutat* **15**:7–12.

den Dunnen JT, Antonarakis SE (2001) Nomenclature for the description of human sequence variations. *Hum Genet* **109**:121–124.

Department of Health and Human Services (2004) Best practices for the licensing of genomic inventions. *Fed Reg* **69**(233): Friday November 19, FR Doc. 04–25647, online: GPO. <http://a257.g.akamaitech.net/7/257/2422/06jun20041800/edocket.access.gpo.gov/2004/04-25671.htm>.

Dequeker E, Ramsden S, Grody WW, Stenzel TT, Barton DE (2001) Quality control in molecular genetic testing. *Nat Rev Genet* **2**(9):717–723.

Deren O, Mahoney MJ, Copel JA, Bahado-Singh RO (1998) Subtle ultrasonographic anomalies: Do they improve the Down syndrome detection rate? *Am J Obstet Gynecol* **178**:441–445.

Derogatis LR (1993) *The Brief Symptom Inventory (BSI): Administration, Scoring and Procedures Manual*, 3rd ed. Minneapolis, MN: *National Computer Systems*.

Derrick JH (2002) Annotation, medical malpractice: Liability for failure of physician to inform patient of alternative modes of diagnosis or treatment, 38 A.L.R.4th 900 (1985 & Supp. 2002).

Descant v. Administrators of the Tulane Educational Fund (1988) 639 So.2d. 246; 706 So.2d. 618 (La. App. 4th Cir.). "To establish consent to a risk, it must be shown both that the patient was aware of the risk and that she consented to encounter it. Therefore, it is obvious that a risk must have been understandably communicated before the element of awareness can be established. A physician's duty to disclose material information, including reasonable alternative therapy must be communicated in terms that a reasonable doctor would believe a reasonable patient in Mrs. Descant's position would understand. Technical language should not be used to inform an untutored lay person in order for a reasonable patient to have awareness of a risk."

deSerres FJ (2003) Alpha-1 antitrypsin deficiency is not a rare disease but a disease that is rarely diagnosed. *Environ Health Perspect* **111**:1851–1854.

Desgeorges M, Kjellberg P, Demaille J, Claustres M (1994) A healthy male with compound and double heterozygosities for DeltaF508, F508C, and M47OV in exon 10 of the cystic fibrosis gene. *Am J Hum Genet* **54**:384–385.

DeVaro E (1998) Consideration of context in the case of disability rights activism and selective abortion. *Health L Rev* **6**:12–19.

Devi G, Ottman R, Tang MX, Marder K, Stern Y, Mayeux R (2000) Familial aggregation of Alzheimer disease among whites, African Americans, and Caribbean Hispanics in northern Manhattan. *Arch Neurol* **57**:72–77.

Devine PC, Malone FD (1999) First trimester screening for structural abnormalities: Nuchal translucency sonography. *Semin Perinatol* **23**:382–392.

Dewald GW (2002) Cytogenetic and FISH studies in myelodysplasia, acute myeloid leukemia, chronic lymphocytic leukemia and lymphoma. *Int J Hematol* **76** (Suppl 2):65–74.

Dewald GW, et al (2002) Cytogenetic studies in neoplastic hematologic disorders. In: McClatchey, KD, ed. *clinical Laboratory Medicine*. Philadelphia: *Lippincott Williams & Wilkins*.

De Wert G (1992) Predictive testing for Huntington disease and the right not to know; Some ethical reflections. In: Evers-Kiebooms G, Fryns JP, Cassiman JJ, Van den Berghe H, eds. *Psychological Aspects of Genetic Counseling*. New York: Wiley-Liss, for the March of Dimes Birth Defects Foundation. *Birth Defects: Original Articles Series* **28**(1):133–138. Also see Ost D (1984) The right not to know. *J Med Philos* **9**:310–312.

Di Marchi B, Ravetz J (1999) Risk management and governance: A post-normal science approach. *Futures* **31**:743–757.

Diabetes Prevention Program Research Group (2002) Reduction in the incidence of type 2 diabetes with lifestyle intervention or metformin. *N Engl J Med* **346**:393–403.

Diack v. Bardsley (1983) 25 C.C.L.T. 159 (B.C.S.C.); 46 B.C.L.R. 240 (S.Ct); aff'd (1984) 31 C.C.L.T. 388 (B.C.C.A.).

Diagnostic and Statistical Manual of Mental Disorders (2000) 4th ed Text Revised. Washington, D.C: American Psychiatric Association.

Dichter MA, Buchhalter JR (2003) The genetic epilepsies. In: Rosenberg RN, Prusiner SB, DiMauro S, Barchi RL, Nestler EJ, eds. *The Molecular and Genetic Basis of Neurologic and Psychiatric Disease*, Philadelphia: Elsevier, pp. 399–419.

Dick DM, Foroud T, Flury L, Bowman ES, Miller MJ, Rau NL, Moe PR, Samavedy N, El-Mallakh R, Manji H, Glitz DA, Meyer ET, Smiley C, Hahn R, Widmark C, McKinney R, Sutton L, Ballas C, Grice D, Berrettini W, Byerley W, Coryell W, DePaulo R, MacKinnon DF, Gershon ES, Kelsoe JR, McMahon FJ, McInnis M, Murphy DL, Reich T, Scheftner W, Nurnberger JI Jr (2003) Genomewide linkage analyses of bipolar disorder: A new sample of 250 pedigrees from the National Institute of Mental Health Genetics Initiative. *Am J Hum Genet* **73**(1):107–114.

Dick DM, Li TK, Edenberg HJ, Hesselbrock V, Kramer J, Kuperman S, Porjesz B, Bucholz K, Goate A, Nurnberger J, Foroud T (2004) A genome-wide screen for genes influencing conduct disorder. *Mol. Psychiatry* **9**(1):81–86.

Dickens BM (1985) The doctrine of "informed consent". In: Abella RS, Rothman ML, eds. *Informed Choice in Medical Care Justice Beyond Orwell*. Montreal: Les Editions Yvon Blais Inc, Canadian Institute for the Administration of Justice, p. 243.

Dickens BM (2002) Informed consent. In: Downie J, Caulfield T, Flood C, eds, *Canadian Health Law and Policy*, 2nd ed. Butterworths, Markham pp. 129–156.

Didato v. Strehler (2001) 262 Va. 617. 554 SE 2d.42.

Dillard JP, Carson CL, Bernard CJ, Laxova A, Farrell PM (2000) An analysis of communication following newborn screening for cystic fibrosis. *Psychol Health* **15**(1):1–12.

Dimopoulos P, Bagaric M (2003) The moral status of wrongful life claims. *Common Law World Rev* **32**:35–64.

Dipple KM, McCabe ERB (2000) Phenotypes of patients with "simple" mendelian disorders are complex traits: Thresholds, modifiers, and systems dynamics. *Am J Hum Genet* **66**:1729–1735.

Ditto v. McCurdy (1997) 86 Haw. 93, 947 P. 2d 961 (Haw. Ct. App.), as amended (June 20, 1997), cert. Granted, (July 3, 1997), aff'd in part, rev'd in part on other grounds, 86 Haw. 84, 947 P. 2d 952 (Haw. 1997).

Diver CS, Cohen JM (2001) Genophobia: What is wrong with genetic discrimination? *Univ Penn Law Rev* **149**:1439–1482.

Djurdjinovic L (1998) Psychosocial counseling. In: Baker DL, Schuette JL, Uhlmann WR, eds. *A Guide to Genetic Counseling*. New York: Wiley-Liss, pp. 127–166.

Doe v. Roe (1977) 400 N.Y.Supp.2d 668.

Dolan SM (2004) New approaches to screening for Down syndrome. *Medscape Ob/Gyn Women's Health* **9**(1). Online.

Doll JJ (1998) The patenting of DNA. *Science* **280**:689–690.

Donat R, McNeil AS, Fitzpatrick DR, Hargreave TB (1997) The incidence of cystic fibrosis gene mutations in patients with congenital bilateral absence of the vas deferens in Scotland, *Br J Urol* **79**:74–77.

Done ML, Ellingham JG, Faunce TA (1996) Disclosure of material risks. The practice of Australian anaesthetists subsequent to Rogers v Whitaker. *Aust Anaesth* **69**.

Dorval M, Farkas Patenaude A, Schneider KA, Kieffer SA, DiGianni L, Kalkbrenner KJ, Bromberg JI, Basili LA, Calzone K, Stopfer J, Weber BL, Garber JE (2000) Anticipated versus actual emotional reactions to disclosure of results of genetic tests for cancer susceptibility: Findings from p53 and BRCA1 testing programs. *J Clin Oncol* **18**:2135–2142.

Dorval M, Gauthier G, Maunsell E, Simard J, members of the INHERIT BRCAs (2003) Are women with an inconclusive BRCA1/2 genetic test result falsely reassured? *Psycho-Oncol* **12**(4 suppl):166.

Doust J, del Mar C (2004) Why do doctors use treatments that do not work? *BMJ* **328**:474–475.

Down JL (1866) Observations on an ethnic classification of idiots. *Clin Lecture Reports, London Hospital* **3**:259.

Dresser R (1994) Freedom of conscience, professional responsibility and access to a bortion. *J Law Med Ethics* **22**:280–285.

Driscoll DA (2004) ACMG policy statement: Second trimester maternal serum screening for fetal open neural tube defects. *Genet Med* **6**(6):in press.

D'Souza G, McCann CL, Hedvick J, Fairley C, Nagel HL, Kushner JD, Kessel R (2000) Tay-Sachs disease carrier screening: A 21-year experience. *Genet Test* **4**(3):257–263.

DuBose ER (1995) The fiduciary focus of the medical province of meaning. In: *The Illusion of Trust*. Boston: Kluwer Academic, at 64–80.

Dubovsky J, Weber JL, Orr HT, et al (1996) A second gene for familial febrile convulsions maps on chromosome 19p. *Am J Hum Genet* **59** (suppl. 1):A223.

DudokdeWit AC (1997) Psychological distress in applicants for predictive DNA testing of autosomal dominant, heritable and late onset disorders. The Rotterdam/Lieden Genetics Workgroup. *J Med Genet* **35**:382–390.

DudokdeWit AC, Tibben A, Frets PG, Meijers-Heijboer EJ, Devilee P, Klijn JG, Oosterwijk JC, Niermeijer MF (1997) BRCA1 in the family: A case description of the psychological implications. *Am J Med Genet* **71**(1):63–71.

DudokedeWit AC, Tibben A, Dulvenvoorden HJ, Niermeijer MF, Passchier J, Trijsburg RW, The Rotterdam/Leiden Genetics Workgroup (1998) Distress in individuals facing predictive DNA testing for autosomal dominant late-onset disorders: Comparing questionnaire results with in-depth interviews. *Am J Med Genet*, **75**:62–74.

Dugan RB, Wiesner GL, Juengst ET, O'Riordan M, Matthews AL, et al (2003) Duty to warn at-risk relatives for genetic disease: Genetic counselors' clinical experience. *Am J Med Genet* **119C**:27–34.

Dumars KW, et al (1996) Practical guide to the diagnosis of thalassemia. Council of Regional Networks for Genetic Services (CORN). *Am J Med Genet* **62**(1):29–37.

Duncan RE (2004) Predictive genetic testing in young people: when is it appropriate? *J Paediatr Child Health* **40**(11):593–5.

Dunne C, Warren C (1998) Legal autonomy: The malfunction of the informed consent mechanism within the context of prenatal diagnosis of genetic variants. *Issues Law Med* **14**(2):1870–2000, at pages 195–96 citing the dissent in *Berman v. Allen* (1979) 80 N.J. 421, 429, 404 A.

2d 8, where the dissenting judge used terms such as "gravely handcapped," "tragedy," and "misfortune" to describe a child both with Down syndrome.

Duplan v. Harper (1999) 188 F.3d 1195, 1198 (10th Cir.). http://caselaw.lp.findlaw.com/scripts/getcase.pl?court=10th&navby=case&no=976344&exact=1.

Dupont WD, Page DL, Rogers LW, Parl FF (1989) Influence of exogenous estrogens, proliferative breast disease, and other variables on breast cancer risk. *Cancer* **63**:948–957.

Dupont WD, Page DL, Pari FF (1999) Estrogen replacement therapy in women with a history of proliferative breast disease. *Cancer* **85**:1277–1283.

Durfy SJ, Buchanan TE, Burke W (1998) Testing for inherited susceptibility to breast cancer: A survey of informed consent forms for BRCA1 and BRCA2 mutation testing. *Am J Med Genet* **75**(1):82–87.

Durfy SJ, Bowen DJ, McTiernan A, Sporleder J, Burke W (1999) Attitudes and interests in genetic testing for breast and ovarian cancer susceptibility in diverse groups of women in western Washington. *Ca Epi Bio Prev* **8**:369–375.

Durner M, Keddache MA, Tomasini L, Shinnar S, Resor SR, Cohen J, Harden C, Moshe SL, Rosenbaum D, Kang H, Ballaban-Gil K, Hertz S, Labar DR, Luciano D, Wallace S, Yohai D, Klotz I, Dicker E, Greenberg DA (2001) Genome scan of idiopathic generalized epilepsy: Evidence for major susceptibility gene and modifying genes influencing the seizure type. *Ann Neurol* **49**:328–335.

Dusseault JH, Laberge C (1973). Thyroxine (T4) determination in dried blood by radioimmunoassay: A screening method for neonatal hypothyroidism. *Union Med Can* **102**:2062–2064.

Duster T (2003) *Backdoor to Eugenics*, 2nd ed. New York: Routledge.

Duttry v. Patterson (2001) 771 A.2d 1255, 1258–59 (Pa.).

Early Breast Cancer Trialists' Collaborative Group (1998) *Lancet* **351**:1451–1467.

Easton DF, Ford D, Bishop DT (1995) Breast and ovarian cancer incidence in BRCA1-mutation carriers. *Am J Hum Genet* **56**:265–271.

Edwards AGK, Pill RM, Stott, NCH (1996) Communicating risk: Use of standard terms is unlikely to result in standard communication. *BMJ* **313**:1483.

Edwards AGK, Hood K, Matthews EJ, Russell D, Russell IT, Barker J, et al (2000) The effectiveness of one-to-one risk communication interventions in health care: A systematic review. *Med Decis Making* **20**:290–297.

Edwards A, Elwyn G, Covey J, Matthews E, Pill R (2001) Presenting risk information—a review of the effects of "framing" and other manipulations on patient outcomes. *J Health Commun* **6**:61–82: A treatment that "saves eight lives out of 10" is perceived to be better than one that "fails to save two in every 10." Also see Greenhalgh T, Kostopoulou O, Harries C (2004) Making decisions about benefits and harms of medicines. *BMJ* **329**:47–50.

Edwards A, Elwyn G, Mulley A (2002) Explaining risks: Turning numerical data into meaningful pictures. *BMJ* **324**(7341):827–830.

Eerola H, Pukkala E, Pyrhonen S, Blomqvist C, Sankila R, Nevanlinna H (2001) Risk of cancer in BRCA1 and BRCA2 mutation-positive and -negative breast cancer families (Finland). *Cancer Causes Control* **8**:739–746.

Egeland JA, Gerhard DS, Pauls DL, Sussex JN, Kidd KK, Allen CR, Hostetter AM, Housman DE (1987) Bipolar affective disorders linked to DNA markers on chromosome 11. *Nature* **325**(6107):783–787.

Eikelboom JW, Lonn E, Genest J, Hankey G, Yusuf S (1999) Homocyste(e)ine and cardiovascular disease: A critical review of the epidemiologic evidence. *Ann Intern Med* **131**:363–375.

Eisbrenner v. Stanley (1981) 106 Mich. App. 357, 308 N.W.2d 209.

Eisen A, Rebbeck TR, Wood WC, Weber BL (2000) Prophylactic surgery in women with a hereditary predisposition to breast and ovarian cancer. *J Clin Oncol* **18**:1980–1995.

Eisen A, Weber BL (2003) Prophylactic mastectomy for women with BRCA1 and BRCA2 mutations – facts and controversy. *N Engl J Med* 345:207–8.

Elderkin-Thompson V, Silver RC, Waitzkin H (2001) When nurses double as interpreters: A study of Spanish-speaking patients in a US primary care setting. *Soc Sci Med* **52**(9):1343–1358.

Elias S, Annas G (1994) Generic consent for genetic screening. *New Engl J Med* **330**:1611.

Elit L (2001) Familial ovarian cancer. *Can Fam Physician* **47**:778–784.

Elit L, Esplen MJ, Butler K, Narod S (2001) Quality of life and psychosexual adjustment after prophylactic oophorectomy for a family history of ovarian cancer. *Familial Ca* **1**:149–156.

Ellis v. Sherman (1986) 515 A. 2d. 1327 (Pa.).

El-Tamer M, Russo D, Troxel A, Bernardino LP, Mazziotta R, Estabrook A, Ditkoff BA, Schnabel F, Mansukhani M (2004) Survival and recurrence after breast cancer in BRCA1/2 mutation carriers. *Ann Surg Oncol* **11**:157–164.

Elting LS, Martin CG, Cantor SB, Rubenstein EB (1999) Influence of data display formats on physician investigators' decisions to stop clinical trials: Prospective trial with repeated measures. *BMJ* **318**:1527–1531.

Emery AEH (1984) In: Emery AEH, Pullen I, eds. *Psychological Aspects of Genetic Counseling*, London: Academic at page 4. "There is often a gap between the counsellor's expectations, and the actual consequences of counselling. For example, in a two-year follow-up study of 200 consecutive couples seen in a genetic counselling clinic, over a third of those who were told they were at high risk of having a child with a serious genetic disease were undeterred and actually planned further pregnancies. In the past, such behaviour has often been regarded as 'irresponsible', a failure on the part of the counsellor and an indictment of counselling in general. But when couples in this study were carefully questioned their reasons for planning further children were often very understandable . . . Thus, a course of action which might seem irresponsible to one person may be eminently reasonable to another. In a free society this choice is the individual's prerogative provided it is made in full knowledge of all the facts and appreciation of the possible consequences. Since the counsellor's role is to help couples reach decisions which are the best for themselves, genetic counselling should always be non-directive." Also see *Royal Commission* (1993) infra, Vol. 2, Ottawa: Minister of Government Services, 1993. As to the issue of prenatal diagnosis, the Commission reported [at pp. 766–767] that "disturbing proportion of referring physicians who do not accept the principle that patients should make their own informed choice about whether to have PND and when to have an abortion after the diagnosis of a fetal disorder . . . This is of great concern to the Commissioners, respect for the pregnant woman's autonomy requires that it be her values and priorities, not the doctor's, that determine her decision to accept or decline PND testing."

Emery J (2001) Is informed choice in genetic testing a different breed of informed decision-making? A discussion paper. *Health Expect* **4**(2):81–86.

Eng C, Brody LC, Wagner TM, Devilee P, Vijg J, Szabo C, Tavtigian SV, Nathanson KL, Ostrander E, Frank TS (2001) Steering Committee of the Breast Cancer Information Core (BIC) Consortium. Interpreting epidemiological research: Blinded comparison of methods used to estimate the prevalence of inherited mutations in BRCA1. *J Med Genet* **38**(12):824–833.

Engel GL (1991) The need for a new medical model. In: Caplan A, ed. *Concepts of Health and Disease: Interdisciplinary Perspectives*. Reading, MA: Addison-Wesley, p. 591.

Englehardt HT (1986) *The Foundation of Bioethics*. New York: Oxford University Press.

English DC (2002) Valid informed consent: A process, not a signature. *Am Surg* **68**(1):45–48.

Enriquez J (2001) *As the Future Catches You*. New York: Crown Business.

Ensenauer RE, et al (in press) Genetic testing: Practical, ethical, and counseling considerations. *Mayo Clinic Proceedings*.

Epps PG (2003) Policy before practice: Genetic discrimination reviewed. *Am J Pharmacogenomics* **3**:405–418.

Epstein RM, Morse DS, Frankel RM, Frarey L, Anderson K, Beckman HB (1998) Awkward moments in patient-physician communication about HIV risk. *Ann Intern Med* **128**(6):435–442.

Eramian D (2004) Patents save lives, speech to the Global Public Policy Institute and Ecole de Science Politique Paris, June 24, 2004, online: BIO <http://www.bio.org/speeches/speeches/20040624.asp>.

Erde EL (1999) Informed consent to Septoplasty: An anecdote from the field. *J Med Philosophy* **24**(1):11–17.

Erickson v. Walter (1977) 116 Ariz. 476, 569 P. 2d. 1374 (App.) The physician delayed two and a half days before referring the patient whose hand required immediate surgery following an industrial accident.

Escayg A, MacDonald BT, Meisler MH, Baulac S, Huberfeld G, An-Gourfinkel I, Brice A, LeGuern E, Moulard B, Chaigne D, Buresi C, Malafosse A (2000) Mutations of SCN1A, encoding a neuronal sodium channel, in two families with GEFS+2. *Nat Genet* **24**:343–345.

Escher M, Sappino AP (2000) Primary care physicians' knowledge and attitudes towards genetic testing for breast-ovarian cancer predisposition. *Ann Oncol* **11**(9):1131–1135.

ESHRE Ethics Task Force, Shenfield F, Pennings G, Devroey P, Sureau C, Tarlatzis B, Cohen J (2003) Taskforce 5: Preimplantation genetic diagnosis. *Hum Reprod* **18**(3):649–651.

Esplen MJ, Hunter J (2002) Grief in women with a family history of breast cancer primary psychiatry: Cancer care therapy. *Psychosoc Issues Cancer Genet* **9**(5):57–64.

Esplen MJ, Toner B, Hunter J, Glendon G, Liede A, Narod S, Stuckless N, Butler K, Field B (2000) A supportive-expressive group intervention for women with a family history of breast cancer: Results of a Phase II trial. *Psycho-Onc* **9**:243–252.

Esplen MJ, Madlensky L, Butler K, McKinnon W, Bapat B, Wong J, Aronson M, Gallinger S (2001) Motivations and psychosocial impact of genetic testing for HNPCC. *Am J Med Genet* **103**(1):9–15.

Esplen MJ, Urquhart C, Butler K, Gallinger S, Aronson M, Wong J (2003a) The experience of loss and anticipation of distress in colorectal cancer patients undergoing genetic testing. *J Psychosom Rsch* **55**(5):427–435.

Esplen MJ, Ardiles P, Stuckless N, Hunter J, Glendon G, Liede A, Metcalfe K, Aronson M, Butler K, Rothenmund H, Berk T, Lefebvre A (2003b) Self-concept instrument development among individuals testing positive for genetic mutations in cancer. *Psycho-Onc* **12**(4)Supp:S230.

Esplen MJ, Hunter J, Leszcz M, Warner E, Narod S, Metcalfe K, Glendon G, Butler K, Liede A, Young MA, Kieffer S, DiProspero L, Irwin E (2004) A multi-centre Phase II study of supportive-expressive group therapy for women with BRCA1 and BRCA2 mutations. *Cancer* **101**(10):2327–2340.

Esplen MJ, Madlensky L, Aronson M, Lesczc, Gallinger (in press) A supportive-expressive group therapy intervention for hereditary non-polyposis colorectal cancer.

Estate of Tranor v. Bloomsburg Hosp (1999) 60 F. Supp. 2d. 412 (M.D. Pa.).

ETC Group (1999) The gene giants: Masters of the universe? Communique, March 30, 1999, online <www.etcgroup.org/article.asp?newsid=180>.

Ethics and genetics in medicine. *CMAJ* **158**(10):1313.

Etzkowitz H, Leydesdorff L, eds. (2001) *Universities and the Global Knowledge Economy*. New York: Continuum.

European Community Respiratory Health Survey Group (1997) Genes for asthma? An analysis of the European Community Respiratory Health Survey. *Am J Respir Crit Care Med* **156**:1773–1780.

European Commission Expert Group on the Ethical, Legal, and Social Aspects of Genetic Testing (2004) 25 recommendations on the ethical, legal and social implications of genetic testing. http://europa.eu.int/comm/research/conferences/2004/genetic/pdf/recommendations_en.pdf.

European Directory of DNA Diagnostic Laboratories (EDDNAL) (2002–2004) Centre de Genetique Humaine, Institut de Pathologie et de Genetique (Loverval, Belgium). http://www.eddnal.com.

European Society of Human Reproduction and Embryology (ESHRE) PGD Consortium Steering Committee (2002) ESHRE Preimplantation Genetic Diagnosis Consortium: Data Collection III. *Hum Reprod* **17**:233–246.

Evans W, Relling MV, R. (2004) Moving towards individualized medicine with pharmacogenomics. *Nature* **429**:464–468.

Evans D, Maher E, MacLead R, Davies DR, Craufurd D (1997) Uptake of genetic testing for cancer predisposition—ethical issues. *J Med Gen* **34**:746–749.

Evers-Kiebooms G, Welkenhuysen M, Claes E, Decruyenaere M, Denayer L (2000) The psychological complexity of predictive testing for late onset neurogenetic diseases and hereditary cancers: Implications for multidisciplinary counselling and for genetic education. *Soc Sci Med* **51**:831–841.

Expert Panel on Detection, Evaluation, and Treatment of High Blood Cholesterol in Adults (2001). Executuve summary of the third report of the national cholesterol education program (NCEP) expert panel on detection, evaluation, and treatment of high blood cholesterol in adults (adult treatment panel III). *JAMA* **285**:2486–2497.

Eysenbach G, Diepgen TL (1998) Responses to unsolicited patient e-mail requests for medical advice on the world wide web. *JAMA* **280**:1333–1335.

Falek A (1984) Sequential aspects of coping and other issues is decision making in genetic counselling. In: Emery AEH, Pullen I, eds. *Psychological Aspects of Genetic Counselling*. London: Academic, pp. 23–36.

Falk MJ, Dugan RB, O'Riordan MA, Matthews AL, Robin NH (2003) Medical geneticists' duty to warn at-risk relatives for genetic disease. *Am J Med Genet* **120A**(3):374–380.

Fallowfield L, Hall A, Maguire G, Baum M (1990) Psychological outcomes of different treatment policies in women with early breast cancer outside a clinical trial. *BMJ* **301**:575–580.

Famy C, Streissguth A, Unis A (1998) Mental illness in adults with fetal alcohol syndrome or fetal alcohol effects. *Am J Psychiatry* **155**:552–554.

Fanos JH, Johnson JP (1995) Perception of carrier status by cystic fibrosis siblings. *Am J Hum Genet* **57**(2):431–438.

Faraone SV, Tsuang MT, Tsuang DW (1999) *Genetics of Mental Disorders*. New York: Guilford.

Farid C (1997) Access to Abortion in Ontario: From Morgentaler 1988 to the Savings and Restructuring Act. *Health Law Journal* **5:**119–145.

Farrell M, Certain L, Farrell P (2001) Genetic counseling and risk communication services of newborn screening programs. *Arch Pediatr Adolesc Med* **155**(2):120–126.

Farrer LA (2000) Familial risk for Alzheimer's disease in ethnic minorities: Nondiscriminating genes. *Arch Neurol* **57**:28–29.

Farrer LA, Conneally PM (1987) Predictability of phenotype in Huntington's disease. *Arch Neurol* **44**(1):109–113.

Farrer LA, Cupples LA, Haines JL, Hyman B, Kukull WA, Mayeux R, Myers RH, Pericak-Vance MA, Risch N, van Duijn CM (1997) Effects of age, sex, and ethnicity on the association between apolipoprotein E genotype and Alzheimer's disease. *JAMA* **278**:1349–1356.

Farrer M, Maraganore DM, Lockhart P, Singleton A, Lesnick TG, de Andrade M, West A, de Silva R, Hardy J, Hernandez D (2001) Alpha-synuclein gene haplotypes are associated with Parkinson's disease. *Hum Mol Genet* **10**(17):1847–1851.

Fearnhead NS, Britton MP, Bodmer WF (2001) The ABC or APC. *Hum Molec Genet* **10**:721–733.

Feinberg KS, Peters JD, Willson JR (1984) *Obstetrics/Gynecology and the Law*. Ann Arbor, MI: Health Administration Press.

Felde v. Vein & Laser Medical Centre (2002) 14 C.C.L.T. (3d) 246: "It is clear that the time which Dr. Castillo spent with Ms Felde prior to the surgery was very limited. In the 15 minute period, Dr. Castillo reviewed her medical history, did a physical examination, which included an assessment of the eyes, and engaged in a risk/consent discussion. He says that he spent about 5 minutes on the discussion of risk and consent, although he was frustrated with putting times on things, since he apparently mixes his assessment of the patient with the explanation of the procedure and, apparently, the discussion of risks. Moreover, even if Dr. Castillo made mention of ectropion and unsatisfactory results, along with other risks, the way in which he carried on this discussion of risk and consent was not conducive to the patient having a reasonable understanding of the material risks of the surgery. Dr. Tarshis and Dr. Taylor [expert witnesses at trial] both indicated that the risk/consent discussion is a dialogue with the patient. The fact that Dr. Castillo mixed his examination, assessment and explanations gives me concern about whether the patient would really be given an adequate understanding of risks and benefits so that a proper dialogue could occur. My concern is significantly heightened by the way in which Dr. Castillo presented as a witness. He speaks in a rapid manner and, through no fault of his, his English is accented, which makes it difficult to understand him at times. The way in which he explained how he proceeded in the discussion and examination of Ms Felde suggested that the discussion consisted of a series of rapid-fire comments and questions from him. In five minutes, he covered the nature of the operation, what to do after the operation, and the possible complications. This would not be easy to follow at the best of times. However, it would be particularly difficult to do so in a time of stress, which was the position in which Ms Felde found herself, since she was booked for surgery that day and was feeling nervous and anxious . . . Dr. Charlebois commented on how nervous and talkative she was that day, and how she pulled back at first from having the procedure. His operative notes describes her as 'nervous, excited, demanding, perfectionist . . . wants procedure.' material risks of the surgery . . . In addition, Dr. Castillo, in my view, is not a good listener, which was evident by the times he interrupted questions during testimony or failed to focus on what was asked. Nor does he give the impression of being particularly empathetic. For example, he denied Dr. Charlebois had a brusque manner—a characteristic that was quite obvious in the courtroom. This does not bode well for an effective dialogue with the patient."

Ferguson T (1998) Digital doctoring—Opportunities and challenges in electronic patient-physician communication. *JAMA* **280**:1361–1362.

Ferguson v. Hamilton Civic Hospital (1983) 40 O.R. (2d) 577; 23 C.C.L.T. 254; aff'd. 33 C.C.L.T. 56 (Ont. C.A.).

Fernandez v. U.S. (1980) 636 F. 2d 636 at 704–08.

FIGO (1991) Ethical Issues Concerning Prenatal Diagnosis of Disease in the Conceptus. In Recommendations on Ethical Issues in Obstetrics and Gynecology by the FIGO Committee for the Ethical Aspects of Human Reproduction and Women's Health, FIGO, London, 2003, pp. 37–38. Available at: <http://www.figo.org/content/PDF/ethics-guidelines-text_2003. pdf>. Accessed November 9, 2004.

FIGO (1995) Guidelines Regarding Informed Consent. In Recommendations on Ethical Issues in Obstetrics and Gynecology by the FIGO Committee for the Ethical Aspects of Human Reproduction and Women's Health, FIGO, London, 2003, pp. 11–12. Available at: <http://www.figo.org/content/PDF/ethics-guidelines-text_2003.pdf>. Accessed November 9, 2004.

FIGO (2004) Recommendations on Ethical Issues in Obstetrics and Gynecology by the FIGO Committee for the Ethical Aspects of Human Reproduction and Women's Health, FIGO, London. Available at: <http://www.figo.org/content/PDF/ethics-guidelines-text_2003.pdf>. Accessed November 9, 2004.

Finckh U (2003) The future of genetic association studies in Alzheimer disease. *J Neural Transm* **110**:253–266.

Fine B (1993) The evolution of nondirectiveness in genetic counseling and implications for the Human Genome Project. In: Bartels DM, LeRoy BS, Caplan AL, eds. *Prescribing Our Future: Ethical Challenges in Genetic Counseling*. New York: Walter deGruyter, pp. 101–118.

Fineman RM, Walton MT (2000) Should genetic health care providers attempt to influence reproductive outcome using directive counseling techniques? A public health prospective. *Women Health* **30**(3):39–47.

Finer LB, Henshaw SK (2003) Abortion incidence and services in the United States in 2000. *Perspect Sex Reprod Health* **35**(1):6–15.

Finkelstein D, Smith MK, Faden R (1993) Informed consent and medical ethics. *Arch Ophthalmol* **111**(3):324–326.

Finucane B (1998) Genetic counseling. In: *Working with Women Who Have Mental Retardation: A Genetic Counselor's Guide*. Elwyn P.A.: Elwyn, Inc., pp. 59–66 at 63.

Fischhoff B, Bostrom A, Quadrel MJ (1993) Risk perception and communication. *Annu Rev Public Health* **14**:183–203, at 198–199. "By causing undue alarm or complacency, poor communication can have great public health impact that the risk that they attempt to describe. It may be no more acceptable to release an untested communication than an untested drug."

Fisher B, Costantino JP, Wickerham DL, Redmond CK, Kavanah M, Cronin WM, Vogel V, Robidoux A, Dimitrov N, Atkins J, Daly M, Wieand S, Tan-Chiu E, Ford L, Wolmark N. (1998) Tamoxifen for prevention of breast cancer: Repoprt of the National Surgical Adjuvant Breast and Bowel Project P-1 Study. *J Natl Cancer Inst* **90**:1371–1388.

Fitzpatrick JL, Hahn C, Costa T, Huggins MJ (1999) The duty to recontact: Attitudes of genetics service providers. *Am J Hum Genet* **64**:852–860. The results of a study by Fitzpatrick et al. (1999) may have limited application because only about 25% of those surveyed responded, and many of the respondents were not involved in direct patient care. However, 46% of respondents thought that geneticists should recontact clients.

Flamm N (1986) Cases of failure to diagnose cancer. *Trial* **22**:82–86.

Fletcher LM, Halliday JW (2002) Haemochromatosis: Understanding the mechanisms of disease and implications for diagnosis and patient management following the recent cloning of novel genes involved in iron metabolism. *J Intern Med* **251**:181–192.

Fletcher JC, Berg K, Tranoy KE (1985) Ethical aspects of medical genetics: A proposal for guidelines in genetic counseling, prenatal diagnosis and screening. *Clin Genet* **27**:199–205, at 201.

Flores G, Laws MB, Mayo SJ, Zuckerman B, Abreu M, Medina L, Hardt EJ (2003) Errors in medical interpretation and their potential clinical consequences in pediatric encounters. *Pediatrics* **111**(1):6–14.

Fodde R (2002) The APC gene in colorectal cancer. *Eur J Cancer* **38**:867–871.

Foley DL, Eaves LJ, Wormley B, Silberg JL, Maes HH, Kuhn J, Riley B (2004) Childhood adversity, monoamine oxidase a genotype, and risk for conduct disorder. *Arch Gen Psychiatry* **61**(7):738–744.

Følling A (1934) Über Ausscheidung von Phenylbrenztraubensäure in den Harn als Stoffwechselanomalie in Verbindung mit Imbezillintät. *Hoppe-Seylers Z Physiol Chem* **227**:169–76.

Følling I (1994) The discovery of phenylketonuria. *Acta Paediatr Suppl* **407**:4–10.

Ford D, Easton DF, Bishop DT, Narod SA, Goldgar DE (1994) Risks of cancer in BRCA1-mutation carriers. *Lancet* **343**:692–695.

Ford D, Easton DF, Stratton M, Narod S, Goldgar D, Devilee P, Bishop DT, Weber B, Lenoir G, Chang-Claude J, Sobol H, Teare MD, Struewing J, Arason A, Scherneck S, Peto J, Rebbeck TR, Tonin P, Neuhausen S, Barkardottir R, Eyfjord J, Lynch H, Ponder BA, Gayther SA, Zelada-Hedman M, et al (1998) Genetic heterogeneity and penetrance analysis of the BRCA1 and BRCA2 genes in breast cancer families. The Breast Cancer Linkage Consortium. *Am J Hum Genet* **62**:676–689.

Forrest CB, Glade GB, Baker AE, Bocian A, von Schrader S, Starfield B (2000) Coordination of specialty referrals and physician satisfaction with referral care. *Arch Pediatr Adolesc Med* **154**(5):499–506.

Forrest K, Simpson SA, Wilson BJ, van Teijlingen ER, McKee L, Haites N, Matthews E (2003) To tell or not to tell: Barriers and facilitators in family communication about genetic risk. *Clin Genet* **64**(4):317–326.

Forrestal v. Magendantz (1988) 848 F. 2d. 303 (1st Circ. R.I.).

Fortado L (2004) Genetic testing maps new legal turf; Doctors' liability grows as tests are more widely used. *National Law J* **26**(50):1.

Fox JL (2002) Eugenics Concerns Rekindle with Application of Gene Therapy and Genetic Counseling. *Nature Biotechnology* **20:**531–532.

Fox KM, Cummings SR, Powell-Threets K, Stone K (1998) Family history and risk of osteoporotic fracture: Study of Osteoporotic Fractures Research Group. *Osteoporos Int* **8**:557–562.

Frank TS, Critchfield GC (2002) Hereditary risk of women's cancer. *Best Pract Res Clin Obstet Gynaecol* **16**:703–13.

Frank TS, Manley SA, Olopade O, Cummings S, Garber JE, Bernhardt B, Antman K, Russo D, Wood ME, Mullineau L, Isaacs C, Peshkin B, Buys S, Venne V, Rowley PT, Loader S, Offit K, Robson M, Hampel H, Brener D, Winer EP, Clark S, Weber B, Strong LC, Rieger P, McClure M, Ward BE, Shattuck-Eidens D, Oliphant A, Skolnick MH, Thomas A (1998) Sequence analysis of BRCA1 and BRCA2: Correlation of mutations with family history and ovarian cancer risk. *J Clin Oncol* **16**:2417–2425.

Frank-Stromberg M (2003) They're real and they're here: The new federally regulated privacy rules under the HIPAA. *Medsurg Nursing* **12**(6):380–414.

Frankenburg WK (1974) Selection of diseases and tests in pediatric screening. *Pediatrics* **54**:612–616.

Fratiglioni L, Ahlborn A, Viitanen M, Winblad B (1993) Risk factors for late-onset Alzheimer's disease: A population-based, case-control study. *Ann Neurol* **33**:258–266.

Freedman AN, Wideroff L, Olson L, Davis W, Klabunde C, Srinath KP, Reeve BB, Croyle RT, Ballard-Barbash R (2003) US physicians' attitudes toward genetic testing for cancer susceptibility. *Am J Med Genet* **12A**:63–71.

Freese v. Lemmon (1973) 210 N.W. 2d. 576.

Frets PG, Niermeijer MF (1990) Reproductive planning after genetic counseling: A perspective from the last decade. *Clin Gen* **38**:295–306.

Frets PG, Verhage F, Niermeijer MF (1991) Characteristics of the postcounseling reproductive decision-making process: An explorative study. *Am J Med Genet* **40**(3):298–303.

Friedman KJ (1997–2004) Neurofibromatosis 1. University of Washington, Seattle. Medical Genetics Information Resource (Online database). Available from http://www.genetests.org.

Fuchs CS, Giovannucci EL, Colditz GA, Hunter DJ, Speizer FE, Willett WC (1994) A prospective study of family history and the risk of colorectal cancer. *N Engl J Med* **331**:1669–1674.

Fukuyama F (2002) Our posthuman future: consequences of the biotechnology revolution. Farrar, Strauss and Giroux, 2002, 256 pp.

Funayama M, Hasegawa K, Kowa H, Saito M, Tsuji S, Obata F (2002) A new locus for Parkinson's disease (PARK8) maps to chromosome 12p11.2-q13.1. *Ann Neurol* **51**:296–265.

Funtowicz SO, Ravetz JR (1993) Science for the post-normal age. *Futures* **25**:739–755.

Furr LA (2002) Perceptions of genetics research as harmful to society: Differences among samples of African-Americans and European-Americans. *Genet Test* **6**(1):25–30.

Gad S, Scheuner M, Pages-Berhouet S, Caux-Moncoutier V, Bensimon A, Aurias A, Pinto M, Stoppa-Lyonnet D (2001) Identification of a large rearrangement of the BRCA1 gene using colour bar code on combed DNA in an American breast-ovarian cancer family previously studied by direct sequencing. *J Med Genet* **38**:388–391.

Gadow S (1984) Medical ethics. In: Buchner AN, ed., *Medical Ethics: A Clinical Textbook and Reference for the Health Care Professional.* Cambridge, MA: MIT Press, p. 358.

Gaffney DK, Brohet RM, Lewis CM, Holden JA, Buys SS, Neuhausen SL, Steele L, Avizonis V, Stewart JR, Cannon-Albright LA (1998) Response to radiation therapy and prognosis in breast cancer patients with BRCA1 and BRCA2 mutations. *Radiother Oncol* **47**:129–136.

Gail MH, Brinton LA, Byar DP, Corle DK, Green SB, Schairer C, Mulvihill JJ (1989) Projecting individualized probabilities of developing breast cancer for white females who are being examined annually. *J Natl Cancer Inst* **81**(24):1879–1886. Online: http://brca.nci.nih.gov/brc/start.htm.

Gallini N, Scotchmer S (2002) Intellectual property: When is it the best incentive system? In: Jaffe A, Lerner J, Stern S, eds *Innovation, Policy and the Economy*, Vol. 2, Cambridge, MA: MIT Press. Online: Socrates <http://ist-socrates.berkeley.edu/~scotch/>.

Gambardella A, Manna I, Labate A, Chifari R, La Russa A, Serra P, Cittadella R, Bonavita S, Andreoli V, LePiane E, Sasanelli F, Di Costanzo A, Zappia M, Tedeschi G, Aguglia U, Quattrone A (2003) GABA (B) receptor 1 polymorphism (G1465A) is associated with temporal lobe epilepsy. *Neurology* **60**:560–563.

Gandhi TK, Sittig DF, Franklin M, Sussman AJ, Fairchild DG, Bates DW (2000) Communication breakdown in the outpatient referral process. *J Gen Intern Med* **15**(9):626–631.

Gardner RJM, Sutherland GR (2004) *Chromosome Abnormalities and Genetic Counseling*, New York: Oxford.

Gardner v. Pawliw (1997) 150 N.J. 359.

Garel M, Gosme-Seguret S, Kaminski M, Cuttini M (2002) Ethical decision-making in prenatal diagnosis and termination of pregnancy: A qualitative survey among physicians and midwives. *Prenat Diag* **22**:811–817.

Garrison A (2003) Between a rock and a hard place. *JAMA* **290**:1217–1218.

Gasser T, Muller-Myhsok B, Wszolek ZK, Oehlmann R, Calne DB, Bonifati V, Bereznai B, Fabrizio E, Vieregge P, Horstmann RD (1998) A susceptibility locus for Parkinson's disease maps to chromosome 2 p13. *Nat Genet* **18**:262–265.

Gaster B, Knight CL, DeWitt DE, Sheffield JV, Assefi NP, Buchwald D (2003) Physicians' use of attitudes toward electronic mail for patient communication. *J Gen Intern Med* **18**(5):385–389.

Gaston v. Hunter (1978) 588 P.2d 326, 351 (Ariz. Ct. App.).

Gates v. Jensen (1979) 92 Wn.2d 246, 595 P.2d 919.

Gattas MR, Mac Millan JC, Meinecke I, Loane M, Wootton R (2001) *Telemed. Telecare* **7**(Suppl 2):68–70

Geer KP, Ropka ME, Cohn WF, Jones SM, Miesfeldt S (2001) Factors influencing patients' decisions to decline cancer genetic counseling services. *J Genet Couns* **10**(1):25–40. A majority of those who refused to proceed with cancer genetic counseling cited concern for insurability for the patient and the family.

Geler v. Akawie (2003) 358 N.J. Super. 437 (App. Div.). 816 A.2d 402.

Geller G, Tambor ES, Chase GA, Hofman KJ, Faden RR, Holtzman NA (1993) Incorporation of genetics in primary care practice. Will physicians do the counseling and will they be directive? *Arch Fam Med* **2**(11):1119–1125.

Geller LN, Alper JS, Billings PR, Barash CI, Beckwith J, Natowicz M (1996) Individual, Family, and Societal Dimensions of Genetic Discrimination: A Case Study Analysis. *Sci. Eng Ethics* **2**(1):71–88. The authors report case studies where physicians have tried to influence, if not coerce, a patient into undergoing genetic testing and agreeing to terminate a pregnancy if test results for a genetic disorder were positive (e.g., Huntington disease and PKU).

Geller G, Botkin JR, Green MJ, Press N, Biesecker BB, Wilfond B, Grana G, Daly MB, Schneider K, Kahn MJ (1997) Genetic testing for susceptibility to adult-onset cancer. The process and content of informed consent. *JAMA* **277**(18):1467–1474.

Geller G, Bernhardt BA, Doksum T, Helzlsouer KJ, Wilcox P, Holtzman NA (1998) Decision-making about breast cancer susceptibility testing: How similar are the attitudes of physicians, nurse practitioners, and at-risk women? *J Clin Oncol* **16**(8):2868–2876.

Gemme v. Goldberg (1993) 626 A.2d 318, 326 (Conn. App. Ct.). Alternative treatments should be disclosed even if the alternative is more hazardous.

General Accounting Office (2003). Newborn Screening Characteristics of State Programs. www.gao.gov/cgi-bin/getrpt?GAO-03-449.

GeneReviews at GeneTests (1997–2004) Medical Genetics Information Resource (database online), copyright, University of Washington, Seattle. Available at http://www.genetests.org. Accessed December 7, 2004.

GeneTests (2004) Medical Genetics Information Resource (database online), copyright, University of Washington, Seattle. Updated weekly. Available at http://www.genetests.org. Accessed December 7, 2004.

Genetic Interest Group (1998) Guidelines for Genetic Services (www.gig.org).

Genetics and Public Policy Center (2004a) Genetic testing of embryos to pick "savior sibling" okay with most Americans. May 3, 2004. http://tools-content.labvelocity.com/pdfs/5/63175.pdf. However, the same study reported disapproval of "selecting an embryo based on whether the baby will be a boy or a girl."

Genetics and Public Policy Centre (2004b) Preimplantation genetic diagnosis: A discussion of challenges, concerns, and preliminary policy options related to the genetic testing of human embryos. Genetics and Public Policy Centre. http://www.dnapolicy.org/

Genetic Information and Health Insurance (1993) *Report of the Task Force on Genetic Information and Health Insurance*. Washington DC: National Institutes of Health.

Genewatch (2001) Patenting genes—stifling research and jeopardising healthcare. Online: Genewatch <http://www.genewatch.org/Patenting/briefs.htm#Special>.

Gershon ES (2002) The challenges of genetic tests for human behavior. *Isr J Psychiatry Relat Sci* **39**(4):206–216.

Gershoni-Baruch R, Dagan E, Fried G (1999) BRCA1 and BRCA2 founder mutations in patients with bilateral breast cancer. *Eur J Hum Gene* **7**:833–836.

Getty v. Hoffman-LaRoche Inc. (1987) 235 Cal. Rptr. 48.

Ghosh AK (2004) Explaining risk. *Minn Med* **87**(2):6.

Giardiello FM, Brensinger JD, Petersen GM, Luce MC, Hylind LM, Bacon JA, Booker SV, Parker RD, Hamilton SR (1997) The use and interpretation of commercial APC gene testing for familial adenomatous polyposis. *N Engl J Med* **336**(12):823–827.

Gibons A (2004) Study results: Employer-based coverage of genetic counseling services. *Benefits Q* **20**(3):48–68.

Gigerenzer G, Edwards A (2003) Simple tools for understanding risks: From innumeracy to insight. *BMJ* **327**:741–744.

Gilbert W (1992) A vision of the grail. In: Kevles DJ, Hood L, eds. *The Code of Codes: Scientific and Social Issues in the Human Genome Project.* Cambridge: Harvard University Press, pp. 83–97.

Gilbert F (2001) Cystic fibrosis carrier screening: Steps in the development of a mutation panel. *Genet Test* **5**(3):223–227.

Gildiner v. Thomas Jefferson University Hospital (1978) 451 F. Supp. 692, (E.D. Pa.).

Gilfix M (1984) Electronic fetal monitoring: Physician liability and informed consent. *Am J Law Med* **10**(1):31–90. The author reviews the difficulty physicians faced, with regard to EFM, in trying to balance standards concerning "customary practice" against a physician's obligations to apply the "best judgment" and "to keep abreast" standards of care.

Gill Estate v. Marriott (1999) O.J. No. 4509.

Gillon R (1986) *Philosophical Medical Ethics.* New York: Wiley.

Giovannucci E, Stampfer MJ, Colditz GA, Hunter DJ, Fuchs C, Rosner BA, Speizer FE, Willett WC (1998) Multivitamin use, folate, and colon cancer in women in the Nurses' Health Study. *Ann Intern Med* **129**:517–524.

Glade GB, Forrest CB, Starfield B, Baker AE, Bocian AB, Wasserman RC (2002) Specialty referrals made during telephone conversations with parents: A study from the pediatric research in office settings network. *Ambul Pediatr* **2**(2):93–98.

Glanz K, Grove J, Lerman C, Gotay C, Le Marchand L (1999) Correlates of intentions to obtain genetic counseling and colorectal cancer gene testing among at-risk relatives from three ethnic groups. *Cancer Epidemiol Biomarkers Prev* **8**(4 Pt 2):329–336.

Glazier AM, Nadeau JH, Aitman TJ (2002) Finding genes that underlie complex traits. *Science* **298**:2345–2349.

Gleitman v. Cosgrove (1967) 49 N.J. 22, 227 A.2d. 689.

Goate A, Chartier-Harlin MC, Mullan M, Brown J, Crawford F, Fidani L, Giuffra L, Haynes A, Irving N, James L, et al (1991) Segregation of a missense mutation in the amyloid precursor protein gene with familial Alzheimer's disease. *Nature* **349**:704–706.

Godard B, Cardinal G (2004) Ethical implications in genetic counseling and family studies of the epilepsies. *Epilepsy Behav* **5**(5):621–626.

Godard B, Simard J (2003) Les enjeux éthiques de l'identification d'une susceptibilité, génétique au cancer du sein dans un contexte de recherche clinique intégrée, In: Hervé C, Knoppers BM, Molinari P eds. *Les pratiques de la recherche médicale visit par la bioéthique.* Paris: Ed. Dalloz, pp. 113–136.

Goel V for Crossroads 99 Group (2001). Appraising organized screening programmes for testing for genetic susceptibility to cancer. *BMJ* **322**:1174–1178.

Gold ER (2000) Moving the gene patent debate forward. *Nature Biotechnol* **18**(12):1319.

Gold ER (2002) Gene patents and medical access. *Intellectual Property Forum* **49**:20.

Gold ER (2003) From theory to practice: Health care and the patent system. *Health Law J,* Special Edition 21.

Gold ER, Adams W, Castle D, Cleret de Langavant G, Cloutier LM, Daar AS, Glass A, Smith PJ, Bernier L (2004) The unexamined assumptions of intellectual property: Adopting an evaluative approach to patenting biotechnological innovation. *Public Affairs Quart* **18**:273.

Goldberg v. Ruskin (1986) 499 N.E. 2d 406 (Ill.). Obstetrician sued by mother for failure to diagnose the presence of Tay-Sachs disease.

Goldman JS, Hou CE (2004) Nat Med **10** (Suppl):S58–62.

Goldman SM, Tanner C (1998) Etiology of Parkinson's disease. In: Jankovic J, Tolosa E, eds. *Parkinson's Disease and Movement Disorders.* Baltimore: Williams and Wilkins, pp. 1333–1358.

Gollust SE, Hull SC, Wilfond BS (2002) Limitations of direct-to-consumer advertising for clinical genetic testing. *JAMA* **288**(14):1762–1767. Also see Gollust SE, Wilfond BS, Hull SC (2003) Direct-to-consumer sales of genetic services on the Internet. *Genet Med* **5**(4):332–337; Hull SC, Prasad K (2001) Reading between the lines: Direct-to-consumer advertising of genetic testing in the USA. *Reprod Health Matters* **9**(18):44–48.

Gomez-Tortosa E, del Barrio A, Garcia Ruiz PJ, Pernaute RS, Benitez J, Barroso A, Jimenez FJ, Garcia Yebenes J (1998) Severity of cognitive impairment in juvenile and late-onset Huntington disease. *Arch Neurol* **55**(6):835–843.

Goodwin FK, Jamison KR (1990) *Manic-Depressive Illness.* New York: Oxford.

Goos LM, Silverman I, Steele L, Stockley T, Ray PN (2004) Providing information at the point of care: Educational diagnostic reports from a genetic testing service provider. *Clin Leadersh Manag Rev* **18**(1):11–24.

Gorab v. Zook (1997) 943 P. 2d. 423 (Colorado).

Gordon C, Walpole I, Zubrick SR, Bower C (2003) Population screening for cystic fibrosis: Knowledge and emotional consequences 18 months later. *Am J Med Genet* **120**A(2):199–208. Also see Axworthy D, Brock DJ, Bobrow M, Marteau TM (1996) Psychological impact of population-based carrier testing for cystic fibrosis: 3-year follow-up. UK Cystic Fibrosis Follow-Up Study Group. *Lancet* **347**(9013):1443–1446.

Gostin L (1991) Genetic discrimination: The use of genetically based diagnostic and prognostic tests by employers and insurers, *Am J Law Med* **17**:109–144.

Gottesman II (1991) *Schizophrenia Genesis: The Origins of Madness.* New York: W.H. Freeman.

Gottesman II, Shields J (1982) *Schizophrenia: The Epigenetic Puzzle.* Cambridge: Cambridge University Press.

Gottlieb S (2003) One in three doctors don't tell patients about services they can't have. *BMJ* **327**:123; http://bmj.bmjjournals.com/cgi/content/full/327/7407/123-a.

Gourfinkel-An I, Baulac S, Nabbout R, Ruberg M, Baulac M, Brice A, LeGuern E (2004) Monogenic idiopathic epilepsies. *Lancet Neurol* **3**:209–218.

Grant, SS (2000) Prenatal genetic screening. *Online J Issues Nurs* **5**(3):2.

Gray J, Brain K, Norman P, Anglim C, France L, Barton G, Branston L, Parsons E, Clarke A, Sampson J, Roberts E, Newcombe R, Cohen D, Rogers C, Mansel R, Harper P (2000) A model protocol evaluating the introduction of genetic assessment for women with a family history of breast cancer. *J Med Genet* **37**(3):192–196.

Greco v. United States (1995) 111 Nev 405, 893 P 2d 345. Also see *Blair v. Hutzel Hospital* (1996) 217 Mich. App. 502, 552 NW2d 507.

Green RM (1997) Parental autonomy and the obligation not to harm one's child genetically. *J Law Med Ethics* **25**(1):5–15, at 16.

Green M, Botkin J (2003) Genetic exceptionalism in medicine: Clarifying the differences between genetic and nongenetic tests. *Ann Internal Med* **137**(7), 571–575.

Green E, Craddock N (2003) Brain-derived neurotrophic factor as a potential risk locus for bipolar disorder: Evidence, limitations, and implications. *Curr Psychiatry Rep* **5**(6):469–476.

Green MJ, Fost N (1997) Who should provide genetic education prior to gene testing? Computers and other methods for improving patient understanding. *Genet Test* **1**(2):131–136.

Green MJ, Biesecker BB, McInerney AM, Mauger D, Fost N (2001a) An interactive computer program can effectively educate patients about genetic testing for breast cancer susceptibility. *Am J Med Genet* **103**:16–23.

Green MJ, McInerney AM, Biesecker BB, Fost N (2001b) Education about genetic testing for breast cancer susceptibility: Patient preferences for a computer program or genetic counselor. *Am J Med Genet* **103**:24–31.

Green MJ, Peterson SK, Baker MW, Harper GR, Friedman LC, Rubinstein WS, Mauger DT (2004) Effect of a computer-based decision aid on knowledge, perceptions, and intentions about genetic testing for breast cancer susceptibility: A randomized controlled trial. *JAMA* **292**(4):442–452.

Greenberg F, James LM, Oakley GP (1983) Estimates of birth prevalence rates of spina bifida in the United States from computer-generated maps. *Am J Obstet Gynecol* **145**:570–573.

Greenberg F, Faucett A, Rose E, Bancalari L, Kardon NB, Mizejewski G, Haddow JE, Alpert E (1992) Congenital deficiency of alpha-fetoprotein. *Am J Obstet Gynecol* **167**:509–511.

Greene MF, Ecker JL (2004) Abortion, health and the law, *New Engl J Med* **350**(2):1–186.

Greenhalgh T, Kostopoulou O, Harries C (2004) Making decisions about benefits and harms of medicines. *BMJ* **329**(7456):47–50.

Greenpeace (2004) The true cost of gene patents: The economic and social consequences of patenting genes and living organisms. Online: Greenpeace <weblog.greenpeace.org/ge/archives/1Study_True_Costs_Gene_Patents.pdf>.

Greeson CJ, McCarthy Veach P, LeRoy BS (2001) A qualitative investigation of Somali immigrant perceptions of disability: Implications for genetic counseling. *J Genet Counsel* **10**(5):359–378.

Grimes DA, Snively GR (1999) Patients' understanding of medical risks: Implications for genetic counseling. *Obstet Gynecol* **1**;93(6):910–914.

Grody WW, Pyeritz RE (1999) Report card on molecular genetic testing: Room for improvement? *JAMA* **281**(9):845–847.

Grody WW, Cutting GR, Klinger KW, Richards CS, Watson MS, Desnick RJ (2001) Laboratory standards and guidelines for population-based cystic fibrosis carrier screening. *Genet Med* **3**(2):149–154.

Grover S, Stoffel EM, Bussone L, Tschoegl E, Syngal S (2004) Physician assessment of family cancer history and referral for genetic evaluation in colorectal cancer patients. *Clin Gastroenterol Hepatol* **2**(9):813–819.

Grubbs ex rel. Grubbs v. Barbourville Family Health Center (2003) P.S.C., 120 S.W. 3d 682 (Ky.).

Grubbs v. Barbourville Family Clinic (2001) Amended; 2003. Ky. S. Ct. SC-0571-DG and SC-0959-DG (aff'd in part; rev'd in part; remanded) consolidated with the case of *Bogan v. Altman, McQuire and Pigg*, PSC. 2001. Ky. App. WL 201848. Quotations from the judgment cited in this text are from pages 11 and 14. In the Grubbs case, the plaintiff alleged that the physicians failed to accurately interpret and/or to report the test results from an ultrasound that indicated spina bifida. In the Bogan case, the plaintiffs alleged that the defendant physician failed to inform them that prenatal genetic tests were available and failed to accurately interpret an ultrasound. The child was born without any eyes, or brain, an underdeveloped brain, a cleft palate, and could not speak. The court in the Bogan case held that the alleged failure to provide the parents with "information necessary to make a decision regarding the continuation of pregnancy stated a viable cause of action for medical negligence. After these cases were consolidated, the Supreme Court of Kentucky declined to recognize a cause of action for wrongful birth or wrongful life: http://162.114.92.72/Opinions/2001-SC-000563-DG.pdf#xml=http://162.114.92.72/dtsearch.asp?cmd=pdfhits&DocId=682&Index=D%3a%5cInetpub%5cwwwroot%5cindices%5cSupremeCourt%5fIndex&HitCount=4&hits=70+71+72+73+&hc=16&req=rendered+w%2F8+august+27%2C+2003+and+not+%22CR+76%22+.

Grumbach K, Selby JV, Damberg C, Bindman AB, Quesenberry C Jr, Truman A, Uratsu C (1999) Resolving the gatekeeper conundrum: What patients value in primary care and referrals to specialists. *JAMA* **282**(3)261–266.

Guadagnoli E, Ward P (1998) Patient participation in decision-making. *Soc Sci Med* **47**(3):329–339.

Guerrini R, Casari G, Marini C (2003) The genetic and molecular basis of epilepsy. *Trend Mol Med* **9**:300–306.

Guibaud S, Robert E, Simplot A, Boisson C, Francannet C, Patouraux MH (1993) Prenatal diagnosis of spina bifida aperta after first-trimester valproate exposure. *Prenat Diagn* **13**:772–773.

Guimond v. Laberge (1956) 4 D.L.R. (2d) 559 at 569 (Ont. C.A.).

Gurm HS, Litaker DG (2000) Framing procedural risks to patients: Is 99% safe the same as a risk of 1 in 100? *Acad Med* **75**(8):840–842.

Guthrie R. (1996) The introduction of newborn screening for phenylketonuria. A personal history. Eur J Pediatr:155 Suppl 1:S4–5.

Gutierrez-Delicado E, Serratosa JM (2004) Genetics of the epilepsies. *Curr Opin Neurol* **17**(2):147–153.

Haddow JE (1998) Antenatal screening for Down's syndrome: Where are we, where next? Commentary on Snijders et al (1998). *Lancet* **351**:336–337.

Haddow JE, Kloza EM, Knight GJ, Smith DE (1981) Relationship between maternal weight and serum alpha-fetoprotein concentration during the second trimester. *Clin Chem* **27**:133–134.

Haddow JE, Palomaki GE, Knight GJ, Foster DL, Neveux LM (1998a) Second trimester screening for Down's syndrome using maternal serum dimeric inhibin A. *J Med Screen* **5**:115–119.

Haddow JE, Palomaki GE, Knight GJ, Williams J, Miller WA, Johnson A (1998b) Screening of maternal serum for fetal Down's syndrome in the first trimester. *N Engl J Med* **338**:955–961.

Hadley DW, Jenkins J, Dimond E, Nakahara K, Grogan L, Liewehr DJ, Steinberg SM, Kirsch I (2003) Genetic counseling and testing in families with hereditary nonpolyposis colorectal cancer. *Arch Intern Med* **163**(5):573–582.

Hadley D, Jenkins J, Dimond E, de Carvalho M, Kirsch I, Palmer C (2004) Colon cancer screening practices after genetic counseling and testing for hereditary nonpolyposis colorectal cancer. *J Clin Oncol* **22**(1):39–44.

Hadlock FP, Deter RL, Harrist RB, Park SK (1982) Fetal biparietal diameter: A critical re-evaluation of the relation to menstrual age by means of real-time ultrasound. *J Ultrasound Med* **1**:97–104.

Hadlock FP, Shah YP, Kanon DJ, Lindsey JV, (1992) Fetal crown-rump length: Reevaluation of relation to menstrual age (5–18 weeks) with high-resolution real-time US. *Radiology* **182**:501–505.

Haffty BG, Harrold E, Khan AJ, Pathare P, Smith TE, Turner BC, Glazer PM, Ward B, Carter D, Matloff E, Bale AE, Alvarez-Franco M (2002) Outcome of conservatively managed early-onset breast cancer by BRCA1/2 status. *Lancet* **359**:1471–1477.

Haga S, Khoury M, Burke W (2003) Genomic profiling to promote a healthy lifestyle: Not ready for prime time. *Nature Genet* **34**(4): 347–350.

Hagerty RG, Butow PN, Ellis PA, Lobb EA, Pendlebury S, Leighl N, Goldstein D, Lo SK, Tattersall MH (2004) Cancer patient preferences for communication of prognosis in the metastatic setting. *J Clin Oncol* **22**(9):1721–1730.

Hahn L (2003) Owning a piece of Jonathan. *Chicago* (May).

Haidet P, Paterniti DA (2003) "Building" a history rather than "taking" one: A perspective on information sharing during the medical interview. *Arch Intern Med* **163**(10):1134–1140.

Hakimian R (2000) Disclosure of Huntington's disease to family members: The dilemma of known but unknowing parties. *Genet Test* **4**(4):359–364.

Halbert CH (2004) Decisions and outcomes of genetic testing for inherited breast cancer risk. *Ann Oncol* **15**(Suppl 1):I35–I39.

Hall v. Hilbun (1985) 466 So. 2d 856, 872–73 (Miss.).

Hall MA, Rich SS (2000a) Laws restricting health insurers' use of genetic information: Impact on genetic discrimination. *Am J Hum Genet* **66**:293–307.

Hall MA, Rich SS (2000b) Patients' fear of genetic discrimination by health insurers: The impact of legal protections. *Genet Med* **2**(4):214–221.

Hall JA, Roter DL, Katz NR (1988) Meta-analysis of correlates of provider behavior in medical encounters. *Med Care* **26**:657–675.

Hall MA, Ellman IM, Strouse DS (1999) *Hall, Ellman and Strouse's Health Care Law and Ethics in a Nutshell*, 2nd ed. West Group Publishing, pp. 117–131.

Hall S, Bobrow M, Marteau TM (2000) Psychological consequences for parents of false negative results on prenatal screening for Down's syndrome: Retrospective interview study. *BMJ* **320**(7232):407–412.

Hall S, Abramsky L, Marteau TM (2003) Health professionals' reports of information given to parents following the prenatal diagnosis of sex chromosome anomalies and outcomes of pregnancies: A pilot study. *Prenat Diagn* **23**(7):535–538.

Hallowell N, Foster C, Eeles R, Ardern-Jones A, Murday V, Watson M (2003) Balancing autonomy and responsibility: the ethics of generating and disclosing genetic information. *J Med Ethics* **29**(2):74–79; discussion 80–83.

Hallowell N, Murton F (1998) The value of written summaries of genetic consultations. *Patient Educ Couns* **35**(1):27–34.

Hallowell N, Murton F, Statham H, Green JM, Richards MPM (1997) Women's need for information before attending genetic counselling for familial breast or ovarian cancer: A questionnaire, interview and observational study. *BMJ* **314**(7076):281–283.

Hanley W (2004) personal communication.

Hansis C, Grifo J (2001) Tay-Sachs disease and preimplantation genetic diagnosis. *Adv Genet* **44**:311–315.

Harbeson v. Parke-Davis, Inc. (1983) 656 P. 2d 483, 494 (Wash.). The court stated that a physician had a duty to conduct a literature search prior to administration of the drug Dilantin.

Harkin LA, Bowser DN, Dibbens LM, Singh R, Phillips F, Wallace RH, Richards MC, Williams DA, Mulley JC, Berkovic SF, Scheffer IE, Petrou S (2002) Truncation of the GABA(A)-receptor gamma2 subunit in a family with generalized epilepsy with febrile seizures plus. *Am J Hum Genet* **70**(2):530–536.

Harmon A (2004a) As gene test menu grows, who gets to choose? *New York Times*, July 21.

Harmon A (2004b) In new tests for fetal defects, agonizing choices. *New York Times*, June 20.

Harris RA (2004) Cost utility of prenatal diagnosis and the risk-based threshold. *Lancet* **363**:276–282.

Harris JM, Fahn S (2003) Genetics of movement disorders. In: Rosenberg RN, Prusiner SB, DiMauro S, Barchi RL, Nestler EJ, eds. *The Molecular and Genetic Basis of Neurologic and Psychiatric Disease*, Philadelphia: Elsevier, pp. 351–368.

Harris JR, Hellman S (1996) Natural history of breast cancer. In: Harris JR, Lipman ME, Morrow MM, et al eds. *Diseases of the Brest*. Philadelphia: Lippincott-Raven, pp. 375–391.

Harris R, Lane B, Harris H, Williamson P, Dodge J, Modell B, Ponder B, Rodeck C, Alberman E (1999) National confidential enquiry into counselling for genetic disorders by non-geneticists: General recommendations and specific standards for improving care. *Br J Obstet Gynaecol* **106**(7):658–663.

Hartlaud PP, Wolkenstein AS, Laufenburg HF (1993) Obtaining informed consent: It is not simply asking "Do You Understand?" *J Fam Pract* **36**(4):383–384.

Hartmann LC, Degnim A, Schaid DJ (2004) Prophylactic mastectomy for BRCA1/2 carriers: progress and more questions *J Clin Oncol* **22**:981–983.

Hartmann LC, Sellers TA, Schaid DJ, Frank TS, Soderberg CL, Sitta DL, Frost MH, Grant CS, Donohue JH, Woods JE, MCDonnell SK, Vockley CW, Deffenbauch A, Cough FJ, Jenkins RB (2001) Efficacy of bilateral prophylactic mastectomy n BRCA1 and BRCA2 gene mutation carriers, *J Natl Cancer Inst* **93**:1633–1637.

Hartmann LC, Schaid DJ, Woods JE, Crotty TP, Myers JL, Arnold PG, Petty PM, Sellers TA, Johnson JL, McDonnell SK, Frost MH, Jenkins RB (1999) Efficacy of bilateral prophylactic mastectomy in women with a family history of breast cancer. *N Engl J Med* **340**:77–84.

Harvard College v. Canada (Commissioner of Patents) (2002) SCC 76; 219 D.L.R. (4th) 577.

Hattori E, Liu C, Badner JA, Bonner TI, Christian SL, Maheshwari M, Detera-Wadleigh SD, Gibbs RA, Gershon ES (2003) Polymorphisms at the G72/G30 gene locus, on 13q33, are associated with bipolar disorder in two independent pedigree series. *Am J Hum Genet* **72**(5):1131–1140.

Haug K, Warnstedt M, Alekov AK, Sander T, Ramirez A, Poser B, Maljevic S, Hebeisen S, Kubisch C, Rebstock J, Horvath S, Hallmann K, Dullinger JS, Rau B, Haverkamp F, Beyenburg S, Schulz H, Janz D, Giese B, Muller-Newen G, Propping P, Elger CE, Fahlke C, Lerche H, Heils A (2003) Mutations in CLCN2 encoding a voltage-gated chloride channel are associated with idiopathic generalized epilepsies. *Nat Genet* **33**:527–532.

Haughian v. Paine (1987) 36 C.C.L.T. 242 (Sask. Q.B.); appeal allowed, 40 C.C.L.T. 13; [1987] 4 W.W.R. 97 (Sask. C.A.) (leave to appeal to S.C.C. refused).

Hauser WA, Hesdorffer DC (1990) *Epilepsy: Frequency, Causes and Consequences*. New York: Demos.

Havens v. Ritchey (1991). 582 N.E. 2d. 792 Ind.

Havighurst CC (1991) Practice guidelines as legal standards governing physician liability. *Law Contemporary Problems* **54**:87–117.

Haydon JH (1999) Legal aspects of health information. *Health Law Canada* **20**(2):1–12.

Hayflick SJ, Eiff MP, Carpenter L, Steinberger J (1998) Primary care physicians' utilization and perceptions of genetics services. *Genet Med* **1**:13–21.

Haymon v. Wilkerson (1987) 535 A. 2d 880 (D.C.). Physician recommended couple not undergo genetic testing of the fetus. The child was diagnosed with Down syndrome.

Health Canada (2002) Congenital Anomalies in Canada: A Perinatal Health Report, 2002, Public Works and Government Services Canada, Ottawa.

Health Services Payment Act Regulations, P.E.I. Reg. EC453/96, s. 1(d)(iv).

Healy DG, Abou-Sleiman PM, Lees AJ, Casas JP, Quinn N, Bhatia K, Hingorani AD, Wood NW (2004) Tau gene and Parkinson's disease: A case-control study and meta-analysis. *J Neurol Neurosurg Psychiatry* **75**(7):962–965.

Hebért PC, Hoffmaster B, Glass KC, Singer PA (1997) Bioethics for clinicians: 7. Truth telling. *CMAJ* **156**(2):225–228.

Hebért PC, Levin AV, Robertson G (2001) Bioethics for clinicians: 23. Disclosure of medical error. *CMAJ* **164**(4):509–513.

Hedenfalk I, Duggan D, Chen Y, Radmacher M, Bittner M, Simon R, Meltzer P, Gusterson B, Esteller M, Kallioniemi OP, Wilfond B, Borg A, Trent J (2001) Gene-expression profiles in hereditary breast cancer. *New Engl J Med* **344**:539–548.

Hedlund v. Superior Ct. (1983) 34 Cal. 3d. 695, 669 P. 2d. 41.

Hegele RA (2001) Genes and environment in type II diabetes and atherosclerosis in aboriginal Canadians. *Curr Atheroscler Rep* **3**(3):216–221.

Heller M, Eisenberg R (1998) Can patents deter innovation? The anticommons in biomedical research. *Science* **280**: 698–701.

Helling v. Carey (1974) 83 Wash2d 514, 519 P2d 981 at 982–983. The court overruled the generally accepted standard of care and found negligent a practice of not routinely testing patients under the age of forty for glaucoma that was sanctioned by professional custom.

Hellman D (2003) What makes genetic discrimination exceptional? *Am J Law Med* **29**:77–116.

Helmes AW, Bowen DJ, Bengel J (2002) Patient preferences of decision-making in the context of genetic testing for breast cancer risk. *Genet Med* **4**(3):150–157.

Henderson R, Orsenigo L, Pisano G (1999) The pharmaceutical industry and the revolution in molecular biology: Interactions among scientific, institutional and organizational change. In: Mowery DC, Nelson RR, eds. *Sources of Industrial Leadership: Studies of Seven Industries*, New York: Cambridge University Press.

Henn W (2000) Consumerism in prenatal diagnosis: A challenge for ethical guidelines. *J Med Ethics* **26**:444–446.

Henriques CU, Damm P, Tabor A, Pedersen JF, Molsted-Pedersen L (1993) Decreased alpha-fetoprotein in amniotic fluid and maternal serum in diabetic pregnancy. *Obstet Gynecol* **82**:960–964.

Henry MR, Cho MK, Weaver MA, Merz JF (2003) A pilot survey on the licensing of DNA inventions. *J Law Med Ethics* **31**(3):442–449.

Henshaw SK, Finer LB (2003) The accessibility of abortion services in the United States, 2001. *Perspect Sex Reprod Health* **35**(1):16–24.

Heron SE, Crossland KM, Andermann E, Phillips HA, Hall AJ, Bleasel A, Shevell M, Mercho S, Seni MH, Guiot MC, Mulley JC, Berkovic SF, Scheffer IE (2002) Sodium-channel defects in benign familial neonatal-infantile seizures. *Lancet* **360**(9336):851–852.

Hicks AA, Petursson H, Jonsson T, Stefansson H, Johannsdottir HS, Sainz J, Frigge ML, Kong A, Gulcher JR, Stefansson K, Sveinbjornsdottir S (2002) A susceptibility gene for late-onset idiopathic Parkinson's disease. *Ann Neurol* **52**:549–555.

Higashi MK, Veenstra DL, del Aguila M, Hujoel P (2002) The cost-effectiveness of interleukin-1 genetic testing for periodontal disease. *J Periodontol* **73**(12):1474–1484.

Hill LM, Fries J, Hecker J, Grzybek P (1994) Second trimester echogenic small bowel: An increased risk of adverse perinatal outcome. *Prenat Diagn* **14**:845–850.

Hilts PJ (1993) Panel reports genetic screening has cost some their health plans. *New York Times*. Nov. 5.

Hino M, Koki Y, Nishi S (1972) Nimpu ketsu na ka noalpha-fetoprotein [Alpha-fetoprotein in pregnant women]. *Igaku No Ayumi* **82**:512.

HIPAA (1996) Health Insurance Portability and Accountability Act of 1996. http://www.aspe .hhs.gov/admnsimp/pl104191.htm.

HIPAA Privacy: Summary of the HIPAA Privacy Rule (2001) http://www.hhs.gov/ocr/ privacysummary.pdf. Federal regulations for the protections of patients' medical records (Privacy Rule or HIPAA–Health Insurance Portability and Accountability Act—regulations) on April 14, 2001. Additional changes were made to the Privacy Rule on August 14, 2002. All covered entities (health care plans, providers, and clearinghouses) were to be in compliance with the regulation by April 14, 2003.

Hirose S, Zenri F, Akiyoshi H, Fukuma G, Iwata H, Inoue T, Yonetani M, Tsutsumi M, Muranaka H, Kurokawa T, Hanai T, Wada K, Kaneko S, Mitsudome A (2000) A novel mutation of KCNQ3 (c.925T->C) in a Japanese family with benign familial neonatal convulsions. *Ann Neurol* **47**:822–826.

Hirschhorn K, Fleisher LD, Godmilow L, Howell RR, Lebel RR, McCabe ERB, McGinniss MJ, Milunsky A, Pelias MZ, Pyeritz RE, Sujansky E, Thompson BH, Zinberg R-E. (1999) Duty to re-contact. Policy Statement: Social Ethical and Legal Issues Committee of the American College of Medical Geneticists. *Genet Med* **1**:171–172.

Hirsh HL (1976) Malpractice liablity of physicians to non patients. *S Med J* **69**:762–763.

Hirsh HL (1995) Failed communication among health care providers about and with the patient. *Med Trial Tech Q* **41**:1–33, at 2–5.

Hitzeroth HW, Niehaus CE, Brill DCS (1995) Phenylketonuria in South Africa. A report on the status quo. *Afr Med J* **85**:33–36.

Hochhauser M (1999) Informed consent and patient's rights documents: A right, a rite, or a rewrite? *Ethics Behav* **9**(1):1–20.

Hodge JG Jr, Gostin LO, Jacobsen PD (1999) Legal issues concerning electronic health information: Privacy, quality and liability. *JAMA* **282**:1466–1471.

Hodge SE (1998) A simple, unified approach to Bayesian risk calculations. *J Genet Couns* **7**:235–261.

Hodge v. Lafayette General Hospital (1981) 399 So. 2d 744 (La. App. 3 Cir.).

Hoeffner v. The Citadel (1993) 311 SC 361, 429 SE.2d 190, 194. A health care professional has an obligation to prevent a suicide where the health care professional has a specific duty to prevent an act of suicide or has specific notice that a patient is likely to commit such an act. Also see *Cunningham v. Helping Hands, Inc.* (2001) 346 S.C. 253, 550 S.E.2d 872 (Ct. App.) certior Opinion No. 25575, 2003 affir'd as modified.

Hoemke v. New York Blood Center (1989) 720 F. Supp. 45 (S.D.N.Y.).

Hoffrage U, Lindsey S, Hertwig R, Gigerenzer G (2000) Communicating statistical information. *Science* **290**:2261–2262.

Hogge WA, Fraer L, Melegari T (2001) Maternal serum screening for fetal trisomy 18: Benefits of patient-specific risk protocol. *Am J Obstet Gynecol* **185**:289–293.

Hoh J, Ott J (2003) Mathematical multi-locus approaches to localizing complex human trait genes. *Nature Rev Genet* **4**:701–709.

Holt PR, Atillasoy EO, Gilman J, Guss J, Moss SF, Newmark H, Fan K, Yang K, Lipkin M (1998) Modulation of abnormal colonic epithelial cell proliferation and differentiation by low-fat dairy foods. *JAMA* **280**:1074–1079.

Holtzman NA (1999) Are genetic tests adequately regulated? *Science* **286**:409.

Holtzman NA (2000) Genetic testing. *Am Fam Physic* **61**(4):950–956.

Holtzman NA, Marteau TM (2000) Will genetics revolutionize medicine? *New Engl J Med* **343**:141–144.

Holtzman N, Watson MS, eds (1997) *Promoting Safe and Effective Genetic Testing in the United States: Final Report of the Task Force on Genetic Testing* Johns Hopkins University

Press, Baltimore: Available at: <http://www.genome.gov/10002393>. Accessed November 8, 2004.

Holtzman S, Ozanne E, Carone B, Goldstein MK, Steinke G, Timbs J (1999) Decision analysis and Alzheimer disease: Three case studies. *Genet Test* 3(1):71–83.

Hood v. Phillips (1977) 554 S.W. 2d. 160, 165 (Tex.).

Hook EB, Cross PK (1987a) Extra structurally abnormal chromosomes (ESAC) detected at amniocentesis: Frequency in approximately 75,000 prenatal cytogenetic diagnoses and associations with maternal and paternal age. *Am J Hum Genet* **40**:83–101.

Hook EB, Cross PK (1987b) Rates of mutant and inherited structural cytogenetic abnormalities detected at amniocentesis: Results on about 63,000 fetuses. *Ann Hum Genet* **51**:27–55.

Hook EB, Cross PK, Scheinemachers DM (1983) Chromosomal abnormality rates at amniocentesis and in live-born infants. *JAMA* **249**:2034–2038.

Hook EB, Topol BB, Cross PK (1989) The natural history of cytogenetically abnormal fetuses detected at midtrimester amniocentesis which are not terminated electively: New data and estimates of the excess and relative risk of later fetal death associated with the 47, +21 and some other abnormal karyotypes. *Am J Hum Genet* **45**:855–861.

Hope v. Landau (1986) 398 Mass. 738, 500 NE 2d 809.

Hopper KD, TenHave TR, Tully DA, Hall TE (1998) The readability of currently used surgical/procedure consent forms in the United States. *Surgery* **123**(5):496–503.

H. (R.) v. Hunter, Rosenbloom, Viner (1996) 32 C.C.L.T. (2d) 44 (Ont. Gen div.). The brother of the plaintiff mother had died of Duchenne muscular dystrophy. She sought counseling in 1981 from the geneticist H, who was held by the court to have acted properly. Seven years later, while under the care of the doctors R and V, she gave birth to her first child and two years later to her second child. Both children suffered from Duchenne muscular dystrophy. The jury found that the doctors R and V should have referred the plaintiff mother for additional genetic counseling.

Horowitz M, Sundin E, Zanko A, Lauer R (2001) Coping with grim news from genetic tests. *Psychosomatics* **42**(2):100–105.

Hoskins KF, Stopfer JE, Calzone KA, Merajver SD, Rebbeck TR, Garber JE, Weber BL (1995) Assessment and counseling for women with a family history of breast cancer. A guide for clinicians. *JAMA* **273**(7):577–585.

Houston TK, Sands DZ, Nash BR, Ford DE (2003) Experiences of physicians who frequently use e-mail with patients. *Health Commun* **15**(4):515–525.

Howse JL, Katz M (2000) The Importance of Newborn Screening. *Pediatrics* **106**:595.

Huang X, Chen PC, Poole C (2004a) APOE epsilon2 allele associated with higher prevalence of sporadic Parkinson disease. *Neurology*. **62**(12):2198–2202.

Huang Y, Cheung L, Rowe D, Halliday G (2004b) Genetic contributions to Parkinson's disease. *Brain Res Rev* **46**(1):44–70.

Hubbard R, Lewontin RC (1996) Pitfalls of genetic testing. *N Engl J Med* **334**(18):1192–1194.

Hudson KL, Rothenberg KH, Andrews LB, Kahn MJE, Collins FS (1995) Genetic discrimination and health insurance: An urgent need for reform. *Science* **270**:391–393.

Huggins M, Bloch M, Wiggins S, Adam S, Suchowersky O, Trew M, Klimek M, Greenberg CR, Eleff M, Thompson LP, et al (1992) Predictive testing for Huntington disease in Canada: Adverse effects and unexpected results in those receiving a decreased risk. *Am J Med Genet* **42**(4):508–515. "Contrary to expectations, approximately 10% of persons with a decreased risk result have had psychological difficulties coping with their new status."

Huggins M, Hahn C, Costa T (1996) Staying informed and recontacting patients about research advances: A study of patient attitudes. *Am J Hum Genet* **59**:335.

Hughes C, Gomez-Caminero A, Benkendorf J, Kerner J, Isaac C, Barter J, Lerman C (1997) Ethnic differences in knowledge and attitudes about BRCA1 testing in women at increased risk. *Pt Edu Counsel* **32**(1–2):51–62.

Hughes C, Lerman C, Schwartz M, Peshkin BN, Wenzel L, Narod S, Corio C, Tercyak KP, Hanna D, Isaacs C, Main D (2002) All in the family: Evaluation of the process and content of sisters' communication about BRCA1 and BRCA2 genetic test results. *Am J Med Genet* **107**(2):143–150.

Huibers AK, Van't Spijker A (1998) The autonomy paradox: Predictive genetic testing and autonomy: Three essential problems. *Pt Edu Counsel* **35**:53–62.

Hull SC, Prasad K (2001a) Reading between the lines: Direct-to-consumer advertising of genetic testing in the USA. *Reprod Health Matters* **9**(18):44–48.

Hull SC, Prasad K (2001b) Reading between the lines: Direct-to-consumer advertising of genetic testing. *Hastings Center Report* **31**(3):33–35.

Hull NE, Hoffer PC, *Roe v. Wade* (2001) *The Abortion Controversy in American History*, Lawrence, Kansas: University of Kansas Press.

Hulley S, Furberg C, Barrett-Connor E, Cauley J, Grady D, Haskell W, Knopp R, Lowery M, Satterfield S, Schrott H, Vittinghoff E, Hunninghake D, HERS Research Group (2002) Non-cardiovascular disease outcomes during 6.8 years of hormone therapy: Heart and Estrogen/Progestin Replacement Study follow-up (HERS II). *JAMA* **288**:58–66.

Human Genetics Commission (2003) Genes Direct. http://www.hgc.gov.uk/genesdirect/#report. The Commission proposed "stricter control" but did not support a statutory ban.

Human Genome Project Information (2004) http://www.ornl.gov/sci/techresources/Human_Genome/home.shtml.

Human Genome Variation Society (Carlton South, VIC, Australia) (2002) http://www.hgvs.org/.

Humana of Kentucky, Inc. v. McKee (1992) 834 S.W.2d 711. Evidence indicated that the hospital did not provide mandatory testing for PKU at birth, or if they did, negative results were incorrectly reported. The jury found the hospital negligent in conducting PKU testing. The appellate court affirmed this judgment.

Humphreys L, Hunter AGW, Zimak A, O'Brien A, Korneluk Y, Cappelli M (2000) Why patients do not show for their appointments at a genetics clinic. *J Med Genet.*

Hunt SC, Williams RR, Barlow GK (1986) A comparison of positive family history definitions for defining risk of future disease. *J Chron Dis* **39**:809–821.

Hunt LM, De Voogd KB, Castaneda H (2004) The routine and the traumatic in prenatal genetic diagnosis: Does clinical information inform patient decision making? *Patient Ed Counsel.* in press.

Hunter A, Wright P, Cappelli M, Kasaboski A, Surh L (1998) Physician knowledge and attitudes towards molecular genetic (DNA) testing of their patients. *Clin Genet* **53**:447–455.

Hunter AG, Sharpe N, Mullen M, Meschino WS (2001) Ethical, legal, and practical concerns about recontacting patients to inform them of new information: The case in medical genetics. *Am J Med Genet* **103**(4):265–276.

Hustead JL, Goldman J (2002) Genetics and privacy. *Am J Law Med* **28**(2–3):285–307.

Hutton JL, Ashcroft RE (2000) Some popular versions of uninformed consent. *Health Care Anal* **8**(1):41–53.

Hutton M, Lendon CL, Rizzu P, Baker M, Froelich S, Houlden H, Pickering-Brown S, Chakraverty S, Isaacs A, Grover A, Hackett J, Adamson J, Lincoln S, Dickson D, Davies P, Petersen RC, Stevens M, de Graaff E, Wauters E, van Baren J, Hillebrand M, Joosse M, Kwon JM, Nowotny P, Che LK, Norton J, Morris JC, Reed LA, Trojanowski J, Basun H, Lannfelt L, Neystat M, Fahn S, Dark F, Tannenberg T, Dodd PR, Hayward N, Kwok JB, Schofield PR,

Andreadis A, Snowden J, Craufurd D, Neary D, Owen F, Oostra BA, Hardy J, Goate A, van Swieten J, Mann D, Lynch T, Heutink P (1998) Association of missense and 5'-splice-site mutations in tau with the inherited dementia FTDP-17. *Nature* **393**(6686):702–705.

Hyer W, Fell JME (2001) Screening for familial adenomatous polyposis. *Arch Dis Child* **84**:377–380.

Hyett JA, Perdu M, Shariand GK, Snijders RS, Nicolaides KH (1997) Increased nuchal translucency at 10–14 weeks of gestation as a marker for major cardiac defects. *Ultrasound Obstet Gynecol* **10**:242–246.

Iafolla AK, Thompson Jr. RJ, Roe CR (1994) Medium-chain acyl-coenzyme A dehydrogenase deficiency: Clinical course in 120 affected children. *J. Pediat 124*:409–415.

ICH (2001) Harmonised Tripartite Guideline: Guideline for good clinical practice.

Imperatore G, Pinsky L, Motulsky A, Reyes M, Bradley L, Burke W (2003) Hereditary hemochromatosis: Perspectives of public health, medical genetics, and primary careenetics in medicine. *Genet Med* **5**(1):1–8.

Infelfinger FJ (1980) Arrogance. *New Engl J Med* **303**:507–511.

Inoue K, Lupski JR (2003) Mendelian, nonmendelian, multigenic inheritance, and complex traits. In: Rosenberg RN, Prusiner SB, DiMauro S, Barchi RL, Nestler EJ, eds. *The Molecular and Genetic Basis of Neurologic and Psychiatric Disease*, Philadelphia: Elsevier, pp. 33–50.

Institute Curie (2004) "The European Patent Office has revoked the Myriad Genetics Patent, press release, online:" Institute Curie <www.curie.fr/upload/presse/190504_gb.pdf>.

Institute of Medicine (2004) Understanding and Eliminating Racial and Ethnic Disparities in Health Care. http://www.iom.edu/IOM/IOMHome.nsf/Pages/Ethnic+disparities.

International Bioethics Committee of the United Nations Educational, Scientific, and Cultural Organization (UNESCO) (2003) Report on pre-implantation genetic diagnosis and germ-line intervention. Available at http://unesdoc.unesco.org/images/0013/001383/138326e.pdf.

International Human Genome Consortium (2004) Finishing the euchromatic sequence of the human genome. *Nature* **431**:931–945.

International Huntington Association and World Federation of Neurology (1994) Guidelines for the molecular genetic predictive test in Huntington's disease. *J Med Genet* **31**:555–559.

International Working Group on Preimplantation Genetics (2001) Report of the 11th Annual Meeting of the International Working Group on Preimplantation Genetics: Preimplantation Genetic Diagnosis: Experience of 3000 Clinical Cycles. *Reprod BioMedicine Online* **3**(1):49–53.

Irvin v. Smith (2001) 31 P. 3d 934 (Kan.). Indirect contact with the patient does not preclude the finding of a physician–patient relationship where advice is given from one medical professional to another; however, the physician's express or implied consent to advise or to treat the patient is required.

ISCN (1995) *An International System for Human Cytogenetic Nomenclature*, Mitelman F, ed. Basel: S. Karger.

Isotalo PA, Wells GA, Donnelly JG (2000) Neonatal and fetal methylenetetrahydrofolate reductatse genetic polymorphisms: An examination of C677T and A1298C mutations. *Am J Hum Genet* **67**:986–990.

Iuculano CK (1987) AIDS and insurance: The rationale for AIDS-related testing. *Harvard Law Rev* **100**:1806–1825.

Ives EJ, Henick P, Levers MI (1979) The malformed newborn-telling the parents. *Birth Defects Orig Artic Ser* **15**(5C):223–231.

Jacobson GM, Veach PM, LeRoy BS (2001) A survey of genetic counselors' use of informed consent documents for prenatal genetic counseling sessions. *J Genet Counsel* **10**(1):3–24.

Jadad AR (1999) Promoting partnerships: Challenges for the Internet age. *BMJ* **319**:761–764.

Jadad AR, Gagliardi A (1998) Rating health information on the Internet: Navigating to knowledge or to Babel? *JAMA* **279**:611–614.

Jadad AR, Haynes RB, Hunt D, Browman GP (2000) The Internet and evidence-based decision-making: A needed synergy for efficient knowledge management in health care. *CMAJ* **162**(3):362–365.

Jamain S, Quach H, Betancur C, Rastam M, Colineaux C, Gillberg IC, Soderstrom H, Giros B, Leboyer M, Gillberg C, Bourgeron T, Paris Autism Research International Sibpair Study (2003) Mutations of the X-linked genes encoding neuroligins NLGN3 and NLGN4 are associated with autism. *Nat Genet* **34**(1):27–29.

James G v. Caserta (1985) 332 SE.2d 872 (W. Va.).

Jansen LA, Ross LF (2001) The ethics of preadoption genetic testing. *Am J Med Genet* **104**:214–220.

Jarvinen HJ, Aarnio M, Mustonen H, Aktan-Collan K, Aaltonen LA, Peltomaki P, De La Chapelle A, Mecklin JP (2000) Controlled 15-year trial on screening for colorectal cancer in families with hereditary nonpolyposis colorectal cancer. *Gastroenterology* **118**:829–834.

Jasanoff S (2003) Technologies of humility: Citizen participation in governing science. *Minerva* **41**:223–244.

Javitt GH (2004) New possibilities in first trimester prenatal screening for fetal chromosomal abnormalities. *Genet Public Policy Centre*. http://www.dnapolicy.org/.

Jedlicka-Kohler I, Gotz M, Eichler I (1994) Utilization of prenatal diagnosis for cystic fibrosis over the past seven years. *Pediatrics* **94**:13 at 15.

Jemal A, Clegg LX, Ward E, Ries LAG, Wu X, Jamison PM, Wingo PA, Howe HL, Anderson RN, Edwards, BK (2004) Annual Report to the Nation on the Status of Cancer, 1975–2001, with a Special Feature Regarding Survival. *Cancer* **101**(1):3–27.

Jimison HB, Sher PP, Appleyard R, LeVernois Y (1998) The use of multimedia in the informed consent process. *J Am Med Informatics Assoc* **5**(3):245–256.

Joffe C (2003) *Roe v. Wade* at 30: What Are the Prospects for Abortion Provision? *Perspect Sex Reprod Health* **35**(1):29–33.

Johnson AM (1981) In: Haddow JE, Wald NJ, eds. Alpha-Fetoprotein Screening: The Current Issues, Scarborough, ME: Foundation for Blood Research, p. 10.

Johnson v. Agoncillo (1994) 183 Wis. 2d 143, 515 N.W. 2d 508 (App.).

Johnson v. Kokemoor (1996) 199 Wis.3d 615, 545 N.W.2d 495, 504–10 (Wis.). Some courts have expanded what constitutes a "material" risk. In *Johnson v. Kokemoor*, the court stated that a surgeon had a duty to disclose a lack of experience and advise that more experienced surgeons in better equipped facilities were available. An analogous line of reasoning, although not within the context of informed consent, was found in the Canadian case of *Kangas v. Parker*, [1976] 5 W.W.R. 25 (Sask. Q.B.): A patient was not given the option of having dental work performed in a hospital as opposed to an office. The Court determined that the failure to give the patient the choice of facilities constituted a failure to meet the standard of care. However, other American courts have held that, " . . . information personal to the physician, whether solicited by the patient or not, is irrelevant to the doctrine of informed consent"—*Duttry v. Patterson*, 741 A. 2d, 1999; 771 A.2d 1255, 1258–1259 (Pa. 2001); also see *Leger v. Louisiana Med. Mut. Ins. Co.*, 732 So. 2d 654, 658 (La. Ct. App. 1999); *Aceto v. Dougherty*, 615 N.E.2d 188, 191 (Mass. 1993); *Whiteside v. Lukson*, 947 P.2d 1263, 1265 (Wash. Ct. App. 1997). In Canada, courts have rejected the requirement that a physician must disclose her or his experience with a procedure: *Hopp v. Lepp* (1980), 112 D.L.R. (3d) 67 (S.C.C.); and see Johnson BJ (2001). Recent decisions: Must doctors disclose their own personal risk factors? *Health Law Rev* **10**: 18–20. The author reviews the contention that a doctor has a duty to disclose her or his medical condition.

Johnson AM, Palomaki GE, Haddow JE (1990a) Maternal serum alpha-fetoprotein levels in pregnancies among black and white women with fetal open spina bifida: A United States collaborative study. *Am J Obstet Gynecol* **162**:328–331.

Johnson AM, Palomaki GE, Haddow JE (1990b) The effect of adjusting maternal serum alpha-fetoprotein levels for maternal weight in pregnancies with fetal open spina bifida: A United states collaborative study. *Am J Obstet Gynecol* **163**:9–11.

Johnson KA, Trimbath JD, Petersen GM, Griffin CA, Giardiello FM (2002) Impact of genetic counseling and testing on colorectal cancer screening behavior. *Genet Test* **6**(4):303–306.

Joint Commission on Accreditation of Healthcare Organizations (JCAHO) (2004) Oakbrook Terrace, IL. http://www.jcaho.org/.

Jolly v. Eli Lilly Co. (1988) 751 P. 2d. 923 (Cal.).

Jones v. Karraker (1982) 109 Ill. App. 3d 363, 440 N.E. 2d 420, aff'd. 98 Ill. 487, 457 N.E. 2d. 23 (3rd. Dist).

Jones v. Rostvig (1999) 44 C.C.L.T. (2D) 313 (B.C.S.C.).

Jones v. United States (1996) 933 F. Supp. 894, 902 (N.D. Cal.).

Jonsen AR, Seigler M, Winslade WJ (1998) *Clinical Ethics*, 4th ed. New York: McGraw-Hill Health Professions Division. Also see Laine C, Davidoff F (1996) Patient-centered medicine: A professional evolution. *JAMA* **275**:152–156.

Jordan v. Bogner (1993) 844 P 2d 664 (Colo.), rehearing denied, 1993 Colo LEXIS 118 (Colo).

Jorde LB, Carey JC, Bamshad MJ and White RL (2003) Medical Genetics, 3rd edition, Mosby, St. Louis.

Jorde LB, Hasstedt SJ, Ritvo ER, Mason-Brothers A, Freeman BJ, Pingree C, McMahon WM, Petersen B, Jenson WR, Mo A (1991) Complex segregation analysis of autism. *Am J Hum Genet* **49**(5):932–938.

Joy v. Eastern Maine Medical Center (1987) 529 A. 2d. 1364 (Me).

Joyal v. Starreveld (1996) 37 Alta. L.R. 3d. 19 (Q.B.).

Jubelirer SJ (1991) Level of reading difficulty in educational pamphlets and informed consent documents for cancer patients. *W V Med J* **87**:554–557.

Julian-Reynier C, Eisinger F, Chabal F, Lasset C, Nogues C, Stoppa-Lyonnet D, Vennin P, Sobol H (2000) Disclosure to the family of breast/ovarian cancer genetic test results: Patient's willingness and associated factors. *Am J Med Genet* **94**(1):13–18.

Julian-Reynier C, Welkenhuysen M, Hagoel L, Decruyenaere M, Hopwood P (2003) Risk communication strategies: State of the art and effectiveness in the context of cancer genetic services. *Eur J Hum Genet* **11**(10):725–736.

Jutras D (1993) Clinical practice guidelines as legal norms. *Can Med Ass'n J* **148**:905; see also Caulfield TA (1994) Health care reform: Can tort law meet the challenge? *Alta Law Rev* **32**:685.

Jorvinen HJ, Aarnio M, Mustonen H, Aktan-Collan K, Aaltonen LA, Peltomaki P, De La Chapelle A, Mecklin JP (2000) Controlled 15-year trial on screening for colorectal cancer in families with hereditary nonpolyposis colorectal cancer. *Gastroenterology* **118**:829–834.

Kaback M, Lim-Steele J, Dabholkar D, Brown D, Levy N, Zeiger K (1993) Tay-Sachs disease—Carrier screening, prenatal diagnosis, and the molecular era. An international perspective, 1970 to 1993. The International TSD Data Collection Network. *JAMA* **270**(19):2307–2315.

Kadouri L, Kote-Jarai Z, Easton DF, Hubert A, Hamoudi R, Glaser B, Abeliovich D, Peretz T, Eeles RA (2004) Polyglutamine repeat length in the A1B1gene modifies breast cancer susceptibility in BRCA1 carriers. *Int J Cancer* **108**:399–403.

Kahnerman D, Slovic P, Tversky A (1982) *Judgment under Uncertainty: Heuristics and Biases.* New York: Cambridge University Press.

Kalet A, Roberts JC, Fletcher R (1994) How do physicians talk with their patients about risks? *J Gen Intern Med* **9**(7):402–404.

Kananura C, Haug K, Sander T, Runge U, Gu W, Hallmann K, Rebstock J, Heils A, Steinlein OK (2002) A splice-site mutation in GABRG2 associated with childhood absence epilepsy and febrile convulsions. *Arch Neurol* **59**:1137–1141.

Kanemoto K, Kawasaki J, Miyamoto T, Obayashi H, Nishimura M (2000) Interleukin (IL)1beta, IL-1alpha, and IL-1 receptor antagonist gene polymorphisms in patients with temporal lobe epilepsy. *Ann Neurol* **47**:571–574.

Kapp MB (2000) Physicians legal duties regarding the use of genetic tests to predict and diagnose Alzheimer disease. *J Legal Med* **21**:445–475.

Karjala D (1992) A legal research agenda for the Human Genome Initiative. *Jurimetrics* **32**:121–311.

Karlin v. Foust (1999) 188 F.3d 446 (7th Cir.). A Wisconsin statute, pertaining to informed consent for an abortion, specified designated topics for discussion. Although the court recognized that a physician can use her or his best medical judgment in determining the exact nature and content of the information, a physician is strictly liable for omitting topics specified in the statute.

Karliner LS, Perez-Stable EJ, Gildengorin G (2004) The language divide. The importance of training in the use of interpreters for outpatient practice. *J Gen Intern Med* **19**(2):175–183.

Kash KM (1995) Psychosocial and ethical implications of defining genetic risk for cancers. *Ann NY Acad Sci* **768**:41–52.

Kash K (1999) Breast Cancer Anxiety Scale (unpublished).

Kash KM, Lerman C (1998) *Psychological, Social, and Ethical Issues in Gene Testing*. Psychooncology. New York, NY: Oxford University Press; Chapter 18, pp. 196–210.

Kash K, Holland J, Halper M, Miller D (1992) Psychological distress and surveillance behaviours of women with a family history of breast cancer. *J Nat'l Ca Inst* **84**:24–30.

Kash KM, Holland JC, Osborne MP, Miller DG (1995) Psychological counseling strategies for women at risk of breast cancer. *J Natl Cancer Inst Monogr* **17**:73–79.

Kash KM, Ortega-Verdejo K, Dabney MK, Holland JC, Miller DG, Osborne MP (2000) Psychosocial aspects of cancer genetics: Women at high risk for breast and ovarian cancer. *Semin Surg Oncol* **18**(4):333–338.

Kass NE, Hull SC, Natowicz MR, Faden RR, Plantinga L, Gostin LO, Slutsman J (2004) Medical privacy and the disclosure of personal medical information: The beliefs and experiences of those with genetic and other clinical conditions. *Am J Med Genet* **128A**(3):261–270.

Kassama v. Magat (2002) 368 Md. 113, 792 A. 2d 1102. Plaintiff claimed that the physician was negligent for failing to communicate in a timely manner the results of an alpha-fetoprotein test showing an increased risk for Down's syndrome. The court denied liability, stating that life could not be considered an "injury."

Kates N, Craven M, Crustolo AM, Nikolaou L, Allen C (1997a) Integrating mental health services within primary care. *A Canadian program Gen Hosp Psychiatry* **19**:324–332.

Kates N, Craven M, Crustolo AM, Nikolaou L, Allen C, Farrar S (1997b) Sharing care: The psychiatrist in the family physicians office. *Cdn J Psychiatry* **42**:960–965.

Kato T (2001) Molecular genetics of bipolar disorder. *Neurosci Res* **40**:105–113.

Katskee v. Blue Cross/Blue Shield of Nebraska (1994) 245 Neb. 808, 515 N.W.2d 645. (Neb.). Also see Glazier AK (1997) Genetic predispositions, prophylactic treatments and private health insurance: Nothing is better than a good pair of genes. *Am J Law Med* **23**:45–68.

Katz J (1977) Informed consent—A fairy tale? *U Pitt L Rev* **39**:137 at 147: "[S]afeguarding self-determination requires asking the patient whether he understands what has been explained to him in order to assess whether his informational needs have been satisfied."

Katz J (1984) *The Silent World of Doctor and Patient.* New York: Free Press, at 28–29.

Katz J (1993) Informed consent—Must it remain a fairy tale? *J Contemp Health Law, Pol'Y* **10**:69; 91, at 73–74. The author argues that informed consent will remain a "charade" unless the physicians adopt cooperative decision making with patients.

Katz SJ, Moyer CA, Cox DT, Stern DT (2003) Effect of a triage-based E-mail system on clinic resource use and patient and physician satisfaction in primary care: A randomized controlled trial. *J Gen Intern Med* **18**(9):736–744.

Kauff ND, Satagopan JM, Robson ME, Scheuer MS, Hensley M, Hudis CA, Ellis NA, Boyd J, Borgen PI, Barakat RR, Norton L, Offit K (2002) Risk-reducing salpingo-oophorectomy in women with a BRCA1 or BRCA2 mutation. *N Engl J Med* **346**:1609–1615.

Kaufman DW, Palmer JR, de Mouzon J, Rosenberg L, Stolley PD, Warshauer ME, Zauber AG, Shapiro S (1991) Estrogen replacement therapy and the risk of breast cancer: Results from the case-control surveillance study, *Am J Epidemiol* **134**:1375–1385.

Kawas CH, Katzman R (1999) Epidemiology of dementia and Alzheimer disease. In: Terry RD, Katzman R, Bick KL, Sisodia SS, eds. Alzheimer Disease, Philadelphia: Lippincott Williams & Wilkins, pp. 95–116.

Kee F, Tiret L, Robo JY, Nicaud V, McCrum E, Evans A, Cambien F (1993) Reliability of reported family history of myocardial infarction. *Br Med J* **307**:1528–1530.

Keel v. Banach (1993) 624 So. 2D 1022.

Keen RW, Hart DJ, Arden NK, Doyle DV, Spector TD (1999) Family history of appendicular fracture and risk of osteoporosis: A population-based study. *Osteoporos Int* **10**:161–166.

Kehoe PG, Katzov H, Feuk L, Bennet AM, Johansson B, Wiman B, de Faire U, Cairns NJ, Wilcock GK, Brookes AJ, Blennow K, Prince JA (2003) Haplotypes extending across ACE are associated with Alzheimer's disease. *Hum Mol Genet* **12**(8):859–867.

Keller M, Jost R, Kadmon M, Wullenweber HP, Haunstetter CM, Willeke F, Jung C, Gebert J, Sutter C, Herfarth C, Buchler MW (2004) Acceptance of and attitude toward genetic testing for hereditary nonpolyposis colorectal cancer: A comparison of participants and nonparticipants in genetic counseling. *Dis Colon Rectum* **47**(2):153–162.

Kelley RI, Hennekam RCM (2000) The Smith-Lemli-Opitz syndrome. *J Med Genet* **37**:321–335.

Kelly P (1987) Risk counselling for relatives of cancer patients: New information, new approaches. *J Psychosoc Onc* **5**:65–79.

Kelly P (1992) Informational needs of individuals and families with hereditary cancers. *Semin Oncol Nurs* **8**:288–292.

Kelly PT (2000) *Assess Your True Risk of Breast Cancer.* New York: Henry Holt.

Kelly PT (2003) Editorial: Women's Health Initiative Results: Breast Cancer Risk in Perspective. The Breast Journal 9: 449–451. Also see: Kelly PT (2002) Breast cancer risks: Some clinically useful approaches. *Current Women's Health Reports* **2**:128–133.

Kennedy D, Chitayat, D, Winsor EJ, Silver M, Toi A (1998) Prenatally diagnosed neural tube defects: Ultrasound, chromosome, and autopsy or postnatal findings in 212 cases. *Am J Med Genet* **26**;77(4):317–321.

Kessler S (1979) The genetic counselor as a psychotherapist. In: Capron AM, Lappe M, Murray RE, Pourledge TM, Twiss SB, Bergsma D, eds. *Genetic Counseling: Facts, Values and Norms.* New York: Alan R. Liss, for National Foundation—March of Dimes, *BD:OAS* **XV**(2):187–200.

Kessler S (1989) Book reviews. *Am J Hum Genet* **42**:392–393.

Kessler S (1992) Process issues in genetic counseling. *Birth Defects Orig Artic Ser* **28**(1):1–10.

Kessler S (1993) Forgotten person in the Huntington disease family. *Am J Med Genet* **48**(3):145–150.

Kessler S (1997a) Psychological aspects of genetic counseling. XI. Nondirectiveness revisited. *Am J Med Genet* **72**:164–171.

Kessler S (1997b) Psychological aspects of genetic counseling. IX. Teaching and counseling. *J Genet Counsel* **6**:287–295.

Kessler S (1999) Psychological aspects of genetic counseling. XIII. Empathy and decency. *J Genet Counsel* **8**:333–344.

Kessler S (2001) Psychological aspects of genetic counseling. XIV. Nondirectiveness and counseling skills. *Genet Test* **5**:187–191.

Kessler S, Bloch M (1989) Social system responses to Huntington disease. *Fam Process* **28**(1):59–68.

Kessler S, Levine EK (1987) Psychological Aspects of Genetic Counseling. IV. The Subjective Assessment of probability. *Amer Jour Med Genet* **28**:361–370.

Kessler RC, McGonagle KA, Zhao S, Nelson CB, Hughes M, Eshleman S, Wittchen HU, Kendler KS (1994) Lifetime and 12-month prevalence of DSM-III-R psychiatric disorders in the United States. Results from the National Comorbidity Survey. *Arch Gen Psychiatry* **51**(1):8–19.

Ketchup v. Howard (2000) 247 Ga. App. 54, 543 S.E.2d 371.

Khoury MJ, Thrasher JF, Burke W, Gettig EA, Fridinger F, Jackson R (2000) Challenges in communicating genetics: A public health approach. *Genet Med* **2**(3):198–202.

Khoury M, McCabe LL, McCabe ER (2003) Population screening in the age of genomic medicine. *New Engl J Med* **348**(1):50–58.

Khoury MJ, Yang Q, Gwinn M, Little J, Dana Flanders W (2004) An epidemiologic assessment of genomic profiling for measuring susceptibility to common diseases and targeting interventions. *Genet Med* **6**(1):38–47.

Kieff FS (2001) Facilitating scientific research: Intellectual property rights and the norms of science—a response to Rai and Eisenberg. *Northwestern L Rev* **95**(2):691–706.

Kievit W, de Bruin JH, Adang EM, Ligtenberg MJ, Nagengast FM, van Krieken JH, Hoogerbrugge N (2004) Current clinical selection strategies for identification of hereditary non-polyposis colorectal cancer families are inadequate: A meta-analysis. *Clin Genet* **65**:308–316.

Kim P, Eng TR, Deering MJ, Maxfield A (1999) Information in practice: Published criteria for evaluating health related web sites. *BMJ* **318**:647–649.

King MC, Marks JH, Mandell JB, New York Breast Cancer Study Group (2003) Breast and ovarian cancer risks due to inherited mutations in BRCA1 and BRCA2. *Science* **302**:643–646.

Kinney A, DeVellis B, Skrzynia C, Millikan R (2001) Genetic testing for colorectal carcinoma susceptibility. *Cancer* **91**:57–65.

Kinsella T (1988) Medical ethics committees: A physician's perspective. *Ann RCPSC* **21**:231–234.

Kip KE, McCreath HE, Roseman JM, Hulley SB, Schreiner PJ (2002) Absence of risk factor change in young adults after family heart attack or stroke. The CARDIA Study. *Am J Prev Med* **22**:258–266.

Kirova YM, Stoppa-Lyonnet D, Savignoni A, Sigal-Zafrani B, Clough KB, Barbier N, Fourquet A (2004) Risk of breast recurrence in relation to BRCA1/2 mutation status following breast-conserving surgery and radiotherapy. *Eur J Cancer Suppl* **2**:150–151 (Abstr.).

Kirschner KL, Ormond KE, Gill CJ (2000) The impact of genetic technologies on perceptions of disability. *Qual Manag Health Care* **8**(3):19–26. The study reports that communication between patients can be biased, at least in part, by the professionals' prior experiences, knowledge, and attitudes toward disability.

Kitada T, Asakawa S, Hattori N, Matsumine H, Yamamura Y, Minoshima S, Yokochi M, Mizuno Y, Shimizu N (1998) Mutations in the parkin gene cause autosomal recessive juvenile parkinsonism. *Nature* **392**:605–608.

Kitch E (1977) The nature and function of the patent system. *J Law Economics* **20**:265–290.

Klanica K (2004) Genetic information and health insurance: A report on the law in the United States. Pages 10–11 http://www.genelaw.info/reports/health.pdf.

Klein BE, Klein R, Moss SE, Cruickshanks KJ (1996) Parental history of diabetes in a population-based study. *Diabetes Care* **19**:827–830.

Klimchuk D, Black V (1997) Negligent failure to warn—Causation: *Arndt v. Smith. Can Bar Rev* **76**:569–588.

Knight GJ, Cole LA (1991) Measurement of choriogonadotropin free beta-subunit: An alternative to choriogonadotropin in screening for fetal Down's syndrome? *Clin Chem* **37**:779–782.

Knight SJL, Flint J (2000) Perfect endings: A review of subtelomeric probes and their use in clinical diagnosis. *J Med Genet* **37**:401–409.

Knight GJ, Palomaki GE, Neveux LM, Haddow JE, Lambert-Messerlian GM (2001) Clinical validation of a new dimeric inhibin-A assay suitable for second trimester Down's syndrome screening. *J Med Screen* **8**:2–7.

Knoppers BM (1999) Status, sale and patenting of human genetic material: An international survey. *Nat Genet* **22**(1):23–26.

Knoppers BM, Isasi RM (2004) Regulatory approaches to reproductive genetic testing. *Hum Reprod* **19**(12):2695–701. Epub 2004 Sept 23.

Knoppers BM, Joly Y (2004) Physicians, genetics, and life insurance. *CMAJ* **170**(9):1421–1423. Also see Knoppers BM, Lemmons T, Godard B, Joly Y, Avard D, Clark T, Hamet P, Hoy M, Lanctot S. Lowden S, et al. (2004) Genetics and life insurance in Canada: Points to consider. *CMAJ* **170**(9); online: 1–3. http://www.cmaj.ca/cgi/data/170/9/1421/DC2/1. Joly Y, Knoppers BM, Godard B (2003) Genetic information and life insurance: A "real" risk? *Eur J Hum Genet* **11**(8):561–564.

Knoppers BM, Strom C, Wright Clayton E, Murray T, Fibison W, Luther L (1998) Professional Disclosure of Familial Genetic Information: Report of the ASHG Social Issues Subcommittee on Familial Disclosure. *Am J Hum Genet* **62**:474–483. http://genetics.faseb.org/genetics/ashg/policy/pol-29.htm.

Knudsen AL, Birgaard ML, Bulow S (2003) Attenuated familial adenomatous polyposis (AFAP). A review of the literature. *Familial Cancer* **2**:43–55.

Knutsen SF, Knutsen R (1991) The Tromso Survey: The family intervention study—The effect of intervention on some coronary risk factors and dietary habits, a 6-year follow-up. *Prev Med* **20**:197–212.

Koch RK (1999) Issues in newborn screening for phenylketonuria. *Am Fam Physician* **60**(5):1462–1466.

Koehly LM, Peterson SK, Watts BG, Kempf KK, Vernon SW, Gritz ER (2003) A social network analysis of communication about hereditary nonpolyposis colorectal cancer genetic testing and family functioning. *Cancer Epidemiol Biomarkers Prev* **12**(4):304–313.

Kohoutek v. Hafner (1986) 383 NW.2d 295, 89 ALR 4th 783 (Minn.).

Kohut R, Dewey D, Love E (2002) Women's knowledge of prenatal ultrasound and informed choice. *J Genet Couns* **11**(4): 265–276.

Kolata G (1997) Advent of testing for breast cancer leads to fears of disclosure and discrimination. *New York Times* February 4: C1–C3.

Kolata G (1998) Genetic testing falls short of public embrace. *New York Times* March 27:A16.

Kolker PL (1997) Patents in the pharmaceutical industry. *Patent World* **88**:34–37.

Koscica KL, Canterino JC, Harrigan JT, Dalaya T, Ananth CV, Vintzileos AM (2001) Assessing genetic risk: Comparison between the referring obstetrician and genetic counselor. *Am J Obstet Gynecol* **185**(5):1032–1034.

Kottow MH (1986) Medical confidentiality: An intransigent and absolute obligation. *J Med Ethics* **12**(3):117–122.

Kottow M (2004) The battering of informed consent. *J Med Ethics* **30**(6):565–569.

Kowdley KV, Tait JF, Bennett RL, Motulsky AG (2004) HFE-associated hereditary hemochromatosis. Genetests (www.genetests.org).

Kramer v. Milner (1994) 265 Ill. App. 3d 875, 639 N.E. 2d 157, 203 Ill. Dec. 118.

Krangle (Guardian ad litem of) v. Brisco (1997) 154 D.L.R. (4th) 707 (B.C.S.C.); [2002] 1 S.C.R. 205. Parents of a child with Down syndrome were not informed of the availability of prenatal genetic testing.

Krauser, J (1989) The future of medicine. *Ont Med Rev* **56**:36 at 36–38, 61–67.

Kriege MK, Brekelmans CTM, Boetes C, Besnard PE, Zonderland HM, Obdeijn IM, Manoliu RA, Kok T, Peterse H, Tilanus-Linthorst MMA, Muller SH, Meijer S, Oosterwijk JC, Beex LVAM, Tollenaar RAEM, deKoning HJ, Rutgers RJT, Klijn JGM, Magnetic Resonance Imaging Screening Study Group (2004) Efficacy of MRI and mammography for breast-cancer screening in women with a familial or genetic predisposition. *N Engl J Med* **351**:427–437.

Kuczewski MG, Marshall P (2002) The decision dynamics of clinical research: The context and process of informed consent. *Med Care* **40**(9 Suppl):V 45–55.

Kugler SL, Stenroos ES, Mandelbaum DE, Lehner T, McKoy VV, Prossick T, Sasvari J, Swannick K, Katz J, Johnson WG (1998) Hereditary febrile seizures: Phenotype and evidence for a chromosome 19p locus. *Am J Med Genet* **79**:354–361.

Kuliev A, Verlinsky Y (2005) Place of preimplantation diagnosis in genetic practice. *Amer J Med Genet* **134A**:105–110.

Kuppermann M, Feeny D, Gates E, Posner SF, Blumberg B, Washington AE (1999) Preferences of women facing a prenatal diagnostic choice: Long-term outcomes matter most. *Prenat Diagn* **19**(8):711–716.

Kurz M, Alves G, Aarsland D, Larsen JP (2003) Familial Parkinson's disease: A community-based study. *Eur J Neurol* **10**:159–163.

Kus v. Sherman Hospital (1995) 268 Ill App 3d 771, 644 NE2d1214, 1216.

Kuszler PC (1999) Telemedicine and integrated health care delivery: Compounding medical malpractice liability. *Am J Law Med Ethics* **25**:297–326. (With respect to creation of the physician–patient relationship by e-mail, see page 310).

LaCaze v. Collier (1983) 434 So. 2d 1039, 1048 (La.).

Lafayette D, Abuelo D, Passero MA, Tantravahl U (1999) Attitudes towards cystic fibrosis carrier and prenatal testing and utilization of carrier testing among relatives of individuals with cystic fibrosis. *J Genet Counsel* **8**(1):17–36.

Lambert-Messerlian GM, Saller DN Jr, Tumber MB, French CA, Peterson CJ, Canick JA (1998) Second trimester maternal serum inhibin A levels in fetal trisomy 18 and Turner syndrome with and without hydrops. *Prenat Diagn* **18**:1061–1067.

Langbehn D, Brinkman R, Falush D, Paulsen J, Hayden M (2004) A new model for prediction of the age of onset and penetrance for Huntington's disease based on GAG length. *Clin Genet* **65**:267–277.

Langenier C, Moschini G (2002) The economics of patents. In: Rothschild M, Newman Sr, eds. *Intellectual Property Rights in Animal Breeding and Genetics*. CAB, pp. 31–50.

Langston JW, Ballard P, Tetrud JW, Irwin I (1983) Chronic Parkinsonism in humans due to a product of meperidine-analog synthesis. *Science* **219**(4587):979–980.

Lanie AD, Jayaratne TE, Sheldon JP, Kardia SLR, Anderson ES, Feldbaum M, Petty EM (2004) Exploring the public understanding of basic genetic concepts. *J Genet Counsel* **13**(4):305–320.

Lankenau v. Dutton (1986) 37 C.C.L.T. 213, at 226 (B.C.S.C.); additional reasons at (1988) 27 B.C.L.R. (2d) 234, 6 W.W.R. 337; aff'd (1991) 7 C.C.L.T. (2d) 421, 79 D.L.R. (4th) 705 (C.A.), 5 W.W.R. 71.

Lapham EV, Kozma C, Weiss JO (1996) Genetic discrimination: Perspective of consumers. *Science* **274**:621–624.

Lapointe v. Hôpital le Gardeur (1992) 1 S.C.R. 351 at 363.

Largey v. Rothman (1988) 110 NJ 204, 213; 540 A.2d 504 (NJ).

Larsen v. Yelle (1976) 310 Minn. 21, 246 N.W. 2d 841.

Larson, EJ (2002) The meaning of human gene testing for disability rights. *U Cin Law Rev* **70**:913.

Larsson C (1978) Natural history and life expectancy in severe alpha1-antitrypsin deficiency, PiZ. *Acta Med Scand* **204**:345–351.

Lau v. Nichols (1974) 414 US 563.

Laurie GT (2001) Challenging medical-legal norms. The role of autonomy, confidentiality, and privacy in protecting individual and familial group rights in genetic information. *J Legal Med* **22**:1–54.

Lavedan C, Buchholtz S, Nussbaum RL, Albin RL, Polymeropoulos MH (2002) A mutation in the human neurofilament M gene in Parkinson's disease that suggests a role for the cytoskeleton in neuronal degeneration. *Neurosci Lett* **322**(1):57–61.

Laws MB, Heckscher R, Mayo SJ, Li W, Wilson IB (2004) A new method for evaluating the quality of medical interpretation. *Med Care* **42**(1):71–80.

Le WD, Xu P, Jankovic J, Jiang H, Appel SH, Smith RG, Vassilatis DK (2003) Mutations in NR4A2 associated with familial Parkinson disease. *Nat Genet* **33**(1):85–89.

Leadbetter v. Brand (1980) 37 N.S.R. (2d) 581, 67 A.P.R. 581, 107 D.L.R. (3d) 252, 3 L. Med. Q. 263.

Learman LA, Kuppermann M, Gates E, Nease RF Jr, Gildengorin V, Washington AE (2003) Social and familial context of prenatal genetic testing decisions: Are there racial/ethnic differences? *Am J Med Genet Part C* (Semin Med Genet) **119C**:19–26.

LeBlang TR (1994) Medical malpractice and physician accountability: Trends in the courts and legislative responses. *Ann Health Law* **3**:105 at 118–120.

LeBlang TR (1995) Informed consent and disclosure in the physician-patient relationship: Expanding obligations for physicians in the United States. *Med Law* **14**(5–6):429–444.

Lee v. O'Farrell (1988) 43 C.C.L.T. 269; additional reasons at (1988) 30 B.C.L.R. (2d) 130 (B.C.S.C.).

Lee LJ, Batal HA, Maselli JH, Kutner JS (2002a) Effect of Spanish interpretation method on patient satisfaction in an urban walk-in clinic. *J Gen Intern Med* **17**(8):641–645.

Lee SJ, Back AL, Block SD, Stewart SK (2002b) Enhancing physician-patient communication. *Hematology (Am Soc Hematol Educ Program)*, 464–483.

Leedy v. Hartnett (1981) 510 F. Supp.1125 (Penn MDDC 1981).

Leek AE, Ruoss, CF, Kitau MJ, Chard T (1973) Raised α-fetoprotein in maternal serum with anencephalic pregnancy. *Lancet* **2**:385.

Lehmann LS, Weeks JC, Klar N, Biener L, Garber JE (2000) Disclosure of familial genetic information: Perceptions of the duty to inform. *Am J Med* **15**:109(9):705–711, and see Wilcke JTR, Seersholm N, Kok-Jensen, Dirksen A (1999) Transmitting genetic risk information in families: Attitudes about disclosing the identity of relatives. *Am J Hum Genet* **65**:902–909.

Lejeune J, Gautier M, Turpin R (1959) Etude des chromosomes somatiques de neuf enfants mongoliens. *CR Acad Sci Paris* **248**:1721–1722.

Lemmens T (2000) Selective justice, genetic discrimination and insurance: Should we single out genes in our laws? *McGill Law J* **45**:347–412.

Lemmens T, Bahamin P (1998) Genetics in life, disability and additional health insurance in Canada: A comparative legal and ethical analysis. In: Knoppers BM, ed. *Socio-Ethical Issues in Human Genetics*. Cowansville, Quebec: Les Editions Yvon Blais.

Leong T, Whitty J, Keilar M, Mifsud S, Ramsay J, Birrell G, Venter D, Southey M, McKay M (2000) Mutation analysis of BRCA1 and BRCA2 carrier predisposition genes in radiation hypersensitive cancer patients. *Int J Radiat Oncol Biol Phys* **48**:959–965.

Lerman C, Croyle R (1994) Psychological issues in genetic testing for breast cancer suscepti-bility. *Arch Intern Med* **154**:609–616.

Lerman C, Croyle R (1996) Emotional and behavioral responses to genetic testing for suscep-tibility to cancer. *Oncology* **10**:191–202.

Lerman C, Daly M, Sands C, Balshem A, Lustbader E, Heggan T, Goldstein L, James J, Engstrom P (1993) Mammography adherence and psychological distress among women at risk for breast cancer. *J Nat'l Ca Inst* **85**:1074–1080.

Lerman C, Daly M, Masny A, Balshem A (1994b) Attitudes about genetic testing for breast-ovarian cancer susceptibility, *J Clin Onc* **12**:843–850.

Lerman C, Lustbader E, Rimer B, Daly M, Miller S, Sands C, Balshem A (1995) Effects of individualized breast cancer risk counseling: A randomized trial. *J Nat'l Ca Inst* **87**:286–292.

Lerman C, Narod S, Schulman K, Hughes C, Gomez-Caminero A, Bonney G, Gold K, Trock B, Main D, Lynch J, Fulmore C, Snyder C, Lemon SJ, Conway T, Tonin P, Lenoir G, Lynch H (1996a) BRCA1 testing in families with hereditary breast-ovarian cancer: A prospective study of patient decision making and outcomes. *J Am Med Assoc* **275**:1885–1892.

Lerman C, Shields AE (2004) Genetic testing for cancer susceptibility: The promise and the pitfalls. *Nature Rev/Cancer* **4**:235–241.

Lerman C, Marshall J, Audrain J, Gomez-Caminero A (1996b) Genetic testing for colon cancer susceptibility: Anticipated reactions of patients and challenges to providers. *Int J Cancer* (*Pred Oncol*) **69**:58–61.

Lerman C, Schwartz MD, Miller SM, Daly M, Sands C, Rimer BK (1996c) A randomized trial of breast cancer risk counselling: Interacting effects of counseling, educational level, and coping style. *Hlth Psychol* **15**:75–83.

Lerman C, Biesecker B, Benkendorf JL, Kerner J, Gomez-Caminero A, Hughes C, Reed MM (1997) Controlled trial of pretest education approaches to enhance informed decision-making for BRCA1 gene testing. *J Natl Cancer Inst* **89**(2):148–157.

Lerman C, Peshkin BN, Hughes C, Isaacs C (1998a) Family disclosure in genetic testing for cancer susceptibility: Determinants and consequences. *J Health Care Law Policy* **1**(2):353–372.

Lerman C, Hughes C, Lemon SJ, Main D, Snyder C, Durham C, Narod S, Lynch HT (1998b) What you don't know can hurt you: Adverse psychological effects in members of BRCA1-linked and BRCA2-linked families who decline genetic testing. *J Clin Oncol* **16**:1650–1654.

Lerman C, Hughes C, Trock BJ, Myers RE, Main D, Bonney A, Abbaszadegan MR, Harty AE, Franklin BA, Lynch JF, Lynch HT (1999) Genetic testing in families with hereditary nonpolyposis colon cancer. *JAMA* **281**(17):1618–1622.

Lerman C, Hughes C, Croyle RT, Main D, Durham C, Snyder C, Bonney A, Lynch JF, Narod SA, Lynch HT (2000) Prophylactic surgery and surveillance practices one year following BRCA1/2 genetic testing. *Prev Med* **1**:75–80.

Lerman C, Croyle RT, Tercyak KP, Hamann H (2002) Genetic testing: Psychological aspects and implications. *J Consult Clin Psychol* **70**(3):784–797.

Leroy E, Boyer R, Auburger G, Leube B, Ulm G, Mezey E, Harta G, Brownstein MJ, Jonnalagada S, Chernova T, Dehejia A, Lavedan C, Gasser T, Steinbach PJ, Wilkinson KD, Polymeropoulos MH (1998) The ubiquitin pathway in Parkinson's disease. *Nature* **395**:451–452.

Letendre M, Godard B (2004) Expending the physician's duty of care: A duty to recontact? *J Med Law* **23**:3.

Leventhal H, Kelly K, Leventhal E (1995) Population risk, actual risk, perceived risk, and cancer control: A discussion. *J Nat'l Can Inst Monogra* **25**:81–85.

Levine EG, Brandt LJ, Plumeri P (1995) Informed consent: A survey of physician outcomes and practices. *Gastrointest Endosc* **41**(5):448–452.

Levinson W (1997) Doctor-patient communication and medical malpractice: Implications for pediatricians. *Pediatr Ann* **26**(3):186–193.

Levinson DF, Mowry BJ (2000) Genetics of schizophrenia. In: Pfaff DW, Berrettini WH, Maxson SC, Joh TH, eds. *Genetic Influences on Neural and Behavioral Functions*, New York: CRC Press.

Levinson W, Roter D (1995) Physicians' psychosocial beliefs correlate with their patient communication skills. *J Gen Intern Med* **10**(7):375–379.

Levinson W, Roter DL, Mullooly JP, Dull VT, Frankel RM (1997) Physician-patient communication. The relationship with malpractice claims among primary care physicians and surgeons. *JAMA* **277**(7):553–559.

Levinson W, Gorawara-Bhat R, Lamb J (2000) A study of patient clues and physician responses in primary care and surgical settings. *JAMA* **284**(8):1021–1027.

Levy N (2002) Deafness, culture, and choice. *J Med Ethics* **28**:284–285.

Levy-Lahad E, Wasco W, Poorkaj P, Romano DM, Oshima J, Pettingell WH, Yu CE, Jondro PD, Schmidt SD, Wang K, et al (1995) Candidate gene for the chromosome 1 familial Alzheimer's disease locus. *Science* **269**:973–977.

Lewis LJ (2002) Models of genetic counseling and their effects on multicultural genetic counseling. *J Genet Counsel* **11**:193–212.

Lewis CM, Levinson DF, Wise LH, DeLisi LE, Straub RE, Hovatta I, Williams NM, Schwab SG, Pulver AE, Faraone SV, Brzustowicz LM, Kaufmann CA, Garver DL, Gurling HM, Lindholm E, Coon H, Moises HW, Byerley W, Shaw SH, Mesen A, Sherrington R, O'Neill FA, Walsh D, Kendler KS, Ekelund J, Paunio T, Lonnqvist J, Peltonen L, O'Donovan MC, Owen MJ, Wildenauer DB, Maier W, Nestadt G, Blouin JL, Antonarakis SE, Mowry BJ, Silverman JM, Crowe RR, Cloninger CR, Tsuang MT, Malaspina D, Harkavy-Friedman JM, Svrakic DM, Bassett AS, Holcomb J, Kalsi G, McQuillin A, Brynjolfson J, Sigmundsson T, Petursson H, Jazin E, Zoega T, Helgason T (2003) Genome scan meta-analysis of schizophrenia and bipolar disorder, part II: Schizophrenia. *Am J Hum Genet* **73**(1):34–48.

Lewis RV (1996) 139 D.L.R. 4th 480 (B.C.S.C.).

Ley P (1979) Memory for medical information. *Br J Soc Clin Psychol* **18**(2):245–255.

Ley P (1982) Satisfaction, compliance and communication. *Br J Clin Psychol* **21**(Pt 4):241–254.

Li T, Ma X, Sham PC, Sun X, Hu X, Wang Q, Meng H, Deng W, Liu X, Murray RM, Collier DA (2004) Evidence for association between novel polymorphisms in the PRODH gene and schizophrenia in a Chinese population. *Am J Med Genet* **129B**(1):13–15.

Liang A (1998) The arguments against a physician's duty to warn for genetic diseases: The conflicts created by Safer v. Estate of Pack. *J Health Care Law Policy* **1**:437, 453.

Lichtenstein P, Holm NV, Verkasalo PA, Iliadou A, Kaprio J, Koskenvuo M, Pukkala E, Skytthe A, Hemminki K (2000) Environmental and heritable factors in the causation of cancer. *New Engl J Med* **343**:78–85.

Liddington v. Burns (1995) 916 F. Supp. 1127, 1132 (W.D. Okla.).

Lidz CW, Meisel A, Osterweis M, Holden JL, Marx JH, Munetz MR (1983) Barriers to informed consent. *Ann Intern Med* **99**(4):539–543.

Lidz CW, Appelbaum PS, Meisel A (1988) Two models of implementing informed consent. *Arch Int Med* **148**:1385–1389.

Lim J, MacIuran M, Price M, Bennett B, Butow P, kConFab Psychosocial Group (2004) Short- and long-term impact of receiving genetic mutation results in women at increased risk for hereditary breast cancer. *J Genet Counsel* **13**(2):115–133.

Lindberg NM, Wellisch D (2004) Identification of traumatic stress reactions in women at increased risk for breast cancer. *Psychosomatics* **45**:7–16.

Lindhout D, Omtzigt JG, Cornel MC (1992) Spectrum of neural-tube defects in 34 infants prenatally exposed to antiepileptic drugs. *Neurology* **42**:111–118.

Lindor NM, Burgart LJ, Leontovich O, Goldberg RM, Cunningham JM, Sargent DJ, Walsh-Vockley C, Petersen GM, Walsh MD, Leggett BA, Young JP, Barker MA, Jass JR, Hopper J, Gallinger S, Bapat B, Redston M, Thibodeau SN (2002) Immunohistochemistry versus microsatellite instability testing in phenotyping colorectal tumors. *J Clin Oncol* **20**:1043–1048.

Lininger v. Eisenbaum (1988) 764 P. 2d. 1202 (Colo.).

Lion R, Meertens RM, Bot I (2002) Priorities in information desire about unknown risks. *Risk Anal* **22**(4):765–776.

Lipari v. Sears-Roebuck and Co. (1980) 497 F. Supp. 185 (D. Neb.).

Lippman A (1998) The politics of health: Geneticization versus health promotion. In: Sherwin S, ed. *The Politics of Women's Health*. Philadelphia: Temple University Press, pp. 64–82.

Lippman-Hand A, Fraser FC (1979) Genetic counseling: Provision and reception of information. *Am J Med Genet* **3**:113–127.

Liu H, Heath SC, Sobin C, Roos JL, Galke BL, Blundell ML, Lenane M, Robertson B, Wijsman EM, Rapoport JL, Gogos JA, Karayiorgou M (2002) Genetic variation at the 22q11 PRODH2/DGCR6 locus presents an unusual pattern and increases susceptibility to schizophrenia. *Proc Natl Acad Sci USA* **99**(6):3717–3722.

Lloyd AJ (2001) The extent of patients' understanding of the risk of treatments. *Qual Health Care* **10**:114–118.

Lloyd SM, Watson M, Oaker G, Sacks N, Querci Della Rovere U, Gui G (2000) Understanding the experience of prophylactic bilateral mastectomy: A qualitative study of ten women. *Psycho-Onc* **9**:473–485.

Lobb E, Butow P, Meiser B, Tucker K, Barratt A (2001) How do geneticists and genetic counselors counsel women from high-risk breast cancer families? *J Genet Counsel* **10**:185–199.

Lobb EA, Butow PN, Meiser B, Barratt A, Gaff C, Young MA, Kirk J, Suthers GK, Tucker K (2002) Tailoring communication in consultations with women from high risk breast cancer families. *Br J Cancer* **87**(5):502–508.

Lobb EA, Butow PN, Barratt A, Meiser B, Gaff C, Young MA, Haan E, Suthers G, Gattas M, Tucker K (2004) Communication and information-giving in high-risk breast cancer consultations: Influence on patient outcomes. *Br J Cancer* **90**(2):321–327.

Lock M (1998) Breast cancer: Reading the omens. *Anthropol Today* **14**(4):7–16.

Lodder L, Frets PG, Trijsburg RW, Meijers-Heijboer EJ, Klijn JG, Duivenvoorden HJ, Tibben A, Wagner A, van der Meer CA, van den Ouweland AM, Niermeijer MF (2001) Psychological impact of receiving a BRCA1/BRCA2 test result. *Am J Med Genet* **98**(1):15–24.

Loeben GL, Marteau TM, Wilfond BS (1998) Mixed messages: Presentation of information in cystic fibrosis-screening pamphlets. *Am J Hum Genet* **63**(4):1181–1189.

Logan v. Greenwich Hosp. Ass'n. (1983) 191 Conn. 282, 299; 465 A. 2d 294, 301–302.

Lohmueller KE, Pearce CL, Pike M, Lander ES, Hirschhorn JN (2003) Meta-analysis of genetic association studies supports a contribution of common variants to susceptibility to common disease. *Nat Genet* **33**(2):177–182.

Longman v. Jasiek (1980) 414 N.E. 2d. 520 (Ill. App.).

Lopez AD, Murray CCJL (1998) The global burden of disease, 1990–2020. *Nat Med* **4**:1241–1243.

Louis ED, Anderson KE, Moskowitz C, Thorne DZ, Marder K (2000) Dystonia-predominant adult-onset Huntington disease: Association between motor phenotype and age of onset in adults. *Arch Neurol* **57**(9):1326–1330.

Louisell DW, Williams H (2001) *Medical Malpractice* New York: Matthew Bender & Company, p. 17G–42.

Love RR, Evans AM, Josten DM (1985) The accuracy of patient reports of a family history of cancer. *J Chron Dis* **38**:289–293.

Lubinski J, Phelan CM, Ghadirian P, Lynch HT, Garber J, Weber B, Tung N, Horsman D, Isaacs C, Monteiro AN, Sun P, Narod SA (2004) Cancer variation associated with the position of the mutation in the BRCA2 gene. *Familial Cancer* **3**:1–10.

Lucassen A, Parker M (2004) Confidentiality and serious harm in genetics-preserving the confidentiality of one patient and preventing harm to relatives. *Eur J Hum Genet* **12**:93–97.

Lucking CB, Durr A, Bonifati V, Vaughan J, De Michele G, Gasser T, Harhangi BS, Meco G, Denefle P, Wood NW, Agid Y, Brice A (2000) Association between early-onset Parkinson's disease and mutations in the parkin gene. *N Engl J Med* **342**:1560–1567.

Lum RG (1987) The patient-counselor relationship in a cross-cultural context. In: Biesecker B, et al, eds. *Strategies in Genetic Counseling II; Religious, Cultural, and Ethnic Influences on the Counselling Process*, vol. 23(6) *Birth Defects Original Articles Series*. New York: Wiley-Liss for the March of Dimes Birth Defects Foundation.

Lundahl v. Rockford Memorial Hospital (1968) 93 Ill. App. 2d 461, 235 N.E. 671 (2d Dist.). See also *Vassos v. Roussalis* (1983) 658 P. 2d. 1284 (Wyo.).

Lynch JJ (1985) The human dialogue. In: *The Language of the Heart: The Body's Response to Human Dialogue*. New York: Basic Books, at 281.

Lynch HT (2002) Cancer family history and genetic testing: Are malpractice adjudications waiting to happen? *Am J Gastroenterol* **97**(3):518–520.

Lynch HT, Watson P (1992) Genetic counselling and hereditary breast ovarian cancer. *Lancet* **339**(8802):1181. The authors report that "several of the 11 individuals, who were told that they were not gene carriers, expressed disbelief... Some indicated that they still wish to continue with intensive surveillance and were still considering prophylactic surgery and much of the counseling process focused on persuading them to think again."

Lynch A, Czeizel A, Salzano FM, Berg K (1990) Case studies: The price of silence. *Hastings Cent Report* **May/June**:31–35.

Lynch HT, Watson P, Tinley S, Snyder C, Durham C, Lynch J, Kirnarsky Y, Serova O, Lenoir G, Lerman C, Narod SA (1999) An update on DNA-based BRCA1/BRCA2 genetic counseling in hereditary breast cancer. *Cancer Genet Cytogenet* **109**(2):91–98.

Maat-Kievit A, Helderman-van den Enden P, Losekoot M, de Knijff P, Belfroid R, Vegter-van der Vlis M, Roos R, Breuning M (2001) Using a roster and haplotyping is useful in risk assessment for persons with intermediate and reduced penetrance alleles in Huntington disease. *Am J Med Genet* **105**(8):737–744.

MacDonald v. Clinger (1982) 84 AD.2d 482, 482 (4th Dep't). Also see *Tighe v. Ginsberg*, 146 AD2d 268, 271, 540 NYS 2d 99 (4th Dept.) 1989.

Macdonald v. York Hospital (1973) 41 D.L.R. (3d) 321 (Ont. C.A.), varying [1972] O.R. 469, 28 D.L.R. (3d) 521 (H.Ct.); aff'd *Vail v. MacDonald* (1976) 66 D.L.R. (3d) 530, 2 S.C.R. 825 (S.C.C.).

Machlup F (1958) An Economic Review of the Patent System, Study No. 15 of the Subcommittee on Patents, Trademarks and Copyrights of the Committee on the Judiciary United States Sen., 85th Cong., 2nd Sess. Washington, DC: U.S. GPO.

MacKey v. Greenview Hospital (1979) 587 SW 2d 249 (Ky. App.).

Mackie R (2001) Harris battles firm over gene patenting. *Globe & Mail* [Toronto], 20 September, A:14.

Macklin R (1992) Privacy and control of genetic information. In: Annas GJ, Elias S, eds. *Gene Mapping Using Law & Ethics as Guides*. New York: Oxford University Press, pp. 160–163.

MacRae AR, Allen LC, Lepage N, Tokmakjian S (2001) Validation of an automated dimeric inhibin A assay for use in maternal serum screening for Down syndrome. *Clin Chem* **47**: A150.

Maddalena A, Richards CS, McGinniss MJ, Brothman A, Desnick RJ, Grier RE, Hirsch B, Jacky P, McDowell GA, Popovich B, Watson M, Wolff DJ (2001) Technical standards and guidelines for Fragile X. *Genet Med* **3**(3):200–205.

Madey v. Duke University (2002) 307 F.3d 1351 (Fed. Cir.), cert. Denied, 123 S. Ct. 2369 (2003).

Maebius SB, Rutt S (2004) Nanotechnology and Intellectual Property: Impacting Laboratories. Online: Foley & Lardner. <http://www.foley.com/publications/pub_detail.aspx?pubid=1918>.

Magnusson C, Colditz G, Rosner B, Bergstrom R, Persson I (1998) Association of family history and other risk factors with breast cancer risk (Sweden). *Cancer Causes Control* **9**:259–267.

Maher NE, Currie LJ, Lazzarini AM, Wilk JB, Taylor CA, Saint-Hilaire MH, Feldman RG, Golbe LI, Wooten GF, Myers RH (2002) Segregation analysis of Parkinson disease revealing evidence for a major causative gene. *Am J Med Genet* **109**:191–197.

Mahowald MB (2003) Aren't we all eugenicists? Commentary on Paul Lombardo's "taking eugenics seriously." *Fla St U L Rev* **30**:219–235.

Maier W, Zobel A, Rietschel M (2003) Genetics of schizophrenia and affective disorders. *Pharmacopsychiatry* **36** (Suppl 3):S195–202.

Mains J (1985) Medical abandonment. *Med Tech Trial Q* **31**:306–328.

Malette v. Shulman (1990) 72 O.R. (2d) 417, 423, 430, 432; 67 DLR (4th) 321 (Ont. C.A.).

Maley JA, Ad Hoc Committee on Ethical Codes and Principles, National Society of Genetic Counselors (1994) *An Ethics Casebook for Genetic Counselors*. Charlottesville, VA: University of Virginia.

Mallet J, Christen Y, eds. (2003) *Neurosciences at the Postgenomic Era*. Berlin: Springer.

Mandl KD, Feit S, Larson C, Kohane IS (2002) Newborn screening program practices in the United States: Notification, research, and consent. *Pediatrics* **109**(2):269–273.

Mansoura MK, Collins FS (1998) Medical implications of the genetic revolution. *J Health Care Law Policy* **1**:329–352.

Manthous CA, DeGirolamo A, Haddad C, Amoateng-Adjepong Y (2003) Informed consent for medical procedures: Local and national practices. *Chest* **124**(5):1978–1984.

Maraganore DM, Lesnick TG, Elbaz A, Chartier-Harlin MC, Gasser T, Kruger R, Hattori N, Mellick GD, Quattrone A, Satoh J, Toda T, Wang J, Ioannidis JP, de Andrade M, Rocca WA, Toda T, UCHL1 Global Genetics Consortium (2004) UCHL1 is a Parkinson's disease susceptibility gene. *Ann Neurol* **55**(4):512–521.

March of Dimes (2004) Medical References Fact Sheets. March of Dimes Birth Defects Foundation. Available: http://www.modimes.org/professionals.

Marini C, Harkin LA, Wallace RH, Mulley JC, Scheffer IE, Berkovic SF (2003) Childhood absence epilepsy and febrile seizures: A family with a GABA(A) receptor mutation. *Brain* **126**:230–240.

Marshall M, Von Tigerstrom B (2003) Health information. In: Downie J, Caulfield T, Flood C, eds. *Canadian Health Law and Policy*, 2nd ed. Toronto: Butterworths, pp.157–203.

Marteau T (1989) Framing of information: Its influence upon decisions of doctors and patients. *Br J Soc Psych* **28**:89–94.

Marteau TM (1992) Psychological implications of genetic screening. In: Evers-Kiebooms G et al, eds. *Psychological Aspects of Genetic Counseling*. New York: Wiley-Liss, pp. 185–190.

Marteau TM, Croyle RT (1998) The new genetics. Psychological responses to genetic testing. *BMJ* **316**(7132):693–696.

Marteau TM, Dormandy E (2001) Facilitating informed choice in prenatal testing: How well are we doing? *Am J Med Genet* **106**(3):185–190.

Marteau T, Michie S (1999a) Genetic counseling. In: Marteau T, Richards M, eds. *The Troubled Helix: Social and Psychological Implications of the New Human Genetics.* Cambridge University Press. Knowledge and risk perception is discussed at pages 107–108.

Marteau TM, Senior V (1997) Illness representations after the human genome project: The perceived role of genes in causing illness. In: Petrie K, Weinman J, eds. *Perceptions of Illness and Treatment Current Psychological Research and Implications.* Amsterdam: Harwood Academic, pp. 41–66.

Marteau TM, Slack J, Kidd J, Shaw RW (1992) Presenting a routine screening test in antenatal care: Practice observed. *Public Health* **106**(2):131–141.

Marteau TM, Michie S, Miedzybrodzka ZH, Allanson A (1999a) Incorrect recall of residual risk three years after carrier screening for cystic fibrosis: A comparison of two-step and couple screening. *Am J Obstet Gynecol* **181**(1):165–169.

Martin v. Richards (1993) 500 N.W.2d 691 (Wis. Ct. App.); 531 N.W.2d 70 (Wis. 1995). Some courts have held that physicians need to disclose diagnostic procedures that might help patients make informed decisions about treatment. For example, in *Martin v. Richards*, the physicians failed to inform parents of a CT scan for a child to detect intracranial bleeding and that another hospital would have been more appropriate to conducxt the procedure and provide appropriate specialized care (neurosurgeon). In *Backlund v. University,* 137 Wash.2d 651, 975 P. 2d. 950, 952 (Wash. 1999), although the jury found that a neonatologist did not breach the standard of care in the decision to treat a newborn infant's jaundice with phototherapy treatment rather than a blood transfusion, this lack of negligence did not preclude a statutory claim for informed consent; the neonatologist did not inform the parents of the treatment alternative. And see Derrick HJ (1985; 2002) Annotation, Medical Malpractice: Liability for failure of physician to inform patient of alternative modes of diagnosis or treatment, A.L.R.4th; 38:900 (1985 and Supp. 2002).

Marvel MK, Epstein RM, Flowers K, Beckman HB (1999) Soliciting the patient's agenda: Have we improved? *JAMA* **281**(3):283–287.

Mascarello JT, Cooley LD, Davison K, Dewald GW, Brothman AR, Herrman M, Park JP, Persons DL, Rao KW, Schneider NR, Vance GH (2003) Problems with ISCN FISH nomenclature make it not practical for use in clinical test reports or cytogenetic databases. *Genet Med* **5**(5):370–377.

Mascarello JT, Cooley LD, Davison K, Dewald GW, Brothman AR, Herrman M, Park JP, Persons DL, Rao KW, Schneider NR, Vance GH (2003) Cytogenetics Resource Committee, College of the American Pathologists; Cytogenetics Resource Committee, American College of Medical Genetics.

Matloff ET (1994) Practice variability in prenatal genetic counseling. *J Genet Counsel* **3**:215–231.

Matloff ET, Shappell H, Brierley K, Bernhardt BA, McKinnon W, Peshkin BN (2000) What would you do? Specialists' perspectives on cancer genetic testing, prophylactic surgery, and insurance discrimination. *J Clin Oncol* **18**(12):2484–2492.

Matthews A, Brandenburg D, Cummings S, Olopade O (2002) Incorporating a psychological counselor in a cancer risk assessment program: Necessity, acceptability, and potential roles, *J Genet Counsel* **11** (1):51–64.

Matthies v. Mastromonaco (1998) 310 N.J. Super. 572, 709 A. 2d 238 (App. Div.).

Mattingly C, Lawlor M (2001) The fragility of healing. *Ethos* **29**(1):30–57.

Mavroforou A, Mavrophoros D, Koumantakis E, Michalodimitrakis E (2003) Liability in prenatal ultrasound screening. *Ultrasound Obstet Gynecol* **21**:525–528.

Maymon R, Betser M, Dreazen E, Padoa A, Herman A (2004) A model for disclosing the first trimester part of an integrated Down's syndrome screening test. *Clin Genet* **65**:113–119.

Mazur DJ (2003) Influence of the law on risk and informed consent. *BMJ* **327**:731–735.

Mazzoleni R, Nelson RR (1998) The benefits and costs of strong patent protection: A contribution to the current debate. *Res Policy* **27**:273–284.

McAllister M (2003) Personal theories of inheritance, coping strategies, risk perception and engagement in hereditary non-polyposis colon cancer families offered genetic testing. *Clin Genet* **64**:179–189.

McAllister v. Ha (1998) 347 N.C. 638, 496 S.E. 2d 577. However, the court recognized a claim for "wrongful conception" or "wrongful pregnancy" compared to wrongful birth.

McCabe LI, McCabe ERB (2002) Newborn screening as a model for population screening. *Molec Genet Metab* **75**:299–307.

McCabe LL, McCabe RB (2004) Direct-to-consumer genetic testing: Access and marketing. *Genet Med* **6**:58–59.

McCance KL, Jorde L (1998) Evaluating the genetic risk of breast cancer. *Nurse Pract* **23**(8):14–16, 19–20, 23–27; quiz 28–29.

McCarthy Veach P, Bartels DM, LeRoy BS (2002) Commentary: Genetic counseling—a profession in search of itself. *J Genet Counsel* **11**(3):187–191.

McCarthy Veach P, et al. (2003) Collaborating with clients: Providing information and assisting in client decision making. In: McCarthy Veach P, LeRoy BS, Bartels DM, eds. *Facilitating the Genetic Counseling Process: A Practice Manual.* New York: Springer, pp. 122–149.

McClellan MB (2003) Presentation, BIO 2003, Washington D.C., 23 June 2003; online: BIO 2003 <http://www.bio.org/events/2003/media/mcclellan_0623.asp>.

McConkie-Rosell A, Sullivan JA (1999) Genetic counseling-stress, coping, and the empowerment perspective. *J Genet Counsel* **8**:345–357.

McConkie-Rosell A, Lachiewicz AM, Spiridigliozzi GA, Tarleton J, Schoenwald S, Phelan MC, et al (1993) Evidence that methylation of the FMR-1 locus is responsible for variable phenotypic expression of the fragile X syndrome. *Am J Hum Genet* **53**:800–809.

McConkie-Rosell A, Robinson H, Wake S, Staley LW, Heller K, Cronister A (1995) Dissemination of genetic risk information to relatives in the fragile X syndrome: Guidelines for genetic counselors. *Am J Med Genet* **59**(4):426–430.

McConkie-Rosell A, Spiridigliozzi GA, Rounds K, Dawson DV, Sullivan JA, Burgess D, Lachiewicz AM (1999) Parental attitudes regarding carrier testing in children at risk for fragile X syndrome. *Am J Med Genet* **82**(3):206–211.

McConkie-Rosell A, Spiridigliozzi GA, Sullivan JA, Dawson DV, Lachiewicz AM (2001) Longitudinal study of the carrier testing process for fragile X syndrome: Perceptions and coping. *Am J Med Genet* **98**(1):37–45.

McConkie-Rosell A, Spiridigliozzi GA, Sullivan JA, Dawson DV, Lachiewicz AM (2002) Carrier testing in fragile X syndrome: When to tell and test. *Am J Med Genet* **110**(1):36–44.

McConkie-Rosell A, Spiridigliozzi GA (2004) Family matters: A conceptual framework for genetic testing in children. *J Genet Counsel* **13**(1):9–29.

McConnell LM, Goldstein MK (1999) The application of medical decision analysis to genetic testing: An introduction. *Genet Test* **3**(1):65–70.

McCullagh DR (1932) Dual endocrine activity of the testis. *Science* **76**:19–20.

McEwen JE, McCarty K, Reilly PR (1992) A survey of state insurance commissioners concerning genetic testing and life insurance. *Am J Hum Genet* **51**:785–792.

McEwen J, McCarty K, Reilly PR (1993) A survey of medical directors of life insurance companies concerning use of genetic information. *Am J Hum Genet* **53**:33–45.

McGovern MM, Benach MO, Wallenstein S, Desnick RJ, Keenlyside R (1999) Quality assurance in molecular genetic testing laboratories. *JAMA* **281**(9):835–840.

McGovern MM, Benach M, Wallenstein S, Boone J, Lubin IM (2003a) Personnel standards and quality assurance practices of biochemical genetic testing laboratories in the United States. *Arch Pathol Lab Med* **127**(1):71–76.

McGovern MM, Benach M, Zinberg R (2003b) Interaction of genetic counselors with molecular genetic testing laboratories: Implications for non-geneticist health care providers. *Am J Med Genet* **119A**(3):297–301.

McKeachie v. Alvarez (1970) 17 D.L.R. (3d) 87 (B.C.S.C.).

McKinnon WC, Baty BJ, Bennett RL, Magee M, Neufeld-Kaiser WA, Peters KF, Sawyer JC, Schneider KA (1997) Predisposition genetic testing for late-onset disorders in adults. A position paper of the National Society of Genetic Counselors. *JAMA* **278**(15):1217–1220.

McKusick VA (1998) *Mendelian Inheritance in Man. A Catalog of Human Genes and Genetic Disorders*, 12th ed. Baltimore: Johns Hopkins University Press.

McKusick VA (1990) *Mendelian Inheritance in Man, 9th ed.*, Johns Hopkins University Press, Baltimore; pp. xi–xxix.

McLaughlin v. Hellbusch (1999) 256 Neb. 615, 591 N.W. 2d 569.

McLeod RS, Canadian Task Force on Preventive Health Care (2001) Screening strategies for colorectal cancer: A systematic review of the evidence. *Can J Gastroenterol* **15**:647–660.

Meade CD (1999) Improving understanding of the informed consent process and document. *Semin Oncol Nurs* **15**(2):124–137.

Meade CD, Howser DM (1992) Consent forms: How to determine and improve their readability. *Oncol Nurs Forum* **19**(10):1523–1528.

Meade CD, Calvo A, Cuthbertson DJ (2002) Impact of culturally, linguistically, and literacy relevant cancer information among Hispanic farmworker women. *J Cancer Educ* **17**(1):50–54.

Meade CD, Calvo A, Rivera MA, Baer RD (2003) Focus groups in the design of prostate cancer screening information for Hispanic farmworkers and African American men. *Oncol Nurs Forum* **30**(6):967–975.

Mechanic D (1998) The functions and limitations of trust in the provision of medical care. *J Health Politics, Policy Law* **23**(40):661–686.

Meikle PJ, Ranieri E, Simonsen H, Rozaklis T, Ramsay S, Whitfield PD, Fuller M, Christensen E, Skovby F, Hopwood JJ (2004) Newborn Screening for Lysosomal Storage Disorders: Clinical Evaluation of a Two-Tier Strategy. *Pediatrics* **114**:909–916.

Meisel A, Kuczewski M (1996) Legal and ethical myths about informed consent. *Arch Intern Med* **156**(22):2521–2526. Also see Mark JS, Spiro H (1990) Informed consent for colonoscopy: A prospective study. *Arch Internal Med* **150**:777, at 780: "[A]utonomous decision making by an informed and self-sufficient patient is probably rarer in clinical practice than in philosophical suggestion."

Meiser B, Dunn S (2000) Psychological impact of genetic testing for Huntington's disease: An update of the literature. *J Neurol Neurosurg Psychiatry* **69**(5):574–578.

Meiser M, Halliday J (2002) What is the impact of genetic counseling in women at increased risk of developing hereditary breast cancer? A meta-analytic review. *Soc Sci Med* **54**:1463–1470.

Meiser B, Gleeson M, Tucker K (2000) Psychological impact of genetic testing for adult-onset disorders: An update for clinicians. *Med J Austral* **172**:126–129.

Meiser B, Eisenbruch M, Barlow-Stewart K, Tucker K, Steel Z, Goldstein D (2001) Cultural aspects of cancer genetics: Setting a research agenda. *J Med Genet* **38**:425–429.

Meissen GJ, Mastromauro CA, Kiely DK, McNamara DS, Myers RH (1991) Understanding the decision to take the predictive test for Huntington Disease. *Am J Med Genet* **39**:404–410.

Menahem S (1998) Counselling strategies for parents of infants with congenital heart disease. *Cardiol Young* **8**(3):400–407.

Menasha JD, Schechter C, Willner J (2000) Genetic testing: A physician's perspective. *Mt Sinai J Med* **67**(2):144–151.

Mennie ME, Compton ME, Gilfillan A, Liston WA, Pullen I, Whyte DA, Brock DJ (1993) Prenatal screening for cystic fibrosis: Psychological effects on carriers and their partners. *J Med Genet* **30**(7):543–548.

Menzel v. Morse (1985) 362 N.W. 2d 465 (Iowa). But note *Helling v. Carey* (1974) 519 P. 2d. 981 (Wash.). The court overruled the generally accepted standard of care and found negligent a practice of not routinely testing patients under the age of 40 for glaucoma that was sanctioned by professional custom.

Meracle v. Children's Service Society of Wisconsin (1989) 149 Wis. 2d.19, 437 N.W.2d 532.

Meredith P, Wood C (1998) Inquiry into the potential value of an information pamphlet on consent to surgery to improve surgeon-patient communication. *Qual Health Care* **7**(2):65–69.

Merenstein SA, Sobesky WE, Taylor AK, Riddle JE, Tran HX, Hagerman RJ (1996) Fragile X syndrome in normal IQ males with learning and behavioral problems: Molecular-clinical correlations in males with an expanded FMR-1 mutation. *Am J Med Genet* **64**:388–394.

Merges RP, Nelson RR (1990) On the complex economics of patent scope. *Columbia Law Rev* **90**:839.

Merikangas KR, Risch N (2003) Will the genomics revolution revolutionize psychiatry? *Am J Psychiatry* **160**:625–635.

Merkatz IR, Nitowsky HM, Macri JN, Johnson WE (1984) An association between low maternal serum alpha-fetoprotein and fetal chromosomal abnormalities. *Am J Obstet Gynecol* **148**:886–894.

Merriam-Webster (2004) Merriam-Webster Online Dictionary. Merriam-Webster Inc. Available: www.Merriam-Webster.com.

Merrill SA, Levin RC, Myers MB, eds. (2004) *A Patent System for the 21st Century*, National Academies Press, online: NAP, p. 31. <www.nap.edu/books/0309089107/html/>.

Merz JF (1993) On a decision-making paradigm of medical informed consent. *J Legal Med* **14**:231, 243–264.

Merz JF, Kriss AG, Leonard DGB, Cho MK (2002) Diagnostic testing fails the test. *Nature* **415**:577–579.

Meschede D, Eigel A, Horst J, Nieschlag E (1993) Compound heterozygosity for the deltaF508 and F508C cystic fibrosis transmembrane regulator (CFTR) mutations in a patient with congenital bilateral aplasia of the vas deferens. *Am J Hum Genet* **53**:292–293.

Metcalfe KA, Liede A, Hoodfar E, Scott A, Foulkes WD, Narod SA (2000) An evaluation of needs of female BRCA1 and BRCA2 carriers undergoing genetic counselling. *J Med Genet* **37**(11):866–874.

Metcalfe KA, Liede A, Trinkaus M, Hanna D, Narod SA (2002) Evaluation of the needs of spouses of female carriers of mutations in BRCA1 and BRCA2. *Clin Genet* **62**(6):464–469.

Metcalfe K, Lynch HT, Ghadirian P, Tung N, Olivotto I, Warner E, Olopade OI, Eisen A, Weber B, McLennan J, Sun P, Foulkes WD, Narod SA (2004a) Contralateral breast cancer in BRCA1 and BRCA2 mutation carriers. *J Clin Oncol* **22**:2328–2335.

Metcalfe KA, Esplen MJ, Goel V, Narod SA (2004b) Psychosocial functioning in women who have undergone bilateral prophylactic mastectomy. *Psycho-Onc* **13**(1):14–25.

Meyer M (1992) Patients' duties. *J Med Philosophy* **17**:541–555.

Michie S, Marteau TM (1999a) The choice to have a disabled child. *Am J Hum Genet* **65**(4):1204–1208.

Michie S, Marteau T (1999b) Genetic counseling. In: Marteau T, Richards M, eds. *The Troubled Helix: Social and Psychological Implications of the New Human Genetics*. United Kingdom: Cambridge University Press—Paperback edition, pp. 108–111.

Michie S, Marteau T, Bobrow M (1997a) Genetic counselling: The psychological impact of meeting patients' expectations. *J Med Genet* **34**:237–241.

Michie S, McDonald V, Marteau TM (1997b) Genetic counseling: Information given, recall and satisfaction. *Patient Educ Counsl*, **32**(1–2):101–106.

Michie S, Bron F, Bobrow M, Marteau TM (1997d) Nondirectiveness in genetic counseling: An empirical study. *Am J Hum Genet* **60**(1):40–47.

Michie S, French D, Allanson A, Bobrow M, Marteau TM (1997c) Information recall in genetic counselling: A pilot study of its assessment. *Patient Educ Counsl* **32**(1–2):93–100. Also see Bertakis KD, Roter D, Putnam SM (1991) The relationship of physician medical interview style to patient satisfaction. *J Fam Pract* **32**(2):175–181. With regard to patient preference for discussion and counseling about psychosocial issues, see Street RL Jr (1992) Analyzing communication in medical consultations. Do behavioral measures correspond to patients' perceptions? *Med Care* **30**(11):976–988.

Michie S, Allanson A, Armstrong D, Weinman J, Bobrow M, Marteau T (1998) Objectives of genetic counselling: Differing views of purchasers, providers and users. *J Pub Health Med* **20**:404–408.

Michie S, Weinman J, Miller J, Collins V, Halliday J, Marteau TM (2002a) Predictive genetic testing: High risk expectations in the face of low risk information. *J Behav Med* **25**(1):33–50.

Michie S, Collins V, Halliday J, Marteau T, FAP Collaborative Research Group (2002b) Likelihood of attending bowel screening after a negative genetic test result: The possible influence of health professionals. *Genet Test* **6**(4):307–311.

Michie S, Smith JA, Senior V, Marteau TM (2003) Understanding why negative genetic test results sometimes fail to reassure. *Am J Med Genet* **119A**(3):340–347.

Mickle v. Salvation Army Grace Hospital Windsor Ontario (1998) 166 D.L.R. (4th) 743 (Ont. Gen. Div.).

Miesfeldt S, Jones SM, Cohn WF (2000) Informed consent for BRCA1 and BRCA2 testing: What clinicians should know about the process and content. *J Am Med Womens Assoc* **55**(5):275–279.

Miller ST, Stilerman TV, Rao SP, Abhyankar S, Brown AK (1990) Newborn screening for sickle cell disease. When is an infant "lost to follow-up"? *Am J Dis Child* **144**(12):1343–1345.

Miller v. Sullivan (1995) 625 N.Y.S. 2d 102 (App. Div.).

Milunsky A (1998) Maternal serum screening for neural tube and other defects. In: Milunsky A, ed. *Genetic Disorders of the Fetus: Diagnosis, Prevention and Treatment*, 4th ed. Baltimore: Johns Hopkins Univ. Press, pp. 635–701.

Milunsky A, Alpert E, Kitzmiller JL, Younger MD, Neff RK (1982) Prenatal diagnosis of neural tube defects. VIII. The importance of alpha-fetoprotein screening in diabetic women. *Am J Obstet Gynecol* **142**:1030–1032.

Milunsky A, Jick SS, Bruell CL, MacLaughlin DS, Tsung YK, Jick H, Rothman KJ, Willett W (1989a) Predictive values, relative risks, and overall benefits of high and low maternal serum alpha-fetoprotein screening in singleton pregnancies: New epidemiologic data. *Am J Obstet Gynecol* **161**(2):291–297.

Milunsky A, Jick H, Jick SS, Bruell CL, MacLaughlin DS, Rothman KJ, Willett W (1989b) Multivitamin/folic acid supplementation in early pregnancy reduces the prevalence of neural tube defects. *JAMA* **262**:2847–2852.

Minderer S, Gloning KP, Henrich W, Stoger H (2003) The nasal bone in fetuses with trisomy 21: Sonographic versus pathomorphological findings. *Ultrasound Obstet Gynecol* **22**:16–21.

Mink v. University of Chicago (1978). (N.D. Ill.) 460 F. Supp. 713 at 718–720. This case concerned some 1000 women who were given prenatal diethylstilbestrol (DES) between the years 1950 and 1952 as part of a medical research project. The women were not told that they were part of an experiment, nor were they told that the pills given to them contained DES. The plaintiffs filed suit alleging that because of the prenatal DES exposure, their daughters had an increased risk of vaginal or cervical cancer. The suit further alleged that the relationship between DES and cancer was known as early as 1971, but the defendants made no effort to notify the patients until 1975 or 1976.

Mio JS, Morris DR (1990) Cross-cultural issues in psychology training programs: An invitation for discussion. *Professional Psych: Res Practice* **21**:434–441.

Mirkovic N, Marti-Renom MA, Weber BL, Sali A, Monteiro ANA (2004) Structure-based assessment of missense mutations in human BRCA1: Implication for breast and ovarian cancer predisposition. *Cancer Res* **64**:3790–3797.

Mischler EH, Wilfond BS, Fost N, Laxova A, Reiser C, Sauer CM, Makholm LM, Shen G, Feenan L, McCarthy C, Farrell PM (1998) Cystic fibrosis newborn screening: Impact on reproductive behavior and implications for genetic counseling. *Pediatrics* **102**(1 Pt 1):44–52.

Mish FC, ed. *Webster's Collegiate Dictionary*. Springfield, MA: Merriam-Webster Inc.

Mittman I, Crombleholme WR, Green JR, Golbus MS (1998) Reproductive genetic counseling to Asian-Pacific and Latin American immigrants. *J Genet Counsel* **7**:49–70.

Modell B (1997) Delivering genetic screening to the community. *Ann Med* **29**(6):591–599.

Mohr v. Commonwealth (1995) 421 Mass. 147, 653 N.E.2d 1104.

Mol BW, Lijmer JG, van der Meulen J, Pajkrt E, Bilardo CM, Bossuyt PM (1999) Effect of study design on the association between nuchal translucency measurement and Down syndrome. *Obstet Gynecol* **94**:864–869.

Molloy et al., v. Meier, Backus, et al. (Minn. 2003) 660 N.W.2d 444; (2004) Minn. C9-02: 1821, C9-02-1837; LEXIS 268. Also see *Didato v. Strehler* (2001) 554 S.E. 2d. 42 (Va.). Also see *Jackson v. Isaac* (2002) 76 S.W. 3d 177 (Tex. App. Eastland). http://caselaw.lp.findlaw.com/scripts/getcase.pl?court=MN&vol=sc%5C0405%5Cop021821–0520&invol=1.

Monaghan KG, Wiktor A, Van Dyke DL (2002) Diagnostic testing for Prader-Willi syndrome and Angelman syndrome: A cost comparison. *Genet Med* **2**(4):448–450.

Monahan v. Weichert (1983) 442 N.Y.S. 295; appeal after reversed, 461 N.Y.S. 2d 633 (N.Y.A.D. 4 Dist.).

Monroe AD, Shirazian T (2004) Challenging linguistic barriers to health care: Students as medical interpreters. *Acad Med* **79**(2):118–122.

Montgomery AA, Fahey T (2001) How do patients' treatment preferences compare with those of clinicians? *Qual Health Care* **10**(Suppl. I): i39–43.

Moore PJ, Adler NE, Robertson PA (2000) Medical malpractice: The effect of doctor-patient relations on medical patient perceptions and malpractice intentions. *West J Med* **173**(4): 244–250.

Moore v. Baker (1993) 989 F.2d 1129 (11th Cir.); also see *Teilhaber v. Greene* (1999) 727 A.2d 518 (N.J. Super. Ct. App. Div.).

Moore v. Preventive Medicine Medical Group Inc. (1986) 178 Cal. App. 3d 728, at pgs. 738–739; 223 Cal. Rptr. 859 (Cal. App. Dist. 2 03-11-1986).

Moore v. Webb (1961) 345 S.W.2d 239 (Mo.App.). "A physician occupies a position of trust and confidence as regards his patient—a fiduciary position. It is his duty to act with the utmost good faith. This duty of the physician flows from the relationship with his patient and is fixed by law—not by the contract of employment . . . The law's exaction of good faith extends to all dealings between the physician and the patient. A person in ill health is more subject to the domination and influence of another than is a person of sound body and mind. The physician has unusual opportunity to influence his patient. Hence, all transactions between physician and patient are closely scrutinized by the courts which must be assured of the fairness of those dealings. In regard to any contract between physician and patient, it is the rule that the physician has the burden of proving that the patient entered into it voluntarily and advisedly, and without undue influence."

Morales AJ, Saffold CW, Bernstein J, Ecker JL, Lorenz RP, Lyerly AD, Martin DJ Sr, Morton C, Tauer CA, Wachbroit RS (ACOG Committee on Ethics) (2004) *Ethics in Obstetrics and Gynecology*, 2nd ed. Washington, DC: American College of Obstetricians and Gynecologists.

Morgan v. Sheppard (1963) 91 Ohio L Abs 579, 188 NE 2d 108 (App.; Cuyahoga Co.).

Morgentaler v. Prince Edward Island (Minister of Health and Social Services) (1995) P.E.I.J. No. 20 (P.E.I. S.C. – T.D.), rev'd 1996. P.E.I.J. No. (P.E.I. S.C. – A.D.), online: QL (PEIJ).

Morlino v. Medical Center of Ocean County (1998) 152 N.J. 563, 706 A. 2d. 721. The physician has a duty to take into account the individual needs of each patient. With regard to an allegation of negligence, the good faith or honesty of the physician is not the issue; the question is whether the exercise of judgment fell below the objective standard of care.

Morreim EH (1997) Medicine meets resource limits: Restructuring the legal standard of care. *Univ Pitt Law Rev* **59**:1–86.

Morris JK, Wald NJ, Watt HC (1999) Fetal loss in Down syndrome pregnancies. *Prenat Diagn* **19**:142–145.

Moskowitz E, Jennings B (1996) Directive counseling on long-acting contraception. *Am J Public Health* **86**(6):787–790.

Moskowitz SM, Gibson RL, Sternen DL, Cheng E, Cutting GR (2004) CFTR-related disorders. GeneTests. http://www.genetests.org/profiles/cf/details.html

Mouchawar J, Valentine Goins K, Somkin C, Puleo E, Hensley Alford S, Geiger AM, Taplin S, Gilbert J, Weinmann S, Zapka J (2003) Guidelines for breast and ovarian cancer genetic counseling referral: Adoption and implementation in HMOs. *Genet Med* **5**(6):444–450.

Moulton D (2003) N.B. courts face battle over abortion clinics. *Lawyers' Weekly* **23**(18):4.

Moutou C, Gardes N, Viville S [2004] Duplex, triplex and quadruplex PCR for the preimplantation genetic diagnosis (PGD) of cystic fibrosis (CF), an exhaustive approach. *Prenat Diagn* **24**(7):562–569.

Muhle R, Trentacoste SV, Rapin I (2004) The genetics of autism. *Pediatrics* **113**(5):e472–486.

Mukai J, Liu H, Burt RA, Swor DE, Lai WS, Karayiorgou M, Gogos JA (2004) Evidence that the gene encoding ZDHHC8 contributes to the risk of schizophrenia. *Nat Genet* **36**(7):725–731.

Mulcahy D, Cunningham K, McCormack D, Cassidy N, Walsh M (1997) Informed consent from whom? *J R Coll Surg Edinb* **42**(3):161–164.

Muller F, Doche C, Ngo S, Faina S, Charvin MA, Rebiffe M, Taguel V, Dingeon B (1999) Stability of free beta-subunit in routine practice for trisomy 21 maternal serum screening. *Prenat Diagn* **19**:85–86.

Munro v. Regent of the University of California (1989) 263 Cal. Rptr. 878, 882 (Cal. Ct. App.). A couple consulted a physician for genetic counseling. A family history was taken. When asked about their respective ethnic background, the mother advised that her background was "primarily German," English and Canadian (the evidence indicated that the defendant physician knew of a small, French Canadian community that was associated with a slightly higher prevalence of Tay-Sachs disease than the general population). Her husband was described as Scottish, Norwegian, and a "peculiar kind of French." Given this family history, the physician did not recommend testing for Tay-Sachs disease; the rate of incidence was described as relatively low for members of the general population except for people of Ashkenazi Jewish ancestry. The baby was born and diagnosed with Tay-Sachs disease. The parents claimed that they had sought out genetic counseling services to diagnose any genetic abnormalities, and the defendant had failed to provide a standard of care in neither offering nor discussing the availability of genetic testing services for Tay-Sachs disease. The court held that the test had not been offered because the parents were not members of the group whose genetic profile met the characteristics most highly associated with the disease. Legal commentators have criticized this decision for establishing too "drastic" a standard; they argue for a standard in which the physician's duty to disclose "is measured by the amount of knowledge a patient needs in order to make an informed choice." Compare this case to Humana of Kentucky, Inc. v. McKee. 1992. 834 S.W.2d 711, where it was alleged the defendant hospital did not perform mandatory PKU testing at birth, or if they did conduct testing, the negative results were incorrectly reported. The plaintiff's parents did not take the necessary dietary precautions affecting the plaintiff child's health. The jury found the hospital negligent in conducting PKU testing (mandatory at the time). An appellate court affirmed this judgment. Reutnauer JE (2000) Medical malpractice liability in the era of genetic susceptibility testing. *Quinnipiac L Rev* **19**:539–579. This case presents the issue, "when is a risk sufficient to obligate a health care professional to discuss and offer genetic testing services?" Some courts have adopted a standard of disclosure similar to that required for informed consent where a health care professional is not obligated to disclose every risk but only those considered "material" [e.g., *Largey v. Rothman* (1988) and *Jones v. United States* (1996)]. As to what constitutes a "material" risk, a court generally will take into consideration what practices constitute a prevailing standard of care. This generally will be determined by "what a reasonable person, in what the physician knows and should have known the plaintiff's position to be, would attach significance to the risk in deciding whether to forego the proposed treatment or not." Given that genetic testing technology and services continue to evolve, and new protocols and obligations continue to be debated, ascertaining what constitutes a "standard" of care may be difficult. What if the scenario in *Munro v. Regent of the University of California* occurred today? Would or should the defendant physician have a reasonable obligation to inquire into the mother's ancestry given the knowledge about the incidence of Tay-Sachs in the French Canadian community? In the case of *Gardner v. Pawliw* (150 N.J. 359. 1997) the court stated that where a diagnostic test may constitute a standard of care, but it is unknown whether performing the test will help diagnose or treat a preexisting condition, the plaintiff must demonstrate that the failure to provide the test increased the risk of harm from the preexistent condition. This may be demonstrated, stated the court, even if such tests are helpful in only a small percentage of cases.

Murakami Y, Okamura H, Sugano K, Yoshida T, Kazuma K, Akechi T, Uchitomi Y (2004) Psychologic distress after disclosure of genetic test results regarding hereditary nonpolyposis colorectal carcinoma. *Cancer* **101**:395–403.

Murgatroyd RJ, Cooper RM (1999) Readability of informed consent forms. *Am J Hosp Pharm* **48**(12):2651–2652.

Murray TH (1992) Speaking unsmooth things about the Human Genome Project. In: Annas GJ, Elias S, eds. *Gene Mapping Using Law & Ethics as Guides*. New York: Oxford University Press, p. 252.

Murray T (1996) *The Worth of a Child*. Berkeley: University of California Press.

Murray CJL, Lopez AD (1997) Alternative projections of mortality and disability by cause 1990–2020: Global burden of disease study. *Lancet* **349**:1498–1504.

Myers MF, Doksum T, Holtzman NA (1999) Genetic services for common complex disorders: Surveys of health maintenance organizations and academic genetic centers. *Genet Med* **1**(6):272–285.

Myriad Genetics (2004) BRCA1 and BRCA2 mutation prevalence tables; http://www .myriadtests.com/provider/mutprev.htm.

Naccarato v. Grob (1970) 180 N.W. 2d 788 (Mich.).

Naccash v. Burger (1982) 223 Va. 406, 290 S.E.2d 825.

Nadel AG (1982) Duty of a medical practitioner to warn patient of subsequently discovered danger from treatment previously given. *Am Law Reports* **12**(4):41.

Nakamura L (1999) A trillion dollars a year in intangible investment and the new economy. In: Hand J, Lev B, eds. *Intangible Assets: Values, Measures, and Risks*, Oxford: Oxford University Press, pp. 19–47 [Paper written in 1999; book published in 2003].

Narayan S, Roy D (2003) Role of APC and DNA mismatch repair genes in the development of colorectal cancers. *Mol Cancer* **2**:41 (http://www.molecular-cancer.com/content/2/1/41.

Narod SA, Foulkes WD (2004) BRCA1 and BRCA2: 1994 and beyond. *Nat Rev Cancer* **4**:665–676.

Narod SA, Risch H, Moslehi R, Dorum A, Neuhausen S, Olsson H, Provencher D, Radice P, Evans G, Bishop S, Brunet J-S, Ponder BAJ, for the Hereditary Ovarian Cancer Clinical Study Group (1998) Oral Contraceptives and the risk of hereditary ovarian cancer. *N Engl J Med* **339**:424–428.

Narod SA, Brunet JS, Ghadirian P, Robson M, Heimdal K, Neuhausen SL, Stoppa-Lyonnet D, Lerman C, Pasini B, de los Rios P, Weber B, Lynch H, Hereditary Breast Cancer Clinical Study Group (2000) Tamoxifen and risk of contralateral breast cancer in BRCA1 and BRCA2 mutation carriers: a case-control study. *Lancet* **356**:1876–1881.

Narod SA, Isaacs C, Matloff E, Daly MB, Olopade O, Weber BL (2004) Bilateral prophylactic mastectomy reduces breast cancer risk in BRCA1 and BRCA2 mutation carriers: The PROSE study group. *J Clin Oncol* **22**:1055–1062.

National Human Genome Research Institute (2004) Genetics FAQ: What are genetic counselors and what do they do? Available: http://www.genome.gov/10001191#6.

National Research Council. (1975) Committee for the Study of Inborn Errors of Metabolism, Genetic Screening: Programs, Principles, and Research. Washington, DC: Natl Acad Sci,

National Research Council NRC (1997). *Intellectual Property Rights and Research Tools in Molecular Biology.* Washington, DC: National Academies Press. Online: NAP, <http://books.nap.edu/html/property/>.

National Society of Genetic Counselors (NSGC) (1991) Position statement on disclosure and informed consent. Available at: <http://www.nsgc.org/newsroom/position.asp>. Accessed October 26, 2004.

National Society of Genetic Counselors (1992) Code of Ethics of the National Society of Genetic Counselors. *J Genet Counsl* **1**(1):41–43. http://www.nsgc.org/about/code_of_ethics.asp.

National Society of Genetic Counselors (NSGC) (2004) Code of Ethics. Available at: <http://www.nsgc.org/newsroom/code_of_ethics.asp>. Accessed October 26, 2004.

National Society of Genetic Counselors (1997) Predisposition genetic testing for late-onset disorders in adults. *JAMA* **278**:1217–1220.

Naylor CD, Chen E, Strauss B (1992) Measured enthusiasm: Does the method of reporting trial results alter perceptions of therapeutic effectiveness? *Ann Intern Med* **117**:916–921.

NCHPEG (2001) Core Competencies in Genetics Essential for All Health-Care Professionals.

Nebiolo L, Ozturk M, Brambati B, Miller S, Wands J, Milunsky A (1990) First-trimester maternal serum alpha-fetoprotein and human chorionic gonadotropin screening for chromosome defects. *Prenat Diagn* **10**(9):575–581.

Nedelcu R, Blazer K, Schwerin B, Gambol P, Mantha P, Uman G, Weitzel J (2004) Genetic discrimination: The clinician perspective. *Clin Genet* **66**(4):311–317. Of 271 clinicians surveyed, 13% would not encourage genetic testing, despite a family history of cancer. Concerns about potential genetic discrimination, and lack of information about protective legislation, could influence access to care.

Nelkin D (2002) Patenting genes and public interests. *Am J Bioethics* **2**(3):13–15.

Nelson E, Robertson G (2001) Liability for wrongful birth and wrongful life. *ISUMA* **2**(3): 13–15. http://www.isuma.net/v02n03/nelson/nelson_e.shtml.

Nelson v. Krusen (1984) 678 SW.2d 918 (Tex.).

Neubauer BA, Fiedler B, Himmelein B, Kampfer F, Lassker U, Schwabe G, Spanier I, Tams D, Bretscher C, Moldenhauer K, Kurlemann G, Weise S, Tedroff K, Eeg-Olofsson O, Wadelius C, Stephani U (1998) Centrotemporal spikes in families with rolandic epilepsy: Linkage to chromosome 15q14. *Neurology* **51**:1608–1612.

Neudeck v. Vestal (1931) 3 P. 2d 595, 117 Cal. App. 266.

Neves-Pereira M, Mundo E, Muglia P, King N, Macciardi F, Kennedy JL (2002) The brain-derived neurotrophic factor gene confers susceptibility to bipolar disorder: Evidence from a family-based association study. *Am J Hum Genet* **71**(3):651–655.

Neveux LM, Palomaki GE, Knight GJ, Haddow JE (1996) Multiple marker screening for Down syndrome in twin pregnancies. *Prenat Diagn* **16**:29–34.

Newborn Screening Task Force (1999) Serving the Family from Birth to Medical Home: A Report from the Newborn Screening Task Force convened in Washington D.C., May 10–11, 1999 (August 2000). *Pediatrics* **106**:2(Supp.). *Newborn Screening: Toward a Uniform Screening Panel and System>* Report for Public Comment Maternal and Child Health Burean 2005- http://www.mchb.hrsa.gov/screening/New Born Screening: Characteristics of State Programs March 2003- http:/www.gao.gov/new.items/d03449.pdf.

Newby D, Aitken DA, Crossley JA, Howatson AG, Macri JN, Connor JM (1997) Biochemical markers of trisomy 21 and the pathophysiology of Down's syndrome pregnancies. *Prenat Diagn* **17**:941–951.

Newby D, Aitken DA, Howatson AG, Connor JM, (2000) Placental synthesis of oestriol in Down's syndrome pregnancies. *Placenta* **21**:263–267.

Nicol D, Nielsen J (2003) Patents and medical biotechnology: An empirical analysis of issues facing the Australian industry. Centre for Law & Genetics, Occasional Paper No. 6. Online: IPRIA, <http://www.law.unimelb.edu.au/ipria/publications/reports.html#Patent>.

Nicolaides KH, Azar G, Byrne D, Mansur C, Marks K (1992) Fetal nuchal translucency: Ultrasound screening for chromosomal defects in first trimester of pregnancy. *BMJ* **304**:867–869.

Nicolaides KH, Heath V, Cicero S (2002) Increased fetal nuchal translucency at 11–14 weeks. *Prenat Diagn* **22**(4):308–315.

Nidorf JF, Ngo KY (1993) Cultural and psychosocial considerations in screening for Thalassemia in Southeast Asian refugee population. *Am J Med Genet* **46**:398–402.

Niederau C, Fischer R, Purschel A, Stremmel W, Haussinger D, Strohmeyer G (1996) Long-term survival in patients with hereditary hemochromatosis. *Gastroenterology* **110**(4):1107–1119.

NIH Consensus Development Conference Statement (1997) Genetic testing for cystic fibrosis. April 14–16, 1997. Available at <http://consensus.nih.gov/cons/106/106_statement.htm#4_3._Shou>. Accessed October 27, 2004.

NIH/DOE Working Group on Ethical, Genetic Information and Health Insurance (1993) Legal and social implications of human genome research. *Hum Gene Therapy* **4**:789–808.

Niklasson L, Rasmussen P, Oskarsdottir S, Gillberg C (2001) Neuropsychiatric disorders in the 22q11 deletion syndrome. *Genet Med* **3**(1):79–84.

Noble J (1998) Natural history of Down's syndrome: A brief review for those involved in antenatal screening. *J Med Screen* **5**:172–177.

Nocton v. Lord Ashburton (1914) A.C. 932 at 955 (H.L.).

Nordin K, Lid A, Hansson M, Rosenquist R, Berglund G (2002) Coping style, psychological distress, risk perception, and satisfaction in subjects attending genetic counselling for hereditary cancer. *J Med Genet* **39**:689–694.

Note (1981) Informed consent: From disclosure to patient participation in medical decision-making. *NW Univ Law Rev* **76**:172–173, 190–195.

Novakovic B, Goldstein AM, Tucker MA (1996) Validation of family history of cancer in deceased family members. *J Natl Cancer Inst* **88**(20):1492–1493.

Nowaczyk MJM, Waye JS (2001) The Smith-Lemli-Opitz syndrome: A novel metabolic way of understanding developmental biology, embryogenesis, and dysmorphology. *Clin Genet* **59**:375–386.

Nowatske v. Osterloh (1996) 198 Wisc. 2d 419, 543 NW 2d 265, on remand 201 Wis. 2d 497, 549 NW 2d 256 (App.).

Nowlan WJ (2000) Genetic testing debate may be entering a new era. *National Underwriter* **104**(5):19–21.

Nuffield Council on Bioethics (2002) The ethics of patenting DNA: A discussion paper, Nuffield Council on Bioethics, London. Online: <http://www.nuffieldbioethics.org/go/ourwork/patentingdna/publication_310.html>.

Nuovo J, Melnikow J, Chang D (2002) Reporting number needed to treat and absolute risk reduction in randomized controlled trials. *JAMA* **287**:2813–2814.

Nussbaum MC (1986) *The Fragility of Goodness: Luck and Ethics in Greek Tragedy and Philosophy.* New York: Cambridge University Press, pp. 32–35. This discussion about the ethics of "choice" is in reference to Aeschylus's "Agamemnon":

Nussbaum RL (2003) The human genome project. In: Rosenberg RN, Prusiner SB, DiMauro S, Barchi RL, Nestler EJ, eds. *The Molecular and Genetic Basis of Neurologic and Psychiatric Disease,* Philadelphia: Elsevier, pp. 91–100.

Nyberg DA, Resta RG, Luthy MA, Hickok DE, Williams MA (1993) Humerus and femur length shortening in the detection of Down syndrome. *Am J Obstet Gynecol* **168**:534–538.

Nyberg DA, Souter VL (2003) Use of genetic sonography for adjusting the risk for fetal Down syndrome. *Sem Perinatol* **27**(2):130–144.

Obedian E, Fischer DB, Haffy BG (2000) Second malignancies after treatment of early-stage breast cancer: Lumpectomy and radiation therapy versus mastectomy. *J Clin Oncol* **18**:2406–2412.

O'Cathain A, Thomas KJ (2004) Evaluating decision aids—where next? *Health Expect* **7**(2):98–103.

O'Cathain A, Walters SJ, Nicholl JP, Thomas KJ, Kirkham M (2002) Use of evidence based leaflets to promote informed choice in maternity care: Randomised controlled trial in everyday practice. *BMJ* **324**(7338):643.

O'Connor AM, Drake ER, Fiset V, Graham ID, Laupacis A, Tugwell P (1999) The Ottawa patient decision aids. *Eff Clin Pract* **2**(4):163–170.

O'Daniel JM, Wells D (2002) Approaching complex cases with a crisis intervention model and teamwork. *J Genet Counsel* **11**(5):369–376.

Offit K, Groeger E, Turner S, Wadsworth EA, Weiser MA (2004) The "duty to warn" a patient's family members about hereditary disease risks. *JAMA* **292**:1469–1473.

Oliveira SA, Scott WK, Martin ER, Nance MA, Watts RL, Hubble JP, Koller WC, Pahwa R, Stern MB, Hiner BC, Ondo WG, Allen FH Jr, Scott BL, Goetz CG, Small GW, Mastaglia F, Stajich JM, Zhang F, Booze MW, Winn MP, Middleton LT, Haines JL, Pericak-Vance MA, Vance JM (2003) Parkin mutations and susceptibility alleles in late-onset Parkinson's disease. *Ann Neurol* **53**:624–629.

Olivier RI, van Beurden M, Lubsen MAC, Rookus MA, Mooij TM, van de Vijver MJ, van't Veer LJ (2004) Clinical outcome of prophylactic oophorectomy in BRCA1/BRCA2 mutation carriers and events during follow-up. *Br J Cancer* **90**:1492–1497.

Olsson HL, Ingvar C, Bladstrom A (2003) Hormone replacement therapy containing progestins and given continuously increases breast carcinoma risk in Sweden. *Cancer* **97**:1387–1392.

Online Mendelian Inheritance in Man, OMIM (TM) (2000) McKusick-Nathans Institute for Genetic Medicine, Johns Hopkins University (Baltimore, MD) and National Center for Biotechnology Information, National Library of Medicine (Bethesda, MD). Online at http://www.ncbi.nlm.nih.gov/omim/.

Ontario AG, Dieleman V (1994) 117 D.L.R. 4th 449 *Ont. Gen. Div.*.

Ontario Advisory Committee on New Predictive Genetic Technologies (2001) Genetics in Ontario: Mapping the Future, November 30. http://www.health.gov.on.ca/english/public/pub/ministry_reports/geneticsrep01/genetic_report.pdf.

Ontario Law Reform Commission (1996) *Report on Genetic Testing.* Toronto: Ontario Law Reform Commission.

Ontario Medical Association (2004) Ontario physicians' guide to referral of patients with family history of cancer to a familial cancer genetics clinic or genetics clinic: http://www.oma.org/pcomm/omr/nov/01genetics.htm.

Ontario, Ministry of Health (2002) Genetics, testing and gene patenting: Charting new territory in healthcare. Online: Ontario Ministry of Health <http://www.health.gov.on.ca/english/public/pub/ministry_reports/geneticsrep02/genetics.html>.

Organisation for Economic Cooperation and Development, OECD (2002) *Genetic Inventions, Intellectual Property Rights & Licensing Practices*, p. 11, 77. Cedex, France: OECD. Online: OECD <http://www.oecd.org/dataoecd/42/21/2491084.pdf>.

Ormond KE, Gill CJ, Semik P, Kirschner KL (2003) Attitudes of health care trainees about genetics and disability: Issues of access, health care communication, and decision making. *J Genet Counsel* **12**(4):333–349.

Osborne P (2003) *The Law of Torts*, 2nd ed. Irwin Law (QL), chpt. 3, Toronto.

Overton C, Serhal P, Davies M (2001) Clinical aspects of preimplantation diagnosis. In: Harper JC, Delhanty JDA, Handyside AH, eds. *Preimplantation Genetic Diagnosis.* Chichester, England: Wiley, pp. 123–140.

Owen MJ, O'Donovan M, Gottesman II (2003) *Psychiatric Genetics and Genomics.* Oxford: Oxford University Press.

Paasche-Orlow MK, Taylor HA, Brancati FL (2003) Readability standards for informed-consent forms as compared with actual readability. *N Engl J Med* **348**(8):721–726.

Pagnatarro MA (2001) Genetic discrimination and the workplace: Employees' right to privacy v. employer's need to know. *Am Bus Law J* **39**:139–185.

Pagon RA (2002) Genetic testing for disease susceptibilities: Consequences for genetic counseling. *Trends Mol Med* **8**(6):306–307.

Pajkrt E, Mol BW, Bleker OP, Bilardo CM (1999) Pregnancy outcome and nuchal translucency measurements in fetuses with a normal karyotype. *Prenat Diagn* **19**:1104–1108.

Pal DK, Evgrafov OV, Tabares P, Zhang F, Durner M, Greenberg DA (2003) BRD2 (RING3) is a probable major susceptibility gene for common juvenile myoclonic epilepsy. *Am J Hum Genet* **73**(2):261–270.

Paling J (2003) Strategies to help patients understand risks. *BMJ* **327**:745–748.

Palmer JR, Rosenberg L, Miller DR, Miller DR, Shapiro S (1991) Breast cancer risk after estrogen replacement therapy: Results from the Toronto breast cancer study. *Am J Epidemiol* **134**:1386–1395.

Palomaki GE, Haddow JE, Knight GJ, Wald NJ, Kennard A, Canick JA, Saller DN, Jr, Blitzer MG, Dickerman LH, Fisher R (1995) Risk-based prenatal screening for trisomy 18 using alpha-fetoprotein, unconjugated oestriol and human chorionic gonadotropin. *Prenat Diagn* **15**:713–723.

Palomaki GE, Haddow JE, Beauregard LJ (1996) Prenatal screening for Down's syndrome in Maine, 1980 to 1993. *N Engl J Med* **334**:1409–1410.

Palomaki GE, Neveux LM, Knight GJ, Haddow JE (2003) Maternal serum-integrated screening for trisomy 18 using both first- and second-trimester markers. *Prenat Diagn* **23**:243–247.

Pandor A, Eastham J, Beverley C, Chilcott J, Paisley S (2004). Clinical effectiveness and cost effectiveness of neonatal screening for inborn errors of metabolism using tandem mass spectrometry: A systematic review. Health Technol Assess **8**(12):1–121.

Pandya PP, Brizot ML, Kuhn P, Snijders RJ, Nicolaides KH (1994) First-trimester fetal nuchal translucency thickness and risk for trisomies. *Obstet Gynecol* **84**:420–423.

Pandya PP, Kondylios A, Hilbert KH (1995a) Abnormalities of the heart and great arteries in first trimester chromosomally abnormal fetuses. *Am J Med Genet* **5**:15–19.

Pandya PP, Kondylios A, Hilbert L, Snijders RJ, Nicolaides KH (1995b) Chromosomal defects and outcome in 1015 fetuses with increased nuchal translucency. *Ultrasound Obstet Gynecol* **5**:15–19.

Pandya PP, Snijders RJ, Johnson SP, De Lourdes Brizot M, Nicolaides KH (1995c) Screening for fetal trisomies by maternal age and fetal nuchal translucency thickness at 10 to 14 weeks of gestation. *Br J Obstet Gynaecol* **102**:957–962.

Panegyres PK, Goldblatt J, Walpole I, Connor C, Liebeck T, Harrop K (2000) Genetic testing for Alzheimer's disease. *Med J Aust* **172**:339–343.

Pantilat SZ, Lindenauer PK, Katz PP, Wachter RM (2001) Primary care physician attitudes regarding communication with hospitalists. *Am J Med* **21**;111(9B):15S–20S.

Pape T (1997) Legal and ethical considerations of informed consent. *AORN J* **65**(6):1122–1127.

Papp Z, Toth Z, Torok O, Szabo M (1987) Prenatal diagnosis policy without routine amniocentesis in pregnancies with a positive family history for neural tube defects. *Am J Med Gen* **26**:103–110.

Parad RB (1996) Heterogeneity of phenotype in two cystic fibrosis patients homozygous for the CFTR exon 11 mutation G551D. *J Med Genet* **33**:711–713.

Parens E, Asch A (2003) Disability rights critique of prenatal genetic testing: Reflections and recommendations. *Ment Retard Develop Disabilities Res Rev* **9**:40–47.

Park YW, Zhu S, Palaniappan L, Heshka S, Carnethon MR, Heymsfield SB (2003) The metabolic syndrome: Prevalence and associated risk factor findings in the US population from the Third National Health and Nutrition Examination Survey, 1988–1994. *Arch Intern Med* **163**:427–436.

Parker CM (1998) Camping trips and family trees: Must Tennessee physicians warn their patients' relatives of genetic risks? *Tenn Law Rev* **65**(2):585–618.

Parker LS, Lidz CW (1994) Familiar coercion to participate in genetic family studies: Is there cause for IRB intervention? *Rev Hum Subjects Res* **16**(1–2):6.

Parker M, Lucassen A (2004) Ethics in practice genetic information: A joint account? *BMJ* **329**:165–167.

Parknell v. Fitzporter (1923) 301 Mo. 317, 256 S.W. 2d 239, 29 A.L.R. 1305.

Parlanti AA, Burbidge LA, Shappell HL, Reid JE, Deffenbaugh AM, Frye CA Critchfield GC (2003) Prevalence of BRCA1 and BRCA2 mutations in women diagnosed with bilateral invasive breast cancer. *San Antonio Breast Cancer Symposium*, December 5 (Abstr.)

Parmigiani G, Berry D, Aguilar O (1998) Determining carrier probabilities for breast cancer-susceptibility genes BRCA1 and BRCA2. *Am J Hum Genet* **62**(1):145–158.

Parsons EP, Bradley DM (2003) Psychosocial issues in newborn screening for cystic fibrosis. *Paediatr Respir Rev* **4**(4):285–292.

Parsons EP, Hood K, Lycett E, Bradley DM (2002) Newborn screening for Duchenne muscular dystrophy: A psychosocial study. *Arch Dis Child Fetal Neonatal Ed* **86**(2):F91–95. There was evidence that reproductive patterning had been modified, and four fetuses carrying a mutation causing Duchenne muscular dystrophy were terminated.

Parsons EP, Clarke AJ, Bradley DM (2003) Implications of carrier identification in newborn screening for cystic fibrosis. *Arch Dis Child Fetal Neonatal Ed* **88**(6):F467–471.

Partial Birth Abortion Act of 2003 (2003) S. 3-8, 108th Congress, 1st Session.

Partridge AH, Winer EP (2002) Informing clinical trial participants about study results. *JAMA* **288**(3):363–365.

Pastor P, Roe CM, Villegas A, Bedoya G, Chakraverty S, Garcia G, Tirado V, Norton J, Rios S, Martinez M, Kosik KS, Lopera F, Goate AM (2003) Apolipoprotein E epsilon 4 modifies Alzheimer's disease onset in an E280A PS1 kindred. *Ann Neurol* **54**:163–169.

Pate v. Threlkel (1995) 661 So. 2d 278, 282 (Fla. 1994). Physician's duty to warn third parties can be satisfied by warning the patient. Compare to *Safer v. Pack* (1996) 291 N.J.Super. 619, 677 A.2d 1188, 1192 (N.J. Super. App. Div.). The court found that the physician owes a duty to warn of an "avertable risk" not only to the patient but members of the immediate family who may be adversely affected by a breach of that duty. The court specifically did not hold that the duty to warn would be satisfied by informing the patient.

Patrick v. Sedwick (1964) 391 P 2d 453, appeal after remand, 413 P 2d 169 (Alaska).

Patt MR, Houston TK, Jenckes MW, Sands DZ, Ford DE (2003) Doctors who are using e-mail with their patients: A qualitative exploration. *J Med Internet Res* **5**(2):e9.

Patterson N, Hattangadi N, Lane B, Lohmueller KE, Hafler DA, Oksenberg JR, Hauser SL, Smith MW, O'Brien SJ, Altshuler D, Daley MJ, Reich D (2004) Methods for high-density admixture mapping of disease genes. *Am J Hum Genet* **74**:979–1000.

Pauker SG, Kassirer JP (1987) Decision analysis. *New Engl J Med* **306**:250–258.

Pauker SG, Pauker SP (1987) Prescriptive models to support decision making in genetics. In: Evers-Kiebooms, Cassiman H, Van den Berghe H, d'Ydewalle, eds. *Genetic Risk, Risk Perception and Decision-Making.* New York: Alan R Liss, for the March of Dimes Birth Defects Foundation. *Birth Defects Original Articles Series* **23**(2):279–296.

Peck v. The Counseling Service of Addison County, Inc (1985) 146 Vt. 61, 499 A.2d 422. After rejection by his father, a mental health patient told the counselor that he wanted to burn down his father's barn. Although the patient promised not to proceed with the plan, the barn subsequently was burnt down. On appeal, the Vermont Supreme Court held that the counselor had a duty to protect an identifiable victim's property interests and that this did not differ substantially from the legal obligation to warn in order to protect the public health.

Pedroza v. Bryant (1984) 677 P.2d 166 at 170-171 (Wash.).

Peiro G, Bornstein BA, Connolly JL, Gelman R, Hetelekidis S, Nixon AJ, Recht A, Silver B, Harris JR, Schnitt SJ (2000) The influence of infiltrating lobular carcinoma on the outcome of patients treated with breast-conxerving surgery and radiation therapy. *Breast Cancer Res Treat* **59**:49–54.

Pelias MZ (1991) Duty to disclose in medical genetics: A legal perspective. *Am J Med Genet* **39**:347–354.

Peltomaki P (2003) Role of DNA mismatch repair defects in the pathogenesis of human cancer. *J Clin Oncol* **21**:1174–1179.

Penchaszadeh VB (2001) Genetic counseling issues in Latinos. *Genet Test* **5**(3):193–200.

Penchaszadeh VB, Punales-Morejon D (1998) Genetic services to the latino population in the United States. *Community Genet* **1**(3):134–141.

Penrose LS (1933) The relative effects of paternal and maternal age in mongolism. *J Genet* **27**:219–224.

Penson RT, Seiden MV, Shannon KM, Lubratovich ML, Roche M, Chabner BA, Lynch TJ Jr (2000) Communicating genetic risk: Pros, cons, and counsel. *Oncologist* **5**(2):152–161.

Peshkin BN, DeMarco TA, Brogan BM, Lerman C, Isaacs C (2001) BRCA1/2 testing: Complex themes in result interpretation. *J Clin Oncol* **19**(9):2555–2565.

Peterson K (1996) Abortion laws: Comparative and feminist perspectives in Australia, England and the United States. *Med Law Int* **2**:77–105.

Peterson SK, Watts BG, Koehly LM, Vernon SW, Baile WF, Kohlmann WK, Gritz ER (2003) How families communicate about Hnpcc genetic testing: Findings from a qualitative study. *Am J Med Genet* **119**C(1):78–86.

Peto J, Collins N, Barfoot R, Seal S, Warren W, Rahman N, Easton DF, Evans C, Deacon J, Stratton MR (1999) Prevalence of BRCA1 and BRCA2 gene mutations in patients with early-onset breast cancer. *J Natl Cancer Inst* **91**:943–949.

Petrikovsky B, Smith-Levitin M, Hosten N (1999) Intra-amniotic bleeding and fetal echogenic bowel. *Obstet Gynecol* **93**:684–686.

Petronis A (2004) The origin of schizophrenia: Genetic thesis, epigenetic antithesis, and resolving synthesis. *Biol. Psychiatry* **55**(10):965–970.

Petrucelli N, Walker M, Schorry E (1998) Continuation of a pregnancy following the diagnosis of a fetal sex chromosome abnormality: A study of parents' counseling needs and experiences. *J Genet Counsel* **7**(5):401–415.

Pfeffer NL, Veach PM, LeRoy BS (2003) An investigation of genetic counselors' discussion of genetic discrimination with cancer risk patients. *J Genet Counsl* **12**(5):419–438.

Pharoah PD, Day NE, Duffy S, Easton DF, Ponder BA (1997) Family history and the risk of breast cancer: A systematic review and meta-analysis. *Int J Cancer* **71**:800–809.

Phelan CM, Dapic V, Tice B, Favis R, Kwan E, Barany F, Manoukian S, Radice P, van der Luijt RB, van Nesselrooij BPM, Chenevix-Trench G, kConfab, Caldes T, de La Hoya M, Lindquist S, Tavtigian SV, Goldgar D, Borg A, Narod SA, Monteiro ANA (2005) Classification of BRCA1 missense variants of unknown clinical significance. *J Med Genet* **42**:138–146.

Phelps C, Platt K, France L, Gray J, Iredale R (2004) Delivering information about cancer genetics via letter to patients at low and moderate risk of familial cancer: A pilot study in Wales. *Fam Cancer* **3**(1):55–59.

Phillips HA, Scheffer IE, Crossland KM, Bhatia KP, Fish DR, Marsden CD, Howell SJ, Stephenson JB, Tolmie J, Plazzi G, Eeg-Olofsson O, Singh R, Lopes-Cendes I, Andermann E, Andermann F, Berkovic SF, Mulley JC (1998) Autosomal dominant nocturnal frontal-lobe epilepsy: Genetic heterogeneity and evidence for a second locus at 15q24. *Am J Hum Genet* **63**:1108–1116.

Phillips KA, Glendon G, Knight JA (1999) Putting the risk of breast cancer in perspective. *New Engl J Med* **340**:141–144.

Phillips KA, Warner E, Meschino WS, Hunter J, Abdolell M, Glendon G, Andrulis IL, Goodwin PJ (2000) Perceptions of Ashkenazi Jewish breast cancer patients on genetic testing for mutations in BRCA1 and BRCA2. *Clin Genet* **57**(5):376–383.

Phillips RL Jr, Bartholomew LA, Dovey SM, Fryer GE Jr, Miyoshi TJ, Green LA (2004) Learning from malpractice claims about negligent, adverse events in primary care in the United States. *Qual Saf Health Care* **13**(2):121–126.

Phillips v. Good Samaritan Hospital (1979) 416 NE 2d 646 (Ohio).

Phillips v. U.S. 566 F. Supp. 1 (D.S.C. 1981).

Picard E, Robertson G (1996) Legal Liability of Doctors and Hospitals in Canada, 3rd ed., Carswell, Toronto.

Pierce LJ, Strawderman M, Narod SA, Oliviotto I, Eisen A, Dawson L, Gaffney D, Solin LJ, Nixon A, Garber J, Berg C, Isaacs C, Heimann R, Olopade OI, Haffty B, Weber BL (2000) Effect of radiotherapy after breast-conserving treatment in women with breast cancer and germline BRCA1/2 mutations. *J Clin Oncol* **18**:3360–3369.

Pietrangelo A (2004a) Hereditary hemochromatosis—A new look at an old disease. *N Engl J Med* **350**:2383–2397.

Pietrangelo A (2004b) The ferroportin disease. *Blood Cells Mol Disease* **32**(1):131–138.

Pike v. Honsinger (1898) 155 NY 201, at 209: Physicians should exercise "that reasonable degree of learning and skill that is ordinarily possessed by physicians and surgeons in the locality where [the physician] practices."

Pilu G, Nicolaides KH (1999) *Diagnosis of Fetal Abnormalities: The 18–23-Week Scan.* London: Parthenon Publishing Group.

Pincock S (2003) Patients put their relationship with their doctors as second only to that with their families. *BMJ* Vol. 327:581.

Pitre v. Opelousas General Hosp (1988) 530 So.2d 1151, 74 A.L.R.4th 777 (La.), rehearing denied (Oct 20, 1988). Online at http://biotech.law.lsu.edu/cases/la/medmal/pitre.htm.

Pitz GF (1987) Evaluating decision aiding technologies for genetic counseling. In: Evers-Kiebooms G, Cassiman H, Van den Berghe H, d'Ydewalle G, eds. *Genetic Risk, Risk Perception and Decision-Making.* New York: Alan R Liss, for the March of Dimes Birth Defects Foundation—Birth Defects Original Articles Series **23**(2):251–278.

Pizzuti A, Flex E, Di Bonaventura C, Dottorini T, Egeo G, Manfredi M, Dallapiccola B, Giallonardo AT (2003) Epilepsy with auditory features: A LGI1 gene mutation suggests a loss-of-function mechanism. *Ann Neurol* **53**:396–399.

Plantinga L, Natowicz MR, Kass NE, Hull SC, Gostin LO, Faden RR (2003) Disclosure, confidentiality, and families: Experiences and attitudes of those with genetic versus nongenetic medical conditions. *Am J Med Genet* **119**(1):51–59.

Pokorski R (1992) Use of genetic information by private insurers—genetic advances: The perspective of an insurance medical director. *J Ins Med* **24**(1):60–68.

Pokorski RJ (1997) Insurance underwriting in the genetic era, *Am J Hum Genet* **60**:205–216.

Polymeropoulos MH, Lavedan C, Leroy E, Ide SE, Dehejia A, Dutra A, Pike B, Root H, Rubenstein J, Boyer R, Stenroos ES, Chandrasekharappa S, Athanassiadou A, Papapetropoulos T, Johnson WG, Lazzarini AM, Duvoisin RC, Di Iorio G, Golbe LI, Nussbaum RL (1997) Mutation in the alpha-synuclein gene identified in families with Parkinson's disease. *Science* **276**:2045–2047.

Poole v. Morgan (1987) 3 W.W.R. 217, 50 Alta. L.R. (2d) 120, 1987, A.J. No. 1414 (Alta. Q.B. Jan 14); [1988] 6 W.W.R. 481, 68 Sask. R. 104: "The highest standard of care is expected of the doctor using a new or experimental procedure or treatment."

Porch JV, Lee IM, Cook NR, Rexrode KM, Burin JE (2002) Estrogen-progestin replacement therapy and breast cancer risk: The Women's Health Study (United States). *Cancer Causes Control* **13**:847–854.

Porteous M, Dunckley M, Appleton S, Catt S, Dunlop M, Campbell H, Cull A (2003) Is it acceptable to approach colorectal cancer patients at diagnosis to discuss genetic testing? A pilot study. *Br J Cancer* **89**(8):1400–1402.

Post SG, Whitehouse PJ, Binstock RH, Bird TD, Eckert SK, Farrer LA, Fleck LM, Gaines AD, Juengst ET, Karlinsky H, Miles S, Murray TH, Quaid KA, Relkin NR, Roses AD, St George-Hyslop PH, Sachs GA, Steinbock B, Truschke EF, Zinn AB (1997) The clinical introduction of genetic testing for Alzheimer disease: An ethical perspective. *JAMA* **277**:832–836.

Powledge TM, Fletcher J (1979) Guidelines for the ethical, social, and legal issues in prenatal diagnosis. *New Engl J Med* **300**:168–172.

Pratt by Pratt v. University of Minnesota Affiliated Hospitals and Clinics (1987) 403 N.W. 2d 865, at 868; 414 NW.2d 399 (Minn.). "Treatment" has been recognized by courts to include diagnostic options and choice of hospital in which a procedure will be performed: *Martin v. Richards* (1993) 500 N.W.2d 691 (Wis. Ct. App.); 531 N.W.2d 70 (Wis. 1995)—the physicians failed to inform the parents of a CT scan for a child.

This case concerned an allegation for "negligent nondisclosure" during genetic counseling. Parents of an afflicted child wanted to know whether the condition "was genetic in origin," and whether future offspring would have an increased risk. The consulting physicians reviewed the parents' respective medical histories, took blood samples, and performed a "chromosome study." Although the physicians were unable to diagnose the child's anomalies, both consulting physicians were aware of the possibility that the disorder may be an autosomal-recessive condition. However, neither physician felt that "the possibility was significant enough" to inform the parents. At trial, the physicians argued that genetic counseling did not constitute treatment and therefore the doctrine of negligent nondisclosure did not apply. The trail court held that it was vital for patients in the situation of having to make a choice to be able to base their decision on as much information as possible. "This is especially true in the area of genetic counseling." The court concluded that the physicians had replied to the parents' request for information "in order to make a major health care decision. In providing that information, we conclude that the respondents [the physicians] did render treatment."

Preloran HM, Browner CH, Lieber E (2001) Strategies for motivating Mexican-origin couples' participation in qualitative research. *Am J Public Health* **91**(11):1832–1841.

Prence EM (1999) A practical guide for the validation of genetic tests. *Genet Test* **3**(2):201–205.

President's Commission for the Study of Ethical Problems in Medicine and Biomedical and Behavioral Research (1983) *Summing Up: Final Report on Studies of the Ethical and Legal Problems in Medicine and Biomedical and Behavioral Research*. Washington, DC: The Commission, pp. 36–37.

Press N, Browner CH (1998) Characteristics of women who refuse an offer of prenatal diagnosis: Data from the California maternal serum alpha fetoprotein blood test experience. *Am J Med Genet* **78**(5):433–445.

Press NA, Fishman JR, Koenig BA (2000) Collective fear, individualized risk: The social and cultural context of genetic testing for breast cancer. *Nurs Ethics* **7**:237–248.

Press N, Yasui Y, Reynolds S, Durfy S, Burke W (2001). Women's interest in genetic testing for breast cancer susceptibility may be based on unrealistic expectations. *Am J Med Genet* **99**:99–110.

Pridham v. Nash (1986) 57 O.R. (2d) 347; (1987) 33 D.L.R. (4th) 304 (H.Ct.)

Prillaman HL (1990) A physician's duty to inform of newly developed therapy. *J Contemp Health L Pol'y* **6**:43, 52–58.

Procanik by Procanik v. Cillo (1984) 97 N.J. 339, 478 A.2d 755, on remand, 206 N.J.Super. 270, 502 A.2d 94 (1985); (1988) 543 A.2d 985. N.J. Super. A.D.

Proffitt v. Bartolo (1987) 412 N.W.2d 232 (Mich. Ct. App.), appeal denied, 430 Mich. 860 (1988).

Prouser N (2000) Case report: Genetic susceptibility testing for breast and ovarian cancer: A patient's perspective. *Genet Counsel* **9**(2):153–159.

Prusiner SB (2003) Degenerative diseases and protein processing. In: Rosenberg RN, Prusiner SB, DiMauro S, Barchi RL, Nestler EJ, eds. *The Molecular and Genetic Basis of Neurologic and Psychiatric Disease*, Philadelphia: Elsevier, pp. 51–64.

Punales-Morejon D (1997) Genetic counseling and prenatal diagnosis: A multicultural perspective. *J Am Med Womens Assoc* **52**(1):30–32.

Punales-Morejon D, Penchaszadeh VB (1992) Psychosocial aspects of genetic counselling: Cross-cultural issues. In: Evers-Kiebooms G, et al., eds. *Psychological Aspects of Genetic Counseling*. New York: Wiley-Liss.

Pyke SDM, Wood DA, Kinmonth AL, Thompson SG (1997) Change in coronary risk and coronary risk factor levels in couples following lifestyle intervention. *Arch Fam Med* **6**:354–360.

Quaid KA, Wesson MK (1995) Exploration of the effects of predictive testing for Huntington disease on intimate relationships. *Am J Med Genet* **57**:46–51.

Quebec Network of Applied Genetic Medicine (2000) Statement of principles: Human genomic research, Montreal, April. Online at http://www.rmga.qc.ca/doc/principes_en_2000.html. Accessed April 8, 2002, Ch. II.

Quill TE, Brody H (1996) Physician recommendations and patient autonomy: Finding a balance between physician power and patient choice. *Ann Internal Med* **125**:763–769.

Quillin JM, Jackson-Cook C, Bodurtha J (2003) The link between providers and patients: How laboratories can ensure quality results with genetic testing. *Clin Leadersh Manag Rev* **17**(6):351–357.

Quinn v. Blau (1997) WL 781874 (Conn. Super.). In *Siemieniec v. Lutheran General Hospital* (1987) 512 BN.E. 2d. 691 (Ill.), a couple consulted with a physician regarding the risk that a prospective child would be born a hemophiliac. Two of the wife's cousins had the disease. The couple was advised that the risk was "very low." The child was born afflicted with hemophilia. The legal remedies of wrongful birth and wrongful life are discussed in Chapter 10. Also see Northern KS (1998) Procreative Torts: Enhancing the Common-Law

Protection for Reproductive Autonomy. *U Ill L Rev* **489**:533: "A physician or counselor could reasonably be expected to ascertain enough of a patient's background, [and] her reasons for seeking pregnancy counseling, . . . to assess what information might be useful to the patient's deliberative process and then to discuss that information with her. The standard only requires physicians and counselors to make reasonable efforts. They need not be mind readers, as long as they reasonably attend to the circumstances of their individual patients. . . . The reasonably prudent patient standard thus takes into account each woman's unique circumstances."

R v Lewis (1996) 139 D.C.R. 4th 480 (B.C.B.C.).

R v. Morgentaler (1988) 1 SCR 30.

Rabino I (2001) How human geneticists in the US view commercialization of the human genome project. *Nat Genet* **29**:15–16.

Rabinowitz I, Luzzatti R, Tamir A, Reis S (2004) Length of patient's monologue, rate of completion, and relation to other components of the clinical encounter: Observational intervention study in primary care. *BMJ* **328**:501–502: "Completed monologues doubled when doctors were told not to interrupt."

Radloff LS (1977) The CES-D Scale: A self-report depression scale for research in the general population. *Appl Psychol Measurement* **1**:385–401.

Raeburn S (2001) Genetic counseling. In: Harper JC, Delhanty JDA, Handyside AH, eds. *Preimplantation Genetic Diagnosis*. Chichester, England: Wiley, pp. 45–51.

Rai AK (2002) Locating gene patents within the patent system. *Am J Bioethics* **2**(3):18–19.

Rapp R (1997) Communicating about chromosomes: Patients, providers, and cultural assumptions. *J Am Med Womens Assoc* **52**(1):28–29, 32.

Rapp R (1999) *Testing Women, Testing the Fetus: The Social Impact of Amniocentesis in America*. New York: Routledge.

Rappaport EB, Knapp M (1989) Isotretinoin embryopathy—a continuing problem. *J Clin Pharmacol* **29**(5):463–465.

Rasmussen SA, Mulinare J, Khoury MJ, Maloney EK (1990) Evaluation of birth defect histories obtained through material interviews. *Am J Hum Genet* **46**(3):478–485.

Ravetz, JR (1971) *Scientific Knowledge and Its Social Problems*. Oxford: Clarendon Press.

Ravine D, McGregor LR, Walker RG, Sheffield LJ (1991) Perceptions of genetic risk in individuals with a one in two chance of developing autosomal dominant polycystic kidney disease. *Med J Aust* **154**:689–691.

Razavi D, Delvaux N, Farvacques C, Robaye E (1990) Screening for adjustment disorders and major depressive disorders in cancer in-patients. *Br J Psych* **156**:79–83.

Razavi D, Delvaux N, Bredart A, Paesmans M, Debusscher L, Bron D, Stryckmans P (1992) Screening for psychiatric disorders in a lymphoma out-patient population. *Eur J Ca* **28A**(11):1869–1872.

Rebbeck TR (2000) Prophylactic oophorectomy in BRCA1 and BRCA2 mutation carriers. *J Clin Oncol* **18**(November Suppl):100s–103s.

Rebbeck T, Weber BL (1997) BRCA1 mutations in women attending clinics that evaluate the risk of breast cancer. *N Engl J Med* **336**:1409–1415.

Rebbeck TR, Levin AM, Eisen A, Snyder C, Watson P, Cannon-Albright L, Isaacs C, Olopade O, Garber JE, Godwin AK, Daly MB, Narod SA, Neuhausen SL, Lynch HT, Weber BL (1999) Breast cancer risk after bilateral prophylactic oophorectomy in BRCA1 mutation carriers. *J Natl Cancer Inst* **91**:1475–1479.

Rebbeck TR, Friebel T, Lynch HT, Neuhausen SL, van't Veer L, Garber JE, Evans GR, Narod SA, Isaacs C, Matloff E, Daly MB, Olopade OI, Weber BL (2004) Bilateral prophylactic mastectomy reduces breast cancer risk in BRCA1 and BRCA2 mutation carriers: the PROSE study group. *J Clin oncol* **22**:1055–1062.

Rebbeck TR, Lynch HT, Neuhausen SL, Narod SA, van't Veer L, Garber JE, Evans G, Isaacs C, Daly MB, Matloff E, Olopade OI, Weber BL (2002) Prophylactic oophorectomy in carriers of BRCA1 or BRCA2 mutations. *N Engl J Med* **346**:1616–1622.

Redelmeier D, Rozin P, Kahneman D (1993) Understanding patients' decisions: Cognitive and emotional perspectives. *JAMA* **270**:72–76.

Redelmeier DA, Tu JV, Schull MJ, Ferris LE, Hux JE (2001) Problems for clinical judgement: 2. Obtaining a reliable past medical history. *CMAJ* **164**(6):809–813.

Reed SC (1955) *Counseling in Medical Genetics*. Philadelphia: WB Saunders.

Reed v. Campagnolo (1993) 332 Md. 226, 630 A. 2d. 1145 (Md.).

Rees G, Fry A, Cull A (2001) A family history of breast cancer: Women's experiences from a theoretical perspective. *Soc Sci Med* **52**:1433–1440.

Reibl v. Hughes (1980) 2 S.C.R. 880.

Reilly P (1977) *Genetics, Law, and Social Policy*. Cambridge, MA: Harvard University Press, p. 151.

Reilly PR (1993) Public policy and legal issues raised by advances in genetic screening and testing. *Suffolk Univ Law Rev* **27**(4):1327–1357.

Reilly PR (1995) Physician Responsibility in Conducting Genetic Testing. *J Nat'l Cancer Inst Monograms*: 59.

Reilly PR (1999) Genetic discrimination. In: Long C, ed. *Genetic Testing and the Use of Information*, Washington DC: AEI Press, pp. 106–133.

Reisner v. Regents of University of California (1995) (2 District) 37 Cal. Rptr. 2d. 518, 31 Cal. App. 4th 1195, review denied (Cal. App. 2d. District).

Rentmeester CA (2001) Value neutrality in genetic counseling: An unattained ideal. *Med. Health Care Phil* **41**:47–51.

Report of the Collaborative Acetylcholinesterase Study (1981) Amniotic fluid acetyl-cholinesterase electrophoresis as a secondary test in the diagnosis of anencephaly and spina bifida in early pregnancy. *Lancet* **2**:321–324.

Resnik DB (2003) Are DNA patents bad for medicine? *Health Policy* **65**:181–197.

Resnik DB (2004) *Owning the Genome: A Moral Analysis of Gene Patenting*. New York: State University of New York Press.

Resta RG (1997) Eugenics and nondirectiveness in genetic counseling. *J Genet Counsl* **6**(2):244–257.

Resta RG (1999) Just watching. *Am J Med Genet* **83**(1):1–2.

Resta RG (2000) *Psyche and Helix, Psychological Aspects of Genetic Counseling*. Wiley-Liss. New York.

Resta RG, Kessler S (2004) Commentary on Robin's A smile, and the need for counseling skills in the clinic. *Am J Med Genet A* **126**(4):437–438; author reply 439.

Reutenauer JE (2000) Medical malpractice liability in the era of genetic susceptibility testing. *Quinnipiac L Rev* **19**:539–579. See page 555 with regard to cases that have recognized the action of wrongful birth.

Rex DK, Johnson DA, Lieberman DA, Burt DW, Sonnenberg A (2000) Colorectal cancer prevention 2000: Screening recommendations of the American College of Gastrenterology. *Am J Gastroenterol* **95**:868–877.

Reyes v. Anka Research Ltd. (1981) 443 N.Y.S. 2d 595.

Rich SS, Annegers JF, Hauser WA, Anderson VE (1987) Complex segregation analysis of febrile convulsions. *Am J Hum Genet* **41**:249–257.

Richards M (1999) Families, kinship and genetics. In: Marteau T, Richards M, eds. *The Troubled Helix: Social and Psychological Implications of the New Human Genetics*. United Kingdom: Cambridge University Press, paperback edition, pp. 249–273.

Richards CS, Bradley LA, Amos J, Allitto B, Grody WW, Maddalena A, McGinnis MJ, Prior TW, Popovich BW, Watson MS, Palomaki GE (2002) Standards and guidelines for CFTR mutation testing. *Genet Med* **4**(5):379–391. Erratum in: *Genet Med* **4**(6):471.

Ricks v. Budge (1937) 91 Utah 307, 64 P 2d 208.

Riegel M, Castellan C, Balmer D, Brecevic L, Schinzel A (1999) Terminal deletion, del(1)(p36.3), detected through screening for terminal deletions in patients with unclassified malformation syndromes. *Am J Med Genet* **82**:249–253.

Ries LAG, Eisner MP, Kosary CL, et al (2004) *SEER Cancer Statistics Review*, 1975–2001. Bethesda, MD, National Cancer Institute. Online at http://seer.cancer.gov/csr/1975_2001/.

Rietze v. Bruser (1979) 1 W.W.R. 31, at 45–50 (Man. Q.B.).

Riggins v. Mauriello (1992) 603 A 2d 827 (Del. Supr.).

Ringberg A, Palmer B, Linell F, Rychterova V, Ljungberg O (1991) Bilateral and multifocal breast carcinoma. A clinical and autopsy study with special emphasis on carcinoma in situ. *Eur J Surg Oncol* **17**:20–29.

Risch HA, McLaughlin JR, Cole DE, Rosen B, Bradley L, Kwan E, Jack E, Vesprini DJ, Kuperstein G, Abrahamson JL, Fan I, Wong B, Narod SA (2001) Prevalence and penetrance of germline BRCA1 and BRCA2 mutations in a population series of 649 women with ovarian cancer. *Am J Hum Genet* **68**:700–710.

Robak v. United States (1981) 658 F.2d 471 (7th Cir.).

Robbins DH, Itzkowitz SH (2002) The molecular and genetic basis of colon cancer. *Med Clin North Am* **86**:1467–1495.

Robert E, Guibaud P (1982) Maternal valproic acid and congenital neural tube defects. *Lancet* **2**:937.

Roberts CS, Cox CE, Reintgen DS, Baile WF, Gibertini M (1994) Influence of physician communication on newly diagnosed breast patients' psychologic adjustment and decision-making. *Cancer* **74**(1 Suppl):336–341.

Roberts LJ, Bewley S, Mackinson AM, Rodeck CH (1995) First trimester fetal nuchal translucency: Problems with screening the general population. 1. *Br J Obstet Gynaecol* **102**:381–385.

Robertson G (1991) Informed consent ten years later. *Canadian Bar Rev* **423**, at 435.

Robertson JA (2001) Preconception gender selection. *Am J Bioethics* **1**(1):2–9.

Robertson JA (2003) Extending preimplantation genetic diagnosis: The ethical debate. *Human Reprod* **18**:465–471.

Robertson JA (2004) Gender variety as a valid choice: A comment on the HFEA—response to Edgar Dahl's "The presumption in favour of liberty." *Reprod Biomed Online* **8**(3):270–271.

Robinson A, Thomson R (2001) Variability in patient preferences for participating in medical decision making: Implication for the use of decision support tools. *Qual Health Care* **10** (Suppl 1):i34–38.

Rocchi A, Pellegrini S, Siciliano G, Murri L (2003) Causative and susceptibility genes for Alzheimer's disease: A review. *Brain Res Bull* **61**:1–24.

Rockhill B (2001) The privatization of risk. *Am J Public Health* **91**:365–368.

Rodgers S (2002) The Legal Regulation of Women's Reproductive Capacity in Canada. In: Downie J, Caulfied T, Flood CM, eds. *Canadian Health Law and Policy*, 2nd ed. Toronto: Butterworths, pp. 331–365.

Rodriguez-Bigas MA, Vasen HFA, O'Malley L, Rosenblatt MJT, Farrell C, Weber TK, Petrelli NJ (1998) Health, life, and disability insurance and hereditary nonpolyposis colorectal cancer. *Am J Hum Genet* **62**:736–737.

Rodych v. Krasey (1971) 4 W.W.R. 358 (Man. Q.B.).

Roe v. Wade (1973) 93 S.Ct. 705, 410 US 113 U.S.S.C.

Rogaev EI, Sherrington R, Rogaeva EA, Levesque G, Ikeda M, Liang Y, Chi H, Lin C, Holman K, Tsuda T, et al (1995) Familial Alzheimer's disease in kindreds with missense mutations in a gene on chromosome 1 related to the Alzheimer's disease type 3 gene. *Nature* **376**:775–778.

Rogers CR (1951/1995) *Client-Centered Therapy: Its Current Practice, Implications and Theory.* Edited by L. Carmichael. La Jolla. Trans-Atlantic Publications (Reprint 1951 original).

Roggenbuck J, Olson JE, Sellers TA, Ludowese C (2000) Perception of genetic risk among genetic counselors. *J Genet Couns* **9**(1):47–59.

Roop W (2000) Not in my womb: Compelled prenatal genetic testing. *Hastings Constitutional Law Quart* **27**:397–421.

Rosa FW (1991) Spina bifida in infants of women treated with carbamazepine during pregnancy. *N Engl J Med* **324**:674–677.

Rose P, Humm E, Hey K, Jones L, Huson SM (1999) Family history taking and genetic counselling in primary care. *Fam Pract* **16**(1):78–83.

Rose PW, Watson E, Yudkin P, Emery J, Murphy M, Fuller A, Lucassen A (2001) Referral of patients with a family history of breast/ovarian cancer—GPs' knowledge and expectations. *Fam Pract* **18**(5):487–490.

Rosen P, Groshen S, Saigo PE, Kinne DW, Hellman S (1989) A long-term follow-up study of survival in stage I (T1 No Mo) and stage II (T1 N1 Mo) breast carcinoma. *J Clin Oncol* **7**:355–366.

Rosen v. Greifenberger (1999) 257 Va. 373, 513 S.E. 2d 861.

Rosenberg RN, Prusiner SB, DiMauro S, Barchi RL, Nestler EJ (2003) *The Molecular and Genetic Basis of Neurologic and Psychiatric Disease,* 3rd ed. Philadelphia: Elsevier.

Rosoff AJ (1999) Informed consent in the electronic age. *Am J L Med* **25**:367, at 383–385. The author reviews interactive computer-based patient education systems, and warns of the danger that such technology may "stifle the dynamic human interaction" between the physician and the patient. Also see Hatdon JH (1999) Legal aspects of health information. *Health Law Canada* **20**:112.

Rostron v. Klein, 23 Mich., App. 288, 178 N.W. 2d 675.

Roter D (2004) Patient centered communication: More than a string of words. *BMJ* **328**:E303–304.

Roter DL, Hall JA (1992) *Doctor Talking with Patients/Patients Talking with Doctors: Improving Communication in Medical Visits.* Westport, CT: Auburn House.

Roter DL, Stewart M, Putnam SM, Lipkin M Jr, Stiles W, Inui TS (1997) Communication patterns of primary care physicians. *JAMA* **277**(4):350–356.

Rothstein MA (1999) *Genetic Secrets: Protecting Privacy and Confidentiality in the Genetic Era.* New Haven, CT: Yale University Press.

Rothstein MA (2004) *Genetics and Life Insurance.* Cambridge, MA: MIT Press.

Rothstein MA, Hoffman S (1999) Genetic testing, genetic medicine, and managed care. *Wake Forest Law Rev* **34**:849–888, at pp. 857–865 with respect to the role of primary care physicians and nurses.

Rottem S, Bronshtein M, Thaler I, Brandes JM (1989) First trimester transvaginal sonographic diagnosis of fetal anomalies. *Lancet* **1**(8635):444–445.

Rowley JD (1993) Can we meet the challenge? *Am J Hum Genet* **54**:403, at 412.

Rowley PT (2004) Screening for an inherited susceptibility to colorectal cancer. *Genetic Testing* **8**:421–430.

Royal College of Physicians and Surgeons of Canada (1998) *Specialty Training Requirements in Medical Genetics.*

Royal Commission (1993) *Proceed with Care: Final Report of The Royal Commission on New Reproductive Technologies,* Vol. 2 (Ottawa: Minister of Government Services, 1993) (Chair:

Dr. Patricia Baird): pp. 788–789. As to the issue of patient autonomy in decision making in the context of prenatal diagnosis, the Commission reported [at 766–767] that a: "disturbing proportion of referring physicians who do not accept the principle that patients should make their own informed choice about whether to have PND and when to have an abortion after the diagnosis of a fetal disorder . . . This is of great concern to the Commissioners, respect for the pregnant woman's autonomy requires that it be her values and priorities, not the doctor's, that determine her decision to accept or decline PND testing." See p. 781: With regard to the issue of nondirectiveness, the Commission reported: [at pgs. 780–782] that patients may not have sufficient knowledge and experience to make decisions and "are almost entirely dependent on their counsellors for information about disabilities any may have difficulty imagining the various possibilities and options. "Counsellors at genetics centres are required by CCMG [Canadian College of Medical Geneticists] guidelines to be non-directive. Non-directiveness is not always welcomed by those receiving counselling. Some women and couples find the information complex and overwhelming and ask the counsellor what he or she would do in their place. Some are frustrated when the counsellor insists that it is their decision . . . In view of this, we believe that the counsellor should provide written summaries of genetic counselling sessions not only to the referring physician but also to the women and couples counselled (which we found was done by some centres)." Editorial Note: Does a refusal to provide direction represent more of a failure to respect autonomy than to promote it? For a discussion, please see: Michie S, Bron F, Bobrow M, Marteau TM (1997) Nondirectiveness in genetic counseling: an empirical study. *Am J Hum Genet* **60**(1):40–47. p. 783: "However, just as in medical care, the genetics counsellor may not be sufficiently expert in recognition or management of complex psychological and emotional problems that may arise, particularly when a fetus is found to be affected."

Royal Society (2003) Keeping the Science Open: The Effect of Intellectual Property on the Conduct of Science. April; online: Royal Society <http://www.royalsoc.ac.uk/document.asp?tip=0&id=1374>.

Rubinsztein D, Leggo J, Coles R, Almqvist E, Biancalana V, Cassiman J, Chotai K, Connarty M, Crauford D, Curtis A, Curtis D, Davidson M, Differ A, Dode C, Dodge A, Frontali M, Ranen N, Stine O, Sherr M, Abbott M, Franz M, Graham C, Harper P, Hedreen J, Hayden M, (1996) Phenotypic characterization of individuals with 30–40 CAG repeats in the Huntington disease (HD) gene reveals HD cases with 36 repeats and apparently normal elderly individuals with 36–39 repeats. *Am J Hum Genet* **59**:16–22.

Rutqvist LE, Johansson H, Signomklao T, Johansson U, Fornander T, Wilking N. (1995) Adjuvant tamoxifen therapy for early stage breast cancer and second primary malignancies. *J Natl Cancer Inst* **87**:645–651.

Rutter M (1995) Relationships between mental disorders in childhood and adulthood. *Acta Psychiatr Scand* **91**(2):73–85.

Ryan M, Miedzybrodzka Z, Fraser L, Hall M (2003) Genetic information but not termination: Pregnant women's attitudes and willingness to pay for carrier screening for deafness genes. *J Med Genet* **40**(6):e80.

Sabeh v. Khosla (2003) No. L-6754-00. Bergen Co., N.J., Super.Ct.; and see Fortado L (2004) Genetic testing maps new legal turf; Doctors' liability grows as tests are more widely used. *National Law J* **26**(50):1.

Safer v. Pack (1996) 291 N.J.Super. 619, 677 A.2d 1188, 1192 (N.J.Super.App.Div.). The court found that the physician owes a duty to warn of an "avertable risk" not only to the patient but members of the immediate family who may be adversely affected by a breach of that duty. The court specifically did not hold that the duty to warn would be satisfied by informing the patient.

Saldov M, Kakai H, McLaughlin L, Thomas A (1998) Cultural barriers in oncology: Issues in obtaining medical informed consent from Japanese-American elders in Hawaii. *J Cross Cult Gerontol* **13**(3): 265–279.

Salgo v. Leland Stanford Jr. University Board of Trustees (1957) 154 Cal.App.2d 560, 317 P.2d 170.

Saller DN Jr, Canick JA (1996) Maternal serum screening for fetal Down syndrome: The detection of other pathologies. *Clin Obstet Gynecol* **39**(4):793–800.

Saltus R (1994) Fear of insurers leading to gene testing in secret. *Boston Globe* September 12:1–10.

Sander T, Berlin W, Gscheidel N, Wendel B, Janz D, Hoehe MR (2000a) Genetic variation of the human mu-opioid receptor and susceptibility to idiopathic absence epilepsy. *Epilepsy Res* **39**(1):57–61.

Sander T, Schulz H, Saar K, Gennaro E, Riggio MC, Bianchi A, Zara F, Luna D, Bulteau C, Kaminska A, Ville D, Cieuta C, Picard F, Prud'homme JF, Bate L, Sundquist A, Gardiner RM, Janssen GA, de Haan GJ, Kasteleijn-Nolst-Trenite DG, Bader A, Lindhout D, Riess O, Wienker TF, Janz D, Reis A (2000b) Genome search for susceptibility loci of common generalized epilepsies. *Hum Mol Genet* **9**:1465–1472.

Sander T, Windemuth C, Schulz H, Saar K, Gennaro E, Bianchi A, Zara F, Bulteau C, Kaminska A, Ville D, Cieuta C, Prud'homme JF, Dulac O, Bate L, Gardiner RM, de Haan GJ, Janssen GA, Witte J, Halley DJ, Lindhout D, Wienker TF, Janz D, European Consortium on the Genetics of Idiopathic Generalized Epilepsy (2002) No evidence for a susceptibility locus for idiopathic generalized epilepsy on chromosome 18q21.1. *Am J Med Genet* **114**:673–678.

Sandhaus LM, Singer ME, Dawson NV, Wiesner GL (2001) Reporting BRCA test results to primary care physicians. *Genet Med* **3**(5):327–334.

Sandhus MS, Luben R, Khaw KT (2001) Prevalence and family history of colorectal cancer: Implications for screening. *J Med Screen* **8**:69–72.

Sandler RS, Halabi S, Baron JA, Budinger S, Paskett E, Keresztes R, Petrelli N, Pipas JM, Karp DD, Loprinzi CL, Steinbach G, Schilsky R (2003) A randomized trial of aspirin to prevent colorectal adenomas in patients with previous colorectal cancer. *N Eng J Med* **348**(10):883–890.

Sangha KK, Dircks A, Langlois S (2003) Assessment of effectiveness of genetic counseling by telephone compared to a clinic visit. *J Genet Counsel* **12**(2):171–184.

Sard v. Hardy (1977) 281 Md. 432.

Sastre-Garau X, Jouve M, Asselain B, Vincent-Salomon A, Beuzeboc P, Dorval T, Durand JC, Fourquet A, Pouillart P (1996) Infiltrating lobular carcinoma of the breast. *Cancer* **77**:113–120.

Saunders T (2003) Renting space on the shoulders of giants: Madey and the future of the experimental use doctrine. *Yale Law Journal* **113**:261.

Savulescu J (2002) Deaf lesbians, "designer disability," and the future of medicine. *BMJ* **325**:771–773.

Say RE, Thomson R (2003) The importance of patient preferences in treatment decisions—challenges for doctors. *BMJ* **327**:542–545.

Schaedel C, Hjelte L, de Monestrol I, Johannesson M, Kollberg H, Kornfalt R, Holmberg L (1999) Three common CFTR mutations should be included in a neonatal screening programme for cystic fibrosis in Sweden. *Clin Genet* **56**(4):318–322.

Scheffer IE, Berkovic SF (1997) Generalized epilepsy with febrile seizures plus: A genetic disorder with heterogeneous clinical phenotypes. *Brain* **120**:479–490.

Scheffer IE, Berkovic SF (2003) The genetics of human epilepsy. *Trends Pharm Sci* **24**:428–433.

Scheffer IE, Bhatia KP, Lopes-Cendes I, Fish DR, Marsden CD, Andermann E, Andermann F, Desbiens R, Keene D, Cendes F, et al (1995) Autosomal dominant nocturnal frontal lobe epilepsy. A distinctive clinical disorder. *Brain* **118**:61–73.

Scheuner MT (2003) Genetic evaluation for coronary artery disease. *Genet Med* **5**:269–285.

Scheuner MT, Raffel LJ, Wang S-J, Rotter JI (1997) The family history: A comprehensive genetic risk assessment method for the chronic conditions of adulthood. *Am J Med Genet* **71**:315–324.

Scheuner MT, Yoon P, Khoury MJ (2004) Contribution of Mendelian disorders to common chronic disease: Opportunities for recognition, intervention and prevention. *Am J Med Genet* **125C**:50–65.

Scheyett A (2002) Approaching complex cases with a crisis intervention model and teamwork: A commentary. *J Genet Counsel* **11**(5): 377–382.

Schieszer J (2001) Genetic underclass emerging. *Internal Med World Report* July 1:1,22.

Schirmer v. Mt. Auburn Obstetrics and Gynecological Assoc (2003) 802 N.E.2d 723 (Oh.); appeal pending to Ohio Supreme Court, *Schirmer v. Mt. Auburn Obstetrics & Gynecologic Assocs.* (2004) Ohio 3069. A mother requested testing to determine if her fetus carried a genetic defect. Incorrect results were reported and the child was born mentally and physically disabled.

Schloendorff v. The Society of the New York Hospital (1914) 211 NY 125; 105 NE 92 (N.Y. Court of Appeals). With regard to the principle of authority; also see Gatter KM (1999) Protecting patient-doctor discourse: Informed consent and deliberative authority. *Or L Rev* **78**(4):941–993 at 961–982.

Schoen RE (2000) Families at risk for colorectal cancer: Risk assessment and genetic testing. *J Clin Gastroenterol* **31**:114–120.

Schrag D, Kuntz KM, Garber JE, Weeks JC (2000) Life expectancy gains from cancer prevention strategies for women with breast cancer and BRCA1 or BRCA2 mutations. *JAMA* **283**:617–624.

Schroeder v. Perkel (1981) 87 NJ 53, 432 A.2d 834 (S.Ct. NJ). The court held that liability could extend to the patient's family where a doctor's failure to diagnose a child with cystic fibrosis led to the birth of a second child with cystic fibrosis, and it was foreseeable that the parents would rely on the diagnosis.

Schuchter K, Wald N, Hackshaw AK, Hafner E, Liebhart E (1998) The distribution of nuchal translucency at 10–13 weeks of pregnancy. *Prenat Diagn* **18**:281–286.

Schuck PH (1994) Rethinking informed consent. *Yale L J* **103**:899–959.

Schuster v. Altenburg (1988) 424 N.W.2d 159 (Wis.).

Schwartz LM, Woloshin S, Welch HG (1999) Risk communication in clinical practice: Putting cancer in context. *J Nat Cancer Inst Monographs* **25**:124–133.

Schwartz MD, Hughes C, Rother J, Main D, Peshkin BN, Isaacs C, Kavanagh C, Lerman C (2000) Spiritual faith and genetic testing decisions among high-risk breast cancer probands. *Ca Epi Bio Prevent* **9**:381–385.

Schwartz M, Peshkin B, Hughes C, Main D, Isaacs C, Lerman C (2002) Impact of BRCA1/BRCA2 mutation testing on psychologic distress in a clinic-based sample. *J Clin Onc* **20**:514–520.

Science Council of Canada (1991) *Genetics in Canadian Health Care*. Ottawa: Minister of Supply and Services, pp. 72–73.

Scotchmer S (Winter 1991) Standing on the shoulders of giants: Cumulative research and the patent law. *J Econ Perspect* **5**(1):29–41.

Scott WK, Nance MA, Watts RL, Hubble JP, Koller WC, Lyons K, Pahwa R, Stern MB, Colcher A, Hiner BC, Jankovic J, Ondo WG, Allen FH Jr, Goetz CG, Small GW, Masterman D,

Mastaglia F, Laing NG, Stajich JM, Slotterbeck B, Booze MW, Ribble RC, Rampersaud E, West SG, Gibson RA, Middleton LT, Roses AD, Haines JL, Scott BL, Vance JM, Pericak-Vance MA (2001) Complete genomic screen in Parkinson disease: Evidence for multiple genes. *JAMA* **286**(18):2239–2244.

Scriver CR, Beaudet AL, Sly WS, Valle D, eds (2001) *The Metabolic and Molecular Bases of Inherited Disease* 8th ed. New York: McGraw-Hill.

Searle J (1997) Routine antenatal screening: Not a case of informed choice. *Australian New Zealand J Public Health* **21**(3):268–274.

Second Report of the UK Collaborative Study on Alpha-Fetoprotein in Relation to Neural-Tube Defects (1979) Amniotic-fluid alpha-fetoprotein measurement in antenatal diagnosis of anencephaly and spina bifida in early pregnancy. *Lancet* **2**:651–662.

Secretary's Advisory Committee on Genetic Testing (2003) http://www4.od.nih.gov/oba/sacgt.htm.

Secretary's Advisory Committee on Genetics, Health and Society (2003) http://www4.od.nih.gov/oba/sacghs/SACGHSPressReleases.html.

Segurado R, Detera-Wadleigh SD, Levinson DF, Lewis CM, Gill M, Nurnberger JI Jr, Craddock N, DePaulo JR, Baron M, Gershon ES, Ekholm J, Cichon S, Turecki G, Claes S, Kelsoe JR, Schofield PR, Badenhop RF, Morissette J, Coon H, Blackwood D, McInnes LA, Foroud T, Edenberg HJ, Reich T, Rice JP, Goate A, McInnis MG, McMahon FJ, Badner JA, Goldin LR, Bennett P, Willour VL, Zandi PP, Liu J, Gilliam C, Juo SH, Berrettini WH, Yoshikawa T, Peltonen L, Lonnqvist J, Nothen MM, Schumacher J, Windemuth C, Rietschel M, Propping P, Maier W, Alda M, Grof P, Rouleau GA, Del-Favero J, Van Broeckhoven C, Mendlewicz J, Adolfsson R, Spence MA, Luebbert H, Adams LJ, Donald JA, Mitchell PB, Barden N, Shink E, Byerley W, Muir W, Visscher PM, Macgregor S, Gurling H, Kalsi G, McQuillin A, Escamilla MA, Reus VI, Leon P, Freimer NB, Ewald H, Kruse TA, Mors O, Radhakrishna U, Blouin JL, Antonarakis SE, Akarsu N (2003) Genome scan meta-analysis of schizophrenia and bipolar disorder, part III: Bipolar disorder. *Am J Hum Genet* **73**:49–62.

Seldin D (1981) Presidential address: The boundaries of medicine. *Trans. Ass Am Phys* **94**:lxxv–lxxxvi.

Selkoe DJ (2001) Alzheimer's disease: genes, proteins, and therapy. *Physiol Rev* **81**:741–766.

Selkoe DJ (2003) Alzheimer's disease. In: Rosenberg RN, Prusiner SB, DiMauro S, Barchi RL, Nestler EJ, eds. *The Molecular and Genetic Basis of Neurologic and Psychiatric Disease* Philadelphia: Elsevier, pp. 331–342.

Sellers TA, Mink PJ, Cerhan JR, Zheng W, Anderson KE, Kushi LH, Folsom AR (1997) The role of hormone replacement therapy in the risk for breast cancer and total mortality in women with a family history of breast cancer. *Ann Intern Med* **127**:973–980.

Seney v. Crooks (1996) 30 C.C.L.T. (2d) 66 at 81.

Sepulveda W (1996) Fetal echogenic bowel. *Lancet* **347**(9007):1043.

Sepulveda W, Sebire NJ (2000) Fetal echogenic bowel: A complex scenario. *Ultrasound Obstet Gynecol* **16**:510–514.

Sepulveda W, Leung KY, Robertson ME, Kay E, Mayall ES, Fisk NM (1996a) Prevalence of cystic fibrosis mutations in pregnancies with fetal echogenic bowel. *Obstet Gynecol* **87**(1):103–106.

Sepulveda W, Reid R, Nicolaidis P, Prendiville O, Chapman RS, Fisk NM (1996b) Second-trimester echogenic bowel and intraamniotic bleeding: Association between fetal bowel echogenicity and amniotic fluid spectrophotometry at 410 nm. *Am J Obstet Gynecol* **174**(3):839–842.

Sermon K, Van Steirteghem A, Liebaers I (2004) Preimplantation genetic diagnosis. *Lancet* **363**(9421):1633–1641.

Severin MJ (1999) Genetic susceptibility for specific cancers. Medical liability of the clinician. *Cancer* **86**(Suppl):2564–2569.

Sevilla C, et al (2002) Testing for BRCA1 mutations: A cost effectiveness analysis. *Eur J Hum Genet* **10**:599–606.

Sevilla C, Moatti JP, Julian-Reynier C, Eisinger F, Stoppa-Lyonnet D, Bressac-de Pailerets B, Sobol H (2002) Testing for BRCA1 mutations: A cost effectiveness analysis. European Journal of Human Genetics 10:599–606.

Shaffer LG, Agan N, Goldberg JD, Ledbetter DH, Longshore JW, Cassidy SB (2001) ACMG statement on diagnostic testing for uniparental disomy. *Genet Med* **3**(3):206–211.

Shafir MS, Silversides C, Waters I, MacRury K, Frank JW, Becker LA (1995) Patient consent to observation. Responses to requests for written consent in an academic family practice unit. *Can Fam Physician* **41**:1367–1372.

Shaheen NJ, Lawrence LB, Bacon BR, Barton JC, Barton NH, Galanko J, Martin CF, Burnett CK, Sandler RS. (2003) Insurance, employment and psychosocial consequences of a diagnosis of hereditary hemochromatosis in subjects without end organ damage. *Am J Gastroent* **98**:1175–1180.

Shapira A (1998) "Wrongful life" lawsuits for faulty genetic counselling: Should the impaired newborn be entitled to sue? *J Med Ethics* **24**(6):369–375.

Shapiro C (2001) Navigating the Patent Thicket: Cross Licenses, Patent Pools and Standard Setting. In: Jaffe A, Lerner J, Stern S, eds. *Innovation Policy and the Economy*, Vol I, MIT Press, prepublication copy, online: Haas School of Business <haas.Berkeley.edu/~shapiro/thicket.pdf>.

Shapiro DE, Boggs SR, Melamed BG, Graham-Pole J (1992) The effect of varied physician affect on recall, anxiety, and perceptions in women at risk for breast cancer: An analogue study. *Health Psychol* **11**(1):61–66. Also see Roberts CS, Cox CE, Reintgen DS, Baile WF, Gibertini M (1994) Influence of physician communication on newly diagnosed breast patients' psychologic adjustment and decision-making. *Cancer* **74**(1 Suppl):336–341.

Sharp PS, Mohan V, Levy JC, Mather HM, Kohner EM (1987) Insulin resistance in patients of Asian Indian and European origin with non-insulin dependent diabetes. *Horm Metab Res* **19**:84–85.

Sharpe NF (1993) Presymptomatic testing for Huntington disease: Is there a duty to test those under the age of eighteen years? *Am J Med Genet* **46**: 250–253.

Sharpe NF (1994a) Informed consent and huntington disease; A model for communication. *Am J Med Genet* **50**:239–246. Also see Katz J (1994) Informed consent—Must it remain a fairy tale? 10 *J Contemp Health L Pol'Y* **69**:81, at 84: The author argues that informed consent will remain a "charade" unless the physicians adopt cooperative decision-making with patients.

Sharpe NF (1994b) Psychological aspects of genetic counseling: A legal perspective. *Am J Med Genet* **50**:234–238.

Sharpe NF (1996) Genetic testing and screening in Canada: A model duty of care. *Health Law J* **4**:199–134.

Sharpe NF (1997) Reinventing the wheel? Informed consent and genetic testing for breast cancer, cystic fibrosis, and Huntington disease. *Queen's Law J* **22**:389–452. This paper reviews the conceptual counseling models of "Decision Analysis" and "Psychotherapeutic," pp. 401–411, compares medicine's dominant Cartesian biomedical model with that of the therapeutic model for genetic counseling pp. 438–442, and discusses the potential implication for judicial standards regarding the evaluation of informed consent.

Sharpe NF (1999) The duty to recontact: Benefit and harm. *Am J Hum Genet* **65**:1201–1204.

Sharpe NF, Carter RF (2002) Human genetics: Research on ethical, legal and social impact of genetic testing. A649. *Encyclopaedia of the Human Genome*. London, England: Nature Publishing Group Reference.

Shaw MW (1986) Avoiding wrongful birth and wrongful life suits. *Am J Med Genet* **25**(1):81–84.

Shaw MW (1987) Invited Editorial Comment: Testing for the Huntington gene: A right to know, a right not to know, or a duty to know. *Am J Med Genet* **26**:243.

Shaw GM, Velie EM, Schaffer D (1996) Risk of neural tube defect-affected pregnancies among obese women. *JAMA* **275**:1093–1096.

Shenfield F, Pennings G, Devroey P, Sureau C, Tarlatzis B, Cohen J, ESHRE Ethics Task Force (2003) Taskforce 5: Preimplantation Genetic Diagnosis. Human Reproduction **18**(3):649–651.

Sheremeta L, Gold ER, Caulfield T (2003) Harmonizing commercialisation and gene patent policy with other social goals. In: Knoppers BM, ed. *Populations & Genetics: Legal and Socio-Ethical Perspectives*, Leiden: Martinus Nijhoff, pp. 423–452.

Sherod C, Sebire NJ, Soares W, Snijders RJM, Nicolaides KH (1997) Prenatal diagnosis of trisomy 18 at the 10–14 week ultrasound scan. *Ultrasound Obstet Gynecol* **10**:387–390.

Sherrington R, Rogaev EI, Liang Y, Rogaeva EA, Levesque G, Ikeda M, Chi H, Lin C, Li G, Holman K, et al. (1995) Cloning of a gene bearing missense mutations in early-onset familial Alzheimer's disease. *Nature* **375**:754–760.

Shifman S, Bronstein M, Sternfeld M, Pisante-Shalom A, Lev-Lehman E, Weizman A, Reznik I, Spivak B, Grisaru N, Karp L, Schiffer R, Kotler M, Strous RD, Swartz-Vanetik M, Knobler HY, Shinar E, Beckmann JS, Yakir B, Risch N, Zak NB, Darvasi A (2002) A highly significant association between a COMT haplotype and schizophrenia. *Am J Hum Genet* **71**(6):1296–1302.

Shilkret v. Annapolis Emergency Hosp. Ass'n (1975) 276 Md. 187, 349 A2d 245.

Shiloh S, Berkenstadt M (1992) Lay conceptions of genetic disorders. *Birth Defects: Original Article Series* **28**(1):191–200.

Shiloh S, Sagi M (1989) Effect of framing on the perception of genetic recurrence risks. *Am J Med Genet* **33**:130–135.

Shiloh S, Saxe L (1989) Perception of risk in genetic counseling. *Psych Health* **3**:45–61.

Shiloh S, Enini NJ, Ben-Neria Z, Sagi M (2001) Framing of prenatal screening test results and women's health-illness orientations as determinants of perceptions of fetal health and approval of amniocentesis. *Psych Health* **16**:313–325.

Shinya M, Okamoto A, Sago H, Saito M, Akiyama Y, Kitagawa M, Tanaka T (2004) Analysis of fetal DNA from maternal peripheral blood by lactim-polymerase chain reaction—single strand conformation polymorphism. *Congenit Anom (Kyoto)* **44**(3):142–146.

Shipp TD, Benacerraf BR (2002) Second trimester ultrasound screening for chromosomal abnormalities. *Prenat Diagn* **22**:296–307.

Shipp TD, Bromley B, Lieberman E, Benacerraf BR (2000) The frequency of the detection of fetal echogenic intracardiac foci with respect to maternal race. *Ultrasound Obstet Gynecol* **15**(6):460–462.

Shirts BH, Nimgaonkar V (2004) The genes for schizophrenia: Finally a breakthrough? *Curr Psychiatry Rep.* **6**(4):303–312.

Shiva V (1997) *Biopiracy: The Plunder of Nature and Knowledge*. Toronto: Between the Lines.

Shprintzen RJ (2000) Velo-cardio-facial syndrome: A distinctive behavioral phenotype. *Ment Retard Dev Disabil Res Rev* **6**(2):142–147.

Shroeder v. Perkel (1981) **432** A 2d. 834 (N.J.).

Siderous M, Lamonthe E (1995) Norms and standards of practice in genetic counselling. *Health L J* **3**:153.

Sidransky E, Ginns EI (1993) Clinical heterogeneity among patients with Gaucher's Disease. *JAMA* **269**:1154–1157.

Siemieniec v. Lutheran General Hosptial (1987) 480 N.E. 2d. 1227; appeal allow'd in part 512 N.E. 2d. 691 (Ill. App. 1st Dist.).

Silberberg JS, Wlodarczyk J, Fryer J, Robertson R, Hensley MJ (1998) Risk associated with various definitions of family history of coronary heart disease: The Newcastle Family History Study II. *Am J Epidemiol* **147**:1133–1139.

Simmons v. W. Covina Medical Clinic (1989) 260 Cal. Rptr. 772 (Cal. Ct. App.) 212 Col. App. 696.

Simoncelli T (2003) Preimplementation diagnosis and selection: From disease prevention to customized conception. Population and Development Program at Hampshire College. **24**: Spring. http://genetics-and-society.org/resources/cgs/200303_difftakes_simoncelli.pdf.

Simonsen v. Swenson (1920) 104 Neb. 224, 177 N.W. 831–832.

Simpson JL and Elias S (2003) Genetics in Obstetrics and Gynecology, 3rd ed., Saunders, Philadelphia.

Singer E, Antonucci T, Van Hoewyk J (2004) Racial and ethnic variations in knowledge and attitudes about genetic testing. *Genet Test* **8**(1):31–43.

Singh NA, Charlier C, Stauffer D, DuPont BR, Leach RJ, Melis R, Ronen GM, Bjerre I, Quattlebaum T, Murphy JV, McHarg ML, Gagnon D, Rosales TO, Peiffer A, Anderson VE, Leppert M (1998) A novel potassium channel gene, KCNQ2, is mutated in an inherited epilepsy of newborns. *Nature Genet* **18**:25–29.

Singleton AB, Farrer M, Johnson J, Singleton A, Hague S, Kachergus J, Hulihan M, Peuralinna T, Dutra A, Nussbaum R, Lincoln S, Crawley A, Hanson M, Maraganore D, Adler C, Cookson MR, Muenter M, Baptista M, Miller D, Blancato J, Hardy J, Gwinn-Hardy K (2003) Alpha-synuclein locus triplication causes Parkinson's disease. *Science* **302**: 841.

Skene L (1998) Patients' rights or family responsibilities? Two approaches to genetic testing. *Med Law Rev* **6**:1–41.

Skillings v. Allen (1919) 143 Minn. 323, 325–26, 173 N.W. 663, 664. A child was hospitalized with scarlet fever. When the parents asked about the nature of the disease and the risk of infection, the physician said that they could visit the child in the hospital and take her home, even though the disease was in its most contagious stage.

Skinner EA (1996) Personality processes and individual differences: A guide to constructs of control. *J Pers Soc Psychol* **71**:549–570.

Skirton H (2001) The client's perspective of genetic counseling—A grounded theory study. *J Genet Counsl* **10**:311–329.

Sklar P, Gabriel SB, McInnis MG, Bennett P, Lim YM, Tsan G, Schaffner S, Kirov G, Jones I, Owen M, Craddock N, DePaulo JR, Lander ES (2002) Family-based association study of 76 candidate genes in bipolar disorder: BDNF is a potential risk locus. Brain-derived neutrophic factor. *Mol Psychiatry* **7**(6):579–593.

Slattery ML, Levin TR, Ma K, Goldgar D, Holubkov R, Edwards S (2003) Family history and colorectal cancer: Predictors of risk. *Cancer Causes Control* **14**:879–887.

Slovic P (1987) Perception of risk. *Science* **236**(4799):280–285.

Smith ACM (1998) Patient education. In: Baker DL, Schuette JL, Uhlmann WR eds. *A Guide to Genetic Counseling* New York: Wiley, pp. 99–155.

Smith v. Ardnt (1994) 21 C.C.L.T. 2d 66, rev'd 25 C.C.L.T. 2d. 262, rev'd, 1997. 148 D.L.R. (4th) 48, 2 S.C.R. 539 (S.C.C.).

Smith v. Rae (1919) 46 O.R. 518, (1920) 51 D.L.R. 323 (C.A.).

Smith AD, Wald NJ, Cuckle HS, Stirrat GM, Bobrow M, Lagercrantz H (1979) Amniotic fluid acetylcholinesterase as a possible diagnostic test for neural-tube defects in early pregnancy. *Lancet* **1**:685–688.

Smith AD, Johnston C, Sim E, Nagy Z, Jobst KA, Hindley N, King E (1994) Protective effect of apoE e2 in Alzheimer's disease. *Lancet* **344**:473–474.

Smith DW, Lemli L, Opitz JM (1964) A newly recognized syndrome of multiple congenital anomalies. *J Pedatr* **64**:210–217.

Smith KR, West JA, Croyle RT, Botkin JR (1999) Familial context of genetic testing for cancer susceptibility: Moderating effect of siblings' test results on psychological distress one to two weeks after BRCA1 mutation testing. *Ca Epi Bio Prevent* **8**:385–392.

Smith MW, Patterson N, Lautenberger JA, Truelove AL, McDonald GJ, Waliszewska A, Kessing BD, Malasky MJ, Scafe C, Le E, De Jager PL, et al (2004) A high-density admixture map for disease gene discovery in african Americans. *Amer J Hum Genet* **74**:1001–1013.

Smith SC, Greenland P, Grundy SM (2000) AHA Conference Proceedings. Prevention Conference V. Beyond secondary prevention: Identifying the high-risk patient for primary prevention. Executive Summary. *Circulation* **101**:111–116.

Smith RA, Cokkinides V, von Eschenbach AC, Levin B, Cohen C, Runowicz CD, Sener S, Saslow D, Eyre HJ (2002) American Cancer Society guidelines for the early detection of cancer. *CA Cancer J Clin* **52**:8–22.

Smith-Bindman R, Feldstein VA, Goldberg JD (2001a) The genetic sonogram in screening for down syndrome. *J Ultrasound Med* **20**:1153–1158.

Smith-Bindman R, Hosmer W, Feldstein VA, Deeks JJ, Goldberg JD (2001b) Second-trimester ultrasound to detect fetuses with Down syndrome. *JAMA* **285**:1044–1055.

Smoller JW, Finn CT (2003) Family, twin, and adoption studies of bipolar disorder. *Am J Med Genet* **123C**(1):48–58.

Sneiderman B, Irvine J, Osborne P (2003) Canadian Medical Law *An Introduction for Physicians, Nurses and Other Health Care Professionals*. Scarborough: Thomson Carswell.

Snijders R, Smith E (2002) The role of fetal nuchal translucency in prenatal screening. *Curr Opin Obstet Gynecol* **14**:577–585.

Snijders RJ, Sebire NJ, Faria M, Patel F, Nicolaides KH (1995) Fetal mild hydronephrosis and chromosomal defects: Relation to maternal age and gestation. *Fetal Diagn Ther* **10**(6):349–355.

Snijders RJM, Johnson S, Sebire NJ, Noble PL, Nicolaides KH (1996) First trimester ultrasound screening for chromosomal defects. *Ultrasound Obstet Gynaecol* **7**:216–226.

Snijders RJM, Noble P, Sebire N, Souka A, Nicolaides KH (1998) UK multicentre project on assessment of risk of trisomy 21 by maternal age and fetal nuchal-translucency thickness at 10–14 weeks of gestation. *Lancet* **351**:343–346.

Snow v. AH Robins Company Inc (1985) 211 Cal. Rptr. 271, 165 Ca. 3d. 720 Ca. App. 3 Dist.

Sobel SK, Cowan DB (2000) Impact of genetic testing for Huntington disease on the family system. *Am J Med Gen* **90**(1):49–59.

Social Ethics and Legal Committee of the American College of Medical Genetics (1999) Policy statement: Duty to re-contact. **1**(4):171. http://www.acmg.net/resources/policies/pol-007.pdf.

Society of Obstetricians and Gynecologists of Canada (2002) Cystic fibrosis carrier testing in pregnancy in Canada. http://sogc.medical.org/SOGCnet/sogc_docs/common/guide/pdfs/ps118.pdf.

SOGC Clinical Practice Guidelines (1998) Guidelines for health care providers involved in prenatal screening and diagnosis. *J Soc Obstet Gynaecol Can* **20**(9):865–870.

SOGC Clinical Practice Guidelines (1999) Guidelines for ultrasound as part of routine prenatal care. *J SOGC* **21**(9):874–879.

Sohl BD, Scioscia AL, Budorick NE, Moore TR (1999) Utility of minor ultrasonographic markers in the prediction of abnormal fetal karyotype at a prenatal diagnostic center. *Am J Obstet Gynecol* **181**(4):898–903.

Sonek JD (2003) Nasal bone evaluation with ultrasonography: A marker for fetal aneuploidy. *Ultrasound Obstet Gynecol* **22**:11–15.

Sorbi S, Nacmias B, Forleo P, Piacentini S, Latorraca S, Amaducci L (1995) Epistatic effect of APP717 mutation and apolipoprotein E genotype in familial Alzheimer's disease. *Ann Neurol* **38**:124–127.

Sorenson JR (1993) Genetic counseling: Values that have mattered. In: Bartels, DM, LeRoy BS, Caplan AL, eds. *Prescribing Our Future: Ethical Challenges in Genetic Counseling.* New York: Walter deGruyter, pp. 3–14.

Souter VL, Nyberg DA, El-Bastawissi A, Zebelman A, Luthhardt F, Luthy DA (2002) Correlation of ultrasound findings and biochemical markers in the second trimester of pregnancy in fetuses with trisomy 21. *Prenat Diagn* **22**(3):175–182.

Speck v. Finegold (1981) 497 Pa. 77, 439 A.2d 110.

Speice J, McDaniel SH, Rowley PT, Loader S (2002) Family issues in a psychoeducation group for women with a BRCA mutation. *Clin Genet* **62**(2):121–127.

Spencer K, Aitken DA, Crossley JA, McCaw G, Berry E, Anderson R, Connor JM, Macri JN (1994) First trimester biochemical screening for trisomy 21: The role for free beta hCG, alpha fetoprotein and pregnancy associated plasma protein A. *Ann Clin Biochem* **31**:447–454.

Spielberg AR (1999) On-line without a net: Physician-patient communication by electronic mail. *Am J Law Med Ethics* **25**:267–295.

Spriggs M (2002) Lesbian couple create a child. *J Med Ethics* **28**:283.

Srivastava A, McKinnon W, Wood ME (2001) Risk of breast and ovarian cancer in women with strong family histories. *Oncology (Huntingt)* **15**:889–902.

Stanford JL, Weiss NS, Voigt LF, Daling JR, Habel LA, Rossing MA (1995) Combined estrogen and progestin hormone replacement therapy in relation to risk of breast cancer in middle-aged women. *JAMA* **274**:137–142.

Stanley v. McCarver (2003) 204 Ariz. 339, 63 P. 3d 1076 (Ct. App. Div.1).

Stapleton H, Kirkham M, Thomas G (2002) Qualitative study of evidence based leaflets in maternity care. *BMJ* **324**:639.

State Genetic Discrimination in Health Insurance Laws (2005) http://www.ncsl.org/programs/health/genetics/ndishlth.htm charts depicting states and their respective legislation.

State Genetic Employment Laws (2005) http://www.ncsl.org/programs/health/genetics/ndiscrim.htm-.

State Genetic Nondiscrimination Laws in Life, Disability, and Long-Term Insurance (2005) http://www.ncsl.org/programs/health/ndislife.htm.

Statham H, Solomou W, Green JM (2003) Communication of prenatal screening and diagnosis results to primary-care health professionals. *Public Health* **117**(5):348–357.

Stauffer v. Karabin (1971) 30 Colo. App. 357, 492 P. 2d 862.

Stefansson H, Steinthorsdottir V, Thorgeirsson TE, Gulcher JR, Stefansson K (2004) Neuregulin 1 and schizophrenia. *Ann Med* **36**(1):62–71.

Steinberg GD, Carter BS, Beaty TH, Childs B, Walsh PC (1990) Family history and the risk of prostate cancer. *Prostate* **17**:337–347.

Steinbock B (2002) Sex selection: Not obviously wrong. *Hastings Center Report* **32**(1):23–28.

Stephenson J (1999) Genetic test information fears unfounded. *JAMA* **282**(23):2197–2198.

Stern RC (1997) The diagnosis of cystic fibrosis. *N Engl J Med* **336**(7):487–491.

Stern JE, Cramer CP, Green RM, Garrod A, DeVries KO (2003) Determining access to assisted reproductive technology: Reactions of clinic directors to ethically complex case scenarios. *Human Reprod* **18**(6):1343–1352.

Stiefel F, Lehmann A, Guex P (1997) Genetic detection: The need for psychosocial support in modern cancer prevention. *Support Care Ca* **5**:461–465.

Stierman L (1995) *Birth Defects in California: 1983–1990.* Sacramento, CA: California Department of Health Services.

St John DJ, McDermott FT, Hopper JL, Debney EA, Johnson WR, Hughes ES (1993) Cancer risk in relatives of patients with common colorectal cancer. *Ann Intern Med* **118**:785–790.

Stocker, AM, Snijders RJ, Carlson DE, Greene N, Gregory KD, Walla CA, Platt LC (2000) Fetal echogenic bowel: Parameters to be considered in differential diagnosis. *Ultrasound Obstet Gynecol* **16**:519–523.

Stogmann E, Zimprich A, Baumgartner C, Aull-Watschinger S, Hollt V, Zimprich F (2002) A functional polymorphism in the prodynorphin gene promoter is associated with temporal lobe epilepsy. *Ann Neurol* **51**:260–263.

Stokes v. Dailey (1957) 85 N.W. 2d 745 (ND), 97 N.W. 2d 676 (ND 1959).

Stolberg S (1994) Genetic bias: Held hostage by heredity. *Los Angeles Times*, March 27: A1–A20.

Stopfer JE (2000) Genetic counseling and clinical cancer genetics services. *Semin. Surg. Oncol.* **18**(4):347–357.

Straub RE, Jiang Y, MacLean CJ, Ma Y, Webb BT, Myakishev MV, Harris-Kerr C, Wormley B, Sadek H, Kadambi B, Cesare AJ, Gibberman A, Wang X, O'Neill FA, Walsh D, Kendler KS (2002) Genetic variation in the 6p22.3 gene DTNBP1, the human ortholog of the mouse dysbindin gene, is associated with schizophrenia. *Am J Hum Genet* **71**(2):337–348.

Straus J (1998) Bargaining around the TRIPS agreement: The case for ongoing public-private initiatives to facilitate worldwide intellectual property transactions. *Duke J Comp Internat'l L* **9**:91–107.

Straus J (2002) Genetic Inventions and Patents—A German Empirical Study presentation to the BMBF & OECD Workshop entitled Genetic Inventions, Intellectual Property Rights and Licensing Practices Berlin, January 24–25, 2002, online: OECD; <www.oecd.org/document/57/0,2340,en_2649_34537_2743225_1_1_1_1,00.html>.

Straus SE, Majumdar SR, McAlister FA (2002) New evidence for stroke prevention. *JAMA* **288**:1388–1395.

Strauss, RP (2002) Beyond easy answers: Prenatal diagnosis and counselling during pregnancy. *Cleft Palate Craniofac J* **39**(2):164–168.

Street RL Jr (1992) Analyzing communication in medical consultations. Do behavioral measures correspond to patients' perceptions? *Med Care* **30**(11):976–988.

Strong C (1997) *Ethics in Reproductive and Perinatal Medicine: A New Framework*, New Haven, CT: Yale University Press.

Strong C (2003) Fetal anomalies: Ethical and legal considerations in screening, detection, and management. *Clin Perinatol* **30**:113–126.

Struewing JP, Lerman C, Kase RG, Giambarresi TR, Tucker MA (1995) Anticipated uptake and impact of genetic testing in hereditary breast and ovarian cancer families. *Ca Epi, Bio Prevent* **4**:169–173.

Struewing JP, Hartge P, Wacholder S, Baker SM, Berlin M, McAdams M, Timmerman MM, Brody LC, Tucker MA (1997) The risk of cancer associated with specific mutations of BRCA1 and BRCA2 among Ashkenazi Jews. *N Engl J Med* **336**:1401–1408.

Suarez BK, Hampe CL, Van Eerdewegh PV (1994) Problems of replicating linkage claims in psychiatry. In: Gershon ES, Cloninger CR, eds. *Genetic Approaches to Mental Disorders*, Washington, DC: American Psychiatric.

Suchman AL, Roter D, Green M, Lipkin M Jr (1993) Physician satisfaction with primary care office visits. Collaborative Study Group of the American Academy on Physician and Patient. *Med Care* **12**:1083–1092.

Suchman AL, Markakis K, Beckman HB, Frankel R (1997) A model of empathic communication in the medical interview. *JAMA* **277**(8):678–682.

Sudell A (2001) To tell or not to tell: The scope of physician-patient confidentiality when relatives are at risk of genetic disease. *J Contemp Health Law Policy* **18**(1):273–295.

Sue D, Arredondo P, McDavis RJ (1995) Multicultural Counseling Competencies and Standards, in Handbook of Multicultural Counseling, edited by Ponterotto JG, Casas JM, Suzuki LA, and Alexander CA. Sage Publications Inc.

Sue D (1990) Cultural specific issues in counseling: A conceptual framework. *Prof Psych Res Prac* **21**:424–433.

Sue DW, Sue D (1990) Counseling the Culturally Different: Theory and Practice. Wiley.

Sugawara T, Tsurubuchi Y, Agarwala KL, Ito M, Fukuma G, Mazaki-Miyazaki E, Nagafuji H, Noda M, Imoto K, Wada K, Mitsudome A, Kaneko S, Montal M, Nagata K, Hirose S, Yamakawa K (2001) A missense mutation of the Na+ channel alpha II subunit gene Na(v)1.2 in a patient with febrile and afebrile seizures causes channel dysfunction. *Proc Natl Acad Sci USA* **98**:10515.

Sugerman PR, Yarisbus VA (1999) Admissibility of managed care financial incentives in medical malpractice cases. *Tort Insur Law J* **34**:735–760.

Sulmasy DP, Lehmann LS, Levine DM, Raden RR (1994) Patients' perceptions of the quality of informed consent for common medical procedures. *J Clin Ethics* **5**(3):189–194.

Suls J, Fletcher B (1985) The relative efficacy of avoidant and nonavoidant coping strategies: A meta-analysis. *Health Psych* **1985**(4):249–288.

Surgical Consultants, P.C. v. Ball (1989) 447 NW 2d 676 (Iowa App.).

Suslak L, Price DM, Desposito F (1985) Transmitting balanced translocation carrier information within families: A follow-up study. *Am J Med Genet* **20**(2):227–232.

Suter SM (2002) The routinization of prenatal testing. *Am J Law Med* **28**:233–270.

Sveinbjornsdottir S, Hicks AA, Jonsson T, Petursson H, Gugmundsson G, Frigge ML, Kong A, Gulcher JR, Stefansson K (2000) Familial aggregation of Parkinson's disease in Iceland. *N Engl J Med* **343**:1765–1770.

Sweet KM, Phelan MC, Tarleton JC, Crawford EC, Christensen B, Schroer RJ, Taylor HA (1989) *Counseling Aids for Geneticists*. Greenwood Genetic Center. Greenwood.

Sweet KM, Willis SK, Ashida S, Westman JA (2003) Use of fear-appeal techniques in the design of tailored cancer risk communication messages: Implications for healthcare providers. *J Clin Oncol* 3375–3376.

Szabo J, Gellen J (1990) Nuchal fluid accumulation in trisomy-21 detected by vaginosonography in first trimester. *Lancet* **336**(8723):1133.

Szabo v. Bryn Mawr Hospital (1994) 432 Pa. Super 409, 638 A 2d 1004.

Szabo CI, Worley T, Monteiro ANA (2004) Understanding germ-line mutations in BRCA1. *Cancer Biol* **3**:515–520.

Szekely AP (2002) Railroad to pay $2.2 million in DNA test case illegally testing workers for genetic defects. *Reuters* May 8.

Tabar L, Vitak B, Chen HH, Duffy SW, Yen MF, Chiang CF, Krusemo UB, Tot T, Smith RA (2000) The Swedish Two-County Trial twenty years later. *Radiol Clin North America* **38**:625–651.

Tacknyk v. Lake of the Woods Clinic (1982) 17 A.C.W.S. (2d) 154 (Ont. C.A.).

Takala T, Gylling HA (2000) Who should know about our genetic makeup and why? *J Med Ethics* **26**:171–174.

Talbot C, Lendon C, Craddock N, Shears S, Morris JC, Goate A (1994) Protection against Alzheimer's disease with apoe e2. *Lancet* **343**:1432–1433.

Tan NC, Mulley JC, Berkovic SF (2004) *Epilepsia* **45**(11):1429–1442. Genetic association studies in epilepsy: "the truth is out there."

Tancredi M, Sensi E, Cipollini G, Aretini P, Lombardi G, Cristofano CD, Presciuttini S, Bevilacqua G, Caligo MA (2004) Haplotype analysis of BRCA1 gene reveals a new gene rearrangement: Characterization of a 19.9 KBP deletion. *Eur J Hum Genet* **12**(9):775–777.

Taneja PR, Pandya A, Foley DL, Nicely LV, Arnos KS (2004) Attitudes of deaf individuals towards genetic testing. *Am J Med Genet* **130A**(1):17–21.

Tang YP, Gershon ES (2003) Genetic studies in Alzheimer's disease. *Dialogues Clin Neurosci* **4**:17–28.

Taniv v. Taub (1998) 683 N.Y.S. 2d. 35 (App. Div. Ist. Dep't.). A physician who was responsible for supervision of the office staff, implementing office policy, transmitting radiology reports, and was a principal of the professional services corporation, was held liable when a patient's X-ray was not sent to the treating physician, resulting in the failure to diagnose a patient's cancer.

Tanner CM, Goldman SM (1996) Epidemiology of Parkinson's disease. *Neuroepidemiology* **14**:317–335.

Tanner v. Norys (1980) 4 W.W.R. 33, 31 A.R. 372; reversing in part, 1979. 21 A.R. 410; leave to appeal to S.C.S. refused 33 N.R. 354n (S.C.C.). Also see *Wenden v. Trikha* (1991) 8 C.C.L.T. (2d.) 138, aff'd, 1993. 14 C.C.L.T. (2d.) 225, leave to appeal refused, 1993. 17 C.C.L.T. (2d.) 285n in *Kines Estate v. Lychak Estate* (1997) 4 W.W. R. 585 (C.A.)

Tanuz v. Carlberg (1996) 921 P. 2d. 309 N.M.App.

Tarasoff v. Regents of University of California (1976) 17 Cal. 3d. 425, 131 Cal. Rptr. 14, 551 P. 2d. 334. With respect to application to HIV antibodies see *Reisner v. Regents of University of California infra*; also see Basanta WE, Foran JM, Cepelewicz BB, Apple GJ, Levin LJ et al (1996) Physician liability to persons other than patients. *Tort Insur Law J* **31**:357–361. In Canada, see *Walker Estate v. York-Finch General Hosp* (1997) 39 C.C.L.T. 1, 1999. 43 O.R. (3d) 461, 129 D.L.R. (4th) 689 (C.A.), leave to appeal to S.C.C. granted Oct. 14, 1999 S.C.C. Bulletin of Proceedings, Oct. 15, 1999 at 1525, [2001] S.C.R. 647, [2001] SCC 23.

Tariq SM, Matthews SM, Hakim EA, Stevens M, Arshad SH, Hide DW (1998) The prevalence of and the risk factors for atopy in early childhood: A whole population birth cohort study. *J Allergy Clin Immunol* **101**:587–593.

Tasca RJ, McClure ME (1998) The emerging technology and application of preimplantation genetic diagnosis. *J Law Med Ethics* **26**(1):7–16; Robertson JA (2003) Extending preimplantation genetic diagnosis: Medical and non-medical uses. *J Med Ethics* **29**:213–216; Caulfield T, Knowles L, Meslin EM (2004) Law and policy in the era of reproductive genetics. *J Med Ethics* **30**(4):414–417.

Task Force (1997) *Promoting Safe and Effective Genetic Testing in the United States: Final Report of the Task Force on Genetic Testing.* Baltimore: Johns Hopkins University Press, 1998. http://www.genome.gov/10001733. With regard to the communication of information, the Task Force stated: "A non-directive approach is of the utmost importance when reproductive decisions are a consequence of testing or when the safety and effectiveness of interventions following a positive test result have not been established." Absolute nondirective counseling may not be a practical possibility (p. 64).

Tatsugawa Z (1998) Personal communication. Genetic Counselor, Fetal Assessment Department, Olive View-UCLA Medical Center. Sylmar, CA.

Taubes G (1995) Epidemiology faces its limits. *Science* **269**:164–169.

Tavani A, Augustin L, Bosetti C, Giordano L, Gallus S, Jenkins DJA, et al (2004) Influence of selected lifestyle factors on risk of acute myocardial infarction in subjects with familial predisposition for the disease. *Prev Med* **38**(4):468–72.

Tavani A, Bosetti C, Dal Maso L, Giordano L, Franceschi S, La Vecchia C (2004) Influence of selected hormonal and lifestyle factors on familial propensity to ovarian cancer. *Gynecol Oncol* **92**:922–926.

Tavakoli-Nouri v. Gunther (2000) 745 A. 2d 939 (D.C.).

Taylor HA (1999) Barriers to informed consent. *Semin Oncol Nurs* **15**(2):89–95.

Taylor MRG (2001) Genetic testing for inherited breast and ovarian cancer syndromes: Important concepts for the primary care physician. *Postgrad Med J* **77**:11–15. The Human Genetics Commission, http://www.hgc.gov.uk/.

Taylor SD (2004) Predictive genetic test decisions for Huntington's disease: Context, appraisal and new moral imperatives. *Soc Sci Med* **58**(1):137–149.

Taylor K, Kelner M (1987) Informed consent: The physicians' perspective. *Soc Sci Med* **24**(2):135–143.

Taylor K, Kelner M (1996) The emerging role of the physician in genetic counseling and testing for heritable breast, ovarian and colon cancer. *CMAJ* **154**:1155–1158 at 1158.

Taylor v. Kurapati (1999) 600 N.W.2d 670 (Mich.).

Taylor v. Wilmington Medical Center (1983) 577 F. Supp. 309, 315.

Teff H (1985) Consent to medical procedures: paternalism, self-determination or therapeutic alliance? *L Q Rev* **101**:432 at 451. Delegation to a junior doctor or intern "is apparently common practice in English hospitals." *ter Neuzen v. Korn* (1993) 16 C.C.L.T. (2d) 65 at 66.

Tennstedt C, Chaoui R, Vogel M, Goldner B, Dietel M (2000) Pathologic correlation of sonographic echogenic foci in the fetal heart. *Prenat Diagn* **20**:287–292.

Tenuto v. Lederle Laboratories (1997) 90 NY 2d 606 (NY).

ter Neuzen v. Korn (1993) 16 C.C.L.T. (2d) 65; (1995), 127 D.L.R. (4th) 577, 3 S.C.R. 674 (S.C.C.).

Tercyak KP, Johnson SB, Roberts SF, Cruz AC (2001a) Psychological response to prenatal genetic counseling and amniocentesis. *Patient Educ Counsl* **43**(1):73–84.

Tercyak KP, Peshkin BN, Streisand R, Lerman C (2001b) Psychological issues among children of hereditary breast cancer gene (BRCA1/2) testing participants. *Psychooncology* **10**(4):336–346.

Terry NP (1999) Cyber-malpractice: Legal exposure for cybermedicine. *Am J Law Med Ethics* **25**:327–366.

Therrell Jr, BL (2001) U.S. Newborn Screening Policy Dilemmas for the Twenty-First Century. Molecular Genetics and Metabolism **74**:64–74.

Thewes B, Meiser B, Tucker K, Schnieden V (2003) Screening for psychological distress and vulnerability factors in women at increased risk for breast cancer: A review of the literature. *Psychol Hlth Med* **8**(3):289–303.

Thirlaway K, Fallowfield L (1993) The psychological consequences of being at risk of developing breast cancer. *Eur J Cancer Prev* **2**:467–471.

Thomas NS, Sharp AJ, Browne CE, Skuse D, Hardie C, Dennis NR (1999) Xp deletions associated with autism in three females. *Hum Genet* **104**(1):43–48.

Thompson D, Easton D (2001) Variation in cancer risks, by mutation position, in BRCA2 mutation carriers. *Am J Hum Genet* **68**:410–419.

Thompson and Thompson (2001) Genetic counseling and risk assessment. In: Nussbaum RL, McInnes RR, Willard HF, eds. *Genetics in Medicine*, 6th ed. Philadelphia: W.B. Saunders Company, pp. 375–389.

Thorevska N, Tilluckdharry L, Ticko S, Havasi A, Amoateng-Adjepong Y, Manthous CA (2004) Informed consent for invasive medical procedures from the patient's perspective. *Conn Med* **68**(2):101–105.

Thornhill v. Midwest Physician Center of Orland Park (2003) 787 NE.2d 247; Ill App. The plaintiff was denied recovery for a wrongful birth action on the factual finding that she would not have terminated the pregnancy if the correct results had been disclosed. http://www.state.il.us/court/Opinions/AppellateCourt/2003/1stDistrict/March/Html/1013050.htm. Accessed December 19, 2004.

Thull DL, Vogel VG (2004) Recognition and management of hereditary breast cancer syndromes. *Oncologist* **9**:13–24.

Tibben A, Vegter-van der Vlis M, Skraastad MI, Frets PG, van der Kamp JJ, Niermeijer MF, van Ommen GJ, Roos RA, Rooijmans HG, Stronks D, Rooymans HG, Verhage F (1992) DNA-testing for Huntington's disease in The Netherlands: A retrospective study on psychosocial effects. *Am J Med Genet* **44**(1):94–99.

Tibben A, Duivenvoorden HJ, Vegter-van der Vlis M, Niermeijer MF, Frets PG, van de Kamp JJ, Roos RA, Rooijmans HG, Verhage F (1993a) Presymptomatic DNA testing for Huntington disease: Identifying the need for psychological intervention. *Am J Med Genet* **48**(3):137–144.

Tibben A, Frets PG, van de Kamp JJ, Niermeijer MF, Vegtervan der Vlis M, Roos RA, Rooymans HG, van Ommen GJ, Verhage F (1993b) On attitudes and appreciation 6 months after predictive DNA testing for Huntington disease in the Dutch program. *Am J Med Genet* **48**(2):103–111.

Tibben A, Timman R, Bannink E, Duivenvoorden H (1997) Three-year follow-up after presymptomatic testing for Huntington's disease in tested individuals and partners. *Hlth Psychol* **16**:20–35.

Tibben A, Vegter-van der Vlis M, Niermeijer M, Frets PG, van de Kamp JJP, Roos RA, Rooymans HG, van Ommen GJ, Verhage F (1999) Testing for Huntington's disease with support for all parties (letter). *Lancet* **335**:553.

Tjio JH, Levan A (1956) The chromosome number of man. *Hereditas*; **42**:1–16.

Tluczek A, Mischler EH, Bowers B, Peterson NM, Morris ME, Farrell PM, Bruns WT, Colby H, McCarthy C, Fost N, Tibben A, Frets PG, van de Kamp JJ, Niermeijer MF, Vegter-van der Vlis M, Roos RA, Rooymans HG, van Ommen GJ, Verhage F (1991) Psychological impact of false-positive results when screening for cystic fibrosis. *Pediatr Pulmonol Suppl* **7**:29–37. "When test results were communicated by telephone rather than with a face to face encounter, and when a brochure was used as a method of informing parents, many parents had misconceptions and adverse emotional responses."

Tluczek A, Mischler EH, Farrell PM, Fost N, Peterson NM, Carey P, Bruns WT, McCarthy C (1992) Parents' knowledge of neonatal screening and response to false-positive cystic fibrosis testing. *J Dev Behav Pediatr* **13**(3):181–186.

Tluczek A, Koscik RL, Forrell PM, Rock MJ (2005) Psychosocial risk with newborn screening for cystic fibrosis: Parents expertence while awaiting the sweet-test appointments. *Padiatrics* **115**(6):1692–1703.

Toledano-Alhadef H, Basel-Vanagaite L, Magal N, Davidov B, Ehrlich S, Drasinover V, Taub E, Halpern GJ, Ginott N, Shohat M (2001) Fragile-X carrier screening and the prevalence of premutation and full-mutation carriers in Israel. *Am J Hum Genet* **69**(2):351–360.

Tonin P, Lenoir G, Lynch H. (1996) BRCA1 testing in families with hereditary breast-ovarian cancer. A prospective study of patient decision making and outcomes. JAMA; **275**:1885–1892.

Torfs CP, Christianson RE (1998) Anomalies in Down syndrome individuals in a large population-based registry. *Am J Med Genet* **77**:431–438.

Torres A, Wagner R (1993) Establishing the physician-patient relationship. *J Dermatol Surg Oncol* **19**:147–149.

Torresani T (2003) Quality control requirements in neonatal screening. *Eur J Pediatr* **162** (Suppl 1):S54–56.

Toth v. Community Hospital (1968) 292 N.Y.S.2d 440, 239 N.E.2d 368, 369 (N.Y.).

Towner D, Loewy RS (2002) Ethics of preimplantation diagnosis for a woman destined to develop early-onset Alzheimer disease. *JAMA* **283**:1038–1040.

Trask P, Paterson A, Wang C, Hayasaka S, Milliron K, Blumberg L, Gonzalez R, Murray S, Merajver S (2001) Cancer-specific worry interference in women attending a breast and ovarian cancer risk evaluation program: Impact on emotional distress and health functioning. *Psycho-Onc* **10**:349–360.

Trefethen A (2000) The emerging tort of wrongful adoption. *J Contemp Legal Issues* **11**:620–624.

Trepanier A, Ahrens M, McKinnon W, Peters J, Stopfer J, Grumet S, Manley S, Culver J, Acton R, Larson-Haidle J, Correia L, Bennett R, Pettersen B, Ferlita T, Costalas J, Hunt K, Donlon S, Skrzynia C, Farrell C, Callif-Daley F,Vockley C (2004). Genetic cancer risk assessment and counseling: Recommendations of the National Society of Genetic Counselors. *J Genet Counsel* **13**(2): 83–114.

Tresemer v. Barke (1976) 17 Cal. 3d. 425, 131 Cal. Rptr. 14, 551 P. 2d 334; 1978. 86 Cal. App. 3d. 656, 150 Cal. Rptr. 384. The physician had provided a patient with a Dalkon Intrauterine Device that, at the time of the original service, was considered safe and effective. The patient had no further contact with the physician, but 3 years later the device was removed from the market because of the serious risk of physical injury. The plaintiff–patient alleged that the defendant physician had not contacted her about this risk, and that the physician had a continuing duty to disclose subsequently discovered risks. The court found for the plaintiff, stating that the duty to warn was created by "the imposed continuing status of physician-patient relationship." The court recognized that a "duty to take action" could arise where prior innocent conduct has created an unreasonable risk of harm to the patient and "where it already has injured him." Also see *Gorab v. Zook* (1997) 943 P.2d 423, 430 (Colo.); *Tanuz v. Carlberg* (1996) 921 P.2d 309, 313, 316 (N.M. Ct. App.).

Tri-Council Policy Statement (1998) Ethical Conduct for Research Involving Humans (and updates) http://www.pre.ethics.gc.ca/english/policystatement/policystatement.cfm.

Trimbath JD, Giardiello FM (2002) Review article: Genetic testing and counselling for hereditary colorectal cancer. *Alimet Pharmacol Ther* **16**:1843–1857.

Troxel v. A.I. Dupont Institute (1996) 675 A. 2d. 314, appeal den'd 685 A. 2d. 547 Pa. Super.

Truman v. Thomas (1980) 611 P.2d 902, 905 (Cal.). "Material information is that which the physician knows or should know would be regarded as significant by a reasonable person in the patient's position when deciding to accept or reject the recommended medical procedure." However, given the lack of effective treatment and cure for many genetic disorders, what standards are to be applied in the determination of what is "reasonable" and from whose's perspective, the physician or the patient? See *Wilkinson v. Vesey* (1972) 295 A.2d 676, 688–90 (R.I.); *Harnish v. Children's Hosp. Med. Ctr.*(1982) 439 N.E.2d 240, 242–44 (Mass.); *Rook v. Trout* (1987) 747 P.2d 61, 66–67 (Idaho); *Carr v. Strode* (1995) 904 P.2d 489, 494–99 (Haw.).

Tsao A (2004) Genetic testing meets Mad. Avenue: Myriad Genetics' groundbreaking ads hawking predictive breast cancer tests are themselves a test of the thorny issues involved. *Business Week*, July. http://www.businessweek.com/technology/content/jul2004/tc20040728_7693_tc024.htm. Also see Myriad Genetics Launches Direct to Consumer Advertising Campaign for Breast Cancer Test (2002): http://www.corporateir-.net/ireye/ir_site.zhtml?ticker=mygn&script=413&layout=9&item_id=333030.

Tsuang MT, Taylor L, Faraone SV (2004) An overview of the genetics of psychotic mood disorders. *J Psychiatr Res* **38**(1):3–15.

Tudor HJ, Dieppe P (1996) Caring effects. *Lancet* **347**:1606–1608.

Tul N, Spencer K, Noble P, Chan C, Nicolaides K (1999) Screening for trisomy 18 by fetal nuchal translucency and maternal serum free á-hCG and PAPP-A at 10–14 weeks of gestation. *Prenat Diagn* **19**:1035–1042.

Turner BC, Harrold E, Matloff E, Smith T, Gumbs AA, Beinfield M, Ward B, Skolnick M, Glazer PM, Thomas A, Haffty BG (1999) BRCA1/BRCA2 germline mutations in locally recurrent breast cancer patients after lumpectomy and radiation therapy: Implications for breast-conserving management in patients with BRCA1/BRCA2 mutations. *J Clin Oncol* **17**:3017–3024.

Turner v. Children's Hosp. Inc. (1991) 76 Ohio App. 3d 541, 602 NE 2d 423 (Franklin Co.). Motion overruled 63 Ohio St. 3d 1469, 590 NE 2d 1268: This court found that the duty to inform arises only when the physician proposes some form of treatment giving rise to a patient decision. Genetic counseling has been recognized as a "treatment": *Pratt by Pratt v. University of Minnesota Affiliated Hospitals and Clinics* (1987) 403 N.W. 2d 865, at 868; 414 NW.2d 399 (Minn.). This case concerned an allegation for "negligent non-disclosure" during genetic counseling. Parents of an afflicted child wanted to know whether the condition "was genetic in origin," and whether future offspring would have an increased risk. The consulting physicians reviewed the parents' respective medical histories, took blood samples, and performed a "chromosome study." Although the physicians were unable to diagnose the child's anomalies, both consulting physicians were aware of the possibility that the disorder may be an autosomal-recessive condition. However, neither physician felt that "the possibility was significant enough" to inform the parents. At trial, the physicians argued that genetic counseling did not constitute treatment and therefore the doctrine of negligent nondisclosure did not apply. The trial court held that it was vital for patients in the situation of having to make a choice to be able to base their decision on as much information as possible. "This is especially true in the area of genetic counseling." The court concluded that the physicians had replied to the parents' request for information "in order to make a major health care decision. In providing that information, we conclude that the respondents [the physicians] did render treatment ... "

Turpin v. Sortini (1982). 31 Cal.3d 220, 231; 643 P.2d 954, 182 Cal.Rptr. 337. "If defendants had performed their jobs properly, [plaintiff] ... would not have been born at all" [31 Cal. 3d 220, 231]. Also see *Gami v. Mullikin Medical Center* (1993) 18 Cal.App.4th 870, 881 with respect to the basis for the claim for compensation.

Tuzmen S, Schechter AN (2001) Genetic diseases of hemoglobin: Diagnostic methods for elucidating beta-thalassemia mutations. *Blood Rev* **15**(1):19–29.

Tymstra T (1991) Prenatal diagnosis, prenatal screening and the rise of the tentative pregnancy. *Int J Tech Assess Hlth Care* **7**:(4):509–516.

Uhlmann WR (1998) A guide to case management. In: Baker DL, Schuette JL, Uhlmann WR, eds. *A Guide to Genetic Counseling*. New York: Wiley-Liss, pp. 199–229.

Umar A, Boland CR, Terdiman JP, Syngal S, de la Chapelle A, Ruschoff J, Fishel R, Lindor NM, Burgart LJ, Hamelin R, Hamilton SR, Hiatt RA, Jass J, Lindblom A, Lynch HT,

Peltomaki P, Ramsey SD, Rodriguez-Bigas MA, Vasen HF, Hawk ET, Barrett JC, Freedman AN, Srivastava S (2004a) Revised Bethesda guidelines for hereditary nonpolyposis colorectal cancer (Lynch syndrome) and microsatellite instability. *J Natl Cancer Inst* **96**:261–268.

Umar A, Risinger JI, Hawk ET, Barrett JC (2004b) Testing guidelines for hereditary nonpoloposis colorectal cancer. *Nat Rev Cancer* **4**:153–158.

United States Preventive Services Task Force (2002b) Aspirin for primary prevention for cardiovascular events: preventive medication. http://www.ahrq.gov/clinic/uspstf/uspsasmi.htm. Accessed March 1, 2004.

United States Preventive Services Task Force (2002a) Clinical guidelines: Screening for colorectal cancer: Recommendation and rationale. *Ann Intern Med* **137**:129–131.

United States Preventive Services Task Force (2001) Lipid disorders: Screening. http://www.ahcpr.gov/clinic/uspstf/uspschol.htm. Accessed March 1, 2004.

USA Congress, Office of Technology Assessment (1992) *Genetic Tests and Health Insurance: Results of a Survey*. Washingon DC: US GPO.

U.S. Department of Health and Human Services (2000) *Fed Reg* **65**(236):76643–76645. http://www4.od.nih.gov/oba/sacgt/reports/Public%20Consultation%20Summary.pdf.

U.S. National Screening Status Report (2004) This web resource lists the status of newborn screening in the United States: http://genes-r-us.uthscsa.edu/resources/newborn/screenstatus.htm.

U.S.-Venezuela Collaborative Research Project, Wexler NS (2004) Venezuelan kindreds reveal that genetic and environmental factors modulate Huntington's disease age of onset. *Proc Natl Acad Sci USA* **101**:3498–3503.

Vadlamudi L, Harvey AS, Connellan MM, Milne RL, Hopper JL, Scheffer IE, Berkovic SF (2004) Is benign rolandic epilepsy genetically determined? *Ann Neurol* **56**(1):129–132.

Vale W, Rivier C, Hsueh A, Campen C, Meunier H, Bicsak T, Vaughan J, Corrigan A, Bardin W, Sawchenko P, et al (1988) Chemical and biological characterization of the inhibin family of protein hormones. *Rec Progress Horm Res* **44**:1–34.

Valente EM, Abou-Sleiman PM, Caputo V, Muqit MM, Harvey K, Gispert S, Ali Z, Del Turco D, Bentivoglio AR, Healy DG, Albanese A, Nussbaum R, Gonzalez-Maldonado R, Deller T, Salvi S, Cortelli P, Gilks WP, Latchman DS, Harvey RJ, Dallapiccola B, Auburger G, Wood NW (2004) Hereditary early-onset Parkinson's disease caused by mutations in PINK1. *Science* **304**(5674):1158–1160.

Valenti C, Schutta EJ, Kehaty T (1968) Prenatal diagnosis of Down's syndrome. *Lancet* **ii**:220.

van der Walt JM, Nicodemus KK, Martin ER, Scott WK, Nance MA, Watts RL, Hubble JP, Haines JL, Koller WC, Lyons K, Pahwa R, Stern MB, Colcher A, Hiner BC, Jankovic J, Ondo WG, Allen FH Jr, Goetz CG, Small GW, Mastaglia F, Stajich JM, McLaurin AC, Middleton LT, Scott BL, Schmechel DE, Pericak-Vance MA, Vance JM (2003) Mitochondrial polymorphisms significantly reduce the risk of Parkinson disease. *Am J Hum Genet* **72**(4):804–811.

van Duijn CM, Farrer LA, Cupples LA, Hofman A (1993) Genetic transmission of Alzheimer's disease among families in a Dutch population-based study. *J Med Genet* **30**:640–646.

van Duijn CM, de Knijff P, Wehnert A, De Voecht J, Bronzova JB, Havekes LM, Hofman A, Van Broeckhoven C (1995) The apolipoprotein E epsilon-2 allele is associated with an increased risk of early-onset Alzheimer's disease and a reduced survival. *Ann Neurol* **37**:605–610.

van Duijn CM, Dekker MC, Bonifati V, Galjaard RJ, Houwing-Duistermaat JJ, Snijders PJ, Testers L, Breedveld GJ, Horstink M, Sandkuijl LA, van Swieten JC, Oostra BA, Heutink P (2001) PARK7, a novel locus for autosomal recessive early-onset parkinsonism, on chromosome 1p36. *Am J Hum Genet* **69**:629–634.

Van Hove JLK, Zhang W, Kahler SG, Roe CR, Chen Y-T, Terada N, Chace DH, Iafolla AK, Ding J-H, Millington DS (1993) Medium-chain acyl-CoA dehydrogenase (MCAD) deficiency: Diagnosis by acylcarnitine analysis in blood. *Am J Hum Genet* **52**:958–966.

Van Lith JM, Pratt JJ, Beekhuis JR, Mantingh A (1992) Second-trimester maternal serum immunoreactive inhibin as a marker for fetal Down's syndrome. *Prenat Diagn* **12**:801–806.

Van Oostrom I, Meijers-Heijboer H, Lodder L, Duivenvoorden H, van Gool A, Seynaeve C, vander Meer C, Klijn J, van Geel B, Burger C, Wladimiroff J, Tibben A (2003) Long-term psychological impact of carrying a BRCA1/2 mutation and prophylactic surgery: A 5 year follow-up study. *J Clin Onc* **21**:3867–3874.

van Zuuren FJ (1997) The standard of neutrality during genetic counselling: An empirical investigation. *Patient Educ Counsl* **32**(1–2):69–79.

van Zuuren FJ, van Schie EC, van Baaren NK (1997) Uncertainty in the information provided during genetic counseling. *Patient Educ Counsl* **32**(1–2):129–139.

Vasdani v. Sehmi (1993) O.J. No. 44 (QL) (Gen. Div.).

Vasen HFA (2000) Clinical diagnosis and management of hereditary colorectal cancer syndromes. *J Clin Oncol* **18**:81s–92s.

Vasen HF, Mecklin JP, Khan PM, Lynch HT (1991) The international collaborative group on hereditary non-polyposis colorectal cancer (ICG-HNPCC). *Dis Colon Rectum* **31**:424–425.

Vasen H, Watson P, Mecklin J, Lynch H (1999) New clinical criteria for hereditary nonpolyposis colorectal cancer (HNPCC, Lynch syndrome) proposed by the International Collaborative group on HNPCC. *Gastroenterology* **116**:1453–1456.

Veach PM (2004) Genetic counselors' impact on the genetics revolution: Recommendations of an informed outsider. *Bioethics Examiner* **8**(2):1–4.

Veenstra-VanderWeele J, Cook EH Jr (2004) Molecular genetics of autism spectrum disorder. *Mol Psychiatry* **9**(9):819–832.

Vehmas S (2002) Is it wrong to deliberately conceive or give birth to a child with mental retardation? *J Med Philos* **27**(1):47–63.

Venditti LN, Venditti CP, Berry GT, Kaplan PB, Kaye EM, Glick H, Stanley CA (2003). Newborn screening by tandem mass spectrometry for medium-chain acyl-CoA dehydrogenase deficiency: A cost-effectiveness analysis Pediatrics **112**:1005–1015.

Verhoog LC, Brekelmans CT, Seynaeve C, Dahmen G, van Geel AN, Bartels CC, Tilanus-Linthorst MM, Wagner A, Devilee P, Halley DJ, et al (1999) Survival in hereditary breast cancer associated with germline mutations of BRCA2. *J Clin Oncol* **17**:3396–3402, 1999.

Verlinsky Y, Rechitsky S, Verlinsky O, Masciangelo C, Lederer K, Kuliev A (2002) Preimplantation diagnosis for early-onset Alzheimer disease caused by V717L mutation. *JAMA* **287**(8):1018–1021.

Verlinsky Y, Cohen J, Munne S, Gianaroli L, Simpson JL, Ferraretti AP, Kuliev A (2004) Over a decade of experience with preimplantation genetic diagnosis. *Fertil Steril* **82**(2):302–303. And see: Wells D (2004) Advances in preimplantation genetic diagnosis. *Eur J Obstet Gynecol Reprod Biol* **115** (Suppl)1:S97–101. With regard to the ethical debate about the application of preimplantation genetic diagnosis, see Tasca RJ, McClure ME (1998) The emerging technology and application of preimplantation genetic diagnosis. *J Law Med Ethics* **26**(1):7–16; Robertson JA (2003) Extending preimplantation genetic diagnosis: Medical and non-medical uses. *J Med Ethics* **29**:213–216; Caulfield T, Knowles L, Meslin EM (2004) Law and policy in the era of reproductive genetics. *J Med Ethics* **30**(4):414–417.

Vernon SW (1999) Risk perception and risk communication for cancer screening behaviors: A review. *J Nat'l Ca Inst Monographs* **25**:101–119.

Vernon A, Gritz S, Peterson S, Amos C, Perz C, Baile W, Lynch P (1997) Correlates of psychologic distress in colorectal cancer patients undergoing genetic testing for hereditary colon cancer. *Hlth Psychol* **16**(1):73–86.

Vibhakar NI, Budorick NE, Scioscia AL, Harby LD, Mullen ML, Sklansky MS (1999) Prevalence of aneuploidy with a cardiac intraventricular echogenic focus in an at-risk patient population. *J Ultrasound Med* **18**(4):265–268.

Vila M, Przedborski S (2004) Genetic contributions to Parkinson's disease. *Brain Res Brain Res Rev* **46**(1):44–70.

Villeneuve v. Sisters of Joseph (1975) 1 S.C.R. 285.

Vincent JB, Kolozsvari D, Roberts WS, Bolton PF, Gurling HM, Scherer SW (2004) Mutation screening of X-chromosomal neuroligin genes: No mutations in 196 autism probands. *Am J Med Genet* **129B**(1):82–84.

Violandi v. New York (1992) 184 App. 2d. 364, 584 NYS 2d 842 (1st. Dept.). But see *Stanley v. McCarver* (2003) 204 Ariz. 339, 63 P. 3d. 1076 (Ct.App. Div. 1). A patient was referred to a radiologist by an employer. An X-ray revealed potential problems. The patient later was diagnosed with lung cancer. The court held that the radiologist had a duty to inform the patient about the X-ray due to the lack of a referring physician.

Virta M, Hurme M, Helminen M (2002) Increased frequency of interleukin-1beta (-511) allele 2 in febrile seizures. *Pediatr Neurol* **26**(3):192–195.

Vlek C (1987) Risk assessment, risk perception and decision making about course of action involving genetic risk: An overview of concepts and methods. In: Evers-Kiebooms, Cassiman H, Van den Berghe H, d'Ydewalle G (eds.) *Genetic Risk, Risk Perception and Decision-Making.* New York: Alan R Liss Inc. for the March of Dimes Birth Defects Foundation—*Birth Defects Original Articles Series* **23**:171–207.

Vogel F, Motulsky AG (1997) *Human Genetics*, New York Springer.

Von Bergen CW, Evers P, Soper (2001) Legal and ethical considerations in genetic testing in the workplace. *Southern Law J* **54**:53–75.

Waalen J, Felitti V, Gelbart T, Ho NJ, Beutler E (2002) Penetrance of hemochromatosis. *Blood Cells, Molecules, Diseases* **29**(3), 418–432.

Wadman M (2001) Testing time for gene patent as Europe rebels. *Nature* **413**:443.

Wagner A, Barrows A, Wijnen J, van der Klift H, Franken P, Verkuijlen P, Nakagawa H, Geugien M, Jaghmohan-Changur S, Breukel C, Meijers-Heijboer H, Morreau H, van Puijenbroek M, Burn J, Coronel S, Kinarski Y, Okimoto R, Watson P, Lynch J, de la Chapelle A, Lynch H, Fodde R (2003) Molecular analysis of hereditary nonpolyposis colorectal cancer in the United States: High mutation detection rate among clinically selected families and characterization of an American founder genomic deletion of the MSH2 gene. *Am J Hum Genet* **72**:1088–1100.

Wainwright v. Leary (1993) 623 So. 2d 233, 237 (La. Ct. App.).

Waisbren SE, Albers S, Amato S, Ampola M, Brewster TG, Demmer L, Eaton RB, Greenstein R, Korson M, Larson C, Marsden D, Msall M, Naylor EW, Pueschel S, Seashore M, Shih VE, Levy HL (2003) Effect of expanded newborn screening for biochemical genetic disorders on child outcomes and parental stress. *JAMA* **290**(19):2564–2572.

Waisbren SE, Albers S, Amato S, Ampola M, Brewster TG, Demmer L, Eaton RB, Greenstein R, Korson M, Larson C, Marsden D, Msall M, Naylor EW, Pueschel S, Seashore M, Shih VE, Levy HL (2004) Effect of expanded newborn screening for biochemical genetic disorders on child outcomes and parental stress. *Obstet Gynecol Surv* **59**(6):415–417.

Waitzkin H (1985) Information giving in medical care. *J Hlth Soc Behav* **26**(2):81–101.

Wald NJ (1994) Guidance on terminology. *J Med Screen* **1**: 76.

Wald NJ (2001) Guidance on terminology. *J Med Screen* **8**(1):56.

Wald N (2000) Down's syndrome. In: Wald N, Leck I, eds. *Antenatal and Neonatal Screening*. Oxford: Oxford University Press, p. 85

Wald NJ, Hackshaw AK (1997) Combining ultrasound and biochemistry in first-trimester screening for Down's syndrome. *Prenat Diagn* **17**:821–829.

Wald NJ, Hackshaw AK (2000) Advances in antenatal screening for Down syndrome. *Baillieres Best Pract Res Clin Obstet Gynaecol* **14**(4):563–580.

Wald NJ, Brock DJH, Bonnar J (1974) Prenatal diagnosis of spina bifida and anencephaly by maternal serum-alpha-fetoprotein measurement. *Lancet* **1**:765–767.

Wald NJ, Cuckle H, Brock JH, Peto R, Polani PE, Woodford FP (1977) Maternal serum-alpha-fetoprotein measurement in antenatal screening for anencephaly and spina bifida in early pregnancy. Report of the UK collaborative study on alpha-fetoprotein in relation to neural-tube defects. *Lancet* **1**:1323–1332.

Wald NJ, Cuckle HS, Boreham J, Stirrat GM, Turnbull AC (1979a) Maternal serum alpha-fetoprotein and diabetes mellitus. *Br J Obstet Gynaecol* **86**:101–105.

Wald NJ, Cuckle HS, Peck S, Stirrat GM, Turnbull AC (1979b) Maternal serum alpha-fetoprotein in relation to zygosity. *Br Med J* **1**:455.

Wald NJ, Cuckle HS, Boreham J, Terzian E, Redman C (1981) The effect of maternal weight on maternal serum alpha-fetoprotein levels. *Br J Obstet Gynaecol* **88**:1094–1096.

Wald N, Cuckle H, Nanchahal K (1989) Amniotic fluid acetylcholinesterase measurement in the prenatal diagnosis of open neural tube defects: Second report of the collaborative acetylcholinesterase study. *Prenat Diagn* **9**:813–829.

Wald NJ, Cuckle HS, Haddow JE, Doherty RA, Knight GJ, Palomaki GE (1991a) Sensitivity of ultrasound in detecting spina bifida. *N Engl J Med* **324**:769–772.

Wald NJ, Cuckle H, Wu T, George L (1991b) Maternal serum unconjugated oestriol and human chorionic gonadotrophin levels in twin pregnancies: Implications for Down's syndrome. *Br J Obstet Gynaecol* **98**:905–908.

Wald NJ, Sneddon J, Densem J, Frost C, Stone R (1991c) Prevention of neural tube defects: Results of the MRC vitamin study. *Lancet* **338**:131–137.

Wald NJ, Stone R, Cuckle HS, Grudzinskas JG, Barkai G, Brambati B, Teisner B, Fuhrmann W (1992) First trimester concentrations of pregnancy associated plasma protein A and placental protein 14 in Down's syndrome. *BMJ* **305**:28.

Wald NJ, Densem JW, Smith D, Klee GG (1994) Four-marker serum screening for Down's syndrome. *Prenat Diagn* **14**:707–716.

Wald NJ, Densem JW, George L, Muttukrishna S, Knight PG (1996a) Prenatal screening for Down's syndrome using inhibin-A as a serum marker. *Prenat Diagn* **16**:143–153.

Wald NJ, George L, Smith D, Densem JW, Petterson K (1996b) Serum screening for Down's syndrome between 8 and 14 weeks of pregnancy. *Br J Obstet Gynaecol* **103**:407–412.

Wald NJ, Huttly W, Wald K, Kennard A (1996c) Down syndrome screening in UK. *Lancet* **347**(8997):330.

Wald NJ, Densem JW, George L, Muttukrishna S, Knight PG, Watt H, Hacksaw A, Morris JK (1997a) Inhibin-A in Down's syndrome pregnancies: Revised estimate of standard deviation. *Prenat Diagn* **17**:285–290.

Wald NJ, Kennard A, Hackshaw A, McGuire A (1997b) Antenatal screening for Down's syndrome. *J Med Screen* **4**:181–246.

Wald NJ, Densem JW, George L, Muttukrishna S, Knight PG, Watt H, Hacksaw A, Morris JK (1997c) Inhibin-A in Down's syndrome pregnancies: Revised estimate of standard deviation. *Prenat Diagn* **17**:285–290.

Wald NJ, Watt HC, Haddow JE, Knight GJ (1998) Screening for Down syndrome at 14 weeks of pregnancy. *Prenat Diagn* **18**:291–293.

Wald NJ, Watt HC, Hackshaw AK (1999) Integrated screening for Down's syndrome on the basis of tests performed during the first and second trimesters. *N Engl J Med* **341**:461–467.

Wald NJ, Rodeck C, Hackshaw AK, Walters J, Chitty L, Mackinson AM (2003) First second trimester antenatal screening for Down's syndrome: The results of the Serum, Urine and Ultrasound Screening Study (SURUSS). *J Med Screen* **10**:56–104.

Wald NJ, Rodeck C, Hackshaw AK, Rudnicka A (2004) SURUSS in perspective. *BJOG* **111**:521–531.

Walker A (1996) Genetic counseling. In: Rimoin DL, Connor JM, Pyeritz RE, eds. *Emery and Rimoin's Principles and Practice of Medical Genetics*, 3rd ed. New York: Churchill Livingstone, pp. 595–618.

Walker A (1998) The practice of genetic counseling. In: Baker DL, Schuette JL, Uhlmann WR, eds. *A Guide to Genetic Counseling*. New York: Wiley, pp. 1–20.

Wallace RH, Berkovic SF, Howell RA, Sutherland GR, Mulley JC (1996) Suggestion of a major gene for familial convulsions mapping to 8q13–21. *J Med Genet* **33**:308–312.

Wallace RH, Wang DW, Singh R, Scheffer IE, George AL Jr, Phillips HA, Saar K, Reis A, Johnson EW, Sutherland GR, Berkovic SF, Mulley JC (1998) Febrile seizures and generalized epilepsy associated with a mutation in the Na+-channel beta-1 subunit gene SCN1B. *Nat Genet* **19**:366–370.

Wallace RH, Marini C, Petrou S, Harkin LA, Bowser DN, Panchal RG, Williams DA, Sutherland GR, Mulley JC, Scheffer IE, Berkovic SF (2001) Mutant GABAA receptor gamma-2 subunit in childhood absence epilepsy and febrile seizures. *Nat Genet* **28**:49–52.

Waller DK, Mills JL, Simpson JL, Cunningham GC, Conley MR, Lassman MR, Rhoads GG (1994) Are obese women at higher risk for producing malformed offspring? *Am J Obstet Gynecol* **170**:541–548.

Waller DK, Lustig LS, Cunningham GC, Feuchtbaum LB, Hook EB (1996) The association between maternal serum alpha-fetoprotein and preterm birth, small for gestational age infants, preeclampsia, and placental complications. *Obstet Gynecol* **88**:816–822.

Wallerstein R, Starkman A, Jansen V (2001) Carrier screening for Gaucher disease in couples of mixed ethnicity. *Genet Test* **5**(1):61–64.

Walsh JP, Arora A, Cohen WM (2004) Effects of research tool patents and licensing on biomedical innovation. In: Cohen WM, Merrill S, eds. *Patents in the Knowledge-Based Economy*, Washington, DC: National Academies Press, pp. 285–340. Online: NAP <http//:books.nap.edu/catalog/10770.html>.

Walstad v. University of Minnesota Hospitals (1971) 442 F 2d 634 (CA 8th Cir.)

Walters v. Rinker (1988) 520 N.E. 2d. 468 Ind. Ct. App.

Wang ST, Pizzolato S, Demshar HP (1997) Receiver operating characteristic plots to evaluate Guthrie, Wallac, and Isolab phenylalanine kit performance for newborn phenylketonuria screening. *Clinical Chemistry* **43**:1838–1842.

Wang V, Marsh FH (1990) Ethical principles and cultural integrity: An Asian ethnocultural perspective in genetic counseling *Am J Hum Genet (Ann Suppl)* **47**:A125–489.

Wang, V (1993) Handbook of Cross-Cultural Genetic Counseling. Unpublished manual.

Wang C, Gonzalez R, Merajver SD (2004) Assessment of genetic testing and related counseling services: Current research and future directions. *Soc Sci Med* **58**(7):1427–1442.

Wapner R, Thom E, Simpson JL, Pergament E, Silver R, Filkins K, Platt L, Mahoney M, Johnson A, Hogge WA, Wilson RD, Mohide P, Hershey D, Krantz D, Zachary J, Snijders R, Greene N, Sabbagha R, MacGregor S, Hill L, Gagnon A, Hallahan T, Jackson L, for the First Trimester Maternal Serum Biochemistry and Fetal Nuchal Translucency Screening (BUN) Study Group (2003) First-trimester screening for trisomies 21 and 18. *N Engl J Med* **349**:1405–1413.

Warner E, Helsey RE, Goel V, Carroll JC, McCready DR (1999) Hereditary breast cancer: Risk assessment of patients with a family history of breast cancer. *Can Fam Phys* **13**:243–281.

Washington State Genetic Documents (2002) Genetics in Managed Care Report: Plan 2002–2005. http://mchneighborhood.ichp.edu/wagenetics/MCO_FinalReport_Final.pdf.

Wasserman D (2003) A choice of evils in prenatal testing. *Florida State Univ Law Rev* **30**:295–313.

Wasserman LM, Jones OW, Trombold JS, Sadler GR (2000) Attitudes of physicians regarding receiving and storing patients' genetic testing results for cancer susceptibility. *J Comm Hlth* **25**(4):305–313.

Watson EK, Mayall ES, Lamb J, Chapple J, Williamson R (1992) Psychological and social consequences of community carrier screening programme for cystic fibrosis. *Lancet* **340**:217–220.

Watson M, Lloyd S, Davidson J, Meyer L, Eeles R, Ebbs S, Murday V (1999) The impact of genetic counselling on risk perception and mental health in women with a family history of breast cancer. *Br J Cancer* **79**:868–874.

Watson MS, Buchanan PD, Cohen MM, DeWald GW, Ledbetter DH, Goldberg JD, Wapner RJ (2000) Technical and clinical assessment of fluorescence in situ hybridization: An ACMG/ASHG position statement. I: Technical considerations. *Genet Med* **2**(6):356–361.

Watson M, Foster C, Eeles R, Eccles D, Ashley S, Davidson R, Mackay J, Morrison PJ, Hopwood P, Evans DG (2004a) Psychosocial impact of breast/ovarian (BRCA1/2) cancer-predictive genetic testing in a UK multi-centre clinical cohort. *Br J Cancer* **91**(10):1787–1794. 20% of female carriers reported insurance problems out of 261 adults (59 male) from nine UK genetics centers participated; 91 gene mutation carriers and 170 noncarriers.

Watson M, Cutting G, Desnick R, Driscoll D, Klinger K, Mennuti M, Palomaki G, Popovich B, Pratt V, Rohlfs E, Strom C, Richards C, Witt D, Grody W (2004b) Cystic fibrosis population carrier screening: 2004 revision of the American College of Medical Genetics mutation panel. *Genet Med* **6**(5):387–391.

Watts G (1999) Fears that a rise in genetic testing will rule out insurance are "paranoia." *BMJ* **319**(7205):273.

Watt HC, Wald NJ, Smith D, Kennard A, Densem J (1996) Effect of allowing for ethnic group in prenatal screening for Downs syndrome. *Prenat Diagn* **16**:691–698.

Wax JR, Philput C (1998) Fetal intracardiac echogenic foci: Does it matter which ventricle? *J Ultrasound Med* **17**(3):141–144; quiz 145–146.

Wax JR, Lopes AM, Benn PA, Lerer T, Steinfeld JD, Ingardia CJ (2000) Unexplained elevated midtrimester maternal serum levels of alpha fetoprotein, human chorionic gonadotropin, or low unconjugated estriol: Recurrence risk and association with adverse perinatal outcome. *J Matern Fetal Med* **9**:161–164.

Weaver KD (1997) Genetic screening and the right not to know. *Issues Law Med* **13**:243–281.

Weaver by Weaver v. University of Mich (1993) Bd of Regents. 201 Mich. App. 239, 506 N.W. 2d 264.

Wecker v. Amend (1996) 22 Kan App 2d 498, 918 P2d 658.

Weil J (2000) *Psychosocial Genetic Counseling*. Oxford: Oxford University Press.

Weil J (2001) Multicultural education and genetic counseling. *Clin Genet* **59**(3):143–149.

Weil J (2003) Psychosocial genetic counseling in the post-nondirective era: A point of view. *J Genet Counsel* **12**(3):199–211.

Weiss v. Solomon (1989) 44 C.C.L.T. 280; R.J.Q. 731 [1989] R.R.A. 374 (C.S.). Also see *Crossman v. Stewart* (1977), 5 C.C.L.T. 45, 82 D.L.R. (3d) 677. Canadian courts have stated that where a treatment is a new one, the standard of care increases [*LaFleur v. Cornelis* (1979), 28 N.B.R. (2d) 569, 63 A.P.R. 569 (Q.B.)].

Weissman MM, Bruce ML, Leaf PJ, Florio LP, et al (1991) Affective disorders. In: Robins LN, Regier DA, eds. *Psychiatric Disorders in America: The Epidemiologic Catchment Area Study*. New York, Free Press.

Weisbrot D (2004) The human genome: lessons for life, love and the law. *J Law Med* **11**(4):428–445. Reproduced with permission.

Welch HG, Burke W (1998) Uncertainties in genetic testing for chronic disease. *JAMA* **280**(17):1525–1527.

Wellisch D (1992) Psychological functioning of daughters of breast cancer patients II. *Psychosomatics* **33**:171–179.

Wellisch D, Lindberg N (2001) A psychological profile of depressed and nondepressed women at high risk for breast cancer. *Psychosomatics* **42**:330–336.

Wellisch DK, Gritz ER, Schain W, et al (1991) Psychological functioning of daughters of breast cancer patients. *Psychosomatics* **32**:324–336.

Wells D (2003) Advances in Preimplantation Genetic Diagnosis. European Journal of Obstetrics and Gynecology and Reproductive Biology 115S:S97–S101.

Welsh v. McCarthy (1996) 677 A. 2d. 1066 Me.

Werler MM, Louik CL, Shapiro S, Mitchell AA (1996) Prepregnant weight in relation to risk of neural tube defects. *JAMA* **275**:1089–1092.

Wertz DC (1997) Society and the not-so-new genetics: What are we afraid of? *J Contem Hlth Law Policy* **13**:299–346.

Wertz D (1999a) Patient and professional views on autonomy: A survey in the United States and Canada. *Hlth Law Rev* **7**, 9–10. Of those surveyed 60% thought that they were entitled to any genetic service that they could pay for and 69% thought that withholding such services was a denial of the patients' rights.

Wertz DC (1999b) Genetic discrimination: Results of a survey of genetic professionals, primary care physicians, patients and public. *Hlth Law Rev* **7**(3):7–8.

Wertz DC, Fletcher DC (1988) Attitudes of genetic counselors: A multinational survey *Am J Hum Genet* **42**:592–600.

Wertz DC, Knoppers BM (2002) Serious genetic disorders: Can or should they be defined? *Am J Med Genet* **108**(1):29–35.

Wertz D, Sorenson JR, Heeren TC (1986) Clients' interpretation of risks provided in genetic counseling. *Am J Hum Genet* **39**:253–264.

Wertz DC, Fletcher JC, Mulvihill JJ (1990) Medical geneticists confront ethical dilemmas: Cross-cultural comparisons among 18 nations. *Am J Hum Genet* **46**:1200–1213.

Wertz DC Janes SR, Rosenfield JM, Erbe RW (1992) Attitudes toward prenatal diagnosis of cystic fibrosis: Factors in decision making among affected families. *Am J Hum Genet* **50**:1077 at 1083. Similar findings are reported in Frets and Niermeijer [1990]. Jedlicka-Kohler et al. [1994] report that the "availability of PD does not substantially change the reproductive behavior of parents of children with CF," p. 15.

Wertz DC, Fanos JH, Reilly PR (1994) Genetic testing for children and adolescents. Who decides? *JAMA* **272**(11):875–881.

West DS, Greene PG, Kratt PP, Pulley L, Weiss HL, Siegfried N, Gore SA (2003) The impact of a family history of breast cancer on screening practices and attitudes in low-income, rural, African American women. *J Womens Hlth (Larchmt)* **12**:779–787.

Wethers DL (2000a) Sickle cell disease in childhood: Part I. Laboratory diagnosis, pathophysiology and health maintenance. *Am Fam Phys* **62**(5):1013–1020, 1027–1028.

Wethers DL (2000b) Sickle cell disease in childhood: Part II. Diagnosis and treatment of major complications and recent advances in treatment. *Am Fam Phys* **62**(6):1309–1314.

WHI Steering Committee (2004) Effects of conjugated equine estrogen in postmenopausal women with hysterectomy. The women's Health Initiative Randomized Controlled Trial. *JAMA* **291**:1701–1712.

White v. Turner (1981) 31 O.R. (2d) 773 (H.C.J.) aff'd (1982), 47 O.R. (2d) 764.

Whiteside v. Lukson (1997) 947 P.2d 1263, 1265 (Wash. Ct. App.): The court's judgment read: "As he was being taken to the Catheterization Lab ("Cath Lab") where the angioplasty was to be performed, Mr. Wilkerson told the staff members that he had not signed a consent form for the angioplasty procedure because it had not been explained to him. The orderly accompanying him responded that the procedure would probably be explained to him at the Cath Lab. Mr. Wilkerson could not recall if he knew why he was being taken to the Cath Lab. However, he did not think any procedure would be done until he had signed a consent form. Shortly after reaching the Cath Lab, Mr. Wilkerson was given Valium. Mr. Wilkerson testified that Dr. Gary Beauchamp, an interventional, or invasive, cardiologist, introduced himself in the Cath Lab just before the angioplasty was performed. Dr. Beauchamp did not ask Mr. Wilkerson if he had any questions about the procedure. Furthermore, Dr. Beauchamp admitted that he did not advise Mr. Wilkerson of the risks of the angioplasty procedure, because he assumed someone else had already done so. Dr. Rowland testified that she discussed the angioplasty and other possible therapeutic options with Mr. and Mrs. Wilkerson on December 12, 1989, after the angiogram and that she discussed them again on December 13, 1989, with Mr. Wilkerson. While Dr. Rowland stated that her first note in Mr. Wilkerson's medical records for December 12, 1989 indicated that she had discussed the risks and benefits of the angioplasty and surgery with Mr. Wilkerson, the notes, in fact, do not reflect a discussion of the procedure. In addition, the prepared written consent form for the angioplasty was never signed by Mr. Wilkerson or anyone on his behalf."

Whitten CF, Whitten-Shurney W (2001) Sickle cell. *Clin Perinatol* **28** (2):435–448.

Whittle M (2001) Down's syndrome screening: Where to now? *Br J Obstet Gynaecol* **108**:559–561.

WHO Genetic Databases (2003). World Health Organization (1998) Proposed International Guidelines on Ethical Issues in Medical Genetics and Genetic Services. Geneva. World Health Organization. 1998: http://www1.umn.edu/humanrts/instree/guidelineproposal.html

WHO (2003) Genetic Databases: http://www.who.int/ethics/topics/hgdb/en/

Wickline v. State (1990) 239 Cal. Rptr. 810 (Cal. App. 1986); also see Kuzzler (1999) infra; Sugerman PR, Yarisbus VA (1999) infra.

Wickstrom EA, Thangavelu M, Parilla BV, Tamura RK, Sabbagha RE (1996) A prospective study of the association between isolated fetal pyelectasis and chromosomal abnormality. *Obstet Gynecol* **88**(3):379–382.

Wiggins S, Whyte P, Huggins M, Adam S, Theilmann J, Bloch M, Sheps SB, Schechter MT, Hayden MR (1992) The psychological consequences of predictive testing for Huntington's disease. Canadian Collaborative Study of Predictive Testing. *N Engl J Med* **327**(20):1401–1405.

Wilcken B, Wiley V, Hammond J, Carpenter K (2003) Screening Newborns for Inborn Errors of Metabolism by Tandem Mass Spectrometry. *N Engl J Med* **348**:2304–2312.

Wildeman S, Downie J (2001) Genetic and metabolic screening of newborns: Must health care providers seek explicit parental consent? *Hlth Law J* **9**:61–111. The authors argue that routine screening for genetic and metabolic disorders without explicit consent "does not mean that this practice is legally defensible." Mandl KD, Feit S, Larson C, Kohane IS (2002) Newborn screening program practices in the United States: Notification, research, and consent. *Pediatrics* **109**(2):269–273. And see Avard D, Knoppers BM (2001) Screening and children—Policy issues for the new millennium. *Isuma* **2**(3):Autumn.

http://www.isuma.net/v02n03/avard/avard_e.shtml. "However, in some circumstances, such as when a severe disorder such as PKU can be prevented, health care professionals have a duty to the newborn to persuade the parents to consent and to override parental refusal to the child being screened."

Wilkie H, Osei-Lah A, Chioza B, Nashef L, McCormick D, Asherson P, Makoff AJ (2002) Association of mu-opioid receptor subunit gene and idiopathic generalized epilepsy. *Neurology* **59**(5):724–728.

Wilkinson v. Vesey (1972) 110 RI 606, 295 A 2d 676, 69 ALR 3d 1202.

Williams JK, Schutte DL, Holkup PA, Evers C, Muilenburg A (2000) Psychological impact of predictive testing for Huntington disease on support persons. *Am J Med Genet* **96**:353–359.

Williams C, Alderson P, Farsides B (2002a) Is nondirectiveness possible within the context of antenatal screening and testing? *Soc Sci Med* **54**(3):339–347.

Williams LJ, Mai CT, Edmonds LD, Shaw GM, Kirby RS, Hobbs CA, Sever LE, Miller LA, Meaney FJ, Levitt M (2002b) Prevalence of spina bifida and anencephaly during the transition to mandatory folic acid fortification in the United States. *Terat* **66**:33–39.

Williams LO, Cole EC, Lubin IM, Iglesias NI, Jordan RL, Elliott LE (2003) Quality assurance in human molecular genetics testing: status and recommendations. *Arch Pathol Lab Med* **127**(10):1353–1358.

Williams-Jones B (1999) Reframing the discussion: Commercial genetic testing in Canada. *Hlth Law J* 7:50–67.

Williams-Jones B (2002) History of a gene patent: Tracing the development and application of commercial BRCA testing. *Hlth Law J* **10**: 123–146.

Wills v. Saunders (1989) 47 C.C.L.T. 235, 93 A.R. 282, 2 W.W.R. 715 (Alta. Q.B.).

Wilson CJ, Champion MP, Collins JE, Clayton PT, Leonard JV (1999) Outcome of medium chain acyl-CoA dehydrogenase deficiency after diagnosis. *Arch Dis Child* **80**:459–462.

Wilson JMG, Jungner F (1968) Principles and Practice of Screening for Disease (Public Health Papers No. 34). Geneva: World Health Organization. Frankenburg WK. (1974) Selection of diseases and tests in pediatric screening. Pediatrics **54**:612–616.

Wilson PWF, Schaefer EJ, Larson MG, Ordovas JM (1996) Apolipoprotein E alleles and risk of coronary disease: A meta-analysis. *Arterioscler Thromb Vasc Biol* **16**:1250–1255.

Wilson RD, Lynch S, Lessoway VA (1997) Fetal pyelectasis: Comparison of postnatal renal pathology with unilateral and bilateral pyelectasis. *Prenat Diagn* **17**:451–455.

Wilson RD, Davies G, Desilets V, Reid GJ, Shaw D, Summers A, Wyatt P, Young D, Crane J, Armson A, de la Ronde S, Farine D, Leduc L, Van Aerde J, Society of Obstetricians and Gynaecologists of Canada (2002) Cystic fibrosis carrier testing in pregnancy in Canada. *J Obstet Gynaecol* **24**(8):644–651.

Wilson v. Blue Cross of Southern California (1990) 271 Cal. Rptr. 876Cal. Ct. App.

Wilson v. Swanson (1956) 5 D.L.R. (2d) 113 (S.C.C.).

Winawer SJ, Fletcher RH, Miller L, et al (1997) Colorectal cancer screening; Clinical guidelines, evidence and rationale. *Gastroenterology* **112**:594–642.

Winawer SJ, Fletcher RH, Miller L, Godlee F, Stolar MH, Mulrow CD, Woolf SH, Glick SN, Ganiats TG, Bond JH, Rosen L, Zapka JG, Olsen SJ, Giardiello FM, Sisk JE, Van Antweep R, Brown-Davis C, Marciniak DA, Mayer RJ (1997) Colorectal cancer screening: clinical guidelines and rationale. Gastroenterology **112**(2):594–642.

Wingrove KJ, Norris J, Barton PL, Hagerman R (1996) Experiences and attitudes concerning genetic testing and insurance in a Colorado population: A survey of families diagnosed with fragile X syndrome. *Am J Med Genet* **64**:378–381.

Winik LW (1998) When you should consider a genetic test. *Parade* April 19:8–9.

Winter TC, Anderson AM, Cheng EY, Komarniski CA, Souter VL, Uhrich SB, et al (2000) Echogenic intracardiac focus in 2nd-trimester fetuses with trisomy 21: Usefulness as a US marker. *Radiology* **216**(2):450–456.

Wolbring G (2003) Disability rights approach towards bioethics? *J Disability Policy Studies* **14**:174–180.

Woloshin S, Schwartz LM, Welch HG (2002) Risk charts: Putting cancer in context. *J Natl Cancer Inst* **94**:799–804. And see Bach PB, Schrag D (2002) Re: Risk charts: Putting cancer in context. *J Natl Cancer Inst* **94**:1584–1585.

Wolpert CM (2000) Human genomics in clinical practice: Bridging the gap. *Clin Rev* **10**(7):67–86.

Wong, SI (2002) At Home with Down Syndrome and Gender. *Hypatia* **17**(3):89–117.

Wood DA, Kinmouth AL, Davies GA, Yarwood J, Thompson SG, Pyke SDM, Kok Y, Cramb R, LeGuen C, Morteav TM, Durrington PN (1994) Randomised controlled trial evaluating cardiovascular screening and intervention in general practice: Principal results of British family heart study. *Br Med J* **308**:313–320.

Woolf SH, Kuzel AJ, Dovey SM, Phillips RL Jr (2004) Astring of mistakes: the importance of coscode analysis in describing, counting, and preventing medical errors. *Am Fam. Med.* **2**(4):292–293.

World Health Organization (1968) Scientific Group on Screening for Inborn Errors of Metabolism. Screening for inborn errors of metabolism. *World Health Org Techn Rep Ser* **401**:1–57.

World Health Organization (1998) Proposed International Guidelines on Ethical Issues in Medical Genetics and Genetic Services. Geneva, December 15–16. http://www1.umn.edu/humanrts/instree/guidelineproposal.html

World Health Organization (WHO) (1999) Draft guidelines on bioethics. World Conference on Science. Available at <http://www.nature.com/wcs/b23a.html>. Accessed August 30, 2004.

World Health Organization (2003) Review of Ethical Issues in Medical Genetics. Les problŠmes ethiques: Relating to a shorter consensus statement "Proposed International Guidelines on Ethical Issues in Medical Genetics and Genetic Services," 1998). http://www.who.int/genomics/publications/en/.

World Medical Association WMA (1992) Declaration on the Human Genome Project, 44th World Medical Assembly, Marbella, Spain, September.

World Medical Association (1995) Declaration on the Rights of the Patient Bali, Indonesia, September.

Worwood M (2005) Inherited iron loading: Genetic testing in diagnosis and management. *Blood Rev* **19**:69–88.

Wright C, Kerzin-Storrar L, Williamson PR, Fryer A, Njindou A, Quarrell O, Donnai D, Craufurd D (2002) Comparison of genetic services with and without genetic registers: Knowledge, adjustment, and attitudes about genetic cousnelling among probands referred to three genetic clinics. *J Med Genet* **39**:e84. (http://www.jmedgenet.com/cgi/content/full/39/12/e84).

Wright D, Bradbury I, Benn P, Cuckle H, Ritchie K (2004) Contingent screening for Down syndrome is an efficient alternative to non-disclosure sequential screening. *Prenat Diagn* **24**(10):762–766.

Writing Group for the Women's Health Initiative Investigators (2002) Risks and benefits of estrogen plus progestin in healthy postmenopausal women. *JAMA* **288**:321–333.

Wu WC, Pearlman RA (1988) Consent in medical decision making: The role of communication. *J Gen Intern Med* **3**(1): 9–14.

Wynia MK, VanGeest JB, Cummins DS, Wilson IB (2003) Do physicians not offer useful services because of coverage restrictions? *Hlth Affairs* **22**(4):190–197.

Yamagishi K (1997) When a 12.86% mortality is more dangerous than 24.14%: Implications for risk communication. *Appl Cogn Psych* **11**:495–506. Yamagishi reported that death rates of 1286 out of 10,000 were rated as more risky than rates of 24.14 out of 100.

Yan H, Papadopoulos N, Marra G, Perrera C, Jiricny J, Boland CR, Lynch HT, Chadwick RB, de la Chapelle A, Berg K, Eshleman JR, Yuan W, Markovitz S, Laken SJ, Lengauer C, Kinzler KW, Vogelstein B (2000) Conversion of diploidy to haploidy. *Nature* **403**:723–724.

Yang Q, Khoury MJ, Coughlin SS, Sun F, Flanders WD (2000) On the use of population-based registries in the clinical validation of genetic tests for disease susceptibility. *Genet Med* **2**(3):186–192.

Yarborough M, Scott JA, Dixon LK (1989) The role of beneficence in clinical genetics: Non-directive counseling reconsidered. *Theoretical Med* **10**(2):139–149.

Yaron Y, Hassan S, Geva E, Kupferminc MJ, Yavetz H, Evans MI (1999) Evaluation of fetal echogenic bowel in the second trimester. *Fetal Diagn Ther* **14**:176–180.

Yeatman TJ, Lyman GH, Smith SK, Reintgen DS, Cantor AB, Cox CE (1997) Bilaterality and recurrence rates for lobular breat cancer: Considerations for treatment. *Ann Surg Oncol* **4**:198–202.

Yepremian v. Scarborough General Hospital (1978) 20 O.R. (2d) 510 (Ont. H.C.), at 8519, affirmed aff'd (1980), 28 O.R. (2d) 494 (Ont. C.A.).

Yesley MS (1997) Genetic privacy, discrimination, and social policy: Challenges and dilemmas. *Microbial Comp Genomics* **2**:19–35.

Yonan AL, Alarcon M, Cheng R, Magnusson PK, Spence SJ, Palmer AA, Grunn A, Juo SH, Terwilliger JD, Liu J, Cantor RM, Geschwind DH, Gilliam TC (2003) A genomewide screen of 345 families for autism-susceptibility loci. *Am J Hum Genet* **73**(4):886–897.

Yoon PW, Scheuner MT, Khoury MJ (2003) Research priorities for evaluating family history in the prevention of common chronic diseases. *Am J Prev Med* **24**:128–135.

Young I (1999) Introduction to risk calculation in genetic counseling, 2nd edition. Oxford University Press, New York, USA. ISBN 0-19-262962.

Zammatteo N, Hamels S, DeLongueville F, Alexandre I, Gala JL, Brasseur F, Remacle J (2002) New chips for molecular biology and diagnostics. *Biotechnol Annu Rev* **8**:85–101.

Zigmond AS, Snaith RP (1983) The Hospital Anxiety and Depression Scale. *Acta Psych Scand* **67**:361–370.

Zinn W (1993) The empathic physician. *Arch Intern Med* **153**(3):306–312.

Zoppi MA, Ibba RM, Floris M, Monni G (2001) Fetal nuchal translucency screening in 12495 pregnancies in Sardinia. *Ultrasound Obstet Gynecol* **18**:649–651.

Zupancic JA, Kirpalani H, Barrett J, Stewart S, Gafni A, Streiner D, Beecroft ML, Smith P (2002) Characterising doctor-parent communication in counselling for impending preterm delivery. *Arch Dis Child Fetal Neonatal Ed* **87**(2):F113–117. "Overall 27% of parents felt: 'I would prefer to have the doctors advise me, rather than asking me to decide'. In 79% of cases, clinicians believed parents preferred advice rather than to make decisions, but in 45% of these, they misidentified those who wished to make their decisions." The authors state that shared decision making is not the "predominant mode of interaction" and that a literature review indicates that "only 10% of the average interaction time ... is being devoted to 'partnership building' in which the doctor elicits patient involvement or facilitates through synthesis and interpretation of the discussion."

Index

Genetic Testing: Care, Consent, and Liability, by Neil F. Sharpe and Ronald F. Carter
© 2006 John Wiley & Sons, Inc.